Theoretical Physics

Theoretical Physics

Editor: Elisha Franks

NY Research Press
New York

Published by NY Research Press
118-35 Queens Blvd., Suite 400,
Forest Hills, NY 11375, USA
www.nyresearchpress.com

Theoretical Physics
Edited by Elisha Franks

International Standard Book Number: 978-1-63238-546-8 (Hardback)

Cataloging-in-Publication Data

Theoretical physics / edited by Elisha Franks.
p. cm.
Includes bibliographical references and index.
ISBN 978-1-63238-546-8
1. Physics. 2. Quantum theory. I. Franks, Elisha.
QC21.3 .T44 2017
530--dc23

Printed in the United States of America.

Contents

Preface

Every book is initially just a concept; it takes months of research and hard work to give it the final shape in which the readers receive it. In its early stages, this book also went through rigorous reviewing. The notable contributions made by experts from across the globe were first molded into patterned chapters and then arranged in a sensibly sequential manner to bring out the best results.

Theoretical physics is a branch of physics that studies physical objects through mathematical methods. It is applied to study natural phenomena and also to predict and analyse it. This book on theoretical physics seeks to incorporate the various theories that have been developed into well-established systems as well as its implications on other fields. It discusses the fundamentals as well as modern approaches of theoretical physics. As this field is emerging at a rapid pace, the contents of this book will help the readers understand the modern concepts and applications of the subject. This text will be of help to students and teachers in the fields of quantum engineering, fluid theory and general relativity.

It has been my immense pleasure to be a part of this project and to contribute my years of learning in such a meaningful form. I would like to take this opportunity to thank all the people who have been associated with the completion of this book at any step.

Editor

1

A minimal physical model captures the shapes of crawling cells

E. Tjhung[1,*], A. Tiribocchi[1,*], D. Marenduzzo[1] & M.E. Cates[1]

Cell motility in higher organisms (eukaryotes) is crucial to biological functions ranging from wound healing to immune response, and also implicated in diseases such as cancer. For cells crawling on hard surfaces, significant insights into motility have been gained from experiments replicating such motion *in vitro*. Such experiments show that crawling uses a combination of actin treadmilling (polymerization), which pushes the front of a cell forward, and myosin-induced stress (contractility), which retracts the rear. Here we present a simplified physical model of a crawling cell, consisting of a droplet of active polar fluid with contractility throughout, but treadmilling connected to a thin layer near the supporting wall. The model shows a variety of shapes and/or motility regimes, some closely resembling cases seen experimentally. Our work strongly supports the view that cellular motility exploits autonomous physical mechanisms whose operation does not need continuous regulatory effort.

[1] SUPA, School of Physics and Astronomy, University of Edinburgh, JCMB Kings Buildings, Mayfield Road, Edinburgh EH9 3JZ, UK. * These authors contributed equally to this work. Correspondence and requests for materials should be addressed to D.M. (email: dmarendu@ph.ed.ac.uk).

How do living cells move around within their environment, whether *in vitro* (for example, on a glass slide) or *in vivo* (for example, within a tissue)? Is their motion sustained by continuous action of the cell complex biochemical networks, or do these issue instructions to a physically autonomous engine of motility? Experiments that replicate cellular motility with simplified *in vitro* systems have gone some way towards answering these questions[1–4].

Cellular self-propulsion in 3D gels or tissues might involve either 'grabbing' of the extracellular matrix[2] or 'swimming' via actomyosin flow in the cell interior[5–7]. However, as 3D motility is hard to study experimentally, detailed mechanisms remain largely speculative. Much better understood is a case we address here: cells moving on a 2D solid substrate.

The crawling motion of such cells combines two processes[2]. One is actin polymerization, in which monomers add to one end of an actin filament and leave from the other. Given a net polarization of the actin network, this 'treadmilling' pushes the leading edge of the cell forwards. The second process is the 'walking' of myosin motors along filaments. This creates a contractile[8] stress that, alongside interfacial tension and/or membrane elasticity, opposes cellular elongation. Contractility thus assists retraction of the rear part of a crawling cell (which includes the nucleus). Also implicated in crawling are assemblies of intracellular membrane proteins called focal adhesions[1], which enhance mechanical connectivity between actomoyosin and the solid substrate. These increase the effective wall friction and improve crawling efficiency (just as a child can crawl more easily on a pile carpet than a polished floor). Focal adhesions also signal alterations to the balance between treadmilling and contractility[9,10]. Notably though, significant motility is maintained even in laser-ablated or reconstituted cell fragments that lack most or all of the cell regulatory machinery[11].

Our goal here is to set up a minimal model of crawling motility, to see how fully the observed behaviour can be understood using physics alone. To achieve this goal, we intentionally rely on a very simplified description of complex biochemical and biophysical processes such as F-actin polymerization and actomyosin contractility. We also employ only an elementary description of the cell membrane, as well as of its coupling to the interior and exterior of the cell. The advantages of such an approach is that there are arguably as few free parameters as possible within the model, and this allows us to isolate the key physical elements of cell crawling (and/or spreading[12]) at surfaces. In other words, we can directly answer the question posed above: can cellular motility be seen as a physics-based machinery, which is coupled to, and controlled by, a complicated underlying biochemical network, but is nevertheless able to function independently? The alternative is that the biochemical detail is fundamental, and no aspects of it can be disregarded, even for a basic understanding of motility. This is an important question, both to understand the basic mechanism of cell motility and to inform subsequent biomimetic work intending to replicate cellular behaviour with synthetic components. In this respect, our simulation model can be viewed as the *in silico* analogue of *in vitro* experiments with cell extracts and cell fragments[11].

The drawback inherent within our approach is that, from the beginning, some of the important biological detail is lost. One can argue that, if experimental observations are reproduced by the minimal framework, this detail may be inessential to the fundamental biophysics of motility, and this by itself is very useful information. Nevertheless, to check the picture and to prove the previous statement unambiguously, one should eventually insert back realistic details one by one, and explicitly compare the results obtained with those of the fully simplified, minimal model. We initiate this last approach here as well, for a few cell biological features that our baseline model does not capture well. These include the allowance for a relatively high membrane viscosity, and, for wall slip, or equivalently, retrograde flow[13]; see Supplementary Note 1–3 for details of these checks.

In our minimal model, we represent a cell or cell extract as a droplet of active polar fluid (actomyosin) immersed in a second fluid and confined by interfacial tension[7]. (The latter is similar to, but simpler than, the fixed area constraint imposed by the outer membrane of a real cell.) As in real cells, actin treadmilling is confined both to the leading edge of the cell and within a finite thickness λ from the substrate. Our droplet interacts with a solid wall through a controllable friction, represented by a partial-slip boundary condition. In contrast to most previous computer models for cell motility[7,14–21], we treat cells as 3D objects. Previous 2D models often incorporated a number of important biological details, such as variable adhesion with the substrate[15] or the dynamics of signalling proteins[18]. A difference with our framework, besides the general philosophy highlighted above, is that these 2D models typically view the cell from above and cannot resolve the flow within the cell, or the presence or absence of slip at the boundary, whereas our model can resolve these features. Importantly, our 3D framework also allows us to directly address the factors influencing the lamellipodium shape, which is the most commonly observed morphology in crawling cells. We are aware of one previous 3D study[22], which recreated the motion of flat cell fragments. Ref. 22 differs from our approach in its aims, as it focuses on a specific type of cell motility, rather than attempting to describe morphological transitions between non-motile and motile cells or between different kinds of motility, as we do here.

Results

Crawling dynamics in quasi-2D. To begin, we consider a quasi-2D geometry describing a thin slice through a crawling droplet (Fig. 1a). At the wall, we assume parallel anchoring of actin polarization (represented by black arrows in the figure), and also a no-slip boundary condition on fluid flow as expected when focal adhesions are plentiful (the effect of a small slip is considered below). In these quasi-2D simulations, actin polarization (hence treadmilling) is confined both to the leading edge of the cell and within a finite thickness λ from the substrate. This is to model real cells where actin filaments are present, and polarized, mainly in a cortex towards the exterior[23]. Importantly, no qualitative differences in the results are observed if we choose instead a simpler representation of treadmilling, where this is no longer localized at the leading edge, but still confined within a thickness λ from the substrate, see Supplementary Fig. 1. At the droplet's interface with the surrounding fluid, we impose outward normal anchoring of variable strength β. The equations of motion are as previously published[7] and detailed in the Methods section.

Figure 1a shows two snapshots of such a quasi-2D droplet in steady state. On increasing the treadmilling rate w_0, we observe progression from a droplet that moves along the wall with only modest distortion to one that extends a thin protrusion at its leading edge (also see Supplementary Movies 1 and 2). In both regimes the droplet moves with a speed close to w_0. Figure 1b plots the steady-state wall contact area against w_0, in the case of no-slip boundary conditions ($s=0$, red/plusses curve) and partial-slip ($s=0.02$, $s=0.04$, $s=0.06$, green/crosses, blue/asterisks and purple/squares curves, respectively). The parameter s controls the amount of slip of the fluid at the solid wall, and is defined such that $s=0$ corresponds to no-slip while $s=1$ corresponds to full-slip (see Supplementary Note 1 for its mathematical definition). The parameter s is thus inversely related to the density of focal adhesions between the substrate and

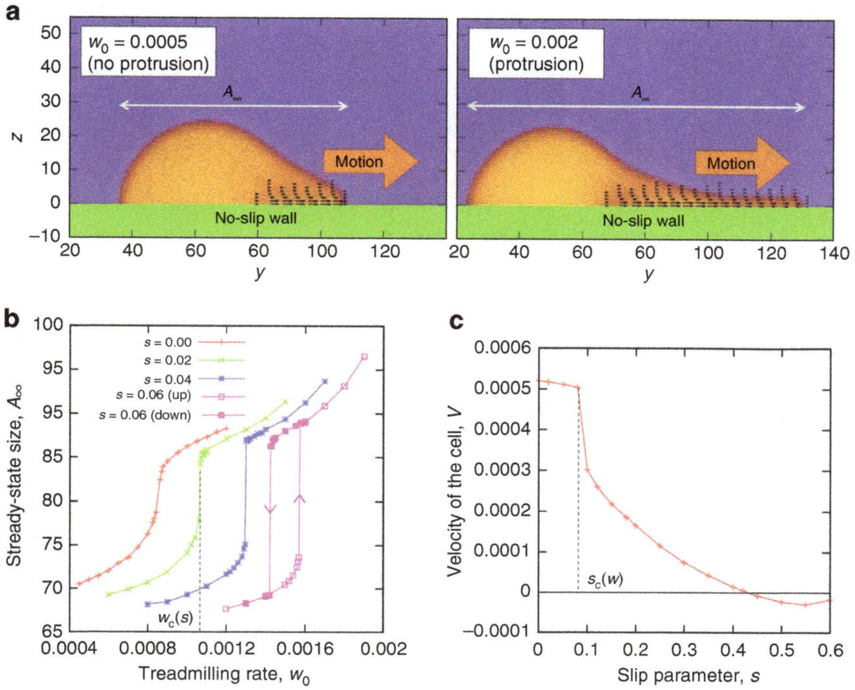

Figure 1 | Quasi-2D crawling dynamics. **(a)** A quasi-2D actomyosin droplet crawling along the treadmilling direction (to the right in the picture). The black arrows here represent actin polarization, which points in the same direction as actin treadmilling. In addition, actin polarization is localized at the leading edge of the droplet to model a crawling eukaryotic cell (also see Supplementary Movies 1 and 2). If the treadmilling rate w_0 is large enough, a thin protrusion form, whose steady-state size (area per unit transverse length) is plotted versus w_0 in **(b)** (red/plus curve), for the case of no-slip boundary condition between the droplet and the substrate. From **(b)**, we can identify the critical treadmilling rate w_c above which a thin layer of protrusion is observed in a crawling cell. This transition becomes sharper as the amount of slip at the substrate increases (green/cross curve) and discontinuous/first order for higher values of s (blue/asterisk and purple/square curves). The hysteresis loop corroborates its discontinuous character. **(c)** A plot of the crawling speed as a function of the slip parameter s (while the treadmilling rate is fixed at $w_0 = 0.002$). This controls the discontinuity in fluid velocity at the solid wall, and is defined such that $s = 0$ corresponds to no-slip while $s = 1$ corresponds to full-slip. An increase of s causes a decrease of the velocity of the cell. This is because the slip velocity at the surface is in the opposite direction to that of the actin polymerization. Again, we observe a discontinuity at $s = s_c$ marking a dynamical phase transition from a round-shaped crawling cell ($s > s_c$) to a protruded shape ($s < s_c$).

the cell. In both cases, with or without wall slip, the contact area increases sharply near a critical value $w_c(s)$, suggestive of a dynamical transition or sharp crossover between two regimes. This transition becomes sharper as the amount of slip on the substrate augments (green/crosses curve), and discontinuous at critical w_c for higher values of s (blue/asterisks and purple/squares curves). The strong hysteresis of the last curve confirms its discontinuous nature. Note that in these quasi-2D simulations, we have set actomyosin contractility to zero. This suggests that actomyosin contraction is not essential to the crawling mechanism; indeed, including ζ leads to almost unchanged results (see Supplementary Fig. 1). Hence the protruded shape observed in crawling cells can indeed be achieved by just a balance of actin treadmilling against interfacial tension at the cell membrane. Finally in Fig. 1c, the crawling speed of our droplet (in steady state) is plotted against the slip parameter s while the treadmilling rate is fixed. As expected, increasing the amount of slip strongly reduces droplet motility via the crawling mechanism. In addition, we still observe a discontinuous transition at critical slip s_c above which no protrusion is observed any longer. This concurs with a previous comparison of crawling with contractility-driven 'cell swimming', in which the wall friction was crudely represented by a depth-integrated drag term[7].

Crawling dynamics in 3D. The quasi-2D simulations in Fig. 1 give intriguing hints to the physics of lamellipodial protrusions, which are the pancake-like structures often seen at the leading edge of crawling cells. However, the structure of such cells is fully three-dimensional, and the lamellipodium is only one of several morphologies attained by live cells, whether crawling or stationary, when attached to a solid support. We have therefore performed full 3D simulations of a slight variation of the same minimal model as we used in 2D. With respect to the 2D case, in the 3D case the effect of actomyosin contractility is much stronger, so we present results only with it switched on from the beginning. Unlike actin treadmilling, actomyosin contractility is known to show highest activity deep inside the cell, close to the cell nucleus, rather than at the cell periphery[23]. To simplify the model in 3D, actin polarization (and thus contractility) is therefore now assumed to be uniform throughout the droplet. In addition, actin polarization is no longer confined to the leading edge of the cell, but we retain a spatially varying treadmilling rate, $w_0 \rightarrow w(z) = w_0 \exp(-z/\lambda)$, to confine actin treadmilling close to the substrate. The use of this simplified model, where the active droplet is homogeneously polarized, is motivated by our finding above that the localization of actin polymerization is not essential to observe the morphological transition between non-motile and motile cells in Fig. 1 (see Supplementary Fig. 1).

As we shall see, our minimal 3D model supports the formation of various dynamical cell shapes, some with surprisingly direct experimental counterparts. Figure 2 shows a selection of these steady-state morphologies, together with a direct comparison with experimental morphologies taken from the literature.

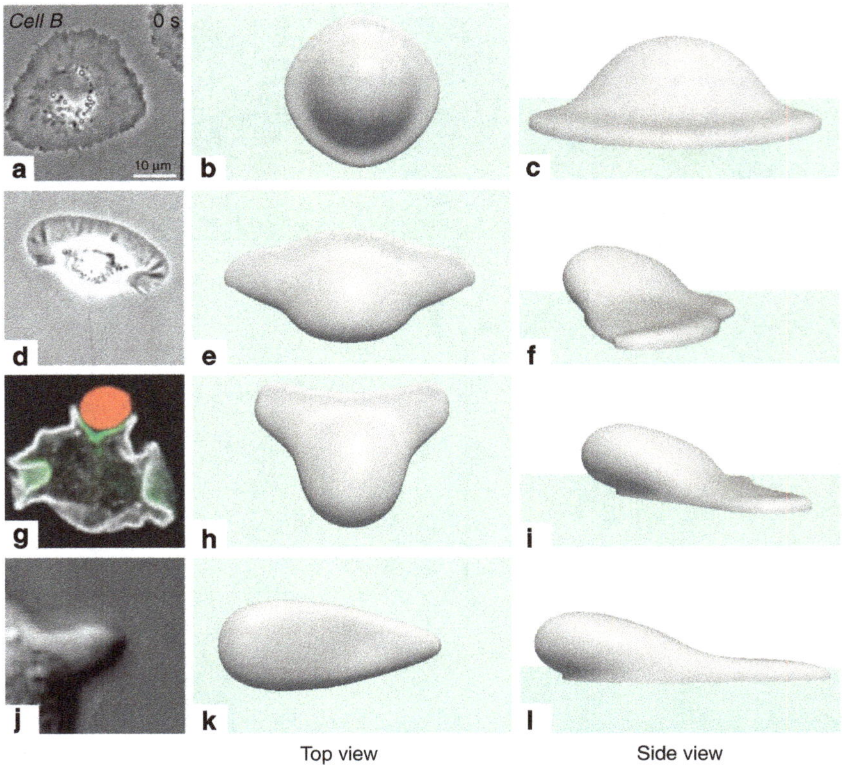

Figure 2 | Spreading and crawling cell morphologies. In this figure, spreading and crawling cell morphologies observed in past experiments (left column) and obtained in the current 3D simulations (middle and right column, giving, respectively, top and side view) are shown. Panels **a–c** show a static cell with a fried-egg shape; the left column panel was originally published in ref. 25. Other frames show moving cells with protrusions resembling, respectively, a lamellipodium (**d–f**; the left column panel was originally published in ref. 27), a phagocytotic cup (**g–i**; left column panel taken from ref. 30 with permission), and a pseudopod (**j–l**; left column panel taken from ref. 32 with permission). Parameters for the 3D modelling (in simulation units; see Supplementary Table 1) are: (**a–c**) $w_0 = 0.035$; $\zeta = -0.001$; $\beta = 0.02$; (**d–f**) $w_0 = 0.04$; $\zeta = -0.0015$; $\beta = 0.001$; (**g–i**) $w_0 = 0.04$; $\zeta = 0$; $\beta = 0.001$; (**j–l**) $w_0 = 0.04$; $\zeta = -0.00015$; $\beta = 0$.

Figure 2a–c shows a droplet subject to strong normal anchoring of the polarity (large β). This boundary condition, which might model the promoted nucleation of outwardly polarized actin by Arp2/3 complexes close to the outer membrane[2], enforces radial polarization with an 'aster' defect at the centre[24]. Treadmilling then causes the droplet to spread, but not to move. In steady state it adopts a symmetric 'fried egg' whose shape resembles that of an inactive platelet[1] or a stationary keratocyte[25]. (We speculate that adding boundary pinning of the cell perimeter, and/or radial modulation of the treadmilling strength w_0, might create shapes more closely resembling mesenchymal stem cells[26].)

Figure 2d–l shows steady-state cell shapes for (rightwards) crawling cells; the morphology is selected by a combination of treadmilling rate w_0, contractility $-\zeta$ and anchoring strength β. Figure 2d–f (see also Supplementary Movie 3) arises when all three effects compete; this closely resembles a motile keratocyte (see for example refs 2,27). The pronounced lamellipodium at the leading edge stems from a balance between contractility-induced splay of the polarity field[7,8] and its soft anchoring at the leading edge. The prediction of this familiar dynamical structure from such minimal physical ingredients is a major finding of this paper (In contrast, the overhang at the cell rear is a detail, and can be eliminated if actin polarization or treadmilling rate $w(x, y, z)$ is set to zero at the back of the cell, as done for example, in Fig. 1a.)

The cell shape in Fig. 2g–i arises in the absence of contractility: the splay is induced by the anchoring and the structure corresponds to an indented structure possibly reminiscent of the 'phagocytic cup', which forms when a cell needs to engulf a solid particle[28]. (Such cups are typically not planar but have been reported for crawling as well as free cells[29,30].) Finally, Fig. 2j–l shows a droplet where treadmilling is dominant: the anchoring is zero and the contractility is much weaker than in Fig. 2d–f. Importantly, without those contributions the cell shape no longer shows the keratocyte-like lamellipodium morphology. Instead, the leading-edge forms a pointed protrusion, qualitatively resembling a 'pseudopod', as often reported in ameoboid cell motion both on surfaces and in 3D geometries[31,32].

Phase diagram. The results shown in Fig. 2 were obtained by varying the treadmilling parameter w_0, the contractile activity $-\zeta$, and also the anchoring β, which controls the orientation of the polarization at the cell boundary. Although a systematic study of the phase diagram in this multi-dimensional parameter space is extremely demanding computationally, and hence outside the scope of the current work, we observe (see Fig. 3) that similar morphologies can also be obtained by keeping the value of w_0 constant, varying β and adjusting ζ to be as large as possible avoiding at the same time droplet breakup (which occurs when ζ is too large; see also ref. 33 where a distinct but related study of contractility-driven breakup 2D active droplets is presented).

For very small values of β, a finger-like structure (the 'pseudopodium' in Fig. 2) is dominant, while fan-shaped geometries (such as the 'lamellipodium' in Fig. 2) emerge when β is larger, due to the higher splay distortion of the actin filaments arising at the leading edge of the cell. For very strong anchoring, the lamellipodium morphology becomes unstable, and the cell attains a fried-egg shape, which is non-motile. Experiments (see for example, ref. 27) have studied how cell shape is affected by adhesion to the substrate. In our model an increase in the adhesion would affect both the treadmilling and the anchoring, decreasing the first and increasing the second. Interestingly, our simulations suggest that lamellipodia are found for intermediate values of the anchoring strength and are disfavoured when treadmilling decreases, qualitatively confirming what has been experimentally observed by Barnhart *et al.*[27]

Evolution of polarization and velocity fields. Besides the concentration field whose structure is shown in Fig. 2, our model also tracks the evolution of the polarization and intracellular flow fields, **P** and **u**, respectively.

In Fig. 4 we show the polarization field close to the substrate (left column) and the flow field (right column) corresponding to the 3D steady-state morphologies reported in Fig. 2. The polarization field shows a transition between the aster corresponding to the 'fried-egg' state (Fig. 4a) and polarized uniform (Fig. 4d) and splayed (Fig. 4b,c) pattern; these are qualitatively in broad agreement with the experimental data for keratocytes

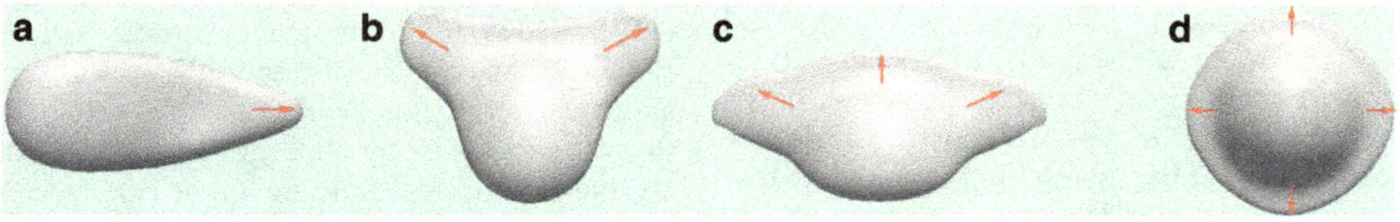

Figure 3 | Phase diagram. 3D steady-state shapes for (**a**) $\beta=0$, $\zeta=-0.00015$, $w_0=0.04$; (**b**) $\beta=0.001$, $\zeta=-0.0001$, $w_0=0.04$; (**c**) $\beta=0.001$, $\zeta=-0.0015$, $w_0=0.04$; (**d**) $\beta=0.02$, $\zeta=-0.001$, $w_0=0.04$. Red arrows indicate the direction of the polarization in the cytoplasmic projections (psedudopod (**a**), phagocytinc cup (**b**), lamellipodium (**c**) and fried-egg crown (**d**)). Notice that w_0 is constant; while ζ varies (both ζ and β affect droplet elasticity, increasing ζ for small β leads to further instabilities such as droplet breakup, which are not relevant for the current work).

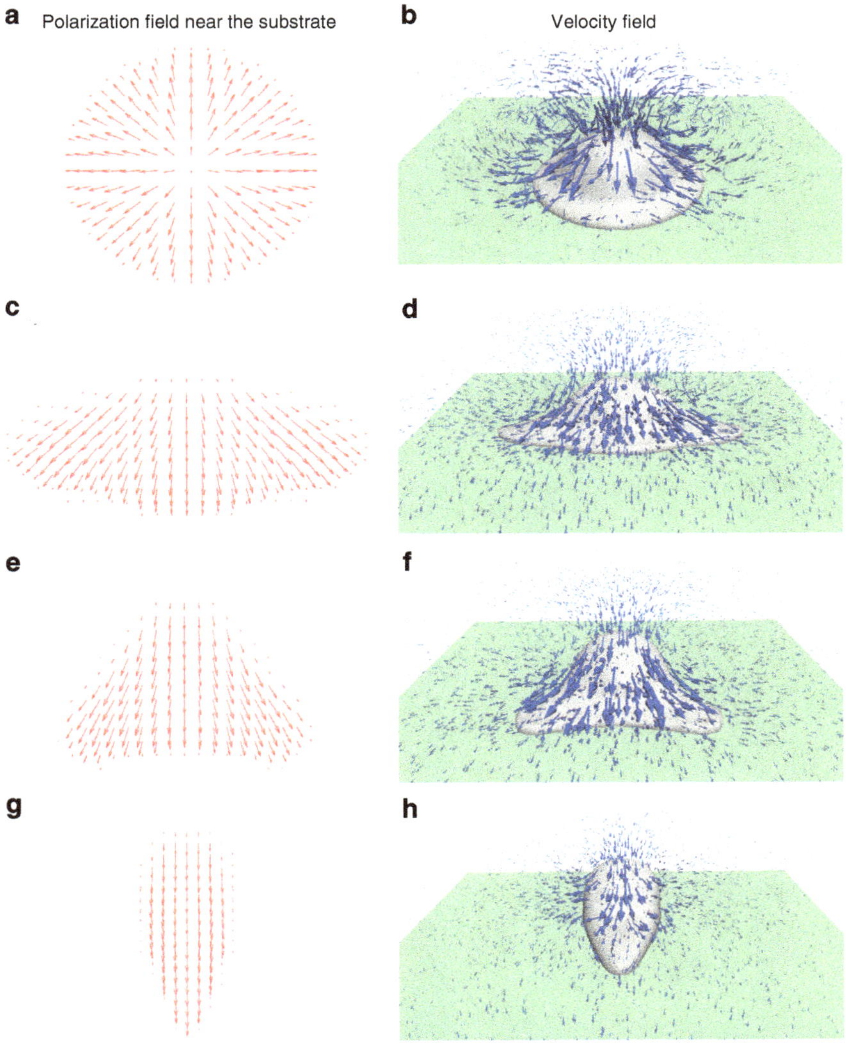

Figure 4 | Polarization and velocity fields. This figure shows the section of the polarization field inside the droplet very close to the substrate (left column) and the velocity field (right column) for (**a,b**) 'fried egg' morphology, (**c,d**) lamellipodium, (**e,f**) phagocytic cup, (**g,h**) pseudopodium, respectively.

presented in ref. 25, where it was established that motility requires symmetry breaking of the actin fibre orientation field, from an aster like conformation as in our 'fried-egg' morphology (Fig. 4a) to an asymmetric conformation in a lamellipodium, where the barbed ends of actin fibres are approximately normal to the cell membrane at the leading edge (as in our simulations in Fig. 4c).

Coming to the flow field, we first note that **u** (plotted in Fig. 4) should be thought of as an intracellular solvent flow inside the cell, which has not been measured often in experiments (an exception is ref. 34 which is discussed in the Supplementary Note 2). This intracellular flow is quite distinct from the flow of F-actin, which is usually tracked in cell biology experiments, and which is commonly referred to as 'retrograde flow'[13]. F-actin flows inwards within the cell, from the leading edge to the bulk. The velocity field for the 'fried-egg' cell is characterized by a quadrupolar velocity profile (in a fully 3D view), more intense close to the surface of the droplet where the curvature is higher than near the substrate (see Fig. 4a). The flow in the vicinity of the membrane is not unlike that measured in the case of an activity-driven motile tumour cell 'swimming' within a 3D matrigel[5], and even closer correspondence is found between that experiment and our contractility-driven droplets (which move in the absence of treadmilling, see Fig. 5). The symmetric structure of the flow field in the 'fried-egg' morphology explains its non-motile character. The flow field for the keratocyte, lamellipodium-like shape and for the phagocytic cup (Fig. 4b,c) is instead unidirectional (along the direction of motion) in the laboratory frame of reference, and gently follows the fan-out structure of the surface cell profile, being more intense close to both cytoplasmic projections. A similar structure is seen for the 'pseudopodium' (Fig. 4d), with the difference that the flow is now almost entirely localized around the finger-like protrusion. Interestingly, the highest magnitude of the fluid flow for the last three cases is twice larger than the one for the fried-egg cell.

The flow fields of these structures in are consistent with previous computer simulations[35] and with the flow fields mapped experimentally in ref. 34 with decreased contractility compared with the wild type (see also Supplementary Fig. 2). Experiments suggest that contractility can propel intracellular flow significantly, and 2D simulations with variable ζ in our model confirm this qualitatively (see Supplementary Fig. 3 and Supplementary Note 2 for further discussion).

Oscillatory dynamics and medium-dependent viscosity. Not all parameters in our model lead to steady-state shapes; we also find spontaneous oscillations in shape and motility (see Supplementary Fig. 4, Supplementary Note 3 and Supplementary Movie 4). These or similar oscillations have long been reported experimentally[36], and it is remarkable that they can emerge in principle from a model that is devoid of all biochemical feedbacks.

All our simulations address the case where the active droplet is surrounded by a fluid of similar viscosity to its interior. This is an adequate model of wall-bound cells moving through an aqueous medium as is normally the case *in vitro* and *in vivo*, where cells are surrounded by at least a thin film of fluid. Accounting for the higher viscosity of the membrane also leads to similar results; we have performed selective numerical checks of this (Supplementary Fig. 5).

Crawling in the absence of treadmilling. Motility of the crawling droplets in Fig. 2d–l is powered by treadmilling, although contractility was found to play a crucial role in determining cell shape. What happens if treadmilling is absent altogether? It was shown in ref. 7 that when suspended in bulk fluid a contractile droplet can 'swim' by undergoing spontaneous splay. This creates an internal toroidal fluid flow that entrains the external fluid and leads to self-propulsion. Our simulations show that the same physical mechanism is still available for a fluid-immersed droplet attached to a wall (Fig. 4). The flow associated with such contractility-driven active droplets is similar to that observed experimentally in cells 'swimming' in 3D matrigel (Fig. 5).

Discussion

In conclusion, we present a minimal physical model of cell crawling, based on an active fluid droplet, attached to a solid wall and immersed in a second fluid. Key ingredients in the model are actin treadmilling (confined to a thin layer of depth λ close to the substrate), actomyosin contractility (throughout the droplet), full anchoring of the actin polarity parallel to the wall and its partial anchoring normal to the fluid–fluid interface. Hydrodynamic fluid flow is properly treated throughout. We showed that on increasing the treadmilling rate, such a droplet undergoes a nonequilibrium morphological transition to form a protrusion layer whose thickness is set by λ (Fig. 6). Variation of the treadmilling, contractility and anchoring parameters leads in 3D to a striking series of cell morphologies, resembling structures seen in real eukaryotic cells including both lamellipodia and pseudopods. We also find regimes where oscillatory modulations of shape and motility arise. Our results show that surprisingly many of the observed features of motile and spreading cells (though of course not all of them) can emerge from relatively simple physical ingredients. This suggests that some aspects of cellular motility might fruitfully be thought of as driven by a physics-based 'motility engine' whose function, although controlled by the cell's complex biochemical feedback networks, does not directly depend on these for its basic operational principles. It

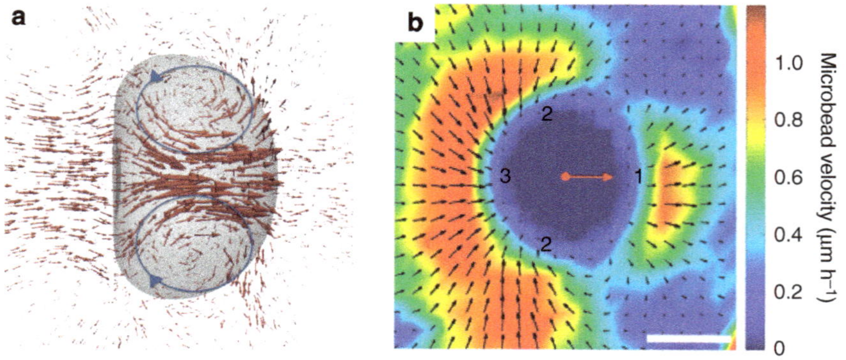

Figure 5 | Flow field of a contractile active droplet. The figure shows the flow field (in the laboratory frame of reference) for a contractility-driven motile active droplet (panel **a**, from ref. 7; for simplicity in the bulk), and for a cell moving in 3D inside matrigel (panel **b**, from ref. 5, with permission).

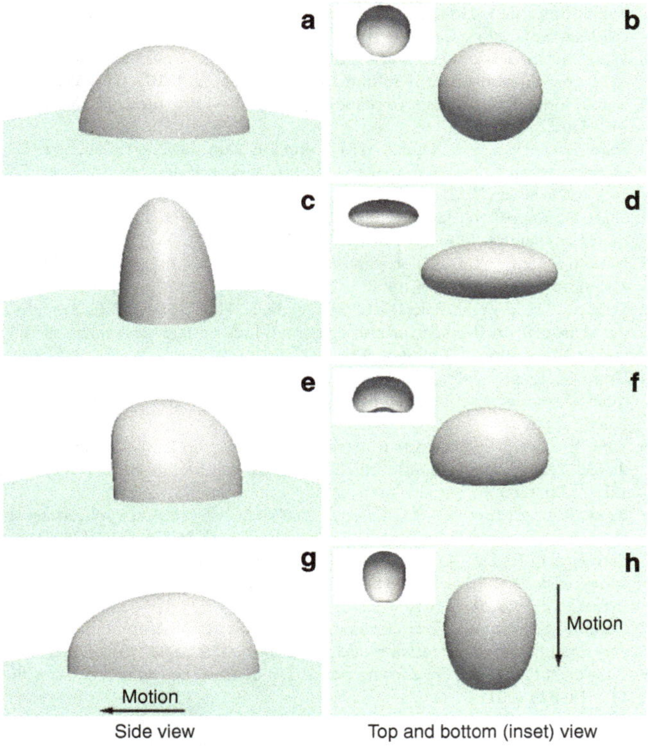

Figure 6 | 3D dynamics in the absence of treadmilling. Snapshots corresponding to the motion of a contractile cell on a substrate, in the absence of treadmilling (parameters were $w=0$, $\zeta=-0.005$, $\beta=0$; **a**,**c**,**e**,**g** show side views, while **b**,**d**,**f**,**h** show top and bottom views of the droplets at different times). The active droplet first elongates perpendicular to the wall due to the active stress; the polarization then splays in 3D and this leads to internal flow and a motile concave shape, resembling half of a free-swimming 3D droplet as investigated in ref. 7.

would be of interest to extend the current model to study cell motility inside complex media, such as the extracellular matrix, or within tissues[37].

Methods

Basic idea of the model. We briefly outline here the hydrodynamic model used in this work. We consider a droplet of cellular matter sitting on a solid substrate as a simple representation of a crawling cell. The droplet phase is represented by a scalar field $\phi(\mathbf{r}, t)>0$. Outside the droplet, we have a passive isotropic fluid with $\phi(\mathbf{r}, t)=0$ to represent the wet environment around the cell and the substrate. Second, we define a polarization field $\mathbf{P}(\mathbf{r}, t)$ to represent the actin filaments, which are polarized and abundant mainly at the cortex of the cell. More specifically, $\mathbf{P}$ is the coarse grained average of all the orientations of the actin filaments: $\mathbf{P}=\langle\mathbf{p}\rangle$ where $\mathbf{p}$ is a unit tangent oriented from the end at which the actin filaments depolymerize to the opposite end at which they polymerize. Finally, we also have the average velocity field $\mathbf{v}(\mathbf{r}, t)$ of both the cellular matter and the solvent.

Underlying free energy. Following the phenomenological approach used in refs 7,38, we first introduce a free-energy functional that controls the equilibrium physics of a polar liquid-crystalline passive droplet

$$F[\phi,\mathbf{P}]=\int d^3r\{\frac{a}{4\phi_{\rm cr}^4}\phi^2(\phi-\phi_0)^2+\frac{k}{2}|\nabla\phi|^2 -\frac{\alpha}{2}\frac{(\phi t(\mathbf{r})-\phi_{\rm cr})}{\phi_{\rm cr}}|\mathbf{P}|^2+\frac{\alpha}{4}|\mathbf{P}|^4+\frac{\kappa}{2}(\nabla\mathbf{P})^2 +\beta\mathbf{P}\cdot\nabla\phi\}. \quad (1)$$

The first two terms of equation (1) control bulk and interfacial properties of the droplet. In particular, $k>0$ determines the droplet interfacial tension and $a>0$ is a phenomenological constant. The bulk term is chosen to stabilize a droplet phase ($\phi\simeq\phi_0$) immersed in a second fluid ($\phi=0$) and $\phi_{\rm cr}=\phi_0/2$. The parameter $t(\mathbf{r})$ controls a second order transition from isotropic ($|\mathbf{P}|=0$) to polar phase ($|\mathbf{P}|>0$) inside the droplet as the value of t changes from 0 to 1. Note that for all our 3D work, and for selected cases in 2D (see Supplementary Fig. 1), we have taken $t(\mathbf{r})=1$ everywhere for simplicity; we discuss some implications of this choice below. The functional form of $t(\mathbf{r})$ for the remaining quasi-2D simulations is:

$$t(y,z)=\frac{1}{4}(1+\tanh(y-y_{back}-\Lambda))(1-\tanh(z-\lambda)), \quad (2)$$

where z is the vertical distance from the substrate, y is the coordinate parallel to the substrate and y_{back} is the position of the cell rear. Basically, the value of t is equal to 1 when z is less than some thickness λ from the substrate and at least a distance Λ from the cell rear. This defines the polarized (hence active) region at the leading edge of the droplet where actin treadmilling takes place (see black arrows in Fig. 1a). Incidentally, λ also sets the lengthscale for the thickness of the protrusion layer. Finally, elsewhere in the droplet t is equal to 0, which gives rise to an isotropic phase (and hence no actin treadmilling).

In summary, there are three equilibrium phases that can be obtained by minimizing the free energy $F[\phi, \mathbf{P}]$ in a state of uniform ϕ and $\mathbf{P}$, setting $\frac{\delta F}{\delta\phi}=0$ and $\frac{\delta F}{\delta\mathbf{P}}=0$. These three phases are as follows: the isotropic passive fluid, external to the droplet ($\phi=0$ and $\mathbf{P}=0$); the polarized region of the droplet, corresponding to the cortical actin–myosin network ($\phi=\phi_{\rm eq1}$ and $\mathbf{P}=\mathbf{P}_{\rm eq}$); and the isotropic region of the droplet, corresponding to the cell nucleus and other cellular materials, which do not play a role in cell motility ($\phi=\phi_{\rm eq2}$ and $\mathbf{P}=0$).

The other three terms in the free energy describe bulk and elastic properties of the polar field, where α is a positive (for stability) constant and κ is the elastic constant (in the one-elastic constant approximation[39]) that describes the energetic cost due to elastic deformation in a liquid-crystalline environment. The surface tension γ will depend on a, k and on the elastic constant κ. Across the droplet interface, the values of the concentration and the polarization fields vary smoothly from $\phi=\phi_{\rm eq2}$ and $\mathbf{P}=\mathbf{P}_{\rm eq}$ to $\phi=0$ and $\mathbf{P}=0$. In general, the width of the diffuse interface also depends on the same parameters. In our simulations, the interfacial width is typically much smaller than the size of the droplet. Finally the last term $\beta\mathbf{P}\cdot\nabla\phi$ takes into account the anchoring of $\mathbf{P}$ to the droplet surface (so that for $\beta>0$ the polar field $\mathbf{P}$ preferentially points perpendicularly outwards at the droplet perimeter, while for $\beta<0$ it preferentially points perpedicularly inwards).

Equations of motion. The velocity field $\mathbf{v}(\mathbf{r}, t)$ obeys the continuity and the Navier-Stokes equations, which in the incompressible limit are

$$\nabla\cdot\mathbf{v}=0, \quad (3)$$

$$\rho\left(\frac{\partial}{\partial t}+\mathbf{v}\cdot\nabla\right)\mathbf{v}=-\nabla P+\nabla\cdot(\underline{\underline{\sigma}}^{\rm active}+\underline{\underline{\sigma}}^{\rm passive}). \quad (4)$$

Here P is the isotropic pressure and $\underline{\underline{\sigma}}^{\rm active}+\underline{\underline{\sigma}}^{\rm passive}$ is the total stress, given by the sum of active and passive contributions. The first of them is

$$\sigma_{\alpha\beta}^{\rm active}=-\zeta\phi P_\alpha P_\beta, \quad (5)$$

that can be derived by summing the contributions from each force dipole and coarse graining[38]. The constant ζ is the activity parameter that is negative for contractile particles and positive for extensile particles. Its magnitude $|\zeta|$ is proportional to the strength of the force dipole. For the case of actomyosin networks, ζ is negative signifying the tendency to contract along the direction of actin filaments. The strength of actomyosin contraction is thus given by its magnitude $|\zeta|$. As we have seen from Fig. 2, actomyosin contractility is responsible for the formation of a variety of complex 3D structures in crawling cells. Also note that in our calculations above, we neglect the contributions of higher order terms (such as $\partial_\alpha P_\beta$) in the active stress, although these are in principle allowed by symmetry[40]. This assumption is appropriate for actomyosin solutions, which are made up by elongated fibres of uniform width whose degree of asymmetry between the particle's head and tail is supposed to be small.

We note that actomyosin contraction shows highest activity at the cell rear[23] rather than at the leading edge. This is in contrast to actin treadmilling that shows highest activity at the leading edge of the cell. Thus, to study the global effect of actomyosin contractility in 3D, we make a further simplification by assuming actin polarization to be homogenous and uniform inside the droplet, which amounts to setting $t(\mathbf{r})=1$ throughout. As actin polarization is now no longer localized at the leading edge, we instead localize the treadmilling parameter w to be large near the substrate to produce a protrusion as in the quasi-2D systems. While these approximations are useful to make progress in 3D, as the numerics is more demanding there, we note that the protrusion transition found in Fig. 1 is still preserved, both in 3D and in 2D when similar approximations are made (Supplementary Fig. 1).

The passive stress consists of a viscous term, written as $\sigma_{\alpha\beta}^{\rm viscous}=\eta(\partial_\alpha v_\beta+\partial_\beta v_\alpha)$, where η is the shear viscosity (the greek indices denote cartesian components), summed with an elastic stress $\underline{\underline{\sigma}}^{elastic}$, due to elastic deformations in the liquid crystal phase, and a surface tension term $\underline{\underline{\sigma}}^{\rm interface}$. The elastic stress is given by

$$\sigma_{\alpha\beta}^{\rm elastic}=\frac{1}{2}(P_\alpha h_\beta-P_\beta h_\alpha)-\frac{\xi}{2}(P_\alpha h_\beta+P_\beta h_\alpha)-\kappa\partial_\alpha P_\gamma\partial_\beta P_\gamma, \quad (6)$$

similar to that used in nematic liquid crystal dynamics. Here the constant ξ depends on the geometry of the active particles: $\xi>0$ describes rod-like particles,

$\xi < 0$ oblate ones. We have set it positive, as in ref. 7. In addition ξ controls whether particles are flow aligning in shear flow ($|\xi| > 1$), hence creating a stable response, or flow tumbling ($|\xi| < 1$), which gives an unstable response. Here we assume $\xi > 1$. Note that the elastic stress also depends on the molecular field h, defined as $\delta F/\delta \mathbf{P}$. The surface tension term takes into account interfacial contribution between the passive and the active phase, and is given by

$$\sigma_{\alpha\beta}^{\text{interface}} = \left(f - \phi\frac{\delta F}{\delta\phi}\right)\delta_{\alpha\beta} - \frac{\partial f}{\partial(\partial_\beta\phi)}\partial_\alpha\phi. \quad (7)$$

This term is borrowed from binary fluids[41], with f being the free-energy density.

The dynamics of the concentration of active material $\phi(\mathbf{r}, t)$ is governed by a convective-diffusion equation

$$\frac{\partial\phi}{\partial t} + \nabla\cdot(\phi(\mathbf{v} + w_0\mathbf{P})) = \nabla\left(M\nabla\frac{\delta F}{\delta\phi}\right), \quad (8)$$

where M is a thermodynamic mobility parameter, related to the diffusion constant $D \simeq Ma$ (see equation (1)). $\delta F/\delta\phi$ is the chemical potential of the system and w_0 is the self-advection parameter. This w_0 can be interpreted as the speed, relative to the bulk fluid velocity $\mathbf{v}(\mathbf{r}, t)$, at which each filament is self propelled along its own tangent[7]. Thus w_0 is also proportional to the rate of actin treadmilling. In our 3D minimal model, the treadmilling rate is a spatially varying function that is large close to the substrate: $w_0 \to w(\mathbf{r}) = w_0 \exp(-z/\lambda)$ where λ is the lengthscale for protrusion thickness.

The last ingredient to complete the description of the physics of the system is an evolution equation for the polarization field $\mathbf{P}(\mathbf{r}, t)$. This is given by

$$\frac{\partial\mathbf{P}}{\partial t} + ((\mathbf{v} + w_0\mathbf{P})\cdot\nabla)\mathbf{P} = -\underline{\underline{\Omega}}\cdot\mathbf{P} + \xi\underline{\underline{D}}\cdot\mathbf{P} - \frac{1}{\Gamma}\frac{\delta F}{\delta\mathbf{P}}, \quad (9)$$

where Γ is the rotational viscosity and $\underline{\underline{D}} = (\underline{\underline{W}} + \underline{\underline{W}}^T)/2$ and $\underline{\underline{\Omega}} = (\underline{\underline{W}} - \underline{\underline{W}}^T)/2$ are the symmetric and the anti-symmetric part of the velocity gradient tensor $W_{\alpha\beta} = \partial_\alpha v_\beta$. Note that this equation is the one usually employed in polar liquid crystal theory[42].

To solve these equations we use a 3D hybrid lattice Boltzmann algorithm already successfully tested in other systems, such as binary fluids[43], liquid crystals[44] and active matter[7]. It consists in solving equations (8) and (9) via a finite difference predictor-corrector algorithm while the integration of the Navier-Stokes equation is taken care of by a standard Lattice Boltzmann approach. This numerical approach is necessary due to the couplings between velocity field and polarization field in equation (9) and velocity field and concentration in equation (8).

References

1. Alberts, B. *et al. Molecular Biology of the Cell,* 4th edn (Garland Science, 2002).
2. Bray, D. *Cell Movements: From Molecules to Motility*. 2nd edn (Garland Science, 2000).
3. Köhler, S., Schaller, V. & Bausch, A. R. Structure formation in active networks. *Nat. Mater.* **10,** 462–468 (2011).
4. Loisel, T., Boujemaa, R., Pantaloni, D. & Carlier, M.-F. Reconstitution of actin-based movement using pure proteins. *Nature* **401,** 613–616 (1999).
5. Poincloux, R. *et al.* Contractility of the cell rear drives invasion of breast tumor cells in 3D Matrigel. *Proc. Natl Acad. Sci. USA* **108,** 1943–1948 (2011).
6. Hawkins, R. J., Poincloux, R., Benichou, O., Piel, M. & Voituriez, R. Spontaneous contractility-mediated cortical flows generates cell migration in three-dimensional environments. *Biophys. J.* **101,** 1041–1045 (2011).
7. Tjhung, E., Marenduzzo, D. & Cates, M. E. Spontaneous symmetry breaking in active droplets provides a generic route to motility. *Proc. Natl Acad. Sci. USA* **109,** 12381–12386 (2012).
8. Simha, R. J. & Ramaswamy, S. Hydrodynamic fluctuations and instabilities in ordered suspensions of self-propelled particles. *Phys. Rev. Lett.* **85,** 058101 (2002).
9. deMali, K. A., Wennerberg, K. & Burridge, K. Integrin signaling to the actin cytoskeleton. *Curr. Opin. Cell Biol.* **15,** 572–589 (2003).
10. Cameron, A. R., Frith, J. E. & Cooper-White, J. J. The influence of substrate creep on mesenchymal stem cell behaviour and phenotype. *Biomaterials* **32,** 5879–5903 (2011).
11. Verkhovsky, A. B., Svitkina, T. M. & Borisy, G. G. Self-polarization and directional motility of cytoplasm. *Curr. Biol.* **9,** 11–20 (1999).
12. Joanny, J.-F. & Ramswamy, S. A drop of active matter. *J. Fluid Mech.* **75,** 46–57 (2012).
13. Vallotton, P., Danuser, G., Bohnet, S., Meister, J. J. & Verkhovsky, A. Tracking retrogade flow in keratocytes: news from the front. *Mol. Biol. Cell* **16,** 1223 (2005).
14. Ziebert, F., Swaminathan, S. & Aranson, I. S. Model for self-polarization and motility of keratocyte fragments. *J. R. Soc. Interface* **9,** 1084–1092 (2011).
15. Ziebert, F. & Aranson, I. S. Effects of adhesion dynamics and substrate compliance, on the shape and motility of crawling cells. *PLoS ONE* **8,** e64511 (2013).
16. Blanch-Mercader, C. & Casademunt, J. Spontaneous motility of actin lamellar fragments. *Phys. Rev. Lett.* **110,** 078102 (2013).
17. Callan-Jones, A. C., Joanny, J. F. & Prost, J. Viscous-fingering-like instability of cell fragments. *Phys. Rev. Lett.* **100,** 258106 (2008).
18. Wolgemuth, C. W., Stajic, J. & Mogilner, A. Redundant mechanisms for stable cell locomotion revealed by minimal models. *Biophys. J.* **101,** 545–553 (2011).
19. Keren, K. *et al.* Mechanism of shape determination in motile cells. *Nature* **453,** 475–480 (2008).
20. Shao, D., Levine, F. & Rappel, W.-J. Coupling actin flow, adhesion, and morphology in a computational cell motility model. *Proc. Natl Acad. Sci. USA* **109,** 6851–6856 (2012).
21. Shao, D., Rappel, W.-J. & Levine, H. Computational model for cell morphodynamics. *Phys. Rev. Lett.* **105,** 108104 (2010).
22. Herant, M. & Dembo, M. Form and function in cell motility: from fibroblasts to keratocytes. *Biophys. J.* **98,** 1408–1417 (2010).
23. Svitkina, T. M., Verkhovsky, A. B., McQuade, K. M. & Borisy, G. G. Analysis of the actin-myosin II system in fish epidermal keratocytes: mechanism of cell body translocation. *J. Cell Biol.* **139,** 397–415 (1997).
24. Kruse, K., Joanny, J.-F., Julicher, F., Prost, J. & Sekimoto, K. Asters, vortices and rotating spirals in active gels of polar filaments. *Phys. Rev. Lett.* **92,** 078101 (2004).
25. Yam, P. T. *et al.* Actin-myosin network reorganization breaks symmetry at the cell rear to spontaneously initiate polarized cell motility. *J. Cell Biol.* **178,** 1207–1221 (2007).
26. Barry, F. P. & Murphy, J. M. Mesenchymal stem cells: clinical applications and biological characterization. *Int. J. Biochem. Cell Biol.* **36,** 568–584 (2004).
27. Barnhart, E. L., Lee, K. C., Keren, K., Mogilner, A. & Theriot, J. A. An adhesion-dependent switch between mechanisms that determine motile cell shape. *PLoS Biol.* **9,** e1001059 (2011).
28. Herant, M., Heinrich, V. & Dembo, M. Mechanics of neutrophil phagocytosis: experiments and quantitative models. *J. Cell Sci.* **119,** 1903–1913 (2006).
29. Tuxworth, R. I. *et al.* A role for myosin VII in dynamic cell adhesion. *Curr. Biol.* **11,** 318–329 (2011).
30. Mercanti, V., Charette, S. J., Bennett, N., Ryckewaert, J.-J. & Letourner, F. Selective membrane exclusion in phagocytic and macropinocytic cups. *J. Cell Sci.* **119,** 4079–4087 (2006).
31. Friedl, P., Borgmann, S. & Brocker, E. B. Amoeboid leukocyte crawling through extracellular matrix: lessons from the Dictostelium paradigm of cell movement. *J. Leukoc. Biol.* **70,** 491–509 (2001).
32. Zhang, H. *et al.* Phosphorylation of the myosin regulatory light chain plays a role in motility and polarity during Dictyostelium chemotaxis. *J. Cell Sci.* **115,** 1733–1747 (2002).
33. Giomi, L. & De Simone, A. Spontaneous division and motility in active nematic droplets. *Phys. Rev. Lett.* **112,** 147802 (2014).
34. Keren, K., Yam, P. T., Kinkhabwala, A., Mogilner, A. & Theriot, J. A. Intracellular fluid flow in rapidly moving cells. *Nat. Cell Biol.* **11,** 1219 (2009).
35. Rubinstein, B., Fournier, M. F., Jacobson, K., Verkhovsky, A. B. & Mogilner, A. Actin-myosin viscoelastic flow in the keratocyte lamellipod. *Biophys. J.* **97,** 1853–1863 (2009).
36. Ehrengruber, M. U., Deranleau, D. A. & Coates, T. D. Shape oscillations of human neutrophil leukocytes: characterization and relationship to cell motility. *J. Exp. Biol.* **199,** 741–747 (1996).
37. Basan, M., Elgeti, J., Hennezo, E., Rappel, W.-J. & Levine, H. Alignment of cellular motility forces with tissue flow as a mechanism for efficient wound healing. *Proc. Natl Acad. Sci. USA* **110,** 2452–2459 (2013).
38. Hatwalne, Y., Ramaswamy, S., Rao, M. & Simha, R. A. Rheology of active-particle suspensions. *Phys. Rev. Lett.* **92,** 118101 (2007).
39. de Gennes, P.-G. & Prost, J. *The Physics of Liquid Crystals* (Clarendon Press, 1993).
40. Marchetti, M. C. & Liverpool, T. B. *Hydrodynamics and Rheology of Active Polar Filaments, Cell Motility* (ed. Lenz, P.) (Springer-Verlag, 2008).
41. de Groot, S. R. & Mazur, P. *Non-Equilibrium Thermodynamics* (Dover Publications, 2011).
42. Kung, W., Marchetti, M. C. & Saunders, K. Hydrodynamics of polar liquid crystals. *Phys. Rev. E* **73,** 031708 (2006).
43. Tiribocchi, A., Stella, N., Gonnella, G. & Lamura, A. Hybrid lattice Boltzmann method for binary fluid mixtures. *Phys. Rev. E* **80,** 026701 (2009).
44. Cates, M. E., Henrich, O., Marenduzzo, D. & Stratford, K. Lattice Boltzmann simulations of liquid crystalline fluids: active gels and blue phases. *Soft Matter* **5,** 37913800 (2009).

Acknowledgements

This work is funded in part by EPSRC Grant EP/J007404/1. We thank R.J. Hawkins for very helpful discussions, and K. Stratford for help with coding. E.T. thanks SUPA for a prize studentship. M.E.C. holds a Royal Society Research Professorship.

Author contributions

All authors designed and performed the research and analysed the data.

Additional information

Competing financial interests: The authors declare no competing financial interests.

2

Topological phase transitions and chiral inelastic transport induced by the squeezing of light

Vittorio Peano[1], Martin Houde[2], Christian Brendel[1], Florian Marquardt[1,3] & Aashish A. Clerk[2]

There is enormous interest in engineering topological photonic systems. Despite intense activity, most works on topological photonic states (and more generally bosonic states) amount in the end to replicating a well-known fermionic single-particle Hamiltonian. Here we show how the squeezing of light can lead to the formation of qualitatively new kinds of topological states. Such states are characterized by non-trivial Chern numbers, and exhibit protected edge modes, which give rise to chiral elastic and inelastic photon transport. These topological bosonic states are not equivalent to their fermionic (topological superconductor) counterparts and, in addition, cannot be mapped by a local transformation onto topological states found in particle-conserving models. They thus represent a new type of topological system. We study this physics in detail in the case of a kagome lattice model, and discuss possible realizations using nonlinear photonic crystals or superconducting circuits.

[1]Institute for Theoretical Physics, University of Erlangen-Nürnberg, Staudtstr. 7, 91058 Erlangen, Germany. [2]Department of Physics, McGill University, 3600 rue University, Montreal, Quebec, Canada H3A 2T8. [3]Max Planck Institute for the Science of Light, Günther-Scharowsky-Straße 1/Bau 24, 91058 Erlangen, Germany. Correspondence and requests for materials should be addressed to V.P. (email: Vittorio.Peano@fau.de).

Waves are not only ubiquitous in physics, but the behaviour of linear waves is also known to be very generic, with many features that are independent of the specific physical realization. This has traditionally allowed us to transfer insights gained in one system (for example, sound waves) to other systems (for example, matter waves). That strategy has even been successful for more advanced concepts in the field of wave transport. One important recent example of this kind is the physics of topological wave transport, where waves can propagate along the boundaries of a sample, in a one-way chiral manner that is robust against disorder scattering. While first discovered for electron waves, this phenomenon has by now also been explored for a variety of other waves in a diverse set of systems, including cold atoms[1], photonic systems[2] and more recently phononic systems[3–9].

In the case of topological wave transport, the connection between waves in different physical implementations can actually be so close that the calculations turn out to be the same. In particular, if we are dealing with matter waves moving in a periodic potential, the results do not depend on whether they are bosons or fermions, as long as interactions do not matter. The single-particle wave equation to be solved happens to be exactly the same. This has allowed to envision and realize photonic analogues of quantum-Hall effect[10–18], the spin Hall effect[19–22], Floquet topological insulators[23,24] and even Majorana-like modes[25]. More generally, the well-known classification of electronic band structures based on the dimensionality and certain generalized symmetries[26] directly applies to photonic systems provided that the particle number is conserved. As we now discuss, this simple correspondence will fail in the presence of squeezing.

Consider the most general quadratic Hamiltonian describing photons in a periodic potential in the presence of parametric driving:

$$\hat{H} = \sum_{\mathbf{k},n} \varepsilon_n[\mathbf{k}]\hat{b}^\dagger_{\mathbf{k},n}\hat{b}_{\mathbf{k},n} + \sum_{\mathbf{k},n,n'}\left(\lambda_{nn'}[\mathbf{k}]\hat{b}^\dagger_{\mathbf{k},n}\hat{b}^\dagger_{-\mathbf{k},n'} + \text{h.c.}\right). \quad (1)$$

The first term describes a non-interacting photonic band structure, where $\hat{b}_{\mathbf{k},n}$ annihilates a photon with quasimomentum $\mathbf{k}$ in the n-th band. The remaining two-mode squeezing terms are induced by parametric driving and do not conserve the excitation number. As we discuss below, they can be controllably realized in a number of different photonic settings. While superficially similar to pairing terms in a superconductor, these two-mode squeezing terms have a profoundly different effect in a bosonic system, as there is no limit to the occupancy of a particular single-particle state. They can give rise to highly entangled ground states, and even to instabilities.

Given these differences, it is natural to ask how anomalous pairing terms can directly lead to topological phases of light. In this work, we study the topological properties of two-dimensional photonic systems described by Equation (1), in the case where the underlying particle-conserving band structure has no topological structure, and where the parametric driving terms do not make the system unstable. We show that the introduction of particle non-conserving terms can break time-reversal symmetry (TRS) in a manner that is distinct from having introduced a synthetic gauge field, and can lead to the formation of bands having a non-trivial pattern of (suitably defined) quantized Chern numbers. This in turn leads to the formation of protected chiral edge modes: unlike the particle-conserving case, these modes can mediate a protected inelastic (but still coherent) scattering mechanism along the edge (that is, a probe field injected into the edge of the sample will travel along the edge, but emerge at a different frequency). In general, the topological phases we find here are distinct both from those obtained in the particle-conserving case, and from those found in topological superconductors. We also discuss possible realizations of this model using a nonlinear photonic crystal or superconducting microwave circuits. Finally, we discuss the formal analogies and crucial differences between the topological phases of light investigated here and those recently proposed for other kinds of Bogoliubov quasiparticles[27–31] (see Discussion section).

Results

Kagome lattice model. For concreteness, we start with a system of bosons on a kagome lattice (Fig.1),

$$\hat{H}_0 = \sum_{\mathbf{j}} \omega_0 \hat{a}^\dagger_{\mathbf{j}}\hat{a}_{\mathbf{j}} - J\sum_{\langle \mathbf{j},\mathbf{j}'\rangle}\hat{a}^\dagger_{\mathbf{j}}\hat{a}_{\mathbf{j}'} \quad (2)$$

(we set $\hbar = 1$). Here we denote by $\hat{a}_{\mathbf{j}}$ the photon annihilation operator associated with lattice site $\mathbf{j}$, where the vector site index has the form $\mathbf{j} = (j_1, j_2, s)$. $j_1, j_2 \in Z$ labels a particular unit cell of the lattice, while the index $s = A, B, C$ labels the element of the sublattice. $\langle \mathbf{j},\mathbf{j}'\rangle$ indicates the sum over nearest neighbours, and J is the (real valued) nearest-neighbour hopping rate; ω_0 plays the role of an onsite energy. As there are no phases associated with the hopping terms, this Hamiltonian is time-reversal symmetric and topologically trivial. We chose the kagome lattice because it is directly realizable both in quantum optomechanics[5] and in arrays of superconducting cavity arrays[13,16]; it is also the simplest model where purely local parametric driving can result in a topological phase.

We next introduce quadratic squeezing terms to this Hamiltonian that preserve the translational symmetry of the lattice and that are no more non-local than our original,

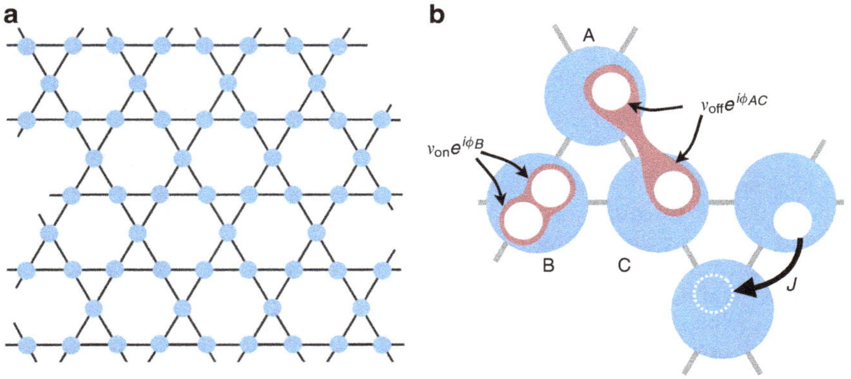

Figure 1 | Setup figure. (**a**) An array of nonlinear cavities forming a kagome lattice. (**b**) Photons hop between nearest-neighbour sites with rate J. Each cavity is driven parametrically leading to the creation of photon pairs on the same lattice site (rate ν_{on}) and on nearest-neighbour sites (rate ν_{off}). A spatial pattern of the driving phase is imprinted on the parametric interactions, breaking the time-reversal symmetry (but preserving the $C3$ rotational symmetry).

nearest-neighbour hopping Hamiltonian:

$$\hat{H}_{\rm L} = -\frac{1}{2}\left[\nu_{\rm on}\sum_{\rm j} e^{i\phi_s}\hat{a}_{\rm j}^{\dagger}\hat{a}_{\rm j}^{\dagger} + \nu_{\rm off}\sum_{\langle {\rm j,j'}\rangle} e^{i\phi_{ss'}}\hat{a}_{\rm j}^{\dagger}\hat{a}_{\rm j'}^{\dagger}\right] + {\rm h.c.} \quad (3)$$

Such terms generically arise from having a nonlinear interaction with a driven auxiliary pump mode (which can be treated classically) on each site, see for example, ref. 32. As we discuss below, the variation in phases in $\hat{H}_{\rm L}$ from site to site could be achieved by a corresponding variation of the driving phase of the pump. Note that we are working in a rotating frame where this interaction is time independent, and thus ω_0 should be interpreted as the detuning between the parametric driving and the true onsite (cavity) frequency $\omega_{\rm cav}$ (that is, $\omega_0 = \omega_{\rm cav} - \omega_{\rm L}/2$, where the parametric driving is at a frequency $\omega_{\rm L}$). The parametric driving can cause the system to become unstable; we will thus require that the onsite energy (that is, parametric drive detuning) ω_0 be sufficiently large that each parametric driving term is non-resonant enough to ensure stability. If one keeps ω_0 fixed, this means that the parametric driving amplitudes $\nu_{\rm on}$, $\nu_{\rm off}$ will be limited to some fraction of ω_0 (the particular value of which depends on J, Supplementary Note 1).

For a generic choice of phases in the parametric driving Hamiltonian of Equation (3), it is no longer possible to find a gauge where $\hat{H} = \hat{H}_0 + \hat{H}_{\rm L}$ is purely real when expressed in terms of real-space annihilation operators: hence, even though the hopping Hamiltonian $\hat{H}_0$ corresponds to strictly zero flux, the parametric driving can itself break TRS. In what follows, we will focus for simplicity on situations where time reversal and particle conservation are the only symmetries broken by the parametric driving: they will maintain the inversion and $\mathcal{C}3$ rotational symmetry of the kagome lattice. We will also make a global gauge transformation so that $\nu_{\rm off}$ is purely real, while $\nu_{\rm on} = |\nu_{\rm on}|e^{i\varphi_\nu}$. In this case, the only possible choices for the ϕ phases have the form $(\phi_A,\phi_B,\phi_C) = (\phi_{AB},\phi_{BC},\phi_{CA}) = \pm(0,\delta,2\delta)$ with $\delta = 2\pi m_\nu/3$, where m_ν is an integer and is the vorticity of the parametric driving phases. We stress that these phases (and hence the sign of the TRS breaking) are determined by the phases of the pump modes used to generate the parametric interaction.

Gap opening and non-trivial topology. $\hat{H}_0$ is the standard tight-binding kagome Hamiltonian for zero magnetic field, and does not have band gaps: the upper and middle bands touch at the symmetry point $\boldsymbol{\Gamma} \equiv (0,0)$, whereas the middle and lower bands touch at the symmetry points $\boldsymbol{K} = (2\pi/3,0)$ and $\boldsymbol{K}' = (\pi/3,\pi/(3)^{1/2})$ where they form Dirac cones (Fig. 2a).

Turning on the pairing terms, the Hamiltonian $\hat{H} = \hat{H}_0 + \hat{H}_{\rm L}$ can be diagonalized in the standard manner as $\hat{H} = \sum_{n,\mathbf{k}} E_n[\mathbf{k}]\hat{\beta}_{n,\mathbf{k}}^{\dagger}\hat{\beta}_{n,\mathbf{k}}$, where the $\hat{\beta}_{n,\mathbf{k}}$ are canonical bosonic annihilation operators determined by a Bogoliubov transformation of the form (see Methods section):

$$\hat{\beta}_{n,\mathbf{k}}^{\dagger} = \sum_{s=A,B,C} u_{n,\mathbf{k}}[s]\hat{a}_{\mathbf{k},s}^{\dagger} - \nu_{n,\mathbf{k}}[s]\hat{a}_{-\mathbf{k},s}. \quad (4)$$

Here $\hat{a}_{\mathbf{k},s}$ are the annihilation operators in quasimomentum space, and $n = 1,2,3$ is a band index; we count the bands by increasing energy. The photonic single-particle spectral function now shows resonances at both positive and negative frequencies, $\pm E_n[k]$, corresponding to particle- and hole-type bands, Fig. 2d. Because of the TRS breaking induced by the squeezing terms, the band structure described by $E_n[\mathbf{k}]$ now exhibits gaps, Fig. 2b; furthermore, for a finite sized system, one also finds edge modes in the gap, Fig. 2d.

The above behaviour suggests that the parametric terms have induced a non-trivial topological structure in the wavefunctions of the band eigenstates. To quantify this, we first need to properly identify the Berry phase associated with a bosonic band eigenstate in the presence of particle non-conserving terms. For each $\mathbf{k}$, the Bloch Hamiltonian $\hat{H}_{\mathbf{k}}$ corresponds to the Hamiltonian of a multi-mode parametric amplifier. Unlike the particle-conserving case, the ground state of such a Hamiltonian is a multi-mode squeezed state with non-zero photon number; it can thus have a non-trivial Berry's phase associated with it when $\mathbf{k}$ is varied, Supplementary Note 2. The Berry phase of interest for us will be the difference of this ground state Berry phase and that associated with a single quasiparticle excitation. One finds that the resulting Berry connection takes the form

$$\boldsymbol{\mathcal{A}}_n = i\langle \mathbf{k}, n|\hat{\sigma}_z \boldsymbol{\nabla}_k|\mathbf{k}, n\rangle. \quad (5)$$

Here the six vector of Bogoliubov coefficients $|\mathbf{k}, n\rangle \equiv (u_{n,\mathbf{k}}[A], u_{n,\mathbf{k}}[B], u_{n,\mathbf{k}}[C], \nu_{n,\mathbf{k}}[A], \nu_{n,\mathbf{k}}[B], \nu_{n,\mathbf{k}}[C])$ plays the role of a singe-particle wavefunction, and $\hat{\sigma}_z$ acts in the particle-hole space, associating $+1$ to the u components and -1 to the ν components, see Methods section. These effective wavefunctions obey the symplectic normalization condition

$$\langle \mathbf{k}, n|\hat{\sigma}_z|\mathbf{k}, n'\rangle = \sum_s u_{n,\mathbf{k}}^*[s]u_{n',\mathbf{k}}[s] - \nu_{n,\mathbf{k}}^*[s]\nu_{n',\mathbf{k}}[s] = \delta_{n,n'}. \quad (6)$$

Having identified the appropriate Berry connection for a band eigenstate, the Chern number for a band n is then defined in the usual manner:

$$C_n = \frac{1}{2\pi}\int_{BZ} (\boldsymbol{\nabla}\times\boldsymbol{\mathcal{A}}_n)\cdot\hat{z}. \quad (7)$$

The definition in Eq. (5) agrees with that presented in ref. 27 and (in one-dimension) ref. 29; standard arguments[27] show that the C_n are integers with the usual properties. We note that, as for superconductors, breaking the $U(1)$ (particle-conservation) symmetry remains compatible with a first-quantized picture after doubling the number of bands. The additional hole bands are connected to the standard particle bands by a particle–hole symmetry; see Methods section. In bosonic systems, the requirement of stability generally implies that particle and hole bands can not touch; this is true for our system. Thus, the sum of the Chern numbers over the particle bands (with $E > 0$) must be zero, and there cannot be any edge states with energies below the lowest particle bulk band (or in particular, at zero energy); Supplementary Note 1.

In the special case where we only have onsite parametric driving (that is, $\nu_{\rm off} = 0, \nu_{\rm on} \neq 0$), the Chern numbers can be calculated analytically (Supplementary Note 3). They are uniquely fixed by the pump vorticity. If $m_\nu = 0$, we have TRS and the band structure is gapless, while for $m_\nu = \pm 1$, $\mathbf{C} = (\mp 1, 0, \pm 1)$. This set of topological phases also occurs in a particle-number conserving model on the kagome lattice with a staggered magnetic field, that is, the Oghushi–Murakami–Nagaosa (OMN) model of the anomalous quantum-Hall effect[33,34].

In the general case, where we include offsite parametric driving, entirely new phases appear. We have computed the Chern numbers here numerically, using the approach of ref. 35. In Fig. 3a, we show the topological phase diagram of our system, where J/ω_0 and m_ν are held fixed, while the parametric drive strengths $\nu_{\rm on}$,$\nu_{\rm off}$ are varied. Different colours correspond to different triplets $\mathbf{C} \equiv (C_1,C_2,C_3)$ of the band Chern numbers, with grey and dark-grey corresponding to the two phases already present in the OMN model. Strikingly, a finite off-diagonal coupling $\nu_{\rm off}$ generates a large variety of phases which are not present in the OMN model, including phases having bands with $|C_n| > 1$. The border between different topological phases represent topological phase transitions, and correspond to parameter values where a pair of bands touch at a particular symmetry

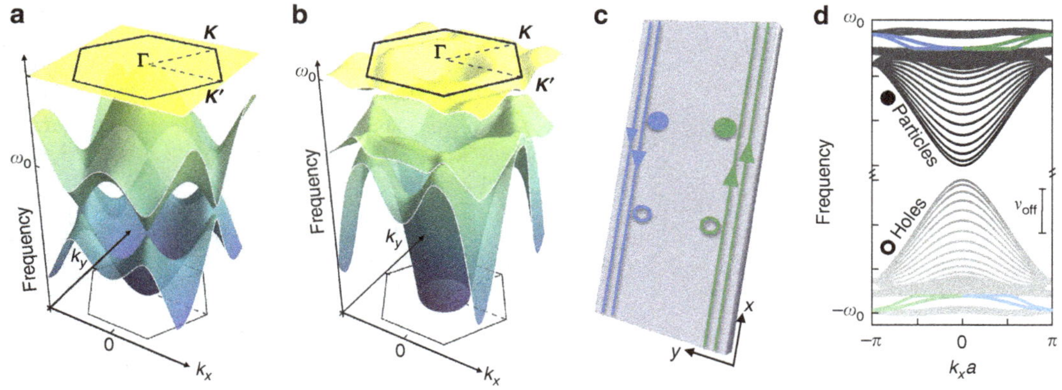

Figure 2 | Topological Band structure. (**a**,**b**) 3D plots of the bulk band structure. The hexagonal Brillouin zone is also shown. (**a**) In the absence of parametric driving, neighbouring bands touch at the rotational symmetry points ***K***, ***K'*** and **Γ**. (**b**) The parametric driving opens a gap between subsequent bands. For the chosen parameters, there is a global band gap between the second and third band. (**d**) Hole and particle bands, $\pm E_m[k_x]$, in a strip geometry (sketched in **c**). The line intensity is proportional to the weight of the corresponding resonance in the photon spectral function, Supplementary Note 1. The edge states localized on the right (left) edge, plotted in green (blue), have positive (negative) velocity. Parameters: Hopping rate $J=0.02\omega_0$ (ω_0 is the onsite frequency); (**b**,**d**), the parametric couplings are $\nu_{on}=-0.085\omega_0$ and $\nu_{off}=0.22\omega_0$.

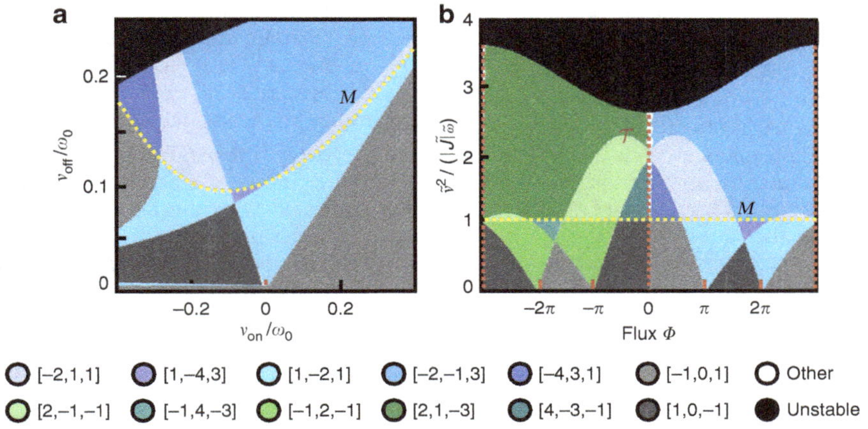

Figure 3 | Symplectic Topological phase diagrams. (**a**)Topological phase diagram for the parametrically driven kagome lattice model. The y (x) axis corresponds to the strength of the onsite parametric drive ν_{on} (offsite parametric drive ν_{off}), and different colours correspond to different triplets $\mathbf{C}=(C_1,C_2,C_3)$ of Chern numbers for the three bands of the model. Note that only the grey and dark-grey phases are found in the particle-conserving version of our model with a staggered field. We have fixed the hopping rate $J/\omega_0=0.02$, and the vorticity of the pump $m_\nu=1$. (**b**) Same phase diagram, but now plotted in terms of the effective flux Φ and effective parametric drive $\tilde{\nu}$ experienced by α quasiparticles.

point; we discuss this further below. Via a standard bulk-boundary correspondence (Supplementary Note 4), the band Chern numbers for a particular phase determine the number of protected edge states that will be present in a system with a boundary; as usual, the number of edge states in a particular bandgap is obtained by summing the Chern numbers of lower-lying bands. We discuss these edge states in greater detail in a following subsection. Finally, the black regions in the phase diagram indicate regimes of instability, which occur when the parametric driving strength becomes too strong.

Dressed-state picture. To gain further insight into the structure of the topological phases found above, it is useful to work in a dressed-state basis that eliminates the local parametric driving terms from our Hamiltonian. We thus first diagonalize the purely local terms in the Hamiltonian; for each lattice site $\mathbf{j}$ we have

$$\hat{H}_{\mathrm{j}}=\omega_0\hat{a}_{\mathrm{j}}^{\dagger}\hat{a}_{\mathrm{j}}-\frac{1}{2}\left[\nu_{\mathrm{on}}e^{i\phi_{\mathrm{j}}}\hat{a}_{\mathrm{j}}^{\dagger}\hat{a}_{\mathrm{j}}^{\dagger}+\mathrm{h.c.}\right]=\tilde{\omega}\hat{\alpha}_{\mathrm{j}}^{\dagger}\hat{\alpha}_{\mathrm{j}}\quad. \tag{8}$$

Here $\tilde{\omega}=\sqrt{\omega_0^2-\nu_{\mathrm{on}}^2}$, and the annihilation operators $\hat{\alpha}_{\mathrm{j}}$ are given by a local Bogoliubov (squeezing) transformation $\hat{\alpha}_{\mathrm{j}}=e^{i\phi_{\mathrm{j}}}e^{-i\varphi_\nu/2}(\cosh(r)a_{\mathrm{j}}-e^{i\phi_{\mathrm{j}}}e^{i\varphi_\nu}\sinh(r)\hat{a}_{\mathrm{j}}^{\dagger})$, where the squeezing factor r is

$$r=\frac{1}{4}\ln\left[\frac{\omega_0+\nu_{\mathrm{on}}}{\omega_0-\nu_{\mathrm{on}}}\right]. \tag{9}$$

On a physical level, the local parametric driving terms attempt to drive each site into a squeezed vacuum state with squeeze parameter r; the $\hat{\alpha}_{\mathrm{j}}$ quasiparticles correspond to excitations above this reference state. Note that we have included an overall phase factor in the definition of the $\hat{\alpha}_{\mathrm{j}}$, which will simplify the final form of the full Hamiltonian.

In this new basis of local quasiparticles, our full Hamiltonian takes the form

$$\hat{H}=\sum_{\mathrm{j}}\tilde{\omega}\hat{\alpha}_{\mathrm{j}}^{\dagger}\hat{\alpha}_{\mathrm{j}}-\sum_{\langle\mathrm{j,l}\rangle}\tilde{J}_{\mathrm{jl}}\hat{\alpha}_{\mathrm{j}}^{\dagger}\hat{\alpha}_{\mathrm{l}}-\left(\frac{\tilde{\nu}}{2}\sum_{\langle\mathrm{j,l}\rangle}\hat{\alpha}_{\mathrm{j}}^{\dagger}\hat{\alpha}_{\mathrm{l}}^{\dagger}+\mathrm{h.c.}\right). \tag{10}$$

The transformation has mixed the hopping terms with the non-local parametric terms: The effective counter-clockwise hopping matrix element is

$$\tilde{J}_{\rm jl} = Je^{i\delta} + e^{3i\delta/2}\left[2J\cos\left(\frac{\delta}{2}\right)\sinh^2 r + v_{\rm off}\sinh 2r\cos\left(\frac{\delta}{2}+\varphi_{\rm v}\right)\right], \tag{11}$$

and the magnitude of the effective non-local parametric driving is

$$|\tilde{v}| = |\, v_{\rm off}e^{-i(\delta/2+\varphi_{\rm v})} + 2v_{\rm off}\cos(\delta/2+\varphi_{\rm v})\sinh^2 r + J\sinh 2r\cos(\delta/2)\,|\,. \tag{12}$$

Note that the phase of $\tilde{v}$ can be eliminated by a global gauge transformation, and hence it plays no role; we thus take $\tilde{v}$ to be real in what follows.

Our model takes on a much simpler form in the new basis: the onsite parametric driving is gone, and the non-local parametric driving is real. Most crucially, the effective hoppings can now have spatially varying phases, which depend both on the vorticity of the parametric driving in $\hat{H}_{\rm L}$ (through δ), and the magnitude of the onsite squeezing (through r). In this transformed basis, the effective hopping phases are the only route to breaking TRS. Our model has thus been mapped onto the standard OMN model for the anomalous quantum-Hall effect, with an additional (purely real) nearest-neighbour two-mode squeezing interaction. In the regime where the parametric interactions between the $\hat{\alpha}$ quasiparticles are negligible (Supplementary Note 3), the complex phases correspond in the usual manner to a synthetic gauge field (that is, the effective flux Φ piercing a triangular plaquette would be $\Phi = 3\arg\tilde{J}$). In other words, the squeezing creates a synthetic gauge field for Bogoliubov quasiparticles. However, in the presence of substantial parametric interaction between $\hat{\alpha}$ quasiparticles, the parameter Φ can not be interpreted anymore as a flux: a flux of 2π can not be eliminated by a gauge transformation because the complex phases reappear in the parametric terms. In that case, only a periodicity of 6π in Φ is retained, since that corresponds to having trivial hopping phases of 2π.

Understanding the topological structure of this transformed Hamiltonian is completely sufficient for our purposes: one can easily show that the Chern number of a band is invariant under any local Bogoliubov transformation, hence the Chern numbers obtained from the transformed Hamiltonian in Equation (8) will coincide exactly with those obtained from the original Hamiltonian in Equation (3). We thus see that the topological structure of our system is controlled completely by only three dimensionless parameters: the flux Φ (associated with the hopping phases), the ratio $|\,\tilde{v}/\tilde{J}\,|$, and the ratio $\tilde{\omega}/\,|\,\tilde{J}\,|$.

The topological phase diagram for the effective model is shown in Fig. 3b. Again, one sees that as soon as the effective non-local parametric drive $\tilde{v}$ is non-zero, topological phases distinct from the standard (particle-conserving) OMN model are possible. The sign of the parametric pump vorticity $m_{\rm v}$ determines the sign of the effective flux Φ, c.f. Equation (11). As such, the right half of Fig. 3b (corresponding to $\Phi > 0$) is a deformed version of the phase diagram of the original model for pump vorticity $m_{\rm v} = 1$, as plotted in Fig. 3a. Changing the sign of $m_{\rm v}$ (and hence Φ) simply flips the sign of all Chern numbers, Supplementary Note 3.

Our effective model provides a more direct means for understanding the boundaries between different topological phases. Most of these are associated with the crossing of bands at one or more high-symmetry points in the Brillouin zone; this allows an analytic calculation of the phase boundary (Supplementary Note 3). Perhaps most striking in Fig. 3b is the horizontal boundary (labelled $\mathcal{M}$), occurring at a finite value of the effective offsite parametric drive, $\tilde{v} \approx \sqrt{\tilde{J}\tilde{\omega}}$. This boundary is set by the closing of a band gap at the $\boldsymbol{M}$ points; as these points are associated with the decoupling of one sublattice from the other two, this boundary is insensitive to the flux Φ. Similarly, the vertical line labelled $\mathcal{T}$ denotes a line where the system has TRS, and all bands cross at the symmetry points $\boldsymbol{K}$, $\boldsymbol{K'}$ and $\boldsymbol{\Gamma}$. The case of zero pump vorticity $m_{\rm v} = 0$ (not shown) is also interesting. Here the effective flux Φ depends on the strength of the parametric drivings, but is always constrained to be 0 or 3π. This implies that the effective Hamiltonian has TRS, even though the original Hamiltonian may not (that is, if Im $v_{\rm off} \neq 0$, the original Hamiltonian does not have TRS). For $m_{\rm v} = 0$, the parametric

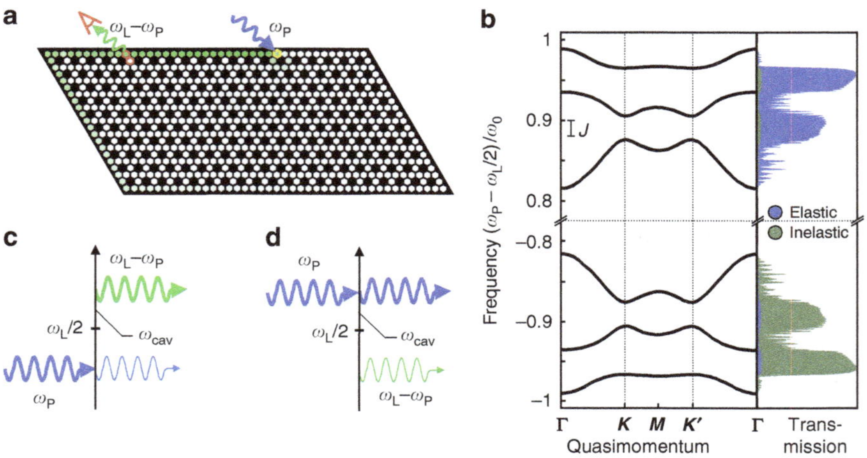

Figure 4 | Topologically protected transport in a finite system. (**a**) A probe beam at frequency $\omega_{\rm p}$ inside the bulk band gap is focused on a site (marked in yellow) at the edge of a finite sample. The probability map of the light transmitted inelastically at frequency $\omega_{\rm L} - \omega_{\rm p}$ (where $\omega_{\rm L}$ is the frequency of the drive tone applied to the auxiliary pump modes) clearly shows that the transport is chiral. (**b**) The elastic and inelastic transmission probability to a pair of sites along the edges (indicated in red in **a**) is plotted in blue and green, respectively. A cut through the bulk bands is shown to the left. (**c**,**d**) Sketch of the relevant scattering processes and energy scales. The inelastic (elastic) transmission has a larger rate when the light is injected in the hole (particle) band gap. Parameters: Hopping rate $J = 0.02\omega_0$ (ω_0 is the onsite frequency), parametric couplings $v_{\rm on} = 0.4\omega_0$ and $v_{\rm off} = 0.02\omega_0$, optical decay rate $\kappa = 0.001\omega_0$. (**a**) $\omega_{\rm p} - \omega_{\rm L}/2 = 0.95\omega_0$.

drivings do not open any band gap and the Chern numbers are not well defined.

Edge states and transport. Despite their modified definition, the Chern numbers associated with our Bogoliubov bands still guarantee the existence of protected chiral edge modes in a system with boundaries via a standard bulk-boundary correspondence, see Supplementary Note 4. These states can be used to transport photons by exciting them with an auxiliary probe laser beam, which is focused on an edge site and at the correct frequency. The lack of particle-number conservation manifests itself directly in the properties of the edge states: along with the standard elastic transmission they can also mediate inelastic scattering processes. In terms of the original lab frame, light injected at a frequency ω_p can emerge on the edge at frequency $\omega_L - \omega_p$, where ω_L is the frequency of the laser parametrically driving the system. This is analogous to the idler output of a parametric amplifier. Here both signal and idler have a topologically protected chirality.

Shown in Fig. 4 are the results of a linear response calculation describing such an experiment, applied to a finite system with corners. We incorporate a finite photon decay rate κ in the standard input–output formalism, see Methods section. Narrow-band probe light inside a topological band gap is applied to a site on the edge, and the resulting inelastic transmission probabilities to each site on the lattice are plotted, Fig. 4a. One clearly sees that the probe light is transmitted in a unidirectional way along the edge of the sample, and is even able to turn the corner without significant backscatter. The corresponding elastic transmission (not shown) is also chiral and shows the same spatial dependence. In Fig. 4b we show the elastic and inelastic transmissions to the sites indicated in red (rescaled by the overall transmission, $1-R$ where R is the reflection probability at the injection site) as a function of the probe frequency ω_p. By scanning the laser probe frequency, one can separately address particle and hole band gaps. The relative intensity of the inelastic scattering component is highly enhanced when the probe beam is inside a hole band gap, see also the sketches in Fig. 4c,d. When the parametric interaction between the $\hat{\alpha}$ quasiparticles is negligible, the ratio of elastic and inelastic transmissions depends only on the squeezing factor r, (c.f. Equation (9)), see Methods section.

Physical realization. Systems of this type could be implemented in photonic crystal coupled cavity arrays[36] fabricated from nonlinear optical χ^2 materials[37–39]. The array of optical modes participating in the transport would be supplemented by pump modes (resonant with the pump laser at twice the frequency). One type of pump mode could be engineered to be spatially co-localized with the transport modes (ν_{on} processes), while others could be located in-between (ν_{off}). The required periodic phase pattern of the pump laser can be implemented using spatial light modulators or a suitable superposition of several laser beams impinging on the plane of the crystal. One method for realizing the required kagome lattice of defect cavities was discussed in ref. 5. Optomechanical systems offer another route towards generating optical squeezing terms[40,41], via the mechanically induced Kerr interaction, and this could be exploited to create an optomechanical array with a photon Hamiltonian of the type discussed here. Alternatively, these systems can be driven by two laser beams to create phononic squeezing terms[42]. A fourth alternative consists in superconducting microwave circuits of coupled resonators, where Josephson junctions can be embedded to introduce χ^2 and higher order nonlinearities, as demonstrated in refs 43,44. Kagome lattices of superconducting resonators have recently been implemented[45].

Discussion

Before concluding, it is worthwhile to discuss the connections between our work and other recent studies. A Hamiltonian of the general form of Equation (1) arises naturally in the mean-field description of a Bose-condensed phase. In this setting, the anomalous pairing terms describe the interactions with the condensate treated at the mean-field level. A few recent studies have proposed to take advantage of these interactions to selectively populate topological edge states[28,30] or, closer to our study, to induce novel topological phases. These include a study of a magnonic crystal[27], as well as general Bose–Einstein condensates in one-dimension[29] and in two-dimensions[31].

There are some crucial differences between the above studies and our work. In our case, Equation (1) describes the real particles of our system, not quasiparticles defined above some background. This difference is not just a question of semantics: in our case, topological effects can directly be seen by detecting photons, whereas in refs 29,31, one would need to isolate the contribution of a small number of Bogoliubov quasiparticles sitting on a much larger background of condensed particles. In addition, in our work the pairing terms in Equation (1) are achieved by driving the system, implying that negative and positive frequencies are clearly physically distinguished (that is, they are defined relative to a non-zero pump frequency). This is at the heart of the topologically protected inelastic scattering mechanism we describe, and is something that is not present in previous studies.

Our work opens the door to a number of interesting new directions. On the more practical side, one could attempt to exploit the unique edge states in our system to facilitate directional, quantum-limited amplification. On the more fundamental level, one could use insights from the corresponding disorder problem[46] and attempt to develop a full characterization of particle non-conserving bosonic topological states that are described by quadratic Hamiltonians. This would then be a counterpart to the classification already developed for fermionic systems[26].

Methods

Bogoliubov transformation and first-quantized picture. We find the normal mode decompositions leading to the band structures in Fig. 2 and the topological phase diagrams in Fig. 3 by introducing a first-quantized picture. Since the relevant Hamiltonians do not conserve the excitation number, this is only possible after doubling the degrees of freedom. This is achieved by grouping all annihilation operators with quasimomentum $\mathbf{k}$ and the creation operators with quasimomentum $-\mathbf{k}$ in the $2N$ vector of operators $\hat{\Psi}_{\mathbf{k}} = (\hat{a}_{\mathbf{k}1}, \ldots, \hat{a}_{\mathbf{k}N}, \hat{a}^{\dagger}_{-\mathbf{k}1}, \ldots, \hat{a}^{\dagger}_{-\mathbf{k}N})$ (where N is the unit cell dimension), and by casting the second quantized Hamiltonian $\hat{H}$ in the form

$$\hat{H} = \frac{1}{2}\sum_{\mathbf{k}} \hat{\Psi}^{\dagger}_{\mathbf{k}} \hat{h}(\mathbf{k}) \hat{\Psi}_{\mathbf{k}}. \tag{13}$$

The $2N \times 2N$ hermitian matrix $\hat{h}(\mathbf{k})$ plays the role of a single-particle Hamiltonian and is referred to as the Bogoliubov de Gennes Hamiltonian. By definition of the normal modes $\hat{H} = \sum_{\mathbf{k},n} E_n[\mathbf{k}]\hat{\beta}^{\dagger}_{n,\mathbf{k}}\hat{\beta}_{n,\mathbf{k}}$, we have $[\hat{H}, \hat{\beta}^{\dagger}_{n,\mathbf{k}}] = E_n[\mathbf{k}]\hat{\beta}^{\dagger}_{n,\mathbf{k}}$. By plugging into the above equation the Bogoliubov ansatz Equation (4) one immediately finds

$$\hat{h}(\mathbf{k})|\mathbf{k}_n\rangle = E_n[\mathbf{k}]\hat{\sigma}_z|\mathbf{k}_n\rangle. \tag{14}$$

Likewise, from $[\hat{H}, \beta_{n,-\mathbf{k}}] = -E_n[-\mathbf{k}]\hat{\beta}_{n,-\mathbf{k}}$ one finds

$$\hat{h}(\mathbf{k})(\mathcal{K}\hat{\sigma}_x|-\mathbf{k}_n\rangle) = -E_n[-\mathbf{k}]\hat{\sigma}_z(\mathcal{K}\hat{\sigma}_x|-\mathbf{k}_n\rangle). \tag{15}$$

Here $\mathcal{K}$ denotes the complex conjugation and the matrix $\hat{\sigma}_x$ exchanges the u's and the v's Bogoliubov coefficients,

$$\hat{\sigma}_x = \begin{pmatrix} 0 & \mathbb{1}_N \\ \mathbb{1}_N & 0 \end{pmatrix}.$$

Thus, the spectrum of the $2N$ matrix $\hat{\sigma}_z\hat{h}(\mathbf{k})$ is formed by the set of $2N$ eigenenergies $E_n[\mathbf{k}]$ (belonging to the particle bands) and $-E_n[-\mathbf{k}]$ (belonging to the hole bands). *Vice versa*, to calculate the eigenenergies $E_n[\mathbf{k}]$ and $-E_n[-\mathbf{k}]$ and the vector of Bogoliubov coefficients in Equation (4), we have to solve the eigenvalue problem

$$\hat{\sigma}_z\hat{h}(\mathbf{k})|m\rangle = \lambda_m|m\rangle. \quad (16)$$

The solutions we are interested in should also display the symplectic orthonormality relations Equation (6).

We note in passing that so far we have implicitly assumed that the normal mode decomposition is possible. However, this is not always the case. When the matrix $\hat{\sigma}_z\hat{h}(\mathbf{k})$ has any complex eigenvalue, the Hamiltonian is unstable. Moreover, at the border of the stable and unstable parameter regions, the matrix $\hat{\sigma}_z\hat{h}(\mathbf{k})$ is not diagonalizable. The Supplementary Note 1 contains a stability analysis of our specific model.

In the stable regime of interest here the matrix $\hat{\sigma}_z\hat{h}(\mathbf{k})$ is diagonalizable and all its eigenvalues are real. In this case, its eigenvectors $|m\rangle$ can be chosen to be mutually $\hat{\sigma}_z$ orthogonal. In addition, there are exactly N positive (negative) norm eigenvectors. Thus, it is always possible to enforce the symplectic orthonormality relations Equation (6) by identifying the (appropriately normalized) positive and negative norm solutions with $|\mathbf{k}_n\rangle$ and $\mathcal{K}\hat{\sigma}_x|-\mathbf{k}_n\rangle$, respectively. The corresponding eigenvalues are then to be identified with $E_n[\mathbf{k}]$ (particle band structure) and $-E_n[-\mathbf{k}]$ (hole band structure), respectively

Particle-hole symmetry. The Bogoliubov de Gennes Hamiltonian has the generalized symmetry $C^\dagger\hat{h}(\mathbf{k})C = -\hat{h}(-\mathbf{k})$ where the charge conjugation operator C is anti-unitary and $C^2 = \mathbb{1}_{2N}$. Thus, our system represents the bosonic analogue of a superconductor in the Class D of the standard topological classification. This is a simple consequence of the doubling of the degrees of freedom in the single-particle picture. It simply reflects that the set of ladder operators $\hat{\beta}^\dagger_{n,\mathbf{k}}$ and $\hat{\beta}_{n,-\mathbf{k}}$ calculated from $\hat{h}(\mathbf{k})$ are the adjoint of the set of operators $\hat{\beta}_{n,\mathbf{k}}$ and $\hat{\beta}^\dagger_{n,-\mathbf{k}}$ calculated from $\hat{h}(-\mathbf{k})$.

Details of the transport calculations. In our transport calculations we have included photon decay. We adopt the standard description of the dissipative dynamics of photonic systems in terms of the Langevin equation and the input–output theory[47], for each site:

$$\dot{\hat{a}}_j = i[\hat{H}, \hat{a}_j] - \kappa\hat{a}_j/2 + \sqrt{\kappa}\hat{a}_j^{(\mathrm{in})}. \quad (17)$$

In practice, we consider an array of detuned parametric amplifiers with intensity decay rate κ and add to the standard description of each parametric amplifier the inter-cell coherent coupling described in the main text. The last term describes the influence of the input field $\hat{a}_j^{(\mathrm{in})}$ injected by an additional probe drive including also the environment vacuum fluctuations. The field $\hat{a}_j^{(\mathrm{out})}$ leaking out of each cavity at site j is given by the input–output relations

$$\hat{a}_j^{(\mathrm{out})} = \hat{a}_j^{(\mathrm{in})} - \sqrt{\kappa}\hat{a}_j. \quad (18)$$

The above formulas give an accurate description of a photonic system where the intrinsic losses during injection and inside the system are negligible. Intrinsic photon absorption can be incorporated by adding another decay channel to the equation for the light field. It reduces the propagation length but does not change qualitatively the dynamics.

In Fig. 3, we show the probabilities $T_{\mathrm{E}}(\omega, l, j)$ and $T_{\mathrm{I}}(\omega, l, j)$ that a photon injected on site j with frequency $\omega_{\mathrm{in}} = \omega + \omega_{\mathrm{L}}/2$ is transmitted elastically (at frequency $\omega + \omega_{\mathrm{L}}/2$) or inelastically (at frequency $\omega_{\mathrm{L}}/2 - \omega$) to site l where it is detected. From the Kubo formula and the input–output relations we find

$$T_{\mathrm{E}}(\omega, l, j) = \left|\delta_{lj} - i\kappa\tilde{G}_{\mathrm{E}}(\omega, l, j)\right|^2, \quad (19)$$

$$T_{\mathrm{I}}(\omega, l, j) = \kappa^2\left|\tilde{G}_{\mathrm{I}}(\omega, l, j)\right|^2. \quad (20)$$

Here $\tilde{G}_{\mathrm{E/I}}(\omega, l, j)$ are the elastic and inelastic components of the Green's function in frequency space,

$$\tilde{G}_{\mathrm{E}}(\omega, l, j) = -i\int_{-\infty}^{\infty} dt\Theta(t)\left\langle\left[\hat{a}_l(t), \hat{a}_j^\dagger(0)\right]\right\rangle e^{i\omega t}, \quad (21)$$

$$\tilde{G}_{\mathrm{I}}(\omega, l, j) = -i\int_{-\infty}^{\infty} dt\Theta(t)\left\langle\left[\hat{a}_l^\dagger(t), \hat{a}_j^\dagger(0)\right]\right\rangle e^{i\omega t}. \quad (22)$$

In a N site array with single-particle eigenstates $|n\rangle = (u_n[1], \ldots, u_n[N], v_n[1], \ldots, v_n[N])^T$, the Green's functions read

$$G_{\mathrm{E}}(\omega, l, j) = \sum_n \frac{u_n[l]u_n^*[j]}{\omega - E[n] + i\kappa/2} - \frac{v_n^*[l]v_n[j]}{\omega + E[n] + i\kappa/2}, \quad (23)$$

$$G_{\mathrm{I}}(\omega, l, j) = \sum_n \frac{v_n[l]u_n^*[j]}{\omega - E[n] + i\kappa/2} - \frac{u_n^*[l]v_n[j]}{\omega + E[n] + i\kappa/2}. \quad (24)$$

We note that for a probe field inside the bandwidth of the particle (hole) sector but far detuned from the hole (particle) sector, only the first (second) term of the summand in Equations (23) and (24) is resonant. Thus, as expected, the inelastic scattering is comparatively larger when the probe field is in the hole band gap.

It is easy to estimate quantitatively the relative intensities of elastically and inelastically transmitted light when the parametric interaction of the $\hat{\alpha}$ Bogoliubov quasiparticles is small (the regime where Φ can be interpreted as a synthetic gauge field experienced by the Bogoliubov quasiparticles). In this case, it is straightforward to show that $|v_n[j]/u_n[j]| \approx \tanh r$ independent of the eigenstate n and the site j. By putting together Equations (20, 23, 24) and neglecting the off-resonant terms we find that for $\omega, \tilde{\omega} \gg |\tilde{J}|, \kappa, |\omega - \tilde{\omega}|$,

$$T_{\mathrm{I}}(\omega, l, j) \approx (\tanh r)^2 T_{\mathrm{E}}(\omega, l, j) \approx T_{\mathrm{I}}(-\omega, l, j) \approx (\coth r)^2 T_{\mathrm{E}}(-\omega, l, j).$$

These analytical formulas agree quantitatively with the numerical results shown in Fig. 4b (note that in Fig. 4b the transmission at the output sites is rescaled by the overall transmission, $\sum_{l \neq j} T_{\mathrm{I}}(\omega, l, j) + T_{\mathrm{E}}(\omega, l, j)$).

References

1. Goldman, N., Juzeliunas, G., Öhberg, P. & Spielman, I. B. Light-induced gauge fields for ultracold atoms. *Rep. Prog. Phys.* **77,** 126401 (2014).
2. Lu, L., Joannopoulos, J. D. & Soljacic, M. Topological photonics. *Nat. Photon.* **8,** 821–829 (2014).
3. Prodan, E. & Prodan, C. Topological phonon modes and their role in dynamic instability of microtubules. *Phys. Rev. Lett.* **103,** 248101 (2009).
4. Kane, C. L. & Lubensky, T. C. Topological boundary modes in isostatic lattices. *Nat. Phys.* **10,** 39–45 (2013).
5. Peano, V., Brendel, C., Schmidt, M. & Marquardt, F. Topological phases of sound and light. *Phys. Rev. X* **5,** 031011 (2015).
6. Yang, Z. *et al.* Topological acoustics. *Phys. Rev. Lett.* **114,** 114301 (2015).
7. Süsstrunk, R. & Huber, S. D. Observation of phononic helical edge states in a mechanical topological insulator. *Science* **349,** 47–50 (2015).
8. Paulose, J., Chen, B. G. & Vitelli, V. Topological modes bound to dislocations in mechanical metamaterials. *Nat. Phys.* **11,** 153–156 (2015).
9. Nash, L. M. *et al.* Topological mechanics of gyroscopic metamaterials. *Proc. Natl Acad. Sci. USA* **112,** 14495–14500 (2015).
10. Haldane, F. D. M. & Raghu, S. Possible realization of directional optical waveguides in photonic crystals with broken time-reversal symmetry. *Phys. Rev. Lett.* **100,** 013904 (2008).
11. Raghu, S. & Haldane, F. D. M. Analogs of quantum-hall-effect edge states in photonic crystals. *Phys. Rev. A* **78,** 033834 (2008).
12. Wang, Z., Chong, Y., Joannopoulos, J. D. & Soljacic, M. Observation of unidirectional backscattering-immune topological electromagnetic states. *Nature* **461,** 772–775 (2009).
13. Koch, J., Houck, A. A., Le Hur, K. & Girvin, S. M. Time-reversal-symmetry breaking in circuit-QED-based photon lattices. *Phys. Rev. A* **82,** 043811 (2010).
14. Umucallar, R. O. & Carusotto, I. Artificial gauge field for photons in coupled cavity arrays. *Phys. Rev. A* **84,** 043804 (2011).
15. Fang, K., Yu, Z. & Fan, S. Realizing effective magnetic field for photons by controlling the phase of dynamic modulation. *Nat. Photon.* **6,** 782–787 (2012).
16. Petrescu, A., Houck, A. A. & Le Hur, K. Anomalous Hall effects of light and chiral edge modes on the Kagomé lattice. *Phys. Rev. A* **86,** 053804 (2012).
17. Tzuang, L. D., Fang, K., Nussenzveig, P., Fan, S. & Lipson, M. Non-reciprocal phase shift induced by an effective magnetic flux for light. *Nat. Photon.* **8,** 701–705 (2014).
18. Schmidt, M., Kessler, S., Peano, V., Painter, O. & Marquardt, F. Optomechanical creation of magnetic fields for photons on a lattice. *Optica* **2,** 635–641 (2015).
19. Hafezi, M., Demler, E. A., Lukin, M. D. & Taylor, J. M. Robust optical delay lines with topological protection. *Nat. Phys.* **7,** 907–912 (2011).
20. Khanikaev, A. B. *et al.* Photonic topological insulators. *Nat. Mater.* **12,** 233–239 (2012).
21. Hafezi, M., Mittal, S., Fan, J., Migdall, A. & Taylor, J. M. Imaging topological edge states in silicon photonics. *Nat. Photon.* **7,** 1001–1005 (2013).
22. Mittal, S. *et al.* Topologically robust transport of photons in a synthetic gauge field. *Phys. Rev. Lett.* **113,** 087403 (2014).
23. Kitagawa, T. *et al.* Observation of topologically protected bound states in photonic quantum walks. *Nat. Commun.* **3,** 882 (2012).
24. Rechtsman, M. C. *et al.* Topological creation and destruction of edge states in photonic graphene. *Phys. Rev. Lett.* **111,** 103901 (2013).
25. Bardyn, C.-E. & Imamoğlu, A. Majorana-like modes of light in a one-dimensional array of nonlinear cavities. *Phys. Rev. Lett.* **109,** 253606 (2012).
26. Ryu, S., Schnyder, A. P., Furusaki, A. & Ludwig, A. W. W. Topological insulators and superconductors: tenfold way and dimensional hierarchy. *New J. Phys.* **12,** 065010 (2010).
27. Shindou, R., Matsumoto, R., Murakami, S. & Ohe, J. Topological chiral magnonic edge mode in a magnonic crystal. *Phys. Rev. B* **87,** 174427 (2013).
28. Barnett, R. Edge-state instabilities of bosons in a topological band. *Phys. Rev. A* **88,** 063631 (2013).
29. Engelhardt, G. & Brandes, T. Topological bogoliubov excitations in inversion-symmetric systems of interacting bosons. *Phys. Rev. A* **91,** 053621 (2015).

30. Galilo, B., Lee, D. K. K. & Barnett, R. Selective population of edge states in a 2d topological band system. *Phys. Rev. Lett.* **115,** 245302 (2015).
31. Bardyn, C.-E., Karzig, T., Refael, G. & Liew, T. C. H. Chiral bogoliubov excitations in nonlinear bosonic systems. *Phys. Rev. B* **93,** 020502 (2016).
32. Gerry, C. C. & Knight, P. L. *Introductory Quantum Optics* (Cambridge University Press (2005).
33. Ohgushi, K., Murakami, S. & Nagaosa, N. Spin anisotropy and quantum hall effect in the *kagomé* lattice: chiral spin state based on a ferromagnet. *Phys. Rev. B* **62,** R6065–R6068 (2000).
34. Green, D., Santos, L. & Chamon, C. Isolated flat bands and spin-1 conical bands in two-dimensional lattices. *Phys. Rev. B* **82,** 075104 (2010).
35. Fukui, T., Hatsugai, Y. & Suzuki, H. Chern numbers in discretized brillouin zone: Efficient method of computing (spin) hall conductances. *J. Phys. Soc. Jpn.* **74,** 1674–1677 (2005).
36. Notomi, M., Kuramochi, E. & Tanabe, T. Large-scale arrays of ultrahigh-q coupled nanocavities. *Nat. Photon.* **2,** 741–747 (2008).
37. Mookherjea, S. & Yariv, A. Coupled resonator optical waveguides. *IEEE J. Quantum Elec.* **8,** 448 (2002).
38. Eggleton, B. J., Luther-Davies, B. & Richardson, K. Chalcogenide photonics. *Nat. Photon.* **5,** 141–148 (2011).
39. Dahdah, J., Pilar-Bernal, M., Courjal, N., Ulliac, G. & Baida, F. Near-field observations of light confinement in a two dimensional lithium niobate photonic crystal cavity. *J. Appl. Phys.* **110,** 074318 (2011).
40. Safavi-Naeini, A. H. *et al.* Squeezed light from a silicon micromechanical resonator. *Nature* **500,** 185–189 (2013).
41. Purdy, T. P., Yu, P. L., Peterson, R. W., Kampel, N. S. & Regal, C. A. Strong Optomechanical Squeezing of Light. *Phys. Rev. X* **3,** 031012 (2013).
42. Kronwald, A., Marquardt, F. & Clerk, A. A. Arbitrarily large steady-state bosonic squeezing via dissipation. *Phys. Rev. A* **88,** 063833 (2013).
43. Bergeal, N. *et al.* Analog information processing at the quantum limit with a Josephson ring modulator. *Nat. Phys.* **6,** 296–302 (2010).
44. Abdo, B., Kamal, A. & Devoret, M. Nondegenerate three-wave mixing with the Josephson ring modulator. *Phys. Rev. B* **87,** 014508 (2013).
45. Underwood, D. L., Shanks, W. E., Koch, J. & Houck, A. A. Low-disorder microwave cavity lattices for quantum simulation with photons. *Phys. Rev. A* **86,** 023837 (2012).
46. Gurarie, V. & Chalker, J. T. Bosonic excitations in random media. *Phys. Rev. B* **68,** 134207 (2003).
47. Clerk, A. A., Devoret, M. H., Girvin, S. M., Marquardt, F. & Schoelkopf, R. J. Introduction to quantum noise, measurement, and amplification. *Rev. Mod. Phys.* **82,** 1155–1208 (2010).

Acknowledgements

V.P., C.B., and F.M. acknowledge support by an ERC Starting Grant OPTOMECH, by the DARPA project ORCHID, and by the European Marie-Curie ITN network cQOM. M.H. and A.A.C. acknowledge support from NSERC.

Author contributions

V.P., A.A.C. and F.M. contributed to the conceptual development of the project and interpretation of results. Calculations and simulations were done by V.P., M.H. and C.B.

Additional information

Competing financial interests: The authors declare no competing financial interests.

3

Description of quantum coherence in thermodynamic processes requires constraints beyond free energy

Matteo Lostaglio[1], David Jennings[1] & Terry Rudolph[1]

Recent studies have developed fundamental limitations on nanoscale thermodynamics, in terms of a set of independent free energy relations. Here we show that free energy relations cannot properly describe quantum coherence in thermodynamic processes. By casting time-asymmetry as a quantifiable, fundamental resource of a quantum state, we arrive at an additional, independent set of thermodynamic constraints that naturally extend the existing ones. These asymmetry relations reveal that the traditional Szilárd engine argument does not extend automatically to quantum coherences, but instead only relational coherences in a multipartite scenario can contribute to thermodynamic work. We find that coherence transformations are always irreversible. Our results also reveal additional structural parallels between thermodynamics and the theory of entanglement.

[1] Department of Physics, Imperial College London, London SW7 2AZ, UK. Correspondence and requests for materials should be addressed to M.L. (email: lostaglio@gmail.com).

We are increasingly able to probe and manipulate the physics of micro- and nanoscale systems. This has led to an explosion of work in the field of nanotechnology, with a myriad of applications to areas in industry, information technology, medicine and energy technologies. With operating scales between 1 and 10^2 nm, there has been remarkable progress in the development of molecular information ratchets, molecular motors, optical thermal ratchets and artificial bipedal nanowalkers inspired by naturally occurring biomolecular walkers[1–5]. There is also increasing evidence for the role of quantum effects within biological systems[6–8].

Towards the lower-end of the nanoscale, quantum mechanical effects such as quantum coherence and entanglement increasingly make their presence felt. Electrical conductance of molecular-scale components no longer obey Kirchhoff's laws and phase coherence can provide both destructive as well as constructive interference effects on electrical transport[9]. Such coherence has been shown to play important roles in thermal to electrical power conversion, heat dissipation in atomic-scale junctions and the engineering toolkit of quantum dots[10]. Conversely, dissipative quantum thermodynamics offers the possibility of on-demand generation of quantum information resources essential for future quantum technologies (communication, encryption, metrology and computing)[11]. Within quantum information science, the question of thermodynamically robust quantum memories, and thermodynamic constraints on quantum computation are still only partially understood and provide deep questions in the overlap between thermodynamics and quantum theory[12,13]. In a similar way, the phenomenon of thermality due to entanglement and the thermodynamics of area laws reveal deep connections between thermodynamics and the theory of entanglement[14,15].

The physics of these remarkable small-scale systems, displaying coherence or entanglement, constitute extreme quantum regimes. As such, a crucial question is: to what degree do traditional thermodynamic formulations and techniques encapsulate this regime? This is a broad, foundational question about thermodynamics. It is increasingly apparent that the traditional entropic formulation that emerges as an essentially unique description of the irreversibility of classical, macroscopic systems, will only place necessary, but not sufficient, constraints on the physics of small-scale systems manifesting coherence or quantum correlations.

The textbook treatments of classical, macroscopic equilibrium thermodynamics are typically based on notions such as Carnot cycles, with the entropy function generically defined via an integral in terms of heat flow[16]. This thermodynamic entropy function is then assumed (but often not proved) to completely describe the irreversible constraints on the system at hand. Alternative approaches follow a statistical mechanical treatment of the system based on underlying microstates, and provide an explanation of the thermodynamics in terms of microscopic degrees of freedom.

However, more rigorous derivations of the entropic form of the second law exist, such as by Carathéodory[17], Giles[18] and more recently by Lieb and Yngvason[19]. Of central importance is the partial order of thermodynamic states, from which an entropy function can then be derived in a rigorous manner. The existence of an essentially unique entropic form of the second law is found to be equivalent to assumptions that fail to hold in small-scale systems or high correlation quantum environments. For example, a scaling hypothesis is required, which is no longer valid for small systems. In addition a 'Comparison Hypothesis'[18,19] is required to hold (or derived from other axioms), which in itself makes a highly non-trivial assumption on the structure of the thermodynamic partial order. Outside of the macroscopic classical regime, quantum systems will generically possess coherence or entanglement, and the ordering of states typically displays a much richer structure[20].

A unique additive entropic function implies that such assumptions must hold[19]. Therefore, their inapplicability in the quantum realm means that no single entropic function can suffice. To fully describe the thermodynamic directionality of nanoscale, non-equilibrium systems, more than one entropy function is required. The results of ref. 21 provide a clean characterization of non-asymptotic, thermodynamic inter-conversions of quantum states with zero coherence between energy eigenspaces. The necessary and sufficient conditions for such state inter-conversions are in terms of a set of entropic free energy functions (here denoted $\Delta F_\alpha \leq 0$).

The present work goes beyond these conditions, showing that even these fail to be sufficient for thermodynamic transformations involving non-zero quantum coherence. Exploiting recent results in asymmetry theory[22,23], we show that thermodynamics can be viewed as being determined by at least two independent resources: the first is quantified by known free energies and measures how far a state is from being thermal; the second, a missing ingredient of previous treatments, measures how much a quantum state breaks time-translation invariance, that is, the degree of coherence in the system. This removes the 'zero coherence' assumption made in numerous recent works, for example see refs 21,24–26. This shift in perspective allows us to extend the free energy relations to a parallel set of thermodynamic constraints for quantum coherence, which take the form $\Delta A_\alpha \leq 0$, where A_α are measures of time-translation asymmetry. These constraints characterize the tendency of any quantum system to 'equilibrate' towards a time-symmetric state. The new laws, irrelevant for a system composed of many, uncorrelated bodies, become essential for the thermodynamics of small/correlated quantum systems. As an application, we show that in certain regimes the free energy splits into two components, one measuring the amount of classical free energy and the other measuring the quantum contribution coming from coherence. We show that coherence is not directly distillable as work, but does admit activation as a relational degree of freedom. We uncover a second form of fundamental irreversibility that parallels the one stressed in ref. 25, but involves coherence transformations. Finally, we shed light on new connections between thermodynamics and entanglement theory.

Results

Free energy second laws. The approach most suited to our needs in this work is the one followed in refs 21,25,27,28, which has emerged from the theory of entanglement[20]. Thermodynamic transformations (also called thermal operations) are defined as the set of all energy-preserving interactions between an arbitrary quantum system and a Gibbsian bath at a fixed temperature (see Methods).

One can allow additional, auxiliary systems to be used catalytically and consider thermodynamic transformations $\rho \otimes \chi_{aux} \rightarrow \sigma \otimes \chi_{aux}$, where an auxiliary system begins and ends in the same state χ_{aux}, yet enables the otherwise forbidden thermodynamic transformation $\rho \rightarrow \sigma$. For this broad setting, it was recently proven[21] that a continuum of quantum second laws govern the allowed thermodynamic transformations. Specifically, the generalized free energies given by $F_\alpha(\rho) = kTS_\alpha(\rho||\gamma) - kT \log Z_H$, $Z_H = \mathrm{Tr}[e^{-\beta H}]$, must all decrease:

$$\Delta F_\alpha \leq 0, \quad \forall \alpha \geq 0. \tag{1}$$

Here γ is the thermal state of the system with Hamiltonian H, $\gamma = e^{-\beta H}/Z_H$, $\beta = (kT)^{-1}$ and S_α (sometimes denoted D_α) are information-theoretic generalizations of the standard relative entropy, called α-Rényi divergences[29] (see Methods). For $\alpha \rightarrow 1$,

$S_\alpha(\rho||\sigma)$ is simply the quantum relative entropy[30] and the constraints of equation (1) reduce to $\Delta F \leq 0$, where $F(\rho) = \mathrm{Tr}[\rho H] - kTS(\rho)$. When applied to isothermal transformations between equilibrium states these conditions reproduce the traditional bound on work extraction[21,25] (see Methods). However, these conditions turn out to be also sufficient for characterizing the states accessible through thermodynamic transformations with the aid of a catalyst, when no coherence is present. For a system of many, uncorrelated particles only $\alpha \to 1$ matters, so that the family of second laws collapse to the traditional constraint of non-increasing free energy[21,28].

Previous work considered either an asymptotic scenario or assumed the states to be block-diagonal in the eigenbasis of the Hamiltonian. Both these assumptions are insensitive to the role of coherence. Indeed, the free energy relations are no longer sufficient for the single-shot thermodynamics of correlated and coherent quantum systems. As we shall see now, additional conditions are required due to the breakdown of time-translation invariance.

Beyond conservation laws. The idea of symmetry is powerful and wide-reaching, and finds countless applications across physics. However, recent work has brought the concept of asymmetry to the fore, and shown it to be a valuable, consumable resource[23,31–35]. An evolution is said to be symmetric if it commutes with the action of a symmetry group, that is, it does not matter if the symmetry transformation is applied before or after the dynamics takes place. Similarly, a state is symmetric if it is invariant under symmetry transformations and asymmetric otherwise (see Methods). Asymmetric states, in analogy with entangled states, constitute a resource that makes possible transformations otherwise impossible under the constraint of a symmetry group.

It has been found that symmetry constraints for closed system dynamics of pure quantum states (not mixed) are encoded by the conservation of all moments of the generators of the symmetry transformations. However, this is not the case for open quantum system dynamics, or for mixed quantum states, and asymmetry monotones, that is, functions that do not increase under symmetric evolution[32], can impose further, non-trivial constraints on the dynamics[22] (see Methods for a brief discussion of the connections between the present approach and fluctuation theorems).

Time-asymmetry and thermodynamics. Noether's theorem tells us that if a system has time-translation invariance then its energy is conserved. However, in general thermodynamic scenarios, we have no time-translation invariance, either for the thermodynamic process on the system or for the quantum state of the system. The thermodynamics of a system generally involves irreversible dynamics and mixed quantum states out of equilibrium, and heat can flow into and out of the thermal reservoir.

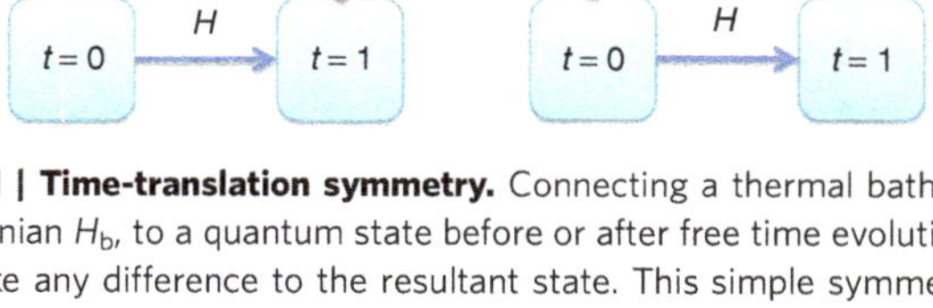

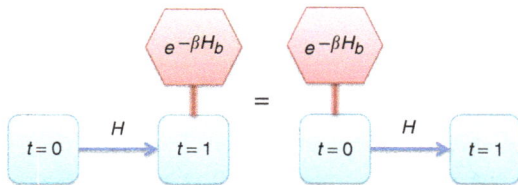

Figure 1 | Time-translation symmetry. Connecting a thermal bath, with Hamiltonian H_b, to a quantum state before or after free time evolution does not make any difference to the resultant state. This simple symmetry implies laws that constrain the approach of a state to time-translation invariance.

One might therefore think that the unitary group generated by the free Hamiltonian H of the system should not play any particular role. However, this is not the case, and from a perspective of asymmetry we find that:

Theorem 1: The set of thermal operations on a quantum system is a strict subset of the set of symmetric quantum operations with respect to time-translations.

The proof of this is provided in the Methods. See also Fig. 1. The implication of this result is that no thermodynamic process can generate additional time-translation asymmetry in the quantum system. A general picture emerges, where thermodynamics is governed by distinct abstract resources. The 'thermodynamic purity' resource component, p, quantifies how ordered the state of the system is in the presence of a thermal bath, and its evolution is constrained by a set of free energy differences[21] (see Methods). If no quantum coherence is present then consideration of p suffices, however more generally, quantum thermodynamics is governed by the interplay of at least two fundamental resources, denoted by (p, a). Free energy relations quantify the former, whereas asymmetry theory provides the tools to quantify the latter.

Coherence second laws. We now present thermodynamic constraints that go beyond free energy relations. In particular, we find that the core measures, used to define the generalized free energy relations[21], can be extended in a natural way that provides asymmetry measures. We introduce the following:

Definition 1: for any $\alpha \geq 0$, the free coherence of a state ρ with respect to a Hamiltonian H is

$$A_\alpha(\rho) := S_\alpha(\rho \,||\, \mathcal{D}_H(\rho)),$$

where $\mathcal{D}_H$ is the operation that removes all coherence between energy eigenspaces. S_α are the quantum Rényi divergences as defined in the Methods.

In the same way in which free energies measure 'how far' a state is from being thermal, free coherences measure 'how far' a state is from being incoherent in energy, that is, time-translation invariant (see Fig. 2). For $\alpha \to 1$, we have $A_1(\rho) \equiv A(\rho)$, which is the asymmetry measure introduced in refs 33,36,37. With these definitions on board, and from *Theorem 1*, we immediately have the following result:

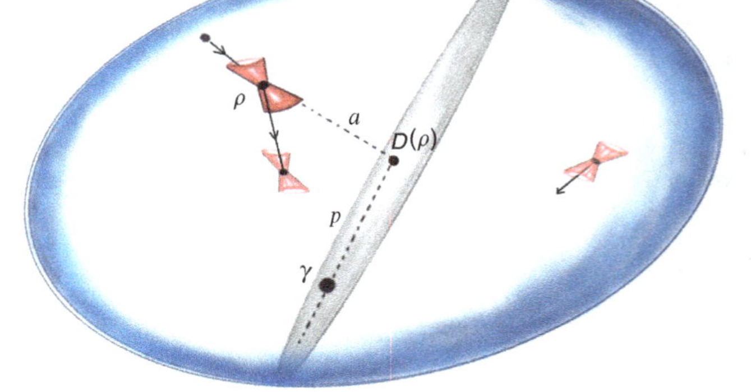

Figure 2 | Quantum thermodynamics as the combination of asymmetry and thermodynamic purity. The blue oval represents the convex set of all quantum states. To any state ρ, we can associate a 'thermal cone' (in red), the convex set of states thermally accessible from it. Any state ρ contributes in terms of thermodynamic purity p, which corresponds to the deviation of $\mathcal{D}_H(\rho)$ from the thermal state γ—as measured by $\{F_\alpha\}$—and asymmetry a, which corresponds to the deviation of ρ from the manifold of time-symmetric states (the grey region)—as measured by $\{A_\alpha\}$.

Theorem 2: For all $\alpha \geq 0$, we necessarily have $\Delta A_\alpha \leq 0$ for any thermal operation.

These laws characterize the depletion of coherence and the tendency to equilibrate onto the manifold of time-translation invariant states. In particular, they also hold for catalytic thermal operations where the catalyst is block-diagonal in the energy eigenbasis and can be extended to time-dependent Hamiltonians (see Methods). Importantly, these provide constraints that are independent of any free energy relations.

The free energy for $\alpha \to 1$, the relevant measure of average work yield[38], naturally splits into a classical and a quantum contribution

$$F(\rho) = F_c(\rho) + kTA(\rho) \tag{2}$$

where $A(\rho) = S(\rho || D_H(\rho))$ measures the amount of coherence in the system and $F_c(\rho) = F(\mathcal{D}_H(\rho))$ is the classical free energy. These results, together with the existing free energy relations, allow us to say that for $\alpha \to 1$ the classical and quantum contributions to the quantum free energy must independently decrease under any thermodynamic process. Notice that a similar result, although differently interpreted, was found in the context of quantum reference frames[39].

The incompleteness of existing second laws. We now establish that the above asymmetry relations are both independent of the free energy relations, and provide additional non-trivial constraints that must be obeyed in any thermodynamic process $\rho \to \sigma$.

To this end, it suffices to consider a qubit system with Hamiltonian $H = |1\rangle\langle 1|$, and choose an initial state $\rho = |1\rangle\langle 1|$, together with the target final state

$$\sigma = (1-\epsilon)\gamma + \epsilon|+\rangle\langle +|.$$

As S_α is monotonically decreasing in α, it suffices to choose $\epsilon > 0$ sufficiently small so that $S_\infty(\sigma||\gamma) \leq S_0(\rho||\gamma)$ to ensure all of the free energy conditions are obeyed. However, as the initial state is a symmetric state, and $A_\alpha(\sigma) > 0$ for any $\epsilon > 0$, it follows that such a transformation is impossible to achieve via a thermodynamic transformation. Thus, the free energy relations are necessarily incomplete.

Another way of seeing that the free energy relations only provide an incomplete description of thermodynamics is through the notion of work. Specifically, work is taken to be an ordered state of elevated energy. This idealized 'work bit' is a two-level system with Hamiltonian $H_w = w|w\rangle\langle w|$ (ref. 25). In its simplest form, it can be thought of as a perfectly controlled atom that gets excited (de-excited) when energy is extracted from (pumped into) a quantum system through a thermodynamic operation, for example,

$$\rho \otimes |w\rangle\langle w| \to \sigma \otimes |0\rangle\langle 0|. \tag{3}$$

Given any two states, ρ and σ, one can readily show there exists a $w > 0$ such that for all $\alpha \geq 0$ the free energy conditions $\Delta F_\alpha \leq 0$ are satisfied by equation (3). Thus, adding enough work, any state transformation is possible (at least catalytically) between block-diagonal states. In this sense, work is a universal resource classically. However, it is easy to see that *Theorem* 2 implies that for the transformation of equation (3) to be possible we need $A_\alpha(\rho) \geq A_\alpha(\sigma)$, for all $\alpha \geq 0$. In quantum thermodynamics, both the energetic and the coherent properties must be considered together.

Emergence of classicality. The constraints of *Theorem* 2 are not only relevant for nanoscale thermodynamics, but also at the macroscopic scale in the presence of correlated quantum systems able to sustain coherence. The regime in which the coherence second laws may be neglected is for systems composed of many, non-interacting bodies. We formalize this question and answer it, by showing that the free coherences per particle in a system of n non-interacting qubits vanish in the $n \to \infty$ limit:

$$\lim_{n\to\infty} A_\alpha(\rho^{\otimes n})/n = 0, \quad \forall \alpha \geq 0. \tag{4}$$

This generalizes the result found in ref. 33 for $\alpha \to 1$ and describes an emergent classical scenario in which states become effectively time-symmetric. This is the reason why only the free energy governs the asymptotic behaviour studied in ref. 28. In particular, the following bound holds (see Methods):

$$0 \leq A_\alpha(\rho^{\otimes n}) \leq \log(n+1). \tag{5}$$

We will use equation (5) shortly to study work extraction at the classical-quantum boundary.

Quantum Szilárd. The notions of work and heat are the primary concerns of thermodynamics, and with the advent of nanoscale technologies it has been necessary to revisit these time-honoured concepts (see for example refs 24,40–44 and references therein). The analysis of Szilárd[45] showed that the information one has about a system has an energetic value in terms of the ordered work one can obtain from a disordered thermal reservoir[3]. Specifically, the possession of a single bit of information can be 'burnt' to obtain $kT\log 2$ Joules. More generally, standard thermodynamic arguments imply that given a state ρ of a d-dimensional system, in contact with a thermal reservoir at a fixed temperature, we can obtain an amount of work $W(\rho) = kT(\log d - S(\rho))$. Previous works[24,25] have shown how to extend this result to deterministic and probabilistic work extraction from single-quantum systems with zero coherence across energy eigenspaces. However, when we encounter quantum states containing coherences we must necessarily take into account the asymmetry constraints. One might think that the work relations extend without alteration, but this is not the case—quantum coherences cannot be simply converted into ordered energy, and so the standard Szilárd result must be modified.

Theorem 3: for a general work extraction:

$$\rho \otimes |0\rangle\langle 0| \to \sigma \otimes \sum_w p(w)|w\rangle\langle w|,$$

the work distributions $p(w)$ that can be obtained from the states ρ and $\mathcal{D}_H(\rho)$ through time-translation symmetric operations coincide.

See Methods for a proof. This phenomenon may be called work-locking, because coherence contributes to the free energy (see equation (2)), but cannot be extracted as work (see also refs 38 and 25). This also sheds light on the origin of the irreversibility noticed in ref. 25. On one hand, the work necessary to form a state, measured by F_∞, is bigger than the work that we can draw from it, given by F_0, because $F_0 < F_\infty$. This first irreversibility is not an intrinsically quantum phenomenon, as it is a sole consequence of the free energy constraints of ref. 21. Indeed, this same irreversibility is present even for diagonal states (probability distributions) undergoing thermal operations (particular stochastic processes), a classical—albeit not deterministic—theory. However, quantum coherence adds another layer of irreversibility, as the work necessary to generate the coherent part of a quantum state cannot be extracted later, due to the fact that thermodynamic operations are time-translation symmetric quantum maps.

In the thermodynamic limit, the work-locking phenomenon is undetectable. From equations (2) and (5), we have $F(\rho^{\otimes n}) \approx F(\mathcal{D}_{\sum_i H_i}(\rho^{\otimes n}))$, when $n \gg \log n$. The free energy is effectively classical, and the maximum extractable work per system approaches the quantum free energy. Equation (5) provides a bound on the rate of this suppression, and we find for n qubits,

$$\frac{A(\rho^{\otimes n})}{F(\rho^{\otimes n})} \leq \frac{\log(n+1)}{n\log 2},$$

independently of the temperature. For example, a naive application of this result to the case of 5 qubits shows that up to 50% of the free energy could be locked in coherences, whereas this number falls to 1% for a system of 1,000 qubits.

Coherent activation of work. We have established that one must associate to a state both purity and asymmetry, abstractly denoted (p, a), and have shown that coherences in isolation do not contribute to thermodynamic work. Schematically, if $(p, a)\rightarrow W$, then $(p, 0)\rightarrow W$ too. It might appear that quantum coherences have no effect on the work output of a thermodynamic process, but this is not the case.

In the case of states block-diagonal in energy eigenspaces, any state that cannot be prepared under thermal operations can be converted into mechanical work. In the fully quantum-mechanical setting, this is no longer the case. There are states that cannot be prepared through thermal operations from which it is impossible to draw any useful work. These are precisely the states ρ with coherence for which $\mathcal{D}_{\mathrm{H}}(\rho)=\gamma$. An extreme case is the pure state

$$|\psi_c\rangle = Z_{\mathrm{H}}^{-1/2}\sum_k e^{-\beta E_k/2}|E_{\mathrm{k}}\rangle, \tag{6}$$

where $|E_k\rangle$ are the eigenstates of H and Z_H is the partition function of H. However, although $(0, a)\rightarrow(W=0)$, it turns out coherence can be activated in the presence of other quantum systems with coherence:

$$(0, a_1)+(0, a_2)\rightarrow(W\neq 0). \tag{7}$$

By $A+B\rightarrow C$, we mean it is possible to transform A and B jointly into C, using thermal operations only. One might expect, following Szilárd, that any pure state should yield $kT\log d$ of work, but if this pure state has $a\neq 0$, this is impossible. Equation (7) tells us the only way to get $kT\log d$ is to smuggle in coherent resources. Only if we allow the use of an external source of coherence does this extraction of work become possible[46].

The way in which coherence in a state ρ can be utilized to obtain mechanical work is readily seen from asymmetry theory and the theory of quantum reference frames[31]. Having shown that thermal operations commute with time-translation, all the results concerning work extraction under the presence of a superselection rule (for example, see ref. 47) can be immediately applied to thermodynamics. If we have *two* quantum systems in states ρ_1 and ρ_2, respectively, for which $\mathcal{D}_{H_1}(\rho_1)=\gamma$ and $\mathcal{D}_{H_2}(\rho_2)=\gamma$, then individually no mechanical work can be obtained in the presence of a thermal reservoir. However, the two systems can instead encode relational coherence that is accessible. Specifically, the introduction of the second system gives $\mathcal{D}_{\bar{H}}(\rho_1\otimes\rho_2)=\sigma_{12}\neq\gamma\otimes\gamma$, where $\bar{H}=H_1\otimes\mathbb{1}+\mathbb{1}\otimes H_2$. This is also why collective actions on multiple copies can extract work in a situation in which operations on single copies would be useless[38].

Alternatively, we can distinguish one of the systems as being the dominant reference. This perspective admits a different physical interpretation. We take the dimension of $\mathcal{H}_2$ to be much larger than $\mathcal{H}_1$, and the state ρ_2 to be highly asymmetric compared with ρ_1. The function of ρ_2 is now to allow the simulation of a non-symmetric operation $\tilde{\mathcal{E}}$ on the first system:

$$\tilde{\mathcal{E}}(\rho_1)=\mathrm{Tr}_2[\mathcal{E}(\rho_1\otimes\rho_2)] \tag{8}$$

A catalytic property in the use of the reference has been recently pointed out in ref. 46 and shown to be a consequence of the fact that time-translations are an abelian group.

Discussion

In refs 21,25, the authors showed that the work needed to create a state ρ is measured by $F_\infty(\rho)$ and the work extractable is given by $F_0(\rho)$. This revealed an inherent irreversibility of thermodynamic transformations. We can now show that a similar irreversibility characterizes the thermodynamic processing of coherence. Although normally one wishes to distill out ordered energy via a thermodynamic process, we could equally ask to obtain a high degree of coherence in the final output state under the allowed quantum operations. One could wish to obtain a d-dimensional uniform superposition of energy states, $|\mathbb{1}(d)\rangle := d^{-1/2}\sum_k|k\rangle$. Conversely, we may want to know how much coherence is needed to create a quantum state. If σ_{sym} is some incoherent quantum state, *Theorem* 2 requires

$$\rho\rightarrow\sigma_{\mathrm{sym}}\otimes|\mathbb{1}(d_{\mathrm{out}})\rangle\langle\mathbb{1}(d_{\mathrm{out}})|\Rightarrow\log d_{\mathrm{out}}\leq A_0(\rho)$$
$$|\mathbb{1}(d_{\mathrm{in}})\rangle\langle\mathbb{1}(d_{\mathrm{in}})|\otimes\sigma_{\mathrm{sym}}\rightarrow\rho\Rightarrow\log d_{\mathrm{in}}\geq A_\infty(\rho),$$

which shows that a further, fundamental irreversibility affects coherence processing as at least $A_\infty(\rho)-A_0(\rho)$ amount of coherence is lost in a cycle.

Shortly after the present work, results appeared[48] on the reduction of quantum coherence under thermal maps, including tight bounds for qubits. Going beyond this, the work in ref. 49 applies the framework developed here to obtain both upper and lower bounds on coherence evolution for general quantum systems. In particular, it highlights that the structure of the bounds in ref. 48 is symmetry based, and that coherence in thermodynamics admits a broader mode-decomposition in terms of spectral analysis.

Our results also shed light on the structural relationships between entanglement theory and thermodynamics[14,15,50] (see Table 1). Beyond structural parallels, this work paves the way for

Table 1 | Structural parallels.

	Quantum thermodynamics	Entanglement theory
Asymptotic conversion	Rel. entropy[28]	Rel. entropy[15]
$\rho^{\otimes n}\rightarrow\sigma^{\otimes m}$	$F(\rho)=kTS(\rho\|\gamma)-kT\log Z_H$	$\inf_{\sigma\in\mathcal{S}} S(\rho\|\sigma)$
$W\rightarrow(p,0)\rightarrow W'<W$	Non-cyclicity[25]	Ent. formation $\neq$ Ent. distillation
$(p, a)\rightarrow W\leftarrow(p, 0)$	Work locking	Bound entanglement[65]
$(0, a_1)+(0, a_2)\rightarrow W$	Coherence activation	Entanglement activation[66]

Quantum thermodynamics and entanglement (Ent.) manipulations present many structural parallels[67]. The asymptotic interconversion of states are governed by relative (Rel.) entropy to the Gibbs states γ and the relative entropy to the manifold of separable states $\mathcal{S}$, respectively. The work necessary to create a state is bigger than the work extractable from it; this similarly happens with entangled state creation and distillation. There are states that cannot be created under thermal (LOCC) operations from which no work (entanglement) can be extracted, but the resource can be activated.

an explicit unification of the resource theories presented here, and of the now well-developed theory of entanglement.

The resource-theoretic perspective is just one recent approach to the thermodynamics of quantum systems, however, we argue that this framework presents an elegant and compact perspective on quantum thermodynamics in terms of the interconversion and quantification of two abstract properties: thermodynamic purity and time-asymmetry. These seem to be necessary components in any unified framework that seeks to describe coherent processes and generic quantum thermodynamic phenomena with no classical counterpart.

Methods

Thermal operations. They are all quantum operations $\mathcal{E}$ of the form[27,28]:

$$\mathcal{E}(\rho) = \mathrm{Tr}_2\left[U(\rho \otimes \gamma_b)U^\dagger\right] \tag{9}$$

where $\gamma_b = e^{-\beta H_b}/\mathrm{Tr}[e^{-\beta H_b}]$, $\beta = (kT)^{-1}$, $[U, H \otimes \mathbb{1} + \mathbb{1} \otimes H_b] = 0$ and U is a joint unitary on system and environment. H is the Hamiltonian of the system and H_b the Hamiltonian of the environment. The more traditional formulation of thermodynamic processes involves time-dependent Hamiltonians. However, as already noted in refs 21,25, this framework can encompass such scenarios through the inclusion of a clock degree of freedom. No restrictions are imposed on the initial and final state of the system, which are in general far from equilibrium, or on the properties of bath or its final state. The interaction is required to preserve total energy, so differently from traditional treatments all external sources of energy (for example, a work source) must be included in the picture and described quantum-mechanically.

Quantum Rényi divergences. There are two non-commutative extensions of the notion of α – Rényi divergence[30,51]. They enjoy operational significance in the regimes $\alpha > 1$ and $\alpha < 1$, respectively, and coincide with the traditional quantum relative entropy for $\alpha \to 1$ (ref. 52). This suggests to follow ref. 52 and define

$$S_\alpha(\rho \,\|\, \sigma) = \begin{cases} \frac{1}{\alpha-1}\log\mathrm{Tr}[\rho^\alpha\sigma^{1-\alpha}], & \alpha \in [0,1) \\ \frac{1}{\alpha-1}\log\mathrm{Tr}\left[\left(\sigma^{\frac{1-\alpha}{2\alpha}}\rho\sigma^{\frac{1-\alpha}{2\alpha}}\right)^\alpha\right], & \alpha > 1 \end{cases}$$

The limit for $\alpha \to 1$ is given by $S_1(\rho\|\sigma) = \mathrm{Tr}[\rho(\log\rho - \log\sigma)]$.

Consistency with equilibrium thermodynamics. When a system undergoes an isothermal transformation from an equilibrium state with respect to Hamiltonian H_1 to an equilibrium state with respect to Hamiltonian H_2, in absence of work, then

$$F(H_1) \geq F(H_2), \tag{10}$$

where $F(H) = -kT\log Z_H$ is the thermodynamic free energy and Z_H is the partition function. The above relation is recovered within the present framework, which shows consistency with the traditional account.

In ref. 21, it is shown that the necessary and sufficient condition for a transformation to be possible between two incoherent quantum states ρ and σ while the Hamiltonian is changed from H_1 to H_2 is

$$F_\alpha(\rho, H_1) \geq F_\alpha(\sigma, H_2) \quad \forall \alpha \geq 0, \tag{11}$$

where $F_\alpha(\rho,H) = kTS_\alpha(\rho\|\gamma_H) - kT\log Z_H$, $\gamma_H = e^{-\beta H}/Z_H$ and $Z_H = \mathrm{Tr}[e^{-\beta H}]$. For any α, $F_\alpha(\gamma_H,H) = -kT\log Z_H$ and therefore if the initial and final states are thermal, $\rho = \gamma_{H_1}$ and $\sigma = \gamma_{H_2}$, all the conditions of equation (11) are equivalent to equation (10), which fully characterizes the transformation $\gamma_{H_1} \to \gamma_{H_2}$ between equilibrium states. However, under a broader class of non-equilibrium operations, the conditions of equation (11) are necessary and sufficient to characterize thermodynamic transformations between two non-equilibrium quantum states, provided that no coherence is present.

The notion of work in the present approach is given by the notion of a work bit[25], as explained in 'The incompleteness of existing second laws'. We can recover another traditional bound by looking at how much work one can extract in the transformation between two equilibrium states:

$$\gamma_{H_1} \otimes |0\rangle\langle 0| \to \gamma_{H_2} \otimes |w\rangle\langle w| \tag{12}$$

Then all equation (11) collapse to the condition

$$w \leq F(H_1) - F(H_2) \tag{13}$$

as expected from traditional treatments.

Symmetric operations. Let G be a Lie group representing a symmetry, and consider a representation of G on a Hilbert space $\mathcal{H}$ given by $U: g \mapsto U(g)$, where $g \in G$ and $U(g)$ is a unitary on $\mathcal{H}$. A quantum operation $\mathcal{E}_G : \mathcal{B}(\mathcal{H}) \to \mathcal{B}(\mathcal{H})$ is called symmetric if [22,23,31]:

$$\mathcal{E}_G\left(U(g)\rho U^\dagger(g)\right) = U(g)\mathcal{E}_G(\rho)U^\dagger(g), \quad \forall \rho, \forall g \in G. \tag{14}$$

A state is called symmetric if it is invariant under symmetry transformations, $U(g)\rho U^\dagger(g) = \rho$. An intuitive example is for the $SU(2)$ representation of the rotation group in three dimensions. The group action defines rotations of quantum states, and those that are invariant (such as the singlet state on two spins) are rotationally symmetric, while all others are asymmetric.

Connection to fluctuation theorems. We can compare our framework with well-established results in non-equilibrium thermodynamics. Specifically, we compare with fluctuation theorem approaches[53–56] that supply powerful descriptions of systems far from equilibrium. The approach of such fluctuation theorems has significant limitations that are not present in the approach taken here. First, although fluctuation theorems can be written down for quantum systems, they only capture stochastic effects, which are 'effectively classical' in nature. More specifically, the requirement of destructive measurements on the initial state unavoidably kills any coherence between energy eigenspaces, and unavoidably kills entanglement between systems. Attempts to generalize to positive operator valued measures (POVMs) quickly hit obstacles when it comes to the pairing of time-reversed trajectories. As such, only a limited set of quantum mechanical features can ever be addressed through fluctuation theorems.

Another issue is the focus on the expectation values of random variables—for example, the moments of work-gain. For small systems, the distributions involved can be quite broad and structured and so it is arguably more natural to analyse it in finer terms, such as those developed within 'single-shot' regimes[42,57]. More significantly, it has been shown recently that even if you knew all the moments $\langle \hat{O}^k \rangle$ of a quantum observable $\hat{O}$, this is *insufficient* to describe the mixed state quantum mechanics of a system in the presence of a conservation law on $\hat{O}$ (ref. 22). In our case, the consequences of energy conservation are not fully captured by energy measurements, the reason being that coherence properties must be also taken into account. Our work is a first step in this direction.

The common feature of these points is that the traditional approaches, when applied to more and more extreme quantum systems, hit against a range of obstacles. The single-shot thermodynamics that has recently emerged has been shown to be consistent with existing thermodynamics, but nevertheless does not suffer from any of the above points. Indeed, as it has been developed from entanglement theory and the theory of quantum information, the framework is ideally suited to describe such phenomena.

Proof of *Theorem* 1. We need to prove (see equation (14)):

$$\forall t, \quad \mathcal{E}\left(e^{-iHt}\rho e^{iHt}\right) = e^{-iHt}\mathcal{E}(\rho)e^{iHt}. \tag{15}$$

For any bath system, $\gamma_b \propto e^{-\beta H_b}$. From equations (9) and (15) follows using $[H_b, \gamma_b] = 0$ and $[U, H + H_b] = 0$. That these operations form a proper subset is seen from the fact that transforming an energy eigenstate into any other energy eigenstate is a symmetric operation, but not a thermally allowed operation.

Thermodynamic purity. Our main result shows that a fundamental resource for the thermodynamics of coherent quantum states is time-translation asymmetry. Previous work[28] has already identified the 'thermodynamic purity' p of a quantum state as a resource for thermodynamics. We speak of 'thermodynamic purity' because, as we shall see, purity in the thermodynamic framework appears within an embedding that takes the Gibbs state to the maximally mixed state. The mapping is effectively the same as that between the canonical and microcanonical ensembles in textbook treatments. While our main results show that thermodynamics is a special resource theory of asymmetry, the ideas that we briefly summarize here show that thermodynamics is a special theory of purity (see Appendix D of ref. 21 for details). The need for two sets of second laws arises from this duality.

The problem solved in ref. 21 is to give necessary and sufficient conditions for the existence of a stochastic operation Λ_{th} that maps a probability distribution p to p' through the aid of a catalyst and leaves the thermal state unchanged:

$$\Lambda_{th}(p \otimes q) = p' \otimes q, \quad \Lambda_{th}(\gamma \otimes \eta_c) = \gamma \otimes \eta_c, \tag{16}$$

where γ is the thermal state of the system (for simplicity, take initial and final Hamiltonian to coincide). Notice that the catalyst can be taken to have a trivial Hamiltonian[21], so that the thermal state of the catalyst is a uniform distribution, denoted by η_c. Here q acts as a catalyst and as such must be given back unchanged. Notice that this approach is limited to quantum states diagonal in the energy basis. An important observation is that operations that leave the thermal state unchanged are equivalent (in terms of interconversion structure) to thermal operations as defined in refs 25,27,28 only if we limit ourselves to initial and final states diagonal in energy (see also ref. 58).

A problem similar to equation (16) was solved in ref. 59 for stochastic maps having the uniform distribution (rather than the thermal distribution) as a fixed point. This was done through an extended notion of majorization, called trumping. Given two probability distributions p and p' we say that p can be trumped into p' if and only if there exists a probability distribution q (the catalyst) such that $p \otimes q$ majorizes $p' \otimes q$. From the Birkhoff-von Neumann theorem, this is equivalent to the existence of a stochastic map Λ such that

$$\Lambda(p \otimes q) = p' \otimes q, \quad \Lambda(\eta \otimes \eta_c) = \eta \otimes \eta_c, \tag{17}$$

where η and η_c are uniform distributions. Notice that stochastic maps which leave the maximally mixed state unchanged give rise to an interconversion structure that is essentially the same as the resource theory of purity studied previously[60]. A necessary and sufficient condition for equation (17) to hold was given in ref. 61:

$$S_\alpha(p \parallel \eta) \geq S_\alpha(p' \parallel \eta), \quad \forall \alpha \tag{18}$$

Here $S_\alpha(\cdot \,||\eta)$ are relative Rényi divergences w.r.t. the uniform distribution η. They measure how pure (that is, far from uniform) a distribution is. To exploit these results to solve problem (16), we need a map embedding the thermal state into the uniform distribution. Given integers $\mathbf{d} = \{d_1,...,d_n\}$, $\sum_i d_i = N$, the authors of ref. 21 defined an embedding map $\Gamma_\mathbf{d}$ as

$$\Gamma_\mathbf{d}(p) = \oplus_i p_i \eta_{i,} \tag{19}$$

where η_i is the uniform distribution of dimension d_i. $\Gamma_\mathbf{d}$ is a map from a space of n-dimensional distributions, which could be called canonical space, to a space of N-dimensional probability distributions, which could be called microcanonical space. The reason for these names is as follows. Let us assume for simplicity that the thermal distribution γ is rational. Then it is easy to see that there exists $\mathbf{d}$:

$$\Gamma_\mathbf{d}(\gamma) = \eta, \tag{20}$$

where η is the N-dimensional uniform distribution.

Crucially, it is possible to show that

$$S_\alpha(p\,||\,\gamma) = S_\alpha(\Gamma_\mathbf{d}(p)\,||\,\eta), \tag{21}$$

Because $F_\alpha(\rho) = kTS_\alpha(\rho||\gamma) - kT \log Z_H$, this shows that the free energy and the purity measures are mapped one into the other by $\Gamma_\mathbf{d}$ and its inverse $\Gamma^*_\mathbf{d}$. For this reason, the generalized free energy differences can be considered measures of thermodynamic purity within the thermodynamic setting. As a consequence of equation (21) and the definition of F_α, if all F_α decrease the purity measures in the embedding space will decrease as well:

$$F_\alpha(p) \leq F_\alpha(p') \Leftrightarrow S_\alpha(\Gamma_\mathbf{d}(p) \parallel \eta) \leq S_\alpha(\Gamma_\mathbf{d}(p') \parallel \eta)$$

Using the necessary and sufficient conditions for trumping (that is, that equations (17) and (18) are equivalent), this implies that the condition of decreasing F_α is equivalent to the existence of a stochastic map Λ that preserves $\eta \otimes \eta_c$ and maps $\Gamma_\mathbf{d}(p)$ to $\Gamma_\mathbf{d}(p')$ through a catalyst q. The map $\Lambda_{\text{th}} = (\Gamma^*_\mathbf{d} \otimes \mathbb{1})\Lambda(\Gamma_\mathbf{d} \otimes \mathbb{1})$ is a stochastic map from the canonical space to itself. As required,

$$\begin{aligned}\Lambda_{\text{th}}(p \otimes q) &= (\Gamma^*_\mathbf{d} \otimes \mathbb{1})\Lambda(\Gamma_\mathbf{d}(p) \otimes q) = \\ &= (\Gamma^*_\mathbf{d} \otimes \mathbb{1})(\Gamma_\mathbf{d}(p') \otimes q) = p' \otimes q.\end{aligned}$$

Moreover, $\Lambda_{\text{th}}(\gamma \otimes \eta_c) = \gamma \otimes \eta_c$, so it is thermal. The embedding map shows a duality between a theory of purity in the microcanonical space and thermodynamics in the canonical space. This proves the decreasing of all generalized free energies F_α is equivalent to the existence of Λ_{th} and q satisfying equation (16) (with the catalyst having a trivial Hamiltonian).

The embedding maps carry zero coherence deviation from equilibrium into the consideration of the purity resource theory on a larger space. Hence, it is not possible to handle coherence in this construction. As a consequence, it is necessary to go beyond free energy relations to capture the role of quantum coherence in thermodynamics.

Beyond free energy constraints. *Proof of Theorem 2*: By assumption, there exists some thermal operation $\mathcal{E}$ such that $\sigma = \mathcal{E}(\rho)$. Since $\mathcal{E}$ is a thermal operation, then it is symmetric (*Theorem* 1). Integrating equation (15) over t gives[31]

$$[\mathcal{E}, \mathcal{D}_H] = 0. \tag{22}$$

Using equation (22) and the data processing inequality for quantum Rényi divergences[52,62–64], we deduce the coherence second laws.

The coherence second laws presented in *Theorem* 2 hold also for a broader set of operations, which allow the aid of a catalyst block-diagonal in energy, as in ref. 21:

Definition 2: We say that a state ρ in $\mathcal{H}$ is transformed into state σ through a catalytic thermal operation

$$\rho \xrightarrow{\text{cat}} \sigma, \tag{23}$$

if there are a quantum state ρ_c in a Hilbert space $\mathcal{H}_c$ with Hamiltonian H_c and a thermal operation $\mathcal{E}$ on $\mathcal{H} \otimes \mathcal{H}_c$:

$$\mathcal{E} = (\rho \otimes \rho_c) = \sigma \otimes \rho_c. \tag{24}$$

Theorem 4: Catalytic thermal operations with a block-diagonal catalyst are symmetric operations, that is, if H is the system's Hamiltonian and $\mathcal{C}$ is a catalytic thermal operation,

$$\mathcal{C}(e^{-iHt}\rho e^{iHt}) = e^{-iHt}\mathcal{C}(\rho)e^{iHt} \tag{25}$$

Proof: A state ρ is sent to ρ' through a catalytic thermal operation with diagonal catalyst if there exists a state σ, s.t. $[\sigma, H_c] = 0$, and a thermal operation $\mathcal{E} : \mathcal{E}(\rho \otimes \sigma) = \rho' \otimes \sigma$. We show that the quantum map $\mathcal{C}(\rho) = \text{Tr}_2\mathcal{E}(\rho \otimes \sigma) = \rho'$ is symmetric. Define $H_{\text{tot}} = H + H_c + H_b$, sum of the Hamiltonians of system, catalyst and bath. Notice that $\sigma = e^{-iH_ct}\sigma e^{-iH_ct}, \gamma_b = e^{-iH_bt}\gamma_b e^{-iH_bt}$. It follows

$$\begin{aligned}\mathcal{C}(e^{-iHt}\rho e^{iHt}) &= \text{Tr}_2\mathcal{E}(e^{-iHt}\rho e^{iHt} \otimes \sigma) \\ &= \text{Tr}_{23}\left[Ue^{-iHt}\rho e^{iHt} \otimes \sigma \otimes \gamma_b U^\dagger\right] \\ &= \text{Tr}_{23}\left[e^{-iH_{\text{tot}}t}U\rho \otimes \sigma \otimes \gamma_b U^\dagger e^{-iH_{\text{tot}}t}\right] \\ &= e^{-iHt}\text{Tr}_2[\mathcal{E}(\rho \otimes \sigma)]e^{iHt} \\ &= e^{-iHt}\mathcal{C}(\rho)e^{iHt}.\end{aligned}$$

Theorem 5: If $[\rho_c, H_c] = 0$,

$$\rho \xrightarrow{\text{cat}} \sigma \Rightarrow A_\alpha(\sigma) \leq A_\alpha(\rho), \quad \forall \alpha \geq 0. \tag{26}$$

Proof: Follows from *Theorem* 4 in the same way in which *Theorem* 2 follows from *Theorem* 1.

Coherence second laws for time-dependent Hamiltonians. In many thermodynamic applications, Hamiltonians are time-dependent. As shown in refs 21,25, we can deal with these situations introducing a classical degree of freedom representing a clock system. This classical degree of freedom can be thought of as a switch for changing the Hamiltonian (for example, the knob tuning a magnetic field). Hence, we might interpret the transformation

$$\rho \otimes |0\rangle\langle 0| \rightarrow \sigma \otimes |1\rangle\langle 1| \tag{27}$$

as the transformation sending ρ with initial Hamiltonian H_0 to σ with Hamiltonian H_1, once we formally define the Hamiltonian

$$H = H_0 \otimes |0\rangle\langle 0| + H_1 \otimes |1\rangle\langle 1| \tag{28}$$

which can be interpreted as a Hamiltonian H_0 that changes into H_1 as the switch goes from 0 to 1. Then *Theorem* 2 admits the following natural extension:

Theorem 6: for all $\alpha \geq 0$, we necessarily have $\Delta A_\alpha \leq 0$ for any thermal operation between ρ and σ in which the Hamiltonian is switched from H_0 to H_1. Here we defined

$$\Delta A_\alpha = A_\alpha(\sigma, H_1) - A_\alpha(\rho, H_0), \tag{29}$$

where $A_\alpha(\cdot, H_i) = S_\alpha(\cdot \parallel \mathcal{D}_{H_i}(\cdot))$.

Proof: From *Theorem* 2 applied to the Hamiltonian of equation (28), we get

$$A_\alpha(\rho \otimes |0\rangle\langle 0|, H) \geq A_\alpha(\sigma \otimes |1\rangle\langle 1|, H).$$

But $\mathcal{D}_H(\rho \otimes |0\rangle\langle 0|) = \mathcal{D}_{H_0 \otimes |0\rangle\langle 0|}(\rho \otimes |0\rangle\langle 0|) = \mathcal{D}_{H_0}(\rho) \otimes |0\rangle\langle 0|$. Hence, $A_\alpha(\rho \otimes |0\rangle\langle 0|, H) = A_\alpha(\rho, H_0)$, and similarly for H_1. The result follows.

Proof of equation (5). We show that for any qubit state ρ and $\alpha \geq 0$,

$$0 \leq A_\alpha(\rho^{\otimes n}) \leq \log(n+1). \tag{30}$$

Proof: without loss of generality we can fix the Hamiltonian of the system to be the Pauli Z. Assume we are able to prove the result for every pure qubit state $|\psi\rangle$. Then for every ρ there exists p and $|\psi\rangle$such that

$$\rho = p|\psi\rangle\langle\psi| + (1-p)\mathbb{1}/2 := \mathcal{E}_{\text{mix}}(|\psi\rangle\langle\psi|) \tag{31}$$

Mixing with the identity is a time-translation symmetric operation:

$$\begin{aligned}\mathcal{E}_{\text{mix}}(e^{-iHt}\sigma e^{iHt}) &= pe^{-iHt}\sigma e^{iHt} + (1-p)\mathbb{1}/2 \\ &= e^{-iHt}\mathcal{E}_{\text{mix}}(\sigma)e^{iHt}\end{aligned}$$

Hence, we can map $|\psi\rangle\langle\psi|^{\otimes n} \rightarrow \rho^{\otimes n}$ by means of symmetric operations. However, it is easy to see that *Theorem* 2 holds, more generally, for any symmetric operation, so

$$A_\alpha(|\psi\rangle\langle\psi|^{\otimes n}) \geq A_\alpha(\rho^{\otimes n}), \quad \forall \alpha \geq 0. \tag{32}$$

We conclude that we need to prove the bound only for pure states and from equation (32) the result will follow for any state. Because rotations about Z are symmetric operations, we can assume $|\psi\rangle$ to lie on the xz plane of the Bloch sphere:

$$\rho = |\psi\rangle\langle\psi| = \begin{bmatrix} p & \sqrt{p(1-p)} \\ \sqrt{p(1-p)} & p \end{bmatrix} \tag{33}$$

We will use the notation $\mathcal{D}_{\sum_{i=1}^n Z_i} \equiv \mathcal{D}$. Expanding in the computational basis,

$$\mathcal{D}(\rho^{\otimes n}) = \oplus_{h=0}^{n} p^{n-h}(1-p)^h|\mathbb{1}_h\rangle\langle\mathbb{1}_h|, \quad |\mathbb{1}_h\rangle = \underbrace{(1,...,1)}_{\binom{n}{h}\text{ elements}}. \tag{34}$$

It is useful to introduce a vector $|v_n\rangle$ whose components are grouped in blocks as follows: 1 component equal to $\sqrt{p^n}$, $\binom{n}{1}$ components equal to $\sqrt{p^{n-h}(1-p)}$, ... , $\binom{n}{h}$ components equal to $\sqrt{p^{n-h}(1-p)^h}$, ... , 1 component equal to $\sqrt{(1-p)^n}$. Then we can compactly rewrite

$$\rho^{\otimes n} = |v_n\rangle\langle v_n| \tag{35}$$

Define

$$P = \mathcal{D}(\rho^{\otimes n})^{\frac{1-\alpha}{2\alpha}}|v_n\rangle\langle v_n|\mathcal{D}(\rho^{\otimes n})^{\frac{1-\alpha}{2\alpha}}.$$

From the definition of A_α,

$$A_\alpha(\rho^{\otimes n}) = \frac{1}{\alpha-1}\log \mathrm{Tr}[P^\alpha], \tag{36}$$

From $\langle \mathbb{1}_h \mid \mathbb{1}_h\rangle = \binom{n}{h}$, for all $\alpha > 0$ we have

$$(|\mathbb{1}_h\rangle\langle\mathbb{1}_h|)^{\frac{1-\alpha}{2\alpha}} = \binom{n}{h}^{\left(\frac{1-\alpha}{2\alpha}-1\right)}|\mathbb{1}_h\rangle\langle\mathbb{1}_h|,$$

so that from equation (34),

$$\mathcal{D}(\rho^{\otimes n})^{\frac{1-\alpha}{2\alpha}} = \oplus_{h=0}^{n}\binom{n}{h}^{\frac{1-3\alpha}{2\alpha}}\left[p^{n-h}(1-p)^h\right]^{\frac{1-\alpha}{2\alpha}}|\mathbb{1}_h\rangle\langle\mathbb{1}_h|.$$

Define $|w_n\rangle = \mathcal{D}(\rho^{\otimes n})^{\frac{1-\alpha}{2\alpha}}|v_n\rangle$. Then $P = |w_n\rangle\langle w_n|$ and

$$|w_n\rangle = \oplus_{h=0}^{n}\binom{n}{h}^{\frac{1-3\alpha}{2\alpha}}\left[p^{n-h}(1-p)^h\right]^{\frac{1-\alpha}{2\alpha}}|\mathbb{1}_h\rangle\langle\mathbb{1}_h||v_n\rangle.$$

The vector $|w_n\rangle$ is also grouped in blocks $h = 0,1,\ldots, n$, each of $\binom{n}{h}$ equal elements. One of the elements of the h-block can be found as follows: the block of ones $|\mathbb{1}_h\rangle\langle\mathbb{1}_h|$ sums the elements in the h-block of $|v_n\rangle$ (which are $\binom{n}{h}$ and identical), getting $\binom{n}{h}\sqrt{p^{n-h}(1-p)^h}$). Adding the prefactors, we see that the elements of the h-block of $|w_n\rangle$ look like:

$$\binom{n}{h}^{\frac{1-3\alpha}{2\alpha}}\left[p^{n-h}(1-p)^h\right]^{\frac{1-\alpha}{2\alpha}}\binom{n}{h}p^{\frac{n-h}{2}}(1-p)^{\frac{h}{2}}.$$

We conclude

$$\langle w_n| = \left(1,\ldots,\underbrace{\binom{n}{h}^{\frac{1-\alpha}{2\alpha}}\left[p^{n-h}(1-p)^h\right]^{\frac{1}{2\alpha}},\ldots}_{\binom{n}{h}\text{ elements}},1\right).$$

Assume $\alpha \in \mathbb{N}$. Then

$$\mathrm{Tr}[P^\alpha] = (\langle w_n|w_n\rangle)^\alpha. \tag{37}$$

But

$$\langle w_n \mid w_n\rangle = \sum_{h=0}^{n}\binom{n}{h}^{1/\alpha}p^{\frac{n-h}{\alpha}}(1-p)^{\frac{h}{\alpha}}. \tag{38}$$

Combining this and equation (37) we obtain

$$\mathrm{Tr}[P^\alpha] = \|x(n)\|_{1/\alpha}, \tag{39}$$

with

$$x(n) := \left\{\binom{n}{0}p^n,\ldots,\binom{n}{h}p^{n-h}(1-p)^h,\ldots,\binom{n}{n}(1-p)^n\right\},$$

and we used the usual definition of ℓ^p-norm

$$\|x\|_p = \left(\sum_i |x_i|^p\right)^{\frac{1}{p}}.$$

Assume $\alpha > 1$. The monotonicity of ℓ^p-norms

$$q > r > 0 \Rightarrow \|x(n)\|_q \le \|x(n)\|_r \tag{40}$$

implies

$$\|x(n)\|_{1/\alpha} \ge \|x(n)\|_1 = \sum_{h=0}^{n}\binom{n}{h}p^{n-h}(1-p)^h = 1.$$

Now, because

$$A_\alpha(\rho^{\otimes n}) = \frac{1}{\alpha-1}\log\|x(n)\|_{1/\alpha} \tag{41}$$

this means that for $\alpha > 1$ any upper bound on $\|x(n)\|_{1/\alpha}$ gives an upper bound on A_α.

Fix $\alpha > 1$. We can now use the following identity concerning p-norms (that follows from Hölder inequality): for all $p > r > 0$, if y is a sequence of k elements,

$$\|y\|_r \le k^{\left(\frac{1}{r}-\frac{1}{p}\right)}\|y\|_p.$$

Choose $p = 1$, $r = 1/\alpha$:

$$\|x(n)\|_{1/\alpha} \le (n+1)^{\alpha-1}. \tag{42}$$

Hence, substituting in equation (41),

$$A_\alpha(\rho^{\otimes n}) \le \frac{1}{\alpha-1}\log\left[(n+1)^{\alpha-1}\right] = \log(n+1),$$

for all integers $\alpha > 1$. From the monotonicity in α of A_α (ref. 30), it is easy to see that this implies the result for every $\alpha \ge 0$, as required. The other inequality, $A_\alpha \ge 0$, follows immediately from the properties of S_α.

Work-locking. *Proof of Theorem 3:* The state $\mathcal{D}_H(\rho)$ can be obtained from ρ through dephasing in energy, which is easily shown to be a time-translation symmetric operation. Hence, any work distribution that can be obtained from $\mathcal{D}_H(\rho)$ can be obtained from ρ as well. Conversely, suppose it is possible to obtain from ρ a work distribution $p(w)$ through a time-symmetric operation $\mathcal{E}$:

$$\mathcal{E}(\rho \otimes |0\rangle\langle 0|) = \sigma \otimes \sum_w p(w)|w\rangle\langle w|. \tag{43}$$

If we apply $\mathcal{E}$ to $\mathcal{D}_H(\rho)$, it is easy to see from equation (22) and (43)

$$\mathcal{E}(\mathcal{D}_H(\rho) \otimes |0\rangle\langle 0|) = \mathcal{D}_H(\sigma) \otimes \sum_w p(w)|w\rangle\langle w|. \tag{44}$$

Hence, any work distribution that can be extracted from ρ can be obtained from $\mathcal{D}_H(\rho)$ as well through the same $\mathcal{E}$.

References

1. Collin, D. *et al.* Verification of the Crooks fluctuation theorem and recovery of RNA folding free energies. *Nature* **437,** 231–234 (2005).
2. Serreli, V., Lee, C., Kay, E. & Leigh, D. A molecular information ratchet. *Nature* **445,** 523–527 (2006).
3. Toyabe, S., Sagawa, T., Ueda, M., Muneyuki, E. & Sano, M. Experimental demonstration of information-to-energy conversion and validation of the generalized Jarzynski equality. *Nat. Phys.* **6,** 988–992 (2010).
4. Alemany, A. & Ritort, F. Fluctuation theorems in small systems: extending thermodynamics to the nanoscale. *Europhys. News* **41,** 27–30 (2010).
5. Cheng, J. *et al.* Bipedal nanowalker by pure physical mechanisms. *Phys. Rev. Lett.* **109,** 238104 (2012).
6. Lloyd, S. Quantum coherence in biological systems. *J. Phys. Conf. Ser.* **302,** 012037 (2011).
7. Lambert, N. *et al.* Quantum biology. *Nat. Phys.* **9,** 10–18 (2013).
8. Gauger, E. M., Rieper, E., Morton, J. J., Benjamin, S. C. & Vedral, V. Sustained quantum coherence and entanglement in the avian compass. *Phys. Rev. Lett.* **106,** 040503 (2011).
9. Vazquez, H. *et al.* Probing the conductance superposition law in single-molecule circuits with parallel paths. *Nat. Nanotech.* **10,** 663–667 (2012).
10. Karlström, O., Linke, H., Karlström, G. & Wacker, A. Increasing thermoelectric performance using coherent transport. *Phys. Rev. B* **84,** 113415 (2011).
11. Lin, Y. *et al.* Dissipative production of a maximally entangled steady state of two quantum bits. *Nature* **504,** 415–418 (2013).
12. Bravyi, S. & Haah, J. Quantum self-correction in the 3d cubic code model. *Phys. Rev. Lett.* **111,** 200501 (2013).
13. Pachos, J. K. *Introduction to Topological Quantum Computation* (Cambridge Univ., 2012).
14. Popescu, S., Short, A. J. & Winter, A. Entanglement and the foundations of statistical mechanics. *Nat. Phys.* **2,** 754–758 (2006).
15. Brandão, F. G. S. L. & Plenio, M. B. Entanglement theory and thermodynamics. *Nat. Phys.* **4,** 873–877 (2008).
16. Fermi, E. *Thermodynamics* (Dover, 1956).
17. Carathéodory, C. *Über die Bestimmung der Energie und der absoluten Temperatur mit Hilfe von reversiblen Prozessen* (Sitzungsber. Akad. Wiss., Phys. Math, 1925).
18. Giles, R. *Mathematical Foundations of Thermodynamics* (Pergamon, Oxford, 1964).
19. Lieb, E. H. & Yngvason, J. The physics and mathematics of the second law of thermodynamics. *Phys. Rept.* **310,** 1–96 (1999).
20. Horodecki, R., Horodecki, P., Horodecki, M. & Horodecki, K. Quantum entanglement. *Rev. Mod. Phys.* **81,** 865–942 (2009).
21. Brandao, F. G. S. L., Horodecki, M., Ng, N. H. Y., Oppenheim, J. & Wehner, S. The second laws of quantum thermodynamics. Preprint at http://www.pnas.org/content/early/2015/02/05/1411728112.abstract (2013).
22. Marvian, I. & Spekkens, R. W. Extending Noether's theorem by quantifying the asymmetry of quantum states. *Nat. Commun.* **5,** 3821 (2014).

23. Marvian, I. & Spekkens, R. W. The theory of manipulations of pure state asymmetry: I. Basic tools, equivalence classes and single copy transformations. *New J. Phys.* **15,** 033001 (2013).
24. Egloff, D., Dahlsten, O. C. O., Renner, R. & Vedral, V. Laws of thermodynamics beyond the von Neumann regime. Preprint at http://arxiv.org/abs/1207.0434 (2012).
25. Horodecki, M. & Oppenheim, J. Fundamental limitations for quantum and nanoscale thermodynamics. *Nat. Commun.* **4,** 2059 (2013).
26. Åberg, J. Truly work-like work extraction via a single-shot analysis. *Nat. Commun* **4,** 1925 (2013).
27. Janzing, D., Wocjan, P., Zeier, R., Geiss, R. & Beth, T. Thermodynamic cost of reliability and low temperatures: tightening landauer's principle and the second law. *Int. J. Theor. Phys.* **39,** 2717–2753 (2000).
28. Brandão, F. G. S. L., Horodecki, M., Oppenheim, J., Renes, J. M. & Spekkens, R. W. Resource theory of quantum states out of thermal equilibrium. *Phys. Rev. Lett.* **111,** 250404 (2013).
29. Renyi, A. in *Fourth Berkeley Symp. Math. Statist. Probability*. 547–561 (University of California Press, 1961).
30. Müller-Lennert, M., Dupuis, F., Szehr, O., Fehr, S. & Tomamichel, M. On quantum Rényi entropies: A new generalization and some properties. *J. Math. Phys.* **54,** 122203 (2013).
31. Bartlett, S. D., Rudolph, T. & Spekkens, R. W. Reference frames, superselection rules, and quantum information. *Rev. Mod. Phys.* **79,** 555 (2007).
32. Gour, G. & Spekkens, R. W. The resource theory of quantum reference frames: manipulations and monotones. *New J. Phys.* **10,** 033023 (2008).
33. Gour, G., Marvian, I. & Spekkens, R. W. Measuring the quality of a quantum reference frame: The relative entropy of frameness. *Phys. Rev. A* **80,** 012307 (2009).
34. Ahmadi, M. *et al.* The WAY theorem and the quantum resource theory of asymmetry. *New. J. Phys.* **15,** 013057 (2013).
35. Marvian, I. & Spekkens, R. W. Modes of asymmetry: the application of harmonic analysis to symmetric quantum dynamics and quantum reference frames. *Phys. Rev. A* **90,** 062110 (2014).
36. Aberg, J. Quantifying Superposition. Preprint at http://arxiv.org/abs/quant-ph/0612146 (2006).
37. Baumgratz, T., Cramer, M. & Plenio, M. B. Quantifying coherence. *Phys. Rev. Lett.* **113,** 140401 (2014).
38. Skrzypczyk, P., Short, A. J. & Popescu, S. Extracting work from quantum systems. Preprint at http://arxiv.org/abs/1302.2811 (2013).
39. Janzing, D. Quantum thermodynamics with missing reference frames: decompositions of free energy into non-increasing components. *J. Statist. Phys.* **125,** 757–772 (2006).
40. Allahverdyan, A. E., Balian, R. & Nieuwenhuizen, T. M. Maximal work extraction from finite quantum systems. *Europhys. Lett.* **67,** 565–571 (2004).
41. Alicki, R., Horodecki, M., Horodecki, P. & Horodecki, R. Thermodynamics of quantum information systems â€" hamiltonian description. *Open Syst. Inf. Dyn.* **11,** 205–217 (2004).
42. Dahlsten, O., Renner, R., Rieper, E. & Vedral, V. Inadequacy of von neumann entropy for characterizing extractable work. *New J. Phys.* **13,** 053015 (2011).
43. Alicki, R. & Fannes, M. Entanglement boost for extractable work from ensembles of quantum batteries. *Phys. Rev. E* **87,** 042123 (2013).
44. Frenzel, M., Jennings, D. & Rudolph, T. Reexamination of pure qubit work extraction. *Phys. Rev. E Stat. Nonlin. Soft Matter Phys.* **90,** 052136 (2014).
45. Maruyama, K., Nori, F. & Vedral, V. *Colloquium*: The physics of Maxwell's demon and information. *Rev. Mod. Phys.* **81,** 1–23 (2009).
46. Åberg, J. Catalytic coherence. *Phys. Rev. Lett.* **113,** 150402 (2014).
47. Vaccaro, J. A., Anselmi, F., Wiseman, H. M. & Jacobs, K. Tradeoff between extractable mechanical work, accessible entanglement, and ability to act as a reference system, under arbitrary superselection rules. *Phys. Rev. A* **77,** 032114 (2008).
48. Ćwikliński, P., Studziński, M., Horodecki, M. & Oppenheim, J. Limitations for thermodynamical processing of coherences. Preprint at http://arxiv.org/abs/1405.5029 (2014).
49. Lostaglio, M., Korzekwa, K., Jennings, D. & Rudolph, T. Quantum coherence, time-translation symmetry and thermodynamics. Preprint at http://arxiv.org/abs/1410.4572 (2014).
50. Jennings, D. & Rudolph, T. Entanglement and the thermodynamic arrow of time. *Phys. Rev. E Stat. Nonlin. Soft Matter Phys.* **81,** 061130 (2010).
51. Wilde, M. M., Winter, A. & Yang, D. Strong converse for the classical capacity of entanglement-breaking and hadamard channels via a sandwiched Rényi relative entropy. *Comm. Math. Phys.* **331,** 593–622 (2014).
52. Mosonyi, M. & Ogawa, T. Quantum hypothesis testing and the operational interpretation of the quantum Rényi relative entropies. *Comm. Math. Phys.* **332,** 1–32 (2014).
53. Jarzynski, C. Nonequilibrium equality for free energy differences. *Phys. Rev. Lett.* **78,** 2690 (1997).
54. Crooks, G. E. Entropy production fluctuation theorem and the nonequilibrium work relation for free energy differences. *Phys. Rev. E* **60,** 2721 (1999).
55. Jarzynski, C. & Wójcik, D. K. Classical and quantum fluctuation theorems for heat exchange. *Phys. Rev. Lett.* **92,** 230602 (2004).
56. Jarzynski, C. Equalities and inequalities: irreversibility and the second law of thermodynamics at the nanoscale. *Annu. Rev. Condens. Matter Phys.* **2,** 329 (2011).
57. Yunger Halpern, N., Garner, A. J. P., Dahlsten, O. C. O. & Vedral, V. Unification of fluctuation theorems and one-shot statistical mechanics. Preprint at http://arxiv.org/abs/1409.3878 (2014).
58. Faist, P., Oppenheim, J. & Renner, R. Gibbs-Preserving Maps outperform Thermal Operations in the quantum regime. Preprint at http://arxiv.org/abs/1406.3618 (2014).
59. Ruch, E., Schranner, R. & Seligman, T. H. The mixing distance. *J. Chem. Phys.* **69,** 386–392 (1978).
60. Horodecki, M., Horodecki, P. & Oppenheim, J. Reversible transformations from pure to mixed states and the unique measure of information. *Phys. Rev. A* **67,** 062104 (2003).
61. Klimesh, M. Inequalities that Collectively Completely Characterize the Catalytic Majorization Relation. Preprint at http://arxiv.org/abs/0709.3680 (2007).
62. Frank, R. L. & Lieb, E. H. Monotonicity of a relative Rényi entropy. *J. Math. Phys.* **54,** 122201 (2013).
63. Beigi, S. Sandwiched Renyi divergence satisfies data processing inequality. *J. Math. Phys.* **54,** 122202 (2013).
64. Datta, N. & Leditzky, F. A limit of the quantum Rényi divergence. *J. Phys. A* **47,** 045304 (2014).
65. Horodecki, M., Horodecki, P. & Horodecki, R. Mixed-state entanglement and distillation: Is there a 'bound' entanglement in nature? *Phys. Rev. Lett.* **80,** 5239–5242 (1998).
66. Horodecki, P., Horodecki, M. & Horodecki, R. Bound entanglement can be activated. *Phys. Rev. Lett.* **82,** 1056–1059 (1999).
67. Skrzypczyk, P., Short, A. J. & Popescu, S. Work extraction and thermodynamics for individual quantum systems. *Nat. Commun.* **5,** 4185 (2014).

Acknowledgements

M.L. thanks K. Korzekwa, J. Oppenheim, F. Mintert, T. Tufarelli and A. Milne for useful discussions. D.J. is supported by the Royal Society. T.R. is supported by the Leverhulme Trust. M.L. is supported in part by EPSRC, COST Action MP1209 and Fondazione Angelo Della Riccia.

Author contributions

M.L., D.J. and T.R. designed the research, M.L. performed the calculations/developed the results, M.L. and D.J. wrote the paper.

Additional information

Competing financial interests: There are no competing financial interests.

4

The H-index of a network node and its relation to degree and coreness

Linyuan Lü[1], Tao Zhou[2,3], Qian-Ming Zhang[2,4] & H. Eugene Stanley[1,4]

Identifying influential nodes in dynamical processes is crucial in understanding network structure and function. Degree, H-index and coreness are widely used metrics, but previously treated as unrelated. Here we show their relation by constructing an operator $\mathcal{H}$, in terms of which degree, H-index and coreness are the initial, intermediate and steady states of the sequences, respectively. We obtain a family of H-indices that can be used to measure a node's importance. We also prove that the convergence to coreness can be guaranteed even under an asynchronous updating process, allowing a decentralized local method of calculating a node's coreness in large-scale evolving networks. Numerical analyses of the susceptible-infected-removed spreading dynamics on disparate real networks suggest that the H-index is a good tradeoff that in many cases can better quantify node influence than either degree or coreness.

[1] Alibaba Research Center for Complexity Sciences, Alibaba Business College, Hangzhou Normal University, Hangzhou 311121, China. [2] CompleX Lab, Web Sciences Center, University of Electronic Science and Technology of China, Chengdu 611731, China. [3] Big Data Research Center, Univesrsity of Electronic Science and Technology of China, Chengdu 611731, China. [4] Department of Physics and Center for Polymer Studies, Boston University, Boston, Massachusetts 02215, USA. Correspondence and requests for materials should be addressed to L.L. (email: linyuan.lv@hznu.edu.cn) or to H.E.S. (email: hes@bu.edu).

The focus of network science research has been shifting from discovering macroscopic statistical regularities[1–4] to uncovering the role played by such microscopic elements as nodes, links and motifs in the structure and dynamics of the system[5–10]. Being able to effectively and efficiently identify the critical nodes associated with the specific dynamics of large-scale networks[11] will allow us to better control the outbreak of epidemics[12], conduct successful advertisements for e-commercial products[13], prevent catastrophic outages in power grids or the Internet[14], optimize the use of limited resources to facilitate information propagation[15], discover drug target candidates and essential proteins[16], and design strategies for communication breakdowns in human and telecommunication networks[17].

The simplest way to measure the importance of a node is to determine its degree, that is, to count the number of its linked neighbours. Previous studies have shown that protecting, immunizing and regulating large-degree nodes can maintain network connectivity[18], halt infectious disease propagation[12], enhance synchronizability[19], improve transport capacity[20] and promote cooperation in evolutionary games[21].

Recently, Kitsak *et al.*[22] argued that the location of a node is more significant than the number of its linked neighbours, and they suggested that coreness is a better indicator of a node's influence on spreading dynamics than degree. The coreness of a node is measured by k-core decomposition[23], and a larger coreness value indicates that the node is more centrally located in the network. The k-core decomposition process is initiated by removing all nodes with degree $k=1$. This causes new nodes with degree $k\leq 1$ to appear. These are also removed and the process is continued until the only nodes remaining are those of degree $k>1$. The removed nodes and their associated links form the 1-shell. This pruning process is repeated for the nodes of degree $k=2$ to extract the 2-shell, that is, in each stage the nodes with degree $k\leq 2$ are removed. The process is continued until all higher-layer shells have been identified and all network nodes have been removed. Then each node i is assigned a shell layer c_i, called the coreness of node i. Recent studies suggest that coreness is a good measure of a node's influence[22,24].

Calculating coreness requires global topological information, and its implementation is usually centralized, which can hinder its application to very large-scale dynamical networks. In contrast, degree is a simple local index, but of lower utility. As a tradeoff, Chen *et al.*[25,26] proposed indices using the local neighbourhood information of individual nodes, which perform well but lack an underlying mathematical structure. This approach brings to mind the Hirsch index (also called the H-index)[27], which was originally used to measure the citation impact of a scholar or a journal[27–29]. For a scholar or journal, the H-index is defined as the maximum value h such that there exists at least h papers, each with citation count $\geq h$.

Here we discuss the extension of the H-index concept to quantify how important a node is to its network[30]. The H-index of a node is defined to be the maximum value h such that there exists at least h neighbours of degree no less than h. In the Supplementary Fig. 1, we compare calculating the H-index of a scholar and the H-index of a node. Degree, H-index and coreness seem to be independent but are actually interrelated. We construct an operator $\mathcal{H}$ on a group of reals that returns a node's H-index when acting on its neighbours' degrees. By sequentially and synchronously applying the $\mathcal{H}$ operator to each node, the returned value soon converges to coreness, that is, in terms of this operator, degree, H-index and coreness are its initial state, intermediate state and steady state, and all other intermediate states can also be treated as centrality measures. We further show that the convergence to coreness can be guaranteed even under an asynchronous updating process, thus allowing a distributed computing algorithm to deal with large-scale dynamical networks. To see whether these centralities can measure a node's influence, we apply the standard susceptible-infected-removed (SIR) spreading model[31] on eight real networks from disparate fields and calculate the correlation between each node's influence and its centrality values. Simulation results show that the H-index outperforms both degree and coreness in several cases, and thus can be considered a good tradeoff between degree and coreness.

Results

Mathematical relationship. We construct an operator, $\mathcal{H}$, which acts on a finite number of reals $(x_1, x_2, \cdots, x_n)$ and returns an integer $y=\mathcal{H}(x_1, x_2, \cdots, x_n)>0$, where y is the maximum integer such that there exist at least y elements in $(x_1, x_2, \cdots, x_n)$, each of which is no less than y. For a scholar with n publications, $x_1, x_2, \cdots, x_n$ is the number of citations to these publications and $\mathcal{H}(x_1, x_2, \cdots, x_n)$ is the scholar's H-index.

Denote $G(V, E)$ an undirected simple network, where V is the set of nodes and E is the set of links. The degree of an arbitrary node i is denoted by k_i and its neighbours' degrees are $k_{j_1}, k_{j_2}, \cdots, k_{j_{k_i}}$. Then, we define the H-index of node i

$$h_i = \mathcal{H}\left(k_{j_1}, k_{j_2}, \cdots, k_{j_{k_i}}\right). \tag{1}$$

We define $h_i^{(0)}=k_i$ to be the zero-order H-index of node i, and define the n-order H-index $(n>0)$ iteratively as

$$h_i^{(n)} = \mathcal{H}\left(h_{j_1}^{(n-1)}, h_{j_2}^{(n-1)}, \cdots, h_{j_{k_i}}^{(n-1)}\right). \tag{2}$$

The H-index of node i is equal to the first-order H-index, namely $h_i^{(1)}=h_i$. A more detailed illustration can be found in Supplementary Note 1.

Theorem 1: for every node $i\in V$ of an undirected simple network $G(V, E)$, its H-index sequence $h_i^{(0)}, h_i^{(1)}, h_i^{(2)}, \cdots$ will converge to the coreness of node i,

$$c_i = \lim_{n\to\infty} h_i^{(n)}. \tag{3}$$

The proof is given in the Methods section. We give an example of iterative process from degree to coreness in the Supplementary Fig. 2.

This theorem shows that the degree, H-index and coreness are respectively the initial, intermediate and steady states under successive operations by $\mathcal{H}$. Given a network $G(V, E)$, the convergence time n_∞ is defined as the minimum number of iterations required to reach coreness from degree using the operator $\mathcal{H}$, that is, n_∞ is the minimum integer such that $h_i^{(n_\infty)}=h_i^\infty=c_i$ for $\forall i\in V$.

Using equation (2) and Theorem 1, we implement the synchronous updating in eight representative real networks drawn from disparate fields, including two social networks (Sex and Facebook), two collaboration networks (Jazz and NS), one communication network (Email), one information network (PB), one transportation network (USAir) and one technological network (Router). In brief, Jazz[32] is a collaboration network of jazz musicians and consists of 198 nodes and 2,742 interactions, NS[33] is a co-authorship network of scientists working on network science, our Email[34] network is of e-mail interchanges between members of the Rovira i Virgili University (Tarragona), our Sex[35] network is of a bipartite sexual activity web community in which nodes are female (sex sellers) and male (sex buyers) and links between them are established when males write posts indicating sexual encounters with females, Facebook[36] is a sample of the friendship network of Facebook users, PB[37] is a network of US political blogs (the original links are directed, but here we treat them as undirected), USAir[38] is the US air transportation network and Router[39] is a symmetrized snapshot of the structure of the Internet at the level of autonomous systems. Their topological features are shown in Table 1. Experiments show that the sequence of H-indices quickly converges to the coreness (n_∞ in Table 1). In addition to the degree, H-index and coreness, all other intermediate states $h^{(2)}, h^{(3)}, \cdots$ can also be considered as centrality measures.

The resolution rate of the $h^{(n)}$-index is the probability that two randomly chosen nodes will have different $h^{(n)}$. It is also a useful index for measuring the degree to which a network is coarse grained. Note that degree is the most distinguishable index and coreness is the least distinguishable, and that the resolution rate decreases as the index order increases (Supplementary Fig. 3; Supplementary Note 2). On the other hand, the calculation of degree requires less information, while coreness requires the most information. For a given node i, the information required to calculate $h_i^{(n)}$ can be measured by information coverage, which is defined as the ratio of the number of nodes with a distance no more than n from i to the network size. The coverage rate increases as the index order increases (Supplementary Fig. 4). Surprisingly, we find that in some networks, such as NS and Router, the coverage rate is <1 even for $h^{(n_\infty)}$, indicating that only partial information is required when calculating the coreness of a node in these networks (Supplementary Note 3).

Figure 1a shows the H-indices in different orders for a typical network Router. From left to right, we see the coarse graining process from degree to coreness. Figure 1b shows the probability distribution $p(h)$, defined as the probability that a randomly selected node's $h^{(n)}$ value is equal to h for the cases $n=0$, $n=1$ and $n=n_\infty$ of the network Router. Note that as the order n increases from 0 to 6 ($n_\infty=6$ for Router), the distribution narrows (see Supplementary Figs. 5–12 for the distributions of all H-family indices for the eight networks under study). Nevertheless, the distribution of values of high-order H-indices is still relatively broad, suggesting its potential as a good indicator of a node's importance[22].

To show the different roles of different H-family indices, we iteratively construct a hierarchical tree (see ref. 40 for a similar method). As shown in Fig. 1c, the initialized network is of two levels, isomorphic to a star with L leaves. Here we set $L=4$. In each step, every leaf node becomes a star with L leaves, and the central node is connected with its original neighbours. After each step, the number of levels is increased by one, and the nodes in the more central positions (that is, of smaller levels) have a higher influence that is not reflected by their degree if the number of levels is high. Note that the three trees in Fig. 1c–e have 2, 3 and 4 levels, respectively, and in a hierarchical tree of D levels we need the index $h^{(0)}, h^{(1)}, \cdots, h^{(D-2)}$ to quantify node influence. Such example clearly shows that a few low-order H-indices are not always enough to distinguish different nodes' influences.

Asynchronous updating. The updating driven by $\mathcal{H}$ uses only local information, and it rapidly converges to coreness. However, the updating from $h^{(n-1)}$ to $h^{(n)}$ is implemented synchronously according to equation (2), and thus in principle requires a centralized controller to set up a global clock that records the order n. In particular, if the target network is evolving, the addition of a single link will require the recalculation of the entire sequence of H-indices. This limits the application of H-indices to large-scale dynamical networks. Fortunately, the asynchronous updating can still guarantee a convergence to coreness, as shown in the following theorem.

Theorem 2: given an undirected simple network $G(V, E)$, for every node $j \in V$, we define $g_j = k_j$. In each iteration of the asynchronous updating process, a node i is randomly selected and its g value updated, that is,

$$\mathcal{H}\left(g_{j_1}, g_{j_2}, \cdots, g_{j_{k_i}}\right) \to g_i, \tag{4}$$

where $j_1, j_2, \cdots, j_{k_i}$ are the neighbouring nodes of i. If $|V|$ is finite, this updating process will reach a steady state $(g_1^\infty, g_2^\infty, \cdots, g_{|V|}^\infty)$ after a finite number of iterations such that the updating at any node will not change its g value, namely,

$$\forall i \in V, \quad g_i^\infty = \mathcal{H}\left(g_{j_1}^\infty, g_{j_2}^\infty, \cdots, g_{j_{k_i}}^\infty\right). \tag{5}$$

In the steady state, for every node i we have $g_i^\infty = c_i$. The proof is given in the Methods section.

Note that in the updating process of equation (4) the g values are not associated with a temporal superscript. In fact, at a certain updating step the values $g_{j_1}, g_{j_2}, \cdots, g_{j_{k_i}}$ could lie in different stages, some updated dozens of times and others never updated. Thus, for any node i, before it reaches the steady state $g_i^\infty = c_i$, all the intermediate values are not necessarily equal to any order of H-indices. Theorem 2 makes a significant step towards making feasible applications to large-scale dynamical networks possible: it guarantees that a decentralized and localized method can be used to calculate the coreness, and even if the network evolves in time, equation (4) can be used to calculate the coreness when new links and nodes are periodically added, and all the previously obtained g values are still usable.

Quantifying spreading influences. Epidemic spreading, one of the most significant dynamic behaviours in complex networks[41], can be used to characterize such real processes as the spreading of infectious diseases[42], the diffusion of microfinance[43] and the propagation of traffic congestion[44]. To see whether the H-indices

Table 1 | The basic topological features and the convergence time of the eight real networks.

Networks	$\lvert V\rvert$	$\lvert E\rvert$	$\langle k\rangle$	$\langle d\rangle$	C	r	n_∞
Jazz	198	2,742	27.7	2.24	0.633	0.02	13
NS	379	914	4.82	6.04	0.798	−0.082	4
Email	1,133	5,451	9.62	3.61	0.254	0.078	16
Sex	15,810	38,540	4.88	5.79	0	−0.115	19
Facebook	63,731	817,090	25.64	8.09	0.253	0.177	63
PB	1,222	16,714	27.36	2.74	0.360	−0.221	18
USAir	332	2,126	12.81	2.74	0.749	−0.208	5
Router	5,022	6,258	2.49	6.45	0.033	−0.138	6

$|V|$ and $|E|$ are the number of nodes and links, respectively. $\langle k\rangle$ and $\langle d\rangle$ are the average degree and the average distance, respectively. C and r are the clustering coefficient[1] and assortative coefficient[3], respectively. Nodes with degree 1 are excluded from the calculation of clustering coefficient. Sex is a bipartite network and thus is characterized by (or 'has') clustering coefficient zero. n_∞ is the convergence time to coreness, defined as the minimum steps required to reach coreness from degree by the operator $\mathcal{H}$.

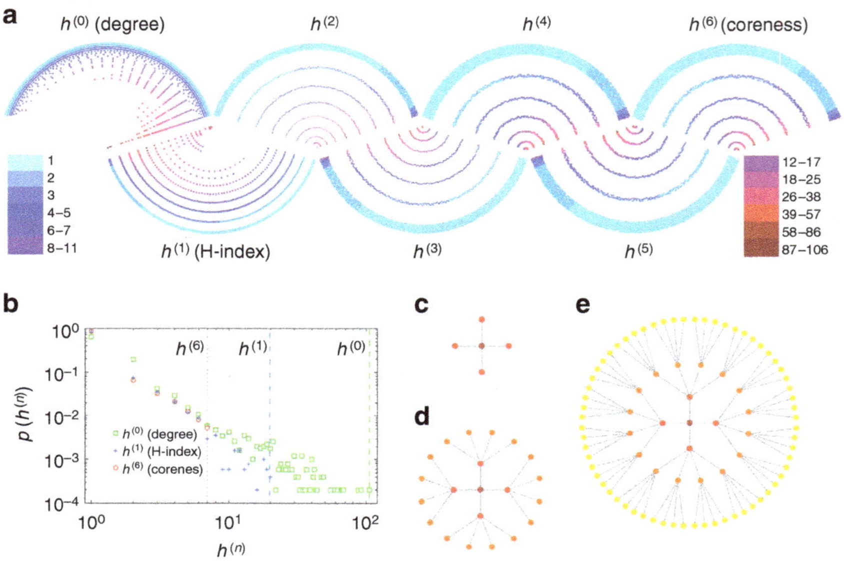

Figure 1 | Comparisons among H-indices in different orders for the network Router. The subplot **a** exhibits a visualized illustration, where the colour represents the node degree (from 1 to 106). The node location represents the $h^{(n)}$-index. Nodes located at a 'fan' closer to the centre of the fan have higher $h^{(n)}$ values, and nodes located at the same layer of a fan have the same $h^{(n)}$ values. The subplot **b** shows the distributions of values of H-indices in different orders, where the green squares, blue crosses and red circles represent the cases for $n=0$, $n=1$ and $n=6$, respectively. The dash lines in different colours mark the largest values for the corresponding indices. The subplots **c**, **d** and **e** show an illustration of the hierarchical trees with 2, 3 and 4 levels, respectively.

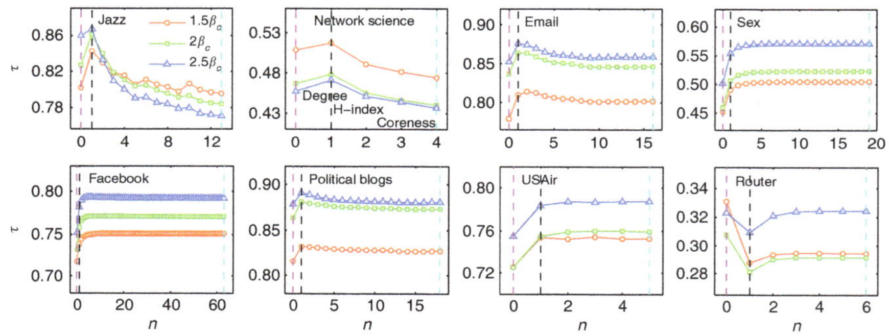

Figure 2 | The Kendall Tau between $h^{(n)}$-index and the node influence index R for undirected networks. The value of n ranges from 0 to n_∞. The red circles, green squares and blue triangles represent the case of $\beta=1.5\beta_c$, $2\beta_c$ and $2.5\beta_c$, respectively. The dash lines in purple, black and cyan colours emphasize the τ values for degree (that is, $h^{(0)}$), H-index (that is, $h^{(1)}$) and coreness (that is, h^{∞}), respectively. The influence R of a node is quantified using the average number of removed nodes after the dynamics over 1,000 independent runs.

can quantify the spreading of node influence, we study the standard SIR spreading model[41] in which the influence R_i of node i is quantified using the average number of removed nodes after the dynamics over 1,000 independent runs, each of which begins with node i as the sole infected seed (see details in Methods).

Given the order n $(0\leq n\leq n_\infty)$, we have two sequences associated with the $|V|$ nodes: the $h^{(n)}$-index $h_1^{(n)}, h_2^{(n)}, \cdots, h_{|V|}^{(n)}$ and the influences $R_1, R_2, \cdots, R_{|V|}$. To quantify to what extent the $h^{(n)}$-index resembles node influence values, we apply the Kendall Tau (τ) coefficient[45], which lies in the range $-1\leq\tau\leq 1$. The larger value of τ means a stronger correlation between the two sequences (see Methods for the definition of τ).

Figure 2 shows that the H-index of node i is highly correlated with the influence value R_i. In many cases (Jazz, NS, Email and PB), the H-index (that is, $h^{(1)}$-index) outperforms both degree and coreness. In other cases (Sex, USAir and Facebook), coreness outperforms degree and the performance of low-order H-indices rapidly approaches that of coreness as the order increases from zero. Router is an exception, where degree performs the best and the H-index the worst. Note that because Router is the most sparse network it may hinder spreading and make predicting influences more difficult (as indicated by the smallest τ value). Thus, the sequence of H-indices, starting from degree and driven by the operator $\mathcal{H}$, provides more alternative centralities in characterizing the importance of nodes, and the low-order H-indices are a good tradeoff between degree and coreness. We further compare the three best known H-indices (degree, H-index and coreness) with two well-known centrality indices (closeness and betweenness) for undirected networks. The definitions of closeness and betweenness are given in Methods. As shown in Table 2, all the H-family indices are competitive, and the H-index and closeness are the overall best performers, but the computational cost in calculating closeness is huge and thus the H-index is the better choice.

We further test two well-known dynamical processes: the susceptible-infected-susceptible (SIS) spreading model[46] and bond percolation[47]. In the SIS model, the node influence index R is defined as the probability that this node will remain infected

Table 2 | The Kendall Tau between the node influence index *R* of SIR model and five centrality indices.

Networks	Degree	H-index	Coreness	Closeness	Betweenness
Jazz	0.8021	**0.8431**	0.7958	0.6961	0.4629
NS	0.5092	**0.5178**	0.4747	0.3510	0.3392
Email	0.7794	**0.8103**	0.8021	0.7747	0.6195
Sex	0.4525	0.4905	0.5049	**0.7029**	0.3834
Facebook	0.7173	0.7381	**0.7513**	0.6716	0.4851
PB	0.8159	**0.8321**	0.8274	0.7375	0.6589
USAir	0.7256	**0.7540**	0.7529	0.7453	0.5442
Router	0.3309	0.2877	0.2946	**0.5975**	0.3228

SIR, susceptible-infected-removed.
They are degree (that is, $h^{(0)}$), H-index (that is, $h^{(1)}$), coreness (that is, h^{∞}), closeness and betweenness. The spreading rate β is set as $\beta = 1.5\beta_c$, and for other values of β, the results are very similar and can be found in the Supplementary Table 1 ($\beta = 2\beta_c$) and Supplementary Table 2 ($\beta = 2.5\beta_c$). In each row, the largest τ is highlighted in bold.

in the steady state. In bond percolation, the node influence index R is defined as either the probability that the target node belongs to the giant component or the size of the connected component that encompasses the target node. The results (Supplementary Tables 3–11) suggest that the H-family indices are competitive, especially for the H-index and coreness. Detailed information about the dynamical processes and the simulation results can be found in Supplementary Note 4.

Discussion

We discover an important relation among degree, H-index and coreness—centrality measures that have previously been treated as unrelated. We construct an operator $\mathcal{H}$ that functions as a 'necklace' stringing together degree, H-index, coreness and other intermediate indices. All these indices are centralities that characterize each node's importance. Using the operator $\mathcal{H}$ to achieve the coreness looks like an inverse way to the iterative removal of nodes with degree less than k that is widely used to determine the k-core of a network. Indeed, they are different, as the iterative removal method cannot result in H-index, or any other H-family indices except for degree and coreness, and the steps required to achieve the final coreness for the two methods are also different.

Although the importance of a given node strongly depends on the type of dynamical processes under consideration and thus there is no single best centrality measure, we need effective and elegant centralities in practice. For example, although we know that degree is not an accurate centrality measure in quantifying node influence in many dynamical processes[9,22], it is still a useful estimation of node importance, even without the specification of dynamics. In despite of its bias and disadvantages[48,49,50] the H-index is now becoming the most widely applied index for academic performances, ranging from individual scientists, scientific journals to universities and even countries. As indicated by the simulations on the SIR model, the SIS model and bond percolation, the H-family indices are effective in quantifying the spreading influences of nodes.

The asynchronous updating by $\mathcal{H}$ can still guarantee the convergence to coreness, and thus one can use a decentralized local algorithm to calculate coreness, which is able to deal with evolving networks. However, randomly selecting nodes to update in each iteration may greatly extend the time required before arriving at the steady state, even in static networks. Thus, the process for selecting which node to update is a nontrivial issue. For example, we can shorten the convergence time by reducing the selection probability of nodes that have been updated many times but have g values seldom changed. In addition, the change of g value of a node will enhance the updating probabilities of its neighbouring nodes, making this issue more complicated and thus more interesting.

The methodologies and results presented here are also applicable to directed networks and weighted networks, in which degree is replaced by in-degree, out-degree or node strength. In this way we can also define the H_{in}-index, H_{out}-index and weighted H-index, as well as in-coreness, out-coreness and weighted coreness. An example of how to calculate the directed H-family indices is presented in Supplementary Fig. 13. To test the performance, we compare the directed H-family indices with PageRank[51] and HITs[52] on seven directed networks. The basic statistics of these seven directed networks are summarized in Supplementary Table 12 and a more detailed introduction given in Supplementary Note 5. The results (Supplementary Tables 13–24) suggest that the directed H-family indices are still very competitive (in-coreness performs overall best).

Methods

Proof of Theorem 1. From the definition of $\mathcal{H}$-function, for any node i and integer $n \geq 0$, we have $h_i^{(n)} \geq 0$ and $h_i^{(0)} \geq h_i^{(1)}$. Applying mathematical induction, we prove that $h_i^{(n)} \geq h_i^{(n+1)}$ for any node i and integer $n \geq 0$. If $h_i^{(n)} \geq h_i^{(n+1)}$ is valid for all $n \leq m$, we then prove it to also be valid for $n = m+1$. From the definition, $h_i^{(m+2)} = \mathcal{H}(h_{j_1}^{(m+1)}, h_{j_2}^{(m+1)}, \cdots, h_{j_{k_i}}^{(m+1)})$, according to the induction hypothesis, $h_j^{(m+1)} \leq h_j^{(m)}$ for any node j. Therefore, $h_i^{(m+2)} = \mathcal{H}(h_{j_1}^{(m+1)}, h_{j_2}^{(m+1)}, \cdots, h_{j_{k_i}}^{(m+1)}) \leq \mathcal{H}(h_{j_1}^{(m)}, h_{j_2}^{(m)}, \cdots, h_{j_{k_i}}^{(m)}) = h_i^{(m+1)}$, namely $h_i^{(m+1)} \geq h_i^{(m+2)}$. Since $h_i^{(0)}, h_i^{(1)}, h_i^{(2)}, \cdots$ is a monotonously nonincreasing sequence, and each element is nonnegative, it has a nonnegative limitation. Thus, we can define h_i^{∞} as the limitation of the sequence $h_i^{(0)}, h_i^{(1)}, h_i^{(2)}, \cdots$.

We then introduce two simple relations. First, if $G'(V', E') \subseteq G(V, E)$, then from the definition of an $\mathcal{H}$-function it is obvious that for any node $i \in V'$ and any integer $n \geq 0$, $h_{i,G'}^{(n)} \leq h_{i,G}^{(n)}$, where the subscript G' indicates that the corresponding index is defined on the subgraph G'. Second, if we denote $k_{\min}$ the minimal degree of G, then for any node i and any integer $n \geq 0$, $h_i^{(n)} \geq k_{min}$. This second relation can also be proven by mathematical induction. It is clear that it is valid for $n = 0$. We next prove that if this relation is valid for all $n \leq m$, then it is also valid for $n = m+1$. For any node i, $h_i^{(m+1)} = \mathcal{H}(h_{j_1}^{(m)}, h_{j_2}^{(m)}, \cdots, h_{j_{k_i}}^{(m)})$, and according to the induction hypothesis, all elements in $(h_{j_1}^{(m)}, h_{j_2}^{(m)}, \cdots, h_{j_{k_i}}^{(m)})$ are no less than $k_{\min}$ and the number of elements in $(h_{j_1}^{(m)}, h_{j_2}^{(m)}, \cdots, h_{j_{k_i}}^{(m)})$ is $k_i \geq k_{\min}$. Therefore, according to the definition of $\mathcal{H}$-function, $h_i^{(m+1)} \geq k_{min}$.

If we denote G' the c_i-core of G, it is clear that $G' \subseteq G$ and in G', $k_{\min} \geq c_i$. Therefore, $h_i^{\infty} \geq k_{min} \geq c_i$. We denote $G''(V'', E'')$ the induced subgraph containing all nodes j such that $h_j^{\infty} \geq h_i^{\infty}$. Note that the node i itself also belongs to G''. For any node $l \in V$, we find $h_l^{\infty} = \mathcal{H}(h_{j_1}^{\infty}, h_{j_2}^{\infty}, \cdots, h_{j_{k_l}}^{\infty})$, where $j_1, j_2, \cdots, j_{k_l}$ are the k_l neighbours of node l. For any node $j \in V''$, since in G, $h_j^{\infty} > h_i^{\infty}$, there are at least h_i^{∞} neighbours of node j with h^{∞} values no less than h_i^{∞}. Thus, these neighbours also belong to V''. Therefore, in G'' the degrees of all the nodes are no less then h_i^{∞}, that is, G'' is a subgraph of G's h_i^{∞}-core. Because c_i is the coreness of i, $h_i^{\infty} \leq c_i$. Combining the two inequalities, we arrive at Theorem 1.

Proof of Theorem 2. For convenience, we introduce the systematic time step t. Initially we set $t = 0$, and for every node $j \in V$ we define $g_j^{(0)} = k_j$. Then, at each time step, we randomly select a node and perform the $\mathcal{H}$ operator on it. If at time step $t > 0$ the node i is selected, then $g_i^{(t)} = \mathcal{H}(g_{j_1}, g_{j_2}, \cdots, g_{j_{k_i}})$. The g value without a temporal superscript indicates the most recently updated value, since only the current value is meaningful in the asynchronous updating procedure. Note that all neglected superscripts are smaller than t. Note also that for an arbitrary pair (t, j), $g_j^{(t)}$ may not exist unless the node j is selected at time step t.

We first prove that if any node $j \in V$ has been selected at time steps t_1 and t_2, and $t_2 > t_1 \geq 0$, then $g_j^{(t_1)} \geq g_j^{(t_2)}$. Note that for any node j selected at time step $t = 1$ we have $g_j^{(0)} \geq g_j^{(1)}$. We apply mathematical induction and assume that the above inequality holds when $t_1 \leq n$ and $t_2 \leq n$, and we next prove this also holds for $t_1 \leq n+1$ and $t_2 \leq n+1$. If node i is the one selected at time step $t = n+1$, then the aforementioned inequality holds for all other nodes $j \neq i$. We denote t' $(0 \leq t' \leq n)$ an arbitrary earlier updating time step of node i, and record two updates, that is, $g_i^{(t')} = \mathcal{H}(g_{j_1}^{(\phi_1)}, g_{j_2}^{(\phi_2)}, \cdots, g_{j_{k_i}}^{(\phi_{k_i})})$ and $g_i^{(n+1)} = \mathcal{H}(g_{j_1}^{(\varphi_1)}, g_{j_2}^{(\varphi_2)}, \cdots, g_{j_{k_i}}^{(\varphi_{k_i})})$. Note that for any m $(1 \leq m \leq k_i)$ we have $\phi_m \leq \varphi_m \leq n$. According to the induction hypothesis, $g_{j_m}^{(\phi_m)} \geq g_{j_m}^{(\varphi_m)}$, together with the definition of $\mathcal{H}$ function, we have $g_i^{(t')} \geq g_i^{(n+1)}$. We denote the updating time steps of any node $i \in V$ to be $0 = t_0 < t_1 < t_2 < \cdots$, then $g_i^{(t_0)}, g_i^{(t_1)}, g_i^{(t_2)}, \cdots$ is a monotonously nonincreasing sequence and each element is nonnegative, and therefore it has a nonnegative limitation. At this point, we can define g_i^{∞} as the limitation of the sequence $g_i^{(t_0)}, g_i^{(t_1)}, g_i^{(t_2)}, \cdots$.

We first prove that for any node $j \in V$, $g_j^{\infty} \geq c_j$. Proving by contradiction, when this inequality does not hold we denote i to be the first node to reach a g value smaller than c_i, and the corresponding updating time step is t, that is, $g_i^{(t)} < c_i$ and before t for all nodes $j \in V$, $g_j \geq c_j$. Note that g_j without a superscript indicates the last updated value before t. Therefore, $g_i^{(t)} = \mathcal{H}(g_{j_1}, g_{j_2}, \cdots, g_{j_{k_i}}) \geq \mathcal{H}(c_{j_1}, c_{j_2}, \cdots, c_{j_{k_i}})$. According to Theorem 1, $\mathcal{H}(c_{j_1}, c_{j_2}, \cdots, c_{j_{k_i}}) = c_i$, namely $g_i^{(t)} \geq c_i$. This leads to a contradiction and thus the inequality $g_j^{\infty} \geq c_j$ is validated.

Analogous to the proof of Theorem 1, after convergence, for any node $i \in V$, all nodes j such that $g_j^{\infty} \geq g_i^{\infty}$, including i itself, constitute an induced subgraph of G's g_i^{∞}-core. Since c_i is the coreness of i, $g_i^{\infty} \leq c_i$. Combining the two inequalities, we arrive at Theorem 2.

Spreading models. The standard SIR model, also referred to as the susceptible-infected-recovered model, is usually applied to analyse the propagation of opinions or news[41]. In the SIR model, there is a group of infected seed nodes and all other nodes are initially susceptible. At each time step, each infected node makes contact with its neighbours and each susceptible neighbour is infected with a probability β. Then, each infected node enters the removed state with a probability λ. For simplicity, we set $\lambda = 1$. To quantify the spreading influence of a target node i, we begin the spreading process with i being the sole infected seed. When there are no longer any infected nodes and the dynamic process ends, the number of removed nodes R_i is a measurement of the influence of node i. Because we use small β values in our simulations, the infected percentage of the population is also small. When β values are high the disease infects a large percentage of the population, irrespective of where it originated, and the influence of individual nodes cannot be measured. According to the heterogeneous mean-field theory[47,53,54], the epidemic threshold of SIR model is approximate to $\beta_c \approx \langle k \rangle / (\langle k^2 \rangle - \langle k \rangle)$. To be more precise, we determine the epidemic threshold β_c by simulation on real networks[55]. We set $\beta = 1.5\beta_c$, $2\beta_c$ and $2.5\beta_c$ in this paper, and we have checked that the choice of theoretical or simulation threshold will not affect the conclusion. Because the fluctuation of R_i is large when β values are small, we use 1,000 independent implementations for averaging.

Benchmark centralities. We also compare two benchmark centrality indices (that is, closeness and betweenness) for undirected networks. Betweenness is one of the most popular geodesic-path-based ranking measures. It is defined as the fraction of shortest paths between all node pairs that pass through the node of interest. Betweenness is, in some sense, a measure of the influence of a node in terms of its role in spreading information[20,56]. For a network $G = (V, E)$, the betweenness centrality of node i, denoted by $B(i)$, is defined as[57,58]

$$B(i) = \sum_{s \neq i, s \neq t, i \neq t} \frac{g_{st}(i)}{g_{st}}, \tag{6}$$

where g_{st} is the number of shortest paths between nodes s and t, and $g_{st}(i)$ denotes the number of shortest paths between nodes s and t that pass through node i.

Closeness of node i is defined as the reciprocal of the sum of geodesic distances to all other nodes of V[59,60],

$$C(i) = \frac{1}{\sum_{j \in V \setminus i} d(i,j)}, \tag{7}$$

where $d(i, j)$ is the geodesic distance between i and j. Closeness can be used to measure of how far information will be able to spread from a given node to other reachable nodes in the network.

The Kendall Tau. We consider two sequences associated with $|V|$ nodes, $X = (x_1, x_2, \cdots, x_{|V|})$ and $Y = (y_1, y_2, \cdots, y_{|V|})$, as well as the $|V|$ two-tuples $(x_1, y_1), (x_2, y_2), \cdots, (x_{|V|}, y_{|V|})$. Any pair of two-tuples (x_i, y_i) and (x_j, y_j) $(i \neq j)$ are concordant if the ranks for both elements agree, that is, if both $x_i > x_j$ and $y_i > y_j$ or if both $x_i < x_j$ and $y_i < y_j$. They are discordant if $x_i > x_j$ and $y_i < y_j$ or if $x_i < x_j$ and $y_i > y_j$. If $x_i = x_j$ or $y_i = y_j$, the pair is neither concordant nor discordant. Comparing all $\frac{1}{2}|V|(|V|-1)$ pairs of two-tuples, the Kendall Tau is defined as $\tau = 2(n_+ - n_-)/[|V|(|V|-1)]$, where n_+ and n_- are the number of concordant and discordant pairs, respectively. The coefficient must be in the range $-1 \leq \tau \leq 1$, and if X and Y are independent, τ should be approximately zero and thus the extent to which τ exceeds zero indicates the strength of the correlation.

Notations. We summarize the notations used in this paper in Supplementary Table 25.

References

1. Watts, D. J. & Strogatz, S. H. Collective dynamics of 'small-world' networks. *Nature* **393**, 440–442 (1998).
2. Barabási, A.-L. & Albert, R. Emergence of scaling in random networks. *Science* **286**, 509–512 (1999).
3. Newman, M. E. J. Assortative mixing in networks. *Phys. Rev. Lett.* **89**, 208701 (2002).
4. Newman, M. E. J. & Girvan, M. Finding and evaluating community structure in networks. *Phys. Rev. E* **69**, 026113 (2004).
5. Csermely, P. *Weak Links: Stabilizers of Complex Systems from Proteins to Social Networks* (Springer, 2006).
6. Alon, U. Network motifs: theory and experimental approaches. *Nat. Rev. Genet.* **8**, 450–461 (2007).
7. Cheng, X.-Q., Ren, F.-X., Shen, H.-W., Zhang, Z.-K. & Zhou, T. Bridgeness: a local index on edge significance in maintaining global connectivity. *J. Stat. Mech.* **2010**, P10011 (2010).
8. Lü, L., Zhang, Y.-C., Yeung, C. H. & Zhou, T. Leaders in social networks: the delicious case. *PLoS ONE* **6**, e21202 (2011).
9. Klemm, K., Serrano, M. Á., Equíluz, V. M. & Miguel, M. S. A measure of individual role in collective dynamics. *Sci. Rep.* **2**, 292 (2012).
10. Wang, P., Lü, J. & Yu, X. Identification of important nodes in directed biological networks: a network motif approach. *PLoS ONE* **9**, e106132 (2014).
11. Pei, S. & Makse, H. A. Spreading dynamics in complex networks. *J. Stat. Mech.* **2013**, P12002 (2013).
12. Pastor-Satorras, R. & Vespignani, A. Immunization of complex networks. *Phys. Rev. E* **65**, 036104 (2002).
13. Leskovec, J., Adamic, L. A. & Huberman, B. A. The dynamics of viral marketing. *ACM Trans. Web* **1**, 5 (2007).
14. Albert, R., Albert, I. & Nakarado, G. L. Structural vulnerability of the North American power grid. *Phys. Rev. E* **69**, 025103 (2004).
15. Morone, F. & Makse, H. A. Influence maximization in complex networks through optimal percolation. *Nature* **524**, 65–68 (2015).
16. Csermely, P., Korcsmros, T., Kiss, H. J., London, G. & Nussinov, R. Structure and dynamics of molecular networks: a novel paradigm of drug discovery: a comprehensive review. *Pharmacol. Ther.* **138**, 333-408 (2013).
17. Resende, M. G. C. & Pardalos, P. M. *Handbook of Optimization in Telecommunications* (Springer, 2006).
18. Albert, R., Jeong, H. & Barabási, A.-L. Error and attack tolerance of complex networks. *Nature* **406**, 378–382 (2000).
19. Motter, A. E., Zhou, C. & Kurths, J. Network synchronization, diffusion, and the paradox of heterogeneity. *Phys. Rev. E* **71**, 016116 (2005).
20. Yan, G., Zhou, T., Hu, B., Fu, Z.-Q. & Wang, B.-H. Efficient routing on complex networks. *Phys. Rev. E* **73**, 046108 (2006).
21. Santos, F. C., Pacheco, J. M. & Lenaerts, T. Evolutionary dynamics of social dilemmas in structured heterogeneous populations. *Proc. Natl Acad. Sci. USA* **103**, 3490–3494 (2006).
22. Kitsak, M. *et al.* Identification of influential spreaders in complex networks. *Nat. Phys.* **6**, 888–893 (2010).
23. Dorogovtsev, S. N., Goltsev, A. V. & Mendes, J. F. F. K-core organization of complex networks. *Phys. Rev. Lett.* **96**, 040601 (2006).
24. Pei, S., Muchnik, L., Andrade, J. S., Zheng, Z. M. & Makse, H. A. Searching for superspreaders of information in real-world social media. *Sci. Rep.* **4**, 5547 (2014).
25. Chen, D.-B., Lü, L., Shang, M.-S., Zhang, Y.-C. & Zhou, T. Identifying influential nodes in complex networks. *Physica A* **391**, 1777–1787 (2012).
26. Chen, D.-B., Gao, H., Lü, L. & Zhou, T. Identifying influential nodes in large-scale directed network: the role of clustering. *PLoS ONE* **8**, e77455 (2013).
27. Hirsch, J. E. An index to quantify an individual's scientific research output. *Proc. Natl Acad. Sci. USA* **102**, 16569–16572 (2005).
28. Braun, T., Glänzel, W. & Schubert, A. A Hirsch-type index for journals. *Scientometrics* **69**, 169–173 (2006).
29. Hirsch, J. E. Does the *h* index have predictive power? *Proc. Natl Acad. Sci. USA* **104**, 19193–19198 (2007).

30. Korn, A., Schubert, A. & Telcs, A. Lobby index in networks. *Physica A* **388,** 2221–2226 (2009).
31. Hethcote, H. W. The mathematics of infectious diseases. *SIAM Rev.* **42,** 599–653 (2000).
32. Gleiser, P. & Danon, L. Community structure in Jazz. *Adv. Complex Syst.* **6,** 565 (2003).
33. Newman, M. E. J. Finding community structure in networks using the eigenvectors of matrices. *Phys. Rev. E* **74,** 036104 (2006).
34. Guimerà, R., Danon, L., Diaz-Guilera, A., Giralt, F. & Arenas, A. Self-similar community structure in a network of human interactions. *Phys. Rev. E* **68,** 065103 (2003).
35. Rocha, L. E., Liljeros, F. & Holme, P. Simulated epidemics in an empirical spatiotemporal network of 50,185 sexual contacts. *PLoS Comput. Biol.* **7,** e1001109 (2011).
36. Viswanath, B., Mislove, A., Cha, M. & Gummadi, K. P. in: *Proceedings of the 2nd Workshop on Online Social Networks.* 37–42 (ACM, 2009).
37. Adamic, L. A. & Glance, N. in: *Proceedings of the 3rd International Workshop on Link Discovery.* 36–43 (ACM, 2004).
38. Batageli, V. & Mrvar, A. Pajek Datasets. Available at http://vlado.fmf.uni-lj.si/pub/networks/data/. 2007.
39. Spring, N., Mahajan, R., Wetherall, D. & Anderson, T. Measuring ISP topologies with Rocketfuel. *IEEE/ACM Trans. Networking* **12,** 2–16 (2004).
40. Ravasz, E. B. & Barabasi, A.-L. Hierarchical organization in complex networks. *Phys. Rev. E* **67,** 026112 (2003).
41. Pastor-Satorras, R., Castellano, C., Van Mieghem, P. & Vespignani, A. Epidemic processes in complex networks. *Rev. Mod. Phys.* **87,** 925 (2015).
42. Keeling, M. J. & Rohani, P. *Modeling Infectious Diseases in Humans and Animals* (Princeton Univ. Press, 2008).
43. Banerjee, A., Chandrasekhar, A. G., Duflo, E. & Jackson, M. O. The diffusion of microfinance. *Science* **341,** 1236498 (2013).
44. Li, D. *et al.* Percolation transition in dynamical traffic network with evolving critical bottlenecks. *Proc. Natl Acad. Sci. USA* **112,** 669–672 (2015).
45. Kendall, M. A new measure of rank correlation. *Biometrika* **30,** 81–89 (1938).
46. Pastor-Satorras, R. & Vespignani, A. Epidemic spreading in scale-free networks. *Phys. Rev. Lett.* **86,** 3200 (2001).
47. Newman, M. E. J. Spread of epidemic disease on networks. *Phys. Rev. E* **66,** 016128 (2002).
48. Egghe, L. Theory and practise of the g-index. *Scientometrics* **69,** 131–152 (2006).
49. Jin, B., Liang, L., Rousseau, R. & Egghe, L. The R-and AR-indices: complementing the h-index. *Chinese Sci. Bull.* **52,** 855–863 (2007).
50. Dorogovtsev, S. N. & Mendes, J. F. F. Ranking Scientists. *Nature Physics* **11,** 882–883 (2015).
51. Brin, S. & Page, L. The anatomy of a large-scale hypertextual Web search engine. *Comput. Networks ISDN Syst.* **30,** 107–117 (1998).
52. Kleinberg, J. Authoritative sources in a hyperlinked environment. *J. ACM* **46,** 604–632 (1999).
53. Cohen, R., Erez, K., Ben-Avraham, D. & Havlin, S. Resilience of the Internet to random breakdowns. *Phys. Rev. Lett.* **85,** 4626 (2000).
54. Castellano, C. & Pastor-Satorras, R. Thresholds for epidemic spreading in networks. *Phys. Rev. Lett.* **105,** 218701 (2010).
55. Shu, P., Wang, W., Tang, M. & Do, Y. Numerical identification of epidemic thresholds for susceptible-infected-recovered model on finite-size networks. *Chaos* **25,** 063104 (2015).
56. Guimerà, R., Das-Guilera, A., Vega-Redondo, F., Cabrales, A. & Arenas, A. Optimal network topologies for local search with congestion. *Phys. Rev. Lett.* **89,** 248701 (2002).
57. Freeman, L. C. A set of measures of centrality based on betweenness. *Sociometry* **40,** 35–41 (1977).
58. Freeman, L. C. Centrality in social networks conceptual clarification. *Soc. Networks* **1,** 215–239 (1979).
59. Bavelas, A. Communication patterns in task-oriented groups. *J. Acoust. Soc. Am.* **22,** 725–730 (1950).
60. Sabidussi, G. The centrality index of a graph. *Psychometrika* **31,** 581–603 (1966).

Acknowledgements

This work was partially supported by the National Natural Science Foundation of China (Grant Nos. 11205042, 11222543, 11075031, 61433014). L.L. acknowledges the research start-up fund of Hangzhou Normal University under Grant No. PE13002004039 and the EU FP7 Grant 611272 (project GROWTHCOM). The Boston University work was supported by NSF Grants CMMI 1125290,CHE 1213217 and PHY 1505000.

Author contributions

L.L. and T.Z. planned and performed this research, contributed equally to this work, and wrote this paper. Q.M.Z. performed research. H.E.S. planned the research and edited this paper. All authors discussed the results and reviewed the manuscript.

Additional information

Competing financial interests: The authors declare no competing financial interests.

5

Direct measurement of large-scale quantum states via expectation values of non-Hermitian matrices

Eliot Bolduc[1], Genevieve Gariepy[1] & Jonathan Leach[1]

In quantum mechanics, predictions are made by way of calculating expectation values of observables, which take the form of Hermitian operators. Non-Hermitian operators, however, are not necessarily devoid of physical significance, and they can play a crucial role in the characterization of quantum states. Here we show that the expectation values of a particular set of non-Hermitian matrices, which we call column operators, directly yield the complex coefficients of a quantum state vector. We provide a definition of the state vector in terms of measurable quantities by decomposing these column operators into observables. The technique we propose renders very-large-scale quantum states significantly more accessible in the laboratory, as we demonstrate by experimentally characterizing a 100,000-dimensional entangled state. This represents an improvement of two orders of magnitude with respect to previous phase-and-amplitude characterizations of discrete entangled states.

[1] Institute of Photonics and Quantum Sciences, School of Engineering & Physical Sciences, Heriot-Watt University, David Brewster Building, Edinburgh EH14 4AS, UK. Correspondence and requests for materials should be addressed to E.B. (email: eliot.bolduc@gmail.com).

One of the current challenges in the field of computing is harnessing the potential processing power provided by quantum devices that exploit entanglement. Experimental research aimed at overcoming this challenge is driven by the production, control and detection of larger and larger entangled quantum states[1–4]. However, the task of characterizing these entangled states quickly become intractable as the number of parameters that define a many-body system scales exponentially with the system size. To keep up with the ever-growing quantum state dimensionality, much effort is put into developing efficient characterization methods[5–19].

Quantum state tomography is the process of retrieving the values that define a quantum system. The process typically involves two steps: (i) gathering an informationally complete set of data and (ii) finding the quantum state most consistent with the data set using post-measurement processing such as the algorithm for maximum-likelihood estimation[20]. Many efficient tomographic methods capitalize on the first step by making simplifying assumptions about the state[11–19], thus reducing the number of measurements required to uniquely identify it. In particular, tomography via compressed sensing allows one to efficiently reconstruct quantum states based on the fact that low-rank density matrices, that is, quasi-pure states are sparse in a particular basis[15–17,21]. Compared with assumption-free tomography, compressive sensing provides a square-root improvement on the required number of measurements[10]. This improvement enabled the reconstruction of the density matrices of a six-qubit state[16] and a (17×17)-dimensional state[17], the largest phase-and-amplitude measurement of an entangled state reported to date. Although compressed sensing does not make use of maximum-likelihood estimation, it does require non-trivial post-measurement processing.

Recently, Lundeen *et al.*[19] reported on the direct measurement of a wavefunction using a method that, for the first time, required no complicated post-measurement processing. Their method is based on weak measurements, whereby one weakly couples a quantum system to a pointer state and subsequently performs a few standard strong measurements on the pointer state. The outcome of a weak measurement is known as the 'weak value', and in the conditions exposed in ref. 19, the weak value is proportional to a given state vector coefficient. The method of Lundeen *et al.* can be used in combination with the assumption that the quantum state at hand is pure, providing the same square-root improvement as compressed sensing. Variations on the original scheme allow measurements of mixed states and increased detection efficiency[22–24].

An important contribution of the work by Lundeen *et al.* was to link the state vector elements to the expectation value of weak measurements. We take a different approach, and point out that the enabling feature that allows access to the complex state vector is not weak measurement but the use of particular non-Hermitian operators. Although weak measurements provide a way to decompose these non-Hermitian operators, it is not the only suitable approach. Moreover, the introduction of weak values in the measurement procedure adds complexity to the experiment and the formalism that links weak values to measurement outcomes involves an approximation that breaks down in a variety of circumstances[24–26].

In this paper, we propose an alternative approach to the direct measurement of quantum states that is exact in the case of pure states, proves to be reliable in the presence of noise, and is consistent with results obtained with well-established tomographic techniques. The key principle of our formalism is to decompose the particular non-Hermitian matrices that yield the complex state vector coefficients using only observables. Our method therefore only requires strong measurements, as in standard tomography, while maintaining the directness of weak-value-assisted tomography. The simplicity in both the experimental procedure and post-measurement processing renders our method ideally suited for the characterization of large-scale systems, which can be high-dimensional, many-body or both. We begin by developing the theory on which our method is based and then demonstrate the potential of this scheme by experimentally retrieving the complex coefficients of a (341×341)-dimensional entangled state.

Results

Theory. Consider a quantum system in a d-dimensional Hilbert space, whose state vector

$$|\Psi\rangle = \sum_{j=0}^{d-1} c_j |j\rangle \tag{1}$$

is expanded in the basis $\{|j\rangle\}$ and where c_j are unknown complex expansion coefficients. To retrieve these coefficients, we introduce the column operators $\widehat{C}_j = |a\rangle\langle j|$, where $|a\rangle$ is an arbitrary reference vector. Each column operator has an expectation value

$$\langle \widehat{C}_j \rangle = \langle \Psi | a \rangle c_j, \tag{2}$$

which is proportional to a complex state vector expansion coefficient. Since the value of $\langle \Psi | a \rangle$ is independent of j, we can express the state vector in terms of the column operators up to a phase factor:

$$|\Psi\rangle = \frac{e^{i\phi}}{v} \sum_{j=0}^{d-1} \langle \widehat{C}_j \rangle |j\rangle, \tag{3}$$

where $v = |\langle \Psi | a \rangle|$ is a normalization constant. We can ignore the phase factor $e^{i\phi}$ since it bears no physical significance.

Most column operators $\widehat{C}_j$ are not Hermitian matrices and are thus not observables. To overcome this apparent constraint, we recognize that any non-Hermitian matrix can be constructed from a complex-weighted sum of Hermitian matrices. Hence, the crucial step to our method is to construct the column operators in terms of measurable quantities: $\widehat{C}_j = \sum_q w_{jq} \widehat{O}_{jq}$, where w_{jq} are complex weights and $\widehat{O}_{jq}$ are observables. As a result, this allows us to retrieve any state vector element with a complex-weighted sum of measurement outcomes:

$$c_j = \frac{1}{v} \sum_q w_{jq} \langle \widehat{O}_{jq} \rangle. \tag{4}$$

Equation 4 is an exact definition of the pure state vector that is provided in terms of measurable quantities. The above formalism readily applies to a general class of quantum states, including high-dimensional and many-body systems.

As an example, consider the case of a qubit $|\Psi\rangle = c_0|0\rangle + c_1|1\rangle$ with $|a\rangle = |0\rangle$ as the reference vector. The first column operator $\widehat{C}_0$ is Hermitian and given by the projector $|0\rangle\langle 0|$. The second column operator $\widehat{C}_1 = |0\rangle\langle 1|$ is not Hermitian but can be constructed a number of ways. The first construction—which, as pointed out earlier, is a key part of the weak value formalism—is the complex-weighted sum of Pauli matrices: $\widehat{C}_1 = (\hat{\sigma}_x + i\hat{\sigma}_y)/2$, a decomposition that requires two observables, each of which is made of two projectors or eigenvectors. A second decomposition requiring only three projectors is given by

$$\widehat{C}_1 = \sum_{q=0}^{2} \frac{2}{3} e^{i2\pi q/3} |s_q\rangle\langle s_q|, \tag{5}$$

where $|s_q\rangle = (|0\rangle + e^{i4\pi q/3}|1\rangle)/\sqrt{2}$ are the states onto which the observables $\widehat{O}_{1q}$ project. In both cases, the qubit state vector is exactly given by $|\Psi\rangle = (\langle \widehat{C}_0 \rangle |0\rangle + \langle \widehat{C}_1 \rangle |1\rangle)/\langle \widehat{C}_0 \rangle^{\frac{1}{2}}$.

Experiment. To demonstrate the power and scalability of our scheme, we apply it to the measurement of a state entangled in greater than 100,000 dimensions. We provide a complete characterization of the spatially entangled two-photon field produced through spontaneous parametric downconversion (SPDC). In general, SPDC can give rise to spatial and frequency correlations between two photons[4,27–36]. The purity of the spatial part of the full state can only be guaranteed if the two types of correlations are completely decoupled, which can be achieved in the collinear regime[27]—see Supplementary Note 1 for a theoretical estimation of our system purity. The consequences of applying our scheme to a quantum state with non-unit purity, which is always the case in the presence of noise, will be discussed below.

We express the spatial part of the entangled state in a discrete cylindrical basis of transverse spatial modes. The azimuthal part of the modes is given by $e^{i\ell\phi}$, where ℓ is an integer between $-\infty$ and ∞ and ϕ is the azimuthal angle. This type of phase profile is known to carry ℓ units of orbital angular momentum (OAM). We decompose the radial part of the field with the recently introduced Walsh modes, labelled by the integer k ranging from 0 to ∞ (ref. 34). The Walsh modes all have the same Gaussian amplitude envelope, but different π-steps radial phase profiles. Combining the OAM modes with the Walsh modes yields a complete basis for coherent two-dimensional images. To perform the characterization of the two-photon spatial field, we consider 31 OAM modes and 11 Walsh modes for each photon. The state vector thus takes the form

$$|\Phi\rangle = \sum_{\ell_1=-15}^{15} \sum_{k_1=0}^{10} \sum_{\ell_2=15}^{-15} \sum_{k_2=0}^{10} c_{\ell_1,k_1}^{\ell_2,k_2} |\ell_1, k_1\rangle |\ell_2, k_2\rangle. \quad (6)$$

Using the column-operator decomposition described in the Methods section, we sequentially measure all 116,281 coefficients $c_{\ell_1,k_1}^{\ell_2,k_2}$, which are shown in Fig. 1a,b. The total Hilbert space dimensionality of this measured state is more than two orders of magnitude larger than any previously reported amplitude-and-phase-characterized discrete entangled state[17]. As a simple verification of the accuracy of our method, we calculate the probabilities associated with each joint mode via the Born rule, $|c_{\ell_1,k_1}^{\ell_2,k_2}|^2$, as shown in Fig. 1c. This result is consistent with the directly measured correlation matrix shown in Fig. 1e, showing that we retrieve the correct magnitude of the amplitudes.

To rigorously assess the validity of the directly measured complex quantum state $|\psi\rangle$, that is, both the amplitudes and the phases, we compare it to the results obtained through full tomography (that is, assumption-free tomography); see Supplementary Note 2 for details of the algorithm used to retrieve the density matrix. As full tomography cannot be performed on a (341×341)-dimensional entangled state in a reasonable time, we characterize a (5×5)-dimensional subset of the SPDC two-photon state. We perform the comparison in a basis of various OAM modes ($\ell_1 \in \{1, -1, 2, -2, 3\}$, $\ell_2 \in \{1, -1, 2, -2, -3\}$) and a fixed radial Walsh mode ($k_1 = k_2 = 0$). The total number of unknown parameters in the corresponding density matrix is equal to 624. After performing the direct measurement procedure in this basis, we record 8,000 random projective measurements that we break into eight sets of 1,000. For each set, we recover a density matrix ρ_{exp} and calculate its purity and the fidelity with the directly measured state $|\psi\rangle$; fidelity is defined as $\sqrt{\langle\psi|\rho_{\text{exp}}|\psi\rangle}$. On average, the purity calculation yields (0.96 ± 0.02), and the fidelity gives (0.985 ± 0.004), where the uncertainties correspond to one standard deviation. After reconstruction of a density matrix, we find that the average error between the measured count rates and the count rates predicted by the density matrix is 5.5%. This can be explained by shot noise, the pixelated nature of the spatial light modulator (SLM), and the finite aperture of the optical elements. While we expect unit purity, the 5% noise level accounts for the discrepancy with the measured value.

Simulation with mixed states. The extremely high fidelity between the tomography results ρ_{exp} and the directly measured state $|\psi\rangle$ indicates the validity of our approach for quantum state measurements applied to near pure states. To evaluate our method in the context of mixed states, we perform a series of numerical simulations where we vary the rank, purity and dimension of an unknown state ρ_{sim}, where no sources of noise are added to the simulated measurement outcomes. We apply our direct measurement procedure to these states and calculate the fidelity $|\langle\psi|\psi_{\text{sim}}\rangle|$, where $|\psi_{\text{sim}}\rangle$ is the eigenvector of ρ_{sim} with the largest eigenvalue. For initial states ρ_{sim} with purity >0.81, we measure a fidelity >0.99 in at least 99% of the cases. The dependency of this result on the dimensionality of the state is negligible. This result indicates that our direct method enables the extraction of the density matrix primary eigenvector, even for a partially mixed state. Full details of this analysis and the density matrix reconstruction are presented in the Supplementary Note 1.

Discussion

Knowledge of the amplitude and phase of the state vector elements allows us to perform otherwise inaccessible calculations. As an example, we perform a calculation of the Schmidt decomposition[37]. This is equivalent to the singular value decomposition for the case of optical transfer matrices. The Schmidt decomposition yields a new joint basis, in which the photons are perfectly correlated and where the joint modes have equal phases, as shown in Fig. 1d. When the Schmidt decomposition is applied to the entire state, we calculate a number of Schmidt modes equal to 142; this represents the effective number of independent joint modes contained within the state (the maximum for a (341×341)-dimensional state being 341). The Schmidt decomposed two-photon field is a good candidate for the violation of very high-dimensional Bell inequalities[29]. Further details on the Schmidt decomposition can be found in the Supplementary Note 3.

There are a number of approaches to reducing the necessary cost and effort for measuring large-scale quantum states. These include, but are not limited to, developing technologies for mode sorting[38] and arbitrary unitary transformations[39,40], reducing the required number of measurement settings, and circumventing the requirement for reconstruction procedures. It is clear that there is significant interplay between each of these approaches. The theoretical implementation of an approach that combines the principles of our work with generalized measurements, such as positive operator value measures, is considered in the Supplementary Note 4. The ability to use positive operator value measures in the laboratory relies on the aforementioned technologies. Access to these types of technologies would reduce the overall number of measurement settings to uniquely recover a quantum state. However, such a system requires arbitrary unitary transformations for spatial states, which is in itself an active area of research[38–40]. Given the limitations of mode sorters for very large dimensions, and the practical nature of projective measurements, our scheme provides a simple and elegant method for the characterization of large-scale quantum states.

Our scheme allows direct access to the complex coefficients that define large-scale quantum states. The main result of our work is a

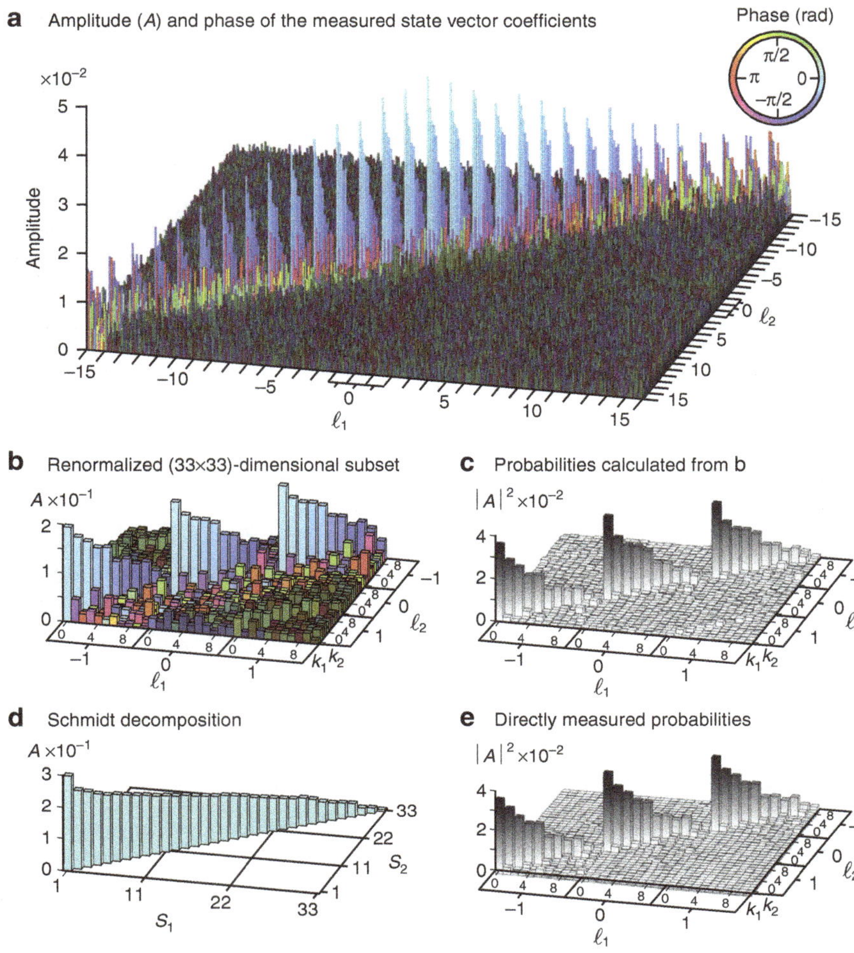

Figure 1 | Measured and calculated properties of the two-photon state. (**a**) We illustrate the complex state vector coefficients in matrix format. This representation is similar to that of an optical transfer matrix, where the lateral axes correspond to input and output modes. Here, each lateral axis corresponds to the spatial state of one photon of the pair. The OAM values ℓ range from -15 to 15. For a fixed OAM value, the radial index k ranges from 0 to 10, thus the combined state has 341×341 dimensions. The amplitude and phase values of a coefficient are given by the height and colour of a bar, respectively. For clarity, we darken the off-diagonal part. The small subset (**b**) of the state shows the phase gradient across the diagonal elements, which is typical of a Gouy phase shift, a common property of light passing through focus[43]. The corresponding calculated probability matrix (**c**) is consistent with the directly measured probabilities (**e**). Finally, we calculate the Schmidt decomposition (**d**) of **b**, which gives the joint basis in which the subset can be expressed with the lowest number of modes. The indices S_1 and S_2 correspond to the states of each photon in this basis.

novel method for retrieving a state vector coefficient with a complex-weighted sum of strong measurement outcomes. One challenge in reconstructing a quantum state from measurement outcomes lies in data processing; our scheme trades the difficulty of data processing for theoretical analysis before the experiment, that is, finding the measurements one has to perform. We anticipate that our work will have an impact on a number of disciplines, for example, quantum parameter estimation, measurement in quantum computing, quantum information and metrology.

Methods

Experiment. The two-photon field is generated via SPDC with a 405-nm laser diode pumping a 1-mm-long periodically poled KTP (PPKTP) crystal with 50 mW of power. The experimental set-up is shown in Fig. 2. We separate the two photons with a right-angle prism and image the plane of the crystal to a Holoeye SLM with a magnification of -10. We simultaneously display two holograms, one on each side of the SLM, to control the amplitude and phase profiles of the two photons independently. To make projective measurements of superposition modes, we make use of intensity masking[41]. We image the plane of the SLM with a magnification of $-1/2{,}500$ to two single-mode fibres. The combination of the SLM and singles mode fibres allows us to make arbitrary projective measurements. All measurements are performed in coincidence with two single-photon avalanche detectors, with a timing window of 25 ns, an integration time of 1 s for modes outside the diagonal and 20 s for the diagonal elements ($\ell_1 = -\ell_2$ and $k_1 = k_2$). We start an automatic alignment procedure with the SLM every 4 hours to compensate for drift. Including the time it takes to calculate and display a hologram (about 1 s), the entire experiment takes 2 weeks; assumption-free tomography would take more than four centuries at the same acquisition rate. We perform no background subtraction and use the fundamental mode ($\ell_1 = -\ell_2 = k_1 = k_2 = 0$) as the reference vector $|a\rangle$. The count rate of the fundamental mode is approximately 900 coincidences per second and varies by 10% over 24-h periods. To correct for long-term drift, we normalize each outcome to the count rate of the fundamental mode, which we measure before the measurement of each column operator. In standard tomography, the calculation of error bounds on the measured state is not a straightforward task[42]. Here, we can calculate the error bound on a given coefficient with a weighted sum of the detector counts used to retrieve it. For a given state vector coefficient, the errors on the amplitude $|c_j|$ and phase $\arg(c_j)$ are both inversely proportional to the overlap v of the reference vector with the quantum state. To minimize the errors, it is important to choose a reference vector that has a high probability of occurrence within the state—the fundamental mode is the most probable one in our case.

Two-body column-operator decomposition. To decompose a given state vector coefficient $c^{\ell_2,k_2}_{\ell_1,k_1}$ into a set of measurement outcomes, we need to find a projector decomposition of the corresponding column operator $\widehat{C}^{\ell_2,k_2}_{\ell_1,k_1} = |0,0\rangle\langle\ell_1,k_1| \otimes |0,0\rangle\langle\ell_2,k_2|$, as in equation 4. We numerically find this column-operator decomposition, that is, the complex weights w_q and the

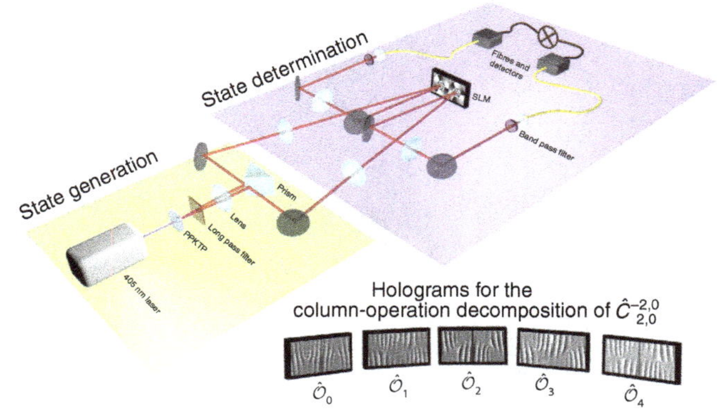

Figure 2 | Generation and characterization of a two-photon field. The entangled state is produced via SPDC in a PPKTP crystal and spatially separated by a prism. For the state determination stage, the crystal plane is imaged onto a SLM, which is in turn imaged to the input facet of two single-mode fibres. To make a given projective measurement, we display the corresponding joint mode on the SLM and measure the coincidence rate between the two single-photon avalanche diode detectors. The inset shows the five joint holograms that correspond to the column-operator decomposition of $\widehat{C}^{-2,0}_{2,0}$. The state vector coefficient $c^{-2,0}_{2,0}$ is given by $\frac{4}{5v}\sum_q \left\langle \widehat{O}_q \right\rangle e^{i4\pi q/5}$, where the expectation value of a given observable is proportional to the measured count rate when displaying the corresponding hologram.

observables $\widehat{O}_q$, using the differential evolution algorithm (see Supplementary Note 5). By inspection, we find that the corresponding analytical form of the state vector coefficients is given by

$$\widehat{C}^{\ell_2,k_2}_{\ell_1,k_1} = \frac{1}{v}\sum_{q=0}^{4}\frac{4}{5}e^{i2\pi q/5}|s_{1,q}\rangle\langle s_{1,q}| \otimes |s_{2,q}\rangle\langle s_{2,q}|, \quad (7)$$

where $\sqrt{2}|s_{m,q}\rangle = |0,0\rangle + e^{i4\pi q/5}|\ell_m, k_m\rangle$ with $m=\{1, 2\}$, and $v = |\langle\Psi|0,0\rangle|$ is a normalization constant. This decomposition is only valid when the state of any photon is different from the reference vector, that is, $|\ell_m, k_m\rangle \neq |0,0\rangle$. Each coefficient measured with the above column-operator decomposition requires five projective measurements, thus explaining the $5D^2$ scaling, where D is the Hilbert space dimensionality of a single particle. The protocol scales much more favourably than assumption-free tomography, which requires D^4 projections.

Here, we briefly explain our protocol for measuring the entire SPDC state vector. We measure more than 99% of the coefficients using the decomposition of equation 7. The remaining column operators are the special cases $|0,0\rangle\langle\ell_1,k_1| \otimes |0,0\rangle\langle 0,0|$ and $|0,0\rangle\langle 0,0| \otimes |0,0\rangle\langle\ell_2,k_2|$, which respectively correspond to a row and a column of the result shown in Fig. 1a. These column operators can be decomposed into only three joint local measurements using the projector $|0,0\rangle\langle 0,0|$ on one system and a column-operator decomposition similar to that of equation 5 on the other system. Finally, the column operator $|0,0\rangle\langle 0,0|\otimes|0,0\rangle\langle 0,0|$ is a projector, and its expectation value can be measured in a single experimental configuration.

Full quantum tomography. We perform full tomography with high count rates in order to achieve high accuracy. We set the magnification between the plane of the SLM and that of the single-mode fibres to 1/400. In this condition, we obtain a count rate of approximately 18,000 counts per second for the fundamental mode and integrate over 1 s for each individual projective measurement. The increase in the count rate of the fundamental mode comes at the price of lower count rates for high order modes. Regarding the full tomography measurements, we take an overcomplete set of 1,000 random projective measurements in a (5×5)-dimensional space. To minimize high-frequency components on the SLM, we limit the random superpositions to two-dimensional subsets of the state space.

References

1. Monz, T. *et al.* 14-Qubit entanglement: creation and coherence. *Phys. Rev. Lett.* **106,** 130506 (2011).
2. Yao, X. C. *et al.* Observation of eight-photon entanglement. *Nat. Photon.* **6,** 225–228 (2012).
3. Yokoyama, S. *et al.* Ultra-large-scale continuous-variable cluster states multiplexed in the time domain. *Nat. Photon.* **7,** 982–986 (2013).
4. Krenn, M. *et al.* Generation and confirmation of a (100 × 100)-dimensional entangled quantum system. *Proc. Natl Acad. Sci. USA* **111,** 6243–6247 (2014).
5. Smith, B. J., Killett, B., Raymer, M. G., Walmsley, I. A. & Banaszek, K. Measurement of the transverse spatial quantum state of light at the single-photon level. *Opt. Lett.* **30,** 3365–3367 (2005).
6. Bogdanov, Y. I. *et al.* Statistical estimation of the efficiency of quantum state tomography protocols. *Phys. Rev. Lett.* **105,** 010404 (2010).
7. Mahler, D. H. *et al.* Adaptive quantum state tomography improves accuracy quadratically. *Phys. Rev. Lett.* **111,** 183601 (2013).
8. Teo, Y. S., Řeháček, J. & Hradil, Z. Informationally incomplete quantum tomography. *Quant. Meas. Quant. Metrol* **1,** 57–83 (2013).
9. Ferrie, C. Self-guided quantum tomography. *Phys. Rev. Lett.* **113,** 190404 (2014).
10. Banaszek, K., Cramer, M. & Gross, D. Focus on quantum tomography. *New J. Phys.* **15,** 125020 (2013).
11. Shabani, A. *et al.* Efficient measurement of quantum dynamics via compressive sensing. *Phys. Rev. Lett.* **106,** 100401 (2011).
12. Flammia, S. T., Silberfarb, A. & Caves, C. M. Minimal informationally complete measurements for pure states. *Found. Phys.* **35,** 1985–2006 (2005).
13. Cramer, M. *et al.* Efficient quantum state tomography. *Nat. Commun.* **1,** 149 (2010).
14. Tóth, G. *et al.* Permutationally invariant quantum tomography. *Phys. Rev. Lett.* **105,** 250403 (2010).
15. Gross, D., Liu, Y.-K., Flammia, S. T., Becker, S. & Eisert, J. Quantum state tomography via compressed sensing. *Phys. Rev. Lett.* **105,** 150401 (2010).
16. Schwemmer, C. *et al.* Experimental comparison of efficient tomography schemes for a six-qubit state. *Phys. Rev. Lett.* **113,** 040503 (2014).
17. Tonolini, F., Chan, S., Agnew, M., Lindsay, A. & Leach, J. Reconstructing high-dimensional two-photon entangled states via compressive sensing. *Sci. Rep.* **4,** 6542 (2014).
18. Lloyd, S., Mohseni, M. & Rebentrost, P. Quantum principal component analysis. *Nat. Phys.* **10,** 631–633 (2014).
19. Lundeen, J. S., Sutherland, B., Patel, A., Stewart, C. & Bamber, C. Direct measurement of the quantum wavefunction. *Nature* **474,** 188–191 (2012).
20. Banaszek, K., D'Ariano, G. M., Paris, M. G. A. & Sacchi, M. F. Maximum-likelihood estimation of the density matrix. *Phys. Rev. A* **61,** 10304 (2000).
21. Liu, W.-T., Zhang, T., Liu, J.-Y., Chen, P.-X. & Yuan, J.-M. Experimental quantum state tomography via compressed sampling. *Phys. Rev. Lett.* **108,** 170403 (2012).
22. Bamber, C. & Lundeen, J. S. Observing Dirac's classical phase space analog to the quantum state. *Phys. Rev. Lett.* **112,** 070405 (2014).
23. Wu, S. State tomography via weak measurements. *Sci. Rep.* **3,** 1193 (2013).
24. Salvail, J. Z. *et al.* Full characterization of polarization states of light via direct measurement. *Nat. Photon.* **7,** 1–6 (2013).
25. Duck, I. M., Stevenson, P. M. & Sudarshan, E. C. G. The sense in which a 'weak measurement' of a spin 1/2 particle's spin component yields a value 100. *Phys. Rev. A* **40,** 2112–2117 (1989).
26. Malik, M. *et al.* Direct measurement of a 27-dimensional orbital-angular-momentum state vector. *Nat. Commun.* **5,** 3115 (2014).
27. Osorio, C. I., Valencia, A. & Torres, J. P. Spatiotemporal correlations in entangled photons generated by spontaneous parametric down conversion. *New J. Phys.* **10,** 113012 (2008).
28. Miatto, F. M., Yao, A. M. & Barnett, S. M. Full characterization of the quantum spiral bandwidth of entangled biphotons. *Phys. Rev. A* **83,** 033816 (2011).
29. Dada, A. C., Leach, J., Buller, G. S., Padgett, M. J. & Andersson, E. Experimental high-dimensional two-photon entanglement and violations of generalized Bell inequalities. *Nat. Phys.* **7,** 1–4 (2011).
30. Agnew, M., Leach, J., McLaren, M., Roux, F. S. & Boyd, R. W. Tomography of the quantum state of photons entangled in high dimensions. *Phys. Rev. A* **84,** 062101 (2011).
31. Leach, J., Bolduc, E., Gauthier, D. J. & Boyd, R. W. Secure information capacity of photons entangled in many dimensions. *Phys. Rev. A* **85,** 060304 (2012).
32. Salakhutdinov, V. D., Eliel, E. R. & Löffler, W. Full-field quantum correlations of spatially entangled photons. *Phys. Rev. Lett.* **108,** 173604 (2012).
33. Tasca, D. S. *et al.* Imaging high-dimensional spatial entanglement with a camera. *Nat. Commun.* **3,** 3:984 (2012).
34. Geelen, D. & Löffler, W. Walsh modes and radial quantum correlations of spatially entangled photons. *Opt. Lett.* **38,** 4108–4111 (2013).
35. Mosley, P. J., Lundeen, J. S., Smith, B. J. & Walmsley, I. A. Conditional preparation of single photons using parametric downconversion: a recipe for purity. *New J. Phys.* **10,** 093011 (2008).
36. Osorio, C. I., Sangouard, N. & Thew, R. T. On the purity and indistinguishability of down-converted photons. *J. Phys. B: At. Mol. Opt. Phys.* **46,** 055501 (2013).

37. Ekert, A. & Knight, P. L. Entangled quantum systems and the Schmidt decomposition. *Am. J. Phys.* **63**, 415–423 (1995).
38. Berkhout, G. C. G. *et al.* Efficient sorting of orbital angular momentum states of light. *Phys. Rev. Lett.* **105,** 153601 (2010).
39. Morizur, J.-F. *et al.* Programmable unitary spatial mode manipulation. *J. Opt. Soc. Am. A Opt. Image Sci. Vis.* **27,** 2524–2531 (2010).
40. Miller, D. Reconfigurable add-drop multiplexer for spatial modes. *Opt. Express* **21,** 20220–20229 (2013).
41. Bolduc, E., Bent, N., Santamato, E., Karimi, E. & Boyd, R. W. Exact solution to simultaneous intensity and phase encryption with a single phase-only hologram. *Opt. Lett.* **38,** 3546–3549 (2013).
42. Christandl, M. & Renner, R. Reliable Quantum State Tomography. *Phys. Rev. Lett.* **109,** 120403 (2012).
43. Boyd, R. W. Intuitive explanation of the phase anomaly of focused light beams. *J. Opt. Soc. Am.* **70,** 877–880 (1980).

Acknowledgements
E.B. and G.G. acknowledge the financial support of the FQRNT, grants #176729 and #173779. J.L. acknowledges the financial support of the Engineering and Physical Sciences Research Council (EPSRC, UK, Grants EP/M006514/1 and EP/M01326X/1). We thank Erik Gauger and George Knee for helpful discussions and feedback.

Author contributions
E.B. conceived the method. E.B. and G.G. developed the theory. E.B. and J.L. designed and performed the experiment and analysed the data. J.L. provided guidance throughout the project. All authors contributed to the manuscript.

Additional information

Competing financial interests: The authors declare no competing financial interests.

6

Mechanical instability at finite temperature

Xiaoming Mao[1], Anton Souslov[2], Carlos I. Mendoza[3] & T.C. Lubensky[4]

Many physical systems including lattices near structural phase transitions, glasses, jammed solids and biopolymer gels have coordination numbers placing them at the edge of mechanical instability. Their properties are determined by an interplay between soft mechanical modes and thermal fluctuations. Here we report our investigation of the mechanical instability in a lattice model at finite temperature T. The model we used is a square lattice with a ϕ^4 potential between next-nearest-neighbour sites, whose quadratic coefficient κ can be tuned from positive to negative. Using analytical techniques and simulations, we obtain a phase diagram characterizing a first-order transition between the square and the rhombic phase and different regimes of elasticity, as well as an 'order-by-disorder' effect that favours the rhombic over other zigzagging configurations. We expect our study to provide a framework for the investigation of finite-T mechanical and phase behaviour of other systems with a large number of floppy modes.

[1] Department of Physics, University of Michigan, Ann Arbor, Michigan 48109, USA. [2] School of Physics, Georgia Institute of Technology, Atlanta, Georgia 30332, USA. [3] Instituto de Investigaciones en Materiales, Universidad Nacional Autónoma de México, Apartado Postal 70-360, 04510 México, D.F., Mexico. [4] Department of Physics and Astronomy, University of Pennsylvania, Philadelphia, Pennsylvania 19104, USA. Correspondence and requests for materials should be addressed to X.M. (email: maox@umich.edu).

Crystalline solids can undergo structural phase transitions in which there is a spontaneous change in the shape or internal geometry of their unit cells[1–4]. These transitions are signalled by the softening of certain elastic moduli or of phonon modes at a discrete set of points in the Brillouin zone. In contrast, lattices with coordination number $z = z_c = 2d$ in d spatial dimensions, which we will call Maxwell lattices[5], exist at the edge of mechanical instability, and they are critical to the understanding of systems as diverse as engineering structures[6,7], diluted lattices near the rigidity threshold[8–10], jammed systems[11–13], biopolymer networks[14–18] and network glasses[19,20]. Hypercubic lattices in d dimensions and the kagome lattice and its generalization to higher dimensions with nearest-neighbour (NN) Hookean springs of spring constant k are a special type of Maxwell lattice whose phonon spectra in a system of N unit cells have harmonic-level zero modes at all $N^{(d-1)/d}$ points on $(d-1)$-dimensional hyperplanes oriented along symmetry directions and passing through the origin[21] of the Brillouin zone. A question that arises naturally is whether these lattices can be viewed as critical lattices at the boundary between phases of different symmetry and, if so, what is the nature of the two phases and the phase transition between them.

Here we introduce a square-lattice model (easily generalized to higher dimensions) in which next-nearest-neighbours (NNNs) are connected via an anharmonic potential consisting of a harmonic term with a spring constant κ tuned from positive to negative, and a quartic stabilizing term. When $\kappa > 0$, the square lattice is stable even at zero temperature. When $\kappa = 0$, NNN springs contribute only at anharmonic order, and the harmonic phonon spectrum is identical to that of the NN lattice. When $\kappa < 0$, the NNN potential has two minima, and the ground state of an individual plaquette is a rhombus that can have any orientation (Fig. 1), leading to a large number of ground states with the same potential energy. The properties of this model, including the subextensive entropy at zero temperature, are very similar to those of colloidal particles confined to a low-height cell[22,23] and to the anti-ferromagnetic Ising model on a deformable triangular lattice[24]. In addition, the scaling of the shear modulus near the zero-temperature critical point is analogous to that observed in finite-temperature simulations of randomly diluted lattices near the rigidity-percolation threshold[25] and to finite-temperature scaling near the jamming transition[26], suggesting that generalizations of our model and approach may provide useful insight into the thermal properties of other systems near the Maxwell rigidity limit.

Using both analytic theory and Monte Carlo (MC) simulations, we discover that (i) among all the equal-energy ground states at $\kappa < 0$, the uniformly sheared rhombic lattice (Fig. 1d) has the lowest free energy through an order-by-disorder effect[24,27–33], (ii) fluctuations stabilize the square phase for a region with $\kappa < 0$ and drive the transition from the square to the rhombic phase to be first order (Fig. 2) and (iii) interesting behaviours, including negative thermal expansion and elastic moduli scalings with fractional exponents, arise near the transition. These intriguing features at finite-temperature mechanical transitions originate from the large but subextensive number of mechanical modes that become floppy at the transition. They may also occur in other transitions at which a large number of modes become floppy simultaneously.

Results

Model. The model we consider is a square lattice with two different types of springs—those connecting NNs and those connecting NNNs, as shown in Fig. 1a. The NN springs are Hookian, with potential

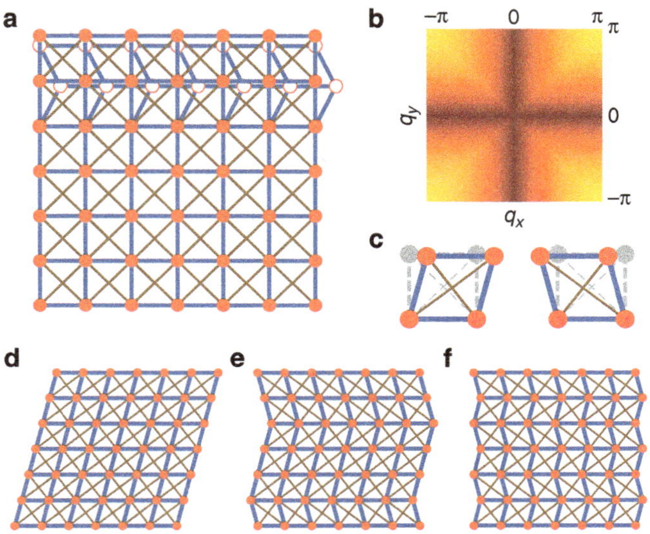

Figure 1 | Square lattice model for mechanical instability. (**a**) The square lattice with NN (blue thick) bonds and NNN (brown thin) bonds. White disks showing a shift of the second row is one example of a floppy modes of the lattice with no NNN bond. (**b**) Density plot of the phonon spectrum of the NN square lattice showing lines of zero modes (darker colour corresponds to lower frequency). (**c**) The $T = 0$ ground states of a plaquette when $\kappa < 0$, with the undeformed reference state shown in grey. (**d-f**) Examples of $T = 0$ ground states of the whole lattice when $\kappa < 0$. (**d**) Uniformly sheared rhombic lattice, which we show to be the preferred configuration at small T in the thermodynamic limit. (**e**) A randomly zigzagging configuration, and **f** is the ordered maximally zigzagging configuration, which has a unit cell consisting of two particles.

$$V_{\mathrm{NN}}(x) = \frac{k}{2}x^2 \tag{1}$$

where $k > 0$. The NNN springs are introduced with an anharmonic potential

$$V_{\mathrm{NNN}}(x) = \frac{\kappa}{2}x^2 + \frac{g}{4!}x^4, \tag{2}$$

where κ can be either positive or negative and g, introduced for stability, is always positive. The Hamiltonian of the whole lattice is thus,

$$H = \sum_{\langle i,j\rangle \in \mathrm{NN}} V_{\mathrm{NN}}\left(|\vec{R}_i - \vec{R}_j| - a\right) + \sum_{\langle i,j\rangle \in \mathrm{NNN}} V_{\mathrm{NNN}}\left(|\vec{R}_i - \vec{R}_j| - \sqrt{2}a\right), \tag{3}$$

where $\vec{R}_i$ is the positions of the node i and a is the lattice constant. In what follows, we will use the reduced variables

$$\tau \equiv \kappa/k \quad \text{and} \quad \lambda \equiv ga^2/k \tag{4}$$

to measure the strength of couplings in V_{NNN}.

For $\kappa > 0$, $V_{\mathrm{NNN}}(x)$ has a unique minimum at $x = 0$, and the ground state of H is the square lattice with lattice spacing a. All elastic moduli of this state are nonzero, and it is stable with respect to thermal fluctuations, although, as we shall see, it does undergo thermal contraction at nonzero temperature with respect to the $T = 0$ equilibrium state. When $\kappa < 0$, $V_{\mathrm{NNN}}(x)$ has two minima at $x = \pm\sqrt{6\kappa/g}$, corresponding to stretch and compression, respectively. This change in length of NNN springs is resisted by the NN springs, and in minimum energy configurations one NNN bond in each plaquette will stretch and the other

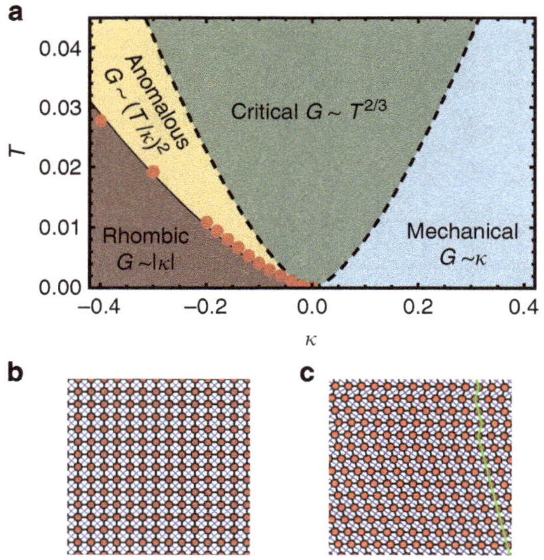

Figure 2 | Phase diagram and Monte Carlo snapshots. (**a**) An example of the phase diagram of the model square lattice at $k=1$, $g=10$, $a=1$. The black solid line (the red dots) shows the boundary obtained from analytic theory (Monte Carlo simulation) between the square phase on the right and the rhombic phase on the left. The square phase is stabilized by thermal fluctuations even for $\kappa<0$, and the shear modulus G of the square lattice exhibits different scaling regimes (separated by black dashed lines) determined by equation (26). (**b**,**c**) Monte Carlo snapshots of the square and rhombic phases, respectively. A small number of zigzags exist in the rhombic phase Monte Carlo snapshots, resulting from finite size effects as discussed in Model in the Results section.

will contract. The alternative of having both stretch or contract would cost too much NN energy. A reasonable assumption, which is verified by our direct calculation, is that the stretching and contraction will occur symmetrically about the centre so that the resulting equilibrium shape is a rhombus rather than a more general quadrilateral (see Supplementary Note 1). The shape of a rhombus is uniquely specified by the lengths d_1 and d_2 of its diagonals (which are perpendicular to each other), whose equilibrium values are obtained by minimizing the sum over plaquettes of

$$\begin{aligned} V_{\mathrm{PL}} =& 2V_{\mathrm{NN}}\left(\frac{1}{2}\sqrt{d_1^2+d_2^2}-a\right)+ \\ & V_{\mathrm{NNN}}\left(d_1-\sqrt{2}a\right)+V_{\mathrm{NNN}}(d_2-\sqrt{2}a). \end{aligned} \quad (5)$$

The ground states of the entire lattice when $\kappa<0$ must correspond to a tiling of the plane by identical rhombi, each of whose vertices are fourfold coordinated. It is clear that zigzag arrangements of rows (or columns) of rhombi in which adjacent rows tilt in either the same or opposite directions constitute a set of ground states. A derivation showing that this is the complete set can be found in ref. 24, which considered packing of isosceles triangles, which make up half of each rhombus. The ground-state energy per site, ϵ_0, is simply V_{PL} evaluated at the equilibrium values of d_1 and d_2.

Each ground-state configuration of a system with N_x vertical columns and N_y horizontal rows has $K=0, \cdots, N_y$ horizontal zigzags or $K=0, \cdots, N_x$ vertical zigzags. Thus, the ground-state entropy is $\ln(2^{N_x}+2^{N_y}) \overset{N_y\to N_x}{\longrightarrow} (N_x+1)\ln 2$, which although subextensive diverges in the thermodynamic limit. Such ground-state configurations are found in other systems, most notably the zigzagging phases seen in suspensions of confined colloidal particles[22,23]. The confined colloidal system has a phase diagram that depends only on the planar density and the height of confinement of the colloids. For sufficiently large heights, the colloids form a phase of two stacked square lattices. In a neighbouring region of the phase diagram, explored in simulations of refs 34,35, this square lattice symmetry is broken through a weakly discontinuous transition and a rhombic phase is observed. This region of the phase diagram of confined colloids thus provides a physical realization of the Hamiltonian (3).

At low but nonzero temperatures, the degeneracy of the ground state is broken by thermal fluctuations through the order-by-disorder mechanism[24,27,29–33]. This splitting of degeneracy due to small phonon fluctuations around the ground state may be calculated using the dynamical matrix in the harmonic approximation to the Hamiltonian (3). For each ground-state configuration $\vec{R}_i$, we write the deformation as $\vec{R}_i \to \vec{R}_i + \vec{u}_i$, and expand to quadratic order in $\vec{u}$, $H=\frac{1}{2}\sum_{\langle ij\rangle}\vec{u}_i\mathbf{D}_{ij}\vec{u}_j$. The Fourier transform of $\mathbf{D}_{ij}$, $\mathbf{D}_q$, is block-diagonal for each configuration, and the associated harmonic free energy is

$$F_{\mathrm{p}}(N_x,N_y,K)=\frac{1}{2}k_BT\sum_{\mathbf{q}}\ln\det\mathbf{D}_{\mathbf{q}} \equiv Nk_BTw_{\mathrm{p}}, \quad (6)$$

where w_{p} is the free energy per site in units of k_BT. In general, F_{p} depends not only on K, but on the particular sequence of zigzags as well. We numerically calculated this free energy for all periodically zigzagged configurations with up to 10 sites per unit cell. We found that the lowest-free-energy state is the uniformly sheared state (Fig. 1c) with $K=0$, and the highest-free-energy state is the maximally zigzagged sate (Fig. 1e) with $K=N_y$ and two sites per unit cell. Both of these energies are extensive in the number of sites N, and we define

$$\begin{aligned} \Delta F_0(N_x,N_y) =& F_{\mathrm{p}}(N_x,N_y,N_y)-F_{\mathrm{p}}(N_x,N_y,0) \\ \equiv& N_xN_yk_BT\Delta w_0 \\ \Delta F(N_x,N_y,K) =& F_{\mathrm{p}}(N_x,N_y,K)-F_{\mathrm{p}}(N_x,N_y,0). \end{aligned} \quad (7)$$

Figure 3a displays our calculation of $\Delta F_0(\tau)$, which vanishes, as expected, at $\tau=0$ and also at large τ at which the rhombus collapses to a line. Figure 3b plots $\Delta F/\Delta F_0$ as a function of $\phi=K/N_y$ for different values of τ. By construction, this function must vanish at $\phi=0$ and be equal to one at $\phi=1$. All of the points lie approximately on a straight line of slope one. Thus, we can approximate $F_{\mathrm{p}}(N_x, N_y, K)$ by

$$F_{\mathrm{p}}(N_x,N_y,K)=F_{\mathrm{p}}(N_x,N_y,0)+N_xKk_BT\Delta w_0. \quad (8)$$

Note that for each K, this energy is extensive in $N=N_xN_y$ as long as $\phi\neq 0$. This approximation treats kinks as independent excitations and effectively ignores interactions between them.

These calculations were carried out in the thermodynamic limit, $N\to\infty$, in which the sum over q is replaced in the continuum limit by an integral. To compare these results with the MC results of the next section, it is necessary to study finite size effects. The first observation is that for any finite N_x, the system will be effectively one dimensional for a sufficiently large N_y, and as a result, we would expect the number of zigzags to fluctuate. To proceed, we continue to use the continuum limit to evaluate equation (6), and we use equation (8) for $F_{\mathrm{p}}(N_x, N_y, K)$. The partition function for this energy is

$$\begin{aligned} Z =& \sum_{K=0}^{N_y} e^{-\beta\left[F_{\mathrm{p}}(N_x,N_y,0)+E_0(N_x,N_y)\right]}\binom{N_y}{K}e^{-N_xK\Delta w_0} \\ =& e^{-\beta\left[F_{\mathrm{p}}(N_x,N_y,0)+E_0(N_x,N_y)\right]}\left(1+e^{-N_x\Delta w_0}\right)^{N_y} \end{aligned} \quad (9)$$

where $E_0(N_x, N_y)$ denotes the potential energy, which is

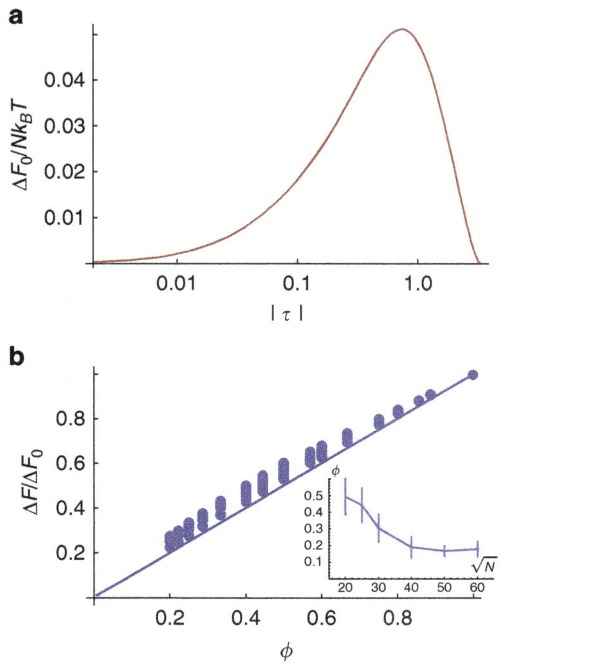

Figure 3 | The phonon contribution to free energy. (**a**) The free-energy difference ΔF_0 between the maximally zigzagging configuration (Fig. 1e) and the uniformly sheared square lattice configuration (Fig. 1c). For sufficiently large τ, the lattice collapses onto a line, at which point the free-energy difference goes to zero. (**b**) The free energy of the phonons as a ratio $\Delta F/\Delta F_0$, where ΔF and ΔF_0 are defined in equation (7), as a function of the zigzag fraction $\phi = K/N_x$. There are multiple data points at each ϕ, corresponding to taking $\tau = -0.02$, -0.1 and -1 ($\lambda = 10$) as well as different zigzag sequences with up to 10 sites per unit cell. Our calculation show that the lattice without zigzags is entropically favoured, and that $\Delta F/\Delta F_0$ is well approximated by the line ϕ, and the differences in τ and the zigzagging sequence only lead to small dispersions, indicating that non-interacting zigzags is a good approximation. The inset of **b** shows Monte Carlo simulation results for the zigzagging fraction ϕ plotted against the linear system size $\sqrt{N}$, with the error bars showing the s.d. obtained from 100 samples.

independent of K, and the full free energy is

$$\begin{aligned} F(N_x, N_y, T) &= -T\ln Z \\ &= N_xN_y(f_0 + e_0) - N_yk_BT\ln\left(1 + e^{-N_x\Delta w_0}\right), \end{aligned} \tag{10}$$

where e_0 and f_0 are the potential energy and phonon free energy per site of the uniformly sheared state. Thus, as expected, when $N_x\Delta w_0 \gg 1$, zigzag configurations make only a very small, subextensive contribution to the free energy. On the other hand, in the opposite limit, they make an extensive contribution of $N_xN_yk_BT\Delta w_0$ to the energy. Therefore, at a given τ, the zigzag configurations are favoured when the system is small, and in thermodynamic limit, the rhombic configuration is always favoured. Our MC simulation verified this (see the inset of Fig. 3b).

MC simulation. We simulate the system using a MC algorithm inside a periodic box whose shape and size are allowed to change to maintain zero pressure. In this version of the Metropolis algorithm, also used in refs 24,36, for each MC step a particle is picked at random and a random trial displacement is performed. The trial displacement is initially uniformly distributed within a radius of $0.1a$, but throughout the simulation the radius is adjusted to keep the acceptance probability between 0.35 and 0.45. Given the initial configuration energy E_i and the trial configuration energy E_j, the trial configuration is accepted with probability $[1 + \exp(E_i - E_j)/T]^{-1}$, that is, Glauber dynamics is used. After initializing the system using the square lattice configuration with lattice constant a, the simulation is first run at a high temperature and is then annealed to the final low temperature. For each intermediate temperature, an equilibration cycle in a sample of N sites consists of at least $N^4 \times 10^2$ MC steps. To accommodate areal and shear distortions in the different phases we encounter, the simulation box area and shape are changed using a similar acceptance algorithm, with the trial deformation adjusted to keep the acceptance probability between 0.35 and 0.45, such that the simulation box retains the shape of a parallelogram[37]. The simulation is thus performed at zero pressure, and a range of temperatures measured in units of ka^2 for up to $N = 3,600$ sites.

We use these simulations to investigate the phase diagram[38] corresponding to the Hamiltonian (3) and to investigate the properties of the phases we encounter, such as ground-state degeneracy, order-by-disorder and negative thermal expansion. As all simulations involve a finite lattice and are run for a finite time, we took care to make sure that the system is sufficiently large to capture the thermodynamic behaviour, and that the simulation time is sufficiently long for the system to relax to equilibrium. To capture the subtlety of the order-by-disorder effect for a finite system, we simulated the model for a range of sizes and times and calculated the average fraction of zigzags n in equilibrium (inset of Fig. 3b). While for small systems, $\phi \approx \frac{1}{2}$, as in a disordered zigzagging configuration, for large systems, ϕ approaches 0, suggesting that the system prefers the configuration of a uniformly sheared square lattice. Thus, we find good agreement with theoretical results from Model subsection in the Results section.

We determined the shape of the boundary between the square and the rhombic phase by calculating the heat capacity of the system as a function of temperature at fixed λ and τ. The location of the peak of the heat capacity corresponds to the location of the phase transition in the thermodynamic limit, and in our simulations, the locations of the peak converge to the values seen in Fig. 2. We characterize this transition using an order parameter $t \equiv \langle|\cot^{-1}\theta|\rangle$ where θ is the angle between the bottom and the left bonds of a plaquette and $\langle\ldots\rangle$ denotes an average over all plaquettes. Obviously in the square phase $\theta = \pi/2$ and $t = 0$, whereas in the rhombic phase $t > 0$. This definition of order parameter is independent of zigzags, because it is extracted from the deformation of each plaquette instead of the overall deformation of the lattice, and thus it is not affected by slow relaxation to the uniformly sheared square lattice. In our simulations, we average over 100 configurations, and the resulting t as a function of κ is shown in Fig. 4a. The behaviour of the order parameter is consistent with a weakly discontinuous transition. We also characterized the thermal expansion $(L - L_0)/L_0$ in simulation, as shown in Fig. 4b.

Analytic theory and the phase diagram. The special feature of our model is its large but subextensive number of soft modes living on the q_x and q_y axes in the first Brillouin zone, as shown in Fig. 1b in the limit $\kappa \to 0)$[21,39]. As we discuss below, these floppy modes provide a divergent fluctuation correction to the rigidity of the square lattice and render the transition from the square to the rhombic phase first order. Thus, this model is analogous to the one introduced by Brazovskii[40] for a liquid–solid transition, in which mode frequencies of the form $\omega = \Delta + (q - q_c)^2/m$ vanish on a $(d-1)$-dimensional hypersphere when $\Delta \to 0$, and contribute terms to the free energy singular in Δ.

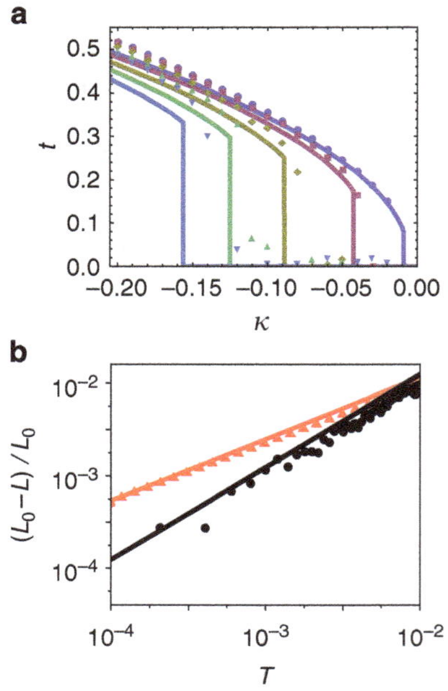

Figure 4 | Order parameter and negative thermal expansion. (**a**) Order parameter t calculated from theory (lines) and simulation (data points) at $\lambda=10$, and from right to left, $T/(ka^2)=0.0001$, 0.001, 0.003, 0.005, 0.007. (**b**) Negative thermal expansion. Shown in the figure are normalized size changes $(L_0-L)/L_0$ as a function of T at $\kappa=0$, $g=10$ (red triangles: MC data; red upper line: theory), and at $\kappa=0.1$, $g=0$ (black circles: MC data; black lower line: theory).

To study the square-to-rhombic transition, we take the $T=0$ square lattice, with site i at position $\vec{R}_{0,i}$, as the reference state. We then represent positions in the distorted lattices as the sum of a part arising from uniform strain characterized by a deformation tensor $\mathbf{\Lambda}$ and deviations $\vec{u}'_i$ from that uniform strain:

$$\vec{R}_i=\mathbf{\Lambda}\cdot\vec{R}_{0,i}+\vec{u}'_i, \tag{11}$$

The deviations $\vec{u}'_i$ are constrained to satisfy periodic boundary conditions, and their average $\langle\vec{u}'_i\rangle$ is constrained to be zero. The former condition ensures that the sum over all bond stretches arising from these deviations vanishes for every configuration. Without loss of generality, we take $\mathbf{\Lambda}_{yx}$ to be zero, leaving three independent parameters to parameterize the three independent strains. As we detail in Supplementary Note 4, the strain parameter characterizing pure shear with axes along the x and y axes of the reference lattice is of order t^2 and can be ignored near $\tau=0$, and we set

$$\mathbf{\Lambda}=\begin{pmatrix}1+s & t\\ 0 & 1+s\end{pmatrix}. \tag{12}$$

Note that $\mathbf{\Lambda}$ is invertible although it is not symmetric. t is the order parameter that distinguishes the rhombic phase from the square phase, as we defined in MC simulation subsection in the Results section. Thermal fluctuations lead to $s<0$ in both phases.

Expanding the Hamiltonian (equation (3)) to second-order $\vec{u}'_i$ about the homogeneously deformed state $\langle\vec{R}_i\rangle=\mathbf{\Lambda}\cdot\vec{R}_{0,i}$, we obtain

$$H=H_0(\mathbf{\Lambda})+\frac{1}{V}\sum_q\vec{u}'_q\cdot\mathbf{D}_q(\mathbf{\Lambda})\cdot\vec{u}'_{-q}+O\left((\vec{u}')^3\right), \tag{13}$$

where H_0 is the energy of the uniformly deformed state, and

$$\mathbf{D}_q(\mathbf{\Lambda})=v_q(\mathbf{\Lambda}^T\mathbf{\Lambda})\mathbf{I}+\mathbf{\Lambda}\cdot\mathbf{M}_q(\mathbf{\Lambda}^T\mathbf{\Lambda})\cdot\mathbf{\Lambda}^T \tag{14}$$

is the $d\times d$ dimensional ($d=2$ being spatial dimension) dynamical matrix with scalar v_q and second-rank tensor $\boldsymbol{M}_q$ determined by the potentials. There is no term linear in $\vec{u}'_i$ in equation (13) because of the periodicity constraint (see Supplementary Note 2).

Integrating out the fluctuations $\vec{u}'$ from the Hamiltonian (13), we obtain the free energy of the deformed state

$$F(\mathbf{\Lambda})=H_0(\mathbf{\Lambda}^T\mathbf{\Lambda})+\frac{T}{2}\ln\mathrm{Det}\tilde{\mathbf{D}}(\mathbf{\Lambda}), \tag{15}$$

where $\tilde{\mathbf{D}}=v_q\mathbf{I}+\boldsymbol{M}_q\mathbf{\Lambda}^T\mathbf{\Lambda}$ depends only on $\mathbf{\Lambda}^T\mathbf{\Lambda}=1+2\epsilon^0$, where $\epsilon^{\mathbf{0}}$ is the full nonlinear strain. (This form is similar to equation (6), except that here we expand around a uniformly deformed lattice characterized by the variable $\mathbf{\Lambda}$, rather than about a zigzagging state.) Thus the one-loop free energy of equation (15) is a function of the nonlinear, rather than the linear strain, so that rotational invariance in the target space is guaranteed and there is a clean distinction between nonlinear terms in linearized deformations arising from nonlinearities in $\epsilon^{\mathbf{0}}$ and from nonlinear terms in the expansion in powers of $\epsilon^{\mathbf{0}}$.

To analyse the transition between the square and the rhombic phases at low temperature, we expand F as a series in $\epsilon^{\mathbf{0}}$, by expanding the transformed dynamical matrix as $\tilde{\mathbf{D}}=\mathbf{D}_0+\mathbf{A}(\epsilon^0)$, where $\mathbf{D}_0=\mathbf{D}|_{\epsilon^0=0}$ is the dynamical matrix of the undeformed state. The free energy is then

$$F(\epsilon^0)=H_0(\epsilon^0)+\frac{T}{2}\mathrm{Tr}\ln\mathbf{D}_0\left[\mathbf{I}+\mathbf{G}_0\cdot\mathbf{A}(\epsilon^0)\right]\equiv Vf(\epsilon^0), \tag{16}$$

where $\mathbf{G}_0\equiv\mathbf{D}_0^{-1}$ is the phonon Green's function in the undeformed state, $V=Na^2$ and f is the free-energy density. The expansion of F at small $\epsilon^{\mathbf{0}}$ follows from this.

Close to the transition, F is dominated by fluctuations coming from the floppy modes, as we discussed above. As $\kappa\to0$, the frequency of these floppy modes vanishes as $\omega\sim\sqrt{\kappa}$, and the corresponding phonon Green's function diverges, leading to divergent fluctuation corrections to the coefficients of ϵ^0 in equation (16), as detailed in Supplementary Note 3. Keeping leading order terms as $\tau\to0$, we can identify the two phases through the equations of state,

$$\frac{\partial f(\epsilon^0)}{\partial s}\simeq2ka^2\left(\frac{\tilde{T}}{\sqrt{\tau}}+s\right)=0 \tag{17}$$

$$\frac{\partial f(\epsilon^0)}{\partial t}\simeq ka^2t\left(\tau+\frac{\lambda\tilde{T}}{\sqrt{\tau}}+\frac{1}{12}\lambda t^2\right)=0, \tag{18}$$

where

$$\tilde{T}=\frac{\pi T}{8ka^2} \tag{19}$$

is a unitless reduced temperature. Equation (18) has three solutions for t: $t=0$ corresponding to the square phase, and two solutions for $t\neq0$ corresponding to the two orientations of the rhombic phase. There is only a single solution for s, with $s<0$, from which we conclude that both phases exhibit negative thermal expansion. The elastic rigidity and thus the stability of the two phases is determined by the second derivatives of F with respect to s and t. In particular, the reduced shear modulus (G/k where G is the shear modulus) is

$$r=\frac{1}{k}\frac{\partial^2f(\epsilon^0)}{\partial t^2}\simeq\tau+\frac{\tilde{T}}{\sqrt{\tau}}+\frac{1}{4}\lambda t^2. \tag{20}$$

To obtain these leading order equations, we (i) assume low T, so we keep only terms singular in τ as $\tau\to0$, such as $t/\sqrt{\tau}$, in the integral of $\frac{T}{2}\ln \mathrm{Det}\,\tilde{\mathbf{D}}(\mathbf{\Lambda})$, and (ii) assume that

$$|s| \sim t^2 \ll \tau \ll 1, \tag{21}$$

the validity of which will be verified below.

As observed in the simulation (Fig. 2), thermal fluctuations at $T>0$ stabilize the square relative to the rhombic phase even for $\kappa<0$. To understand this phenomenon within the analytic approach, we use a self-consistent-field approximation in which τ is replaced by its renormalized value r in the phonon Green's function $\mathbf{G}_0$ and thus in the denominators on the right-hand sides of equations (17, 18 and 20). In this approximation, the shear rigidity of the square ($t=0$) and rhombic ($t\neq 0$) phases satisfy

$$r = \begin{cases} \tau + \frac{\tilde{T}\lambda}{\sqrt{r}} & \text{square} \\ -2\tau - 2\frac{\tilde{T}\lambda}{\sqrt{r}} & \text{rhombic} \end{cases}, \tag{22}$$

where we used the equation of state, equation (18) to eliminate t^2 from equation (20). In the square phase, equation (22) has a solution $r>0$, implying local stability, everywhere except at $\tilde{T}=0$, $\tau<0$ and in the uninteresting limit, $z\to-\infty$. This local stability implies that the transition to the rhombic phase must be first order. In the rhombic phase, solutions $r>0$ only exist for $\tau<\tau_{c1}$ ($z<z_{c1}$), where

$$\tau_{c1} = -\frac{3}{2}(\tilde{T}\lambda)^{2/3}. \tag{23}$$

The solutions to equation (22) can conveniently be expressed as scaling functions in the two phases:

$$\frac{|\tau|}{r} = h_v(|z|), \quad z = \frac{\tau}{(\tilde{T}\lambda)^{2/3}}, \tag{24}$$

where $v=$ s, r for the square and rhombic phases, respectively. The scaling functions $h_s(|z|)$ and $h_r(|z|)$ depicted in Fig. 5 have the following limits

$$h_s(|z|) \sim \begin{cases} 1 & z\to+\infty \\ |z| & z\to 0 \\ |z|^3 & z\to-\infty \end{cases} \tag{25}$$

$$h_r(|z|) \sim \begin{cases} (3/2)-\sqrt{6\,|\,z-z_{c1}\,|} & z\to z_{c1}^- \\ 1/2 & z\to-\infty \end{cases} \tag{26}$$

where $z_{c1}=-3/2$. The z^3 regimes of h_s is in the metastable regime where the rhombic phase is stable. These results yield the scaling phase diagram of Fig. 2.

The phase boundary of this discontinuous transition occurs along the coexistence line (that is, equal free-energy line) of the two phases. Following Brazovskii, we have already calculated the limit of metastability of the rhombic phase, that is, for the value of $\kappa=\tau_{c1}$ (equation (23)) at the local free-energy minimum where that phase first appears. We then calculate the free-energy difference between the two phases, which is evaluated through the following integral for a given τ,

$$\Delta F = \int_0^{t_{\text{rhombic}}} \frac{dF(\epsilon^0)}{dt}dt = \int_{r_0}^{r_1} \frac{dF(\epsilon^0)}{dt}\frac{dt}{dr}dr, \tag{27}$$

where we have substituted $r(t)$ for the integral measure. Here r_0 and r_1 are the values of r at the minima of F, corresponding to the square and the rhombic phases determined by equation (22). Along the path of this integral, equation (20) is valid, but the equations of state (17) and (18) and the equation (equation (22)) for r in the rhombic phase are not satisfied, because they only apply to equilibrium states. The phase boundary corresponds to the curve of $\tau=\tau_{c2}$ along which $\Delta F|_{\tau=\tau c2}=0$. As shown in Supplementary Note 4, an asymptotic solution valid at low τ can be obtained by expanding the equation around $\tau=\tau_{c1}$, assuming that τ_1 and τ_2 are of the same order of magnitude (verified below). This yields

$$\tau_{c2} = -\left(\frac{3}{2}+c\right)(\tilde{T}\lambda)^{2/3} = -1.716\tilde{T}^{2/3}, \tag{28}$$

for $\tilde{T}\ll 1$ where $c\simeq 0.216$ is a constant. This transition line is shown in Fig. 2. Excellent agreement between theory and simulation is obtained without any fitting parameter.

Along the phase boundary, $r_0, r_1 \sim \tilde{T}^{2/3}>0$, so that both phases are locally stable. The order parameter for the transition, t, jumps from 0 to

$$t_{c2} = 3.4\tilde{T}^{1/3}\lambda^{-1/6} \tag{29}$$

at the transition. As $T\to 0$ this discontinuity vanishes, consistent with the continuous nature of the transition at $T=0$. A good agreement between t-values in theory and in simulation is shown in Fig. 4a.

From equation (17), the negative thermal expansion coefficient in both the square and rhombic phases,

$$s = -\frac{\tilde{T}}{\sqrt{r}}, \tag{30}$$

is determined by the equation of state, equation (17). Equation (22) for r then implies the following behaviour for s in different regions of the phase diagram: in the critical region $0<|\tau|<\tilde{T}^{2/3}$ of the square phase,

$$s \simeq -\tilde{T}^{2/3}\lambda^{-1/3}. \tag{31}$$

Deep in the square and rhombic phases, where $0<\tilde{T}^{2/3}\ll|\tau|$,

$$s = \begin{cases} -\tilde{T}/\sqrt{|\tau|} & \text{square} \\ -\tilde{T}/\sqrt{2|\tau|} & \text{rhombic} \end{cases}. \tag{32}$$

Finally, along the coexistence curve in both phases, $s\sim-\tilde{T}/\sqrt{|\tau|}\sim\tilde{T}^{2/3}$. These results agree well with simulation measurement of negative thermal expansion, as shown in Fig. 4b. In this lattice, the negative thermal expansion behaviour results from strong transverse fluctuations associated with soft modes.

These solutions for s and t verify that our assumptions in equation (21) are satisfied, provided that $\lambda\gg 1$.

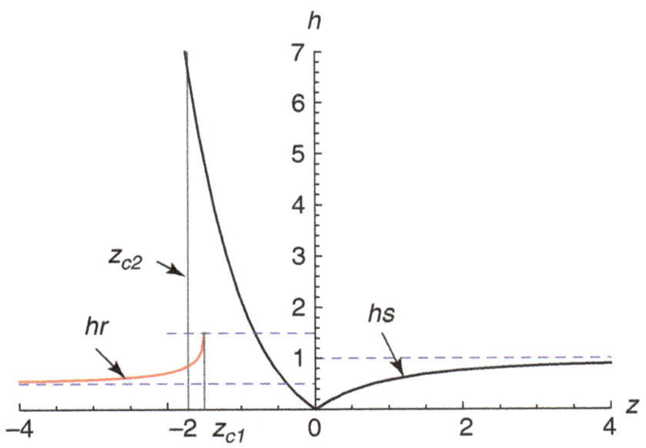

Figure 5 | Plots of $h_s(|z|)$ (black) and $h_r(|z|)$ (red) as a function of z. Note the singular behaviour of $h_r(|z|)$ in the vicinity of z_{c1} and the large difference between h_s and h_r at the first-order transition at $z=z_{c2}$.

Discussion

Our model of the square-to-rhombic transition is very similar to a model, studied by Brazovskii[40], for the transition from an isotropic fluid to a crystal and later applied to the

Rayleigh–Bénard instability[41] and to the nematic-to-smectic-C transitions in liquid crystals[42,43]. In all of these systems, an infinite number of zero modes exist in the disordered phase, leading to a singular contribution to the free energy. These zero modes live on a subextensive but infinite manifold in momentum space, which is a $(d-1=2)$-dimensional spherical shell in the Brazovskii case, a $(d-1=1)$-dimensional circle in the Rayleigh–Bénard case and two $(d-2=1)$-dimensional circles in the liquid-crystal case. We use the Brazovskii theory to calculate the temperature of the first-order square-to-rhombic transition as a function of $\kappa<0$ and negative thermal expansion in the square phase.

In applying the Brazovskii approach to our problem, we develop an expansion of the free energy that maintains rotational invariance of elastic distortions in the target space. Previous treatments[1] of structural transitions tend to mix up nonlinear terms arising from nonlinearities in the strain tensor required to ensure rotational invariance and nonlinear terms in the elastic potential itself. This advance should be useful for the calculation of renormalized free energies and critical exponents in standard structural phase transitions.

The number and nature of zero modes of the critical NN square lattice has a direct impact on the properties of the lattices with $\kappa>0$ and $\kappa<0$, and it is instructive to review their origin. A powerful index theorem[44] relates the number of zero modes of a frame consisting of N points and $N_B \equiv \frac{1}{2}zN$ bonds, where z is the average coordination number, via the relation

$$N_0 = dN - N_B + S, \tag{33}$$

where S is the number of independent states of self-stress, in which bonds are under tension or compression and in which the net force on each point is zero. In his seminal 1864 paper[5], Maxwell considered the case with $S=0$ that yields the Maxwell relation for the critical coordination number at which N_0 is equal to the number $n(d)$ ($=d(d+1)/2$ for free boundary conditions and $=d$ for periodic boundary conditions, which Maxwell did not consider) of zero modes of rigid translation and rotation:

$$z_c^N = 2d - \frac{2n(d)}{N}. \tag{34}$$

In the limit of large N, $z_c^N \to z_c^\infty = 2d$.

There are many small-unit-cell periodic Maxwell lattices with $N_0=S$. The NN square and kagome lattices in two dimensions and the cubic and pyrochlore lattices in three dimensions are special cases of such lattices that have sample-spanning straight lines of bonds that support states of self-stress under periodic boundary conditions. Therefore, they have of order $N^{(d-1)/d}$ states of self-stress and the same number of zero modes, which are indicators of buckling instabilities of the lines when subjected to compression. Geometrical distortions of theses lattices that remove straight lines, as is the case with the twisted kagome lattice[45], remove states of self-stress and associated zero modes. When subjected to free rather than periodic boundary conditions, these distorted lattices continue to have no bulk zero modes (other than those of uniform translation and rotation), and they have of order $N^{(d-1)/d}$ fewer bonds than the periodic lattice of N sites. As a result, they have of order $N^{(d-1)/d}$ zero modes that, except for the modes of rigid translation and rotation, are necessarily surface modes, which can have a topological character[46] or be described in the long-wavelength limit by a conformal field theory[45]. Unlike the infinitesimal zero modes of hypercubic lattices, those of the kagome and pyrochlore do not translate into finite zero modes of the lattices when finite sections are cut from a lattice under periodic boundary conditions. Thus, it is not yet clear whether the ground state of the latter lattices are highly degenerate or not. Nevertheless, the Brasovskii theory should provide a sound description of thermal properties of these lattices in the vicinity of the point $T=0$, $\kappa=0$.

There has been very little research on the thermal properties of Maxwell and related lattices. There is, however, a recent paper[25] that studies the effects of temperature on the shear modulus in the vicinity of the rigidity percolation threshold, $z=z_{cen}\approx 4$, of the randomly diluted triangular lattice. Increasing either τ in our model or $\Delta z \equiv z - z_{cen}$ in the rigidity percolation model leads to mechanically stable $T=0$ lattices ($\kappa \propto (\Delta z)^2$ in kagome and square lattices with randomly added NNN bonds[39,47]), and the phase diagram in the $T-\Delta z$ plane calculated in ref. 25 is qualitatively similar to our phase diagram in the $T-\tau$ plane (Fig. 2), if the ordered phase in our model is ignored. Both diagrams exhibit a mechanical regime in which $G \propto T^{b_m}$, a critical regime in which $G \propto T^{b_c}$ and an anomalous regime in which $G \propto T^{b_a}$, where in both cases $b_a > b_c > b_m = 0$, although the values of b_a and b_c are different. The critical points in both models are at or near the Maxwell limit of $z=4$, and it would seem that it is this feature that is responsible for the similarities in their properties. There are, however, important differences between the two models. In our model, there are mechanically stable $T=0$ phases (although one with high entropy) on both sides of the critical point, whereas in the rigidity percolation model, the phase with $z<z_{cen}$ is not stable at $T=0$. The critical point of our model is the NN square lattice, which has special states of self-stress that give rise to linear manifolds of zero energy that ultimately control the critical and anomalous regimes. The critical point of rigidity percolation on the triangular lattices is less well characterized. At the rigidity threshold of the 'generic' triangular lattice[10] in which sites are randomly displaced to remove straight lines of bonds, there certainly are a large number of zero modes[48]. We speculate that there are a similarly large number of zero modes at the rigidity threshold in the undistorted but diluted triangular lattice, and that they are responsible for the observed temperature regimes in the phase diagram in ref. 25. Clearly there is work to be done to understand more fully the effects of temperature on general Maxwell lattices.

We have presented an analysis of a model based on the square lattice with NN harmonic and NNN anharmonic springs that can be tuned at zero temperature from a stable square lattice through the mechanically unstable NN square lattice to a highly degenerate zigzag state by changing the coefficient κ of the harmonic term in the NNN spring from positive through zero to negative. Using analytic theory, including a generalization of the Brazovskii theory for the liquid-to-crystal transition, we investigated the phase diagram and mechanical properties of this model at $T>0$. The degeneracy of the $T=0$ zigzag state is broken by an order-by-disorder effect, thermal fluctuations drive the square-to-rhombic phase transition first order and the elastic modulus of the square phase exhibits a crossover from mechanical ($\propto \kappa$), to critical ($\propto T^{2/3}$), to anomalous ($\propto T^2$), as κ is tuned from positive to negative at finite T, and this crossover is characterized by the scaling variable $z = \tau/(\tilde{T}\lambda)^{2/3} = (\kappa/k)/(Tg/k^2)^{2/3}$. This behaviour arises because the spectrum of the NN square lattice with N sites exhibits $\sqrt{N}$ zero modes on a one-dimensional manifold in the Brillouin Zone. Other lattices such as the two-dimensional kagome lattice, the three-dimensional simple cubic lattice and the three-dimensional pyrochlore and β-cristobalite[49] lattices have similar spectra, and it is our expectation that generalizations of our model to these lattices will exhibit similar behaviour. It is also likely that our model can inform us about more physically realistic models in which interactions lead to spectra with a large set of modes with small but nonzero frequency[50–53].

References

1. Folk, R., Iro, H. & Schwabl, F. Critical elastic phase transtions. *Z. Physik B* **25,** 69–81 (1976).
2. Cowley, R. A. Structural phase-transitions I. Landau theory. *Adv. Phys.* **29,** 1–110 (1980).
3. Bruce, A. Structural phase transitions II. Static critical behaviour. *Adv. Phys.* **29,** 111–217 (1980).
4. Fujimoto, M. *The Physics of Structural Phase Transitions* 2nd edn (Springer, 2005).
5. Maxwell, J. C. On the calculation of the equilibrium and stiffness of frames. *Philos. Mag.* **27,** 294 (1864).
6. Heyman, J. *The Science of Structural Engineering* (Cengage Learning, 2005).
7. Kassimali, A. *Structural Analysis* (Cengage Learning, 2005).
8. Feng, S. & Sen, P. N. Percolation on elastic networks: New exponent and threshold. *Phys. Rev. Lett.* **52,** 216–219 (1984).
9. Feng, S., Sen, P. N., Halperin, B. I. & Lobb, C. J. Percolation on two-dimensional elastic networks with rotationally invariant bond-bending forces. *Phys. Rev. B* **30,** 5386–5389 (1984).
10. Jacobs, D. J. & Thorpe, M. F. Generic rigidity percolation: the pebble game. *Phys. Rev. Lett.* **75,** 4051–4054 (1995).
11. Liu, A. J. & Nagel, S. R. Jamming is not just cool any more. *Nature* **396,** 21–22 (1998).
12. Wyart, M. On the rigidity of amorphous solids. *Ann. Phys. Fr.* **30,** 1–96 (2005).
13. Liu, A. J. & Nagel, S. R. The jamming transition and the marginally jammed solid. *Annu. Rev. Condens. Matter Phys.* **1,** 347–369 (2010).
14. Elson, E. L. Cellular mechanism as an indicator of cytoskeletal structure and function. *Annu. Rev. Biophys. Biophys. Chem.* **17,** 397–430 (1988).
15. Kasza, K. *et al.* The cell as a material. *Curr. Opin. Cell Biol.* **19,** 101–107 (2007).
16. Alberts, B. *et al. Molecular Biology of the Cell* 4th edn (Garland, 2008).
17. Janmey, P. *et al.* Resemblance of actin-binding protein/actin gels to covalently cross-linked networks. *Nature* **345,** 89–92 (1990).
18. Broedersz, C. P. & MacKintosh, F. C. Modeling semiflexible polymer networks. *Rev. Mod. Phys.* **86,** 995–1036 (2014).
19. Phillips, J. C. Topology of covalent non-crystalline solids II: medium-range order in chalcogenide alloys and A-Si(Ge). *J. Non-Cryst. Solids* **43,** 37–77 (1981).
20. Thorpe, M. Continuous deformations in random networks. *J. Non-Cryst. Solids* **57,** 355–370 (1983).
21. Souslov, A., Liu, A. J. & Lubensky, T. C. Elasticity and response in nearly isostatic periodic lattices. *Phys. Rev. Lett.* **103,** 205503 (2009).
22. Pieranski, P., Strzelecki, L. & Pansu, B. Thin colloidal crystals. *Phys. Rev. Lett.* **50,** 900–903 (1983).
23. Han, Y. *et al.* Geometric frustration in buckled colloidal monolayers. *Nature* **456,** 898–903 (2008).
24. Shokef, Y., Souslov, A. & Lubensky, T. C. Order by disorder in the antiferromagnetic ising model on an elastic triangular lattice. *Proc. Natl Acad. Sci. USA* **108,** 11804–11809 (2011).
25. Dennison, M., Sheinman, M., Storm, C. & MacKintosh, F. C. Fluctuation-stabilized marginal networks and anomalous entropic elasticity. *Phys. Rev. Lett.* **111,** 095503 (2013).
26. Ikeda, A., Berthier, L. & Biroli, G. Dynamic criticality at the jamming transition. *J. Chem. Phys.* **138,** 12a507 (2013).
27. Villain, J., Bidaux, R., Carton, J.-P. & Conte, R. Order as an effect of disorder. *J. Phys.* **41,** 1263–1272 (1980).
28. Shender, E. Anitferromagnetic garnets with fluctuationally interacting sublattices. *Sov. Phys. JETP* **56,** 178–184 (1982).
29. Henley, C. L. Ordering by disorder: ground state selection in fcc vector antiferromagnets. *J. Appl. Phys.* **61,** 3962–3964 (1987).
30. Henley, C. L. Ordering due to disorder in a frustrated vector antiferromagnet. *Phys. Rev. Lett.* **62,** 2056–2059 (1989).
31. Chubukov, A. Order from disorder in a kagome antiferromagnet. *Phys. Rev. Lett.* **69,** 832–835 (1992).
32. Reimers, J. N. & Berlinsky, A. J. Order by disorder in the classical heisenberg kagome antiferromagnet. *Phys. Rev. B* **48,** 9539–9554 (1993).
33. Bergman, D., Alicea, J., Gull, E., Trebst, S. & Balents, L. Order-by-disorder and spiral spin-liquid in frustrated diamond-lattice antiferromagnets. *Nat. Phys.* **3,** 487–491 (2007).
34. Schmidt, M. & Löwen, H. Freezing between two and three dimensions. *Phys. Rev. Lett.* **76,** 4552–4555 (1996).
35. Schmidt, M. & Löwen, H. Phase diagram of hard spheres confined between two parallel plates. *Phys. Rev. E* **55,** 7228–7241 (1997).
36. Shokef, Y. & Lubensky, T. C. Stripes, zigzags, and slow dynamics in buckled hard spheres. *Phys. Rev. Lett.* **102,** 048303 (2009).
37. Frenkel, D. & Smit, B. *Understanding Molecular Simulations* (Academic Press, 2001).
38. Binder, K. Critical properties from Monte Carlo coarse graining and renormalization. *Phys. Rev. Lett.* **47,** 693–696 (1981).
39. Mao, X., Xu, N. & Lubensky, T. C. Soft modes and elasticity of nearly isostatic lattices: Randomness and dissipation. *Phys. Rev. Lett.* **104,** 085504 (2010).
40. Brazovskii, S. A. Phase-transition of an isotropic system to an inhomogeneous state. *Zh. Eksp. Teor. Fiz.* **68,** 175–185 (1975).
41. Swift, J. & Hohenberg, P. C. Hydrodynamic fluctuations at convective instability. *Phys. Rev. A* **15,** 319–328 (1977).
42. Chen, J. H. & Lubensky, T. C. Landau-Ginzburg mean-field theory for nematic to smectic-C and nematic to smectic-A phase transitions. *Phys. Rev. A* **14,** 1202–1207 (1976).
43. Swift, J. Fluctuations near nematic-smectic-C phase transition. *Phys. Rev. A* **14,** 2274–2277 (1976).
44. Calladine, C. R. Buckminster Fuller "Tensegrity" structures and Clerk Maxwell's rules for the construction of stiff frames. *Int. J. Solids Struct.* **14,** 161–172 (1978).
45. Sun, K., Souslov, A., Mao, X. & Lubensky, T. C. Surface phonons, elastic response, and conformal invariance in twisted kagome lattices. *Proc. Natl Acad. Sci. USA* **109,** 12369–12374 (2012).
46. Kane, C. & Lubensky, T. Topological boundary modes in isostatic lattices. *Nat. Phys.* **10,** 39–45 (2014).
47. Mao, X. & Lubensky, T. C. Coherent potential approximation of random nearly isostatic kagome lattice. *Phys. Rev. E* **83,** 011111 (2011).
48. Jacobs, D. J. & Thorpe, M. F. Generic rigidity percolation in two dimensions. *Phys. Rev. E* **53,** 3682–3693 (1996).
49. Hammonds, K. D., Dove, M. T., Giddy, A. P., Heine, V. & Winkler, B. Rigid-unit phonon modes and structural phase transitions in framework silicates. *Am. Mineral.* **81,** 1057–1079 (1996).
50. Rubinstein, M., Leibler, L. & Bastide, J. Giant fluctuations of cross-linked positions in gels. *Phys. Rev. Lett.* **68,** 405–407 (1992).
51. Barriere, B. Elatic moduli of 2d grafted tethered membranes. *J. Phys. I France* **5,** 389–398 (1995).
52. Plischke, M. & Joós, B. Entropic elasticity of diluted central force networks. *Phys. Rev. Lett.* **80,** 4907 (1998).
53. Tessier, F., Boal, D. H. & Discher, D. E. Networks with fourfold connectivity in two dimensions. *Phys. Rev. E* **67,** 011903 (2003).

Acknowledgements

A.S. gratefully acknowledges discussions with P. A. Rikvold, Gregory Brown, Shengnan Huang and Andrea J. Liu. This work was supported in part by the NSF under grants DMR-1104707 and DMR-1120901 (T.C.L.), DMR-1207026 (A.S.), grant DGAPA IN-110613 (C.I.M.) and by the Georgia Institute of Technology (A.S.).

Author contributions

X.M., A.S. and T.C.L. conceived the research. X.M. and T.C.L. carried out analytical calculations, and A.S. and C.I.M. performed simulations. All authors contributed to the manuscript preparation.

Additional information

Competing financial interests: The authors declare no competing financial interests.

7

Phase-selective entrainment of nonlinear oscillator ensembles

Anatoly Zlotnik[1], Raphael Nagao[2], István Z. Kiss[2] & Jr-Shin Li[3]

The ability to organize and finely manipulate the hierarchy and timing of dynamic processes is important for understanding and influencing brain functions, sleep and metabolic cycles, and many other natural phenomena. However, establishing spatiotemporal structures in biological oscillator ensembles is a challenging task that requires controlling large collections of complex nonlinear dynamical units. In this report, we present a method to design entrainment signals that create stable phase patterns in ensembles of heterogeneous nonlinear oscillators without using state feedback information. We demonstrate the approach using experiments with electrochemical reactions on multielectrode arrays, in which we selectively assign ensemble subgroups into spatiotemporal patterns with multiple phase clusters. The experimentally confirmed mechanism elucidates the connection between the phases and natural frequencies of a collection of dynamical elements, the spatial and temporal information that is encoded within this ensemble, and how external signals can be used to retrieve this information.

[1]Center for Nonlinear Studies, MS B258, Los Alamos National Laboratory, Los Alamos, New Mexico 87545, USA. [2]Department of Chemistry, Saint Louis University, 3501 Laclede Ave., St Louis, Missouri 63103, USA. [3]Department of Electrical and Systems Engineering, Washington University in St Louis, CB 1042, 1 Brookings Drive, St Louis, Missouri 63130, USA. Correspondence and requests for materials should be addressed to J.-S.L. (email: jsli@wustl.edu).

Complex interactions among nonlinear periodic phenomena emerge in many natural and engineered systems[1,2]. Numerous instances appear in chemical reactions[3,4] and biological systems[5,6], which exhibit endogenous and emergent multiscale oscillations[7]. There is significant interest in characterizing synchronization in oscillators interconnected in networks[8,9], which is especially important for understanding the highly complex dynamics of man-made systems such as electric power grids[10], and elucidating the functions of neural systems[11–13]. Understanding entrainment of oscillating systems to an exogenous forcing signal is crucial to modelling circadian timekeeping[14], dynamic neural regulation[15] and for the design of synchronizing or desynchronizing treatments of cardiac arrhythmias[16], Parkinson's disease[17], epilepsy[18] and movement disorders[19].

Even the simplest models of interacting oscillators can exhibit highly complex behaviour[20], and individual oscillating units may themselves possess complicated dynamics[21]. These factors are aggravated in practice by model and parameter uncertainty and the impracticality of obtaining feedback information, such as for *in vivo* biological applications, and pose challenges to manipulating or controlling oscillating ensembles. Such tasks require tractable yet accurate simplifications of the complex dynamic interactions involved, and demand suitable mathematical approaches that characterize ensemble-level properties while withstanding experimental uncertainties.

Control-theoretic techniques have been applied to control a single oscillator[22–24]. In contrast, finely manipulating individual subsystems in underactuated ensembles, such as thousands of neurons in the brain affected by one electrode, rather than activating them homogeneously remains a fundamental challenge. Synchronization has been engineered in collections of oscillators using feedback[25–27], or tuning coupling strengths[4,28,29]. Such approaches require certain coupling structures, exact model specification, state feedback information, or precise knowledge of initial conditions, but still are not able to produce a prescribed phase pattern corresponding to frequency clusters of the oscillators.

Versatile open-loop control techniques were developed for simultaneous control of ensembles of quantum spin systems, which motivated the field of ensemble control[30]. Inspired by selective pulse design in nuclear magnetic resonance (NMR)[31], which enabled revolutionary applications including functional magnetic resonance imaging (fMRI), we develop a method for selectively manipulating the subunits of oscillator ensembles using periodic inputs that are robust to parameter uncertainty and disturbances. Specifically, we exploit the slight heterogeneity and high nonlinearity of an ensemble of structurally similar oscillators far past the Hopf bifurcation, rather than relying on a known coupling structure, state feedback or initial condition information.

In this manuscript, we present a methodology for constructing weak, globally applied, open-loop control inputs that synchronize a collection of structurally similar yet heterogeneous nonlinear oscillators while selectively assigning their relative phases on the periodic orbit. Using the technique, the synchronization structure of an oscillating ensemble can be manipulated among diverse phase patterns, seen in relative positions on the limit cycle. Our theory is developed specifically to overcome the challenges of experimental implementation when feedback information is unavailable, initial conditions are unknown and the oscillators are subject to uncertainty in subsystem parameters and stochastic disturbances. The control inputs create and maintain such phase patterns when the coupling between oscillators is weak, while preserving the intrinsic nature of the ensemble to enable nondestructive application to fragile biological and chemical systems. The dynamics of the oscillators may be arbitrary, as long as all are structurally similar and exhibit sufficient nonlinear relaxation for the control design to be realizable. A coherent structure may be established and robustly maintained indefinitely by a single periodic waveform, which can be altered to switch between patterns. We demonstrate the theoretical methodology in practice with experiments to control complex electrochemical ensembles whose dynamics are variable and unknown, and for which state information is unavailable[26,32].

Results

Phase model approximation. We approximate the effect of inputs on periodic dynamical systems using phase models[33], which can be computed for systems with known dynamics[34] or experimentally inferred in practice when the dynamics are unknown[35]. These models are used to characterize circadian cycles[36], cardiac rhythms[16] and phenomena in neural and chemical systems[11,37], and their simplicity has enabled control design for neuron models[22] given initial conditions and exact parameters. Control techniques have recently been developed for individual nonlinear oscillators and finite collections described by phase models that require exactly known initial conditions, parameters, and dynamics[22,38,39]. Many studies on synchronization focus on the network structure of couplings between oscillators, and the nonlinearity in the phase response of each unit is simplified to sinusoidal couplings with its neighbours[9]. However, for the manipulation and desynchronization of electrochemical and neural systems[11,40,41], complex, hierarchical interactions must be established or broken in large collections of nonlinear systems. The dynamics, parameters, and interconnections of these systems are typically problematic to infer, may be noisy, variable, or uncertain, and state observations may be incomplete or unavailable. Such conditions elude tractable formulation, and require an approach where synchronization properties of the systems, that is, asymptotic phase structure, are manipulated rather than steering the system states directly[7,42].

Entrainment of an ensemble. Our method relies on entrainment, which refers to the dynamic synchronization of oscillating systems to periodic inputs. Each system in an ensemble of N structurally similar units exhibits endogenous oscillation along an attractive periodic orbit with period T_j, and is represented by the Winfree phase model

$$\dot{\psi}_j = \omega_j + Z\left(\psi_j\right)u, \qquad (1)$$

where $\omega_j = 2\pi/T_j$ is the natural frequency and Z is the phase response curve (PRC), which quantifies how a weak perturbation u advances or delays the phase ψ_j (refs 2,33). A value of $\psi_j = 0$ (equivalently 2π) corresponds to a measurement of the jth system reaching its maximum. More details on phase coordinate transformation are given in Supplementary Note 1. We demonstrate our phase-selective entrainment techniques using experiments in which nickel electrodes undergo anodic dissolution in sulfuric acid and exhibit electrochemical oscillations[32], for which the experimental apparatus is described in section 'Experimental apparatus'. Before these experiments, the PRCs of the ensemble elements, which are shown in Fig. 1a, were estimated and averaged for use as the nominal PRC in equation (1). This was done using a pulse perturbation procedure for system identification that was previously used for electrochemical oscillators[43], and is described in Supplementary Note 2.

To synchronize oscillation of ensemble elements, each subsystem receives the same weak, periodic forcing input of frequency Ω of the form $u(t) = v(\Omega t)$, where v is 2π-periodic. When the forcing frequency is near the natural frequencies in the

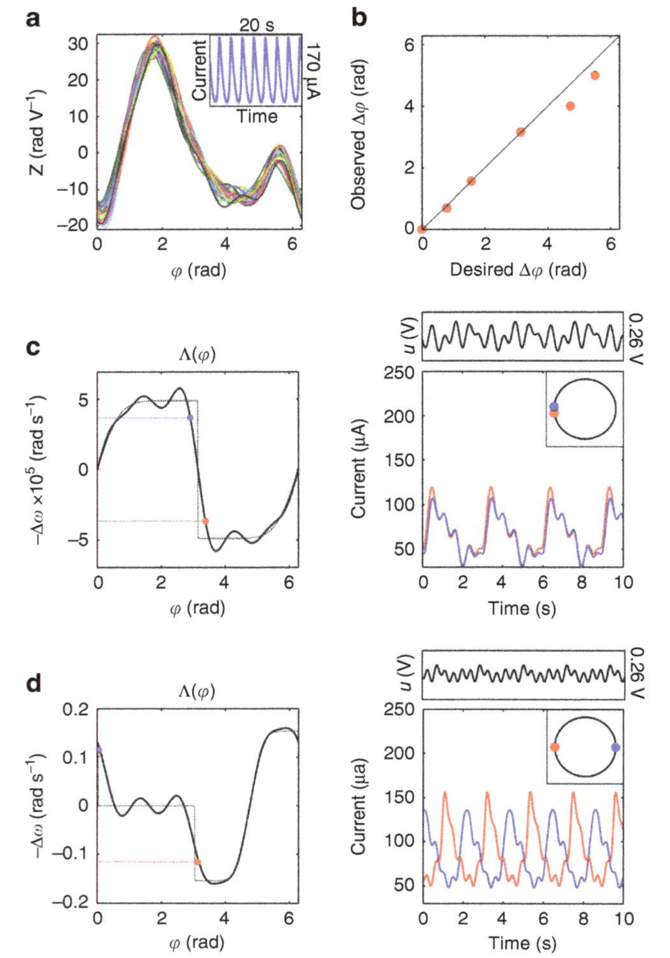

Figure 1 | Tuning the phase difference between two electrochemical oscillators. (**a**) PRCs measured simultaneously for 20 working electrodes, and observed current oscillations (inset). (**b**) Designed versus experimentally measured phase difference and fit (dotted line). The left panels in **c** and **d** show how phase assignment is constructed using an ideal interaction function (dotted line), and the respective best achievable approximation (solid line) for two nonidentical oscillators. The right panel shows the time-series of the entrained oscillators and the periodic control signal (above), and the entrained oscillator phases on the unit circle (inset). (**c**) In-phase phase assignment: $(\varphi_1^*, \varphi_2^*) = (\pi, \pi)$, with natural frequencies $(\omega_1, \omega_2) = (0.330, 0.348)$ Hz shown in (blue,red) with $\Omega = 0.339$ Hz. (**d**) Anti-phase phase assignment: $(\varphi_1^*, \varphi_2^*) = (0, \pi)$ and natural frequencies $(\omega_1, \omega_2) = (0.443, 0.480)$ Hz in (blue,red), with $\Omega = 0.462$ Hz.

ensemble, averaging theory[44] states that the mean difference φ_j between each phase ψ_j and the forcing phase $\theta = \Omega t$ follows the time-independent dynamic equation

$$\dot{\varphi}_j = \Delta\omega_j + \Lambda_v(\varphi_j), \tag{2}$$

where $\Delta\omega_j = \omega_j - \Omega$ is called the frequency detuning, and

$$\Lambda_v(\varphi) = \frac{1}{2\pi}\int_0^{2\pi} Z(\theta + \varphi)v(\theta)\mathrm{d}\theta \tag{3}$$

is a 2π-periodic interaction function that characterizes the average effect of the periodic input on the oscillation phase[43]. Such ergodic averaging is discussed in more detail in Supplementary Note 3 and illustrated in Supplementary Fig. 1. If equation (2) has a unique attractive fixed point φ_j^* that satisfies $\Delta\omega_j + \Lambda_v(\varphi_j^*) = 0$, then the phase of the jth oscillator becomes entrained to the forcing phase θ with an average offset of φ_j^*. This analysis is widely applied to determine the interaction function resulting from a measured PRC and an input waveform, and equation (2) is used to infer the entrained system's stability[27]. We reverse this approach by choosing the desired asymptotic behaviour, constructing a suitable interaction function and using the PRC to obtain the input by circular deconvolution of equation (3).

Interaction functions for phase selection. We design the input $v(\Omega t)$ so that each system in equation (1) is entrained to a forcing frequency Ω, e.g., the mean of natural frequencies $\omega_1 < \omega_2 < \ldots < \omega_N$, and such that the jth oscillator cycles its orbit with a phase offset of φ_j^* relative to the forcing phase θ. The set of pairs (ω_j, φ_j^*) constitutes a pattern for selective entrainment. We require that $\dot{\varphi}_j = \Delta\omega_j + \Lambda_v(\varphi_j^*) = 0$ eventually holds for each oscillator, that is, equation (2) has an attractive fixed point at φ_j^* for all j at which the slope of the interaction function Λ_v is negative[43]. The function that best satisfies these ideal conditions has steep decreases at phase values φ_j^* where entrainment must occur, and crosses (from above) horizontal lines at frequency detuning values $-\Delta\omega_j$. This creates the desired attractive fixed points for equation (2). Because the interaction function is periodic, it must then increase so that $\Lambda_v(2\pi) = \Lambda_v(0)$ holds. Crossings of $-\Delta\omega_j$ from below are unstable fixed points, and do not affect convergence.

The concept is illustrated in Fig. 1, which describes an experiment where a phase difference $\Delta\varphi^* = \varphi_2^* - \varphi_1^*$ is assigned between two entrained oscillators. In Fig. 1c we desire in-phase synchronization at phase offsets of $(\varphi_1^*, \varphi_2^*) = (\pi, \pi)$, so the ideal interaction function has one steep decrease at $\varphi_1^* = \varphi_2^* = \pi$ that intersects horizontal lines through $-\Delta\omega_1$ (blue) and $-\Delta\omega_2$ (red) at π radians. Figure 1d illustrates anti-phase synchronization with $(\varphi_1^*, \varphi_2^*) = (0, \pi)$, where the interaction function has two steep decreases at $\varphi_1^* = 0$ and $\varphi_2^* = \pi$ that intersect horizontal lines at $-\Delta\omega_1$ (blue) and $-\Delta\omega_2$ (red), respectively. The best achievable interaction function (solid line) and the PRC estimate yield the input from equation (3). The right columns of Fig. 1c,d show the observed current of two oscillators entrained in in-phase and anti-phase arrangements by the input waveform (shown above). These configurations are achieved regardless of initial oscillator phases, because the interaction function crosses the line $-\Delta\omega_j$ only once from above, so each system has a globally attractive fixed point. For the electrochemical system, phase differences in nearly the entire 0 to 2π region are achievable, with small deviations as $\Delta\varphi^*$ approaches 2π, as seen in Fig. 1b.

Separation of oscillator ensembles into phase clusters. Uniquely attractive phase patterns are desired, where a common input synchronizes the oscillators to a pattern independently of their initial phases. The fixed point φ_j^* of equation (2) must be unique for each j, which is achieved when the interaction function crosses each horizontal line $\Delta\omega_j$ from above only once at φ_j^*. This is possible when the phase offsets are monotonically ordered as $\varphi_1^* < \varphi_2^* < \ldots < \varphi_N^*$ for $\omega_1 < \omega_2 < \ldots < \omega_N$, as demonstrated by segregation of 20 inhomogeneous electrochemical oscillators into clusters in the experiments described in Fig. 2. An anti-phase configuration $(\varphi_1, \varphi_2) = (0, \pi)$ is achieved for electrodes in balanced $(N_1, N_2) = (10, 10)$ and unbalanced $(N_1, N_2) = (1, 19)$ clusters in Fig. 2a,b, respectively. In these two-cluster examples, the interaction function decreases in two steps, of which the top and bottom correspond to clusters with slower (blue) and faster (red) natural frequencies. Figure 2c shows the formation of four balanced clusters of $(N_1, N_2, N_3, N_4) = (5, 5, 5, 5)$ oscillators with the phase structure $(\varphi_1^*, \varphi_2^*, \varphi_3^*, \varphi_4^*) = (0, 1.1, 2.1, \pi)$ radians. The phase offsets φ_j^* are increasing as $-\Delta\omega_j$ decreases (and ω_j

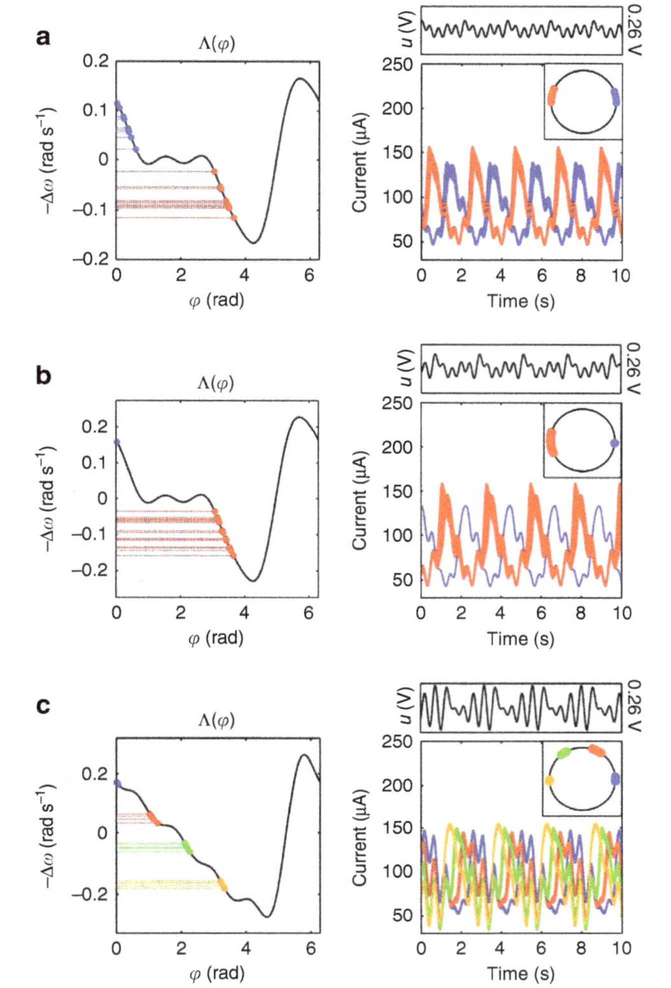

Figure 2 | Phase assignment for 20 electrochemical oscillators. The first column shows the frequency detunings and the intersections with the interaction function at the phase offsets φ_j^*. The second column represents the entrained time-series with the respective periodic waveform of the control signal (above) and a plot of the phase offsets on the unit circle (inset). Phase assignment for: (**a**): $(N_1, N_2) = (10, 10)$ clusters; cluster 1 (blue): $\varphi_1=0$ and $\omega_1 = 0.450$ Hz; cluster 2 (red): $\varphi_2=\pi$ and $\omega_2 = 0.471$ Hz; forcing at $\Omega = 0.463$ Hz. (**b**): $(N_1, N_2) = (1, 19)$ clusters; one electrode (blue): $\varphi_1 = 0$ and $\omega_1 = 0.419$ Hz, cluster 2 (red): $\varphi_2 = \pi$ and $\omega_2 = 0.454$ Hz; forcing at $\Omega = 0.450$ Hz. (**c**): $(N_1, N_2, N_3, N_4) = (4, 4, 4, 4)$ clusters; cluster 1 (blue): $\varphi_1=0$ and $\omega_1 = 0.386$ Hz, cluster 2 (red): $\varphi_2=1.1$ rad and $\omega_2 = 0.404$ Hz; cluster 3 (green): $\varphi_3=2.1$ rad and $\omega_3 = 0.421$ Hz; cluster 4 (yellow): $\varphi_4=\pi$ and $\omega_4 = 0.440$ Hz; forcing at $\Omega = 0.413$ Hz.

increases), and the designed interaction function decreases monotonically as it crosses the required intersections. Observe that the assumption of a common PRC is reasonable, because the functions in Fig. 1a are very similar, yet the oscillator frequencies are sufficiently heterogeneous for our technique to create the patterns in Fig. 2.

Control of pattern transitions in an ensemble. In addition, we establish patterns without monotone phase ordering by designing an interaction function of the form at the bottom of Fig. 3d, which crosses a horizontal line at $-\Delta\omega_j$ from above twice, yielding multiple possible entrained phases and dependence on initial conditions. We apply precursor waveforms to steer subsets of the ensemble into attractive regions for the desired phase

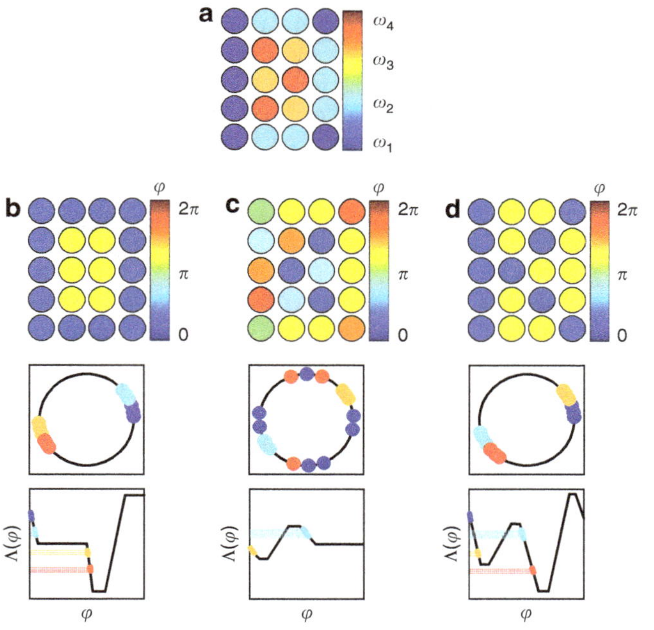

Figure 3 | Schematic of pattern switching 'O→K→O' by phase-selective entrainment. (**a**) Natural frequencies for four spatially distributed oscillator clusters of sizes $(N_1, N_2, N_3, N_4) = (7, 7, 3, 3)$ in an ascending order: $\omega_1 < \omega_2 < \omega_3 < \omega_4$. Cluster 1 contains electrodes $n = 1$ to 7 (blue), cluster 2 contains $n = 8$ to 14 (cyan), cluster 3 contains $n = 15$ to 17 (yellow) and cluster 4 contains $n = 18$ to 20 (red). Items **b** to **d** depict snapshots of phase patterns at the forcing phase $\theta = 0$ (top row), the description of the phases on a unit circle (middle row), and a sketch of the interaction function (bottom row). Phase assignment for: (**b**): Pattern 'O': $(\varphi_1^*, \varphi_2^*, \varphi_3^*, \varphi_4^*) = (0, 0, \pi, \pi)$. (**c**): Precursor of pattern 'K': $(\varphi_2^*, \varphi_3^*) = (\pi, 0)$. (**d**): Pattern 'K': $(\varphi_1^*, \varphi_2^*, \varphi_3^*, \varphi_4^*) = (0, \pi, 0, \pi)$.

offsets φ_j^*, then finalize and hold the pattern with an ultimate input. This procedure is applied to steer an ensemble between spatially associated clusters by alternating selective inputs. Fig. 3 illustrates input design for itinerant formation of letters in the word 'OK' in the array of 20 electrochemical oscillators used in the experiments in Fig. 2. We produce anti-phase assignment between clusters to display the letter 'O', then switch the input to produce the letter 'K'. Rhythmic elements are assigned desired phase offsets of $\varphi_j^* = 0$ or $\varphi_j^* = \pi$, which correspond to 'on' (pattern) or 'off' (background) states, respectively, that are visualized using a colour scale where 0 ('on') is blue and π ('off') is yellow. Switching between two patterns is accomplished using four numbered clusters, where 1 is 'on' for both, 2 switches from 'on' to 'off', 3 switches from 'off' to 'on', and 4 is always 'off'. Electrodes in clusters of $(N_1, N_2, N_3, N_4) = (7, 7, 3, 3)$ elements with mean natural frequencies $(\omega_1, \omega_2, \omega_3, \omega_4) = (0.390, 0.406, 0.427, 0.442)$ Hz are positioned in the spatial arrangement in Fig. 3a. Figure 3b–d each have panels that show, from top to bottom, the spatial distribution of phase offsets, the structure on the unit circle, and a sketch of the ideal interaction function.

The pattern 'O', shown in Fig. 3b, is realized using the phases $(\varphi_1^*, \varphi_2^*, \varphi_3^*, \varphi_4^*) = (0, 0, \pi, \pi)$, which are achieved by an interaction function as in Fig. 2a. The phases $(\varphi_1^*, \varphi_2^*, \varphi_3^*, \varphi_4^*) = (0, \pi, 0, \pi)$ used for 'K' are not monotonically ordered, so a precursor waveform is applied to generate globally attractive phase offsets $\varphi_2^* = \pi$ and $\varphi_3^* = 0$ for clusters 2 and 3. This anti-phase pattern establishes initial conditions for the final input waveform, while clusters 1 and 4 lose their entrainment, as shown in Fig. 3c. Figure 3d illustrates the design of the finalizing control for pattern 'K', where the phase assignments for clusters 1 and 4 are globally attractive, as seen in the bottom panel of Fig. 3d, while clusters 2 and 3 stay at phase offsets established in the precursor stage. The transition from pattern 'K' back to 'O' is accomplished by applying the initial control for the pattern in Fig. 3b. We provide additional descriptions of interaction function construction (Supplementary Note 4; Supplementary Figs 1 and 2), pattern realizability (Supplementary Note 5), control design for monotonically ordered patterns (Supplementary Note 6; Supplementary Fig. 3), and precursor waveform engineering (Supplementary Note 7; Supplementary Figs 2 and 4).

Measurements of the 'O→K→O' pattern switching experiment appear in Fig. 4. Figure 4a shows current oscillations for the reaction units given zero input, when no pattern forms. When the controls (shown above the current) are applied, the ensemble is entrained within several cycles, selectively forming the patterns for 'O', the precursor, and 'K'. These results are shown in Fig. 4b–d, and correspond to Fig. 3b–d, respectively. The ensemble is returned to pattern 'O', as shown in Fig. 4e, to demonstrate switching. The automatically generated interaction functions and control waveforms are presented in section 'Automatic control waveforms generated in experiments'. An animation produced using the experimental current traces and oscillation phases is included as Supplementary Movie 1.

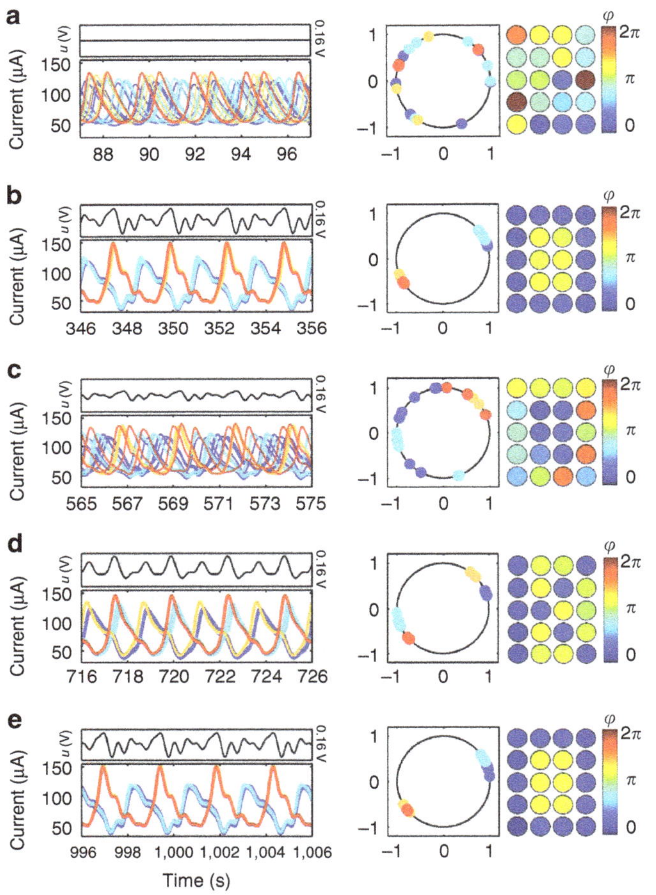

Figure 4 | Pattern switching experiment 'O→K→O' on an array of electrochemical oscillators. Left: current oscillations with the control signal above. Right: oscillator phases on a unit circle and a snapshot of spatial phase assignment at a time t_s. **(a)**: Current oscillations with no control and mean natural frequencies for four electrode clusters $(N_1, N_2, N_3, N_4) = (7, 7, 3, 3)$ of $(\omega_1, \omega_2, \omega_3, \omega_4) = (0.390, 0.406, 0.427, 0.442)$ Hz at $t_s = 92$ s. Phase assignment with forcing at $\Omega = 0.408$ Hz for: **(b)**: Pattern 'O': $(\varphi_1^*, \varphi_2^*, \varphi_3^*, \varphi_4^*) = (0, 0, \pi, \pi)$, at $t_s = 351$ s. **(c)**: Precursor of pattern 'K': $(\varphi_2^*, \varphi_3^*) = (\pi, 0)$, at $t_s = 570$ s. **(d)**: Pattern 'K': $(\varphi_1^*, \varphi_2^*, \varphi_3^*, \varphi_4^*) = (0, \pi, 0, \pi)$, at $t_s = 721$ s. **(e)**: Same as item **(b)** with $t_s = 1{,}001$ s.

Discussion

Phase-selective entrainment enables the use of a single global signal to robustly assign elements of a noisy nonlinear oscillator ensemble to specific phases without using coupling or feedback information. Control design using interaction functions simplifies the creation of complex synchronization patterns to drawing or automatically generating curves through sets of crossing points and computing the resulting controls with a simple formula, which is an accessible technique for experimentalists. Greater relaxation in the oscillation and ensemble heterogeneity increases pattern controllability, and performance is improved as the oscillations move farther away from the Hopf bifurcation. The asymptotic nature of entrainment yields robustness to noise, disturbances and model parameter variability while preserving the intrinsic nature of the ensemble.

Such resilience is required for nondestructive control of underactuated, noisy and uncertain biological and chemical ensembles that cannot be readily observed. For example, an effective technology for neurological treatment of Parkinson's disease[17] is provided by deep brain stimulation, which alleviates pathological synchronization in the brain. Selective entrainment could be extended to ensembles with weak coupling to design robust desynchronization inputs, which would potentially benefit noninvasive neurostimulation technology[41]. The goal could be a target distribution that is found to be optimal for leveraging neuroplasticity to prevent resynchronization after the stimulus is ended. The technique could also improve phase regulation to treat cardiac arrhythmias[16] and sleep irregularities[45]. The formalism could also represent the entrainment that occurs in circadian timekeeping[14].

We note that a simple sinusoidal forcing of the form $v(\Omega t) = \sin(\Omega t)$ results in a sinusoidal interaction function, because of orthogonality of the trigonometric Fourier basis. Sinusoidal forcing can thus be used to create monotone ordered phase patterns, and could also be used for desynchronization. However, because such an interaction function is decreasing on an interval of length π, the maximum achievable distance between extremal phase offsets φ_1^* and φ_N^* is $|\varphi_N^* - \varphi_1^*| < \pi$. Thus, a sinusoidal input cannot produce anti-phase synchronization. Our approach enables more versatile manipulation of phase relationships beyond this limitation. We describe the application to desynchronization in Supplementary Note 8 and Supplementary Fig. 5, and quantify how our approach increases the achievable relative phase desynchronization difference over sinusoidal forcing. More rigorous mathematical characterization of ensemble desynchronization by periodic inputs is a compelling direction for further investigation.

In our methodology, we take advantage of approximations that are possible in the specific experimental setting. In our experiments, the distribution of natural frequencies of ensemble oscillations varies by ±20% from the (non-zero) mean, the oscillators are very weakly coupled, the amplitude of the required external forcing signal is relatively small, and the entrainment process is approximated well by averaged phase models. In addition, although the ensemble subsystems are slightly heterogeneous and noisy, with variation in natural frequencies and dynamic properties, the phase response curves are very similar. We expect the methodology to function well in other experimental settings in which these conditions are satisfied. Moreover, the method holds promise for extension to other scenarios, e.g., in sub- and superhamornic entrainments (the oscillations are locked

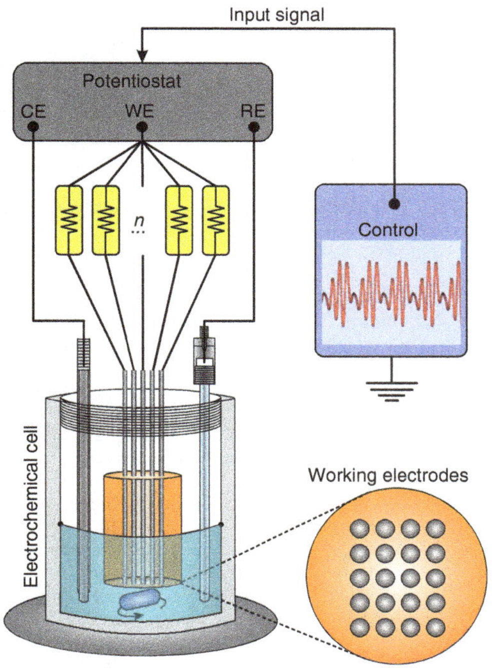

Figure 5 | Electrochemical set-up. The working electrode (WE) is composed of an array of nickel wires, the counter electrode (CE) is a Pt coated Ti wire, and the reference electrode (RE) is Hg/Hg_2SO_4/sat. K_2SO_4. The control waveform is a potential signal generated by the multifunction M Series DAQ PXI-6255 (National Instruments) and implemented in the WE by a potentiostat (Bank Instruments) as a superimposed signal on the applied baseline potential V_0. Each nickel wire is connected in series to an individual resistance $R_{ind} = 2{,}500$ Ohm.

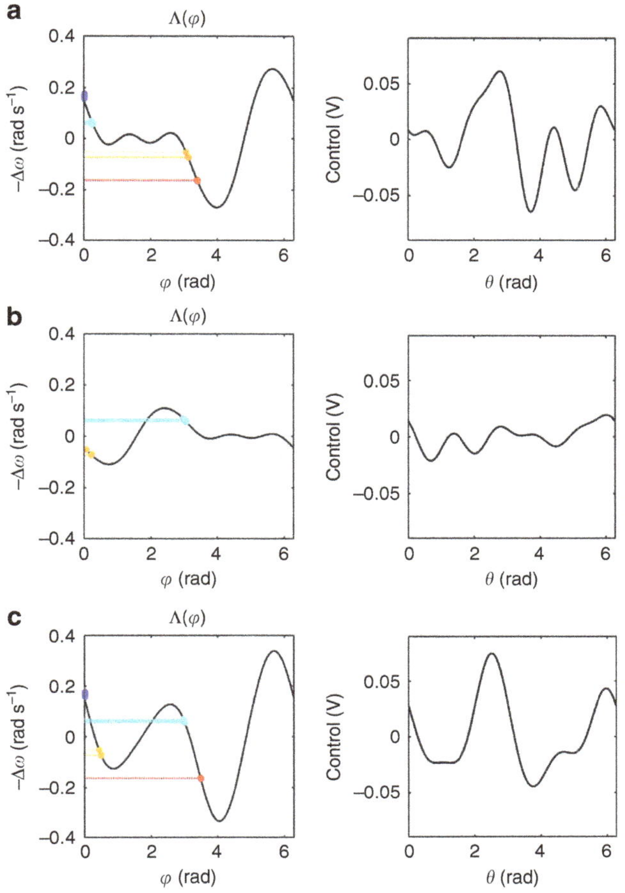

Figure 6 | Interaction function and control waveform for pattern switching 'O→K→O' described in Fig. 4. Left: the best approximation to the desired interaction function; Right: one cycle of the resulting control waveform, obtained using the control design procedure described in section 'Control of pattern transitions in an ensemble', which is applied to entrain the electrochemical oscillators. The ensemble is grouped into four clusters of quantities $(N_1, N_2, N_3, N_4) = (7, 7, 3, 3)$. Cluster 1 for oscillators $j = 1$ to 7 (blue), cluster 2 for $j = 8$ to 14 (cyan), cluster 3 for $j = 15$ to 17 (yellow) and cluster 4 for $j = 18$ to 20 (red). Phase assignment for: (**a**): Pattern 'O': $(\varphi_1^*, \varphi_2^*, \varphi_3^*, \varphi_4^*) = (0, 0, \pi, \pi)$. (**b**): Precursor of pattern 'K': $(\varphi_2^*, \varphi_3^*) = (\pi, 0)$. (**c**): Pattern 'K': $(\varphi_1^*, \varphi_2^*, \varphi_3^*, \varphi_4^*) = (0, \pi, 0, \pi)$.

to different frequency ratios), where the interaction function-based phase description is possible.

The arrangement of frequency clusters in an oscillator ensemble can also be viewed as encoding information within the spatial pattern produced when selective entrainment is applied. One of several encoded messages can then be retrieved using the phase-selective entrainment process, for which the passkey for retrieving the message is the temporal information contained in a periodic input signal. The passkey is constructed using the PRC and natural frequencies of the dynamical subsystems, and after that input signal is applied, the spatial phase pattern produced in the ensemble reveals the message. This approach may be incorporated in neurocomputing architectures[46] that mimic neural systems in nature. Future investigation is required to understand how network coupling could be suppressed or taken advantage of to improve pattern resilience and information capacity, or effectively encrypt the message by preventing estimation of PRCs and natural frequencies of oscillators in the spatial array.

Methods

Experimental apparatus. Our methodology for controlling the phase structure of an ensemble of heterogeneous oscillators is experimentally verified in the electrochemical dissolution of nickel in 3 mol l^{-1} sulfuric acid solution using potential actuation. A schematic description of the experimental set-up is depicted in Fig. 5. The apparatus consists of 20 nickel wires that function as working electrodes (WE), with diameters of 0.69 mm, spaced by 2.0 mm and embedded in epoxy resin. Prior to the electrochemical measurements, the WE were polished with sandpaper in six levels of roughness, ranging from 180 to 4,000 grit. The current of all electrodes is monitored independently. Under such conditions, the spontaneous formation of self-sustained electrochemical oscillations driven by the presence of a negative differential resistance[47] is observed in the anodic dissolution of nickel.

Once the WE are placed in the electrochemical cell, a slow positive sweep of 0.01 V s^{-1} from 0 V to a constantly applied potential V_0 was conducted to form a thin passive layer on the electrode surface. This baseline is set in reference to an Hg/Hg_2SO_4/sat. K_2SO_4 reference electrode (RE) in an electrochemical cell, containing a 1.6 mm diameter Pt coated Ti wire counter electrode at constant temperature of 10 °C. The potential V_0 was initially set using a potentiostat (Bank Instruments) at a value for which the oscillation is close to the Hopf bifurcation, which is ~1.15 V. Inputs to the oscillating system consist of an additional potential u superimposed onto the baseline potential V_0 using the potentiostat.

Soon after, the PRCs were measured simultaneously for the WE by the automatic procedure of applying a pseudorandomly timed potential pulse sequence of −0.20 V with pulse-width of 0.05 s and post-processing the observed current using the pulse perturbation procedure as described in section 'Entrainment of an ensemble'. Measurements of the current oscillation in the electrochemical reactions were carried out by a real-time data acquisition by a high-speed multifunction M Series DAQ PXI-6255 (National Instruments) interface with a sample rate of 200 Hz. Simultaneously, the periodic potential perturbations u were applied in the electrochemical oscillator ensemble using the potentiostat to superimpose the waveform onto the applied baseline voltage V_0. Each control waveform was generated based on the PRC obtained preliminary to the experiment and the targeted interaction function generated from the desired set of phase offsets for the oscillators.

Automatic control waveforms generated in experiments. Figure 6 displays the interaction function (left column) and the respective control waveform (right column) used to entrain the electrochemical oscillators for the pattern switching procedure 'O→K→O' described in Fig. 4 (see main text for additional details). An electrochemical oscillator ensembles with clusters of size (N_1, N_2, N_3, N_4) = (7, 7, 3, 3) were selected by tuning the mean natural frequency of the oscillations in the following order: cluster 1 with $\omega_1 = 0.390$ Hz (electrodes $j = 1$ to 7 in blue), cluster 2 with $\omega_2 = 0.406$ Hz (electrodes $j = 8$ to 14 in cyan), cluster 3 with $\omega_3 = 0.427$ Hz (electrodes $j = 15$ to 17 in yellow) and cluster 4 with $\omega_4 = 0.442$ Hz (electrodes $j = 18$ to 20 in red). The phase assignments for pattern switching 'O→K→O' are listed in Fig. 6.

References

1. Strogatz., S. *Nonlinear Dynamics And Chaos: With Applications To Physics, Biology, Chemistry, And Engineering* (Westview Press, 2001).
2. Pikovsky, A., Rosenblum, M. & Kurths., J. *Synchronization: A Universal Concept in Nonlinear Science* (Cambridge Univ. Press, 2001).
3. Kuramoto., Y. *Chemical Oscillations, Waves, and Turbulence* (Springer, 1984).
4. Epstein, I. R. & Pojman., J. A. *An Introduction to Nonlinear Chemical Dynamics: Oscillations, Waves, Patterns, and Chaos (Topics in Physical Chemistry)* (Oxford Univ. Press, 1998).
5. Glass, L. Synchronization and rhythmic processes in physiology. *Nature* **410,** 277–284 (2001).
6. Glass, L. & Mackey., M. C. *From clocks to chaos: The rhythms of life* (Princeton Univ. Press, 1988).
7. Zlotnik, A. & Li, J.-S. Optimal subharmonic entrainment of weakly forced nonlinear oscillators. *SIAM J. Appl. Dynam. Syst.* **13,** 1654–1693 (2014).
8. Arenas, A., Daz-Guilera, A., Kurths, J., Moreno, Y. & Zhou, C. Synchronization in complex networks. *Phys. Rep.* **469,** 93–153 (2008).
9. Dörfler, F. & Bullo, F. Synchronization in complex networks of phase oscillators: A survey. *Automatica* **50,** 1539–1564 (2014).
10. Dörfler, F., Chertkov, M. & Bullo, F. Synchronization in complex oscillator networks and smart grids. *Proc. Natl Acad. Sci USA* **110,** 2005–2010 (2013).
11. Izhikevich., E. *Dynamical Systems in Neuroscience* (MIT Press, 2007).
12. Uhlhaas, P. J. & Singer, W. Neural synchrony in brain disorders: relevance for cognitive dysfunctions and pathophysiology. *Neuron* **52,** 155–168 (2006).
13. Buzsáki, G. & Draguhn, A. Neuronal oscillations in cortical networks. *Science* **304,** 1926–1929 (2004).
14. Leloup, J.-C. & Goldbeter, A. Toward a detailed computational model for the mammalian circadian clock. *Proc. Natl Acad. Sci. USA* **100,** 7051–7056 (2003).
15. Hunter, J. & Milton, J. Amplitude and Frequency Dependence of Spike Timing: Implications for Dynamic Regulation. *J. Neurophysiol.* **90,** 387–394 (2003).
16. Guevara, M. & Glass., L. Phase locking, period doubling bifurcations and chaos in a mathematical model of a periodically driven oscillator. *J. Math. Biol.* **14,** 1–23 (1982).
17. Hammond, C., Bergman, H. & Brown, P. Pathological synchronization in parkinson's disease: networks, models and treatments. *Trends Neurosci.* **30,** 357–364 (2007).
18. Good, L. Control of synchronization of brain dynamics leads to control of epileptic seizures in rodents. *Int. J. Neural Syst.* **19,** 173–196 (2009).
19. Reinkensmeyer, D. J. *et al.* Tools for understanding and optimizing robotic gait training. *J. Rehab. Res. Dev* **43,** 657–670 (2014).
20. Strogatz., S. H. From Kuramoto to Crawford: exploring the onset of synchronization in populations of coupled oscillators. *Phys. D* **143,** 1–20 (2000).
21. Izhikevich, E. M. Synchronization of elliptic bursters. *SIAM Rev.* **43,** 315–344 (2001).
22. Moehlis, H., Brown, E. & Rabitz, H. Optimal inputs for phase models of spiking neurons. *J. Comput. Nonlin. Dyn.* **1,** 358–367 (2006).
23. Dasanayake, I. & Li, J.-S. Optimal design of minimum-power stimuli for phase models of neuron oscillators. *Phys. Rev. E* **83,** 061916 (2011).
24. Efimov, D., Sacré, P. & Sepulchre, R. *Proceedings of the 48th IEEE Conference on Decision and Control,* 7692–7697 (Shanghai, China, 2009).
25. Sepulchre, R., Paley, D. A. & Leonard, N. E. Stabilization of planar collective motion: All-to-all communication. *IEEE Trans. Autom. Control* **52,** 811–824 (2007).
26. Kiss, I. Z., Rusin, C. G., Kori, H. & Hudson, J. L. Engineering complex dynamical structures: sequential patterns and desynchronization. *Science* **316,** 1886–1889 (2007).
27. Kori, H., Rusin, C. G., Kiss, I. Z. & Hudson, J. L. Synchronization engineering: Theoretical framework and application to dynamical clustering. *Chaos* **18,** 026111 (2008).
28. Rosenblum, M. G. & Pikovsky, A. S. Controlling synchronization in an ensemble of globally coupled oscillators. *Phys. Rev. Lett.* **92,** 114102 (2004).
29. Orosz, G. Decomposition of nonlinear delayed networks around cluster states with applications to neurodynamics. *SIAM J. Appl. Dyn. Syst.* **13,** 1353–1386 (2014).
30. Li, J.-S. & Khaneja, N. Control of inhomogeneous quantum ensembles. *Phys. Rev. A* **73,** 030302 (2006).
31. Kobzar, K., Luy, B., Khaneja, N. & Glaser, S. J. Pattern pulses: design of arbitary excitation profiles as a function of pulse amplitude and offset. *J. Magn. Reson.* **173,** 229–235 (2005).
32. Kiss, I. Z., Zhai, Y. & Hudson, J. Emerging coherence in a population of chemical oscillators. *Science* **296,** 1676–1678 (2002).
33. Winfree., A. T. *The Geometry of Biological Time* (Springer-Verlag, 1980).
34. Ermentrout, B. Type I membranes, phase resetting curves, and synchrony. *Neural Comput.* **8,** 979–1001 (1996).
35. Galan, R. F., Ermentrout, G. B. & Urban, N. N. Efficient estimation of phase-resetting curves in real neurons and its significance for neural-network modeling. *Phys. Rev. Lett.* **94,** 158101 (2005).
36. Strogatz, S. H. Human sleep and circadian rhythms: a simple model based on two coupled oscillators. *J. Math. Biol.* **25,** 327–347 (1987).
37. Nakao, H., Yanagita, T. & Kawamura., Y. Phase-reduction approach to synchronization of spatiotemporal rhythms in reaction-diffusion systems. *Phys. Rev. X* **4,** 021032 (2014).
38. Li, J.-S., Dasanayake, I. & Ruths, J. Control and synchronization of neuron ensembles. *IEEE Trans. Autom. Control* **58,** 1919–1930 (2013).
39. Dasanayake, I. S. & Li, J.-S. Constrained charge-balanced minimum-power controls for spiking neuron oscillators. *Syst. Control Lett.* **75,** 124–130 (2015).
40. Zhai, Y., Kiss, I. Z., Tass, P. A. & Hudson, J. L. Desynchronization of coupled electrochemical oscillators with pulse stimulations. *Phys. Rev. E* **71,** 065202 (2005).
41. Tass., P. A. A model of desynchronizing deep brain stimulation with a demand-controlled coordinated reset of neural subpopulations. *Biol. Cybern.* **89,** 81–88 (2003).
42. Zlotnik, A. & Li, J.-S. Optimal entrainment of neural oscillator ensembles. *J. Neural Eng.* **9,** 046015 (2012).
43. Zlotnik, A., Chen, Y., Kiss, I. Z., Tanaka, H.-A. & Li, J.-S. Optimal waveform for fast entrainment of weakly forced nonlinear oscillators. *Phys. Rev. Lett.* **111,** 024102 (2013).
44. Hoppensteadt, F. & Izhikevich., E. *Weakly connected neural networks* (Springer-Verlag, 1997).
45. Gooley, J. J. *et al.* Spectral responses of the human circadian system depend on the irradiance and duration of exposure to light. *Sci. Trans. Med.* **2,** 31ra33 (2010).
46. Hoppensteadt, F. & Izhikevich, E. Oscillatory Neurocomputers with Dynamic Connectivity. *Phys. Rev. Lett.* **82,** 2983 (1999).
47. Koper., M. T. M. The theory of electrochemical instabilities. *Electrochim. Acta.* **37,** 1771–1778 (1992).

Acknowledgements

This work was supported by the U.S. National Science Foundation under the awards CMMI-1301148, CMMI-1462796, ECCS-1509342, and CHE-1465013, the CNPq Award 229171/2013-3, and the U.S. Department of Energy under contract DE-AC52-06NA25396.

Author contributions

The authors contributed equally to developing the methodology presented here. A.Z. developed computational algorithms and did primary writing for the manuscript and R.N. performed the experiments and created figures.

Additional information

Competing financial interests: The authors declare no competing financial interests.

8

Computational multiqubit tunnelling in programmable quantum annealers

Sergio Boixo[1], Vadim N. Smelyanskiy[1,2], Alireza Shabani[1], Sergei V. Isakov[1], Mark Dykman[3], Vasil S. Denchev[1], Mohammad H. Amin[4,5], Anatoly Yu. Smirnov[4], Masoud Mohseni[1] & Hartmut Neven[1]

Quantum tunnelling is a phenomenon in which a quantum state traverses energy barriers higher than the energy of the state itself. Quantum tunnelling has been hypothesized as an advantageous physical resource for optimization in quantum annealing. However, computational multiqubit tunnelling has not yet been observed, and a theory of co-tunnelling under high- and low-frequency noises is lacking. Here we show that 8-qubit tunnelling plays a computational role in a currently available programmable quantum annealer. We devise a probe for tunnelling, a computational primitive where classical paths are trapped in a false minimum. In support of the design of quantum annealers we develop a nonperturbative theory of open quantum dynamics under realistic noise characteristics. This theory accurately predicts the rate of many-body dissipative quantum tunnelling subject to the polaron effect. Furthermore, we experimentally demonstrate that quantum tunnelling outperforms thermal hopping along classical paths for problems with up to 200 qubits containing the computational primitive.

[1]Google, Venice, California 90291, USA. [2]NASA Ames Research Center, Moffett Field, California 94035, USA. [3]Department of Physics and Astronomy, Michigan State University, East Lansing, Michigan 48824, USA. [4]D-Wave Systems Inc., Burnaby, British Columbia, Canada V5C 6G9. [5]Department of Physics, Simon Fraser University, Burnaby, British Columbia, Canada V5A 1S6. Correspondence and requests for materials should be addressed to S.B. (email: boixo@google.com) or to V.N.S. (smelyan@google.com).

Quantum annealing[1-5] is a technique inspired by classical simulated annealing[6] that aims to take advantage of quantum tunnelling. In classical cooling optimization algorithms such as simulated annealing, the initial temperature must be high to overcome tall energy barriers. As the algorithm progresses, the temperature is gradually lowered to distinguish between local minima with small energy differences. This causes the stochastic process to freeze once the thermal energy is lower than the height of the barriers surrounding the state. In contrast, quantum tunnelling transitions are still present even at zero temperature. Therefore, for some energy landscapes, one might expect that quantum dynamical evolutions can converge to the global minimum faster than the corresponding classical cooling process.

The goal of quantum annealing is to find low-energy states of a 'problem Hamiltonian'

$$H_P = -\sum_{\mu} h_\mu \sigma^z_\mu - \sum_{\mu\nu} J_{\mu\nu}\sigma^z_\mu \sigma^z_\nu, \qquad (1)$$

where the Pauli matrices σ^z_μ correspond to spin variables with values $\{\pm 1\}$. The local fields $\{h_\mu\}$ and couplings $\{J_{\mu\nu}\}$ define the problem instance. Quantum annealing is characterized by evolution under the Hamiltonian

$$H_0(s) = A(s)H_D + B(s)H_P, \qquad (2)$$

where $H_D = -\sum_\mu \sigma^x_\mu$. The annealing parameter s slowly increases from 0 to 1 throughout the annealing time t_{qa}. Initially, $A(0) \gg B(0)$. With increasing s, $A(s)$ monotonically decreases to 0 for $s=1$, whereas $B(s)$ increases.

The performance of chips designed to implement quantum annealing using superconducting electronics has been studied in a number of recent lines of work[7-24]. Given a quantum annealer operating at finite temperature, noise and dissipation strengths, does it utilize tunnelling or thermal activation for computation? Here we address precisely this question. We introduce a 16-qubit probe for tunnelling, a computational primitive where classical paths are trapped in a false minimum. We present a nonperturbative theory of multiqubit tunnelling, which takes into account both high- and low-frequency noises. To distinguish between tunnelling and thermal activation, we study the thermal dependence of the probability of success for the computational primitive. Thermal activation shows an increasing probability of success with increasing temperature, as expected. Multiqubit tunnelling, on the other hand, shows a decreasing probability of success with increasing temperature, both in theory and experiment. Finally, we study a generalization of the computational primitive to a larger number of qubits that contain the same 'motif' multiple times. Quantum tunnelling outperforms thermal hoping for these problems under similar parameters.

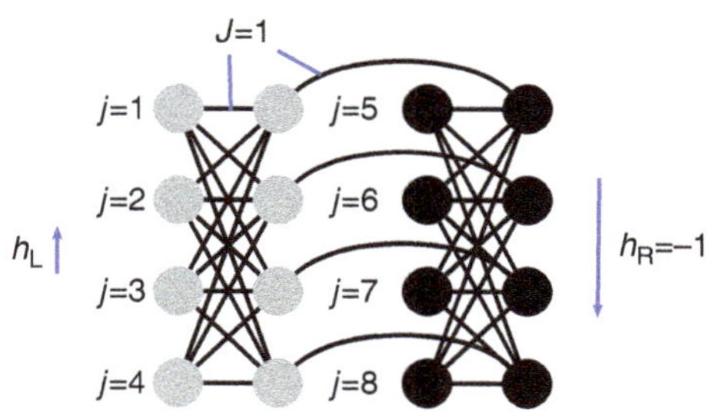

Figure 1 | Graph of the tunnelling probe Hamiltonian. The 16 qubits are coupled ferromagnetically with $J=1$ (lines). The applied fields are $0<h_L<J/2$ ($h_R=-1$) for the left (right) qubit cell. The symmetry and strong intracell ferromagnetic coupling makes each 8-qubit cluster evolve together.

Results

Quantum tunnelling probe. We now describe a probe for computational tunnelling: a non-convex optimization problem consisting of just one global minimum and one false (local) minimum. For concreteness, we use the problem Hamiltonian depicted in Fig. 1, implementable in a D-Wave Two quantum annealer (a description of the flux qubits used in D-Wave Two is given in the Supplementary Note 1). The problem Hamiltonian consists of two cells, left and right, each with $n=8$ qubits. The local fields $0<h_L<0.5$ and $h_R=-1$ are equal for all the spins within each cell, and all the couplings $J=1$ are ferromagnetic. The spins within each cell tend to move together as clusters because of symmetry and the strong intracell ferromagnetic coupling energy. We choose $|h_R|>|h_L|$ so that in the low-energy states of H_P the right cluster is pointing along its own local field as seen in Fig. 1. The difference in energy of the states with opposite polarization in the left cluster is $n(J-2h_L)$. Choosing $h_L<J/2=0.5$, the global minimum corresponds to both clusters having the same orientation, while in the false minimum they have opposite orientations.

We can gain an intuitive understanding of the effective energy landscape if we represent each qubit by a mean field spin vector in the xz plane. Denote by θ_μ the angle of the spin vector for qubit μ with the x quantization axis. We assume that all the qubits in the left (right) cluster have the same angle θ_L (θ_R). This assumption is based on symmetry and the strong intracluster ferromagnetic energy. The resulting energy potential can also be derived using more formal methods, such as the Villain representation (see Supplementary Note 2 and ref. 25). Figure 2 plots the effective energy potential for the left cluster as a function of θ_L with $h_L=0.44$. At the beginning of the annealing process $B(s)/A(s) \ll 1$, and we have $\langle\sigma^z_\mu\rangle \simeq h_\mu B(s)/A(s)$ (the coupling terms are quadratic in the z polarizations $\langle\sigma^z_\mu\rangle$ and therefore negligible at this point). As h_L and h_R have opposite signs, so will the z-projections of spins in the two clusters early in the evolution. To escape this path classically all spins in the left cluster must flip sign, which requires traversing an energy barrier. The barrier peak corresponds to zero total z-polarization of the left cluster. Therefore, the barrier grows with the ferromagnetic energy of the cluster $(n/2)^2 J$. The barrier height is much greater than the residual energy, which grows with $n(J-2h_L)$.

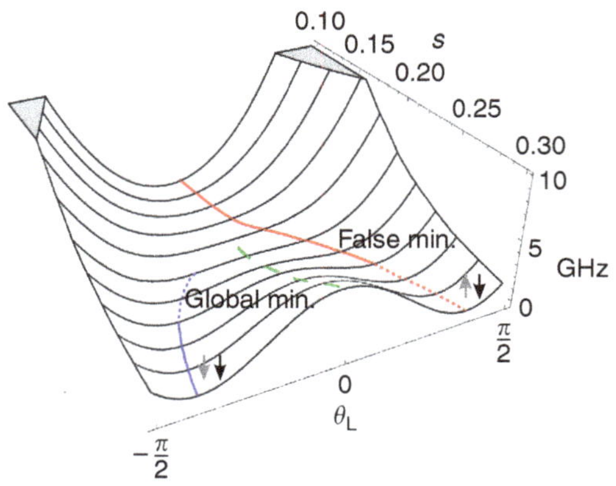

Figure 2 | Effective energy potential using $h_L=0.44$. The mean field potential is plotted versus annealing parameter s and tilt angle θ_L of each spin vector in the left cluster. The red line corresponds to a path that starts in the initial global minimum and follows the instantaneous local energy minimum. A second local minimum (dashed blue line) forms at the bifurcation point $s=0.18$. The global minimum is found in this second path after $s=0.24$ (dashed to continuous blue line). To reach this global minimum the system state has to traverse the energy barrier between them (dashed green line), either by thermal activation or by quantum tunnelling.

Quantum mechanically, if $h_L < J/2$ the system evolution goes through an 'avoided-crossing' where the two lowest eigenstates $E_1(s)$ and $E_0(s)$ approach closely to, and then repel from, each other (see inset in Fig. 3). Higher-energy states remain well separated during the evolution. This level repulsion occurs because of the collective tunnelling of qubits in the left cluster between the opposite z polarizations. At the point where the gap $\hbar\Omega_{10}(s) = E_1(s) - E_0(s)$ reaches its minimum, the corresponding adiabatic eigenstates are formed by the symmetric and antisymmetric superpositions of the cluster orientations. The size of the minimum gap varies with h_L as seen in Fig. 3. The position of the avoided-crossing can be estimated to occur at the point where $|\langle\sigma_\mu\rangle| \sim 2h_L/J$ and moves towards $s = 1$ as h_L approaches $J/2$. Note that for $h_L = J/2$ the residual energy $n(J - 2h_L)$ vanishes and the final ground space is degenerate. There is no avoided-crossing when $h_L > J/2$ (Supplementary Fig. 1).

Characterization of noise and dissipation. Under realistic conditions, a quantum annealer can be strongly influenced by coupling to the environment. We introduce a detailed phenomenological open quantum system model based on single-qubit measurable noise parameters. We shall assume that each qubit is coupled to its own environment with an independent noise source. In the concrete case studied here, this is consistent with experimental data[9], and the coupling of the environment to each flux qubit is proportional to a σ^z qubit operator (flux fluctuations).

In the analysis of the transitions between the states we start from the initial (gapped) stage when the instantaneous energy gap $\hbar\Omega_{10}(s)$ between the two lowest eigenstates $|\psi_0(s)\rangle$, $|\psi_1(s)\rangle$ is sufficiently large compared with the linewidth $\hbar W$. Then, the coupling to the environment can be treated as a perturbation and the transition rate between these states is given by Fermi's golden rule $\Gamma_{1\to 0}(s) \approx a(s)S(\Omega_{10}(s))/\hbar^2$, where $S(\omega)$ is the noise spectral density (see Methods). Here

$$a(s) = \sum_{\mu=1}^{2n} \left| \langle\psi_0(s)|\sigma_\mu^z|\psi_1(s)\rangle \right|^2 \quad (3)$$

is a sum of (squared) transition matrix elements between the two eigenstates.

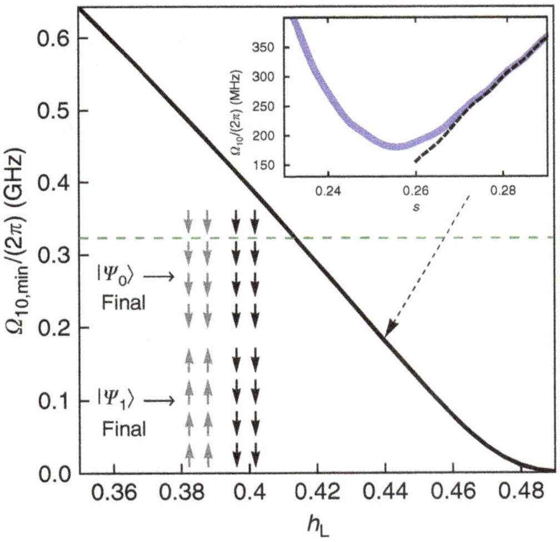

Figure 3 | Quantum energy gap. Inset shows the gap $\hbar\Omega_{10} = E_1(s) - E_0(s)$ versus s, using $h_L = 0.44$. The dashed line is the gap in the diabatic (pointer) basis. In the main plot, the minimum gap decreases with h_L. The horizontal green dashed line (324 MHz) corresponds to 15.5 mK, the lowest temperature in our experiments. The lower inset shows the spin configurations of the two lowest eigenstates at the end of the annealing.

In the minimum gap region, the (squared) matrix element $a(s)$ for the transition rate is large, and the system is thermalized (Fig. 4). More precisely, we have $\Gamma_{1\to 0} \gg 1/t_{qa}$, where the inverse of the annealing time $1/t_{qa}$ is an approximation for the annealing rate. The ground-state population is given by the Boltzmann distribution at the experimental temperature.

After the avoided-crossing region (at $s = 0.255$) we observe a steep exponential fall-off of the matrix element $a(s)$ with s, eventually causing multiqubit freezing (Fig. 4). Multiqubit freezing is quite distinct from single-qubit freezing. Single-qubit tunnelling[10] decays slowly as the magnitude of the transverse field $A(s)$ decreases. The multiqubit transition rate, however, decays exponentially fast (see inset of Fig. 4). This is due to the increasing effective barrier width (Fig. 2), which results in an exponential decrease in quantum tunnelling and a slowdown of the transition rate $\Gamma_{1\to 0}$. Formally, the barrier width corresponds to the Hamming distance

$$h(s) = \sum_{\mu=1}^{2n} \left| \left\langle \psi_0 \middle| \sigma_\mu^z \middle| \psi_0 \right\rangle - \left\langle \psi_1 \middle| \sigma_\mu^z \middle| \psi_1 \right\rangle \right|^2 /4 \quad (4)$$

between the opposite z orientations of the left cluster in the two lowest-energy eigenstates. The exponential sensitivity of multiqubit tunnelling to the width or Hamming distance $h(s)$ is the cause of the exponential decay of the matrix element $a(s)$, and of the multiqubit freezing.

We distinguish a slowdown phase (roughly, $0.1 < t_{qa}\Gamma_{1\to 0} < 10$) and a frozen phase ($t_{qa}\Gamma_{1\to 0} < 0.1$). In the frozen phase, there are no dynamics. Part of the system population remains trapped in the excited state $|\psi_1(s)\rangle$ corresponding to the false minimum of the effective potential until the end of the quantum annealing process (Fig. 4).

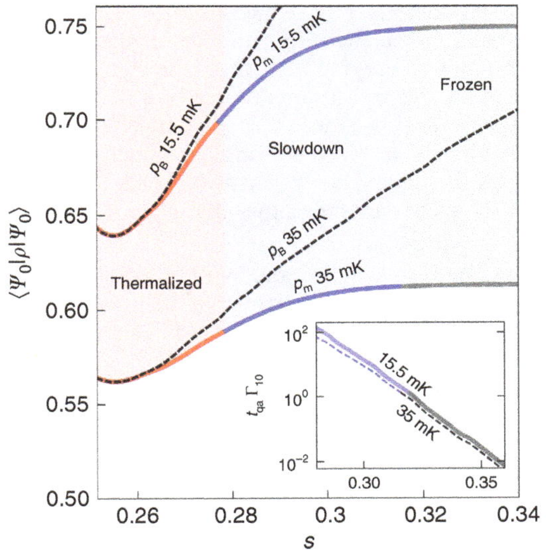

Figure 4 | Multiqubit freezing. Solid lines correspond to the modelled population p_m of the lowest-energy eigenstate along the quantum annealing process using $h_L = 0.44$ at 15.5 mK (top line) and 35 mK (bottom line). Dashed lines correspond to the thermal equilibrium population p_B. In the thermalization phase (red) the transition rate is fast and the population remains close to thermal equilibrium. As the multiqubit energy barrier increases, the transition rates are exponentially reduced with s, as shown in the inset. We define the slowdown regime (blue) as $t_{qa}\Gamma_{1\to 0} < 10$ and the frozen regime (grey) as $t_{qa}\Gamma_{1\to 0} < 0.1$. Comparing the data at 15.5 and 35 mK, we see a small change in the transition rate relative to the larger change in the thermal equilibrium ground-state population. Therefore, the probability of success is lower at higher temperature.

When the energy gap is similar to (or smaller than) the noise linewidth W, the environment cannot be treated as a perturbation. We develop a multiqubit nonperturbative analysis in the spirit of the Non-interacting Blip Approximation (NIBA)[26] that covers all quantum annealing (QA) stages. In the slowdown phase, when the Hamming distance approaches its maximum value, $h \sim n$, the instantaneous decay rate of the first excited state takes the form

$$\Gamma_{1\to 0} = \int_{-\infty}^{\infty} \mathrm{d}\tau \, e^{i\Omega_{10}\tau - h\left(i\epsilon_p\tau + (W\tau)^2/2\right)} \left[\frac{\pi\tau_c}{i\beta} \operatorname{csc} h \frac{(\tau - i\tau_c)}{\beta/\pi}\right]^{\frac{h\eta}{2\pi}} D(\tau), \quad (5)$$

where W and ϵ_p are the linewidth and reconfiguration energy from the low-frequency noise, η and $1/\tau_c$ are the high-frequency Ohmic noise coupling and cutoff and β is the inverse temperature $\hbar/k_B T$. The dependence on the annealing parameter s is implicit. The factor $D(\tau)$ is related to the tunnelling permeability of the potential barrier in Fig. 2 (similar to the coefficient a). It has the expression

$$D(\tau, s) = \left[\left(\epsilon_p(s)c_+(s) - ic_-(s)\partial_\tau\zeta(\tau, s)\right)^2 + a(s)\partial_{\tau\tau}\zeta(\tau, s)\right] e^{-i\epsilon_p(s)\mathrm{d}(s)\tau} \quad (6)$$

where

$$\begin{aligned} Z_\mu^{\gamma\gamma'}(s) &= \left\langle \psi_\gamma(s) \middle| \sigma_\mu^z \middle| \psi_{\gamma'}(s) \right\rangle, \quad \gamma, \gamma' = 0, 1 \\ c_\pm(s) &= \frac{1}{4}\sum_{\mu=1}^{2n} Z_\mu^{10}(s)\left(Z_\mu^{11}(s) \pm Z_\mu^{00}(s)\right), \\ \mathrm{d}(s) &= \frac{1}{4}\sum_{\mu=1}^{2n}\left[\left(Z_\mu^{11}(s)\right)^2 - \left(Z_\mu^{00}(s)\right)^2\right] \\ \zeta(\tau, s) &= i\epsilon_p(s)\tau + \frac{(W(s)\tau)^2}{2} \end{aligned} \quad (7)$$

Equation (5) describes collective tunnelling of the left qubit cluster assisted by the environment. The crucial difference from the single-qubit theory[27,28] is that the parameters of the environment in the transition rate are rescaled by the barrier width or Hamming distance $h(s)$. The effective low-frequency noise linewidth is $h^{1/2}(s)W(s)$, the reconfiguration energy is $h(s)\epsilon_p(s)$ and the Ohmic coefficient is $h(s)\eta(s)$. This is important at the late stages of quantum annealing when $h \sim n \gg 1$.

Comparison of NIBA with data. We observe a very close correspondence between the results of the analysis with the NIBA Quantum Master Equation for the dressed cluster states and the D-Wave Two data displayed in Fig. 5, which shows the dependence of the ground-state population on h_L. We emphasize that for NIBA (and the standard Redfield equation with Ohmic $S_\mathrm{oh}(\omega)$, see Methods) we do not have any parameter fitting: the parameters are obtained from experiments (see Methods). The success probability of quantum annealing is (roughly) determined by the thermal equilibrium ground-state population during the slowdown phase (Fig. 4). When the temperature grows, the ground-state population decreases appreciably, while the transition rate changes little. Consequently, quantum mechanically, the probability of success decreases with increasing temperature, as seen in Fig. 6, for sufficiently big gaps. Figure 6 shows the dependence of the ground-state population with temperature.

For h_L closer to the degeneracy value $h_\mathrm{L} = J/2$, the minimum gap Ω_{10}^{min} becomes smaller, as seen in Fig. 3. Where $\Omega_{10} \ll W$, the adiabatic basis of the instantaneous multiqubit states $\{|\psi_0(s)\rangle, |\psi_1(s)\rangle\}$ loses its physical significance. Because the coupling to the bath is relatively strong here, the system quickly approaches the states corresponding to predominantly opposite cluster orientations, similar to diabatic states (see inset of Fig. 3).

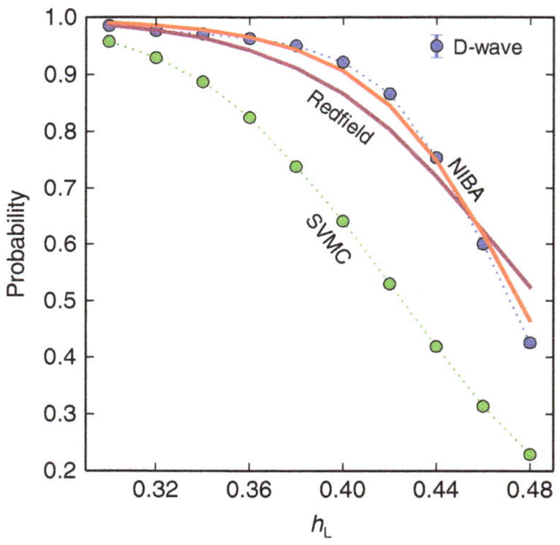

Figure 5 | Probability of success versus h_L. We plot the probability of measuring the final ground state for different values of h_L. The physical temperature for D-Wave is 15.(5) mK, and the annealing time is 20 µs. Both Redfield and NIBA use only measured parameters (no fitting). SVMC uses an algorithmic temperature equal to the physical temperature and 128,000 sweeps, as explained in the text. Error bars (s.e.m.) are smaller than markers.

Transitions between these states, also called pointer states[29], occur at a much slower rate as a consequence of the polaronic effect. As a result, for sufficiently small mininum gaps the multiqubit freezing starts before the avoided-crossing and the success probability increases with temperature[12]. This is captured by the multiqubit NIBA equation, but not by the standard Redfield Quantum Master Equation.

Spin vector Monte Carlo. We want to distinguish quantum tunnelling from thermal activation along classical paths of product states (which preclude multiqubit tunnelling). To give a more precise description of the classical paths of product states, let each qubit be represented by a mean field spin vector in the xz plane and denote by θ_μ the angle of the spin vector for qubit μ with the x quantization axis, as before. A classical path (red line in Fig. 2) that follows the local minimum of the effective energy potential gets trapped in a false minimum and fails to solve the corresponding optimization problem, as explained above. In the absence of quantum tunnelling, the global minimum could be reached through thermal excitations for over-the-barrier escape from the false minimum. This thermal activation results in an increasing probability of success with rising temperature.

This intuition is supported by spin vector Monte Carlo (SVMC), a numerical algorithm consisting of thermal Metropolis updates of the spin vectors. SVMC was introduced recently in ref. 20 and studied in related lines of work[21,23]. The dynamics are constrained to spin–vector product states, with one spin vector per qubit. For a given Hamiltonian $H_0(s)$, we denote the corresponding energy by $E_s(\theta_1,\ldots,\theta n_\mathrm{q})$, where n_q is the number of qubits. The evolution consists of a sequence of sweeps along the Hamiltonian path $\{H_0(s)\}$. In each sweep, a Monte Carlo update is proposed for each qubit in two steps. First, a new angle θ'_μ is drawn from the uniform distribution in $[0, 2\pi]$. Second, the spin vector for qubit μ is updated $\theta_\mu \leftarrow \theta'_\mu$ according to the Metropolis rule for the energy difference

$$D = E_s\left(\ldots, \theta'_\mu, \ldots\right) - E_s\left(\ldots, \theta_\mu, \ldots\right).$$

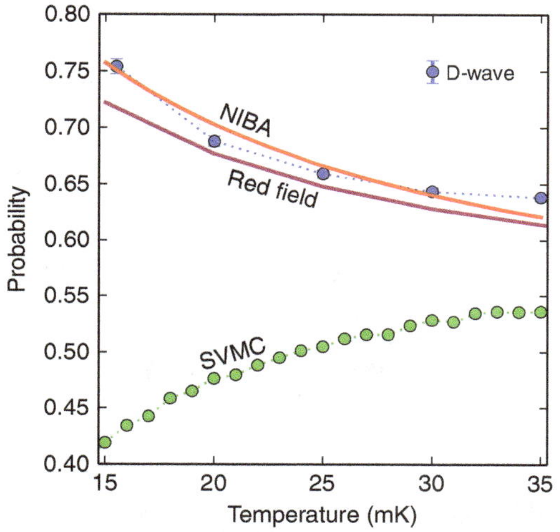

Figure 6 | Probability of success versus temperature for $h_L = 0.44$. The decreasing probability with increasing temperature is only matched with theories based on quantum tunnelling. This is the opposite tendency to thermal activation (SVMC). The annealing time is 20 μs. Both Redfield and NIBA use only measured parameters (no fitting). SVMC uses an algorithmic temperature equal to the physical temperature and 128,000 sweeps. In this temperature range the lowest two states (the double-well potential) account for all the probability (0.9998 for D-Wave, 0.99998 for SVMC). Error bars (s.e.m.) are smaller than markers.

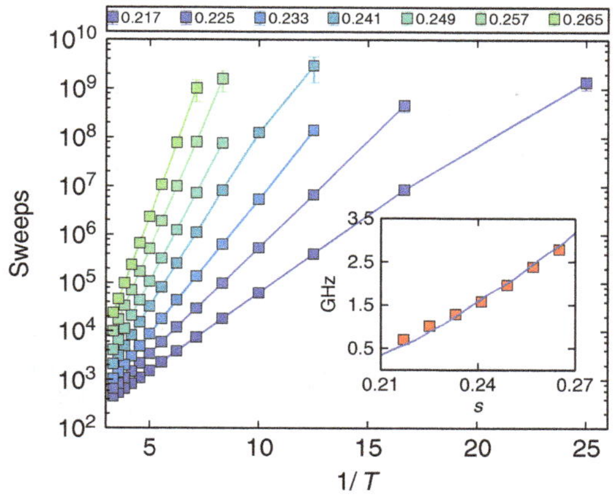

Figure 7 | Analysis of the activation energy for Kramer's scape for SVMC. The main figure shows, in a semilog scale, the average number of sweeps as a function of temperature. We plot lines for different points in the annealing schedule, from $s = 0.217$ (dark blue, see legend) to $s = 0.265$ (green). Error bars correspond to the s.e.m. The embedded figure shows the activation energy (red dots) and the semiclassical energy barrier (blue). There is a good correspondence between SVMC and the effective energy potential.

That is, the move is always accepted if D is negative, and with probability given by the Boltzmann factor $\exp(-D/k_BT)$ if D is positive.

The initial state is chosen to be the global minimum of the transverse field. For low T and sufficient sweeps, the evolution proceeds along the false minima path of Fig. 2. This numerical method allows us to study thermal hopping between the classical paths. To check this correspondence, we studied the height of the energy barrier obtained from Kramers' theory applied to SVMC. For the effective potential at a fixed value of s, we initialized the spin vector state at a local minima, and watch for Kramers events. A Kramers event corresponds to the arrival at the other minima under thermal activation. According to Kramers' theory, the dependence on temperature for the expected number of sweeps necessary for a Kramers event follows the formula $\exp(\Delta U/T)$, where ΔU is the height of the energy barrier. We extract the energy barrier by fitting the curve of the average number of sweeps for different T. We find that this matches almost exactly the energy barrier height in Fig. 2 for different values of s (Fig. 7).

A disadvantage of SVMC as outlined above and introduced in ref. 20 is that there is no natural choice to relate the number of sweeps to the physical evolution time. As in other lines of work, we will choose the number of sweeps to obtain a good correlation with the probability of success of the D-Wave chip for a benchmark of random Ising models with binary couplings $J_{\mu\nu} \in \{1, -1\}$ (refs 15,18,20,23). This will allow us to phenomenologically correlate the number of sweeps to physical time. We set the algorithmic temperature of SVMC to be the same as the physical temperature because we are interested in the dependence of the success probability with temperature. There are no important differences for the correlation with other temperature choices. The correlation with the random Ising benchmark for 128,000 sweeps (Fig. 8) is 0.92, and the residual probabilities $P_{SVMC} - P_{D-Wave}$ have a mean of 0.05 and a s.d. of 0.12. This is consistent with the best values found over a wide range of parameters. We therefore use 128,000 sweeps at 15 mK as our reference parameters for SVMC.

Figure 6 confirms the thermal activation in SVMC. This is opposite to both open quantum system theory and experiments with the D-Wave chip, which show a reduction in the probability of success with rising temperature, as explained above. Furthermore, Fig. 5 shows that the probability of success for SVMC is lower than the probability of success for D-Wave and open system quantum models.

Larger problems. A generalization of the 16-qubit problem to a larger number of qubits is achieved by studying problems that contain the same 'motif' (Fig. 1) multiple times within the connectivity graph (Fig. 9). The success probabilities for up to 200 qubits are shown in Fig. 10. We fit the average success probability as $P(n_q) \propto \exp(-\alpha n_q)$, where n_q is the number of qubits. The fitting exponent α for the D-Wave Two data is $(1.1 \pm 0.05) \times 10^{-2}$, while the fitting exponent for the SVMC numerics is $(2.8 \pm 0.17) \times 10^{-2}$. We conclude that, for instance with multiqubit quantum tunnelling, the D-Wave Two processor returns the solution that minimizes the energy with consistently higher probability than physically plausible models of the hardware that only employ product states and do not allow for multiqubit tunnelling transitions.

Discussion

The role of multiqubit tunnelling as a computational resource is an open problem of active research. Nevertheless, it is instructive to consider some plausible estimates for the case of minimization of an Ising problem that contains pairwise interactions between all qubits. A way to think of multiqubit tunnelling as a computational resource is to regard it as a form of large neighbourhood search. Collective tunnelling transitions involving K qubits explore a K variable neighbourhood, and there is a combinatorial number of such neighbourhoods. Using standard resources, the cost of exhaustively searching on a Hamming ball of binary strings of radius K is $\sum_{i=1}^{K} \binom{n}{i}$, which is bounded from below by $\binom{n}{K}$, where n is the total number of qubits. Therefore, for $K \ll n$, the cost is $\sim n^K/K! \sim \exp(K \log(n))$. As one can see, the exponent in K is very steep ($\log n$). On the other hand, for $K \sim n/2$ the exponent is that of the exhaustive search in the entire n—bit string space (2^n). We now compare this with the tunnelling rate across a barrier of

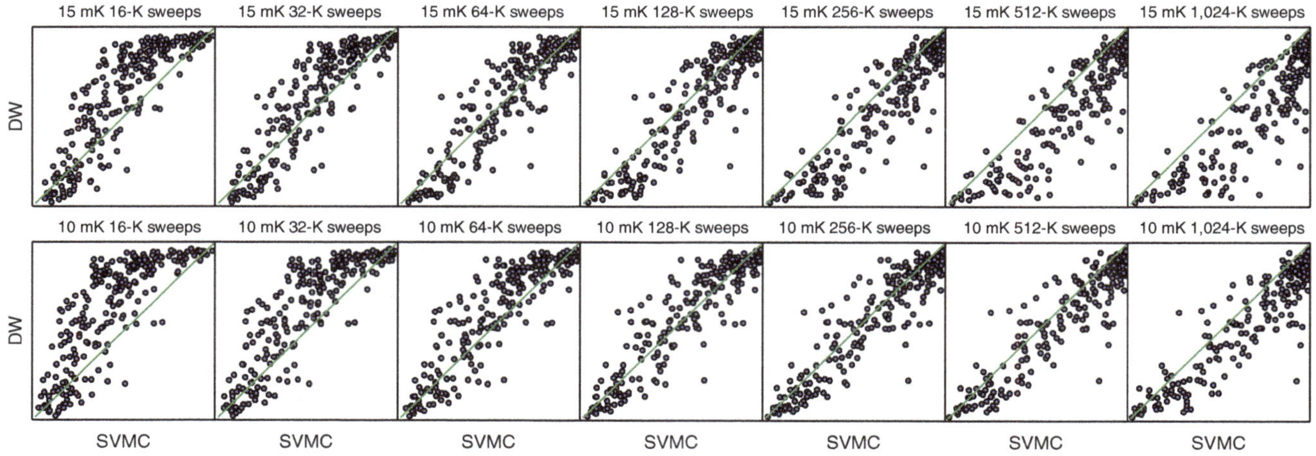

Figure 8 | Scatter plots showing the correlation of D-Wave Two data and SVMC. Correlations for the random Ising benchmark for different algorithmic temperatures (in mK) and number of sweeps. We will use the parameters $T = 15$ mK and sweeps $= 128{,}000$ for SVMC in the rest of the paper.

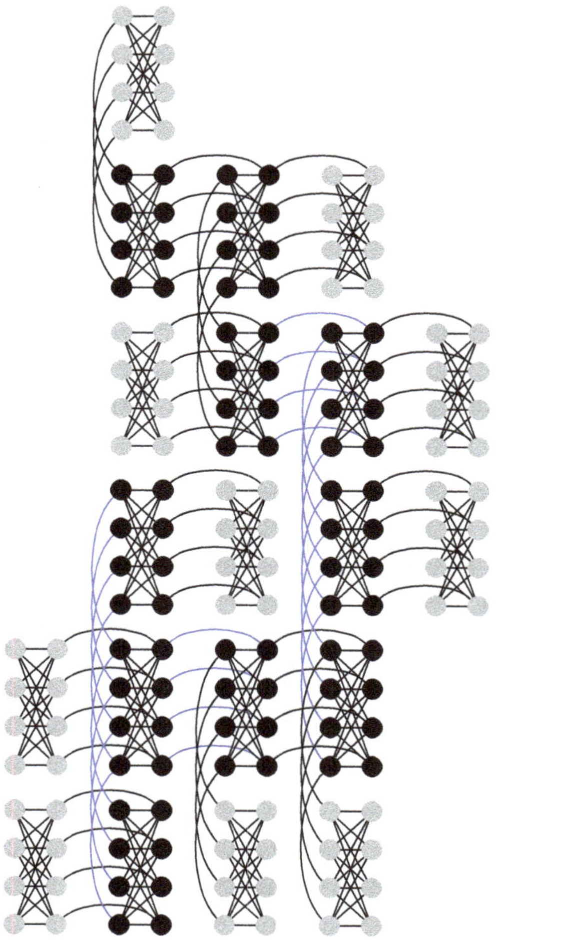

Figure 9 | Larger problems that contain the tunnelling probe 'motif' as subproblems. As in Fig. 1, the black (grey) cluster has a strong $h_R = -1$ (weak $h_L = 0.44$) local field. The black clusters are connected in a glassy manner to make the problem less regular: all connections between any two neighbouring black clusters are set randomly to either -1 or $+1$. The -1 connections are depicted in blue.

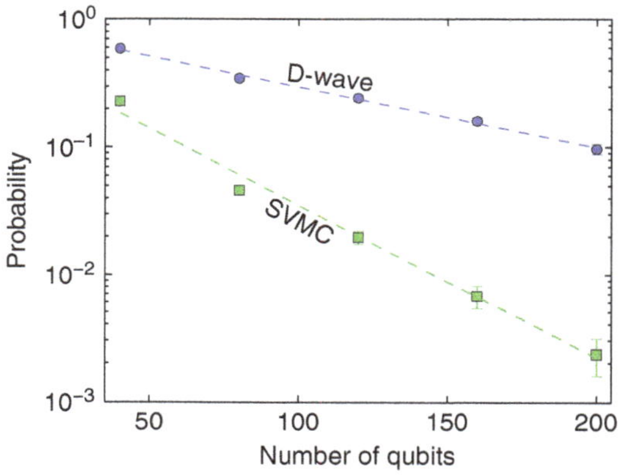

Figure 10 | Success probability for a glass of clusters as a function of the number of qubits. We fit the mean probability of success $P(n_q) \propto \exp(-\alpha n_q)$ as a function of the number of qubits n_q (dashed lines). The fitting exponent α for the D-Wave Two data is $(1.1 \pm 0.05) \times 10^{-2}$, while the fitting exponent for the SVMC numerics is $(2.8 \pm 0.17) \times 10^{-2}$. The error estimates for the exponents (s.e.m.) are obtained by bootstrapping.

width K. It is given by $\exp(-cK)$ for some constant c. If K is smaller than n then, as n increases, $\exp(cK) \ll \exp(K \log(n))$ and tunnelling can be faster than large neighbourhood search. On the other hand, in many problems similar to the two-cluster problem the barrier width is K in $O(n)$. In these cases the tunnelling rate can still be $\exp(cn) \ll 2^n$ (refs 30–35). Therefore, again tunnelling can still provide a dramatically faster search option.

We find that the current-generation D-Wave Two annealer enables tunnelling transitions involving at least 8 qubits. It will be an important future task to determine the maximal K attainable by current technology and how large it can be made in next generations. The multiqubit polaronic quantum master equation presented here lays down the theory to answer this question. It guides the design of next-generation architectures and helps to understand for which computational problems quantum-enhanced optimization may offer an advantage. The larger the K the easier it should be to translate the quantum resource 'K-qubit tunnelling' into a possible computational speedup. We want to emphasize that this paper does not claim to have established a quantum speedup. To this end, one would have to demonstrate that no known classical algorithm finds the optimal solution as fast as the quantum process. To establish such an advantage it will be important to study to what degree collective tunnelling can be emulated in classical algorithms such as Quantum Monte Carlo or by employing cluster update methods. However, the collective tunnelling phenomena demonstrated here present an important step towards what we would like to call a physical speedup: a speedup relative to a hypothetical version of the hardware operated under the laws of classical physics.

Methods

Experimental properties of the noise. The properties of the noise are determined by the noise spectral density $S(\omega)$, which is characterized by single-qubit macroscopic resonant tunnelling (MRT) experiments in a broad range of biases (0.4 MHz–4 GHz) and temperatures (21–38 mK) for tunnelling amplitudes of a single flux qubit below 1 MHz. The MRT data collected are surprisingly well described[28,36] by a phenomenological 'hybrid' thermal noise model $S(\omega) = S_{\mathrm{lf}}(\omega) + S_{\mathrm{oh}}(\omega)$. Here $S_{\mathrm{oh}}(\omega) = \hbar^2 \eta \omega e^{-\omega\tau_c}/(1 - e^{-\hbar\omega/k_B T})$ denotes the high-frequency part and has Ohmic form with dimensionless coupling η and cutoff frequency $1/\tau_c$ (assumed to be very large). The low-frequency part S_{lf} is of the $1/f$ type[37], and in current D-Wave chips this noise is coupled to the flux qubit relatively strongly. Its effect can be described with only two parameters: the width W and the Stokes shift ϵ_p of the MRT line[27]. The experimental shift value is related to the width by the fluctuation-dissipation theorem $(\epsilon_p = \hbar W^2/2k_B T)$ and represents the reorganization energy of the environment. The values of the noise parameters measured at the end of the annealing ($s = 1$) for the D-Wave Two chip are $W/(2\pi) = 0.40(1)$ GHz and $\eta = 0.24(3)$.

References

1. Finnila, A. B., Gomez, M. A., Sebenik, C., Stenson, C. & Doll, J. D. Quantum annealing: a new method for minimizing multidimensional functions. *Chem. Phys. Lett.* **219**, 343–348 (1994).
2. Kadowaki, T. & Nishimori, H. Quantum annealing in the transverse Ising model. *Phys. Rev. E* **58**, 5355–5363 (1998).
3. Farhi, E., Goldstone, J. & Gutmann, S. Quantum adiabatic evolution algorithms versus simulated annealing. Preprint at http://arxiv.org/abs/quant-ph/0201031(2002).
4. Brooke, J., Bitko, D., Rosenbaum, T. F. & Aeppli, G. Quantum annealing of a disordered spin system. *Science* **284**, 779–781 (1999).
5. Santoro, G. E., Martoňák, R., Tosatti, E. & Car, R. Theory of quantum annealing of an Ising spin glass. *Science* **295**, 2427–2430 (2002).
6. Ray, P., Chakrabarti, B. K. & Chakrabarti, A. Sherrington-Kirkpatrick model in a transverse field: absence of replica symmetry breaking due to quantum fluctuations. *Phys. Rev. B* **39**, 11828–11832 (1989).
7. Mooij, J. *et al.* Josephson persistent-current qubit. *Science* **285**, 1036–1039 (1999).
8. Harris, R. *et al.* Experimental investigation of an eight-qubit unit cell in a superconducting optimization processor. *Phys. Rev. B* **82**, 024511 (2010).
9. Lanting, T. *et al.* Cotunneling in pairs of coupled flux qubits. *Phys. Rev. B* **82**, 060512 (2010).
10. Johnson, M. *et al.* Quantum annealing with manufactured spins. *Nature* **473**, 194–198 (2011).
11. Boixo, S., Albash, T., Spedalieri, F. M., Chancellor, N. & Lidar, D. A. Experimental signature of programmable quantum annealing. *Nat. Commun.* **4**, 2067 (2013).
12. Dickson, N. *et al.* Thermally assisted quantum annealing of a 16-qubit problem. *Nat. Commun.* **4**, 1903 (2013).
13. McGeoch, C. C. & Wang, C. in *Proceedings of the ACM International Conference on Computing Frontiers* 23 (ACM, 2013).
14. Dash, S. A note on qubo instances defined on chimera graphs. Preprint at http://arxiv.org/abs/1306.1202 (2013).
15. Boixo, S. *et al.* Evidence for quantum annealing with more than one hundred qubits. *Nat. Phys.* **10**, 218–224 (2014).
16. Lanting, T. *et al.* Entanglement in a quantum annealing processor. *Phys. Rev. X* **4**, 021041 (2014).
17. Santra, S., Quiroz, G., Ver Steeg, G. & Lidar, D. A. MAX 2-SAT with up to 108 qubits. *New J. Phys.* **16**, 045006 (2014).
18. Rønnow, T. F. *et al.* Defining and detecting quantum speedup. *Science* **345**, 420–424 (2014).
19. Vinci, W. *et al.* Hearing the shape of the Ising model with a programmable superconducting-flux annealer. *Sci. Rep.* **4**, 5703 (2014).
20. Shin, S. W., Smith, G., Smolin, J. A. & Vazirani, U. How "quantum" is the D-Wave machine? Preprint at http://arxiv.org/abs/1401.7087 (2014).
21. Albash, T., Vinci, W., Mishra, A., Warburton, P. A. & Lidar, D. A. Consistency tests of classical and quantum models for a quantum annealer. *Phys. Rev. A* **91**, 042314 (2015).
22. Venturelli, D. *et al.* Quantum optimization of fully connected spin glasses. *Phys. Rev. X* **5**, 031040 (2015).
23. Albash, T., Rønnow, T., Troyer, M. & Lidar, D. Reexamining classical and quantum models for the D-Wave One processor: the role of excited states and ground state degeneracy. *Eur. Phys J. Special Top.* **224**, 111–129 (2015).
24. Martin-Mayor, V. & Hen, I. Unraveling quantum annealers using classical hardness. *Sci. Rep.* **5**, 15324 (2015).
25. Boulatov, A. & Smelyanskiy, V. N. Quantum adiabatic algorithm and large spin tunnelling. *Phys. Rev. A* **68**, 062321 (2003).
26. Leggett, A. J. *et al.* Dynamics of the dissipative two-state system. *Rev. Mod. Phys.* **59**, 1–85 (1987).
27. Amin, M. H. S. & Averin, D. V. Macroscopic resonant tunneling in the presence of low frequency noise. *Phys. Rev. Lett.* **100**, 197001 (2008).
28. Lanting, T. *et al.* Probing high-frequency noise with macroscopic resonant tunneling. *Phys. Rev. B* **83**, 180502 (2011).
29. Zurek, W. H. Pointer basis of quantum apparatus: into what mixture does the wave packet collapse? *Phys. Rev. D* **24**, 1516–1525 (1981).
30. Farhi, E., Goldstone, J., Gutmann, S. & Sipser, M. Quantum computation by adiabatic evolution. Preprint at http://arxiv.org/abs/quant-ph/0001106(2000).
31. Roland, J. & Cerf, N. J. Quantum search by local adiabatic evolution. *Phys. Rev. A* **65**, 042308 (2002).
32. Reichardt, B. W. in *Proceedings of the Thirty-Sixth Annual ACM Symposium on Theory of Computing* 502–510 (New York, NY, USA, 2004).
33. Somma, R. D., Boixo, S., Barnum, H. & Knill, E. Quantum simulations of classical annealing processes. *Phys. Rev. Lett.* **101**, 130504 (2008).
34. Somma, R. D. & Boixo, S. Spectral gap amplification. *SIAM J. Comput.* **42**, 593–610 (2013).
35. Kechedzhi, K. & Smelyanskiy, V. N. Open system quantum annealing in mean field models with exponential degeneracy. Preprint at http://arxiv.org/abs/1505.05878(2015).
36. Harris, R. *et al.* Probing noise in flux qubits via macroscopic resonant tunneling. *Phys. Rev. Lett.* **101**, 117003 (2008).
37. Sendelbach, S. *et al.* Decoherence of a superconducting qubit due to bias noise. *Phys. Rev. B* **67**, 094510 (2003).

Acknowledgements

We thank John Martinis, Edward Farhi and Anthony Leggett for useful discussions and reviewing the manuscript. We also thank Ryan Babbush and Bryan O'Gorman for reviewing the manuscript, and Damian Steiger, Daniel Lidar and Tameem Albash for comments about the temperature experiment. The work of V.N.S. was supported in part by the Office of the Director of National Intelligence (ODNI), Intelligence Advanced Research Projects Activity (IARPA) via IAA 145483 and by the AFRL Information Directorate under grant F4HBKC4162G001.

Author contributions

S.B., V.N.S. and H.N. designed the project. H.N., V.N.S. and S.B. proposed frustration patterns for tunnelling. V.N.S., M.D., M.H.A., A.Y.S. and S.B. completed the open quantum system study. S.B., A.S., S.V.I. and V.D. performed numerical studies. All authors contributed to several tasks, such as analysis of the results and discussions of the draft.

Additional information

Competing financial interests: Mohammad H. Amin and Anatoly Yu. Smirnov are employees of D-Wave Systems Inc., a company seeking to develop a processor that uses a computational model known as quantum annealing. The remaining authors declare no competing financial interests.

9

Voltage collapse in complex power grids

John W. Simpson-Porco[1], Florian Dörfler[2] & Francesco Bullo[3]

A large-scale power grid's ability to transfer energy from producers to consumers is constrained by both the network structure and the nonlinear physics of power flow. Violations of these constraints have been observed to result in voltage collapse blackouts, where nodal voltages slowly decline before precipitously falling. However, methods to test for voltage collapse are dominantly simulation-based, offering little theoretical insight into how grid structure influences stability margins. For a simplified power flow model, here we derive a closed-form condition under which a power network is safe from voltage collapse. The condition combines the complex structure of the network with the reactive power demands of loads to produce a node-by-node measure of grid stress, a prediction of the largest nodal voltage deviation, and an estimate of the distance to collapse. We extensively test our predictions on large-scale systems, highlighting how our condition can be leveraged to increase grid stability margins.

[1] Department of Electrical and Computer Engineering, Engineering Building 5, University of Waterloo, Waterloo, Ontario, Canada N2L 3G1. [2] Automatic Control Laboratory, Swiss Federal Institute of Technology (ETH), Physikstrasse 3, CH-8092 Zürich, Switzerland. [3] Department of Mechanical Engineering, Center for Control, Dynamical Systems and Computation, Engineering Building II, University of California at Santa Barbara, Santa Barbara, 93106-9560 California, USA. Correspondence and requests for materials should be addressed to J.W.S.-P. (email: jwsimpson@uwaterloo.ca).

Modern power grids are some of the largest and most complex engineered systems. Currently however, growing consumer demand and the transition to distributed and deregulated small-scale generation are leading to increased system stress, and grid operators have strong economic incentives to operate networks close to their physical limits[1–3]. When these physical limits are approached or breached, power systems can experience a form of network-wide failure termed voltage collapse[4–8]. Voltage collapse and related instabilities have been identified as contributing factors in several recent large-scale blackouts, including Scandinavia (2003), the northeastern United States (2003), Athens (2004) and Brazil (2009) (refs 7–9). An obstacle in predicting voltage collapse is the extensive use of capacitor banks to hold up voltage levels at substations and along transmission lines. This voltage support keeps the system within operational constraints, but conceals the low stability margin of the network, leading to increased blackout risk[7,10]. Voltage fluctuations are presently being further aggravated by the increasing integration of utility-scale wind and photovoltaic sources. A key problem is therefore to develop physically insightful, easily computable stability conditions under which a network is safe from voltage collapse.

Applications of network theory and statistical mechanics to power transmission networks have to this point focused heavily on synchronization[11–19], a phenomenon associated with the self-stabilizing collective behaviour of synchronous generators[20]. Synchronization is primarily controlled by the flow of active power; the real power used by loads to do work[8]. Interest in synchronization has led to a robust theoretical understanding of active power[1,16,21–23], and a plethora of closed-form conditions under which power networks synchronize. In contrast, voltage collapse—a collective nonlinear 'instability'[4,5,7]—has received little attention from a network perspective.

While voltage collapse is a multifaceted phenomena involving generator and transformer limits, the most important fundamental effect is a saddle-node bifurcation of the network equations, resulting in the loss of system equilibrium. Voltage phenomena are driven primarily by 'reactive power', a much less intuitive concept than active power. Reactive power represents the ebb and flow of energy in the electromagnetic fields of system components. This energy is stored and released during each a.c. cycle, allowing system components to function normally and to facilitate the transfer of useful active power with minimal transmission losses[7]. Understanding and controlling reactive power is therefore essential for the efficient and safe operation of the grid.

Theoretical understanding of reactive power flow and voltage collapse in complex networks is poor, however, and numerical simulation is currently the only satisfactory approach to guard against voltage collapse; see refs 4,5,7,24–29 for numerical tests based on sensitivity matrices, and refs 1,10,23,30–32 for approaches based on continuation methods, optimization and energy methods. The network is usually analysed not only under normal conditions, but under a large set of contingencies generated from single-component failures. A broad survey of computational approaches can be found in ref. 33. While effective computational tools in practice, these numerical approaches often offer little theoretical insight into how the underlying parameters and network structure influence voltage stability. An exception is the branch flow monitoring approach in refs 34,35, where voltage collapse and network structure are linked by showing that collapse is preceded by the saturation of transfer paths between sources and sinks of power (Supplementary Note 2).

In contrast with computational methods focused on predicting voltage collapse with great accuracy, here we develop a simple and new analytical framework for analysing voltage collapse, and focus in particular on understanding how the structure of the network influences stability margins. While previous analytic works[36,37] have relied on spectral graph measures such as algebraic connectivity[13,14,16], the closed-form voltage stability condition, we propose below accounts for the grid structure by simultaneously incorporating all eigenvalues of an appropriate system matrix, and combines this information with the sizes and locations of shunt capacitors and loads. To our knowledge, this stability condition is the first to achieve this combination. Our analysis, which is based on a simplified power flow model, yields predictions for the voltage profiles of power grids and provides an explicit stability margin against voltage collapse. The predictions are found to be quite accurate in standard test cases. Our approach is not only mathematically accurate, but also appealing and intuitive to scholars versed in network science and dynamic processes over networks. Since we focus on the influence of grid structure on voltage collapse, we analyse the simplest possible network model that captures the essential bifurcation phenomena; we discuss important extensions involving second-order effects due to active power coupling, as well as component failures in the 'Discussion' section. While our simplified model does not account for active power coupling, we show through extensive numerical experiments that our predictions remain robust when including these effects, and we specifically highlight when they break down.

Results

Power network modelling. We consider a high-voltage power network with $n \geq 1$ load nodes and $m \geq 1$ generator nodes, and in this article we focus on the decoupled reactive power flow equations

$$Q_i = -\sum_{j=1}^{n+m} V_i B_{ij} V_j, \quad i \in \{1, \ldots, n\}, \tag{1}$$

where Q_i (resp. V_i) is the reactive power demanded (resp. voltage magnitude) at load $i \in \{1,\ldots,n\}$. Voltage magnitudes V_j at generators nodes $j \in \{n+1,\ldots,n+m\}$ are regulated by internal controllers to constant values, and the sum in equation (1) therefore contains both quadratic and linear terms in the unknown load voltages $\mathbf{V}_{\mathrm{L}} = (V_1,\ldots,V_n)$. The symmetric coefficients $B_{ij} = B_{ji}$ quantify the effective strength of connection between nodes i and j. These coupling coefficients have the form $B_{ij} = b_{ij} \cos(\theta_i - \theta_j)$, where $b_{ij} \geq 0$ quantifies the strength of the transmission line joining nodes i and j, and $\theta_i - \theta_j$ is the difference between the angles of the voltage phasors at the two nodes. These phase angles may be estimated in advance using a decoupled active power flow model[38], or come from the output of a numerical power flow solver. The diagonal elements are defined by $B_{ii} = -\sum_{j \neq i} b_{ij} + b_{ii}$, where b_{ii} accounts for inductive or capacitive shunts (connections to ground). The sparsity pattern of the matrix B_{ij} therefore encodes both the structure of the physical network and the degree of coupling between nodes after accounting for active power transfers. Equation (1) arises from considering the balance of reactive power at each node in the network while neglecting second-order effects accounting for coupling with active power flows and phase-angle dynamics; more modelling information may be found in (Supplementary Note 3).

A novel mechanical analogy for the power flow (1) is shown in Fig. 1b. The equilibrium configuration of the spring network corresponds to the desirable high-voltage solution of (1), and can be interpreted as a local minimum (Fig. 1c) of the energy function[31]

$$E(V_1, \ldots, V_n) = \frac{1}{2}\sum_{i=1}^{n} \sum_{j=i+1}^{n+m} B_{ij}(V_i - V_j)^2 - \sum_{i=1}^{n} \left(\frac{1}{2}\kappa_i V_i^2 + Q_i \ln(V_i)\right), \tag{2}$$

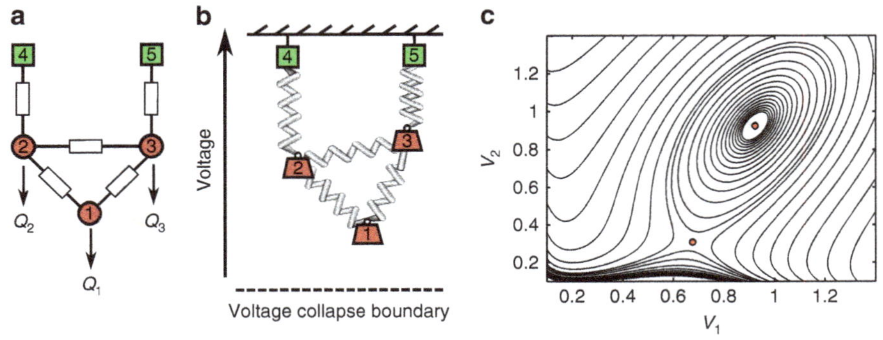

Figure 1 | Mechanical and energy interpretations of power flow. (**a**) An example power network with two generators (green) supplying power to three loads (red). Power demands (Q_1, Q_2, Q_3) are placed on the load nodes; (**b**) a mechanical analogy: a linear spring network placed in a potential field. The generator voltages (green) are 'pinned' at constant values, while the load voltages (red) are masses 'hanging' off the generators, their equilibrium values being determined by their weights (the power demands $\mathbf{Q}_L = (Q_1, Q_2, Q_3)$), the heights of the fixed-generator voltages (V_4, V_5), and by the stiffness of the spring network (the susceptance matrix **B**). Voltage collapse can occur when one of the masses crosses an appropriate collapse boundary curve; (**c**) Contour plot of energy function when $Q_3 = 0$ and node 3 is eliminated via Kron reduction[13]. Since $E(\mathbf{V}_L)$ contains logarithms, it tends to $-\infty$ as either axis is approached. In a normalized system of units, the stable high-voltage equilibrium rests in a local minimum at (0.94, 0.94), while an unstable low-voltage equilibrium sits at the saddle (0.68, 0.30). Voltage collapse occurs when these equilibria coalesce and the system trajectory diverges.

where $\kappa_i \triangleq \sum_{j=1}^{n+m} B_{ij}$ (Supplementary Note 4). Note that the power demands Q_i generate a logarithmic potential, leading to multiple equilibria (Fig. 1c). Standard practice is that for stable and economical network operation with minimal transmission losses, nodal voltages should remain near their open-circuit values as obtained for an unloaded (and thus unstressed) network[8]. Intuitively then, a stable steady-state is characterized by

$$|V_i - V_i^*|/V_i^* \leq \delta, \quad i \in \{1, \dots, n\}, \tag{3}$$

where V_i^* is the open-circuit voltage at the ith node and $\delta > 0$ is a dimensionless variable quantifying an allowable percentage limit on deviations. Intuition from Fig. 1 suggests that a stiff, lightly loaded grid will have a high and uniform voltage profile with small deviation δ, while a weak, heavily loaded grid will result in voltage collapse. The following 'Analytic results' section will make this intuition precise and mathematically accurate.

Analytic results. We suggest that for assessing voltage stability and collapse, one should consider not the underlying electrical network encoded in the susceptance matrix **B**, but a reduced and re-weighted auxiliary network. This auxiliary network shares the same topology as the physical network, but with new edge weights which encode both generator voltage levels and the topology and strength of connections between loads and generators. After potentially reordering the network nodes so that loads and generators are labelled, respectively, $\{1,\dots,n\}$ and $\{n+1,\dots,n+m\}$, we may partition the $(n+m)\times(n+m)$ coupling matrix **B** with elements B_{ij} into four block matrices as

$$\mathbf{B} = \begin{pmatrix} \mathbf{B}_{LL} & \mathbf{B}_{LG} \\ \mathbf{B}_{GL} & \mathbf{B}_{GG} \end{pmatrix}. \tag{4}$$

The $n \times n$ sub-matrix $\mathbf{B}_{LL}$ now describes the interconnections among loads, while the $n \times m$ matrix $\mathbf{B}_{LG}$ specifies the interconnections between loads and generators. This partitioning suggests a natural mapping from generators to loads through the matrix $\mathbf{B}_{LL}^{-1}\mathbf{B}_{LG}$, which we can use to define the open-circuit load voltages $\mathbf{V}_L^* = (V_1^*, \dots, V_n^*)$ by

$$\mathbf{V}_L^* = -\mathbf{B}_{LL}^{-1}\mathbf{B}_{LG}\mathbf{V}_G, \tag{5}$$

where $\mathbf{V}_G = (V_{n+1},\dots,V_{n+m})$ is the vector of fixed-generator voltages. To quantify the stiffness of the spring network in Fig. 1b, we combine the nominal voltages in equation (5) with the sub-matrix $\mathbf{B}_{LL}$ in equation (4) to obtain the symmetric stiffness matrix

$$\mathbf{Q}_{crit} \triangleq \frac{1}{4}\mathrm{diag}(\mathbf{V}_L^*)\cdot\mathbf{B}_{LL}\cdot\mathrm{diag}(\mathbf{V}_L^*), \tag{6}$$

where $\mathrm{diag}(\mathbf{V}_L^*)$ is the matrix with $(V_1^*, \dots, V_n^*)$ on the main diagonal. In other words, $\mathbf{Q}_{crit}$ has units of power and its ijth entry is given by $V_i^* V_j^* B_{ij}/4$. Selected topological features, edge weights, generator voltages, and the relative locations of generators and loads are all concisely encoded in the stiffness matrix $\mathbf{Q}_{crit}$.

Just as the stiffness matrix of a standard spring network relates displacements to spring forces, the matrix $\mathbf{Q}_{crit}$ can be thought of as relating the dimensionless voltage deviations $(V_i - V_i^*)/V_i^*$ to the reactive power demands $\mathbf{Q}_L = (Q_1,\dots,Q_n)$. Indeed, this normalization to dimensionless variables is key to our theoretical analysis. To arrive at small normalized deviations of the form (3), it then seems reasonable that the dimensionless matrix-vector product $\mathbf{Q}_{crit}^{-1}\mathbf{Q}_L$ should be small in some sense. Our main result below shows that this intuition based on linear spring networks can be made precise, leading to guarantees on voltage deviations for the nonlinear network (1). A derivation and a formal proof can be found in the 'Methods' section and in Supplementary Note 5 respectively.

Theorem 1: The power flow equations (1) have a unique, stable, high-voltage solution $(V_1,\dots,V_n)$ if

$$\Delta = \|\mathbf{Q}_{crit}^{-1}\mathbf{Q}_L\|_\infty < 1, \tag{7}$$

where $\|\mathbf{Q}_{crit}^{-1}\mathbf{Q}_L\|_\infty$ is the largest magnitude of the entries of the vector $\mathbf{Q}_{crit}^{-1}\mathbf{Q}_L$. Moreover, each component V_i of the unique high-voltage solution satisfies the bound $|V_i - V_i^*|/V_i^* \leq \delta_-$, where $\delta_- = (1-\sqrt{1-\Delta})/2$.

The matrix-vector product $\mathbf{Q}_{crit}^{-1}\mathbf{Q}_L$ captures the interaction between the auxiliary network structure and the locations of loads, with the infinity norm $\|\cdot\|_\infty$ identifying the maximally stressed node. The scalar δ_- then bounds the largest voltage deviation in the network. No reactive loading corresponds to zero stress $\Delta = 0$ and $\delta_- = 0$; voltages align with their open-circuit values. Conversely, when $\Delta = 1$, the network's guaranteed stability margin has been depleted. Said differently, $\Delta < 1$ guarantees the existence of a stable equilibrium, while $\Delta \geq 1$ is a necessary condition for voltage collapse, where at least one node of the network has become overly stressed. The stability condition (7) can be therefore be interpreted as a dual to previous literature showing that voltage collapse is always preceded by at least one edge of the network becoming overly stressed[34,35]. Moreover, the bound $\Delta < 1$ is the 'tightest' possible general bound, as cases can be constructed where voltage collapse occurs at $\Delta = 1$ (Supplementary Note 5). Note that equation (7) captures the desired intuition of the spring network analogy in Fig. 1b; the network stiffness matrix $\mathbf{Q}_{\text{crit}}$ should be large when compared with the reactive loading $\mathbf{Q}_{\text{L}}$; see (Supplementary Note 5) for complex network, power system and circuit-theoretic interpretations of the stability condition. In terms of Fig. 1c, $\sqrt{1-\Delta}$ lower bounds the distance in voltage-space between the stable and unstable equilibria in the power system energy landscape. In summary, the stability condition (7) concisely and elegantly captures the physical intuition developed in Fig. 1 and in the previous section, and guarantees the existence of a unique equilibrium for the nonlinear network equation (1).

For fixed reactive demands $\mathbf{Q}_{\text{L}}$, the stability test (7) states that the largest stability margins are obtained by making $\mathbf{Q}_{\text{crit}}^{-1}$ small. Since the parameters of the grid are embedded in the stiffness matrix $\mathbf{Q}_{\text{crit}}$ defined in equation (6), the stability test (7) provides insight into how the parameters of the network influence its stability margins. Rigorous statements may be found in Supplementary Note 6, while here we present the key insights. For example, by examining the definitions (5) and (6) one observes that raising generator voltage levels $\mathbf{V}_{\text{G}}$ will weaken (in magnitude) the elements of $\mathbf{Q}_{\text{crit}}^{-1}$ and therefore increase stability margins. In terms of Fig. 1b, this corresponds to 'raising the ceiling', which increases the distance to the stability boundary. Since the coupling weights B_{ij} enter the stiffness matrix (6) both directly and through the open-circuit voltages $\mathbf{V}_{\text{L}}^*$, their effects on stability margins are subtle, and counter-examples can be constructed where increasing the coupling between generators and loads decreases stability margins (Supplementary Note 6). Nonetheless, one may show rigorously that under normal network conditions, strengthening the edge weights B_{ij} between loads and generators and increasing the shunt capacitances b_{ii} at loads are both beneficial to stability margins. The first corresponds to stiffening the springs (4, 2) and (5, 3) of Fig. 1b, while the second can be thought of as extra upward force directly applied to nodes {1, 2, 3}. In summary, the stability condition (7) can be leveraged to provide new qualitative insights into how the network structure and parameters influence stability margins.

Finally, in contrast to standard voltage collapse studies, note that we have made no assumptions about the direction of the reactive power demands $\mathbf{Q}_{\text{L}}$, which appear linearly in equation (7). Therefore, the condition (7) simultaneously accounts for all directions in the space of reactive power demands. This generality may result in the test (7) being conservative for a particular direction in the space of power demands. On the other hand, this generality allows one to assess network stability for an entire set of possible power demands via a single evaluation of the condition (7).

The inverse of the stiffness matrix is the sensitivity matrix relating percentage changes in voltage to changes in reactive demands $\mathbf{Q}_{\text{L}}$, as can be seen from the linearized relationship $(V_i - V_i^*)/V_i^* = -(\mathbf{Q}_{\text{crit}}^{-1}\mathbf{Q}_{\text{L}})_i/4$. A comparison of the stiffness matrix $\mathbf{Q}_{\text{crit}}$ and its inverse is shown in Fig. 2. The stiffness matrix $\mathbf{Q}_{\text{crit}}$ is itself very sparse, mirroring the physical topology of the grid. This sparsity allows the inequality (7) to be rapidly checked by solving a sparse linear system $\mathbf{Q}_{\text{crit}}\mathbf{x} = \mathbf{Q}_{\text{L}}$; the vector $\mathbf{x}$ serves as a linear approximation of (and an upper bound on) the exact voltage deviations $(V_i - V_i^*)/V_i^*$. In contrast, the inverse $\mathbf{Q}_{\text{crit}}^{-1}$ is a dense matrix with significant off-diagonal elements, indicating the importance of not only local but also multi-hop interactions.

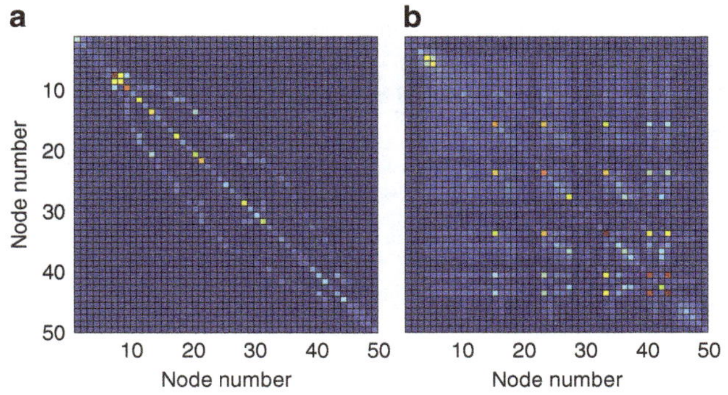

Figure 2 | Sparsity patterns of network matrices for 57 node test case. (**a**) the stiffness matrix $\mathbf{Q}_{\text{crit}}$ representing the auxiliary network. (**b**) The inverse stiffness matrix $\mathbf{Q}_{\text{crit}}^{-1}$. The 57 node network contains 50 loads and 7 generators. Nodes are sorted and grouped by connected components of the subgraph induced by $\mathbf{Q}_{\text{crit}}$, with connected components ordered from largest to smallest; nodes {1,...,48} are part of one large connected component, while nodes {49, 50} each constitute their own component. Colour scale represents normalized values of the matrix elements, with dark blue being zero and red being one. Diagonal elements of $\mathbf{Q}_{\text{crit}}$ are displayed in absolute value for clarity.

While we omit the details here, the stability condition (7) can be extended to additionally guarantee the satisfaction of hard, predefined limits on both voltage magnitudes and the reactive power injections of generators (Supplementary Note 5 and Supplementary Note 6, respectively).

Numerical assessment of voltage stability condition. In this section we provide three numerical studies to assess the accuracy of the stability condition (7) in large-scale power networks, and to determine its predictive limitations. Our first study focuses on the accuracy of the theoretical bound $|V_i - V_i^*|/V_i^* \leq \delta_-$ in typical networks operating in the normal regime far away from voltage collapse. We consider 11 widely established test cases[39], ranging from a small 9 node network to a representation of the Polish grid with nearly 2,400 nodes. To generate a diverse set of sample networks, we construct 1,000 realizations of each network, with up to 30% deviation from forecast conditions in generation and up to 50% deviation in active and reactive power demands, drawn from a normal distribution centred around base conditions; see (Supplementary Methods) for details. For each realization, we solve the more-realistic lossless coupled active/reactive a.c. power flow equations numerically, and we compare the largest nodal voltage deviation $\delta_{\text{exact}} = \max_i |V_i - V_i^*|/V_i^*$ from the numerically determined voltage profile to the analytic bound $\delta_- = \frac{1}{2}(1 - \sqrt{1-\Delta})$ from our main result (7) based on the simplified model (1) with the numerically determined phase angles $\theta_i - \theta_j$ substituted.

Our findings are reported in Table 1. The theoretical prediction of the stability test (7) is that $\delta_{\text{exact}} \leq \delta_-$; the first column indicates that this inequality held for all realizations for which the numerical solver converged. All realizations for which the numerical solver failed to converge were discarded; this occurred in fewer than 1% of all cases. The second and third columns list the average values of these two quantities over all realizations. As can be seen, the voltage deviations range from roughly 1% to 6% from open-circuit conditions. The final column shows the average of the prediction error $(\delta_- - \delta_{\text{exact}})/\delta_{\text{exact}}$ over all realizations. For all networks from 9 to 2,383 nodes (except the 57 and 300 node networks) the prediction error is less than 1%, indicating that prediction accuracy is not directly dependent on system size.

Table 1 | Voltage stability condition applied to 11 test networks.

Numerical testing of theoretical predictions

Test case (1,000 instances)	Condition correctness	Exact deviation (δ_{exact})	Predicted deviation (δ_-)	Condition accuracy
9 bus system	True	$5.50 \cdot 10^{-2}$	$5.52 \cdot 10^{-2}$	$3.56 \cdot 10^{-3}$
14 bus system	True	$2.50 \cdot 10^{-2}$	$2.51 \cdot 10^{-2}$	$1.96 \cdot 10^{-3}$
RTS 24	True	$3.28 \cdot 10^{-2}$	$3.29 \cdot 10^{-2}$	$3.28 \cdot 10^{-3}$
30 bus system	True	$4.72 \cdot 10^{-2}$	$4.75 \cdot 10^{-2}$	$7.64 \cdot 10^{-3}$
New England 39	True	$5.95 \cdot 10^{-2}$	$5.99 \cdot 10^{-2}$	$5.97 \cdot 10^{-3}$
RTS '96 (2 area)	True	$3.44 \cdot 10^{-2}$	$3.45 \cdot 10^{-2}$	$3.81 \cdot 10^{-3}$
57 bus system	True	$0.97 \cdot 10^{-1}$	$0.99 \cdot 10^{-1}$	$2.97 \cdot 10^{-2}$
RTS '96 (3 area)	True	$3.57 \cdot 10^{-2}$	$3.58 \cdot 10^{-2}$	$3.94 \cdot 10^{-3}$
118 bus system	True	$2.68 \cdot 10^{-2}$	$2.69 \cdot 10^{-2}$	$3.63 \cdot 10^{-3}$
300 bus system	True	$1.32 \cdot 10^{-1}$	$1.36 \cdot 10^{-1}$	$3.03 \cdot 10^{-2}$
Polish 2,383 system	True	$4.03 \cdot 10^{-2}$	$4.06 \cdot 10^{-2}$	$8.55 \cdot 10^{-3}$

Condition correctness is whether the implication $\Delta = \|\mathbf{Q}_{crit}^{-1}\mathbf{Q}_L\|_\infty < 1 \Rightarrow \delta_{exact} \le \delta_-$ holds for every network realization, where $\delta_- = \frac{1}{2}\left(1 - \sqrt{1-\Delta}\right)$ and δ_{exact} is determined numerically. Exact and predicted deviations are averaged values of the respective quantities over all realizations. Condition accuracy is calculated as $(\delta_- - \delta_{exact})/\delta_{exact}$, and averaged over 1,000 randomized instances for each network, with 30% of generation (resp. 30% of load) randomized by 30% (resp. 50%) using a normal distribution centred around base conditions.

Perhaps surprisingly, considering the simplicity of the condition (7), the least accurate prediction overestimates voltage deviations by only 3.8%. We conclude that for normally stressed large-scale networks, the bounds predicted by the stability condition (7) hold and are accurate even when tested on more complicated coupled power flow models.

Our second study analyses the predictions of (7) in a highly stressed network, again for the more-realistic lossless coupled active/reactive power flow model. As our focus is on studying bifurcation phenomena for the network equations, we discard generator limitations in this study and assume internal generator controls hold the network-side generator voltages constant; see Supplementary Note 7 for theoretical extensions which include generator limits. As we noted previously, $\Delta \geq 1$ is a necessary condition for voltage collapse, and we now test the gap between this necessary condition and true point of collapse. We consider the 39-node reduced representation of the New England power grid, illustrated in Fig. 5a. Beginning from normal base case loading conditions, the active and reactive power demands and generation are increased continuously along a chosen ray in parameter-space, with the size of the increase parameterized by a scalar λ, until voltage collapse occurred at a value $\lambda = \lambda_{collapse}$. For each $\lambda \in [0, \lambda_{collapse}]$, we determine numerically the system equilibrium and recalculate Δ from equation (7) using the numerically determined phase angles $\theta_i - \theta_j$.

The above testing procedure obviously depends on the choice of direction for increase in the space of power demands and generation. We select two directions and study them separately, to illustrate the strengths and limitations of our analytic approach based on a simplified power flow model. As a first choice, we select a direction where the mean power factor in the network is decreased 20% to a value of 0.7. (The power factor of the ith load is defined as $P_i/\sqrt{(P_i^2 + Q_i^2)}$, where P_i is the active power drawn by the load. If $P_i = Q_i$, then the power factor is 0.707.) This corresponds to a case where loads consume roughly equal amounts of active and reactive power, which in practice is unusually highly reactive power consumption. We therefore expect that instabilities associated with reactive power flow should dominate any unmodeled active power effects, and the simplified model (1) should serve as a good proxy for the coupled active/reactive power flow equations. As a function of λ, Fig. 3 displays the trace of the voltage magnitude at node 4 (solid black), the loading margin Δ (dashed blue), and the bound $V_4^*(1 - \delta_-)$ (dotted red) determined by equation (7). Node 4 was determined through equation (7) to be the most stressed node in the network, and hence the node for which our theoretical bound would be best tested. First, observe that the numerically determined voltage trace is bounded below by the trace of the theoretical bound, as expected. The loading margin Δ increases roughly linearly with λ, with $\Delta = 1$ occurring at $\lambda/\lambda_{collapse} = 0.98$. Our previous conclusions regarding the necessity of $\Delta > 1$ for voltage collapse therefore hold in this highly stressed case for the more complicated coupled active/reactive power flow model, and the gap between the necessary condition $\Delta > 1$ and the true point of collapse is a surprisingly small 2%.

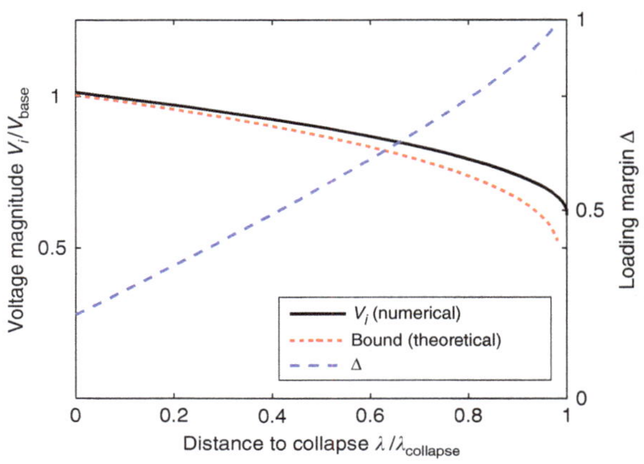

Figure 3 | Stress testing of voltage stability condition for low power factor loading. The horizontal voltage axis is scaled by $V_{base} = 345$ kV. The solid black trace is the numerically computed voltage magnitude at node four, while the dotted red trace is given explicitly by $V_4^*(1 - \delta_-)$, where δ_- is determined as below (7). The stability margin Δ is shown in dashed blue. When $\Delta > 1$, δ_- becomes undefined and the corresponding bound is no longer plotted.

As a second loading direction for testing, we maintain the direction of the base case, for which the average power factor of loads is approximately 0.88. In this regime reactive power transfers will be less prominent, and we expect the unmodeled coupling between active and reactive power flows to induce voltage collapse at a loading level lower than expected from the simplified model (1). Again as a function of λ, Fig. 4 displays the desired traces. While the trace of $V_4^*(1 - \delta_-)$ continues to lower bound the trace of the node voltage V_4, we find in this case that $\Delta = 0.75$ when voltage collapse occurs for the coupled equations at $\lambda/\lambda_{collapse} = 1$. As expected, in this regime the unmodeled coupled power flow effects become crucial and the simplified decoupled model (1), on which our analysis is based, becomes

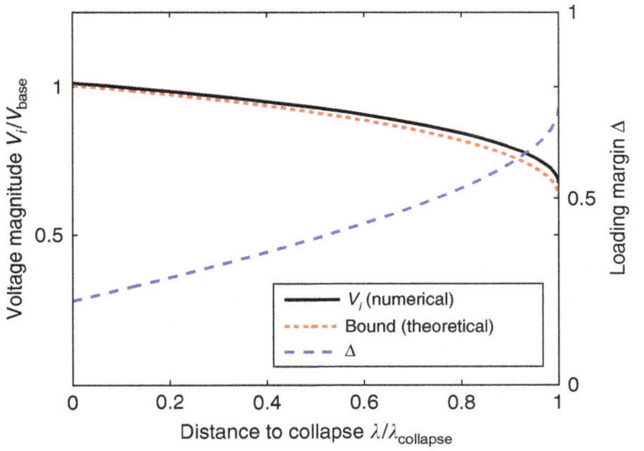

Figure 4 | Stress testing of voltage stability condition for high power factor loading. The horizontal voltage axis is scaled by $V_{base} = 345$ kV. The solid black trace is the numerically computed voltage magnitude at node four, while the dotted red trace is given explicitly by $V_4^*(1-\delta_-)$, where δ_- is determined as below (7). The stability margin Δ is shown in dashed blue.

invalid. Said differently, when reactive power demands in the network are low, our analytic prediction of the point of voltage collapse based on the simplified model (1) is overly optimistic. We comment further on extensions of our analysis to the coupled case in the 'Discussion' section and in Supplementary Note 5.

Our final study illustrates the use of our stability condition (7) for determining corrective actions, with the goal of increasing grid stability margins. The New England grid in Fig. 5a is experiencing peak loading conditions, and shunt capacitors have been switched in at all substations (red nodes) to support voltage magnitudes, keeping the voltage profile (solid black in Fig. 5b) within operational bounds (dotted grey). Node 8 is under particularly heavy loading with a poor power factor of 0.82, and additional shunt capacitors at nodes 7 through 9 have been used to support the voltages in this area. While all voltages are maintained within the operational bounds, we calculate using the condition (7) that $\Delta = 0.64$, indicating the network is actually under significant stress. This stress is also apparent by numerically solving the lossy coupled power flow equations plotting the ratio V_i/V_i^* of the nodal voltage to the open-circuit voltage (solid red in Fig. 5b), as these ratios take into account the effects of shunt compensation; node 8 is experiencing the greatest stress. Consider the possibility of control equipment being present at the ith node of the network, capable of supplying an additional amount of reactive power q_i to the grid. Our goal is to select $\mathbf{q} = (q_1,\ldots,q_n)$ to optimally increase grid stability margins. Such control could be realized actively through power electronic devices, or passively by curtailing local power consumption; in either case it is also desirable to minimize the total control action.

With this additional control capability, the stability metric (7) is modified to $\left\|\mathbf{Q}_{\mathrm{crit}}^{-1}(\mathbf{Q}_{\mathrm{L}}+\mathbf{q})\right\|_\infty < 1$. One immediately observes that the elements of $\mathbf{Q}_{\mathrm{crit}}^{-1}$ are providing information on where control action will be the most effective. For example, suppose that control equipment is present only at nodes seven and nine, but not at node eight (Fig. 5a). One finds for this example that $(\mathbf{Q}_{\mathrm{crit}}^{-1})_{87}/(\mathbf{Q}_{\mathrm{crit}}^{-1})_{89} = 1.98$, indicating that control action at node seven will be nearly twice as effective in reducing stress at node eight as the same control action would be if applied at node nine. From a purely topological viewpoint, this discrepancy in control sensitivity is surprising, as both nodes are neighbours of node eight. The stiffness matrix $\mathbf{Q}_{\mathrm{crit}}$ incorporates not only the topology, but also the strength of connections between nodes, the locations of shunt capacitors and the relative proximity of generation (green nodes). Increasing q_7 and q_9 in this ratio provides the desired control action, allowing capacitor banks to be switched out, and we find that $\Delta = 0.52$ after control. A simple heuristic control has therefore reduced network stress by $(0.64-0.52)/(0.52) \simeq 23\%$, while the voltage profile of the grid (dotted black) is essentially unchanged.

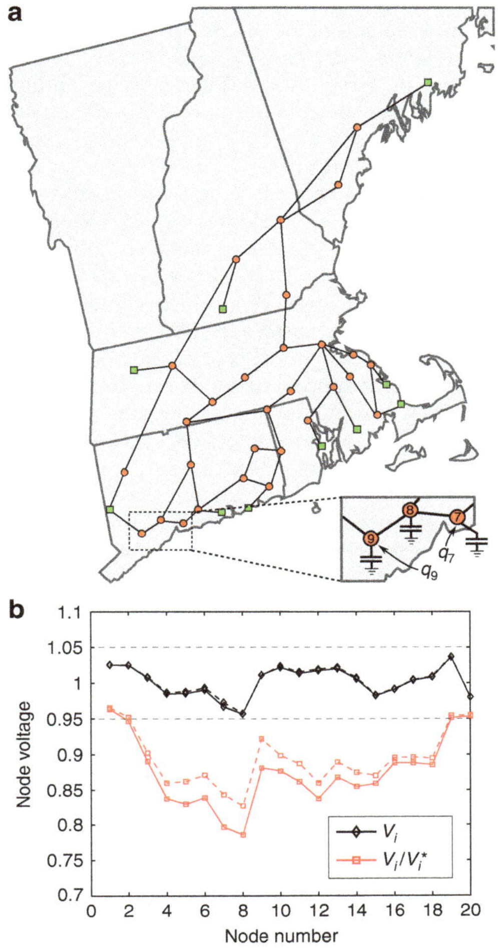

Figure 5 | Corrective action results for the reduced New England 39-node network. (**a**) Depiction of the reduced New England grid. Load nodes {1,…,30} are red circles, while generators {31,…,39} are green squares. Shunt capacitors are present at all load nodes, but shown explicitly at nodes 7, 8 and 9. (**b**) Results of corrective action study. Voltage profile V_i (black) and scaled voltages V_i/V_i^* (red), before (solid) and after (dashed) corrective action. All voltages were scaled by the grid's base voltage $V_{base} = 345$ kV. Horizontal dashed lines are operational limits for V_i of $\pm 5\%$ from base voltage. For clarity only nodes {1,…,20} are plotted. Map by freevectormaps.com.

In summary, the stability condition (7) can be simply and intuitively used to select control policies which increase grid stability margins with minimal control effort; additional details on eigenvector-based control directions[40] and on the simulation setup are available in Supplementary Note 5 and the Supplementary Methods, respectively.

Discussion

The stability condition (7) provides a long sought-after connection between network structure, reactive loading and the resulting voltage profile of the grid. As such, the condition (7) can be used to identify weak network areas and trace geographical origins of voltage instability by examining the entries of the vector $\mathbf{Q}_{\mathrm{crit}}^{-1}\mathbf{Q}_{\mathrm{L}}$.

This allows for the effective placement of voltage control equipment, and the automatic dispatch of generation to mitigate voltage fluctuations, creating a self-healing network. The condition (7) can serve as a bridge between intuition-based heuristics for voltage control and more computational optimization approaches, and the use of (7) for systematic control design is currently under investigation.

The results reported here are a first step towards an analytic approach to assessing and strengthening the voltage stability of power grids. A limitation of the current work is that active power demands are included only implicitly in the condition (7), through the stiffness matrix $\mathbf{Q}_{\text{crit}}$ which contains the effective coupling weights $B_{ij} = b_{ij}\cos(\theta_i - \theta_j)$. While our formal theoretical results hold only for the approximate model, the results of Table 1 show that this approximation is extremely accurate under normal operating conditions, and the results of Fig. 5 indicate that our framework provides effective control guidelines even when this assumption is violated. As can be seen from Fig. 4, however, this decoupling approximation tends to degrade near points of voltage collapse, where second-order effects due to active power flows become crucial, and the predictions of the simplified decoupled model and the coupled active/reactive power flow model diverge (Supplementary Note 5). The key direction for future work is therefore the development of a more advanced analytic test which explicitly includes active power demands and does not require that the stiffness matrix be updated as phase-angle differences $\theta_i - \theta_j$ change. This should allow for the rigorous extension of our theoretical results to coupled active/reactive power flow. Another limitation of model (1) is the assumption that resistances between nodes in the network are negligible. While this assumption is quite reasonable in large high-voltage transmission networks, resistances, nonetheless, generate additional voltage drops, and losses may become sizable due to large current flows as the network becomes stressed. Extending the stability test (7) to lossy power flow models is therefore another key step towards an analytic understanding of power flow. These two extensions are under investigation, and if completed will translate the new theoretical framework presented here into a robust set of analysis and design tools for practical power grids. We expect that a generalized stiffness matrix similar to equation (6) will play a key role in these more general problem setups.

An area where these results may have a major impact is in contingency screening, where system operators computationally assess failure scenarios to determine if the grid remains stable. Due to the low computational overhead of evaluating analytic conditions such as our stability condition (7), further developments of the theory may allow for the fast assessment of many more contingencies than is currently feasible, or a single condition could be derived which guarantees the stability of the system under all contingencies within a certain class. Finally, we note that similar matrix techniques for incorporating network structure should prove relevant in other complex networked systems displaying polynomial nonlinearities, such as ecological population models, chemical reaction networks, and viral epidemic spreading.

Methods

Main result derivation. The key step in deriving equation (7) is recognizing the physical significance of the open-circuit voltages $\mathbf{V}_{\text{L}}^*$ in equation (5). Physically, V_i^* is the voltage one would measure at the ith node of the network when $Q_1 = Q_2 = \cdots = Q_n = 0$. The condition (7) was derived by reformulating the power flow (1) as a fixed-point equation of the form $\mathbf{x} = \mathbf{f}(\mathbf{x})$, where $x_i = (V_i - V_i^*)/V_i^*$ is a shifted and normalized voltage variable. With this notation, the power flow (1) takes the dimensionless form $\mathbf{x} = \mathbf{f}(\mathbf{x}) \triangleq -\frac{1}{4}\mathbf{Q}_{\text{crit}}^{-1}\text{diag}(\mathbf{Q}_{\text{L}}) \cdot \mathbf{r}(\mathbf{x})$, where $\mathbf{r}(\mathbf{x}) = (\frac{1}{1+x_1}, \ldots, \frac{1}{1+x_n})$. Imposing invariance of the set $\{x : |x_i| \leq \delta,\ i = 1,\ldots,n\}$ under the fixed-point map $\mathbf{f}(\mathbf{x})$ leads to condition (7). Existence and uniqueness of the equilibrium was shown by applying the contraction mapping theorem. Finally, stability was confirmed by showing that the Hessian matrix of the energy function is positive definite at the equilibrium (Supplementary Note 3 and 5).

Properties of stiffness matrix. In all publicly available test cases, the sub-matrix $\mathbf{B}_{\text{LL}}$ is a nonsingular Metzler matrix. It follows that its inverse has nonpositive elements[41], that the matrix $-\mathbf{B}_{\text{LL}}^{-1}\mathbf{B}_{\text{LG}}$ is nonnegative, and hence that the open-circuit voltages $\mathbf{V}_{\text{L}}^* = -\mathbf{B}_{\text{LL}}^{-1}\mathbf{B}_{\text{LG}}\mathbf{V}_{\text{G}}$ as defined in equation (5) are strictly positive. The stiffness matrix $\mathbf{Q}_{\text{crit}}$ used in the condition (7) inherits this Metzler property, and also posses an inverse with nonpositive elements. In particular, it holds that $(\mathbf{Q}_{\text{crit}}^{-1})_{ij} < 0$ with strictly inequality if and only if there exists a path in the network between load node i and load node j which does not intersect any generator node. Thus, reactive loading at node j influences the voltage at node i and vice versa, even if nodes i and j are not one-hop neighbours. When there are multiple groups of loads electrically isolated from one another by generators, the stability test (7) therefore decouples into an identical test for each group.

Numerical studies. Extensive details on the construction of our three numerical experiments may be found in the Supplementary Methods. All studies were implemented using the standard power flow solution techniques from the MATPOWER package[39].

References

1. Hiskens, I. A. & Davy, R. J. Exploring the power flow solution space boundary. *IEEE Trans. Power Syst.* **16,** 389–395 (2001).
2. Hiskens, I. A. Analysis tools for power systems—contending with nonlinearities. *Proc. IEEE* **83,** 1573–1587 (2002).
3. Brummitt, C. D., Hines, P. D. H., Dobson, I., Moore, C. & D'Souza, R. M. Transdisciplinary electric power grid science. *Proc. Natl Acad. Sci.* **110,** 12159 (2013).
4. Dobson, I. & Chiang, H. D. Towards a theory of voltage collapse in electric power systems. *Syst. Control Lett.* **13,** 253–262 (1989).
5. Dobson, I. Observations on the geometry of saddle node bifurcation and voltage collapse in electrical power systems. *IEEE Trans. Circuits Syst. I Fundam. Theory Appl.* **39,** 240–243 (1992).
6. Taylor, C. W. *Power System Voltage Stability* (McGraw-Hill, 1994).
7. Van Cutsem, T. & Vournas, C. *Voltage Stability of Electric Power Systems* (Springer, 1998).
8. Machowski, J., Bialek, J. W. & Bumby, J. R. *Power System Dynamics.* 2nd edn (John Wiley & Sons, 2008).
9. Ordacgi Filho, J. M. Brazilian blackout 2009: Blackout watch. Protection, Automation and Control World (2010). Available at: http://www.pacw.org/fileadmin/doc/MarchIssue2010/Brazilian_Blackout_march_2010.pdf (Accessed on 10 October 2015).
10. Van Cutsem, T. Voltage instability: phenomena, countermeasures, and analysis methods. *Proc. IEEE* **88,** 208–227 (2000).
11. Rohden, M., Sorge, A., Timme, M. & Witthaut, D. Self-organized synchronization in decentralized power grids. *Phys. Rev. Lett.* **109,** 064101 (2012).
12. Dobson, I. Complex networks: synchrony and your morning coffee. *Nat. Phys.* **9,** 133–134 (2013).
13. Motter, A. E., Myers, S. A., Anghel, M. & Nishikawa, T. Spontaneous synchrony in power-grid networks. *Nat. Phys.* **9,** 191–197 (2013).
14. Menck, P. J., Heitzig, J., Kurths, J. & Joachim Schellnhuber, H. How dead ends undermine power grid stability. *Nat. Commun.* **5,** 3969 (2014).
15. Rohden, M., Sorge, A., Witthaut, D. & Timme, M. Impact of network topology on synchrony of oscillatory power grids. *Chaos* **24,** 013123 (2014).
16. Dörfler, F., Chertkov, M. & Bullo, F. Synchronization in complex oscillator networks and smart grids. *Proc. Natl Acad. Sci.* **110,** 2005–2010 (2013).
17. Skardal, P. S. & Arenas, A. Control of coupled oscillator networks with application to microgrid technologies. *Sci. Adv.* **1,** e1500339 (2015).
18. Schultz, P., Heitzig, J. & Kurths, J. Detours around basin stability in power networks. *New J. Phys.* **16,** 125001 (2014).
19. Dewenter, T. & Alexander, K. H. Large-deviation properties of resilience of power grids. *New J. Phys.* **17,** 015005 (2015).
20. Nishikawa, T. & Motter, A. E. Comparative analysis of existing models for power-grid synchronization. *New J. Phys.* **17,** 015012 (2015).
21. Wu, F. & Kumagai, S. Steady-state security regions of power systems. *IEEE Trans. Circuits and Syst.* **29,** 703–711 (1982).
22. Kaye, R. & Wu, F. Analysis of linearized decoupled power flow approximations for steady-state security assessment. *IEEE Trans. Circuits Syst.* **31,** 623–636 (1984).
23. Molzahn, D. K., Lesieutre, B. C. & DeMarco, C. L. A sufficient condition for power flow insolvability with applications to voltage stability margins. *IEEE Trans. Power Syst.* **28,** 2592–2601 (2012).
24. Tamura, Y., Mori, H. & Iwamoto, S. Relationship between voltage instability and multiple load flow solutions in electric power systems. *IEEE Trans. Power App. Syst.* **102,** 1115–1125 (1983).

25. Greene, S., Dobson, I. & Alvarado, F. L. Sensitivity of the loading margin to voltage collapse with respect to arbitrary parameters. *IEEE Trans. Power Syst.* **12,** 262–272 (1997).
26. Schlueter, R. A., Costi, A. G., Sekerke, J. E. & Forgey, H. L. *Voltage stability and security assessment. Tech. Rep.* EPRI EL-5967; Project 1999-8; (Division of Engineering Research, Michigan State University; 1988).
27. Lof, P.-A., Andersson, G. & Hill, D. J. Voltage stability indices for stressed power systems. *IEEE Trans. Power Syst.* **8,** 326–335 (1993).
28. Chiang, H.-D. & Jean-Jumeau, R. Toward a practical performance index for predicting voltage collapse in electric power systems. *IEEE Trans. Power Syst.* **10,** 584–592 (1995).
29. Overbye, T. J. & Klump, R. P. Effective calculation of power system low-voltage solutions. *IEEE Trans. Power Syst.* **11,** 75–82 (1996).
30. Ajjarapu, V. & Christy, C. The continuation power flow: a tool for steady state voltage stability analysis. *IEEE Trans. Power Syst.* **7,** 416–423 (1992).
31. Overbye, T. J., Dobson, I. & DeMarco, C. L. Q-V curve interpretations of energy measures for voltage security. *IEEE Trans. Power Syst.* **9,** 331–340 (1994).
32. Cañizares, C. A. Calculating optimal system parameters to maximize the distance to saddle-node bifurcations. *IEEE Trans. Circuits Syst. I, Fundam. Theory Appl.* **45,** 225–237 (1998).
33. Cañizares, C. A. (ed.) *Voltage Stability Assessment: Concepts, Practices and Tools. Tech. Rep.* PES-TR9 (IEEE-PES Power System Stability Subcommittee, 2002).
34. Grijalva, S. & Sauer, P. W. A necessary condition for power flow Jacobian singularity based on branch complex flows. *IEEE Trans. Circuits Syst. I Fundam. Theory Appl.* **52,** 1406–1413 (2005).
35. Grijalva, S. Individual branch and path necessary conditions for saddle-node bifurcation voltage collapse. *IEEE Trans. Power Syst.* **27,** 12–19 (2012).
36. Thorp, J., Schulz, D. & Ilic′-Spong, M. Reactive power-voltage problem: conditions for the existence of solution and localized disturbance propagation. *Int. J. Electr. Power Energy Syst.* **8,** 66–74 (1986).
37. Bolognani, S. & Zampieri, S. On the existence and linear approximation of the power flow solution in power distribution networks. *IEEE Trans. Power Syst.* **31,** 163–172 (2015).
38. Dörfler, F. & Bullo, F. in *IEEE Power & Energy Society General Meeting* (Vancouver, 2013).
39. Zimmerman, R. D., Murillo-Sánchez, C. E. & Thomas, R. J. MATPOWER: Steady-state operations, planning, and analysis tools for power systems research and education. *IEEE Trans. Power Syst.* **26,** 12–19 (2011).
40. Capitanescu, F. & Van Cutsem, T. Unified sensitivity analysis of unstable or low voltages caused by load increases or contingencies. *IEEE Trans. Power Syst.* **20,** 321–329 (2005).
41. Berman, A. & Plemmons, R. J. *Nonnegative Matrices in the Mathematical Sciences* (SIAM, 1994).

Acknowledgements

This work was supported in part by the National Science Foundation NSF CNS-1135819, by ETH Zürich funds, the SNF Assistant Professor Energy Grant #160573, and by the National Science and Engineering Research Council of Canada.

Author contributions

Research design, theoretical results and numerics were performed by J.W.S.-P., with F.D. and F.B. supervising the project. All authors contributed to editing the manuscript.

Additional information

Competing financial interests: The authors declare no competing financial interests.

10

Trainable hardware for dynamical computing using error backpropagation through physical media

Michiel Hermans[1], Michaël Burm[2], Thomas Van Vaerenbergh[3], Joni Dambre[2] & Peter Bienstman[3]

Neural networks are currently implemented on digital Von Neumann machines, which do not fully leverage their intrinsic parallelism. We demonstrate how to use a novel class of reconfigurable dynamical systems for analogue information processing, mitigating this problem. Our generic hardware platform for dynamic, analogue computing consists of a reciprocal linear dynamical system with nonlinear feedback. Thanks to reciprocity, a ubiquitous property of many physical phenomena like the propagation of light and sound, the error backpropagation—a crucial step for tuning such systems towards a specific task—can happen in hardware. This can potentially speed up the optimization process significantly, offering important benefits for the scalability of neuro-inspired hardware. In this paper, we show, using one experimentally validated and one conceptual example, that such systems may provide a straightforward mechanism for constructing highly scalable, fully dynamical analogue computers.

[1]OPERA photonique, Université Libre de Bruxelles, Avenue F. Roosevelt 50, 1050 Brussels, Belgium. [2]ELIS Department, Ghent University, Sint Pietersnieuwstraat 41, 9000 Ghent, Belgium. [3]INTEC Department, Ghent University, Sint Pietersnieuwstraat 41, 9000 Ghent, Belgium. Correspondence and requests for materials should be addressed to M.H. (email: michiel.hermans@ulb.ac.be).

In a variety of forms, neural networks have seen an exponential rise in attention during the last decade. Neural networks trained with gradient descent are currently outperforming other, more classical approaches in a broad number of challenging tasks. One often-quoted result is their state-of-the-art performance in computer vision[1], an example of a problem that is considered easy for humans, and hard for conventional computer algorithms. In combination with a number of semi-heuristic methods, this result has been obtained using the backpropagation algorithm, a method that has been around since the 1960s.

One highly interesting set of neural architectures are so-called recurrent neural networks (RNN), that is, neural networks that have temporal feedback loops. This makes them highly suited for problems that have a natural time element, such as speech recognition[2] and character-based language modelling[3,4]. Conventional neural networks may be limited in solving such tasks, as they can only include a finite temporal context (usually a window of data of a fixed duration). RNNs, on the other hand, can—at least in principle—have indefinitely long memory. In addition, it is believed that temporal feedback loops are primary functional components of the human brain.

Currently, (recurrent) neural networks are most often implemented in software on digital devices, mostly for reasons of convenience and availability. At their core, however, these networks are analogue computers, and they come far closer to mimicking the workings of the brain than classic computation algorithms do. A computer made up of state-of-the-art components that matches the processing power of the human brain is estimated to consume about 0.5 GW of power[5], a full seven orders of magnitude more than the 20 W required by the brain itself. If we ever wish to achieve any degree of scalability for devices performing brain-like computation, we will need to embrace physical realizations of analogue computers, where the information is encoded in hardware by continuous physical variables, and where these data are processed by letting them interact through physical nonlinear dynamical systems.

One line of research that has been partially successful in accomplishing this goal is that of reservoir computing (RC)[6]. This paradigm, which combines several previous lines of research[7,8], essentially employs randomly constructed dynamical systems to process a time-varying signal. If the system is sufficiently varied, nonlinear and high dimensional, it acts as an efficient random feature generator, which expands the input signal into a high-dimensional space in which the time series-processing problem becomes much easier to solve. All that remains to be optimized is a a very small number of meta-parameters, and a linear mapping of these features to a desired output signal, which can be performed efficiently with any linear algebra solver.

The last decade of research into RC has shown a variety of interesting examples of physical implementations of analogue processors. It has been shown to work with water ripples[9], mechanical constructs[10,11], electro-optical devices[12,13], fully optical devices[14] and nanophotonic circuits[15,16]. Despite some remarkable successes, the RC concept still faces the problem of being ineffective for tasks that require a great deal of modelling power (for example, natural language processing, video stream processing and so on). The main reason is that any particular feature that needs to be extracted from the input data has to be present in the randomly constructed nonlinear feature space offered by the reservoir. When the input dimensionality grows, the probability of having the correct features present becomes extremely small. Even for moderately demanding tasks, this means that good performance requires a disproportionally large dimensionality of the system at hand, up to tens of thousands of state variables in practice[17]. By contrast, neural networks trained by the backpropagation algorithm can build the required nonlinear features internally during the training process, which makes them far more scalable.

In recent work, we have shown that it is possible to extend backpropagation (in particular the variant used to train RNNs, called backpropagation through time (BPTT)[18]) to models of physical dynamical systems[19]. We have shown that it can serve as an efficient automated tool to find highly non-trivial solutions for complex dynamical problems. In ref. 20, we have shown that backpropagation can be used on models of existing electro-optical reservoirs, offering a method to optimize their input encoding. So far, this work relies on simulation, however. The backpropagation algorithm operates on a computer model of the dynamical system. If we would use it to train physical analogue computers, we would face the same scaling issues that we encountered with systems that mimic the human brain.

In the following, we offer a definition of a set of specific physical dynamical systems that can act as analogue computing systems. Crucially, we show that—in contrast to the results presented in ref. 19—the backpropagation algorithm can be performed physically on such systems, with only minor additional hardware requirements and without the need for simulation. This approach significantly reduces the external computational demands of the optimization process, hence greatly speeding it up. First, this means that we can overcome the aforementioned limitations of the RC paradigm by building systems that are more optimized for a specific task. Second, it implies that we can break barriers in terms of scalability: an analogue physical computing set-up is inherently massively parallel, and its speed of operation does not directly depend on its complexity. Whereas training systems to perform complex, large-scale machine learning tasks currently take several days to complete, physical hardware as introduced in this paper may offer a way to reduce this to more manageable times.

Results

Theory. In this section, we will describe the general physical system we use to implement machine learning models, and we explain how their reciprocity allows to perform the training process in hardware. We start by introducing an abbreviated notation for a multivariate convolution. If $\mathbf{x}(t)$ is a multivariate signal and $\mathbf{W}(t)$ a matrix with time-varying elements defined for $t > 0$, we define the signal $\mathbf{y}(t)$ as the convolution of $\mathbf{x}(t)$ with $\mathbf{W}(t)$ as follows:

$$\mathbf{y}(t) = \int_0^\infty \mathrm{d}t' \, \mathbf{W}(t')\mathbf{x}(t-t') = [\mathbf{W}*\mathbf{x}](t). \quad (1)$$

We will consider systems as follows. We will assume that there are a set of N signal input sources $s_i(t)$ that excite the linear dynamical system (LDS) and a set of M output receivers that receive an output signal $o_i(t)$. We can write both the sets as a single source and receiver vector $\mathbf{s}(t)$ and $\mathbf{o}(t)$, respectively. The LDS will cause the following transformation between the source and receiver:

$$\mathbf{o}(t) = [\mathbf{W}_{\mathbf{so}}*\mathbf{s}](t), \quad (2)$$

where the impulse response, or first-order Volterra kernel $\mathbf{W}_{\mathbf{so}}(t)$ characterizes the transfer function of the system. Furthermore, we are able to use the LDS reversely, where the receivers now act as sources and the sources act as receivers. In this case, $\mathbf{s}'(t)$ acts as the input of the system and $\mathbf{o}'(t)$ represent the received signal at the places of the original sources. If the LDS is a reciprocal system, the following equation holds:

$$\mathbf{o}'(t) = \left[\mathbf{W}_{\mathbf{so}}^{\mathrm{T}} * \mathbf{s}'\right](t). \qquad (3)$$

This property is crucial, as it will allow us to perform error backpropagation (which we introduce later on) physically. Reciprocal systems are ubiquitous in physics. One important example of such a system, which we will use for the rest of the paper, is the propagation of waves through a linear medium. Suppose for instance we have a chamber in which we place a set of different speakers and a set of microphones, the signals the microphones receive would indeed be described by equation (2), where $\mathbf{W}_{\mathbf{so}}(t)$ would be determined by the shape of the room, the absorption of the walls, the air density and so on. If we then replace each speaker by a microphone and vice versa, the signal we received would be described by equation (3). Another set of examples are systems described by the linear heat (or diffusion) equation. Here the source would have a controllable temperature, and the receivers would be thermometers, where the medium performs an operation as described by the above equations.

Linear systems can only perform linear operations on their input signal. In order for the full system to be able to model nonlinear relationships, we add nonlinear feedback. We provide a set of $M_{\mathbf{a}}$ additional receivers and $N_{\mathbf{a}}$ sources. The signal that is detected at these new receivers gets sent through a nonlinear operator $\mathbf{f} : \mathbb{R}^{M_{\mathrm{a}}} \rightarrow \mathbb{R}^{N_{\mathrm{a}}}$ and is fed back into the system via the new sources. We denote the signal after the function as $\mathbf{a}(t)$. The impulse response matrix for the transition from the input sources to the receivers for the nonlinear feedback we denote as $\mathbf{W}_{\mathbf{sa}}(t)$, and those for the transition from the nonlinear feedback sources to the output receivers and nonlinear feedback receivers with $\mathbf{W}_{\mathbf{ao}}(t)$ and $\mathbf{W}_{\mathbf{aa}}(t)$, respectively. A schematic diagram of the full system is shown in the top of Fig. 1a. The system is described as follows:

$$\begin{aligned}\mathbf{a}(t) &= \mathbf{f}([\mathbf{W}_{\mathbf{sa}} * \mathbf{s}](t) + [\mathbf{W}_{\mathbf{aa}} * \mathbf{a}](t)) \\ \mathbf{o}(t) &= [\mathbf{W}_{\mathbf{so}} * \mathbf{s}](t) + [\mathbf{W}_{\mathbf{ao}} * \mathbf{a}](t).\end{aligned} \qquad (4)$$

As we argue in Supplementary Note 1, for specific choices of the impulse response matrices, these equations reduce to those of neural networks, including all kinds of deep networks, RNNs and so on. Neural networks are usually trained using gradient descent based on backpropagation. Therefore, if we wish to use this system as a trainable model for signal processing, we need to be able to calculate gradients for the parameters we can change (the impulse response matrices and the input and output encoding which we define later). First of all, we define a cost functional $C(\mathbf{o}(t))$, a functional of the whole history of $\mathbf{a}(t)$ in the interval $t \in \{0 \cdots T\}$ that we wish to minimize. This could, for instance, be the time integral of the squared difference between the actual and a desired output signal. Next we define the partial derivative of $C(\mathbf{o}(t))$ with respect to (w.r.t.) $\mathbf{o}(t)$ as $\mathbf{e}_{\mathbf{o}}(t)$. The error backpropagation process is then described by the following equations:

$$\begin{aligned}\mathbf{e}_{\mathbf{a}}(s) &= \mathbf{J}^{\mathrm{T}}(s)\left(\left[\mathbf{W}_{\mathbf{ao}}^{\mathrm{T}} * \mathbf{e}_{\mathbf{o}}\right](s) + \left[\mathbf{W}_{\mathbf{aa}}^{\mathrm{T}} * \mathbf{e}_{\mathbf{a}}\right](s)\right) \\ \mathbf{e}_{\mathbf{s}}(s) &= \left[\mathbf{W}_{\mathbf{so}}^{\mathrm{T}} * \mathbf{e}_{\mathbf{o}}\right](s) + \left[\mathbf{W}_{\mathbf{sa}}^{\mathrm{T}} * \mathbf{e}_{\mathbf{a}}\right](s).\end{aligned}$$

Here $s = T - t$, that is, the equations run backwards in time and $\mathbf{J}(t)$ is the Jacobian of $\mathbf{f}$ w.r.t. its argument. From the variables $\mathbf{e}_{\mathbf{a}}(t)$ and $\mathbf{e}_{\mathbf{s}}(t)$, gradients w.r.t. all impulse response matrices within the system, and w.r.t. $\mathbf{s}(t)$ can be found, which in turn can be used for optimization (see Supplementary Note 2 and Supplementary Fig. 1 for the derivations and for the mathematical interpretation of $\mathbf{e}_{\mathbf{a}}(t)$ and $\mathbf{e}_{\mathbf{s}}(t)$). The crucial property of these equations is that, thanks to the reciprocity of the LDS, they too can be performed physically on the same system. First of all, we consider the time s as physical time running forwards (which in practice means that we need to time-reverse the external input signals $\mathbf{e}_{\mathbf{o}}(t)$ and $\mathbf{J}(t)$). If we then switch the positions of the sources and receivers (leading to transposed impulse response matrices), and instead of providing nonlinear feedback, we modulate the feedback with $\mathbf{J}^{\mathrm{T}}(s)$, we have physically implemented equation (5), and hence can record $\mathbf{e}_{\mathbf{a}}(s)$ and $\mathbf{e}_{\mathbf{s}}(s)$ directly from the system. This principle is depicted in the bottom of Fig. 1a. In summary, the requirements for our system are a reciprocal LDS, a physical implementation of the nonlinear feedback and a physical implementation of time-varying modulation with $\mathbf{J}(s)$.

In many cases, it is difficult to send signals to multiple sources and to record signals from multiple receivers due to the cost and

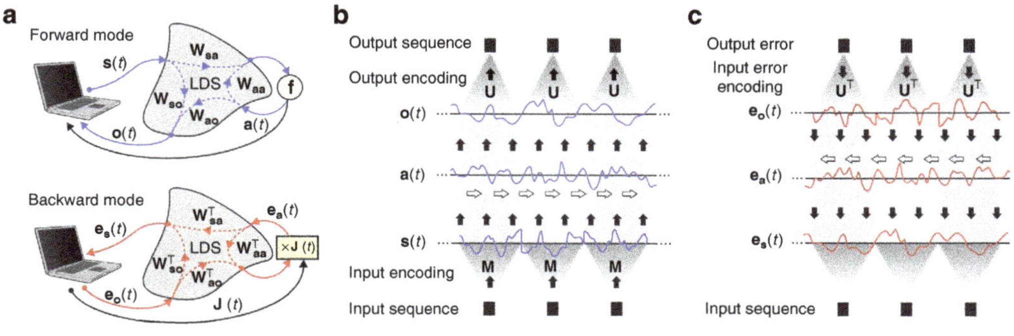

Figure 1 | Schematic of physically implemented backpropagation. (**a**) Illustration of the most general set-up of the physical neural network studied in this paper. The top diagram shows how the signals propagate through the system and the nonlinear feedback with blue arrows during the forward pass. The filter operations are depicted with dashed lines running through the LDS, which is depicted as a grey blob. The bottom diagram shows the error backpropagation phase, where the signal runs backwards through all functional dependencies. Here filter operations run in the opposite direction such that they are represented by the transpose of their impulse response matrices. Note that the computer is not in the loop during the forward or backward pass, but only serves to send out a predefined signal, and to record at the same time. (**b**) Depiction of the masking principle in the forward direction. At the bottom, we see three consecutive instances of an input time series. Each of these is converted into a finite time segment through the masking signals $\mathbf{M}(t)$. These segments are next concatenated in time and serve as the input signal $\mathbf{s}(t)$ for the dynamical system (where time runs according to the white arrows), and which in turn generates an output signal $\mathbf{o}(t)$. The output signal $\mathbf{o}(t)$ is divided into finite length pieces, which are decoded into output instances of an output time series using the output masks $\mathbf{U}(t)$. (**c**) The backpropagation process happens in a completely similar manner as in the forward direction. This time, the transpose of the output masks serve as the encoding masks. Finally, the input error signal $\mathbf{e}_{\mathbf{s}}(t)$ is also segmented in time before it is used to determine the gradients w.r.t. $\mathbf{M}(t)$.

complexity associated with the necessary hardware. Yet a high-dimensional state space is desirable as it increases the complexity and therefore the representational power of the system. To increase the effective dimensionality of the system, we use an input–output multiplexing scheme that was first introduced in ref. 21. This increases the effective dimensionality of the system and its parameters can also be optimized using the physical backpropagation method. Suppose we have an input time series $\mathbf{x}_i$ that we wish to map onto an output time series $\mathbf{y}_i$. First of all, we define an encoding that transforms the vector $\mathbf{x}_i$, the i-th instance of the input data sequence, into a continuous time signal segment $\mathbf{s}_i(t)$:

$$\mathbf{s}_i(t) = \mathbf{s}_b(t) + \mathbf{M}(t)\mathbf{x}_i \quad for \quad t \in [0 \cdots P], \tag{6}$$

where P is the masking period and $\mathbf{s}_b(t)$ is a bias time trace. The matrix $\mathbf{M}(t)$ contains the so-called input masks, defined for a finite time interval of duration P. The input signal $\mathbf{s}(t)$ is now simply the time concatenation of the finite time segments $\mathbf{s}_i(t)$.

The output decoding works in a very similar fashion. If there is a time series $\mathbf{y}_i$ that represents the output associated with the i-th instance of the input time series, we can define an output mask $\mathbf{U}(t)$. We divide the time trace of the system output $\mathbf{o}(t)$ into segments $\mathbf{o}_i(t)$ of duration P. The i-th network output instance is then defined as

$$\mathbf{y}_i = \mathbf{y}_b + \int_0^P dt\, \mathbf{U}(t)\mathbf{o}_i(t), \tag{7}$$

with $\mathbf{y}_b$ a bias vector. Effectively, the whole time trace of $\mathbf{o}_i(t)$ now serves as the immediate state vector of the system, drastically increasing its effective dimensionality.

The process described here is essentially a form of time multiplexing, and is depicted in Fig. 1b. The backpropagation phase happens in a similar fashion. Suppose we have a time series with instances $\mathbf{e}_i$, which are the gradient of the chosen cost function w.r.t. the output instances $\mathbf{y}_i$. Completely equivalent to the input masking, we can now define the error signal $\mathbf{e}_\mathbf{o}(t)$ as a time concatenation of finite time segments $\mathbf{e}_\mathbf{o}^i(t)$:

$$\mathbf{e}_\mathbf{o}^i(t) = \mathbf{U}^\mathrm{T}(t)\mathbf{e}_i. \tag{8}$$

Using this signal as an input to the system during backpropagation will provide us with an error $\mathbf{e}_\mathbf{s}(t)$, which in turn can be used to determine the gradient for the masking signals $\mathbf{M}(t)$ as depicted in Fig. 1c (for the derivations, see Supplementary Note 3). Note that the signal encoding and decoding happens on an external computer as well, and not physically. These operations can be fully parallelized, however, such that we do not lose the advantages of the processing power of the system. In ref. 20, we showed in simulation that optimizing input masks using BPTT on a computer model of an electro-optical set-up as used in ref. 12 had significant benefits. As we will show experimentally, physical backpropagation allows us to largely omit the modelling and simulation, and allow for a more direct optimization.

The full training process works as follows. First, we sample a representative amount of data, encode it to a continuous time input signal, run it through the physical system and record the output signals. Next, we compute the cost corresponding to the output and construct the signal $\mathbf{e}_\mathbf{o}(t)$ (on the computer), which we then also run through the system physically, and record the signal $\mathbf{e}_\mathbf{s}(t)$ (and $\mathbf{e}_\mathbf{a}(t)$ if applicable). Once this is completed, we can compute gradients w.r.t. the relevant parameters on the computer. We then subtract these gradients from the parameters, multiplied with a (small) learning rate. This process is repeated for many iterations until we reach satisfactory performance. All the operations that take place on the computer can be performed fully in parallel, and hence do not significantly slow down the training process.

A real-life acoustic set-up. We have tested the principles described above by building a system that uses the propagation of acoustic waves as an LDS. To reduce the complexity of the set-up, we work with only one signal source (a small computer speaker) and one receiver (a voice microphone). The sound enters a 6-m long plastic tube via a paper funnel, and the microphone receives the signal at the other end of the tube. The tube delays the signal and introduces reflections and resonance frequencies. Note that, due to the fact that the signal in this case is scalar, all impulse response matrices determining the system are scalar too, which means that they are equal to their transpose. This has the advantage that we do not need to switch the speaker and microphone between the forward and backward phase.

The received signal is electronically truncated at 0 V (such that only positive voltages can pass), implementing a so-called linear rectifier function, which acts as the system nonlinearity **f**. The linear rectifier function is currently a popular activation function in neural architectures[22]. Feedback is implemented by adding this signal to the external input signal. One important advantage of the linear rectifier function is that its derivative is a binary signal: equal to one when the signal is transmitted, and equal to zero when it is cutoff. This means that multiplication with the Jacobian is equivalent to either transmitting the feedback signal unchanged, or setting it to zero, which can be easily implemented using an analogue switch. For a more detailed explanation of the relation between the acoustic system and the general case described by equation (4), we refer to the methods section. Note that other types of nonlinearity such as, for example, a sigmoid and its corresponding derivative, can also be implemented in analogue hardware (see for instance ref. 23).

We have used the physical backpropagation set-up to train input and output masks. Note that in principle it could also be used to optimize properties of the acoustic set-up itself, but we have omitted this for reasons of experimental simplicity. We have tested the set-up on an academic task that combines the need for nonlinearity and memory. The input time series q_i is scalar and consists of a series of i.i.d. integers from the set {0,1,2}, which are encoded into an acoustic signal as described above. The desired output time series y_i is defined as

$$y_i = q(i - q(i)), \tag{9}$$

that is, the task consists of retrieving the input with a delay that depends on the current input. The fact that the delay is variable makes that the task is nonlinear, that is, it cannot be solved by any linear filtering operation on the input.

For details concerning the experiments, we refer to the Methods section. A schematic depiction of the set-up and the main results of the experiments are shown in Fig. 2. In Fig. 2e, we show a comparison between the system output and the target, indicating that the system has learned to solve the task successfully. We show the evolution of the normalized root mean square error (NRMSE) during the training process in Fig. 2b. To make sure that the physical backpropagation works as intended, we have run two additional tests in which we trained either only the output masks (the RC approach, which does not require backpropagation) or only the input masks, keeping the other random and fixed. As can be seen in Fig. 2b, training only output or input masks in both cases reduces performance. Note that if the input masks are trained while the output masks are kept fixed and random, all adaptations to the system parameters are exclusively due to the error signal that has been propagated through the system in the form of sound. On top of this, the input mask training needs to find a solution that works with a

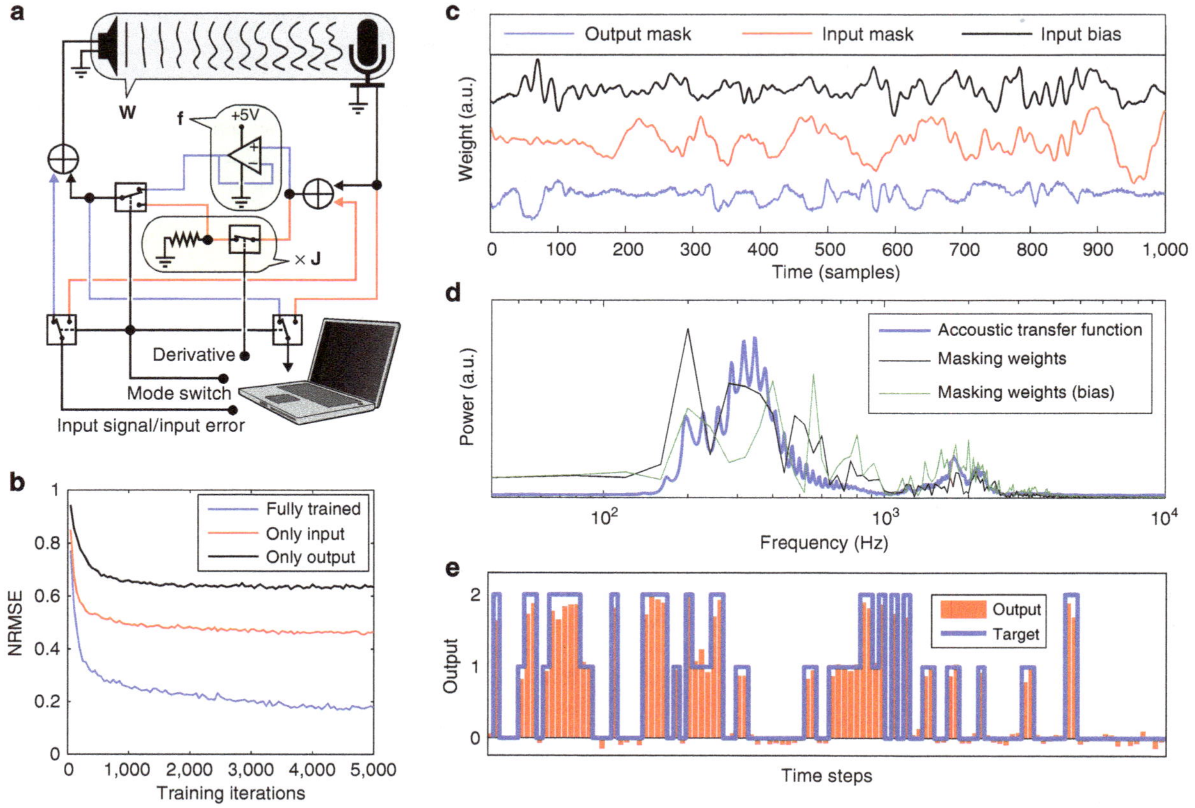

Figure 2 | Results of the acoustic backpropagation experiment. (**a**) Schematic depiction of the electric circuit of the experiment. The grey box at the top represents the speaker, tube and microphone. The blue circuit lines are only in use during the forward propagation and the red lines are in use during the backpropagation phase. The forward or backward mode can be toggled by a logic signal that comes from the PC, and which controls three analogue switches in the circuit. In the forward mode, the nonlinear feedback is implemented by an op-amp voltage follower which cuts off the signal at 0 V (green box), implementing a linear rectifier function. In the backprop mode, the multiplication with the Jacobian is implemented by a fast analogue switch that either outputs zero, or transmits the signal. (**b**) Example of the normalized root mean square error (NRMSE) as a function of the number of training iterations for three cases. That where both the output and input masks are trained, and those where only one of each is trained. (**c**) Example of resulting input, bias and output masks after training, corresponding to $\mathbf{M}(t)$, $\mathbf{s}_b(t)$ (equation (6)) and $\mathbf{U}(t)$ (equation (7)), respectively. (**d**) Example power spectrums of the input masks and the power spectrum of the transmission of the speaker-tube-microphone system. (**e**) Example of the network output versus the target.

completely random instantiation of the output mask. The fact that it can achieve this at least to some degree (NRMSE $\approx$ 0.47) demonstrates that physical BPTT works as intended.

Note that the result for fixed input masks and trained output masks is not meant to be interpreted as a fair comparison with the RC method, but rather as a naive baseline. Indeed, it is likely that the results for the untrained input masks could be improved by, for example, constructing the random input mask such that it has a spectrum matching the transmission of the acoustic system.

The input and output masks after training are shown in Fig. 2c, showing their temporal structure. In Fig. 2d, we show the power spectra of the input masks (and consequently the power spectra of the signals sent into the system) compared with the power spectrum of the system transmission (measured as the squared ratio in amplitude between the output voltage in the microphone and the input voltage in the speaker). Clearly, the acoustic parts of the full system (speaker–tube–microphone) only transmit certain frequency bands (a.o. the resonance frequencies of the tube are visible as a set of peaks.) and the input masks seem to have learned to match this spectrum. Note that at no point during training or testing we ever required a model of the acoustic part of the system. The impulse response matrices of the system do not need to be known for the backpropagation to work. Also important to mention is that no process on the external computer or in the electric circuitry provides the required memory for the system to process the time series. Instead, this happens due to the physical memory inherent to the acoustic system and to the nonlinear feedback, which the training process learns to exploit. This means that at any point in time, information about past inputs exists solely as acoustic waves travelling through the tube.

A conceptual electro-optical implementation. The described acoustic implementation can be extended to a larger and faster system relatively easily. For instance, one could use ultrasound for higher data transmission (combined with a medium with a high speed of sound). Here one could use piezoelectric transducers that can both emit and receive signals. Sound propagation is in that sense an interesting platform for a physical analogue neural network. The most attractive medium, however, but also more technologically challenging, would be light. Light can transport information at a very high speed, and unlike sound waves, it can be easily guided through fibre optics and integrated photonic waveguides. Just like sound waves, light transmission is reciprocal, making it possible to perform error backpropagation physically on the system. Indeed, using light as a medium for neuro-inspired processing has been studied extensively in the past[24,25]. These examples primarily exploit parallel processing that happens when light travels through a (often holographic) medium.

In our case, we would like to exploit not only parallelism, but also the time delays that are inherent to travelling light.

As a proof of concept, we propose a circuit to perform physical backpropagation electro-optically, partially inspired by the systems described in refs 12,13. Delays are physically implemented by means of long optical fibres. For this example, we wish not just to train the input masks, butto control the way in which the signals are mixed as well. Concretely, if we have an N-dimensional state $\mathbf{a}(t)$, we wish to optically implement a mixing matrix $\mathbf{W}$ of size $N \times N$, such that the mixing matrix $\mathbf{W}_{\mathbf{aa}}(t) = \delta(t - D)\mathbf{W}$, where D is the delay introduced by the optical fibres. Set-ups for computing matrix–vector products optically have been experimentally demonstrated in the past[26,27], and here we will assume that it is possible to perform them all-optically (see methods). Note that we do not need very high parameter precision for neural network applications, as the detection of the signal will be inherently noisy.

Similar to the acoustic example, we conceive an electro-optical 'neuron' (see Fig. 3a), which sends and receives optical signals, and either applies a nonlinear function or multiplies with the Jacobian. We encode the state $\mathbf{a}(t)$ as light intensity. Each neuron will have a fixed-power laser source, which can be modulated between the minimal and maximal value. The nonlinear function can be simply implemented by electronically truncating the feedback signal in a range corresponding to a minimum and maximum intensity levels of the laser, conveniently making full use of the signal range. Note that such a behaviour can be implemented relatively easily in high speed electronics, as it is an inherent property of amplifiers used in optical telecommunication[28]. The Jacobian of such a function is again a simple binary function. Finally, we can send the light into the optical circuit either in the forward or backward direction by using a 2×2 optical switch.

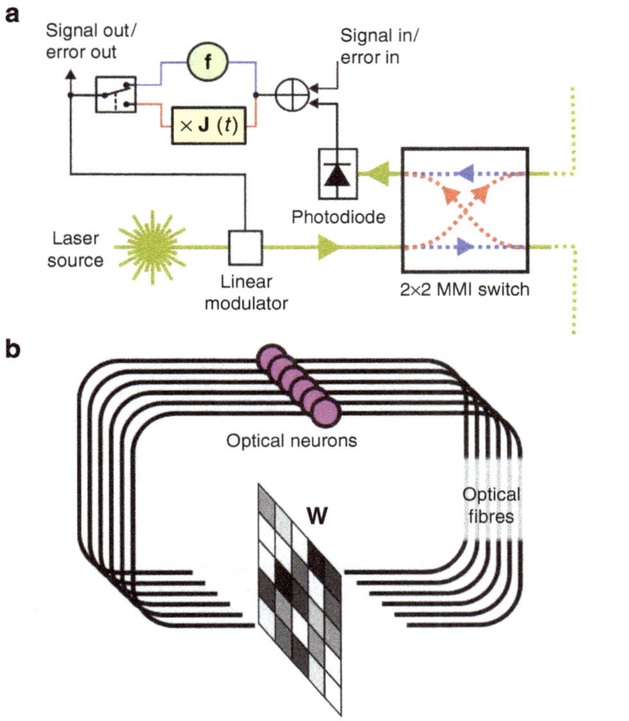

Figure 3 | The electro-optical set-up. (**a**) Schematic depiction of an electro-optical neuron, where pathways that are exclusive in the forward or backward mode are depicted in blue or red, respectively. It is largely similar to the electronic circuit in the acoustic set-up, with the difference that the input now enters the system before the nonlinearity. The 2×2 switch allows light to travel either forwards or backwards through the fibre network (fibres depicted by yellow lines). (**b**) Depiction of a network of electro-optical neurons, each purple circle represents a neuron. They send their output signals through optical fibres (which also incorporate delay) to an optical matrix-vector multiplier which multiplies with a matrix $\mathbf{W}$ in the forward direction and $\mathbf{W}^{\mathrm{T}}$ in the backwards direction. The elements of $\mathbf{W}$ can be set electronically, and are adapted in each training iteration. Note that each neuron also has an incoming and outgoing connection to external hardware that sends input and records output (not depicted).

In the final set-up, we simulate 20 electro-optical nodes. We add different levels of noise to the measured intensity. We applied the simulated version of this system on a realistic phoneme recognition task, which is part of speech processing and hence a typical example of a problem that can be solved using RNNs. We used the often-used TIMIT data set, which is a speech corpus in which the phonemes are directly labelled on the speech signal itself. As an error measure, we use the frame error rate, which is the fraction of frames (time steps) in the input signal that have been mislabeled. More details on the task can be found in the Methods section.

We ran a number of simulations that range from ideally (mathematically correct) implemented backpropagation to simulations which include unavoidable non-ideal behaviour for a real physical set-up. In particular, we included nonlinearity in the backpropagation phase, as well as measurement noise and the case in which the modulation with the Jacobian is only an approximation of the true Jacobian (see Supplementary Fig. 2). Details and the results of these experiments can be found in Supplementary Note 4 and Supplementary Table 1, respectively, as well as a discussion on the expected speed of the conceptual electro-optical set-up. What we found was a remarkable robustness against non-ideality for the task at hand, only increasing the test error to a limited degree compared with the ideal scenario. We obtain a test frame error rate of $\sim$30%. Comparing this with the other results in literature, we find that we perform in the same ballpark as some other established machine learning techniques (see for instance an overview of results on frame error rate in ref. 29, where the results range from 25 to 39%). When we compare this with the RC paradigm, similar results are only obtained using extremely large reservoirs, with up to 20,000 nodes[30].

Discussion

In this paper, we have proposed a framework for using reciprocal physical dynamical systems with nonlinear feedback as analogue RNNs. We have demonstrated that the error backpropagation algorithm, which efficiently optimizes an RNN, can be implemented physically on the same system, thus greatly reducing the necessary computations required for the optimization process. This in turn paves the way to faster, more scalable analogue computing. We have experimentally verified the proposed system using a real-world acoustic set-up as a proof of concept, as well as a simulated electro-optical set-up. In this second, more complex set-up, we explored the impact of expected sources of non-ideal behaviour on the performance of a real-world task and where we demonstrated that good performance does not require a very precise physical implementation of the backpropagation algorithm, and can tolerate reasonable levels of noise and nonlinearity. We have also included a short discussion on what processing speed may be obtained for an electro-optical set-up in the Supplementary Discussion.

The concepts presented in this paper provide a sizeable step forward towards a novel form of processing, which relies far less on software implementations and comes much closer to brain-like computation, especially in the sense that the system can partially internalize its own training procedure.

By using analogue physical processes to compute, we may benefit from inherent massively parallel computing, great potential speed benefits and low-power usage, the properties that were the initial motivation for research into physical RC systems. Specifically, the obtainable speed does not depend on the dimensionality (number of sources and receivers) of the system, offering inherent scalability. The possibility to fully optimize all internal system parameters using the backpropagation algorithm offers great performance improvements, and makes the application domain of the proposed set of systems far greater (as has been evidenced in ref. 20). This also means that such physical set-ups can potentially become competitive with digitally implemented neural networks, which are currently the state-of-the-art for several important signal processing problems. The training process too benefits from being implemented physically, meaning that there is only limited need for external processing. If the physical system under consideration has speed benefits compared with a digitally implemented neural architecture, these benefits are also present for training the system. Finally, while physically implementing the training process comes at an additional complexity cost (modulation of the feedback with the jacobian), the benefits over optimization of the parameters in simulation is paramount. Optimization in simulation might require a very precise model, and it is hard to predict how model-reality discrepancies would manifest themselves once the parameters obtained in simulation are applied to the physical set-up. When instead optimizing on the physical system there can be imperfections like the ones we explored in Supplementary Note 4 but one is certain that the measurements used to base the training on are those from the real system, and no additional system characterization is needed to perform the training.

Important challenges for the large-scale execution of this scheme still remain. The current set-up still requires some level of external processing for computing the gradients. If the analogue part of the system is sufficiently fast, gradient computations may become the bottleneck of training, though this may be partially redeemed by the fact that they are solely matrix–matrix multiplications (without a sequential part), which means that it is fully parallelizable.

Recording the signals in the forward and backward pass still requires digitization. Currently, this is the most important hurdle to scaling up the system in practice, as analogue-digital conversion at high speeds is expensive and consumes a lot of power. This limits the number of sources and receivers that can be practically applied. Time multiplexing of the input, as used in both examples of this paper, partially solves this problem, but at the cost of reducing the obtainable speed of the system. Larger numbers of sources and receivers would also benefit more from the spatial parallelism that is offered by acoustic or optic systems.

With currently available hardware, one could potentially build physical systems that are competitive with digitally implemented neural networks, as we demonstrated using a simulated electro-optical set-up. Truly exploiting the full potential of analogue physical computation, however, very likely requires the design of novel hardware that internalizes all necessary elements into a single device. In particular, future research into this topic should explore ways to develop hardware which has impulse responses determined by large amounts of controllable parameters. This would increase the number of trainable parameters and hence the representational power of the systems. Finally, it is also of importance to relate the results found in this paper to developments in neural network research. For one, it was recently found[31] that random feedback weights for the backpropagation phase can also be used to train feedforward networks. It should be investigated if this has implications for the recurrent systems under considerations in this paper.

Methods

Acoustic experiments. For the acoustic set-up, we used a data acquisition card, which samples the signal at 40 kHz (well above the maximum frequency that is still passed through speaker–tube–microphone system). Input and output masks consist of 1,000 samples, which means that time series are processed at a rate of 40 time steps s^{-1}.

We train the input and output masks over the course of 5,000 iterations. Initial values for the input and output masks are picked by independent, identically distributed sampling from a normal distribution with zero mean and variance equal to 0.2 and 0.1, respectively. Each training batch consists of a newly generated time series of 100 instances. This means that each training iteration (forward and backward pass) takes ~5 s, and the complete training takes about 7 h. We found little to no significant variation in performance for different random initializations of the input and output masks.

As absolute scaling of the error signal does not matter for the backpropagation phase, we always normalize and rescale the error signal before we use it as the input. This ensures that the signal always remains well above the noise levels. Note that this step causes us to lose the absolute magnitude of the gradients. For parameter updates, we therefore normalize the obtained gradients before using them. The learning rate is set to 0.25 at the start of the experiment, and then linearly decays to zero to ensure convergence at the end.

Mathematically, the full system can be described as follows: let $s(t)$ be the input signal after the encoding and $a(t)$ the (scalar) state of the system at time t, then

$$a(t) = f([W*(s+a)](t)). \tag{10}$$

Here $f(x) = max(0,x)$ is the linear rectifier function and $\mathbf{W}(t)$ is the scalar impulse response of the speaker–tube–microphone system, that is, it is the signal that would be received by the microphone for a Dirac delta voltage impulse for the speaker. The output of the system is also the state $a(t)$. This means that we can relate the acoustic system to the general case of equation (4) if $\mathbf{W}_{sa}(t) = \mathbf{W}_{aa}(t) = W(t)$, $\mathbf{W}_{ao}(t) = \delta(t)$ (the Dirac delta function) and $\mathbf{W}_{so}(t) = 0$.

Conceptual photonic set-up. The system is described by the following equation:

$$\mathbf{a}(t) = f(\mathbf{W}\mathbf{a}(t-D) + \mathbf{s}(t))), \tag{11}$$

where $\mathbf{W}$ is the matrix, which is implemented optically (see below). The function $f(x)$ truncates the signal between minimal and maximal intensity (which we define as -1 and 1, respectively):

$$f(x) = \begin{cases} -1 & \text{if } x \leq -1 \\ x & \text{if } -1 < x < 1 \\ 1 & \text{if } x \geq 1. \end{cases} \tag{12}$$

For what follows, we assume that all the nodes' laser sources have different wavelengths, such that light intensities add up linearly. Note that this is feasible within the presented set-up: optical 2×2 switches can have bandwidths that exceed 100 nm (ref. 32, whereas laser bandwidths are usually of the order of 1 nm or less. If we assume laser light with a wavelength of $\approx 1\,\mu m$, and we assume that the different wavelengths differ as little as 1 nm between the nodes, the slowest resulting fluctuations in intensity from adding up these signals are of the order of 100 GHz (derived form the difference in their frequencies), 100 times faster than the assumed speed at which we measure the signal. An alternative approach would be to use light sources that are inherently broadband (for example, not lasers but light-emitting diodes, which have a bandwidth of 50–100 nm).

Several ways to implement optical matrix–vector multiplication have been discussed in literature. One possible method would be to encode the vector $\mathbf{a}$ as light intensities. We can let the light pass through a spatial array of tuneable intensity modulators and focus the light on the other side to perform a matrix–vector product. Conversely, if the light comes from the other direction, the signal is effectively multiplied with the transpose of the matrix. Directly implementing the whole concept using intensities to encode the signal would have the important limitation that all elements of $\mathbf{W}$ would be positive, as light intensities can only be added up. This is a strong drawback, as this would mean that the system can only provide weighted averages of the individual states and cannot use and enhance differences between them. Therefore, we envision a different approach where each element of $\mathbf{a}$ is encoded by two light signals with intensities: $\mathbf{k}+\mathbf{a}$ and $\mathbf{k}-\mathbf{a}$, with $\mathbf{k}$ a vector with all elements equal to one. This ensures that the intensities fall into a positive range between 0 and 2, which can correspond to the minimum and maximum output intensity level of each neuron. Now, these constituents are sent to two separate arrays of intensity modulators $\mathbf{W}_1 = \mathbf{K} + \mathbf{W}/2$ and $\mathbf{W}_2 = \mathbf{K} - \mathbf{W}/2$, where $\mathbf{K}$ has all elements equal to one. As all elements of $\mathbf{W}_1$ and $\mathbf{W}_2$ need to be positive, this means that the range of the elements of $\mathbf{W}$ fall within the range -2 to 2. If we combine the light signals after the intensity modulation and hence add up the intensities we get:

$$(\mathbf{W}_1 + \mathbf{W}_2)\mathbf{k} + (\mathbf{W}_1 - \mathbf{W}_2)\mathbf{a} = \mathbf{K}\mathbf{k} + \mathbf{W}\mathbf{a}. \tag{13}$$

The first term will simply introduce a constant bias value, which we can remove electronically after measuring. The second term now contains the matrix–vector product where the elements of **W** can be both positive and negative.

For the simulations, we use a piecewise constant input signal with a fixed sample period (equal to the sample period of the measurement). We again used the masking scheme, where each masking period consisted of 50 sample periods. We chose the delay D at 51 sample periods and used a network of 20 optical neurons.

The TIMIT data set[33] consists of labelled speech. It has a well-defined training and test set, which makes comparison with results in literature possible. Each time step (frame) needs to be labelled with one out of 39 possible phonemes. The input signal consists of a 39-dimensional time series (the fact that it has the same dimensionality as the output signal is coincidental), which encodes the speech signal using MFCCs. For more details on how the data is encoded, please check, for example, ref. 30.

We trained for 50,000 iterations. In each iteration, we randomly sampled 200 sequences of 50 frames from the training set. We again normalized the gradients. Parameter updates were performed with a learning rate starting at 1, which linearly dropped to zero over the course of the training.

References

1. Krizhevsky, A., Sutskever, I. & Hinton, G. Imagenet classification with deep convolutional neural networks. *Adv. Neural Inf. Process. Syst.* **25,** 1106–1114 (2012).
2. Graves, A., Mohamed, A.-r. & Hinton, G. in *IEEE International Conference on Acoustic Speech and Signal Processing (ICASSP 2013)* 6645–6649 (Vancouver, Canada, 2013).
3. Sutskever, I., Martens, J. & Hinton, G. in *Proc. 28th International Conference on Machine Learning* 1017–1024 (Seattle, WA, USA, 2011).
4. Hermans, M. & Schrauwen, B. Training and analysing deep recurrent neural networks. in *Adv. Neural Inf. Process. Syst.* **26,** 190–198 (2013).
5. Kogge, P. The tops in flops. *Spectrum, IEEE* **48,** 48–54 (2011).
6. Lukosevicius, M. & Jäger, H. Reservoir computing approaches to recurrent neural network training. *Comp. Sci. Rev.* **3,** 127–149 (2009).
7. Jaeger, H. & Haas, H. Harnessing nonlinearity: predicting chaotic systems and saving energy in wireless telecommunication. *Science* **308,** 78–80 (2004).
8. Maass, W., Natschläger, T. & Markram, H. Real-time computing without stable states: a new framework for neural computation based on perturbations. *Neural Comput.* **14,** 2531–2560 (2002).
9. Fernando, C. & Sojakka, S. in *Proc. 7th European Conference on Artificial Life* (eds Banzhaf, W., Christaller, T., Dittrich, P., Kim, J. T. & Ziegler, J.) 588–597 (Springer, 2003).
10. Caluwaerts, K., D'Haene, M., Verstraeten, D. & Schrauwen, B. Locomotion without a brain: physical reservoir computing in tensegrity structures. *Artif. Life* **19,** 35–66 (2013).
11. Hauser, H., Ijspeert, A. J., Füchslin, R. M., Pfeifer, R. & Maass, W. Towards a theoretical foundation for morphological computation with compliant bodies. *Biol. Cybern.* **105,** 355–370 (2011).
12. Larger, L. *et al.* Photonic information processing beyond turing: an opto-electronic implementation of reservoir computing. *Opt. Express* **3,** 20 (2012).
13. Paquot, Y. *et al.* Optoelectronic reservoir computing. *Sci. Rep.* **2,** 1–6 (2012).
14. Brunner, D., Soriano, M. C., Mirasso, C. R. & Fischer, I. Parallel photonic information processing at gigabyte per second data rates using transient states. *Nat. Commun.* **4,** 1364 (2013).
15. Vandoorne, K. *et al.* Toward optical signal processing using photonic reservoir computing. *Opt. Express* **16,** 11182–11192 (2008).
16. Vandoorne, K. *et al.* Experimental demonstration of reservoir computing on a silicon photonics chip. *Nat. Commun.* **5,** 3541 (2014).
17. Triefenbach, F., Demuynck, K. & Martens, J.-P. Large vocabulary continuous speech recognition with reservoir-based acoustic models. *IEEE Signal Proc. Lett.* **21,** 311–315 (2014).
18. Rumelhart, D., Hinton, G. & Williams, R. *Learning Internal Representations by Error Propagation* (MIT Press, 1986).
19. Hermans, M., Schrauwen, B., Bienstman, P. & Dambre, J. Automated design of complex dynamic systems. *PLoS ONE* **9,** e86696 (2014).
20. Hermans, M., Dambre, J. & Bienstman, P. Optoelectronic systems trained with backpropagation through time. *IEEE Trans. Neural Netw. Learn. Syst.* (2014).
21. Appeltant, L. *et al.* Information processing using a single dynamical node as complex system. *Nat. Commun.* **2,** 468 (2011).
22. Glorot, X., Bordes, A. & Bengio, Y. in *Proc. 14th International Conference on Artificial Intelligence and Statistics. JMLR W&CP Volume* **vol. 15,** 315–323 (2011).
23. Shi, B. & Lu, C. Generator of neuron transfer function and its derivative. US Patent 6,429,699 (2002).
24. Caulfield, H. J., Kinser, J. & Rogers, S. K. Optical neural networks. *Proc. IEEE* **77,** 1573–1583 (1989).
25. Yu, F. T. & Jutamulia, S. *Optical Signal Processing, Computing, and Neural Networks* (John Wiley & Sons Inc., 1992).
26. Goodman, J. W., Dias, A. & Woody, L. Fully parallel, high-speed incoherent optical method for performing discrete fourier transforms. *Opt. Lett.* **2,** 1–3 (1978).
27. Yeh, P. & Chiou, A. E. Optical matrix-vector multiplication through four-wave mixing in photorefractive media. *Opt. Lett.* **12,** 138–140 (1987).
28. Säckinger, E. *Broadband CIrcuits for Optical Fiber Communication* (John Wiley & Sons, 2005).
29. Keshet, J., McAllester, D. & Hazan, T. in *International Conference on Acoustics, Speech and Signal Processing (ICASSP)* 2224–2227 (2011).
30. Triefenbach, F., Jalalvand, A., Schrauwen, B. & Martens, J.-P. Phoneme recognition with large hierarchical reservoirs. *Adv. Neural Inf. Process. Syst.* **23,** 2307–2315 (2010).
31. Lillicrap, T., Cownden, D., Tweed, D. & Akerman, C. Random feedback weights support learning in deep neural networks Preprint at http://arxiv.org/abs/1411.0247 (2014).
32. Van Campenhout, J., Green, W. M., Assefa, S. & Vlasov, Y. A. Low-power, 2î2 silicon electro-optic switch with 110-nm bandwidth for broadband reconfigurable optical networks. *Opt. Express* **17,** 24020–24029 (2009).
33. Garofolo, J. *et al.* *TIMIT Acoustic-phonetic Continuous Speech Corpus* (Linguistic Data Consortium, 1993).

Acknowledgements

We thank Tim Waegeman for help with the data acquisition for the acoustic experiment. This work was partially supported by the interuniversity attraction pole (IAP) Photonics@be of the Belgian Science Policy Office and the ERC NaResCo Starting grant and the European Union Seventh Framework Programme under grant agreement no. 604102 (Human Brain Project).

Author contributions

M.H. conceived, designed and conducted the physical and simulated experiments, M.B. contributed in building the electronic hardware for the acoustic experiment, T.V.V. co-designed the simulated experiments and helped conceive the electro-optical concept. J.D. and P.B. supervised the experiments and the project.

Additional information

Competing financial interests: The authors declare no competing financial interests.

11

Information-theoretic implications of quantum causal structures

Rafael Chaves[1], Christian Majenz[2] & David Gross[1]

It is a relatively new insight of classical statistics that empirical data can contain information about causation rather than mere correlation. First algorithms have been proposed that are capable of testing whether a presumed causal relationship is compatible with an observed distribution. However, no systematic method is known for treating such problems in a way that generalizes to quantum systems. Here, we describe a general algorithm for computing information-theoretic constraints on the correlations that can arise from a given causal structure, where we allow for quantum systems as well as classical random variables. The general technique is applied to two relevant cases: first, we show that the principle of information causality appears naturally in our framework and go on to generalize and strengthen it. Second, we derive bounds on the correlations that can occur in a networked architecture, where a set of few-body quantum systems is distributed among some parties.

[1] Institute for Physics, University of Freiburg, Rheinstrasse 10, D-79104 Freiburg, Germany. [2] Department of Mathematical Sciences, University of Copenhagen, Universitetsparken 5, DK-2100 Copenhagen Ø, Denmark. Correspondence and requests for materials should be addressed to R.C. (email: rafael.chaves@physik.uni-freiburg.de).

A causal structure for a set of classical variables is a graph, where every variable is associated with a node and a directed edge denotes functional dependence. Such a causal model offers a means of explaining dependencies between variables, by specifying the process that gave rise to them. More formally, variables $X_1, \ldots, X_n$ form a Bayesian network with respect to a directed, acyclic graph (commonly abbreviated DAG), if every variable X_i depends only on its graph–theoretic parents PA_i. This is the case[1,2] if and only if the distribution factorizes as in

$$p(x_1, \ldots, x_n) = \prod_{i=1}^{n} p(x_i | x_{\mathrm{PA}_i}). \quad (1)$$

One can ask the following fundamental question: given a subset of variables, which correlations between them are compatible with a given causal structure? In this work, we measure 'correlations' in terms of the collection of joint entropies of the variables, which we allow to be quantum systems as well as classical random variables (a precise definition will be given below).

This problem appears in several contexts. In the young field of causal inference, the goal is to learn causal dependencies from empirical data[1,2]. If observed correlations are incompatible with a presumed causal structure, it can be discarded as a possible model. This is close to the reasoning employed in Bell's theorem[3]—a connection that is increasingly appreciated among quantum physicists[4–10]. In the context of communication theory, these joint entropies describe the capacities that can be achieved in network-coding protocols[11].

In this work, we are interested in quantum generalizations of causal structures. Nodes are now allowed to represent either quantum or classical systems, and edges are quantum operations. An important conceptual difference to the purely classical set-up is rooted in the fact that quantum operations disturb their input. Put differently, quantum mechanics does not assign a joint state to the input and the output of an operation. Therefore, there is in general no analogue to (1), that is, a global density operator for all nodes in a quantum causal structure cannot be defined. However, if we pick a set of nodes that do coexist (for example, because they are classical, or because they are created at the same instance of time, that is, they do have a joint density operator), then we can again ask: which joint entropies of coexisting nodes can result from a given quantum causal structure?

Here, we answer this question by introducing a universal framework that generalizes previous results on the classical case[6,7,12–14]. This framework's versatility and practical relevance can be illustrated with two examples. In the context of distributed quantum architectures[15–19], the framework can be employed to systematically compute limits on the correlations that are imposed solely by the topology of the networks. The machinery can also be used to generalize and strengthen information causality (IC)[20], a principle that may explain the 'degree of non-locality' exhibited by quantum mechanics. The details, along with more examples including dense coding schemes[21], are presented in the main text.

Results

Quantum causal structures. Informally, a quantum causal structure specifies the functional dependency between a collection of quantum systems and classical variables. We find it helpful to employ a graphical notation, where we aim to closely follow the conventions of classical graphical models[1,2]. There are two basic building blocks: root nodes are labelled by a set of quantum systems and represent a density operator for these systems

$$\boxed{AB} \quad \triangleq \varrho_{AB} \quad (2)$$

The second type is given by nodes with incoming edges. Again, both the edges and the node carry the labels of quantum systems. Such symbols represent a quantum operation (completely positive, trace-preserving (CPTP) map) from the systems associated with the edges to the ones associated with the node:

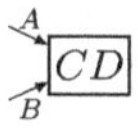
$$\triangleq \Phi_{AB\to CD} : A \otimes B \to C \otimes D \quad (3)$$

These blocks may be combined: a node containing a system X can be connected to an edge with the same label. The interpretation is, of course, that X serves as the input to the associated operation. For example,

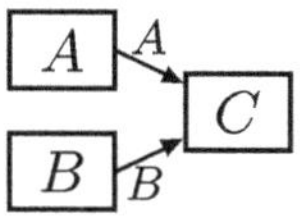
$$\triangleq \varrho_C = \Phi_{AB\to C}(\varrho_A \otimes \varrho_B) \quad (4)$$

says that the state of system C is the result of applying an operation $\Phi_{\mathrm{AB}\to\mathrm{C}}$ to a product state on AB. To avoid ambiguities, we will never use the same label in two different nodes (in particular, we always assume that the output systems of an operation are distinct from the input systems). For a more involved example, note that Fig. 1a gives a fairly readable representation of the following cumbersome algebraic statement:

$$\begin{aligned}\varrho_{ABC} = &[\Phi_{\mathrm{A_1A_2}\to\mathrm{A}} \otimes \Phi_{\mathrm{B_1B_2}\to\mathrm{B}} \otimes \Phi_{\mathrm{C_1C_2}\to\mathrm{C}}] \\ &(\varrho_{\mathrm{A_1B_1}} \otimes \varrho_{\mathrm{A_2C_1}} \otimes \varrho_{\mathrm{B_2C_2}})\end{aligned} \quad (5)$$

(where the operation defined in the first line is acting on the state defined in the second line). The graphical representation does not indicate which input state or which operation to employ. We suppress this information, because we will be interested only in constraints on the resulting correlations that are implied by the topology of the interactions alone, regardless of the choice of states and maps.

We denote the labels of classical variables (equivalently, quantum systems described by states that are diagonal in a given basis) in circles, as opposed to the rectangles we use for quantum systems. In principle, classical variables could have >1 outgoing edge. Of course, the no-cloning principle precludes a quantum system being used as the input to two different operations. Moreover, only graphs that are free of cyclic dependencies can be interpreted as specifying a causal structure. Thus, as is the case in classical Bayesian networks, every quantum causal structure is associated with a DAG.

We note that graphical notations for quantum processes have been used frequently before. The most popular graphical calculus is probably the gate model of quantum computation[22], where,

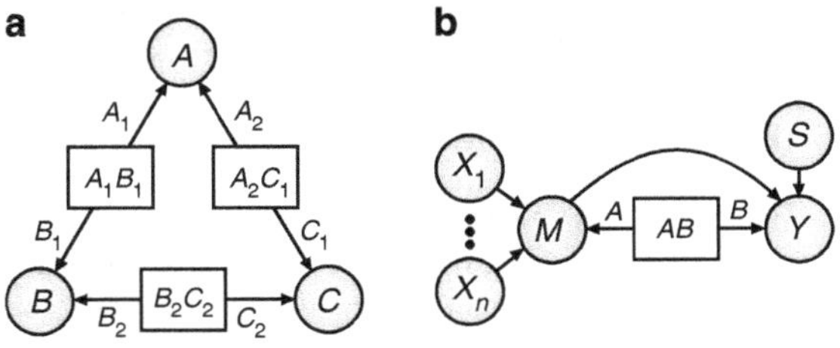

Figure 1 | Causal structures. (**a**) An example of distributed architecture involving bipartite entangled states. Each of the underlying quantum states can connect at most two of the observable variables, which implies a non-trivial monogamy of correlations as captured in (12). (**b**) The quantum causal structure associated with the information causality principle.

directly opposite to our conventions, operations are nodes and systems are edges. That is also the case in the recently introduced generalized Bayesian networks in ref. 9. There, the authors even allow for post-quantum resources. Quantum communication scenarios are often visualized the same way we employ here[23].

We have noted in the introduction that a classical Bayesian network not only defines the functional dependencies between random variables, but also provides a structural formula (1) for the joint distribution of all variables in the graph. Again, such a joint state for all systems that appear in a quantum causal structure is not in general defined. However, other authors have considered quantum versions of distributions that factor as in (1) and have developed graphical notations to this end. Well-known examples include the related constructions that go by the name of finitely correlated states, matrix–product states, tree–tensor networks or projected entangled pairs states (a highly incomplete set of starting points to the literature is given in refs 24–26). Also, certain definitions of quantum Bayesian networks[27,28] fall into that class.

Entropic description of quantum causal structures. The entropic description of classical-quantum DAGs can be seen as a generalization of the framework for the case of purely classical variables[6,7,12–14] that consists of three main steps. Consider a classical DAG consisting of n variables and a subset of them that are observable. In step 1, one needs to construct a description of the unconstrained Shannon cone. As we will see below, this means enumerating all elementary inequalities that entropies of n variables must respect, regardless of their causal relations. In step 2, the causal constraints must be added, which corresponds to listing all conditional independence relations implied by the DAG. In step 3, a marginalization is performed, that is, because some of the variables may not be observable, these need to be eliminated from our description. The final result of this three-step programme is the description of the marginal entropic constraints implied by the model under investigation. To further understand the meaning of these steps and how they need to be modified to cope with quantum causal structures, in the following we briefly discuss each of them.

To begin with, we denote the set of indices of the random variables by $[n]=\{1, \ldots, n\}$ and its power set (that is, the set of subsets) by $2^{[n]}$. For every subset $S\in 2^{[n]}$ of indices, let X_S be the random vector $(X_i)_{i\in S}$ and denote by $H(S):=H(X_S)$ the associated entropy vector (for some, still unspecified entropy function H). Entropy is then a (partial) function $H:2^{[n]}\rightarrow\mathbb{R}$, $S\mapsto H(S)$ on the power set.

Note that as entropies must fulfil some constraints, not all entropy vectors are possible. That is, given the linear space of all set functions denoted by R_n and a function $h\in R_n$ the region of vectors in R_n that correspond to entropies is given by

$$\{h \in R_n \,|\, h(S) = H(S) \text{ for some entropy function } H\}.$$

Clearly, this region will depend on the chosen entropy function.

For classical variables, H is chosen to be the Shannon entropy given by $H(X_S) = -\sum_{x_s} p(x_s) \log_2 p(x_s)$. In this case, an outer approximation to the associated entropy region has been studied extensively in information theory, the so-called Shannon cone Γ_n (ref. 11), which is the basis of the entropic approach in classical causal inference[14]. The Shannon cone is the polyhedral closed convex cone of set functions h that respect two elementary inequalities, known as polymatroidal axioms. The first relation is the sub-modularity (also known as strong subadditivity) condition, which is equivalent to the positivity of the conditional mutual information, for example, $I(A{:}B|C)=H(A,C)+H(B,C)-H(A,B,C)-H(C)\geq 0$. The second inequality—nown as monotonicity—is equivalent to the positivity of the conditional entropy, for example, $H(A|B)=H(A,B)-H(B)\geq 0$. Therefore, the first step of the algorithm corresponds to listing all these elementary inequalities.

The elementary inequalities discussed above encode the constraints that the entropies of any set of classical variables are subject to, regardless of their causal relations. Therefore, in the second step of the algorithm we need to list the causal relationships between the variables. These are encoded in the conditional independences (CIs) implied by the graph and can be algorithmically enumerated using the so-called d-separation criterion[1]. Therefore, if one further demands that classical random variables are a Bayesian network with respect to some given DAG, their entropies will also fulfil the additional CI relations implied by the graph. The CIs, relations of the type $p(x,y|z)=p(x|z)p(y|z)$ that defines nonlinear constraints in terms of probabilities, are equivalent to homogeneous linear constraints on the level of entropies, for example, $I(X{:}Y|Z)=0$.

Finally, we are interested in situations where not all joint distributions are accessible. Most commonly, this is because the variables of a DAG can be divided into observable and not observable ones (for example, the underlying quantum states in Fig. 1). Given the set of observable variables, in the classical case, it is natural to assume that any subset of them can be jointly observed. However, in quantum mechanics that situation is more subtle. For example, position Q and momentum P of a particle are individually measurable, however, there is no way to consistently assign a joint distribution to both position and momentum of the same particle[3]. That is while $H(Q)$ and $H(P)$ are part of the entropic description of classical-quantum DAGs, joint terms such as $H(Q,P)$ cannot be part of it. This motivates the following definition: given a set of variables $X_1, \ldots, X_n$ contained in a DAG, a marginal scenario $\mathcal{M}$ is the collection of those subsets of $X_1, \ldots, X_n$ that are assumed to be jointly measurable. Given the inequality description of the DAG and the marginal scenario $\mathcal{M}$ under consideration, the third and last step of the algorithm consists of eliminating from this inequality description the variables that are not observable, that is the variables that are not contained in $\mathcal{M}$. This is achieved, for example, via a Fourier–Motzkin (FM) elimination (see Methods for further details).

We now turn our attention to the generalization of the algorithm to include quantum systems, that are described in terms of the quantum analogue of the Shannon entropy, the von Neumann entropy $H(\varrho_{A,B}) = -\operatorname{Tr}(\varrho_{A,B}\log\varrho_{A,B})$.

In the first step of the algorithm there are two differences. First, while quantum systems respect sub-modularity, the von Neumann entropy fails to comply with monotonicity. In this case, one needs to resort to the weak version of the monotonicity inequality (for example, $H(\varrho_A)+H(\varrho_B)\leq H(\varrho_{AC})+H(\varrho_{BC})$), a constraint that is fulfilled by the von Neumann entropy. Note, however, that for sets consisting of both classical and quantum systems, monotonicity may still hold. That is because the uncertainty about a classical variable A cannot be negative, even if we condition on an arbitrary quantum system ϱ, following then that $H(A|\varrho)\geq 0$ (ref. 29). Furthermore, for a classical variable A, the entropy $H(A)$ reduces to the Shannon entropy[30].

The second difference is the fact that measurements, or more generally CPTP maps, on a quantum state will generally destroy/disturb the state. To illustrate that, consider the classical-quantum DAG in Fig. 1. Consider the classical and observable variable A. It can without loss of generality be considered a deterministic function of its parents ϱ_{A_1} and ϱ_{A_2}, as any additional local parent can be absorbed in one of the latter. For the variable A to assume a definite outcome, a joint CPTP map is applied to both parents ϱ_{A_1} and ϱ_{A_2} that will in general disturb these variables. The random variable A does not coexist with quantum systems A_1 and A_2. Therefore, no entropy can be associated to these variables simultaneously, that is, $H(A,A_1,A_2)$ cannot be part of the entropic description of the classical-quantum DAG. As a result of that, only elementary inequalities involving coexisting variables can be listed in step 1. Classically, this problem does not arise as the underlying classical hidden variables could be accessed without disturbing them.

In the second step of the algorithm, we need to list all the causal relations as encoded in CIs implied by the graph. All the classical CIs (that is, following from the d-separation criterion) that involve coexisting variables also hold for the quantum causal structures considered here[9]. However, some classically valid CIs may, in the quantum case, involve non-coexisting variables and therefore are not valid for quantum systems. As an example, consider the DAG in Fig. 1b. For the classical analogue of this DAG it follows, for example, that $I(A{:}B|A_1,A_2,B_1,B_2)$. This relation states that the correlations between A and B should be screened off conditioned on their common ancestor. Because this CI involves a term such as $H(A, A_1, A_2)$, this CI cannot be defined in the quantum case. Another example of that is illustrated below for the IC scenario.

Furthermore, because terms such as $H(A, A_1, A_2)$ are not part of our description, we need, together with the CIs implied by the quantum causal structure, a rule telling us how to relate the entropies of underlying quantum systems to the entropies of their classical descendants, for example, how to relate $H(A_1,A_2)\rightarrow H(A)$. This is achieved by the data processing (DP) inequality, another basic property that is valid both for the classical and quantum cases[22]. The DP inequality basically states that the information content of a system cannot be increased by acting locally on it. To exemplify, one DP inequality implied by the DAG in Fig. 1 is given by $I\ (A\ :\ B)\leq I\ (A_1,\ A_2\ :\ B_1,\ B_2)$, that is, the mutual information between the classical variables cannot be larger then the information shared by their underlying quantum parents.

Defined by the marginal scenario of interest, the third step of the algorithm is identical to the classical case, that is, the elimination of variables representing unobservable random variables or quantum systems. In two of the examples below (IC and quantum networks), all the observable quantities correspond to classical variables, corresponding, for example, to the outcomes of measurements performed on quantum states. Therefore, the marginal description will be given in terms of linear inequalities involving Shannon entropies only. This contrasts with another example we will mention: a generalization of super-dense coding. There, the final description does involve a quantum system, and therefore a mixed inequality with Shannon as well as von Neumann entropy terms results.

Information causality. The 'no-signalling principle' alone is insufficient to explain the 'degree of non-locality' exhibited by quantum mechanics[31]. This has motivated the search for stronger, operationally motivated principles, that may single out quantum–mechanical correlations[20,32–39]. One of these is IC principle[20], which can be understood as a game: Alice receives a bit string x of length n, while Bob receives a random number s $(1\leq s\leq n)$. Bob's task is to make a guess Y_s about the sth bit of the bit string x using as resources a m-bit message M sent to him by Alice and some correlations shared between them. It would be expected that the amount of information available to Bob about x should be bounded by the amount of information contained in the message, that is, $H(M)$. IC makes this notion precise, stating that the following inequality is valid in quantum theory[20]

$$\sum_{s=1}^{n} I\,(X_s : Y_s) \leq H\,(M) \tag{6}$$

where $I(X{:}Y)$ is the classical mutual information between the variables X and Y and the input bits of Alice are assumed to be independent. This inequality is valid for quantum correlations but is violated by all nonlocal correlations beyond Tsirelson's bound, as originally shown in ref. 20 and further explored in refs 29,30,40,41.

Consider the case where $X=(X_1,X_2)$ is a 2-bit string. The corresponding causal structure to the IC game is the one shown in Fig. 1b). The only relevant CI is given by $I(X_1,X_2{:}AB)=0$. Note that classically the CIs $I(X_1,X_2{:}Y_s|M,B)=0$ (with $s=1$, 2) would also be part of our entropic description. However, because we cannot assign a joint entropy to Y_s and ϱ_B, this classically valid CI cannot be part of the entropic description in the quantum case. We can now proceed with the general framework. But before doing that, we first need to specify in which marginal scenario we are interested. In ref. 20, the authors implicitly restricted their attention to the marginal scenario defined by $\{X_1, Y_1\}$, $\{X_2, Y_2\}$, $\{M\}$. Proceeding with this marginal scenario, we find that the only non-trivial inequality characterizing this marginal entropic cone is given by

$$I(X_1 : Y_1) + I(X_2 : Y_2) \leq H\,(M) + I(X_1 : X_2), \tag{7}$$

which corresponds exactly to the IC inequality obtained in ref. 30, where the input bits are not assumed to be independent.

Note, however, that using the aforementioned marginal scenario, available information is being discarded. The most general possible marginal scenario is given by $\{X_1, X_2, Y_s, M\}$ (with $s=1$, 2). That is, in this case we are also interested in how much information the guess Y_1 of the bit X_1, together with the message M, may contain about the bit X_2 (similarly for B_2 and X_1). Proceeding with this marginal scenario, we find different classes of non-trivial tight inequalities describing the marginal information causality cone. Of particular relevance is the following tighter version of the original IC inequality

$$\begin{aligned} I(X_1 : Y_1, M) + I(X_2 : Y_2, M) + I(X_1 : X_2\,|\,Y_2, M) \\ \leq H(M) + I(X_1 : X_2). \end{aligned} \tag{8}$$

Two different interpretations can be given to this inequality: as a monogamy of correlations or as a classical quantification of causal influence.

For the first interpretation, consider for simplicity the case where the input bits are independent, that is, $I(X_1{:}X_2)=0$. These independent variables may, however, become correlated conditioned on other variables that depend on them. That is, in general $I(X_1{:}X_2|Y_2,M)\neq 0$. However, the underlying causal relationships between the variables impose constraints on how much we can correlate these variables. In fact, as we can see from (8), the more information the message M and the guess Y_i contain about about the input bit X_i, the smaller is the correlation we can generate between the input bits by conditioning on them. As an extreme example suppose Alice decides to send $M=X_1\oplus X_2$. Then X_1 and X_2 are fully correlated given M, but M does not contain any information about the individual inputs X_1 and X_2.

As for the second interpretation, we need to rely on the classical concept of how to quantify causal influence between two sets of variables X and Y. As shown in ref. 42, a good measure $\mathcal{C}_{X\rightarrow Y}$ of the causal influence of a variable X over a variable Y should be lower bounded as $\mathcal{C}_{X\rightarrow Y}\geq I(X{:}Y\,|\,PA_Y^X)$, where PA_Y^X stands for all the parents of Y but X. That is, excluding the correlations between X and Y that are mediated via PA_Y^X, the remaining correlations give a lower bound to the direct causal influence between the variables. Consider for instance that we allow for an arrow between the input bits X and the guess Y. Therefore, the classical CI $I(X_1, X_2 : Y_1, Y_2|M, B)=0$ that is valid for the DAG in Fig. 1b), does not hold any longer. In this case $I\big(X{:}Y\big|PA_Y^X\big) = I(X_1, X_2{:}Y_1, Y_2|M, B)$, an object that is part of the entropic description in the classical case. Proceeding with the general framework one can prove that

$$\begin{aligned} C_{X\to Y} \geq & I(X_1 : Y_1, M) + I(X_2 : Y_2, M) \\ & + I(X_1 : X_2 \,|\, Y_2, M) - H(M) - I(X_1 : X_2). \end{aligned} \tag{9}$$

That is, the degree of violation of (8) (for example, via a Popescu-Rohrlich(PR)-box) gives a lower bound to the minimum amount of direct causal influence required to obtain the same level of correlations within a classical model.

Inequality (8) refers to the particular case of two input bits for Alice. As we prove in the Methods section, the following generalization is valid within the quantum theory for any number of input bits:

$$\begin{aligned} & \sum_{i=1}^{n} I(X_i : Y_i, M) + \sum_{i=2}^{n} I(X_1 : X_i \,|\, Y_i, M) \\ & \leq H(M) + \sum_{i=1}^{n} H(X_i) - H(X_1, \ldots, X_n). \end{aligned} \tag{10}$$

We further notice that the IC scenario is quite similar to the super-dense coding set-up[21], where, however, in the latter case the message M is quantum. On the level of the entropies, this difference is manifested by the fact that monotonicity $H(M|X_0, X_1, B) \geq 0$ must be replaced by weak monotonicity $H(M|X_0, X_1, B) + H(M) \geq 0$. As proved in the Methods section, this implies that a similar inequality to (10) is a also valid for the super-dense coding scenario if one replaces the Shannon entropy $H(M)$ by twice the von Neumann entropy $2H(M)$. This generalizes the well-known phenomenon of super-dense coding[21], which states that in the presence of shared entanglement, the exchange of one d-level system allows for one of d^2-diferent messages to be communicated.

Finally, to understand how much more powerful inequality (8) may be as a witness of post-quantum correlations, we perform a similar analysis to the one in ref. 43. We consider the following section of the non-signalling polytope

$$p(a, b \,|\, x, y) = \gamma p_{\mathrm{PR}} + \epsilon p_{\mathrm{det}} + (1 - \gamma - \epsilon)\, p_{\mathrm{white}} \tag{11}$$

with $p_{\mathrm{PR}}(a, b|x, y) = (1/2)\delta_{a\oplus b, xy}$, $p_{\mathrm{white}}(a, b|x, y) = 1/4$ and $p_{\mathrm{det}}(a, b|x, y) = \delta_{a,0}\delta_{b,0}$ corresponding, respectively, to the PR-box, white noise and a deterministic box. The results are displayed in Fig. 2, where it can be seen that the new inequality is considerably more powerful then the original one. Our new inequality can witness, already on the single-copy level, the postquantumness of distributions that could not be detected before even in the limit of many copies.

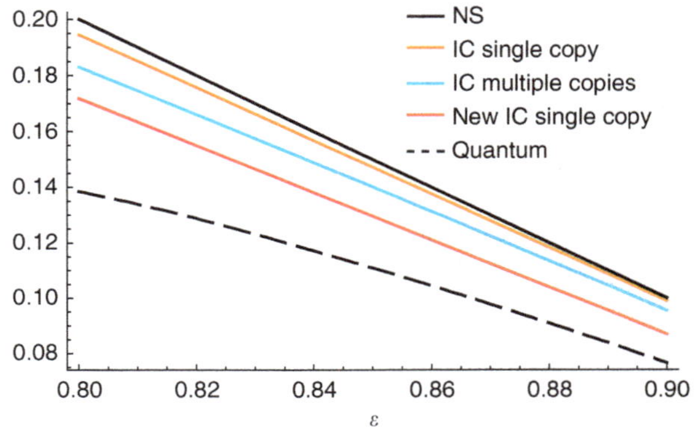

Figure 2 | Comparison of the strengthened and original IC principles. A slice of the non-signalling polytope corresponding to the distribution (11). The lower black-dashed line is an upper limit on quantum correlations obtained via the criterion in ref. 48 while the upper solid black line bounds the set of non-signalling correlations. The solid red, blue and orange curves correspond, respectively, to the boundaries obtained with the IC inequalities (8), (6) and (7). Above each of the curves, the corresponding inequalities are violated. See Supplementary Note 1 for details of how the curves are computed.

Quantum networks. Quantum networks are ubiquitous in quantum information. The basic scenario consists of a collection of entangled states that are distributed among several spatially separated parties to perform some informational task, for example, entanglement percolation[17], entanglement swapping[44] or distributed computing[15,16]. A similar set-up is of relevance in classical causal inference, namely the inference of latent common ancestors[14,45]. As we will show next, just the topology of these quantum networks already imply non-trivial constraints on the correlations that can be obtained between the different parties. We will consider the particular case where all the parties can be connected by at most bipartite states. We note, however, that our framework applies as well to the most general case and results along this line are presented in the Supplementary Notes 2 and 3.

The problem can be restated as follows. Consider n observable variables that may be assumed to have no direct causal influence on each other (as they are space-like separated). Given some observed correlations between them, the basic question is then: can the correlations between these n variables be explained by (hidden) common ancestors each connecting at most two of them? The simplest of such common ancestors scenarios ($n=3$), the so-called triangle scenario[5,45,46], is illustrated in Fig. 1a.

In the case where the underlying hidden variables are classical (for example, separable states), the entropic marginal cone associated with this DAG has been completely characterized in ref. 7. Following the framework delineated before, we can prove that facets of this cone are also obtained if we replace the underlying classical variables by quantum states (Supplementary Note 2). This implies that entropically quantum correlations respect the same type of monogamy relations as classical variables.

The natural question is how to generalize this result to more general common ancestor structures for arbitrary n. With this aim, we prove in the Methods section that the monogamy relation

$$\sum_{\substack{i=1,\cdots,n \\ i\neq j}} I(V_i : V_j) \leq H(V_j), \tag{12}$$

recently derived in ref. 14 is also valid for quantum theory. We also prove in the Supplementary Note 3 that this inequality is valid for general non-signalling theories, generalizing the result obtained in ref. 9 for $n=3$. In addition, we exhibit that for any non-trivial common ancestor structure there are entropic constraints even if we allow for general non-signalling theories.

The inequality (12) can be seen as a kind of monogamy of correlations. Consider for instance the case $n=3$ and label the common ancestors (any non-signalling resource) connecting variables V_i and V_j by $\varrho_{i,j}$. If the correlation between V_1 and V_2 is large, that means that V_1 has a strong causal dependence on their common mutual ancestor $\varrho_{1,2}$. That implies that V_1 should depend only mildly on its ancestor $\varrho_{1,3}$ and therefore its correlation with V_3 should also be small. The inequality (12) makes this intuition precise.

Discussion

In this work, we have introduced a systematic algorithm for computing information-theoretic constraints arising from quantum causal structures. Moreover, we have demonstrated the versatility of the framework by applying it to a set of diverse examples from quantum foundations, quantum communication and the analysis of distributed architectures. In particular, our framework readily allows us to obtain a much stronger version of information causality.

These examples aside, we believe that the main contribution of this work is to highlight the power of systematically analysing entropic marginals. A number of future directions for research immediately suggests themselves. In particular, it will likely be fruitful to consider multipartite versions of information causality or other information-theoretical principles and to further look into the operational meaning of entropy inequality violations.

Methods

A linear programme framework for entropic inequalities. Given the inequality description of the entropic cone describing a causal structure, to obtain the description of an associated marginal scenario $\mathcal{M}$ we need to eliminate from the set of inequalities all variables not contained in $\mathcal{M}$. After this elimination procedure, we obtain a new set of linear inequalities, constraints that correspond to facets of a convex cone, more precisely the marginal entropic cone characterizing the compatibility region of a certain causal structure[7]. This can be achieved via a FM elimination, a standard linear programming algorithm for eliminating variables from systems of inequalities[47]. The problem with the FM elimination is that it is a doubly exponential algorithm in the number of variables to be eliminated. As the number of variables in the causal structure of interest increases, typically this elimination becomes computationally intractable.

While it can be computationally very demanding to obtain the full description of a marginal cone, to check whether a given candidate inequality is respected by a causal structure is relatively easy. More precisely, the algorithm is admittedly exponential in the number of random variables/quantum systems, but in comparison the FM elimination method for finding all the inequalities is doubly exponential in the number of variables participating in the set of linear inequalities, that is, triply exponential in the number of random variables/quantum systems. Consider that a given causal structure leads to a number N of possible entropies. These are organized in a n-dimensional vector $\boldsymbol{h}$. In the purely classical case, the graph consisting of n nodes $(X_1, \ldots, X_n)$ will lead to a $N=2^n$-dimensional entropy vector that can be organized as $\boldsymbol{h}=(H(\phi), H(X_n), H(X_{n-1}), H(X_{n-1}X_n), \ldots, H(X_1, \ldots, X_n))$. In the quantum case, since not all subsets of variables may jointly coexist we will have typically that N is strictly smaller than 2^n.

As explained in detail in the main text, for this entropy vector to be compatible with a given causal structure, a set of linear constraints must be fulfilled. These linear constraints can be cast as a system of inequalities of the form $M\boldsymbol{h}\geq \boldsymbol{0}$, where M is a $m\times N$ matrix with m being the number of inequalities characterizing the causal structure.

Given the entropy vector $\boldsymbol{h}$, any entropic-linear inequality can be written simply as the inner product $\langle \mathcal{I}, \boldsymbol{h}\rangle \geq 0$, where $\mathcal{I}$ is the associated vector to the inequality. A sufficient condition for a given inequality to be valid for a given causal structure is that the associated set of inequalities $M\boldsymbol{h}\geq\boldsymbol{0}$ be true for any entropy vector $\boldsymbol{h}$. That is, to check the validity of a test inequality, one simply needs to solve the following linear programme:

$$\begin{aligned} &\underset{h\in\mathbb{R}^N}{\text{minimize}}\ \langle \mathcal{I}, \boldsymbol{h}\rangle \\ &\text{subject to}\ M\boldsymbol{h}\geq \boldsymbol{0} \end{aligned} \tag{13}$$

In general, this linear programme only provides a sufficient but not necessary condition for the validity of an inequality. The reason for that is the existence of non-Shannon type inequalities, which are briefly discussed in the Supplementary Note 2.

Proving the new IC inequality. We provide in the following an analytical proof of the validity of the generalized IC inequality (10) for the quantum causal structure in Fig. 1b). Further details can be found in the Supplementary Note 1.

Proof: first rewrite the following conditional mutual information as

$$I(X_1 : |X_i|Y_i, M) = I(X_i{:}X_1, Y_i, M) - I(X_i{:}Y_i, M). \tag{14}$$

The left-hand side of the inequality (10) can then be rewritten as

$$I(X_1 : Y_1, M) + \sum_{i=2}^{n} I(X_i : X_1, Y_i, M). \tag{15}$$

This quantity can be upper bounded as

$$\leq I(X_1 : B, M) + \sum_{i=2}^{n} I(X_i : X_1, B, M) \tag{16}$$

$$= \sum_{i=1}^{n} H(X_i) + H(B, M) + (n-2)H(X_1, B, M) - \sum_{i=2}^{n} H(X_1, X_i, B, M) \tag{17}$$

$$\leq \sum_{i=1}^{n} H(X_i) + H(B, M) - H(X_1, \ldots, X_n, B, M) \tag{18}$$

$$\leq \sum_{i=1}^{n} H(X_i) + H(B, M) - H(X_1, \ldots, X_n, B) \tag{19}$$

$$= \sum_{i=1}^{n} H(X_i) + H(B, M) - H(X_1, \ldots, X_n) - H(B) \tag{20}$$

$$\leq \sum_{i=1}^{n} H(X_i) + H(M) - H(X_1, \ldots, X_n) \tag{21}$$

leading exactly to the inequality (10). In the proof above we have used consecutively (i) the DP inequalities $I(X_1 : Y_1, M) \leq I(X_1 : B, M)$ and $I(X_i : X_1, Y_i, M) \leq I(X_i : X_1, B, M)$, (ii) the fact that $-\sum_{i=1}^{N-1} H(X_i, X_1, B, M) \leq -H(X_1, \ldots, X_n, B, M) - (n-2)H(X_1, B, M)$ (as can be easily proved inductively using the strong subadditivity property of entropies), (iii) the monotonicity $H(M|X_1, \ldots, X_n, B) \geq 0$, (iv) the independence relation $I(X_1, \ldots, X_n : B) = 0$ and (v) the positivity of the mutual information $I(B{:}M) \geq 0$. This concludes the proof.

Note that this proof can be easily adapted to the case where the message M sent from Alice to Bob is a quantum state. In this case there are two differences. First, because the message is disturbed to create the guess Y_i, we cannot assign an entropy to M and Y_i simultaneously. That is, in the left-hand side of the inequality (10), we replace $I(X_i : Y_i, M) \rightarrow I(X_i : Y_i)$ and $I(X_1 : X_i|Y_i, M) \rightarrow I(X_1 : X_i|Y_i)$. The second difference is in step iii, because we have used the monotonicity $H(M|X_1, \ldots, X_n, B) \geq 0$ that is not valid for a quantum message. Instead of that, we can use a weak monotonicity inequality, namely $H(M|X_1, \ldots, X_n, B) + H(M) \geq 0$. Therefore, in the final inequality (10), $I(X_i : Y_i, M) \rightarrow I(X_i : Y_i)$ and $I(X_1 : X_i|Y_i, M) \rightarrow I(X_1 : X_i|Y_i)$ and $H(M)$ is replaced by $2H(M)$—where H now stands for the von Neumann entropy.

Proving the monogamy relations of quantum networks. In the following, we provide an analytical proof of the monogamy inequality (12) in the main text. Further details can be found in the Supplementary Note 2.

We start with the case $n=3$. For a Hilbert space $\mathcal{H}$, we denote the set of quantum states, that is, the set of positive semidefinite operators with trace one, on it by $\mathsf{S}(\mathcal{H})$.

Theorem 1. Let $\varrho_{A_1A_2B_1B_2C_1C_2} = \varrho_{A_1B_1} \otimes \varrho_{B_2C_2} \otimes \varrho_{C_1A_2}$ be a six-partite quantum state on $\mathcal{H} = \mathcal{H}_{A_1} \otimes \mathcal{H}_{A_2} \otimes \mathcal{H}_{B_1} \otimes \mathcal{H}_{B_2} \otimes \mathcal{H}_{C_1} \otimes \mathcal{H}_{C_2}$. Let further $\Phi_N : \mathsf{S}(\mathcal{H}_{N_1} \otimes \mathcal{H}_{N_2}) \rightarrow \mathsf{S}(\mathcal{H}_N)$ be an arbitrary measurement for $N = A, B, C$. Then

$$I(A:B) + I(A:C) \leq H(A). \tag{22}$$

Proof: DP yields

$$I(A:B) + I(A:C) \leq I(A:B_1B_2) + I(A:C_1C_2). \tag{23}$$

Then we exploit the chain rule twice and afterward DP again,

$$\begin{aligned} I(A:B_1B_2) &= I(A:B_1) + I(A:B_2|B_1) \\ &= I(A:B_1) + I(AB_1:B_2) - I(B_1:B_2) \\ &\leq I(A:B_1) + I(A_1A_2B_1:B_2) \\ &= I(A:B_1), \end{aligned} \tag{24}$$

where in the last step, we have used the independence relation between the quantum states. We have therefore

$$I(A:B) + I(A:C) \leq I(A:B_1) + I(A:C_1), \tag{25}$$

for which the right-hand side can be bounded as

$$\begin{aligned} &I(A{:}B_1) + I(A:C_1) \\ &= 2H(A) + H(B_1) + H(C_1) - H(AB_1) - H(AC_1) \\ &\leq H(A) + H(B_1) + H(C_1) - H(AB_1C_1) \\ &= H(A) - H(A|B_1C_1) \\ &\leq H(A), \end{aligned} \tag{26}$$

leading to inequality (22). In the third line of (26) we used strong subadditivity, and in the last line we used that the entropy of a classical state conditioned on a quantum state is positive. This concludes the proof.

This proof can easily be generalized to the case of an arbitrary number of random variables resulting from a classical-quantum Bayesian network in which each parent has at most two children.

Corollary. 2 Let

$$\varrho = \bigotimes_{\substack{i,j=1 \\ i<j}}^{n} \varrho_{(ij),(ji)}$$

be an $n(n-1)$-partite quantum state on

$$\mathcal{H} = \bigotimes_{\substack{i,j=1 \\ i\neq j}}^{n} \mathcal{H}_{(ij)},$$

and let

$$\Phi_i : \mathbf{S}\left(\bigotimes_{j\neq i} \mathcal{H}_{(ij)}\right) \rightarrow \mathbf{S}\left(\mathcal{H}_i\right)$$

be an arbitrary measurement for $i = 1, \ldots, n$. Then

$$\sum_{i=2}^{n} I(1:i) \leq H(1) \tag{27}$$

Proof: First, utilize the independences in the same way as in the proof of Theorem 1 to conclude

$$\sum_{i=2}^{n} I(1:i) \leq \sum_{i=2}^{n} I(1:(i1)). \tag{28}$$

Now continue by induction. For $n = 3$ we have, according to the proof of Theorem 1,

$$I(1:(21)) + I(1:(31)) \leq H(1). \tag{29}$$

Now assume

$$\sum_{i=2}^{n-1} I(1:(i1)) \leq H(1). \tag{30}$$

Using the proof of Theorem 1 again and stopping before the last inequality in (26) we get

$$I(1:(n-11)) + I(1:(n1)) \leq I(1:(n-11)(n1)), \tag{31}$$

that is, we get

$$\sum_{i=2}^{n} I(1:(i1)) \leq \sum_{i=2}^{n-1} I(1:(i1')) \leq H(1), \tag{32}$$

where we defined the primed systems by $\mathcal{H}_{(n-11)'} = \mathcal{H}_{(n-11)} \otimes \mathcal{H}_{(n1)'}$, observing that this yields a classical-quantum bayesian network with $n-1$ nodes and connectivity two and used the induction hypothesis. This concludes the proof.

References

1. Pearl, J. *Causality* (Cambridge Univ. Press, 2009).
2. Spirtes, P., Glymour, N. & Scheienes, R. *Causation, Prediction, and Search* 2nd edn (The MIT Press, 2001).
3. Bell, J. S. On the Einstein—Podolsky—Rosen paradox. *Physics* **1,** 195–200 (1964).
4. Wood, C. J. & Spekkens, R. W. The lesson of causal discovery algorithms for quantum correlations: causal explanations of bell-inequality violations require fine-tuning. Preprint at http://arxiv.org/abs/1208.4119 (2012).
5. Fritz, T. Beyond bell's theorem: correlation scenarios. *New J. Phys.* **14,** 103001 (2012).
6. Fritz, T. & Chaves, R. Entropic inequalities and marginal problems. *IEEE Trans. Inform. Theory* **59,** 803–817 (2013).
7. Chaves, R., Luft, L. & Gross, D. Causal structures from entropic information: geometry and novel scenarios. *New J. Phys.* **16,** 043001 (2014).
8. Fritz, T. Beyond bell's theorem ii: scenarios with arbitrary causal structure. Preprint at http://arxiv.org/abs/1404.4812 (2014).
9. Henson, J., Lal, R. & Pusey, M. F. Theory-independent limits on correlations from generalised bayesian networks. Preprint at http://arxiv.org/abs/1405.2572 (2014).
10. Pienaar, J. & Brukner, C. A graph-separation theorem for quantum causal models. Preprint at http://arxiv.org/abs/1406.0430 (2014).
11. Yeung, R. W. *Information technology—transmission, processing, and storage* (Springer, 2008).
12. Chaves, R. & Fritz, T. Entropic approach to local realism and noncontextuality. *Phys. Rev. A* **85,** 032113 (2012).
13. Chaves, R. Entropic inequalities as a necessary and sufficient condition to noncontextuality and locality. *Phy. Rev. A* **87,** 022102 (2013).
14. Chaves, R. *et al.* Inferring latent structures via information inequalities. in *Proceedings of the 30th Conference on Uncertainty in Artificial Intelligence* 112–121 (2014).
15. Van Meter, R. *Quantum Networking* (Wiley, 2014).
16. Buhrman, H. & Röhrig., H. Distributed quantum computing. *In Mathematical Foundations of Computer Science* (Springer, 2003).
17. Acn, A., Cirac, J. I. & Lewenstein, M. Entanglement percolation in quantum networks. *Nat. Phys.* **3,** 256–259 (2007).
18. Brunner, N., Cavalcanti, D., Pironio, S., Scarani, V. & Wehner, S. Bell nonlocality. *Rev. Mod. Phys.* **86,** 419 (2014).
19. Sangouard, N., Simon, C., de Riedmatten, H. & Gisin, N. Quantum repeaters based on atomic ensembles and linear optics. *Rev. Mod. Phys.* **83,** 33 (2011).
20. Pawłowski, M. *et al.* Information causality as a physical principle. *Nature* **461,** 1101 (2009).
21. Bennett, C. H. & Wiesner, S. J. Communication via one- and two-particle operators on einstein-podolsky-rosen states. *Phys. Rev. Lett.* **69,** 2881–2884 (1992).
22. Nielsen, M. A. & Chuang, I. L. *Quantum Computation and Quantum Information* (Cambridge Univ. Press, 2010).
23. Wilde., M. M. *Quantum Information Theory* (Cambridge Univ. Press, 2013).
24. Fannes, M., Nachtergaele, B. & Werner, R. F. Finitely correlated states on quantum spin chains. *Commun. Math. Phys.* **144,** 443–490 (1992).
25. Pérez Garcia, D., Verstraete, F., Wolf, M. M. & Cirac, J. I. Matrix product state representations. *Quantum Inf. Comput.* **7,** 401 (2007).
26. Shi, Y.-Y., Duan, L.-M. & Vidal, G. Classical simulation of quantum many-body systems with a tree tensor network. *Phys. Rev. A* **74,** 022320 (2006).
27. Tucci, R. R. Quantum bayesian nets. *Int. J. Mod. Phys. B* **9,** 295–337 (1995).
28. Leifer, M. S. & Poulin, D. Quantum graphical models and belief propagation. *Ann. Phys.* **323,** 1899 (2008).
29. Barnum, H. *et al.* Entropy and information causality in general probabilistic theories. *New J. Phys.* **12,** 033024 (2010).
30. Al-Safi, S. W. & Short, A. J. Information causality from an entropic and a probabilistic perspective. *Phys. Rev. A* **84,** 042323 (2011).
31. Popescu, S. & Rohrlich, D. Quantum nonlocality as an axiom. *Found. Phys.* **24,** 379–385 (1994).
32. van Dam., W. *Nonlocality & Communication Complexity.* PhD thesisFaculty of Physical Sciences, Univ. Oxford, 1999).
33. Brassard, G. *et al.* Limit on nonlocality in any world in which communication complexity is not trivial. *Phys. Rev. Lett.* **96,** 250401 (2006).
34. Gross, D., Müller, M., Colbeck, R. & Dahlsten, O. C. O. All reversible dynamics in maximally nonlocal theories are trivial. *Phys. Rev. Lett.* **104,** 080402 (2010).
35. de la Torre, G., Masanes, L., Short, A. J. & Müller, M. P. Deriving quantum theory from its local structure and reversibility. *Phys. Rev. Lett.* **109,** 090403 (2012).
36. Navascués, M. & Wunderlich., H. A glance beyond the quantum model. *Proc. Roy. Soc. Lond. A* **466,** 881–890 (2010).
37. Fritz, T. *et al.* Local orthogonality as a multipartite principle for quantum correlations. *Nat. Commun.* **4,** 2263 (2013).
38. Sainz, A. B. *et al.* Exploring the local orthogonality principle. *Phys. Rev. A* **89,** 032117 (2014).
39. Navascués, M., Guryanova, Y., Hoban, M. J. & Acn, A. Almost quantum correlations. Preprint at http://arxiv.org/abs/1403.4621 (2014).
40. Dahlsten, O. C. O., Lercher, D. & Renner., R. Tsirelson's bound from a generalized data processing inequality. *New J. Phys.* **14,** 063024 (2012).
41. Short, A. J. & Wehner, S. Entropy in general physical theories. *New J. Phys.* **12,** 033023 (2010).
42. Janzing, D., Balduzzi, D., Grosse-Wentrup, M. & Schölkopf, B. Quantifying causal influences. *Ann. Statist.* **41,** 2324–2358 (2013).
43. Allcock, J., Brunner, N., Pawlowski, M. & Scarani, V. Recovering part of the boundary between quantum and nonquantum correlations from information causality. *Phys. Rev. A* **80,** 040103 (2009).
44. Żukowski, M. Z., Zeilinger, A., Horne, M. A. & Ekert, A. K. 'Event-ready-detectors' bell experiment via entanglement swapping. *Phys. Rev. Lett.* **71,** 4287–4290 (1993).
45. Steudel, B. & Ay., N. Information-theoretic inference of common ancestors. Preprint at http://arxiv.org/abs/1010.5720 (2010).
46. Branciard, C., Rosset, D., Gisin, N. & Pironio, S. Bilocal versus nonbilocal correlations in entanglement-swapping experiments. *Phys. Rev. A* **85,** 032119 (2012).
47. Williams., H. P. Fourier's method of linear programming and its dual. *Amer. Math. Monthly* **93,** 681–695 (1986).
48. Navascués, M., Pironio, S. & Acin, A. Bounding the set of quantum correlations. *Phys. Rev. Lett.* **98,** 010401 (2007).

Acknowledgements

We would like to thank E. Wolfe and M. Pusey for valuable feedback and comments. We acknowledge support by the Excellence Initiative of the German Federal and State Governments (Grant ZUK 43), the Research Innovation Fund from the University of Freiburg. The research by D.G. is supported by the US Army Research Office under contracts W911NF-14-1-0098 and W911NF-14-1-0133 (quantum characterization, verification and validation). C.M. acknowledges support by the German National Academic Foundation, by a Sapere Aude grant of the Danish Council for Independent Research, the ERC starting grant 'QMULT' and the CHIST-ERA project 'CQC'.

Author contributions

All authors contributed extensively to the work presented in this paper.

Additional information

Competing financial interests: The authors declare no competing financial interests.

12

Understanding congested travel in urban areas

Serdar Çolak[1], Antonio Lima[1,2] & Marta C. González[1,3]

Rapid urbanization and increasing demand for transportation burdens urban road infrastructures. The interplay of number of vehicles and available road capacity on their routes determines the level of congestion. Although approaches to modify demand and capacity exist, the possible limits of congestion alleviation by only modifying route choices have not been systematically studied. Here we couple the road networks of five diverse cities with the travel demand profiles in the morning peak hour obtained from billions of mobile phone traces to comprehensively analyse urban traffic. We present that a dimensionless ratio of the road supply to the travel demand explains the percentage of time lost in congestion. Finally, we examine congestion relief under a centralized routing scheme with varying levels of awareness of social good and quantify the benefits to show that moderate levels are enough to achieve significant collective travel time savings.

[1] Department of Civil and Environmental Engineering, MIT, Cambridge, Massachusetts 02139, USA. [2] School of Computer Science, University of Birmingham, Edgbaston B15 2TT, UK. [3] Engineering Systems Division, MIT, Cambridge, Massachusetts 02139, USA. Correspondence and requests for materials should be addressed to M.C.G. (email: martag@mit.edu).

Cities have a long-standing history cultivating technological innovations that allow citizens to efficiently access goods and opportunities. However, the ease of access has been increasingly difficult to maintain under rapid urbanization[1–7]. As growing population densities create excessive demand for cities' infrastructure, the increasing penetration and advancement of technology generates massive amounts of multidimensional data that can be used to study and mitigate this demand. Specifically, the availability of mobile phone data has led researchers to quantify fundamental spatiotemporal patterns to better understand human mobility in urban areas[8–12]. With the continuous increase in the volume and accuracy of new data sources, new methods that process and distill mobile phone data are consistently refined, and traditional models of mobility such as the gravity-, radiation- or activity-based models are being updated in tandem[13–18]. In the context of travel demand estimation, previous efforts focused on developing models that combine household travel surveys with census and land-use information[19,20]. Despite the robust methodology and meticulous implementation of these models, the high costs associated with obtaining the infrequent and small data have proven to be the bottleneck. To supplement these approaches, traffic simulations and demand estimation models have begun incorporating big data sources into their forecasts, building portable data pipelines to create data-driven decision-making tools for policy makers[21–23].

Understanding of the complex interplay of road infrastructure and travel patterns to model travel times and congestion in not a single city but many at once has been a particular challenge in this line of research[24–26]. Road networks, the circulatory system sustaining a city's accessibility and cultivating its economic prosperity[27–29] are seized with congestion in most large metropolitan areas. In their 2013 report, TomTom, a leading GPS company, states that in cities such as Moscow, Istanbul, Rio de Janeiro, Mexico City and Beijing, people on average spend >75% extra time travelling due to traffic. The resulting loss of time, money and energy are borne by the city's citizens and travellers. Municipalities continually invest in road infrastructure construction and maintenance to increase supply, although controversies on whether more roads alleviate congestion persist[30]. Other efforts to reduce congestion aim to decrease driving demand by promoting alternative travel modes, high-occupancy driving lanes, carpooling, congestion pricing and, in extreme cases, road space rationing. Even with all these measures, congestion remains inherent and drivers are increasingly leveraging real-time information through GPS devices and online routing tools to move faster. With everyone having easy access to traffic information, drivers make decisions without coordination based on near-perfect information, resulting in suboptimal system configuration. This general trend of using raw real-time information in decision-making has significant implications, as it might be also used as a tool to guide drivers to make choices for the benefit of the city, thus creating a more optimal traffic configuration. The extent of the global inefficiency has been of great interest[31–34] in many contexts, ranging from wireless networks to transportation[35–40]. Theoretical approaches to bring the system to optimality generally converge to marginal cost taxation, which essentially forms the basis of congestion pricing schemes today[41,42]. Despite the abundance of research on optimal flow configurations and their implications in the transportation, urban planning and economics literature, there is a shortage of works that use big data sources to understand the role of travel demand and actual travel times in metropolitan regions when comparing cities. This highlights a need to build a framework that can be replicated to systematically generate meaningful travel times to not only understand cities better but also test solutions to urban problems such as congestion or pollution.

In this work, we address this issue by coupling travel demand profiles and travel time estimates to analyse how efficiently people move across cities. We begin by modelling the supply by parsing publicly available OpenStreetMap data to obtain road networks. To model travel demand, we mine massive mobile phone data sets, also referred to as call detail records (CDRs)[43]. This procedure requires home and work location detection for millions of users, mining of their location shifts, and the proper sampling procedures to represent accurately the trip tables for the whole city (see Supplementary Notes 1–3). Using this information of the trip distribution within the city, we estimate morning peak vehicular volumes from origins to destinations and compare the inferred travel times based on demand with the estimates of an online map provider in the respective routes and hour of the day. We then explore the relationship between travel distance and travel time across many cities. We show that the time lost due to congestion in each city can be accounted by a dimensionless parameter Γ that measures the ratio between the vehicular travel demand and the road infrastructure supply for the city. To a lesser extent, the differences in congestion levels depend on the population density and the spatial distribution of population. Next, we calculate the detrimental effects of selfish routing by comparing obtained travel times to those that would be observed if the routes were selected to attain the social optimum. We then explore the bounds of the benefits of leveraging information technologies to influence route choices in ways that would help create a more optimal system configuration for vehicular travel. To do so, we implement a generalized selfish routing model that generates expected travel times for varying levels of consideration of overall social good, or λ. We analyse the system gains of socially aware driver behaviour, as well as exploring the distributions of benefits and losses at the individual level. We present our findings for ;five major cities around the world: Boston and San Francisco Bay Area in the United States, Rio de Janeiro in Brazil, and Lisbon and Porto in Portugal.

Results

Approach. We formalize the traffic problem by modelling route choice as follows: every driver i makes a choice of the route p to their destination. This choice depends on a personal utility $u_i = \sum_{e\in p} c_e(x_e)$, expressed as the sum of the costs c of every road segment e along the chosen route. For simplicity, we assume that the cost of a road segment for driver i is equal to the travel time, $c_e(x_e) = t_e(x_e)$, where $t_e(x_e)$ represents the travel time t observed on road e for vehicle flow x_e. We can then define the total cost incurred by all users as $C = \sum_{e\in E} x_e t_e(x_e)$. The flow configuration that results in the optimal cost is referred to as the socially optimal flows obtained by a typical minimum cost network flow programme[44]:

$$\begin{aligned} \underset{x_e \forall e\in E}{\text{Minimize}} \quad & C \\ \text{subject to} \quad & \sum_p f_p^{st} = f^{st} \\ & x_e = \sum_s \sum_t \sum_p f_p^{st} \delta^{st}(p, e), \\ & x_e \geq 0, f_p^{st} \geq 0. \end{aligned} \tag{1}$$

where x_e refers to the flow on road e, f_p^{st} is the flow between the source s and target t on route p, and $\delta^{st}(p, e) = 1$ when road e lies on route p.

As drivers make selfish choices, the system settles into a suboptimal state. Although driver i only experiences and considers his/her own travel time, the cost the whole system incurs also includes the marginal cost driver i imposes on all

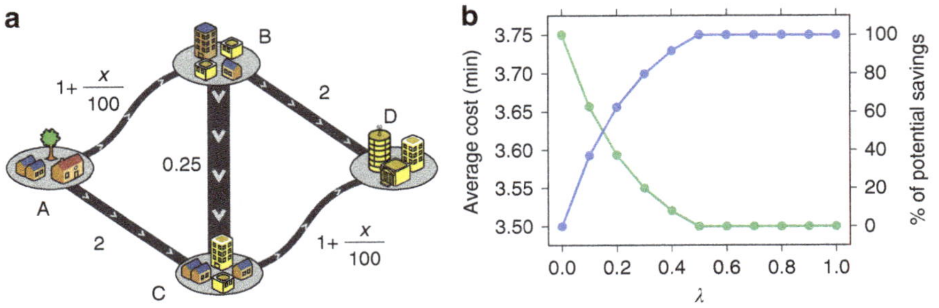

Figure 1 | Illustration of routing equilibrium. (**a**) In this small network, 100 drivers are going from A to D. The road labels represent the costs of travel as a function of vehicle flows. User equilibrium allocates the flows between paths as $f_{ABD}=f_{ACD}=25$ and $f_{ABCD}=50$, and the average travel time is 3.75 min for all drivers. Socially optimal flows decrease total travel time to 3.5 by $f_{ABD}=f_{ACD}=50$ and $f_{ABCD}=0$, with road *BC* remaining unused. (**b**) Achieved percentage of potential savings for increasing values of social good weight λ: 10 and 20% social good weight results in 40 and 60% of potential savings, respectively.

other drivers on the road segments he/she takes. The set of flows that occur when every driver minimizes their own travel time is referred to as the user equilibrium flows. Theoretically, in the resulting system state, no driver can benefit from deviating from their route. This idea, essentially describing a Nash equilibrium in roads, is captured in Wardrop's principles in transportation[36]: the journey times on all the used routes for an origin–destination (OD) pair are equal and are less than those that would be experienced by a single vehicle on any unused route. This routing game is solved through a potential function $\phi_e(x_e)=\int_0^{x_e} t_e(x)dx$ such that $\phi'_e(x_e)=t_e(x_e)$(ref. 45). The convex programme for the user equilibrium problem has been formulated[46] as follows:

$$\begin{array}{ll} \underset{x_e \forall e\in E}{\text{minimize}} & \sum_{e\in E} \phi_e = \sum_{e\in E} \int_0^{x_e} t_e(x)dx \\ \text{subject to} & \text{constraints in Eq. 1.} \end{array} \tag{2}$$

Figure 1a depicts an example that captures solutions for equilibrium and optimal flows for a widely used toy network. For the demand of $d_{AD}=100$, the user equilibrium flows allocate 50 drivers on path ABCD and 25 drivers on paths ABD and ACD each, resulting in a travel time from A to D of 3.75, regardless of the path chosen. The socially optimal configuration avoids allocating too much flow on the path ABCD, as its marginal cost is higher than those of paths ABD and ACD. By minimizing the marginal cost, path ABCD receives no flow and the average cost is minimized at 3.5.

To assess the benefits of different scenarios based on travel demand information, we make use of the formulation proposed in ref. 47. We reconfigure the utility function of a driver as a linear combination of the cost he/she will incur and the total marginal cost his/her choice imposes on everyone else:

$$\begin{aligned} c_e^{\lambda}(x_e) &= (1-\lambda)t_e(x_e) + \lambda \frac{d[x_e t_e(x_e)]}{dx_e} \\ &= t_e(x_e) + \lambda x_e \frac{dt_e(x_e)}{dx_e} \end{aligned} \tag{3}$$

λ defines the weight towards social good; it is a parameter ranging between 0 and 1. A driver with $\lambda=1$ chooses routes with respect to the marginal costs, thus moving the system closer to the system optimum. Conversely, a user with $\lambda=0$ only considers the cost of his route and potentially moves the system away from optimality. The resulting convex programme for the socially aware routing problem is as follows:

$$\begin{array}{ll} \underset{x_e \forall e\in E}{\text{minimize}} & \sum_{e\in E} \int_0^{x_e} c_e^{\lambda}(x_e)\cdot x_e dx_e \\ \text{subject to} & \text{constraints in Eq. 1.} \end{array} \tag{4}$$

For the city depicted in Fig. 1a, the user equilibrium configuration results in an average cost of 3.75 min per driver versus 3.5 min the system optimum, meaning solely by adjusting routing behaviour to $\lambda=1$, a benefit of 0.25 min can be achieved per driver. Figure 1b shows that for $\lambda=0.1$, when the drivers begin valuing social good as well, the average cost drops to $\sim$3.65 and almost 40% of potential savings are realized. In fact, the social optimum is achieved at $\lambda=0.5$.

Travel times. To understand the relationship between travel demand and driving travel times, we begin by comparing our five cities during estimated morning peak period traffic conditions. The areas of analysis are significantly diverse: Rio is very highly populated over its large extensions, whereas Porto's population density considerably decreases after $r>20$ km from the most dense location. Rio de Janeiro, the Bay Area and Lisbon extend across Guanabara Bay, the Bay and Tagus, respectively, and have many inhabitants commuting on few bridges (see Supplementary Fig. 3 and Supplementary Table 1 for more details). As a consequence of their differences, cities demonstrate varying traffic conditions, as shown in Fig. 2. The volume-over-capacity ratio (VOC) measures how successfully a road segment is able to cope with the assigned volume of vehicles, with high VOC values indicating more congestion. High VOCs are generally observed on highways, as they provide faster means of travel due to their wider roads, increased number of lanes and higher speed limits. In addition, bridges and roads that lie central in the network topology are typically congested due to a lack of alternative routes.

We begin by analysing the efficiency of urban mobility for the five regions to understand the mechanisms underlying observed travel times. The main determinant of congestion is travel demand, which is heavily tied to commuting trip distances during weekday peak travel times. In Fig. 3a, we demonstrate that the straight-line (Euclidean) commuting distances, d, follow a lognormal distribution, $f(d)=\frac{1}{\sqrt{2\pi}\sigma d_b}e^{-(\ln(d)-\mu)^2/2\sigma^2}$ with means ranging from 5 to 8 km ($\mu=1.6$–2.1) and s.d. ranging from 2 to 4 km ($\sigma=0.7$–1.2) (see Supplementary Fig. 7). It can be observed that majority of trips span relatively short distances and trips over 25 km are uncommon. However, what makes a city more traversable are the speeds at which drivers can span these distances. In Fig. 3b we investigate the effective speeds in both free and congested traffic conditions. It can be observed that cities exhibit similar free travel-speed distributions, normally distributed with μ fluctuating around 50 km h^{-1} with mean values reported in the legend. The differences in road network supply $S=\sum_{x_e>0,e\in E} l_e C_e$ (km vehicles per hour), where l_e and C_e are the length (km) and the flow capacity (vehicles per hour) of a road segment e, explains the slight differences in free flow speeds, as

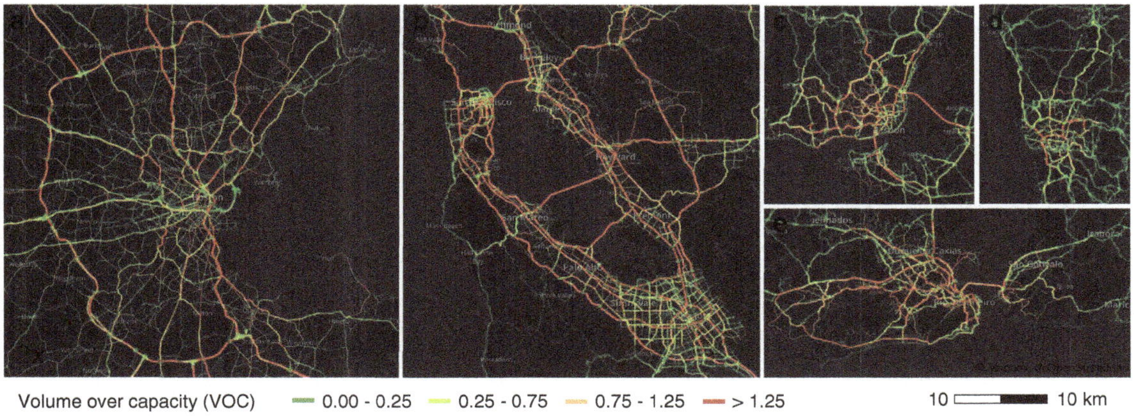

Figure 2 | The maps of VOCs (volume over capacity) of the roads in the user equilibrium configuration. The depicted cities are (**a**) Boston, USA, (**b**) San Francisco Bay Area, USA, (**c**) Lisbon, Portugal, (**d**) Porto, Portugal, and (**e**) Rio de Janeiro, Brazil. Higher VOCs are generally observed in highways, as they provide faster means of travel. (Boston is 2x the distance scale.) Maps under © OpenStreetMap contributors BY-SA.

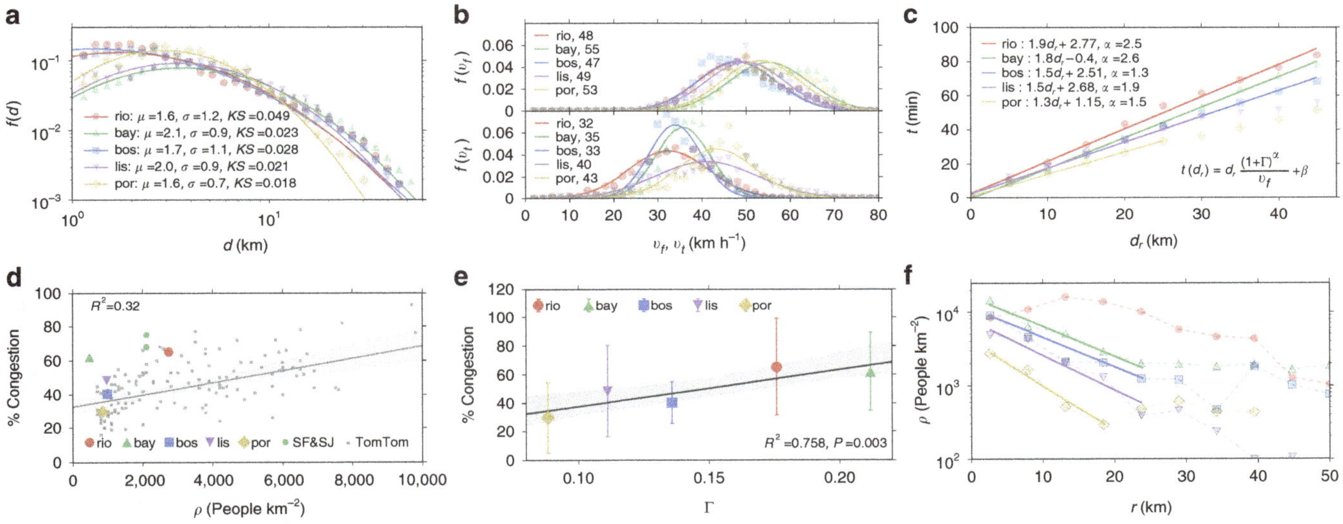

Figure 3 | Comparisons of cities and their congested travel. (**a**) Distributions of commuting trip distances, d, in the morning peak period with parameters of the fitted lognormal distribution depicted in the legend (see Supplementary Fig. 7 and Supplementary Table 2 for more detail). (**b**) Distribution of trip free flow speeds, v_f, and in traffic conditions, v_t. (**c**) Commuting travel times versus route distances of commuters, d_r. (**d**) Estimates of overall mean % of time lost in congestion versus population density p for TomTom Traffic Index estimates and our analysis. (**e**) Relationship of overall mean % congestion to the demand to supply ratio, Γ, for the five subject cities, with error bars specifying the s.d. (see Supplementary Fig. 8). (**f**) Average population density ρ as a function of distance from the most dense area in the region, r.

seen in Table 1. These differences are significantly more apparent in speed distributions under real traffic conditions: the effective OD travel speeds in Rio, the Bay Area and Boston decay considerably compared with those in free traffic conditions, whereas the speeds in Porto and Lisbon change less. We explore further these two different responses given the demand profiles of each city.

To that end, we analyse the experienced travel times per distances travelled in Fig. 3c. We observe a strong yet very simple relationship that pronounces the differences between the subject cities: Rio de Janeiro is the slowest city and is followed next by the Bay Area, and Porto is the fastest. All cities exhibit a linear relationship, with the exception of long-distance trips in Porto and Lisbon where a different regime appears for longer distances. To explain this observation, we model travel times by city-specific parameters describing the demand, the capacity and observed free traffic speeds. In doing so, we define demand-to-supply ratio of a city as

$$\Gamma = \frac{\sum_{e\in E} l_e x_e}{\sum_{x_e>0, e\in E} l_e C_e}. \quad (5)$$

This dimensionless measure is a simple ratio of the total distance travelled by all vehicles to the upper bound of the total vehicle kilometres the road network can support per hour, thus capturing the load on the road infrastructure by bringing together trip distances, trip magnitudes, road capacities and the distances they span as shown in Table 1. Using this measure along with v_f, the average free travel speed of each city, we are able to better explain the linear relationship between travel time and distance by

$$t(d_r) = d_r \frac{(1+\Gamma)^{\alpha}}{v_f} + \beta, \quad (6)$$

Table 1 | A comparison of general properties of the subject cities.

	City				
	Rio	**SF Bay**	**Boston**	**Lisbon**	**Porto**
Population (millions)	12.6	7.15	4.5	2.8	1.7
Area (1,000 km^2)	4.6	18.1	4.6	2.9	2.0
Demand (vehicle km h^{-1})	3.1	9.1	5.4	2.9	1.1
Supply (vehicle km h^{-1})	17.6	43.0	39.7	25.5	11.7
Demand-to-supply (Γ)	0.18	0.21	0.14	0.11	0.09
Expansion factor	890	100	32	96	164
Vehicle usage (vehicle per person)	0.25	0.67	0.67	0.56	0.62

Table 2 | Comparison of cost findings in the subject cities for the morning peak hour.

	City				
(min)	**Rio**	**SF Bay**	**Boston**	**Lisbon**	**Porto**
FTT	20.6	21.1	19.3	22.4	15.3
Loss	**14.1**	**12.5**	**8.2**	**8.0**	**4.0**
UE	34.7	33.6	27.5	30.4	19.3
Benefit	**2.6**	**2.6**	**1.3**	**2.1**	**1.1**
SO	32.1	31.0	26.2	28.3	18.2
%S	18	21	16	27	28

FTT, free travel time; SO, social optimum; UE, user equilibrium; % S, percentage of total congestion attributed to selfish routing, defined as $S = 100$*Benefit/Loss.
Bold rows indicate the loss of travel times from free travel times to socially optimal flows, and from socially optimal flows to user equilibrium flows for commuters, respectively.

where α-values vary between 1.3 and 2.5, essentially describing the sensitivity of the city to the stress imposed by travel demand on its road infrastructure.

To untangle the particular ordering of cities in terms of speed and understand why some cities are more congested than others, we investigate a typical relationship in Fig. 3d, to test the common conception that cities with higher population densities tend to exhibit more heterogeneity in their demand profiles, and therefore tend to be more congested. For this purpose, we measure the ratio of the time lost in traffic to the travel time under free flow conditions, known as the traffic index, along with those measured for many other urban areas by TomTom, a leading GPS company. We consider the percentage of congestion, defined as the percentage of additional travel time due to traffic compared with free flow conditions, for different population densities in these various cities. We observe that Boston, Lisbon and Porto fall on the fit model, whereas the Bay Area and Rio demonstrates a significantly higher level of congestion. The outlier appearance of the Bay Area is a consequence of the arbitrary definitions of urban areas and its influence in population density as pointed out in ref. 4. To account for this, we plot the subdivisions of San Francisco and San Jose, which support the relationship, as they lie closer to the fit. Interestingly, the dimensionless demand-to-supply ratio Γ lacks this problem and presents a better linear trend with congestion for the five analysed urban areas as depicted in Fig. 3e, despite the broad behaviour of the traffic response. The two most congested cities have the highest ratios, the Bay Area closely followed by Rio de Janeiro, whereas Porto and Lisbon, the two least congested cities, have lower ratios.

To finalize our analysis, in Fig. 3f we measure how population densities are spatially distributed from the most densely populated region in each of the subject cities based on the chosen administrative level. The results show different spatial distributions in the population density of the five cities. First, it verifies the expected effect of higher population densities in increasing congestion. It also highlights the importance of the spatial distribution around the highest density point. Lisbon and Porto present densities of population below 500 people per km^2 for distances of $r > 20$ km, whereas the other three cities stabilize in values $> 1{,}000$ people per km^2. These differences can explain the two types of responses in the effective travel speeds presented in Fig. 3b, where Lisbon and Porto belong to a city type of lower density. Taking these results together, we observe that congestion increases with Γ and appears to be influenced by the spatial distribution of population density and its gradient.

Selfish routing. In this section, we compare the travel times for commuters in free flow, socially optimal and user equilibrium flow configurations. Our findings in the five subject cities are outlined in Table 2. Although the estimated free travel time averages are similar, congestion plays a significant role: Lisbon commuters lose 2.1 min on average by selfish routing preferences. Rio de Janeiro exhibits an average loss of 2.6 min on average incurred by selfish routing. The results show that on average 15–30% of total minutes lost in congestion is caused solely by selfish routing.

Although a more nuanced methodology incorporating stochastic traffic assignment and probabilistic OD matrices would probably improve validation, our formulation and central findings would remain robust, as they are based on aggregate and endogenous, albeit simplified, behaviour of our system. Furthermore, a principled and singular validation source does not exist for our cities; we instead use an online map provider as a validation benchmark. Although the validation data are also the product of internal models and estimations, it is of value as they are obtained from an independent data source to ours. In Fig. 4a, we compare the distributions of obtained travel times with those obtained from the map provider in the morning peak hour between 7:30 and 8:30 h for 2,000 OD pairs with the highest commuting flows (see Supplementary Table 3 for statistics related to the regressions). There is an overall overestimation of travel times, which strengthens the notion that route choice in reality might not be a perfect user equilibrium or a social optimum, but somewhere in between. Neither the provider's nor our findings are expected to have accurate travel time variability, as these comparisons are estimates of typical travel times for the given OD pairs and they act as a first step towards the validation of our estimated travel times based on the assigned traffic flows obtained from the phone data.

Weight of social good. In assessing the effects of socially aware routing behaviour for the subject cities, we calculate the average commuting time for various levels of λ. The inset of Fig. 4b depicts the decrease in average commuting travel times for increasing λ in all five cities, ranging from an average of 1–3 min. More importantly, the shape of the curves indicate that even modest social consideration weights can realize a significant portion of the potential savings. Figure 4b collapses these curves to represent realized potential savings as a percentage to exhibit a striking similarity between the five cities in terms of response to socially aware routing. To assess the economies of such routing behaviour, we measure the Gini index of the obtained curves; by definition, higher values of G indicate higher savings for smaller levels of social good weight. Our findings show that G ranges from 30–40%: $G_{\mathrm{rio}} = 41\%$, $G_{\mathrm{bay}} = 42\%$, $G_{\mathrm{bos}} = 33\%$, $G_{\mathrm{lis}} = 30\%$ and $G_{\mathrm{por}} = 34\%$. These findings indicate congested cities benefit

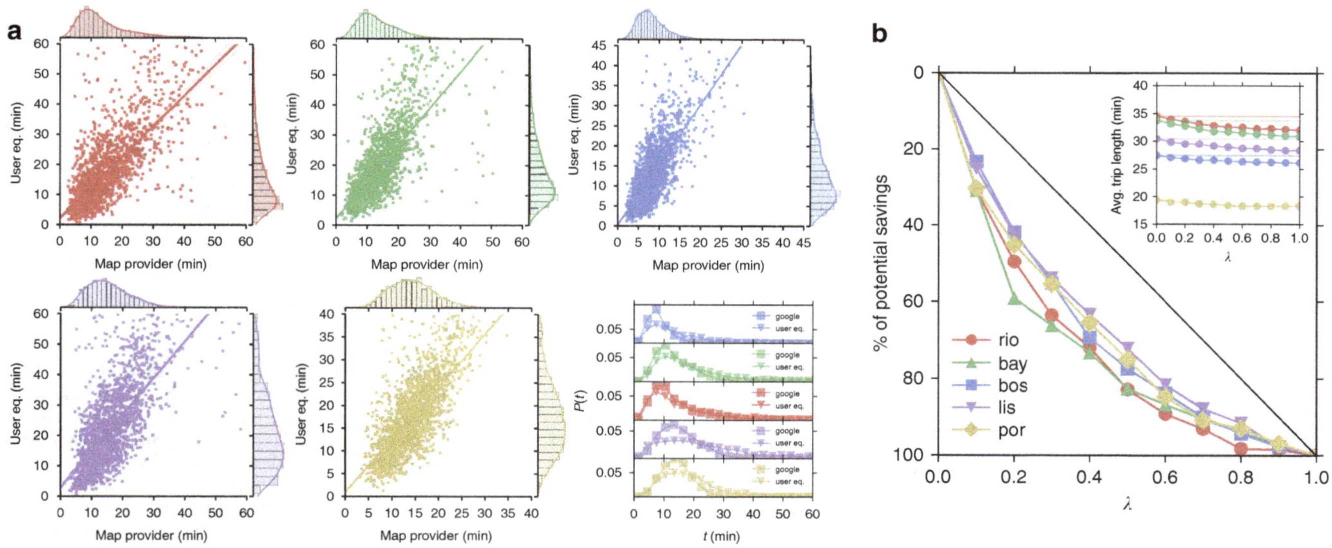

Figure 4 | Travel time comparisons and potential savings. (**a**) Comparison of travel times and their distributions between user equilibrium versus routes obtained from the online map provider. OD samples consist of 2,000 OD pairs with the highest commuting flow magnitudes for each city. (**b**) The percentage of potential savings in average commuting times for the five cities for varying levels of social good weight of routing. (inset: the travel time savings represented in actual minutes).

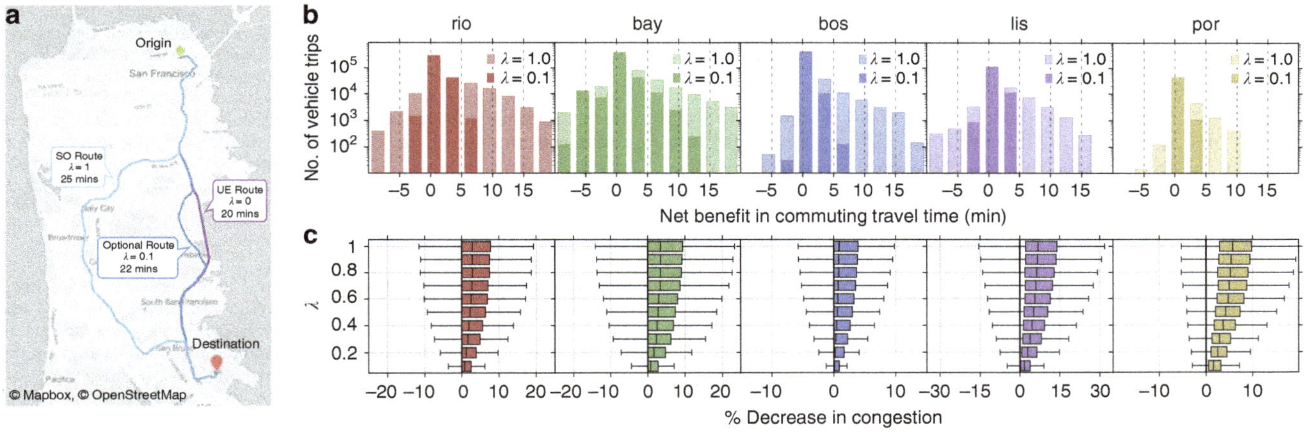

Figure 5 | Benefit and congestion decrease distributions for different weights of social good. (**a**) A depiction of three route alternatives with the corresponding travel times for a trip from Union Square to San Francisco Airport for $\lambda = 0$, $\lambda = 0.2$ and $\lambda = 1$, respectively. (**b**) Counts of vehicle trips and observed travel time benefits for $\lambda = 1$ and $\lambda = 0.1$. Negative benefits refer to increase in travel times for vehicles sacrificing for the social good. The spread of the distributions increase for higher λ. (**c**) The response of distributions of percentage decrease in time lost to congestion to increasing values of λ. The skewness towards positive values of congestion decrease indicate movement towards more optimal configurations. Maps under © OpenStreetMap contributors BY-SA.

more from incorporating social good considerations into routing behaviour.

Travel time benefit distributions. In the previous section we characterized the percentage of potential savings that can be obtained for increasing levels of social consideration. However, these benefits are achieved at the expense of time of drivers who adjust their commute for the benefit of others. The unwillingness to give up time is the defining factor in drivers' failure to reach an optimal state on their own. This highlights the importance of fairness of the distribution of who has to sacrifice versus who benefits in terms of both the success potential of the implementation of policies or a reward/punishment reinforcement schema. Figure 5a demonstrates one such schema, where drivers are shown a route that corresponds to a choice, which might result in a travel time sacrifice.

Our findings, in accordance with the results of the previous sections, indicate a net bias towards benefits, meaning the number of drivers who benefit outnumber those who sacrifice. Figure 5b summarizes the benefit distributions for the five cities for $\lambda = 0.1$ and $\lambda = 1$. The former exhibits a less spread distribution than the latter but the skewness remains inherent to the distributions. Although the average benefits described in the previous sections appear small, it should be noted that 10-min benefits can be observed for tens of thousands of vehicles. Figure 5c describes in more detail how the positive skewness evolves for increasing social consideration. For higher λ, the % decrease in congestion distributions are shifted towards positive values, indicating a net benefit. This result demonstrates the potential of incentive schemes, which could compensate the few drivers who sacrifice under consideration of social good.

Discussion

The economic and social costs of congestion are crippling. In addition to the overall loss of time, congestion underlies many major economic and urban issues such as increased gas consumption, infrastructure deterioration and CO_2 emissions. In this work, we use massive amounts of data to estimate peak hour travel demand and understand travel times. We then explore the power of information-based routing on congestion alleviation.

Our findings suggest very interesting similarities in the behaviour of the five subject cities to explain congestion and potential benefits of social routing. Commuting distances follow a lognormal distribution and free travel speeds are normally distributed. A city's unique congestion fingerprint is strongly related to measurable characteristics. The population density and its spatial distribution together with the Γ parameter of demand-to-supply ratio are the two driving factors of the observed congestion in a diverse range of cities. Further, given the current state of traffic, we then estimate how centralized routing schemes using the power of information would reach possible benefits in travel times. Such information is important, as it allows the assessment of the upper bounds of routing policies; if effective in implementation, it would influence the traffic on a city scale. In practice, this would imply that we could have similar routing applications that we use today with the incorporation of demand profiles, to provide routes that are not necessarily the shortest but also the best for decreasing overall congestion.

We find that routing solutions that mimic socially optimal configurations, that is, $\lambda = 1$, have a limit of decreasing time lost in congestion by up to 30%. This is in contrast with the effectiveness of direct and costly interventions where 1% target decrease in demand can achieve 18% decrease in travel times[18]. Although in both scenarios the collective benefits for the whole city can be significant (15–30% decrease), the observed time benefits the average individual receives are marginal, ranging from 1 to 3 min. Furthermore, these times are below the travel time variability based on events, weather conditions or traffic lights. Our findings indicate that in the best-case scenario, time savings would be imperceptible for the majority of the drivers. From this, it is clear that such routing solutions cannot fix the traffic problem for individual drivers but rather would contribute to the city as a whole. The advantage is that in the context of the implied routing application, the number of vehicles sacrificing their travel time is significantly smaller than the number of those that benefit. Lower levels of weight towards social good will also moderate the magnitude of benefits and losses, consequently making the policies fairer and easier to implement.

Open work in this subject contains, but is not limited to, a more generalized bottom-up approach to comparison of cities that includes various modes of transportation to demonstrate their similarities, differences and their consequences. As the volume, the variety and the resolution of data increase along with the expected disruptions from connected self-driving cars and similar technologies, this front of research will become more relevant to facilitate the study and planning for the future of urban mobility. With more updated demand models extracted from communication technologies, understanding the network effects on congestion will become easier to pinpoint and address. In addition, planning tasks on urban mobility previously difficult to tackle may now be addressed at lower costs and with much larger samples of the population. For example, a thorough analysis of how travel time and congestion is distributed among the population and its split by income and other sociodemographic characteristics remains an open front.

Methods

Mobile phone data. Mobile phone data sets, also referred to as CDRs, used in this study consist of at least 3 weeks of records of all mobile phone users of a particular carrier across each subject city. Each individual CDR consists of a hashed user identification string, a timestamp and the location of the activity. The spatial granularity of the data varies between cell tower level, where calls are mapped to tower locations and distributed uniformly within the Voronoi cell that it forms, and triangulated geographical coordinate pairs, where each call has a unique pair of coordinates accurate to within a few hundred metres. Market shares associated with the carriers that provide the data also vary (see Supplementary Figs 1 and 2, and Supplementary Note 1).

Census and travel survey data. At the census tract (or equivalent) scale, we obtain the population, vehicle usage rate and median income of residents in that area. For US cities, the American Community Survey provides this data on the level of census tracts (each containing roughly 5,000 people). Census data are obtained for Brazil through IBGE (Instituto Brasileiro de Geografia e Estatística) and for Portugal through the Instituto de Nacional de Estatística. All cities analysed in this work have varying spatial resolutions of the census information. Wherever possible, we obtain the most recent travel demand model or survey from the subject city and compare the results with those output by our methods. We use the 2011 Massachusetts Household Travel Survey for Boston, 2,000 Bay Area Transportation Survey for the Bay Area and a recent transportation model output provided by the local government for Rio de Janeiro. For Lisbon, the most recent estimates from the MIT-Portugal UrbanSim LUT model that uses the 1994 Lisbon transportation survey as input are used. We found no recent travel survey or model for Porto (see Supplementary Note 2).

Extraction of validated OD information. Traditional modelling approaches to OD information use data obtained from travel surveys, possibly combined with land-use and point-of-interest information, to generate estimates of trip production and attraction for locations. Although new data sources such as CDRs do not provide the same detailed demographic and contextual information about individuals or trips, they do provide many high-resolution data points over a far longer observation period. Mobile phones offer good, but imperfect measurements of geographic position due to the uncertainty of the location estimates and the non-uniform sampling frequency (see Supplementary Fig. 5 and Supplementary Note 3 for procedures to generate OD matrices and more descriptive information). For further questions and inquiries about the OD data, please contact the corresponding author.

Road networks. For many cities in the United States, detailed road network data are made available by local or state transportation authorities. These data sets generally are well maintained; however, many properties are often incomplete or missing entirely. For this purpose, we infer required road characteristics to build realistic and routable networks using OpenStreetMap, an open-source crowd sourced mapping tool (see Supplementary Note 4).

Traffic flow and travel time. Relating travel performance to traffic conditions has been a long-standing problem in transportation. Many different characterizations exist, ranging from conical volume-delay functions to more complex approaches (see Supplementary Fig. 4 and Supplementary Note 5).

Traffic assignment. Traffic assignment is a mature domain that aims to bring together travel demand with road infrastructure, to better understand traffic, and has been studied extensively by urban and transportation planners. In this work, we follow an efficient, static, origin-based assignment algorithm that focuses on the equilibration of a directed acyclic graph structure emanating from every origin node (see Supplementary Fig. 6 and Supplementary Note 6).

References

1. Glaeser, E. L., Kallal, H. D., Scheinkman, J. A. & Shleifer, A. Growth in cities. *Working Paper 3787* (National Bureau of Economic Research, 1991).
2. Batty, M. The size, scale, and shape of cities. *Science* **319,** 769–771 (2008).
3. Bettencourt, L. M., Lobo, J., Helbing, D., Kühnert, C. & West, G. B. Growth, innovation, scaling, and the pace of life in cities. *Proc. Natl Acad. Sci. USA* **104,** 7301–7306 (2007).
4. Arcaute, E. *et al.* Constructing cities, deconstructing scaling laws. *J. R. Soc. Interface* **12,** 20140745 (2015).
5. Bettencourt, L. M. A. The origins of scaling in cities. *Science* **340,** 1438–1441 (2013).
6. Hernando, A., Hernando, R. & Plastino, A. Space-time correlations in urban sprawl. *J. R. Soc. Interface* **11,** 20130930 (2014).
7. Jacobs, J. *The Death and Life of Great American Cities* (Vintage, 1961).

8. González, M. C., Hidalgo, C. A. & Barabasi, A. -L. Understanding individual human mobility patterns. *Nature* **453,** 779–782 (2008).
9. Brockmann, D., Hufnagel, L. & Geisel, T. The scaling laws of human travel. *Nature* **439,** 462–465 (2006).
10. Song, C., Koren, T., Wang, P. & Barabási, A.-L. Modelling the scaling properties of human mobility. *Nat. Phys.* **6,** 818–823 (2010).
11. Song, C., Qu, Z., Blumm, N. & Barabási, A.-L. Limits of predictability in human mobility. *Science* **327,** 1018–1021 (2010).
12. de Montjoye, Y.-A., Hidalgo, C. A., Verleysen, M. & Blondel, V. D. Unique in the crowd: The privacy bounds of human mobility. *Sci. Rep.* **3,** 1376 (2013).
13. Stouffer, S. A. Intervening opportunities: a theory relating mobility and distance. *Am. Sociol. Rev.* **5,** 845–867 (1940).
14. Ren, Y., Ercsey-Ravasz, M., Wang, P., González, M. C. & Toroczkai, Z. Predicting commuter flows in spatial networks using a radiation model based on temporal ranges. *Nat. Commun.* **5,** 5347 (2014).
15. Simini, F., González, M. C., Maritan, A. & Barabási, A.-L. A universal model for mobility and migration patterns. *Nature* **484,** 96–100 (2012).
16. Yan, X.-Y., Zhao, C., Fan, Y., Di, Z. & Wang, W.-X. Universal predictability of mobility patterns in cities. *J. R. Soc. Interface* **11,** 20140834 (2014).
17. Schneider, C. M., Belik, V., Couronné, T., Smoreda, Z. & González, M. C. Unravelling daily human mobility motifs. *J. R. Soc. Interface* **10,** 20130246 (2013).
18. Wang, P., Hunter, T., Bayen, A. M., Schechtner, K. & González, M. C. Understanding road usage patterns in urban areas. *Sci. Rep.* **2,** 1001 (2012).
19. Ortúzar, J. D. & Willumsen, L. G. *Modelling Transport* (John Wiley & Sons, 1994).
20. Balmer, M. *et al. Agent-Based Simulation of Travel Demand: Structure and Computational Performance of MATSim-T* (ETH, Eidgenössische Technische Hochschule Zürich, IVT Institut für Verkehrsplanung und Transportsysteme, 2008).
21. Toole, J. L. *et al.* The path most traveled: Travel demand estimation using big data resources. *Transport. Res. C Emerg. Technol.* **58,** 162–177 (2015).
22. Alexander, L., Jiang, S., Murga, M. & Gonzlez, M. C. Origin-destination trips by purpose and time of day inferred from mobile phone data. *Transport. Res. C Emerg. Technol.* **58,** Part **B,** 240–250 (2015).
23. Çolak, S., Alexander, L. P., Alvim, B. G., Mehndiretta, S. R. & González, M. C. Analyzing cell phone location data for urban travel: current methods, limitations, and opportunities. *Transport. Res. Rec. J. Transport. Res. Board* **2526,** 126–135 (2015).
24. Louf, R. & Barthelemy, M. How congestion shapes cities: from mobility patterns to scaling. *Sci. Rep.* **4,** 5561 (2014).
25. Noulas, A., Scellato, S., Lambiotte, R., Pontil, M. & Mascolo, C. A tale of many cities: universal patterns in human urban mobility. *PLoS ONE* **7,** e37027 (2012).
26. Louail, T. *et al.* Uncovering the spatial structure of mobility networks. *Nat. Commun.* **6,** 6007 (2015).
27. Lämmer, S., Gehlsen, B. & Helbing, D. Scaling laws in the spatial structure of urban road networks. *Phys. A Stat. Mech. Appl.* **363,** 89–95 (2006).
28. Rosvall, M., Trusina, A., Minnhagen, P. & Sneppen, K. Networks and cities: An information perspective. *Phys. Rev. Lett.* **94,** 028701 (2005).
29. Barthélemy, M. Spatial networks. *Phys. Rep.* **499,** 1–101 (2011).
30. Braess, D., Nagurney, A. & Wakolbinger, T. On a paradox of traffic planning. *Transport. Sci.* **39,** 446–450 (2005).
31. Van Huyck, J. B., Battalio, R. C. & Beil, R. O. Tacit coordination games, strategic uncertainty, and coordination failure. *Am. Econ. Rev.* **80,** 234–248 (1990).
32. Roughgarden, T. & Tardos, É. How bad is selfish routing? *JACM* **49,** 236–259 (2002).
33. Roughgarden, T. *Selfish Routing and the Price of Anarchy* (MIT Press, 2005).
34. Roughgarden, T. & Tardos, É. Bounding the inefficiency of equilibria in nonatomic congestion games. *Games Econ. Behav.* **47,** 389–403 (2004).
35. Vickrey, W. S. Congestion theory and transport investment. *Am. Econ. Rev.* 251–260 (1969).
36. Wardrop, J. G. in *Proceedings of the Institution of Civil Engineers* vol. **1,** 325–378, Part 2 (1952).
37. Boyce, D. E., Mahmassani, H. S. & Nagurney, A. A retrospective on Beckmann, Mcguire and Winsten's Studies in the economics of transportation. *Pap. Reg. Sci.* **84,** 85–103 (2005).
38. Youn, H., Gastner, M. T. & Jeong, H. Price of anarchy in transportation networks: Efficiency and optimality control. *Phys. Rev. Lett.* **101,** 128701 (2008).
39. Sheffi, Y. *Urban Transportation Networks* (Prentice-Hall, Englewood Cliffs, NJ, 1985).
40. Correa, J. R., Schulz, A. S. & Stier-Moses, N. E. A geometric approach to the price of anarchy in nonatomic congestion games. *Games Econ. Behav.* **64,** 457–469 (2008).
41. Pigou, A. C. *The Economics of Welfare* (Palgrave Macmillan, 2013).
42. Smith, M. The marginal cost taxation of a transportation network. *Transport. Res. B Methodol.* **13,** 237–242 (1979).
43. Blondel, V. D., Decuyper, A. & Krings, G. A survey of results on mobile phone datasets analysis. *EPJ Data Sci.* **4,** 10 (2015).
44. Ahuja, R. K., Magnanti, T. L. & Orlin, J. B. *Network Flows: Theory, Algorithms, and Applications* 1st edn (Prentice Hall, 1993).
45. Monderer, D. & Shapley, L. S. Potential games. *Games Econ. Behav.* **14,** 124–143 (1996).
46. Beckmann, M., Mc Guire, C. & Weinstein, C. *Studies in the Economics of Transportation* (Yale Univ. Press, 1956).
47. Chen, P.-A. & Kempe, D. in *Proceedings of the 9th ACM Conference on Electronic Commerce* 140–149 (ACM, 2008).
48. Bertaud, A. The spatial organization of cities. *Deliberate Outcome or Unforeseen Consequence, Background Paper to World Development Report* (2003).

Acknowledgements

We thank Saurabh Amin for stimulating discussions and helpful suggestions and Airsage for the data provided. The research was partly funded by the World Bank, Ford, the Department of Transportation's grant of the New England UTC Y25, the MIT Portugal Program, the MIT-Brazil seed Grants Program and the Center for Complex Engineering Systems at KACST-MIT, and A.L. was funded by the Vest Scholarship.

Author contributions

S.Ç. and A.L. processed and analysed the data. S.Ç. and M.C.G. designed the study and wrote the manuscript. All authors read, commented and approved the final version of the manuscript.

Additional information

Competing financial interests: The authors declare no competing financial interests.

13

Spatio-temporal propagation of cascading overload failures in spatially embedded networks

Jichang Zhao[1], Daqing Li[2,3], Hillel Sanhedrai[4], Reuven Cohen[5] & Shlomo Havlin[4]

Different from the direct contact in epidemics spread, overload failures propagate through hidden functional dependencies. Many studies focused on the critical conditions and catastrophic consequences of cascading failures. However, to understand the network vulnerability and mitigate the cascading overload failures, the knowledge of how the failures propagate in time and space is essential but still missing. Here we study the spatio-temporal propagation behaviour of cascading overload failures analytically and numerically on spatially embedded networks. The cascading overload failures are found to spread radially from the centre of the initial failure with an approximately constant velocity. The propagation velocity decreases with increasing tolerance, and can be well predicted by our theoretical framework with one single correction for all the tolerance values. This propagation velocity is found similar in various model networks and real network structures. Our findings may help to predict the dynamics of cascading overload failures in realistic systems.

[1] School of Economics and Management, Beihang University, Beijing 100191, China. [2] School of Reliability and Systems Engineering, Beihang University, Beijing 100191, China. [3] Science and Technology on Reliability and Environmental Engineering Laboratory, Beijing 100191, China. [4] Department of Physics, Bar-Ilan University, Ramat Gan 5290002, Israel. [5] Department of Mathematics, Bar-Ilan University, Ramat Gan 5290002, Israel. Correspondence and requests for materials should be addressed to D.L. (email: daqingl@buaa.edu.cn).

Resilience of individual components in networks is determined not only by their own intrinsic properties, but also by their functional interactions with other components. For example, a failure of one component in a network may lead to overloads and failures of other components. Starting with a localized failure, such interactions between components can ignite a domino-like cascading failure, which may result in catastrophes such as those observed in many realistic networks[1–5]. The devastating consequences of these cascading failures have stimulated extensive studies[6–9].

Many studies provided deep insight on the conditions and outcome of cascading failures[10–17], which may help to evaluate the system resilience. However, to predict and mitigate the failure spreading in a network, understanding their spatio-temporal propagation properties is essential but still missing. Different from structural cascading failures caused by direct causal dependencies[18–20], overload failures[21–24] usually propagate through invisible paths as a result of cooperative interactions in the system. Actually, a fundamental question has rarely been posed: how overload failures propagate with time in space during the cascading process? Indeed, predicting the spatio-temporal propagation of cascading failures could determine the timing and resource allocation of an effective mitigation strategy in corresponding self-healing technologies.

In this paper, we aim to understand the spatio-temporal propagation of the cascading overload failures in spatially embedded systems[25]. We define two basic quantities to describe the spatio-temporal propagation properties of cascading failures. The first quantity, $r_c(t)$, is the average Euclidean distance of the failures appearing at cascading step t from the centre of the initial failure. The second quantity, $F_r(t)$, is the number of node failures that occur at cascading step t. While r_c can help system regulators to set a 'firewall' at suitable locations before the failure arrives, F_r can suggest the 'height' of the firewall. In our current study, the propagation of cascading overloads is found to follow an approximately constant velocity. This propagation velocity decreases as the system tolerance increases, which can be well predicted by our theoretical analysis with one single constant correction. The propagation velocity is also found here to be similar in various model networks and real network structures.

Results

Propagation properties of cascading overload failures. In this study, we focus on the cascading overloads on spatially embedded networks caused by localized attacks, which are common in natural disaster and malicious attacks[26,27]. In Fig. 1, we show snapshots of the simulated cascading failures, where the failures spread almost radially from the initial attack region and finally spread over the whole system. As can be seen from Fig. 1, the nodes with spatial location closer to the initial failure begin to fail first and form approximately a 'ring of failures'. The ring begins to grow and expand with time until it reaches the system's boundary.

Considering the ring shape of the cascading overload failures (originating from the initial location), it is reasonable to quantify $r_c(t)$ and $F_r(t)$ at cascading step t. Figure 2a shows that the propagation radius of cascading failures is increasing almost linearly with time for different system sizes. This means that the failures spread during the cascade process with an almost constant speed (slope of $r_c(t)$). It can be seen that as the system linear size L increases, the propagation velocity of overload failures increases. The propagation radius of failures is increasing with time and becomes saturated near the boundary. The propagation size of failure F_r (number of new failures at each instant) increases until a certain time step and then decreases (Fig. 2b). The behaviour at large t is due to the finite size of the system, when $r_c(t)$ reaches the order of L the amount of damage can only decrease. Note that for different system sizes, the propagation size of failure F_r reaches the maximum at similar instants, which results from the higher velocity in larger system sizes.

To further understand the effect of system size and tolerance α on failure propagation $r_c(t)$, we rescale it in Fig. 2c by the system linear size L. We find that the curves of rescaled r_c collapse into a single curve for different system sizes at a given tolerance α, suggesting that failures spread in the same relative velocities for different system sizes. When the size of failures, $F_r(t)$, is rescaled by the system size, $L \times L$, the different curves of $F_r(t)/L^2$ also collapse into a single curve (Fig. 2d). It can be seen that the spreading time needed to reach the maximum $F_r(t)/L^2$ is determined by the tolerance α, which implies that large tolerance can postpone the collapse of the system.

In contrast to the dynamics such as epidemics[28] that propagate owing to nearest-neighboring interactions such as contact infection, cascading overloads propagate as a result of the global interaction between all the flows contributed by the whole system. Surprisingly, although the interactions are global, the propagation dynamics in the model network and realistic networks (see Supplementary Figs 1–6 and Supplementary Notes 1 and 2 for more examples) are found rather local. Here the local propagation means that there is a finite characteristic distance ($\Delta(r_c)$) between the successive overloads. This characteristic distance is the value of propagation velocity, which increases non-linearly with decreasing tolerance. The nearest-neighbor propagation of overloads usually assumed in some complex

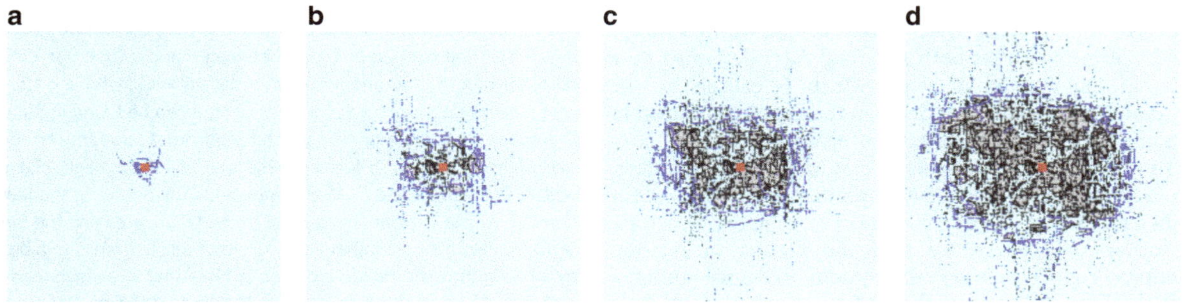

Figure 1 | The propagation of the overload failures in the network. We demonstrate step 1, 3, 5 and 7 of the cascading failures on a 200 × 200 lattice with periodic boundary conditions and a Gaussian distribution of weights. The disorder is $\sigma = 0.01$, the initial attack size (in red) is 6 × 6 and the tolerance of system is set to $\alpha = 0.5$. In each figure, the deep blue dots stands for the overloaded nodes in the current step, while the black ones are the nodes failed in the previous steps. The cyan dots are the functional nodes that did not fail. (**a**) $t = 1$, (**b**) $t = 3$, (**c**) $t = 5$ and (**d**) $t = 7$.

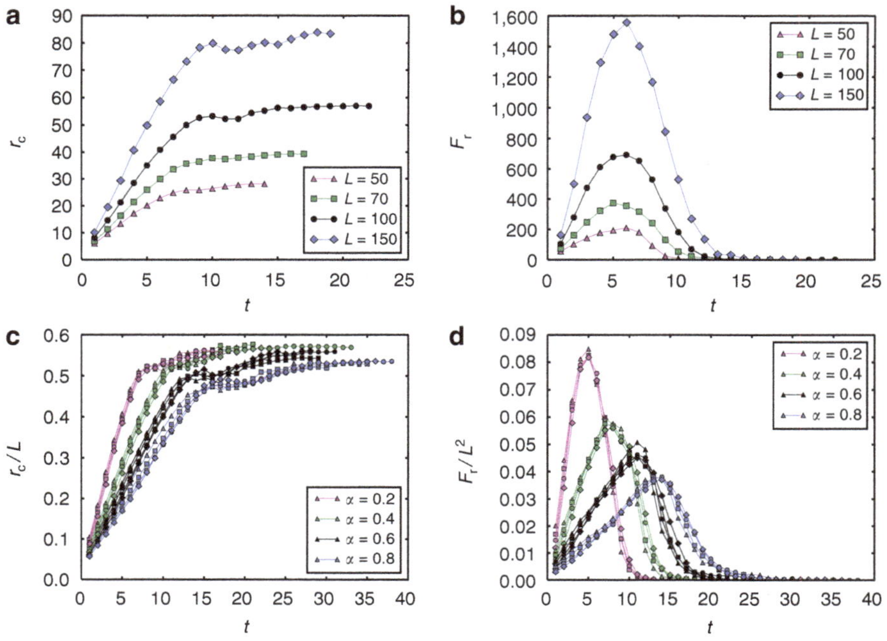

Figure 2 | Spatio-temporal propagation of cascading overload failures in simulations. (**a**,**b**) are the spreading radius $r_c(t)$ of failures and amount of failures at each step, $F_r(t)$, as a function of time for $\alpha=0.25$, $\sigma=0.1$ and $l=6$. (**c**,**d**) are the results of r_c and F_r scaled by the system size, including $L=70$ (triangle), $L=80$ (square), $L=90$ (circle) and $L=100$ (diamond), respectively. The results are obtained by averaging over fifty realizations. Note that simulations are limited by the computational complexity of the most efficient algorithm for calculating node betweenness, which is $O(NM+N^2\log N)$ for weighted networks, where N is the system size and M is the number of edges (NM is the order of N^2 for sparse network).

network models only corresponds to the limiting case of this local propagation ($\Delta(r_c)=1$).

Propagation results based on theoretical analysis. To further explore these propagation behaviours of cascading failures found in simulations and their relations with tolerance, we develop the theory (see Fig. 3, Methods section and Supplementary Methods for more details) to describe the cascading overloads.

As can be seen in Fig. 4, our theoretical analysis reproduces well the spatio-temporal propagation features of cascading overloads found in the simulations (Fig. 2). Specifically, as shown in Fig. 4a,c, the velocity of the failure propagation is almost constant in most of the time t, and decreases with increasing tolerance or decreasing system size. As shown in Fig. 4b,d, the number of failures increases with time and then begins to drop after reaching the peak, which is reduced by increasing the tolerance α. The instant for the maximal failure size is independent of the system size, but increases with the system tolerance. Moreover, similar to simulation results, both the radius and size of failures in the cascades can also be rescaled with system size (L and L^2, respectively) as seen in Fig. 4c,d, where the different curves for different system sizes collapse into a single curve at a given α. Note that both $r_c(t)$ and $F_r(t)$ demonstrate a small slope in the few initial steps, which is caused by the extremely small initial failure considered in the theory. As shown in Supplementary Fig. 7, if the initial size of attack is large, both quantities show a higher slope also in the few initial cascading steps.

As seen in Supplementary Fig. 8 (see Supplementary Note 3 for more details), there exists a critical value of α, α_c, above which no spreading occurs. Moreover, we find the theoretical relation (Supplementary Fig. 9) between the system tolerance and the critical initial failure size, where the system can sustain a larger size of initial failure for a larger α.

Comparison between simulation and theory. To test our theory, we perform a quantitative comparison between the simulation and the theoretical model. As shown in Fig. 5a, the propagation velocity in different model networks and real networks can be well predicted by the theory with the same constant correction for

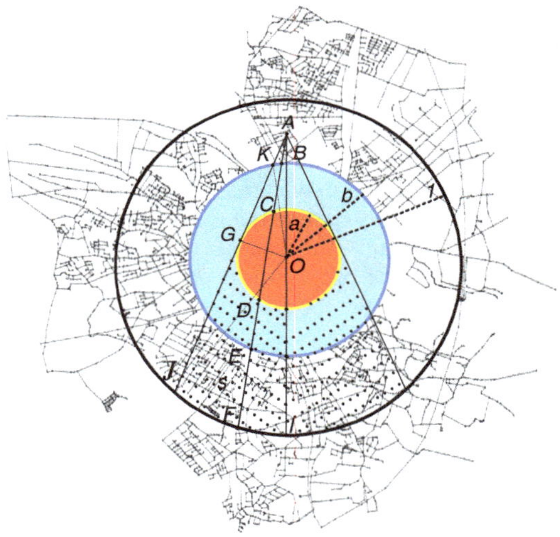

Figure 3 | Theory for overload propagation. The network is embedded in a two-dimensional circular plate centred at O with a radius of 1 unit and the initial failure is located at the centre of the network within a circle of radius $a\ll1$. The ring centred at O and between a and b ($b>a$) is defined as the adjacent ring. A is a random node in the network, whose distance to O is $r\leq1$ (here we assume $r>b$, the case of $r\leq b$ can be found in Supplementary Methods). AF is the original path starting from A to a random node F on the system border. Since $r>b$, the intersection points between AF and the circle with radius b are B and E. AF also intersects the border of the failure area at C and D. AJ is a straight line tangent to the failure area and the tangent point is G. We also define another path AI, which starts from A to a border node I by passing through the plate centre O. Note that a realistic road network is embedded behind the circular plate for demonstration.

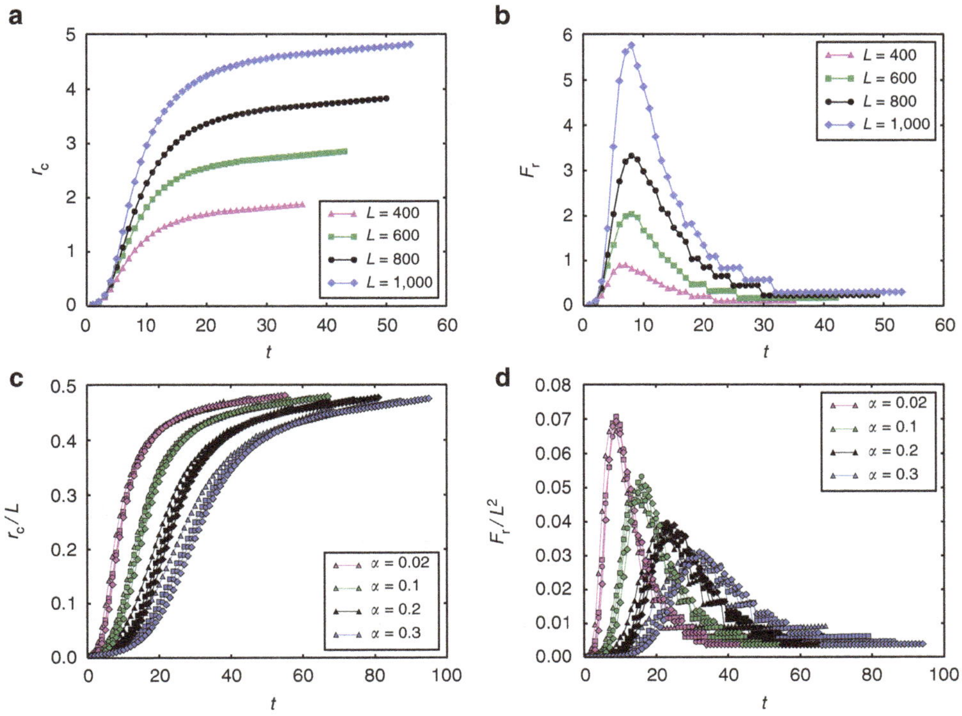

Figure 4 | Theoretical results of overloads propagation. In **a** and **b**, the results from theory with unit distance resolution $1/R$ ($L = 2R$ is the system linear size) are reported for r_c and F_r, where $\alpha = 0.01$. In **c** and **d**, different symbols stand for different resolutions, including 1/200 (triangle), 1/300 (square), 1/400 (circle) and 1/500 (diamond), respectively. Here the absolute radius of the initial damage is set to 3 units. As can be seen from the figure, our theory can predict well the propagation behaviours of cascading overloads in the model simulation. Meanwhile, the numerical calculation of our theory is much less expensive than that of the simulation, which can be applied in the prediction of cascading overloads in large spatially embedded systems.

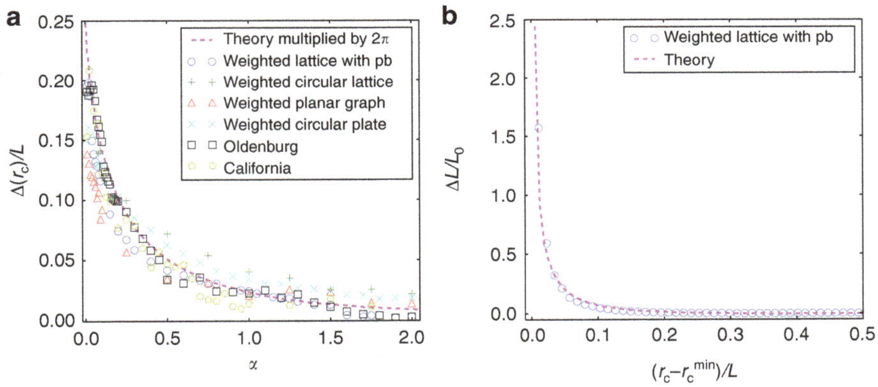

Figure 5 | Comparison between theory and simulation. (**a**) Relative velocity as a function of tolerance in the theory, models as well as in real structures. The relative velocity is calculated in the linear regime of $r_c(t)$, $\Delta\langle r_c\rangle/L$, which decreases with α. The velocity in the theory is multiplied by a constant 2π. We find that the velocity is similar in different model networks (lattice, circular lattice, planar graph and circular plate) and real networks (road networks in Oldenburg and California). (**b**) The average overload as a function of relative distance from the initial attack. The overload in the weighted lattice (circle symbol) and theory (dashed line) after the initial damage is shown. The results are both shifted by the linear size of the initial damage (r_c^{min}), which makes two results comparable from the initial attack. We also multiplied the x axis of theory by 2π, which is consistent with the velocity difference shown in **a**.

all the values of tolerances. This constant correction, close to 2π, is a result of the anisotropic propagation of overloads in the simulation, which is assumed isotropic in the theory. These good agreements between theory and simulations support the validity of our proposed theoretical framework for cascading overloads.

The local propagation found in the simulation and theory is due to the mechanism of overload redistribution. In the simulation of the Motter–Lai model in Fig. 5b, the overloads propagate rather locally (with a characteristic distance) after an initial damage due to the redistribution of optimal paths between existing pairs of nodes. The optimal paths that passed through the previous damaged area will now mainly be redistributed close to this area, leading to the local overloads in spite of global interactions. Similar to our findings in the Motter–Lai model, the overloads in the theory (as shown in Fig. 5b and Supplementary Fig. 10) are found here mainly distributed close to the previous damage area, which causes the local propagation of failures (also as shown in Supplementary Fig. 11 and Supplementary Note 4). Note the agreement in Fig. 5b between theory and simulations (with the same constant correction of 2π, due to the anisotropic nature of the failure spreading, which is assumed isotropic in the theory). The excellent agreement between simulations and theory of the overload (Fig. 5b) as a function of distance from the original failure also supports our theoretical approach.

It is worth noting that the analysis of our theoretical framework is not limited by the computational complexity of calculating optimal paths. This is in contrast to simulations based on betweenness (number of optimal paths passing through a node), which require heavy computations and therefore are limited to relatively small systems. Our theoretical framework thus provides an efficient way to explore and understand the cascading behaviour of failures for any system size.

Universality of propagation velocity. To explore the possibility of universal feature in overload propagation, here we analyse the propagation of cascading overload failures on different network models and realistic networks. Besides the model networks including lattice, circular lattice, planar graph and circular plate, we also study the overload propagation on the realistic road networks[29]. We initiate a local attack in the geographical centre of these spatially embedded networks and analyse the spatio-temporal evolution of r_c and F_r. As shown in Fig. 5a (see Supplementary Figs 1–6 for more examples), we find that propagation velocity of overloads is similar in various model and real network structures, which can be well predicted by our theory. The propagation velocity is independent of detailed network structures, which makes our findings more applicable. This universal propagation of overloads can be attributed to the common mechanism of overload redistribution in different networks, independent of structure difference between these networks. Furthermore, both theory and simulations suggest that for a given system size and a given tolerance, the size of initial failure does not influence the spreading velocity (Supplementary Figs 12 and 13 and Supplementary Notes 5 and 6).

Discussion

Cascading failures represent the manifestation of non-linear butterfly effect in infrastructure networks, which can cause catastrophic damages due to a small local disturbance. The Motter–Lai-type overload cascade models are an important class of cascading failure dynamics, characterized by the nonlocal—in contrast to epidemic spreading-type local cascade models—interactions, and have been studied extensively for the last decade. Given the inherent global interactions in the mechanism of overload formation, it is of interest that the overloads spread rather locally from the initial attack region, at velocity that increases non-linearly with decreasing tolerance. Here the local propagation means that there is a finite characteristic distance between the successive overloads, which is the value of propagation velocity. The nearest-neighboring propagation of overloads usually assumed in some complex network models only corresponds to the limiting case of this local propagation (characteristic distance is 1).

For different model and realistic networks, our results suggest the existence of universal propagation features of cascading overloads, which are characterized by a finite linear propagation velocity. This velocity can be predicted by our theory with the same constant correction for all the values of tolerances. This constant correction, close to 2π, is a result of the anisotropic propagation of overloads in the simulation, which is assumed isotropic in the theory. This universal behaviour comes from the common global mechanisms of overload redistributions in different networks. When certain extreme heterogeneous networks like embedded scale free networks are considered, a revision of the theoretical framework may be needed. While the focus of this manuscript is on the spatio-temporal propagation of the cascading overload failures in spatially embedded systems, the overloads may spread very fast in general non-spatially embedded sparse network due to its small diameter.

The present study may help to bridge the longstanding gap between the overload model[16,22] and the model of dependency links proposed by Buldyrev *et al.*[13] and Parshani *et al.*[18], in particular the lattice version of the model[19,20] where dependency links can have a characteristic length r. Indeed, as can be seen in Figs 2 and 4, overload failures propagate in a nearly constant speed, which suggests a characteristic dependency distance, between successive overload failures. Furthermore, this speed or characteristic distance is found to increase with decreasing tolerance. This suggests a possible mapping between systems with overload failures and networks with dependency links, where networks with different characteristic length of dependency links can serve as a suitable model to describe cascading overload failures. This mapping can be useful since overload models usually require heavy computations and are therefore limited to small systems, while dependency models require significantly less computations, and large systems can be easily analysed.

When a disturbance is detected in networks, the knowledge of spatio-temporal propagation properties of cascading failures is essential for predicting and mitigating the cascading failures. Meanwhile, realistic cascading failures are usually the result of the collective interactions between different processes including overloads and other system operation procedures[30–32]. The universal features of overload propagation found here across different spatially embedded networks may help to better mitigate realistic cascading failure, if combined with the detailed knowledge of other processes including system operations and planning procedures.

Methods

Simulation of model. To study the propagation properties of cascading overload failures, we model the spatially embedded network as a randomly weighted $L \times L$ lattice with periodic boundaries, where L is the linear length of the lattice. The weight of each link is taken from a Gaussian distribution $N(\mu, \sigma^2)$ with mean weight μ and the disorder is represented by the standard deviation σ. In this model[22], the load on a node i, L_i, represents the number of optimal paths between all pairs of nodes passing through this node. A node i will fail when its load L_i is more than $(1+\alpha)$ times its original load, where α represents the system tolerance to overloads. A randomly localized region of the system is initially removed to trigger the cascading overloads. This kind of initial failure is motivated by the fact that natural disasters (like earthquake or floods) or malicious attacks (like electromagnetic pulse (EMP)) usually occur in specific geographical locations and destroy initially localized regions of the network. The initial failure is located in a randomly selected square of $l \times l$ nodes ($l \ll L$), which are removed initially. In addition to being realistic, since failures are usually localized, this configuration can also help us to follow and analyse the spatio-temporal propagation pathway of the cascading failures. This local failure may trigger failures of other nodes if their load value exceeds the tolerance threshold due to the load redistribution across the entire network. Given the periodic boundary conditions, we can position the initial attack region at the centre of the lattice.

Theoretical analysis. Cascading overload failures due to an initial local failure are produced by the redistribution of loads in the network. From the observations of simulations on weighted network, we find (Fig. 1) that failures spread in a ring shape from the centre of the initial damage. This inspires us to assume in our theory that the network is embedded in a two-dimensional circular plate (Fig. 3), the initial failure is within a (red) circle of radius a, and the main overload due to traffic after the initial failure is located in the (cyan) ring adjacent to the initial failure (see the cyan ring in Fig. 3), whose size will be determined by the theory. The overload is reflected by the increase of the number and lengths of shortest paths passing through this ring. If the overload exceeds the capacity tolerance of a node ($(1+\alpha)$ times its original load) within the adjacent ring, the node will fail in the next step, causing the cascading failure to propagate forward. Owing to the initial failed area, shortest paths from a given node A to destinations located in the shadow area s (the dotted area in Fig. 3) would be affected and become longer, since they now have to surround the failed area. Specifically, the shortest paths from A to nodes in s (for example, AF) across the failed area can be separated into two parts in the adjacent ring (for example, BC and DE). For the first part, its length within the ring changes from BC to KG. As for the second part, its overload on the ring can be calculated from symmetry by switching A and F (source and target). Finally, the integration from $r=a$ to $r=1$ covers all the overloads added to the adjacent ring (cyan) due to the initial failure, can be written as

$$\Delta L_r(a,b) = \int_a^b \left[\sqrt{r^2-a^2}\ s(a,r) - v(a,b,r)\left(s(a,r)+\pi a^2\right)\right] 2\pi r dr + \int_b^1 \left[\sqrt{b^2-a^2}\ s(a,r) - v(a,b,r)\left(s(a,r)+\pi a^2\right)\right] 2\pi r dr, \quad (1)$$

where the length of KG is $\sqrt{b^2-a^2}$ for $r>b$ and $\sqrt{r^2-a^2}$ for $r\leq b$, $s(a, r)$ is the area of s, that is, the number of A's destinations in the shadow and $v(a, b, r)$ is the average length of the first part of shortest paths (within the adjacent ring) before the failure. Similarly, we can obtain the initial load of nodes located within the circle centred at O with radius b as $L_{\mathrm{ini}}(b)$ (Supplementary Methods). Then, for a node on the circle centred O with radius b, the overload produced by the failure is

$$\Delta L = \frac{1}{2\pi b}\frac{\partial}{\partial b}\Delta L_r(a,b) \quad (2)$$

and its initial load is

$$L_0 = \frac{1}{2\pi b}\frac{\partial}{\partial b}L_{\mathrm{ini}}(b) \quad (3)$$

For each functional node in the network, the critical condition for failure can be written as $\alpha=\Delta L/L_0$. Specifically, if $\alpha\geq\Delta L/L_0$, it survives, otherwise it fails. More details for the solution of the theoretical model can be found in Supplementary Figs 14–19 and Supplementary Methods.

References

1. Helbing, D. Globally networked risks and how to respond. *Nature* **497**, 51–59 (2013).
2. Dobson, I., Carreras, B. A., Lynch, V. E. & Newman, D. E. Complex systems analysis of series of blackouts: cascading failure, critical points, and self-organization. *Chaos* **17**, 026103 (2007).
3. Toroczkai, Z. & Bassler, K. E. Network dynamics: jamming is limited in scale-free systems. *Nature* **428**, 716 (2004).
4. Caldarelli, G., Chessa, A., Pammolli, F., Gabrielli, A. & Puliga, M. Reconstructing a credit network. *Nat. Phys.* **9**, 125–126 (2013).
5. Frank, K. T., Petrie, B., Choi, J. S. & Leggett, W. C. Trophic cascades in a formerly cod-dominated ecosystem. *Science* **5728**, 1621–1623 (2005).
6. Albert, R. & Barabási, A.-L. Statistical mechanics of complex networks. *Rev. Mod. Phys.* **74**, 47–97 (2002).
7. Boccaletti, S., Latora, V., Moreno, Y., Chavez, M. & Hwang, D.-U. Complex networks: Structure and dynamics. *Phys. Rep.* **424**, 175–308 (2006).
8. Barrat, A., Barthelemy, M. & Vespignani, A. *Dynamical Processes on Complex Networks* (Cambridge University Press, 2008).
9. Cohen, R. & Havlin, S. *Complex Networks: Structure, Robustness and Function* (Cambridge University Press, 2010).
10. Watts, D. J. A simple model of global cascades on random networks. *Proc. Natl Acad. Sci.* **99**, 5766–5771 (2002).
11. Lorenz, J., Battiston, S. & Schweitzer, F. Systemic risk in a unifying framework for cascading processes on networks. *Eur. Phys. J. B* **71**, 441-460 (2009).
12. Araújo, N. A., Andrade, J. S., Ziff, R. M. & Herrmann, H. J. Tricritical point in explosive percolation. *Phys. Rev. Lett.* **106**, 095703 (2011).
13. Buldyrev, S. V., Parshani, R., Paul, G., Stanley, H. E. & Havlin, S. Catastrophic cascade of failures in interdependent networks. *Nature* **464**, 1025–1028 (2010).
14. Zhao, J. H., Zhou, H. J. & Liu, Y. Y. Inducing effect on the percolation transition in complex networks. *Nat. Commun.* **4**, 2412 (2013).
15. Radicchi, F. & Arenas, A. Abrupt transition in the structural formation of interconnected networks. *Nat. Phys.* **9**, 717–720 (2013).
16. Motter, A. E. Cascade control and defense in complex networks. *Phys. Rev. Lett.* **93,** 098701 (2004).
17. Simonsen, I., Buzna, L., Peters, K., Bornholdt, S. & Helbing, D. Transient dynamics increasing network vulnerability to cascading failures. *Phys. Rev. Lett.* **100,** 218701 (2008).
18. Parshani, R., Buldyrev, S. V. & Havlin, S. Critical effect of dependency groups on the function of networks. *Proc. Natl Acad. Sci. USA* **108,** 1007–1010 (2011).
19. Li, W., Bashan, A., Buldyrev, S. V., Stanley, H. E. & Havlin, S. Cascading failures in interdependent lattice networks: the critical role of the length of dependency links. *Phys. Rev. Lett.* **108,** 228702 (2012).
20. Bashan, A., Berezin, Y., Buldyrev, S. V. & Havlin, S. The extreme vulnerability of interdependent spatially embedded networks. *Nat. Phys.* **9,** 667–672 (2013).
21. Goh, K.-I., Kahng, B. & Kim, D. Universal behavior of load distribution in scale-free networks. *Phys. Rev. Lett.* **87,** 278701 (2001).
22. Motter, A. E. & Lai, Y. C. Cascade-based attacks on complex networks. *Phys. Rev. E* **66,** 065102 (2002).
23. Crucitti, P., Latora, V. & Marchiori, M. Model for cascading failures in complex networks. *Phys. Rev. E* **69,** 045104 (2004).
24. Li, D., Jiang, Y., Kang, R. & Havlin, S. Spatial correlation analysis of cascading failures: congestions and blackouts. *Sci. Rep.* **4,** 5381 (2014).
25. Li, D., Kosmidis, K., Bunde, A. & Havlin, S. Dimension of spatially embedded networks. *Nat. Phys.* **7,** 481–484 (2011).
26. Neumayer, S., Zussman, G., Cohen, R. & Modiano, E. Assessing the vulnerability of the fiber infrastructure to disasters. *IEEE/ACM Trans. Netw.* **19,** 1610–1623 (2011).
27. Bernstein, A., Bienstock, D., Hay, D., Uzunoglu, M. & Zussman, G. *Power Grid Vulnerability to Geographically Correlated Failures - Analysis and Control Implications* (Columbia University, Electrical Engineering, Technical Report, 2011).
28. Pastor-Satorras, R. & Vespignani, A. Epidemic spreading in scale-free networks. *Phys. Rev. Lett.* **86,** 3200 (2001).
29. Li, F. Real Datasets for Spatial Databases: Road Networks and Points of Interest. http://www.cs.utah.edu/~lifeifei/SpatialDataset.htm (2013).
30. Menck, P. J., Heitzig, J., Kurths, J. & Schellnhuber, H. J. How dead ends undermine power grid stability. *Nat. Commun.* **5,** 3969 (2014).
31. Cornelius, H. J., Kath, W. L. & Motter, A. E. Realistic control of network dynamics. *Nat. Commun.* **4,** 1942 (2013).
32. Chen, J., Thorp, J. S. & Dobson, I. Cascading dynamics and mitigation assessment in power system disturbances via a hidden failure model. *Int. J. Electr. Power Energy Syst.* **27,** 318–326 (2005).

Acknowledgements

We thank the support from Collaborative Innovation Center for industrial Cyber-Physical System. D.L. is also supported by the National Natural Science Foundation of China (Grant 61104144) and the National Basic Research Program of China (2012CB725404). S.H. thanks DTRA, ONR, the LINC and the Multiplex (No. 317532) EU projects, the DFG, and the Israel Science Foundation for support. J.Z. was partially supported by NSFC (Grant Nos 71501005 and 71531001) and 863 Program (Grant No. SS2014AA012303).

Author contributions

All authors contributed to all aspects of this work.

Additional information

Competing financial interests: The authors declare no competing financial interests.

14

Purely hydrodynamic ordering of rotating disks at a finite Reynolds number

Yusuke Goto[1] & Hajime Tanaka[1]

Self-organization of moving objects in hydrodynamic environments has recently attracted considerable attention in connection to natural phenomena and living systems. However, the underlying physical mechanism is much less clear due to the intrinsically nonequilibrium nature, compared with self-organization of thermal systems. Hydrodynamic interactions are believed to play a crucial role in such phenomena. To elucidate the fundamental physical nature of many-body hydrodynamic interactions at a finite Reynolds number, here we study a system of co-rotating hard disks in a two-dimensional viscous fluid at zero temperature. Despite the absence of thermal noise, this system exhibits rich phase behaviours, including a fluid state with diffusive dynamics, a cluster state, a hexatic state, a glassy state, a plastic crystal state and phase demixing. We reveal that these behaviours are induced by the off-axis and many-body nature of nonlinear hydrodynamic interactions and the finite time required for propagating the interactions by momentum diffusion.

[1] Institute of Industrial Science, University of Tokyo, 4-6-1 Komaba, Meguro-ku, Tokyo 153-8505, Japan. Correspondence and requests for materials should be addressed to H.T. (email: tanaka@iis.u-tokyo.ac.jp).

Self-organization is the autonomous organization of components into patterns or structures and the underlying mechanism can be grouped into two classes: thermal and athermal origin[1,2]. Compared with the understanding of self-assembly of a thermal system, which can be explained in terms of the free energy, that of an athermal system is far behind due to the lack of a firm theoretical background. Typical examples for the latter is active matter in hydrodynamic environments, which shows many interesting unconventional pattern formation with exotic dynamical features[1,3–11]. Directional swimmers undergoing translational motion, like fishes, tend to swim together while forming a cluster of particular shape[6,7,9]. The hexatic ordering of directionally self-propelling particles was also recently studied[12].

Rotation is another important type of motion. Self-organization of passive rotors was first observed in laboratory experiments by Grzybowski, Stone and Whitesides[4]. Such rotation can be induced by applying a torque to a particle by magnetic[4,13,14] and optical fields[15]. Self-organization of active rotors has also been studied intensively. For example, it was found that there are interesting stable bound states of spinning *Volvox* algae[8]. Furthermore, it was shown[16] that microorganisms, like spermatozoa, self-organize into dynamic vortices and they form an array with local hexagonal order. This study indicates that large-scale coordination of cells can be regulated hydrodynamically, and chemical signals are not required. It was also shown recently that self-assembly of rotors is a generic feature of aggregating swimmers[17]. There has also been a theoretical prediction for intriguing self-organization of rotating molecular motors in membranes[18]. This and related problems have also been investigated by numerical simulations[11,19–24].

As described above, there exist two types of rotors[25]: rotors driven by external torques are called passive rotors, while those that are internally driven are called active rotors and many such examples can be seen in biological systems. A realistic realization of a truly active system of self-rotors in biological systems may be one in which the particles are torque dipole with no resultant net torque on the system[18,21,25]. To elucidate the physics of self-organization of flow created by rotors at a finite Reynolds number, a system of hard disks each of which is rotated by an externally applied torque is an ideal model system. Ordering of this system may be regarded as a nonequilibrium counterpart of thermodaynamic ordering of hard disks. We note that, unlike a thermodynamic system, where a state is selected solely by free energy, dynamic factors such as hydrodynamic interactions also affect the selection of a state of an out-of-equilibrium system.

For example, it has been now recognized that nonlinear hydrodynamic interactions play crucial roles in self-organization of rotating disks[4,13,14]. In other words, the phenomena are beyond Stokes approximations, and may even have a link to self-organization of vortices[26], which is observed at a high Reynolds number. Vortex crystals are very interesting examples[27,28]. The nonequilibrium, nonlinear, nonlocal, non-instantaneous nature of hydrodynamic interactions makes analytical approaches to this problem very difficult and thus numerical simulations are expected to play a crucial role. This problem also has a link to self-organization of a point vortex system[29,30], but the finite size and solidity of disks lead to far more rich behaviour. Here we employ a fluid particle dynamics (FPD) method[31], which we developed for studying hydrodynamic interactions between colloidal particles (Methods). This method treats a solid colloid as an undeformable fluid particle inside which the viscosity is considerably higher than the surrounding liquid. This approximation makes us free from solid–fluid boundary conditions, which significantly simplifies the computation. This method can quite naturally deal with many-body hydrodynamic interactions even at a high Reynolds number.

In this Article, we study self-organization of rotating hard disks in a two-dimensional (2D) incompressible liquid by using the FPD method. Like the Ising model for magnetic ordering or the hard sphere system for crystallization, this system may serve as a fundamental model system for studying dynamical phase behaviour caused by hydrodynamic interactions between rotating particles. The situation is similar to the above-mentioned experimental and theoretical works[4,13,14]. We make our simulation in 2D to compare with the behaviour of its thermal counterpart, 2D hard disks, whose thermodynamic behaviour is reasonably understood[32]. Here we report surprisingly rich phase-ordering behaviours, such as aggregation, re-entrant order–disorder transition, glass transition, plastic crystal formation and phase demixing in this class of strongly nonequilibrium systems.

Results

Behaviours of a single and a pair of rotating disks. Before discussing many-body interactions between disks rotating with an angular frequency Ω, first we describe the behaviour of a dilute limit: behaviours of a single rotating disk and a pair of rotating disks (Supplementary Fig. 1; Supplementary Note 1). We characterize the rotation speed of a disk either by the angular frequency Ω or the relevant Reynolds number $\mathrm{Re}=\rho a^2\Omega/\eta_\ell$ (ρ: density; a: particle radius: η_ℓ: liquid viscosity). Figure 1a,b shows the 2D flow field around a rotating disk and the velocity distribution, respectively. The latter shows that the rotating velocity linearly increases with the distance from the centre of mass, r, inside a disk and decays as $1/r$ in its outside. This is the characteristic of the so-called Rankin vortex, that is, a forced vortex in the central core surrounded by a free vortex. We note that the Rankin vortex is known to mimic the 2D flow field of tornado and hurricane[33]. For two co-rotating 'point' vortexes, the analytical solution is known and the two particles rotate around their centre of mass towards the same direction as the rotating direction[29,30]. In this case, the radius of rotation is the half of the initial interparticle distance. For particles with a finite size, on the other hand, the direction of rotation and the rotation centre are the same as the case of point particles, but the radius of the rotation decreases with an increase in the rotation speed of the particles Ω, or Re. This problem has a similarity to viscous interactions of co-rotating vortices[29,34], but the crucial difference arises from the fact that the vortex core is an undeformable solid and not a liquid in our case. There are a few studies on spinning particles in 3D[25,35] however, we note that there is an essential difference between 2D and 3D problems (Supplementary Fig. 2; Supplementary Note 2). We confirm that if we fix the centres of mass of two rotating particles on a fixed line they always repel each other by the Magnus force[4,13,36]. The attraction between particles is thus due to interparticle hydrodynamic interactions, as schematically explained in Fig. 1c. The flow field generated by a rotating disk makes the other disk follow it, and vice versa. Thus, two particles try to follow each other while rotating around the centre of mass of the pair. In 3D, on the other hand, such hydrodynamic interactions weaken due to the presence of the escape dimension, thus the Magnus force wins over the hydrodynamic attractive force, and particles repel each other. In 2D, this hydrodynamic attraction overwhelms the repulsion due to the Magnus force, which leads to the rotating pair of particles (Fig. 1d), whose interparticle distance monotonically decreases with an increase in Ω, or Re (Fig. 1e), as long as the area fraction of disks, Φ, is sufficiently small. In relation to the above dimensionality effect on hydrodynamic interactions between

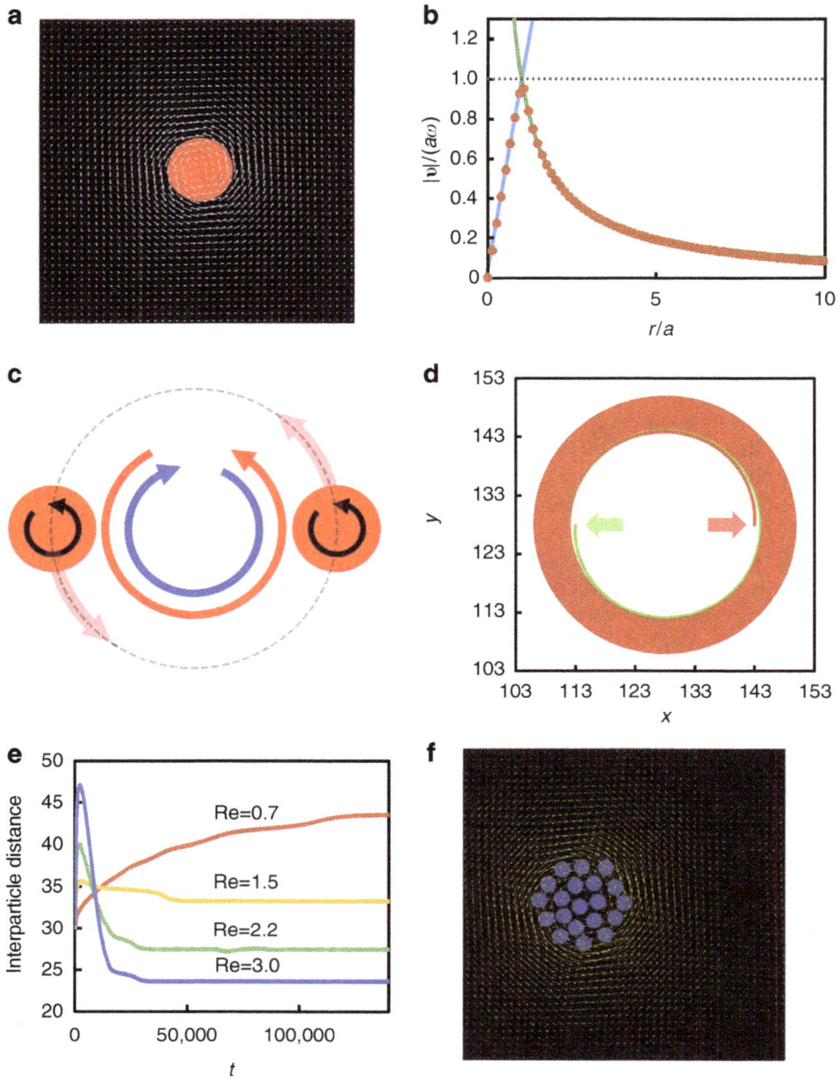

Figure 1 | Behaviours of an isolated rotating disk and co-rotating disks in a dilute suspension. (**a**) 2D flow fields around an isolated rotating disk. (**b**) The velocity profile as a function of r/a. $v/(a\omega)$ increases linearly with r inside the disk and then decays as $1/r$ in its outside, which is the characteristic of the so-called Rankin voltex. (**c**) Schematic picture of a pair of two disks rotating around the centre of mass. The two disks are rotating anticlockwise with the same speed. (**d**) Trajectory of coupled rotating particles (the system size $= 256 \times 256$ lattices). The initial separation of the particles (arrowed) was 30 lattices and each particle rotates anticlockwise with $Re = 0.76$. (**e**) Temporal change in the interparticle separation. We note that for all the cases including $Re = 0.7$ the interparticle distance eventually reaches a final steady-state value of the separation (see **d**), which monotonically decreases with Re. (**f**) A cluster of disks formed at $\Phi = 0.04$ at $Re = 5.94$. We confirm that irrespective of the value of Re a cluster is always formed in the range of Re studied.

rotors, it is worth noting that the hexagonal ordering observed by Grzybowski *et al.*[1,4,13,14] at finite Reynolds numbers is due to the effective hydrodynamic repulsion between the disks floating at an interface in 3D (that is, 2.5D).

Structural ordering due to many-body hydrodynamic interactions. Now we consider the dynamical behaviour of a system of many disks rotating in the anticlockwise direction with Ω. The hydrodynamic attraction between rotating disks leads to the formation of rotating clusters for low Φ (Fig. 1f). The higher rotation speed of individual disks leads to the formation of a more compact rotating cluster. This tendency is basically the same as that for a pair of rotors (Fig. 1e). For this regime, only one cluster is formed in the simulation box and the whole cluster rotates in the anticlockwise direction, as shown in Fig. 1f. A cluster always tends to have a circular shape, but it does not have any particular internal structural order, partly because imperfect matching between the size and the number of particles leads to structural fluctuations: the internal structure is basically controlled by the number of disks in it and Re, which determine the cluster size, but fluctuating with time. With a further increase in Re, however, this cluster state becomes unstable, since the repulsive Magnus force of nonlinear origin eventually wins over the hydrodynamic attraction for high Re (see below). Above a critical Φ ($\Phi_c \sim 0.05$), on the other hand, a system exhibits a re-entrant transition between states as a function of Ω (Fig. 2a): for low Ω, a system is in a disordered liquid state with large fluctuations, but with an increase in Ω it enters into a rather stable hexatic phase where rotating particles are localized on a hexagonal lattice. The border between the cluster to the hexatic state is rather sharp as a function of Φ.

The transition can be characterized by the nature of interparticle interactions. Here we analyse a point pattern to

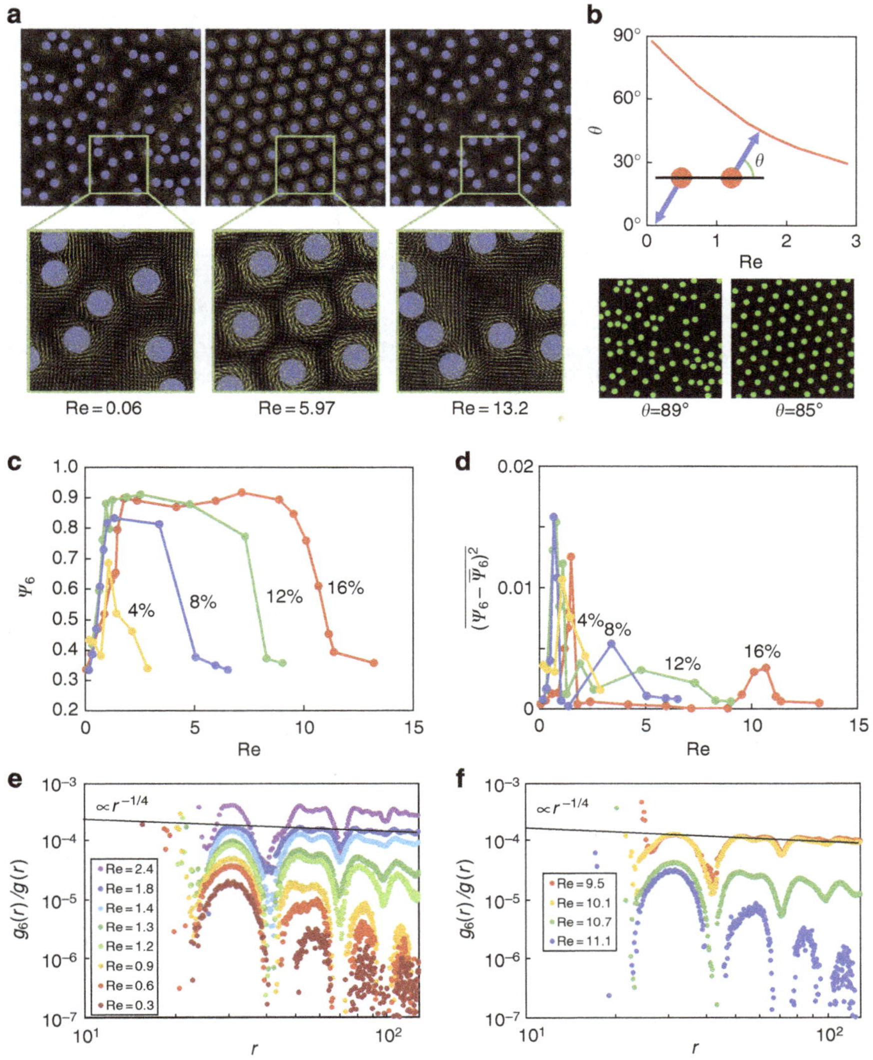

Figure 2 | Re-entrant state transitions observed at a rather high Φ. (**a**) Snapshots of particle configurations together with the velocity fields. The system size is 256 × 256 lattices, the number of particles is 80 and $\Phi \sim 0.157$. We can see sequential transitions from disordered, hexatic ordered to disordered states with an increase in Re. (**b**) Upper panel: the direction of the hydrodynamic force θ measured from the interparticle axis. Lower panels: patterns formed by Brownian simulations for the two off-axis forces: left is a liquid state for $\theta = 89°$, whereas the right panel is a hexatic state for $\theta = 85°$. (**c**) Re-dependence of the hexatic order parameter Ψ_6, which clearly shows the re-entrant behaviour for various Φ's (expressed in %). (**d**) The degree of fluctuations of Ψ_6 as a function of Re for various Φ's (expressed in %). We can see that, as in a thermodynamic hexatic-ordering transition, the order parameter exhibits large amplitude fluctuations near the transition points. (**e**) Re-dependence of the decay of the spatial correlation function of the hexatic order parameter $g_6(r)$ normalized by the radial distribution function $g(r)$ around the transition at low Re for $\Phi = 0.157$. In the hexatic state, $g_6(r)/g(r)$ decays with a power law with the exponent of $-1/4$, as it should be[32]. In the disordered states, it decays almost exponentially. (**f**) The same as **e**, but for around the transition at high Re. In the hexatic state, $g_6(r)/g(r)$ again decays with a power law with the exponent of $-1/4$. In the disordered states, it decays almost exponentially.

extract the nature of interparticle interactions. The point pattern analysis is very useful to determine the overall interparticle interactions[37]. Here we use what we call N function, which is the number of connected regions as a function of the radius of circle whose centre is located at the centre of each disk. The number of connected regions decreases monotonically, with an increase in the circle radius, from the total number of disks N_0 to 1. Here we use N for the one normalized by N_0. By comparing N for a reference system made of randomly distributed disks, that is, Poisson pattern, we can judge whether the interaction is repulsive or attractive. For a system of particles with repulsive interactions, particles form a rather regular pattern and N decays slower than that for the corresponding Poisson pattern. For a system of particles with attractive interactions, on the other hand, particles form a cluster pattern and N decays faster than that for the corresponding Poisson pattern. In Fig. 3a, we show the N function for $\Phi = 0.04$. We can clearly see that for the cluster-forming case at Re = 0.72 the interaction is attractive, whereas for the disordered state at Re = 5.97 the interaction is repulsive. In Fig. 3b, we show the N function for $\Phi = 0.15$. We can see that for all the value of Re the interaction is basically repulsive, but its strength is maximum at Re = 5.9, for which the hexatic order is formed.

The transition can be seen even more clearly by the degree of localization of the flow field: for a hexatic state, each particle has its own localized rotational flow field, whereas for a cluster state a single vortex is always formed and thus the flow field is strongly delocalized. However, the precise nature of the cluster-hexatic

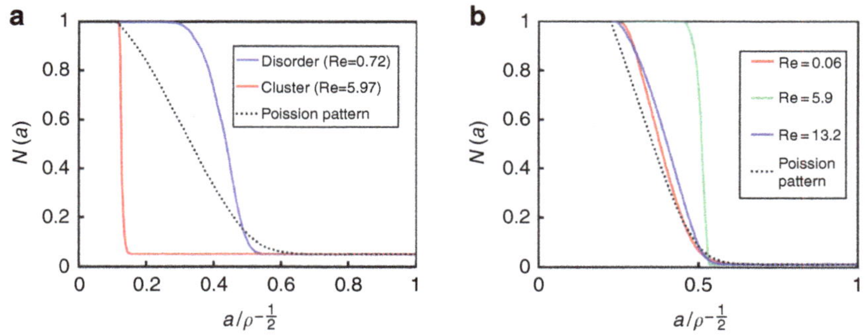

Figure 3 | The analysis of point patterns. (**a**) The decrease in N as a function of the particle radius $a/\rho^{1/2}$ for $\Phi = 0.04$, where ρ is the particle number density and $\rho^{-1/2}$ is the average interparticle distance. (**b**) The decrease in N as a function of the particle radius $a/\rho^{1/2}$ for $\Phi = 0.15$.

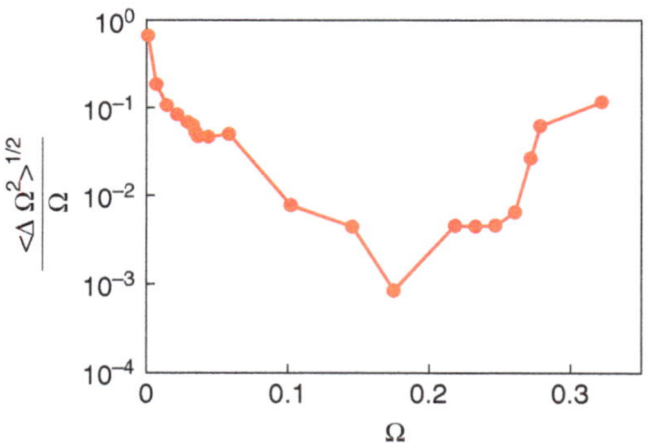

Figure 4 | The normalized variance of the angular frequency Ω of rotating disks as a function of the averaged Ω for $\Phi = 0.15$. We can see the negative correlation between the degree of the hexatic order and the variance.

phase transition, such as whether the transition is continuous or discontinuous and whether a disorder state always exists between the two states or there exists a critical point between the cluster and the hexatic state, is not clear at this moment. To access this problem, we need to survey the border region of a bigger system size with a high resolution of Φ and Re. Although this is a very interesting problem, we leave this for future investigation.

We also find that the hexatic state is eventually destabilized and melts by a further increase in Ω and the system becomes disordered again. We note that for all these states interparticle interactions are basically repulsive, unlike the case of low Φ. Although the torque exerted to each disk is exactly the same, the rotation speed of a disk can in principle depend on the particle configuration around it. Here we show in Fig. 4 the normalized variance of the angular frequency Ω of rotating disks as a function of the averaged Ω. For ordered states, the variance, that is, the fluctuations of rotation speed, becomes very small, indicating all the disks rotating with almost the same frequency. For disordered states, on the other hand, the variance is large, reflecting the large fluctuations of particle environment. This result indicates a strong negative correlation between the degree of fluctuations of rotational speed of particles and the degree of order.

In Fig. 2c, we show the Ω-dependence of the hexatic order parameter Ψ_6, which clearly indicates the re-entrant nature of the state transitions. The transitions can also be characterized by the magnitude of fluctuations of Ψ_6, or the susceptibility (see Fig. 2d), which is also observed in a thermodynamic hexatic ordering in 2D disks. In the hexatic ordered state, we confirm the power law decay of the spatial correlation of the hexatic order that is specific to the hexatic phase (see Fig. 2e,f). We do not see any indication of the positional order since the radial distribution function decays almost exponentially (see below).

Here we show a hexatic phase observed in a large system (the system size = $2{,}048^2$ lattices and the number of particles = 5,120) at $\Phi = 0.15$ and Re = 5.9. We can see grain boundaries between hexatic order with different orientations as shown in Fig. 5a,b. Figure 6a,b shows the decay of the correlation function of the hexatic order normalized by the radial distribution function $g(r)$, $g_6(r)/g(r)$, and that of $g(r) - 1$, respectively. We can see that $g_6(r)/g(r)$ decays algebraically for a short distance $r < 400$, but decays faster for long distance. This is because the size of monodomain regions is finite (see Fig. 5a,b). On the other hand, we can see that $g(r)$ decays faster than an algebraic decay even for $r < 400$, where $g_6(r)/g(r)$ decays algebraically. This clearly indicates the absence of quasi-long-range translational order in the ordered phase. Thus, we conclude that the ordered phase is the hexatic phase. The appearance of the transitions between dynamical states as a function of Φ and Ω in athermal systems is quite striking, which is reminiscent of phase transitions in a thermal system. We stress that the interparticle interaction in our system is of purely hydrodynamic origin.

Here we consider a quite interesting feature of hydrodynamic interparticle interactions, which are absent in a thermal system. Binary interaction potentials that lead to phase ordering in thermal systems always act along the interparticle axis, that is, along the line connecting the centres of mass of two interacting particles. However, this is not the case for our athermal system. The nonlinear Magnus force acts along the line connecting two particles, whereas the Stokes force acts according to the flow direction. Furthermore, linear hydrodynamic interactions are tensorial and act not along the interparticle direction. Thus, the total hydrodynamic force does not act along the interparticle axis. The direction of the hydrodynamic force as a function of Re is shown in the upper panel of Fig. 2b with a schematic explanation. With an increase in Re, both the Magnus and the hydrodynamic force increase. However, the nonlinear Magnus force increases more rapidly than the linear hydrodynamic force. Thus, the direction of the total force, which is repulsive, approaches the interparticle axis. A strong enough repulsive force acting on particles not so far from the interparticle direction leads to the formation of hexatic order. To verify this scenario, we have performed Brownian dynamics simulation (without hydrodynamic interactions) by changing the angle θ between the direction of an artificial repulsive force and the interparticle direction. We find that when θ is continuously decreased a system indeed forms hexatic order below a critical value of θ (see the lower panels of Fig. 2b).

Next we discuss the nature of the translational motion of rotating disks in the disordered states. To see this, we calculate the mean-square displacement of disks $\langle \Delta r^2(t) \rangle$ (see Fig. 7a).

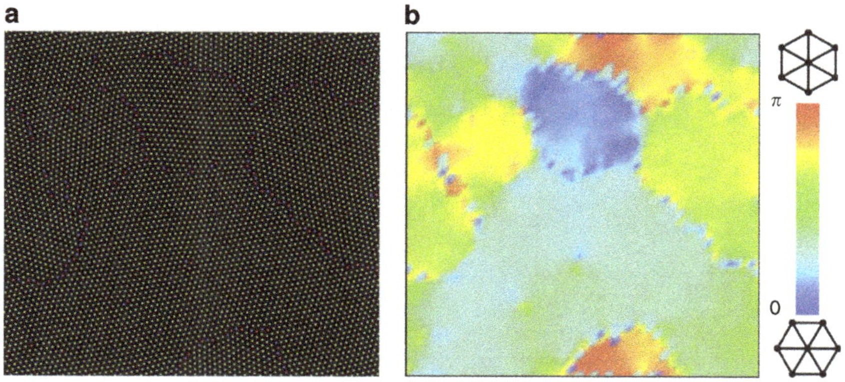

Figure 5 | Hexatic ordering in a large system. Here we show results of large-size simulation (the system size = $2{,}048^2$ lattices and the number of particles = 5,120). The area fraction $\Phi = 0.15$ and Re = 5.9. (**a**) The structure of a hexatic state shown with three colours (yellow for particles with six neighbours; red for particles with more than 7 neighbours; blue for particles with less than 5 neighbours). (**b**) The same as **a**, but with orientations of hexatic order. See the colour bar for the meaning of colour.

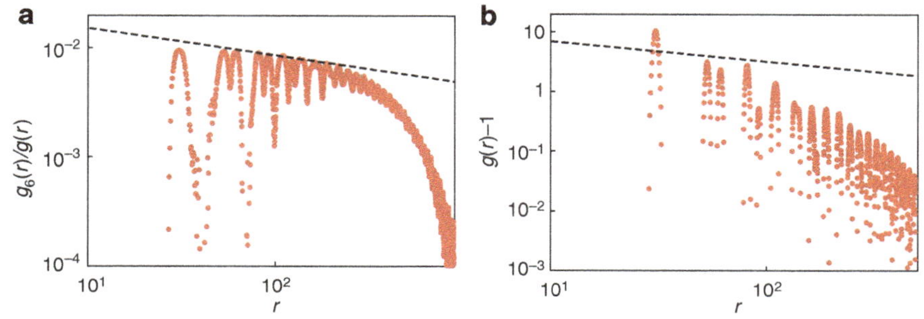

Figure 6 | Spatial correlation of hexatic and positional order. (**a**) The spatial decay of the correlation function of the hexatic order $g_6(r)$ in Fig. 5a normalized by the radial distribution function $g(r)$. We can see that it decays algebraically in a short distance, but decays faster for long distance because of the finiteness of the domain size. The dashed line has a slope of $-1/4$. (**d**) The spatial decay of the radial distribution function $g(r)$ for the pattern shown in Fig. 5a. It decays more quickly than the decay of $g_6(r)$ even for rather short distance ($r<400$). The dashed line has a slope of $-1/3$.

Interestingly, in the two types of disordered states, particle motion is apparently diffusional: in the long-time limit, we observe the relation $\langle \Delta r^2(t) \rangle \sim D_{\mathrm{eff}} t$, where D_{eff} is the effective diffusion constant and t is the time duration. It should be noted that in our system there is no thermal noise and an isolated rotating disk exhibits no motion. For a pair of rotating particles, we also observe that their trajectory is very stable and there is no fluctuation (see Fig. 1d). Thus, the apparently diffusional motion should be the consequence of self-generated force noise due to many-body hydrodynamic interactions of both linear and nonlinear origin. We can also see no diffusion, or non-ergodic behaviour, for the hexatic state. The random nature of fluctuations may be related to the long-range nature of hydrodynamic interactions, which makes a number of surrounding particles affecting a particle large enough to provide strong stochastic spatio-temporal fluctuations.

For high Re, the particle diffusion starts to become comparable or faster than momentum diffusion: $D_{\mathrm{eff}} \geq \nu$, where ν is the momentum diffusion constant or the kinematic viscosity $\nu = \eta_\ell/\rho$ (see Fig. 7a,b). This implies that hydrodynamic interactions induced by the rotation of a particle cannot fully propagate to its neighbouring particles for high Re. This weakens the repulsive interactions of particles and eventually leads to the melting of the hexatic state. Thus, the re-entrant hexatic ordering as a function of Re may be explained as follows: the increase in the Magnus force of nonlinear origin with an increase in Re makes the direction of the interparticle force more aligned along the interparticle direction and also increases the strength of the repulsive force. This leads to stabilization of the hexatic order, as explained above. The ordered state is stable until the repulsive interaction is weakened by the intrinsically kinetic nature of hydrodynamic interactions: Unlike ordinary interparticle interactions that propagate with the speed of light, hydrodynamic interactions propagate much slower in a diffusive manner with the momentum diffusion constant ν. This kinetic weakening of the repulsive force eventually destabilizes the hexatic state and leads to melting into the disordered chaotic state.

Here we summarize what we observed in our system and show the state diagram as a function of Φ and Re (see Fig. 7c). We can see the three states, that is, disordered fluid, cluster and the ordered hexatic state. It is quite striking that particles interacting by hydrodynamic interactions alone exhibit such rich phase behaviours.

Other interesting states formed by rotating disks. Finally, we show other interesting states formed by rotating particles. The introduction of size polydispersity to hard spheres is known to lead to the formation of a glass state for a thermal system[38]. Motivated by this, we introduce the polydispersity in the rotating speed of particles, whose variance is δ, and indeed find a non-ergodic glassy state of the rotating particles (Fig. 8a) for high enough δ (Fig. 8b): liquid–glass transition in an athermal system. Reflecting the nearly continuous nature of the liquid-to-hexatic transition[32,39], there is no sharp transition from a hexatic to a disordered glassy state. Although disks are rotating around their centres of mass, their positions are frozen in a disordered configuration and thus the system can be regarded as a

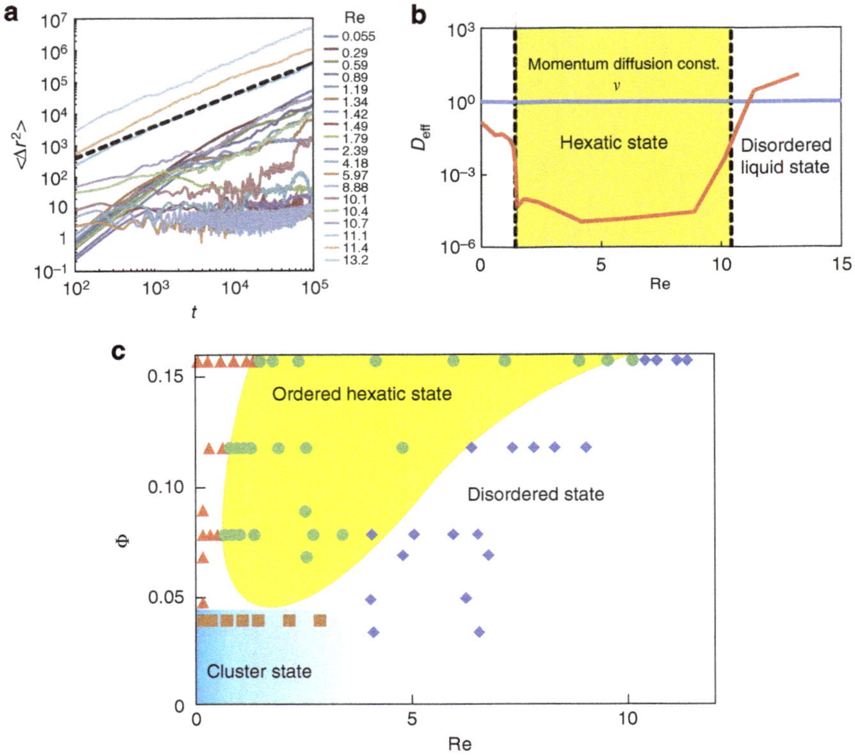

Figure 7 | Dynamical behaviour and state diagram of rotating disks. (**a**) The mean-square displacement $\langle \Delta r^2 \rangle^{1/2}$ versus time t. Despite the absence of thermal noise, particles undergo Brownian-like diffusive motion $\langle \Delta r^2 \rangle^{1/2} = D_{\mathrm{eff}} t$ with an effective diffusion constant D_{eff} in the long-time limit. The black dashed line is the momentum diffusion constant, or the kinematic viscosity $\nu (= \eta_\ell / \rho)$. (**b**) Re-dependence of the effective diffusion constant D_{eff}. In the hexatic ordered state, D_{eff} is very low and there is no diffusive behaviour, as it should be (see **a**). At high Re, D_{eff} exceeds ν (indicated by the blue horizontal line), which means that fluctuations of particles cannot be suppressed by hydrodynamic interactions at high Re (see text). (**c**) State diagram on the Re–Φ plane. At low Φ and low Re, the system forms a cluster due to hydrodynamic attractive interactions. At high Φ, on the other hand, the system exhibits re-entrant transition between a disordered chaotic liquid state and ordered hexatic state. The latter state is basically stabilized by the repulsive interaction due to the Magnus effect.

non-ergodic glassy state. Here we show how the distribution of the rotation speed of particles leads to the loss of hexatic order and results in the formation of a glassy state. As shown in Fig. 9a, the larger deviation from the averaged Re leads to the larger deviation from the average number of nearest neighbours ($=6$). With an increase in the variance of the distribution δ, more defects are produced and the hexatic order is eventually lost above $\delta \geq 0.2$. We can also see that the the interparticle distance is larger for a particle rotating with a faster speed because of stronger hydrodynamic repulsive force, that is, Magnus force (Fig. 9). Both of these disorder effects are responsible for the formation of a glassy state: the number of the nearest neighbours and the average distance to the neighbours are both strongly correlated to the rotational speed of particles, which explains a wide enough distribution of the rotation speed results in the formation of a non-ergodic amorphous state instead of a hexatic ordered state. This glassy state is non-ergodic if we consider the particle configuration, yet maintains strong flow fields, which makes this state very unique. Reflecting the kinetic origin of interparticle interactions, the introduction of disorder in the dynamic quantity, Ω, is essential for avoiding the ordering, which is an interesting point unique to purely kinetic athermal systems.

We also find a plastic crystal-like state in a rotating particle pair (dumbbell) system (Fig. 8c) above a critical Re (Fig. 8d). This state is characterized by hexatic ordering of the centre of mass of disk pairs (dumbbells) without any orientational order in the axis directions of dumbbells. So we find for this type of athermal systems made of rotating particles almost all states seen in its thermodynamic counterparts, including a liquid, a hexatic phase, a glass and a plastic crystal. The only missing state is a liquid-crystalline state, but the lack of this state is natural consequence of the fact that the system is composed of rotating elements. We also note that the modification of the rotational direction leads to complex behaviours; for example, we can introduce phase demixing of particles rotating oppositely (Fig. 8e), which may be used to separate different types of passive and active rotors. In relation to this, it is worth noting that phase separation between particles rotating clockwise and anticlockwise was recently observed even without hydrodynamic interactions[40]. It is interesting that such phase separation is observed in both systems with and without hydrodynamic interactions. It is also worth noting that the presence of regular arrays of vortices, for example, the triangle state, are predicted for an active polar film[41]. So far we have not seen such a regular state, but the similarity in the physics between the two systems implies that it might exist in a certain parameter range. This is an interesting problem for future study.

Discussion

It is remarkable that our athermal system where particles interact only via hydrodynamic interactions exhibits such rich phase (or more strictly, state) behaviour and reproduces almost all the physical states observed in its thermal counterpart. We hope that these phase behaviours will be observed experimentally. For this purpose, a quasi-2D version of the experimental setup used in

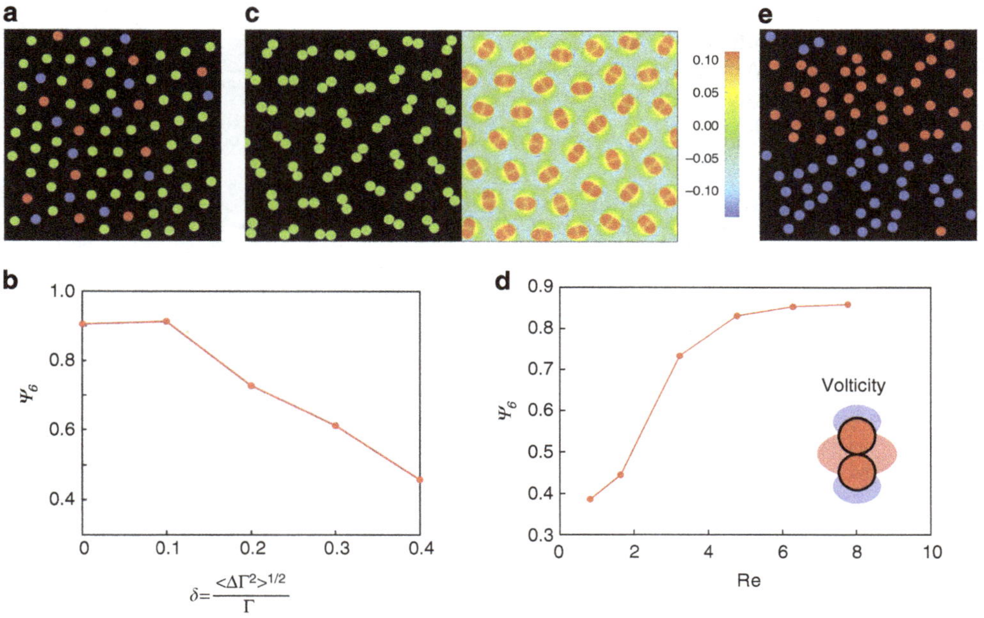

Figure 8 | Other interesting states formed in a system made of rotating particles. (**a**) A glassy non-ergodic state formed in a system of rotating disks where the torque Γ acting on particles has a Gaussian distribution, whose normalized variance is $\delta = \langle \Delta\Gamma^2 \rangle^{1/2}/\Gamma$. Here $\Phi = 0.157$, the average Re ($\bar{\mathrm{Re}}$) = 4.22 and $\delta = 0.2$. The colour of particles is green when the number of nearest neighbour particles NN is 6. For NN > 6, the colour is red and for NN < 6 blue. (**b**) δ-dependence of the hexatic-order parameter Ψ_6. Here $\Phi = 0.157$ and Re = 7.76. With an increase in δ, the hexatic order monotonically decreases and the system eventually enters into a non-ergodic glassy state. (**c**) A plastic phase formed in a system made of rotating dumbbells (anticlockwise). The torque applied is the same for all particles. The rotation direction is anticlockwise. The system exhibits hexatic order at a certain range of Re, but without any orientational order of the axes of dumbbells. (**d**) Re-dependence of Ψ_6 for rotating dumbbells. We note that the transition is rather broad. The schematic image represents the distribution of volticity around a single rotating dumbbell (red: anticlockwise; blue: clockwise). (**e**) Phase separation of particles rotating clockwise (Ω) and anticlockwise ($-\Omega$; $\Omega = 0.155$ (or Re = 6.35)). $\Phi = 0.157$ and the number fraction of particles rotating clockwise is 0.5.

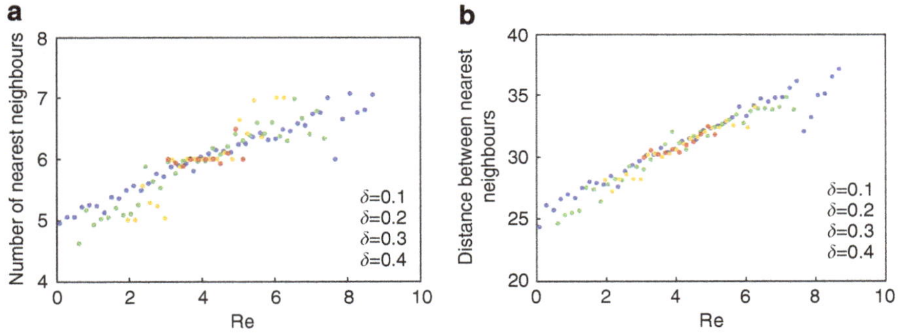

Figure 9 | Correlation between local structure and rotation speed (or, Re) of each particle in a glassy state. (**a**) Dependence of the number of nearest neighbours on Re for individual particles. The system size is 256^2 lattices, $\Phi = 0.15$, and the average value of Re, $\bar{\mathrm{Re}}$, is 4.2. We can see that the larger deviation from $\bar{\mathrm{Re}}$ leads to the larger deviation from the average number of nearest neighbours (= 6). (**b**) Dependence of the distance to nearest neighbour particles on Re for individual particles. The conditions are the same as the above. We can see that the interparticle distance is larger for a particle rotating with a faster speed (or, larger Re) because of stronger hydrodynamic repulsive force, that is, Magnus force.

refs 1,4,13,14 may be suitable, since the 2D nature of hydrodynamic interactions is important. Here we have focused on the bulk behaviour of rotating disks to study the fundamental nature of dynamical phase behaviour from a viewpoint of nonequilibrium statistical physics. The effects of confinement on such a system are also quite interesting not only from a fundamental viewpoint, but also from a viewpoint of applications to microfluidics[42,43]. Such effects on rotors in a fluid have recently been numerically studied by Götze and Gompper[22,23]. It is quite interesting to study how such spatial confinements affect all the dynamical phases we reported and the transition between them.

Here it is worth stressing that even non-ergodic states of particles, such as hexatic, glassy and plastic crystal states, have strong hydrodynamic flow fields, which makes these states quite distinct from their thermal counterparts. The interesting and unique feature of hydrodynamic self-organization is that structural ordering is the consequence of self-organization of flow dissipating energy and thus even non-ergodic states are maintained by dynamical flow. In nature, there are many dynamical systems in which crucial interactions between elements are of purely hydrodynamic origin. The coexistence of linear and nonlinear hydrodynamic interactions, the resulting unconventional off-axis force, the finite propagation speed of the interactions and the significance of hydrodynamic degrees of freedom even for non-ergodic (apparently static) states lead to rich and non-trivial self-organization. We hope that our study sheds new light on hydrodynamic self-organization and stimulate further study on this intriguing problem.

Methods

Simulation method. Treating hydrodynamic interactions between colloids is difficult even for a thermal system. There are several methods such as Stokesian dynamics, lattice Boltzmann, smooth particle and FPD methods. Here we employ our FPD method[31,44], which has an advantage in its theoretical transparency and its applicability to a high Reynolds number (Re) regime. We can access a high Re regime rather easily particularly because a torque applied externally makes the flow field inside a fluid disk almost exactly that for a solid disk even at high Re.

Here we briefly explain the FPD method[31] and the physical concept behind it. A particle whose centre of mass is located at r_i is represented by a smooth viscosity change as $\eta(\boldsymbol{r}) = \eta_\ell + \sum_i^N (\eta_c - \eta_\ell)\phi_i(\boldsymbol{r})$, where η_ℓ is the liquid viscosity and η_c is the viscosity inside a colloidal particle. The summation is taken over all N particles. ϕ_i represents particle i as $\phi_i(\boldsymbol{r}) = \{\tanh[(a - |\boldsymbol{r} - \boldsymbol{r}_i|)/\xi] + 1\}/2$, where a is the particle radius and ξ is the interface thickness. Then, the equation of motion to be solved is

$$\rho\left(\frac{\partial}{\partial t} + \boldsymbol{v}\cdot\nabla\right)\boldsymbol{v} = \boldsymbol{f}_U + \boldsymbol{f}_T - \nabla\cdot\overleftrightarrow{\boldsymbol{\Pi}} \tag{1}$$

with $\overleftrightarrow{\boldsymbol{\Pi}} = p\overleftrightarrow{\boldsymbol{I}} - \eta\left(\nabla\boldsymbol{v}^\dagger + \nabla\boldsymbol{v}\right)$, where $\overleftrightarrow{\boldsymbol{I}}$ is the unit tensor. Here ρ is the mass density and we assume that the density of the liquid is the same as that of particles. $v(\boldsymbol{r})$ is the velocity field, and the pressure p is determined to satisfy the incompressibility condition $\nabla\cdot\boldsymbol{v} = 0$. Here $f_U(\boldsymbol{r})$ is the force density due to the interparticle interaction determined as $\boldsymbol{f}_U(\boldsymbol{r}) = -\sum_i^N (\phi_i(\boldsymbol{r})/A)\sum_{j\neq i}^N \partial U(|\boldsymbol{r}_{ij}|)/\partial \boldsymbol{r}_{ij}$, where $U(r)$ is the interparticle potential, $\boldsymbol{r}_{ij} = \boldsymbol{r}_i - \boldsymbol{r}_j$, and $A = A_i = \int d\boldsymbol{r}\phi_i(\boldsymbol{r})$ is the area of each particle. $f_T(\boldsymbol{r})$ is the force density due to the torque: $f_T(\boldsymbol{r}) = \alpha|\boldsymbol{r}|\phi(\boldsymbol{r})\boldsymbol{e}_\theta$, where $\boldsymbol{e}_\theta$ is the unit angular vector in the anticlockwise direction. α is the strength of the torque and $\alpha > 0$ leads to the anticlockwise rotation of a particle.

In our FPD method, the particle rigidity is approximately expressed by introducing the smooth viscosity profile, $\eta(\boldsymbol{r})$. The approximation is better for a larger viscosity ratio η_c/η_ℓ and a smaller ξ/a. By multiplying both sides of Equation (1) by $\phi_i(\boldsymbol{r})$ and then performing its spatial integration, we can straightforwardly obtain an approximate equation of motion of particle i: $M_i d\boldsymbol{V}_i/dt = \boldsymbol{F}_i + \boldsymbol{K}_i$, where $M_i = \rho A = M$ and $\boldsymbol{V}_i = \int d\boldsymbol{r}\boldsymbol{v}\phi_i/A$ are the mass and the average velocity of particle i, respectively. On the right hand side, $\boldsymbol{F}_i = \int d\boldsymbol{r}\phi_i(\boldsymbol{f}_U + \boldsymbol{f}_T)$ is the force arising from the interparticle interaction and the torque, and $\boldsymbol{K}_i = -\int d\boldsymbol{r}\phi_i\nabla\cdot\overleftrightarrow{\boldsymbol{\Pi}} \cong -\int dS_i\hat{\boldsymbol{n}}_i\cdot\overleftrightarrow{\boldsymbol{\Pi}}$ the force exerted by the fluid. Here we use the following approximate relation $\int d\boldsymbol{r}\nabla\phi_i\cdot\overleftrightarrow{\boldsymbol{Q}} \cong -\int_{S_i}\hat{\boldsymbol{n}}_i dS_i\cdot\overleftrightarrow{\boldsymbol{Q}}$ for an arbitrary tensor $\overleftrightarrow{\boldsymbol{Q}}(\boldsymbol{r})$, where S_i is the surface of particle i and $\hat{\boldsymbol{n}}_i$ is the unit outward normal vector to S_i. In practical numerical calculations, the on-lattice velocity field, $v(\boldsymbol{r}, t+\Delta t)$, is evaluated from the physical quantities at time t by Equation (1). Then, we move particle i off-lattice as a rigid body by $\boldsymbol{r}_i(t+\Delta t) = \boldsymbol{r}_i(t) + \Delta t\boldsymbol{V}_i(t+\Delta t)$, where Δt is the time increment of the numerical integration.

In our simulation, the units of length ℓ and time τ are related as $\tau = \ell^2/(\eta_\ell/\rho)$, which sets both the scaled density and viscosity of the fluid region to unity. This τ is a time required for the fluid momentum to diffuse over a lattice size ℓ. The units of stress and energy are $\bar{\sigma} = \rho(\ell/\tau)^2$ and $\bar{\epsilon} = \bar{\sigma}\ell^3$, respectively. Furthermore, we set $\eta_c/\eta_\ell = 50$, $\Delta t = 0.003$, $\ell = 0.5$, $\xi = 0.5$ and $a = 6.4$. We confirmed that this choice yields reliable results by comparing them wit the analytical solution for a single particle rotation (see, for example, Fig. 1a,b). The simulation box used was typically $L^2 = 256^2$ lattices. To avoid cumbersome expressions, we will use the same characters for the scaled variables below. We solve the equation of motion (Equation (1)) by the marker-and-cell method with a staggered lattice under the periodic boundary condition.

Interparticle potentials. To mimic rotating hard disks, we employ the Weeks–Chandler–Andersen repulsive potential[45]: $U_{jk}(r) = 4\epsilon\{(\sigma_{jk}/r)^{12} - (\sigma_{jk}/r)^6 + 1/4\}$ for $r < 2^{\frac{1}{6}}\sigma_{jk}$, otherwise $U_{jk}(r) = 0$, where ϵ gives the energy scale, $\sigma_{jk} = (\sigma_j + \sigma_k)/2$ and σ_j represent the size of particle j.

Characterization of structures. The 2D radial distribution function $g(r)$ was calculated as

$$g(r) = \frac{1}{2\pi r\Delta r\rho(N-1)}\sum_{j\neq k}\delta(r - |\boldsymbol{r}_{jk}|), \tag{2}$$

which is the ratio of the ensemble average of the number density of particles existing in the region $r \sim r + \Delta r$ to the average number density $\rho = N/L^2$. Here N is the number of particles in the simulation box, whose side length is L, and Δr is the increment of r.

Similarly, the spatial correlation of Ψ_6^j is calculated as[32]

$$g_6^{2D}(r) = \frac{L^2}{2\pi r\Delta r N(N-1)}\sum_{j\neq k}\delta(r - |\boldsymbol{r}_{jk}|)\Psi_6^j\Psi_6^{k*}. \tag{3}$$

The spatial correlation of the bond-orientational order can then be characterized by $g_6^{2D}(r)/g(r)$.

References

1. Whitesides, G. M. & Grzybowski, B. Self-assembly at all scales. *Science* **295,** 2418–2421 (2002).
2. Grzybowski, B. A., Wilmer, C. E., Kim, J., Browne, K. P. & Bishop, K. J. M. Self-assembly: from crystals to cells. *Soft Matter* **5,** 1110–1128 (2009).
3. Marchetti, M. C. *et al.* Hydrodynamics of soft active matter. *Rev. Mod. Phys.* **85,** 1143–1189 (2013).
4. Grzybowski, B. A., Stone, H. A. & Whitesides, G. M. Dynamic self-assembly of magnetized, millimetre-sized objects rotating at a liquid-air interface. *Nature* **405,** 1033–1036 (2000).
5. Ishikawa, T. & Hota, M. Interaction of two swimming Paramecia. *J. Exp. Biol.* **209,** 4452–4463 (2006).
6. Lauga, E. & Powers, T. R. The hydrodynamics of swimming microorganisms. *Rep. Prog. Phys.* **72,** 096601 (2009).
7. Baskaran, A. & Marchetti, M. C. Statistical mechanics and hydrodynamics of bacterial suspensions. *Proc. Natl Acad. Sci. USA* **106,** 15567–15572 (2009).
8. Drescher, K. *et al.* Dancing Volvox: hydrodynamic bound states of swimming algae. *Phys. Rev. Lett.* **102,** 168101 (2009).
9. Ramaswamy, S. The mechanics and statistics of active matter. *Annu. Rev. Condens. Matter Phys.* **1,** 323–345 (2010).
10. Wensink, H. H. *et al.* Meso-scale turbulence in living fluids. *Proc. Natl Acad. Sci. USA* **109,** 14308–14313 (2012).
11. Zöttl, A. & Stark, H. Hydrodynamics determines collective motion and phase behavior of active colloids in quasi-two-dimensional confinement. *Phys. Rev. Lett.* **112,** 118101 (2014).
12. Bialké, J., Speck, T. & Löwen, H. Crystallization in a dense suspension of self-propelled particles. *Phys. Rev. Lett.* **108,** 168301 (2012).
13. Grzybowski, B. A., Jiang, X., Stone, H. A. & Whitesides, G. M. Dynamic, self-assembled aggregates of magnetized, millimeter-sized objects rotating at the liquid-air interface: Macroscopic, two-dimensional classical artificial atoms and molecules. *Phys. Rev. E* **64,** 011603 (2001).
14. Grzybowski, B. A. & Whitesides, G. M. Directed dynamic self-assembly of objects rotating on two parallel fluid interfaces. *J. Chem. Phys.* **116,** 8571–8577 (2002).
15. Friese, M. E. J., Nieminen, T. A., Heckenberg, N. R. & Rubinsztein-Dunlop, H. Optical alignment and spinning of laser-trapped microscopic particles. *Nature* **394,** 348–350 (1998).
16. Riedel, I. H., Kruse, K. & Howard, J. A self-organized vortex array of hydrodynamically entrained sperm cells. *Science* **309,** 300–303 (2005).
17. Schwarz-Linek, J. *et al.* Phase separation and rotor self-assembly in active particle suspensions. *Proc. Natl Acad. Sci. USA* **109,** 4052–4057 (2012).
18. Lenz, P., Joanny, J. F., Jülicher, F. & Prost, J. Membranes with rotating motors. *Phys. Rev. Lett.* **91,** 108104 (2003).
19. Gehrig, E. & Hess, O. Nonlinear dynamics and self-organization of rotary molecular motor ensembles. *Phys. Rev. E* **73,** 051916 (2006).
20. Llopis, I. & Pagonabarraga, I. Hydrodynamic regimes of active rotators at fluid interfaces. *Eur. Phys. J. E Soft Matter* **26,** 103–113 (2008).
21. Leoni, M. & Liverpool, T. B. Dynamics and interactions of active rotors. *Europhys. Lett.* **92,** 64004 (2010).
22. Götze, I. O. & Gompper, G. Flow generation by rotating colloids in planar microchannels. *Europhys. Lett.* **92,** 64003 (2010).
23. Götze, I. O. & Gompper, G. Dynamic self-assembly and directed flow of rotating colloids in microchannels. *Phys. Rev. E* **84,** 031404 (2011).
24. Yang, Y., Qiu, F. & Gompper, G. Self-organized vortices of circling self-propelled particles and curved active flagella. *Phys. Rev. E* **89,** 012720 (2014).
25. Fily, Y., Baskaran, A. & Marchetti, M. C. Cooperative self-propulsion of active and passive rotors. *Soft Matter* **8,** 3002–3009 (2012).
26. Eyink, G. L. & Sreenivasan, K. R. Onsager and the theory of hydrodynamic turbulence. *Rev. Mod. Phys.* **78,** 87–135 (2006).
27. Durkin, D. & Fajans, J. Experiments on two-dimensional vortex patterns. *Phys. Fluids* **12,** 289–293 (2000).
28. Aref, H., Newton, P. K., Stremler, M. A., Tokieda, T. & Vainchtein, D. L. Vortex crystals. *Adv. Appl. Mech.* **39,** 1–79 (2003).
29. Rasmussen, J. J., Nielsen, A. H. & Naulin, V. Dynamics of vortex interactions in two-dimensional flows. *Phys. Scr.* **T98,** 29–33 (2002).
30. Aref, H. Point vortex dynamics: a classical mathematics playground. *J. Math. Phys.* **48,** 065401 (2007).
31. Tanaka, H. & Araki, T. Simulation method of colloidal suspensions with hydrodynamic interactions: fluid particle dynamics. *Phys. Rev. Lett.* **85,** 1338–1341 (2000).

32. Nelson, D. R. *Defects and Geometry in Condensed Matter Physics* (Cambridge Univ. Press, 2002).
33. Montgomery, M. T., Vladimirov, V. A. & Denissenko, P. V. An experimental study on hurricane mesovortices. *J. Fluid Mech.* **471,** 1–32 (2002).
34. Le Dizes, S. & Verga, A. Viscous interactions of two co-rotating vortices before merging. *J. Fluid Mech.* **467,** 389–410 (2002).
35. Yeo, K., Maxey, M. R. & Karniadakis, G. E. Dynamic self-assembly of spinning particles. *J. Fluids Eng.* **129,** 379–387 (2007).
36. Landau, L. D. & Lifshitz, E. M. *Fluid mechanics* Vol. 6, 227–229 (Course of Theoretical Physics, Pergamon Press, Oxford, 1987).
37. Tanaka, H., Hayashi, T. & Nishi, T. Digital image analysis of droplet patterns in polymer systems: point pattern. *J. Appl. Phys.* **65,** 4480–4495 (1989).
38. Kawasaki, T., Araki, T. & Tanaka, H. Correlation between dynamic heterogeneity and medium-range order in two-dimensional glass-forming liquids. *Phys. Rev. Lett.* **99,** 215701 (2007).
39. Bernard, E. P. & Krauth, W. Two-step melting in two dimensions: first-order liquid-hexatic transition. *Phys. Rev. Lett.* **107,** 155704 (2011).
40. Nguyen, N. H. P., Klotsa, D., Engel, M. & Glotzer, S. C. Emergent collective phenomena in a mixture of hard shapes through active rotation. *Phys. Rev. Lett.* **112,** 075701 (2014).
41. Voituriez, R., Joanny, J. F. & Prost, J. Generic phase diagram of active polar films. *Phys. Rev. Lett.* **96,** 028102 (2006).
42. Terray, A., Oakey, J. & Marr, D. W. M. Microfluidic control using colloidal devices. *Science* **296,** 1841–1844 (2002).
43. Bleil, S., Marr, D. W. M. & Bechinger, C. Field-mediated self-assembly and actuation of highly parallel microfluidic devices. *Appl. Phys. Lett.* **88,** 263515 (2006).
44. Furukawa, A. & Tanaka, H. Key role of hydrodynamic interactions in colloidal gelation. *Phys. Rev. Lett.* **104,** 245702 (2010).
45. Weeks, J. D., Chandler, D. & Andersen, H. C. Role of repulsive forces in determining the equilibrium structure of simple liquids. *J. Chem. Phys.* **54,** 5237–5247 (1971).

Acknowledgements

We are grateful to Akira Furukawa for discussion on the origin of hydrodynamic attractive interactions. We also thank Jun Russo for a critical reading of the manuscript. This study was partly supported by Grants-in-Aid for Scientific Research (S) and Specially Promoted Research from the Japan Society for the Promotion of Science (JSPS) and the Aihara Project, the FIRST program from JSPS, initiated by the Council for Science and Technology Policy (CSTP).

Author contributions

H.T. proposed and supervised the study, Y.G. performed simulations and H.T. wrote the manuscript.

Additional information

Competing financial interests: The authors declare no competing financial interests.

15

Suppression law of quantum states in a 3D photonic fast Fourier transform chip

Andrea Crespi[1,2], Roberto Osellame[1,2], Roberta Ramponi[1,2], Marco Bentivegna[3], Fulvio Flamini[3], Nicolò Spagnolo[3], Niko Viggianiello[3], Luca Innocenti[3,4], Paolo Mataloni[3] & Fabio Sciarrino[3]

The identification of phenomena able to pinpoint quantum interference is attracting large interest. Indeed, a generalization of the Hong-Ou-Mandel effect valid for any number of photons and optical modes would represent an important leap ahead both from a fundamental perspective and for practical applications, such as certification of photonic quantum devices, whose computational speedup is expected to depend critically on multi-particle interference. Quantum distinctive features have been predicted for many particles injected into multimode interferometers implementing the Fourier transform over the optical modes. Here we develop a scalable approach for the implementation of the fast Fourier transform algorithm using three-dimensional photonic integrated interferometers, fabricated via femtosecond laser writing technique. We observe the suppression law for a large number of output states with four- and eight-mode optical circuits: the experimental results demonstrate genuine quantum interference between the injected photons, thus offering a powerful tool for diagnostic of photonic platforms.

[1]Istituto di Fotonica e Nanotecnologie, Consiglio Nazionale delle Ricerche (IFN-CNR), Piazza Leonardo da Vinci, 32, I-20133 Milano, Italy. [2]Dipartimento di Fisica, Politecnico di Milano, Piazza Leonardo da Vinci, 32, I-20133 Milano, Italy. [3]Dipartimento di Fisica, Sapienza Università di Roma, Piazzale Aldo Moro 5, I-00185 Roma, Italy. [4]Università di Roma Tor Vergata, Via della ricerca scientifica 1, I-00133 Roma, Italy. Correspondence and requests for materials should be addressed to R.O. (email: roberto.osellame@polimi.it) or to F.S. (email: fabio.sciarrino@uniroma1.it).

The amplitude interference between wavefunctions corresponding to indistinguishable particles lies at the very heart of quantum mechanics. Right after the introduction of laser amplification, the availability of strong coherent pulses allowed to test interference between different light pulses[1,2], while generation of pairs of identical photons through parametric fluorescence[3] led subsequently to the milestone experiment of Hong *et al.*[4–8]. Later on, photonic platforms have been demonstrated to be in principle capable to perform universal quantum computing[9].

Recently, multi-particle interference effects of many photons in large interferometers are attracting a strong interest, as they should be able to show unprecedented evidences of the superior quantum computational power compared with that of classical devices[10–12]. The main example is given by the boson sampling[13] computational problem, which consists in sampling from the probability distribution given by the permanents of the $n \times n$ submatrices of a given Haar random unitary. The problem is computationally hard (in n) for a classical computer, since calculating the permanent of a complex-valued matrix is a #P-hard problem. However, sampling from the output distribution can be efficiently achieved by letting n indistinguishable photons evolve through an optical interferometer implementing the unitary transformation in the Fock space, and by detecting output states with an array of single-photon detectors. The chance to provide evidences of a post-classical computation with this relatively simple set-up has triggered a large experimental effort, leading to small-scale implementations[14–20], as well as theoretical analyses on the effects of experimental imperfections[21,22] and on possible implementations including alternative schemes[23,24].

In the context of searching for experimental evidences against the extended Church–Turing thesis, a boson sampling experiment poses a problem of certification of the result's correctness in the computationally hard regime[25]. The very complexity of the boson-sampling computational problem precludes the use of a brute-force approach, that is, calculating the expected probability distribution at the output and comparing it with the collected data. Efficient statistical techniques able to rule out trivial alternative distributions have been proposed[26] and tested[18,19], but the need for more stringent tests able to rule out less trivial distributions has led, and continues to encourage, additional research efforts in this direction.

In particular, an efficient test able to confirm true n-photon interference in a multimode device has been recently proposed[27]. The protocol is based on the use of an interferometer implementing the transformation described by the n^p-dimensional Fourier matrix, with p being any integer. When feeding this device with multi-photon states of a specific symmetry, suppression of many output configurations is observed[28], due to granular[27] many-particle interference. This effect is able to rule out alternative models requiring only coarse-grained features like the ones present in Bose–Einstein condensates[29–31]. Indeed, the implications of this effect go well beyond the certification of boson sampling devices. As a generalization of the two-photon/two-modes Hong–Ou–Mandel (HOM) effect, the suppression law, also named Zero-Transmission law[28], is important at a fundamental level, while at the practical level it could be used as a diagnostic tool for a wide range of photonic platforms[27,32,33]. During the review process of this work, the implementation of a discrete Fourier transform circuit in a fully reconfigurable chip has been reported[34]. The Zero-Transmission law for three-photon no-bunching events has been demonstrated in this planar six-mode interferometer.

In this article, we report the experimental observation of the recent theoretically proposed[27] suppression law for Fourier matrices, and its use to validate quantum many-body interference against alternative non-trivial hypotheses resulting in similar output probability distributions. The Fourier matrices have been implemented with an efficient and reliable approach by exploiting the quantum version of the fast Fourier transform (qFFT), an algorithm developed by Barak and Ben-Aryeh[35] to optimize the number of optical elements required to build the Fourier transform over the optical modes. Here we implement the qFFT on photonic integrated interferometers by exploiting the three-dimensional (3D) capabilities of femtosecond laser writing[36,37], which makes it possible to fabricate waveguides arranged in 3D structures with arbitrary layouts[38–40], by adopting an architecture scalable to a larger number of modes. The observations have been carried out with two-photon Fock states injected into four-mode and eight-mode qFFT interferometers. The peculiar behaviour of Fock states compared with other kinds of states is investigated, showing in principle the validity of the certification protocol for the identification of true granular n-particle interference, which is the source of a rich landscape of quantum effects such as the computational complexity of boson sampling.

Results

Suppression law in Fourier transform matrices. As a generalization of the HOM effect, it has been pointed out that quantum interference effects in multimode interferometers may determine suppression of a large fraction of the output configurations[28,31,41], depending on the specific unitary transformation being implemented and on the symmetry of the input state. In particular[28,31], let us consider a cyclic input, that is, an n-photon Fock state over $m = n^p$ modes (for some integer p) where the occupied modes j_r^s are determined by the rule $j_r^s = s + (r-1)n^{p-1}$, with $r = 1, \ldots, n$ and $s = 1, \ldots, n^{p-1}$. The index s takes into account the fact that there are n^{p-1} possible n-photon arrangements with periodicity n^{p-1}, which simply differ by a translation of the occupational mode labels. For example, for $n = 2$ and $m = 4$ there are $2^1 = 2$ possible cyclic states, (1,0,1,0) and (0,1,0,1), while for $n = 2$ and $m = 8$ there are $2^2 = 4$ possible (collision-free) cyclic inputs, that is, the states (1,0,0,0,1,0,0,0), (0,1,0,0,0,1,0,0), (0,0,1,0,0,0,1,0) and (0,0,0,1,0,0,0,1).

We consider the evolution of such states through an interferometer implementing the transformation described by the Fourier matrix

$$(U_m^F)_{l,q} = \frac{1}{\sqrt{m}} e^{i\frac{2\pi lq}{m}}. \quad (1)$$

Such evolution results in the suppression of all output configurations not fulfilling the equation

$$\mathrm{mod}\left(\sum_{l=1}^{n} k_l, n\right) = 0, \quad (2)$$

where k_l is the output mode of the lth photon. An interesting application of suppression laws is to certify the presence of true many-body granular interference during the evolution in the interferometer, ruling out alternative hypotheses which would result in similar output probability distributions. In particular, in the case of Fourier matrices, the observation of the suppression law (2) allows to certify that the sampled output distribution is not produced by either distinguishable particles or a mean field state (MF)[27]. The latter is defined as a single-particle state $|\psi^s\rangle$ of the form

$$|\psi^s\rangle = \frac{1}{\sqrt{n}} \sum_{r=1}^{n} e^{i\theta_r} |j_r^s\rangle, \quad (3)$$

with a random set of phases θ_r for each state, being $|j_r^s\rangle$ a single-particle state occupying input mode j_r^s. This state reproduces macroscopic interference effects, such as bunching or bosonic clouding[19], and cannot be distinguished from true multi-particle interference with criteria based on these features. Since it is possible to efficiently simulate the evolution of a MF with a classical algorithm[27], it is of fundamental importance to assess the ability of a validation scheme to discriminate such a state in the context of an untrusted party claiming to perform a boson sampling experiment. Hence, the MF represents an optimal test bed for the certification protocol based on the suppression law (see Fig. 1).

It is possible to quantify the degree of violation $\mathcal{D} = N_{\text{forbidden}}/N_{\text{events}}$ of the suppression law as the number of observed events in forbidden output states divided by the total number of events[27]. If a Fock state is injected in a Fourier interferometer, a violation $\mathcal{D} = 0$ would be observed. In the case of distinguishable photons there is no suppression law, and the violation would be simply the fraction of suppressed outputs, each one weighted with the number of possible arrangements of the n distinguishable particles in that output combination. In the case of two-photon states, the weighting factor is 2 for collision-free outputs and 1 otherwise, and a degree of violation of 1/2 is expected (see Section 'Observation of the suppression law'). In contrast, in the case of two-photon MF, bunching effects occur leading to an expected degree of violation of half the weighted fraction of suppressed outputs (1/4 for two-photon MF). It has been shown that the fraction of forbidden outputs is always large[31]. Hence, a comparison of the observed value of $\mathcal{D}$ with the expected one represents an efficient way, in terms of necessary experimental runs, to discriminate between Fock states, distinguishable particles states and MFs.

Realization of 3D qFFT interferometers. Let us now introduce our experimental implementation of the qFFT. The general method to realize an arbitrary unitary transformation using linear optics was introduced by Reck *et al.*[42], who provided a decomposition of a unitary of dimension m as a sequence of $m(m-1)/2$ beam splitters and phase shifters. However, in the special case of Fourier matrices a more efficient method has been proposed[35,43], which takes advantage of their symmetries to significantly reduce the number of linear optical elements required. On the basis of the classical algorithm of Cooley and Tukey[44], who first introduced the fast Fourier transform algorithm as a more efficient way to calculate the discrete Fourier transform, Barak and Ben-Aryeh developed a quantum analogue in the linear optics domain, leading to the concept of qFFT. This approach, valid for 2^p-dimensional Fourier matrices, requires only $(m/2)\log m$ beam splitters and phase shifters, to be compared with the $O(m^2)$ elements needed for the more general Reck decomposition, thus enhancing the compactness and scalability of the platform for a more reliable experimental realization. The overall linear transformation on the optical modes implemented by the qFFT circuit is naturally equivalent to the transformation described by the Fourier matrix, hence $U_m^{\text{qFFT}} = U_m^{\text{F}}$.

Here we introduce a new methodology for an integrated implementation of the qFFT, which exploits the 3D capabilities of the femtosecond laser writing technique. The sequential structure arising from the decomposition of the m-dimensional Fourier matrix using the Barak and Ben-Aryeh algorithm is reproduced by the consecutive layers shown in Fig. 2. The complex arrangement of pairwise interactions necessary for the qFFT method cannot be easily implemented using a planar architecture. However, femtosecond laser writing technique allows to overcome this issue exploiting the third dimension, arranging the waveguides along the bidimensional sections of the integrated chip.

The strategy can be outlined as follows (see also Supplementary Note 1): the 2^p modes are ideally placed on the vertices of a p-dimensional hypercube; in each step of the algorithm the vertices connected by parallel edges having one specific direction are made to interact by a two-mode Hadamard transformation, with proper phase terms. An optical interferometer implementing this procedure is thus composed of $\log_2 m = p$ sections, each employing $m/2$ balanced beam splitters and phase shifters.

We fabricated waveguide interferometers realizing the Fourier matrix for $m = 4$ and 8 modes in borosilicate glass chips using femtosecond laser micromachining[36,37]. A schematic representation of these two interferometers is given in Fig. 2. According to the scheme outlined above and by exploiting the 3D capabilities of the fabrication technique, the waveguides are placed, for what concerns the cross-section of the device, on the vertices of a two-dimensional projection of the p-dimensional hypercube (see also Supplementary Fig. 1). 3D directional couplers, with proper interaction length and distance to achieve a balanced splitting, connect in each step the required vertices. The insets of Fig. 2 show, at each step i, which modes are connected by directional couplers (L_i) and the amount of phase shift that needs to be introduced in specific modes (P_i). Phase shifters, where needed, are implemented by geometrical deformation of the connecting S-bends. Fan-in and fan-out sections at the input and output of the devices allows interfacing with 127 μm spaced single-mode fibre arrays. Note that in our device geometry, in each step, the vertices to be connected are all at the same relative distance. This means that, unless geometric deformations are designed where needed, light travelling in different modes does not acquire undesired phase delays. It is worth noting that the geometric construction here developed is scalable to an arbitrary number of modes with a number of elements increasing as $m\log_2 m$.

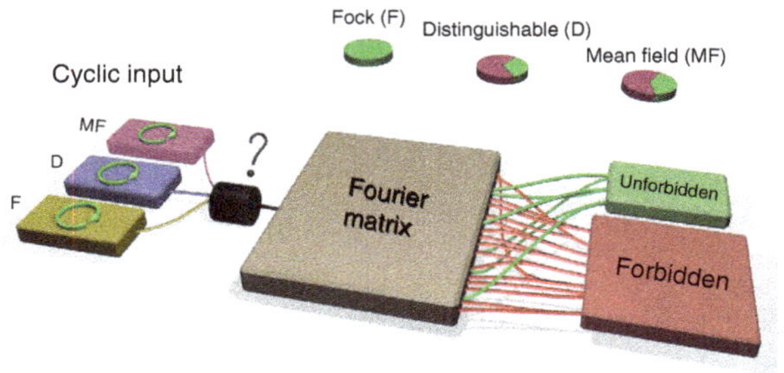

Figure 1 | Suppression law for Fock states in a Fourier interferometer. Conceptual scheme of the protocol: the possible configurations of n photons at the output of an m-mode interferometer can be divided into two categories, unforbidden and forbidden, depending on whether they satisfy or not the suppression condition (2), respectively. The pie charts show the expected output statistics with different classes of particles, where green and red areas represent events with unforbidden and forbidden outputs, respectively. The injection of a cyclic Fock state (beige box) in an m-mode Fourier interferometer results in total suppression of forbidden output states. Cyclic states with distinguishable particles (blue box) show no suppression, so that each output combination is equally likely to occur. A mean field state (purple box), which reproduces some of the features of bosonic statistics, shows suppression with highly reduced contrast. Therefore, with a cyclic input the m-mode Fourier interferometer is able to discriminate, through the measurement of degree of violation $\mathcal{D} = N_{\text{forbidden}}/N_{\text{events}}$, which of these three hypotheses the input state belongs to.

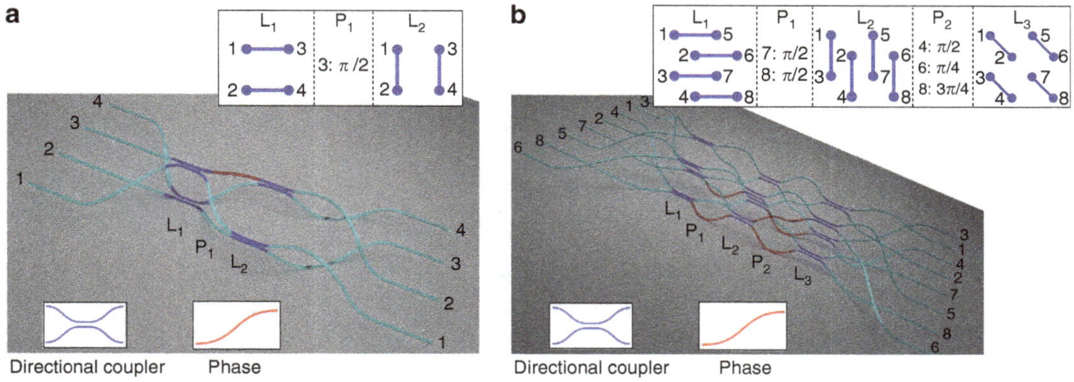

Figure 2 | Schematic representation of the structure of the integrated devices. Internal structure of the four-mode (**a**) and eight-mode (**b**) integrated interferometers implementing the qFFT over the optical modes. In the eight-mode case, the Barak and Ben-Aryeh algorithm requires an additional relabelling of the output modes (not shown in the figure), namely 2↔5 and 4↔7, to obtain the effective Fourier transformation. The mode arrangement has been chosen in a way to minimize bending losses. The insets show the actual disposition of the waveguides in the cross-section of the devices. The modes coupled together in each step (L_i) of the interferometer are joined by segments. The implemented phase shifts in each step (P_i) are also indicated.

One- and two-photon measurements. The two implemented interferometers of $m=4$ and 8 modes are fed with single- and two-photon states. The experimental set-up, preparing a biphoton wave packet to be injected into the devices, is shown in Fig. 3. Further details on the photon generation and detection scheme are described in the Methods section. To test the validity of the suppression law, we measured the number of coincidences at each forbidden output combination injecting cyclic inputs with two indistinguishable photons. The degree of violation $\mathcal{D}$ of the suppression law could simply be evaluated with a counting experiment. Alternatively, the same quantity $\mathcal{D}$ can be expressed as a function of single-photon input–output probabilities and of the HOM visibilities, defined as

$$V_{i,j}=\frac{N^D_{i,j}-N^Q_{i,j}}{N^D_{i,j}} \qquad (4)$$

where $N^D_{i,j}$ is the number of detected coincidences for distinguishable photons and $N^Q_{i,j}$ for indistinguishable photons. The subscripts (i,j) are the indexes of the two output modes, for a given input state. The degree of violation can therefore be expressed as

$$\begin{aligned}\mathcal{D}&=\frac{N_{\text{forbidden}}}{N_{\text{events}}}=P_{\text{forbidden}}=\\&=\sum_{(i,j)_{\text{forbidden}}}P^Q_{i,j}=\sum_{(i,j)_{\text{forbidden}}}P^D_{i,j}(1-V_{i,j})\end{aligned} \qquad (5)$$

where $P^Q_{i,j}$ ($P^D_{i,j}$) are the probabilities of having photons in the outputs i,j in the case of indistinguishable (distinguishable) particles. Here $P^D_{i,j}$ can be obtained from single-particle probabilities. The visibilities are measured by recording the number of coincidences for each output combination as a function of the temporal delay between the two injected photons.

For the four-mode device, we measured the full set of $\binom{4}{2}^2=36$ collision-free input–output combinations, that is, where the two photons exit from different output ports. These contributions have been measured by recording the number of coincidences for each combination of two outputs as a function of the temporal delay between the two input photons. Because of the law given by equation (2), we expect to observe four suppressed outcomes (over six possible output combinations) for the two cyclic input states (1,3) and (2,4). Since distinguishable photons exhibit no interference, HOM dips in the coincidence patterns are expected for the suppressed output states. Conversely, peaks are expected in the non-suppressed output combinations. The experimental results are shown in Fig. 4a, where the expected pattern of four suppressions and two enhancements is reported, with average visibilities of $\overline{V}_{\text{supp}}=0.899\pm0.001$ and $\overline{V}_{\text{enh}}=-0.951\pm0.004$ for suppression and enhancement, respectively.

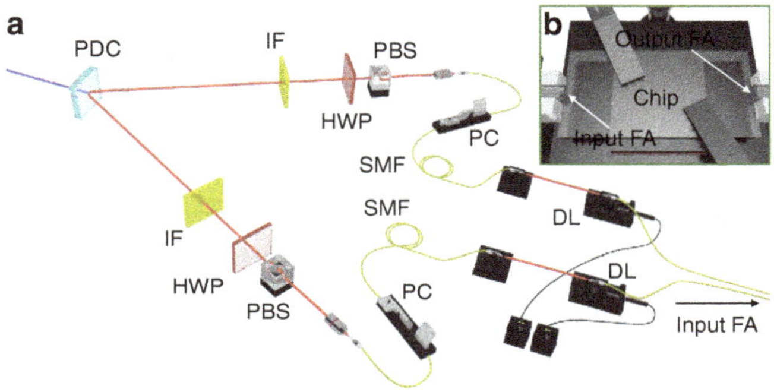

Figure 3 | Experimental apparatus for input state preparation. (**a**) The photon source (IF, HWP, PBS, PC, PDC, DL and SMF). (**b**) Photon injection (extraction) before (after) the evolution through the interferometer. DL, delay lines with motorized stages; FA, fibre array; HWP, half-wave plate; IF, interferential filter; PBS, polarizing beam splitter; PC, polarization compensator; PDC, parametric downconversion; SMF, single-mode fibre.

For the cyclic inputs, we also measured the interference patterns for the output contributions where the two photons exit from the same mode. These terms have been measured by inserting an additional symmetric beam splitter on each output mode, and by connecting each of its two outputs to a single-photon detector. These cases correspond to a full-bunching scenario with $n=2$, and a HOM peak with $V=-1$ visibility is expected independently from the input state and from the unitary operation[45]. This feature has been observed for the tested inputs, where an average visibility of $\overline{V}_{\text{bunch}}=-0.969\pm0.024$ has been obtained over all full-bunching combinations. Note that the measured two-mode correlation matrix is not compatible with classical light (see Supplementary Note 4).

The existence of a general rule for the prediction of suppressed output combinations when injecting a cyclic Fock state in a Fourier interferometer is due to the intrinsic symmetry of the problem, as opposed to the general boson sampling scenario[13]. Suppressed outputs for non-cyclic inputs can be predicted by

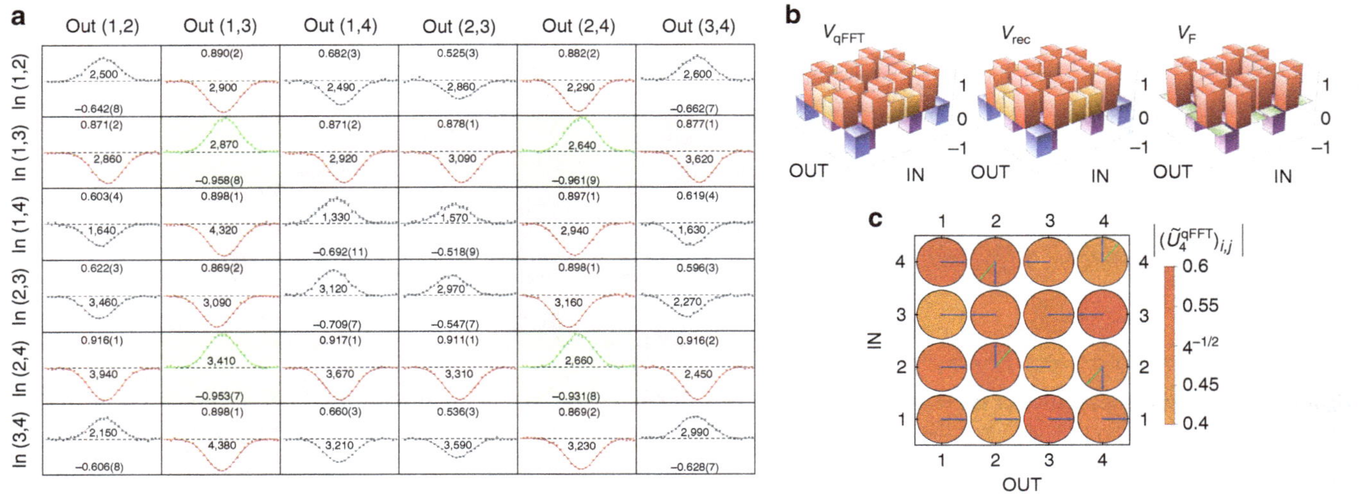

Figure 4 | Suppression law in a four-mode qFFT integrated chip. (**a**) Complete set of 36 measured coincidence patterns (raw experimental data) for all input-output combinations in the four-mode chip. For each input-output combination, the measured coincidence pattern as a function of the time delay is shown (points: experimental data, lines: best-fit curves). Cyclic inputs (1,3) and (2,4) exhibit enhancement (green) and suppression (red) on cyclic and non-cyclic outputs, respectively. For all points, error bars are due to the Poissonian statistics of the events. In each subplot the measured visibility with corresponding error and the sample size are reported. For each visibility, the error is obtained through a Monte Carlo simulation by averaging over 3,000 simulated data sets. In each subplot the zero level coincides with the baseline, while a dashed line represents the number of coincidence events in the distinguishable limit. (**b**) HOM visibilities for all 36 input-output configurations. (left to right) Experimental measured visibilities (V_{qFFT}, obtained from raw experimental data), visibilities calculated from the reconstructed unitary (V_{rec}), and visibilities calculated from the theoretical unitary (V_F). (**c**) Representation of the reconstructed experimental transformation $\tilde{U}_4^{qFFT}$, and comparison with U_4^F. Coloured disks represent the moduli of the reconstructed matrix elements (all equal to $4^{-1/2}$ for U_4^F). Arrows represent the phases of the unitary matrix elements (green: reconstructed unitary, blue: Fourier matrix).

calculating the permanent of the submatrix given by the intersection of the columns and rows of U^F corresponding to the occupied input and output modes, respectively. The complete set of measured dips and peaks is shown in Fig. 4a, highlighting the symmetry in the Fourier transform interference pattern. The injection of the non-cyclic input states has been employed for the complete reconstruction of the chip action $\tilde{U}_4^{qFFT}$, using a data set statistically independent from the one adopted to observe the suppression law. The adopted reconstruction algorithm, which exploits knowledge on the internal structure of the interferometers (specified in Fig. 2), works in two steps. In a first step, the power-splitting ratios measured with classical light are employed to extrapolate the transmissivities of the directional couplers. In a second step, the two-photon visibilities for the non-cyclic inputs are used to retrieve the values of the fabrication phases. In both steps the parameters are obtained by minimizing a suitable χ^2 function. The results are shown in Fig. 4c. The fidelity between the reconstructed unitary $\tilde{U}_4^{qFFT}$ and the theoretical Fourier transform U_4^F is $\mathcal{F} = 0.9822 \pm 0.0001$, thus confirming the high quality of the fabrication process. The error in the estimation of the fidelity is obtained through a Monte Carlo simulation, properly accounting for the degree of distinguishability of the photons with a rescaling factor in the visibilities.

For the eight-mode chip we recorded all the $\binom{8}{2} = 28$ two-photon coincidence patterns, as a function of the relative delay between the input photons, for each of the four collision-free cyclic inputs and for one non-cyclic input. The reconstruction of the actual unitary transformation $\tilde{U}_8^{qFFT}$ implemented has been performed with the same algorithm of the four-modes, by using the power-splitting ratios measured with classical light and the two-photon visibilities for one non-cyclic input. The latter has been chosen in a way to maximize the sensitivity of the measurements with respect to the five fabrication phases. The results are shown in Fig. 5. The fidelity between the reconstructed unitary $\tilde{U}_8^{qFFT}$ and the ideal eight-mode Fourier transform U_8^F is $\mathcal{F} = 0.9527 \pm 0.0006$. More details on the reconstruction algorithm can be found in the Supplementary Note 3.

Observation of the suppression law. The suppression of events which do not satisfy equation (2) is fulfilled only when two perfectly indistinguishable photons are injected in a cyclic input of a perfect Fourier interferometer. In such a case, we would have the suppression of all output states whose sum of the indexes corresponding to the occupied modes is odd. For the four-mode (eight-mode) interferometer, this corresponds to four (16) suppressed and two (12) non-suppressed collision-free outputs (each one given by two possible arrangements of the two distinguishable photons), plus four (8) terms with two photons in the same output, each one corresponding to a single possible two-photon path.

The expected violation for distinguishable particles can be obtained from classical considerations. Let us consider the case with $n = 2$. The two distinguishable photons evolve independently from each other, and the output distribution is obtained by classically mixing single-particle probabilities. All collision-free terms are equally likely to occur with probability $q = 2/m^2$, while full-bunching events occur with probability $q' = q/2 = 1/m^2$. The degree of violation $\mathcal{D}_D$ can then be obtained by multiplying the probability q by the number of forbidden output combinations. As a result, we expect a violation degree of $\mathcal{D}_D = 0.5$ for distinguishable two-photon states. The evaluation of the expected value for a MF state, which is due to single-particle bosonic statistic effects, requires different calculations[27]. It can be shown that for $n = 2$ the degree of violation is $\mathcal{D}_{MF} = 0.25$.

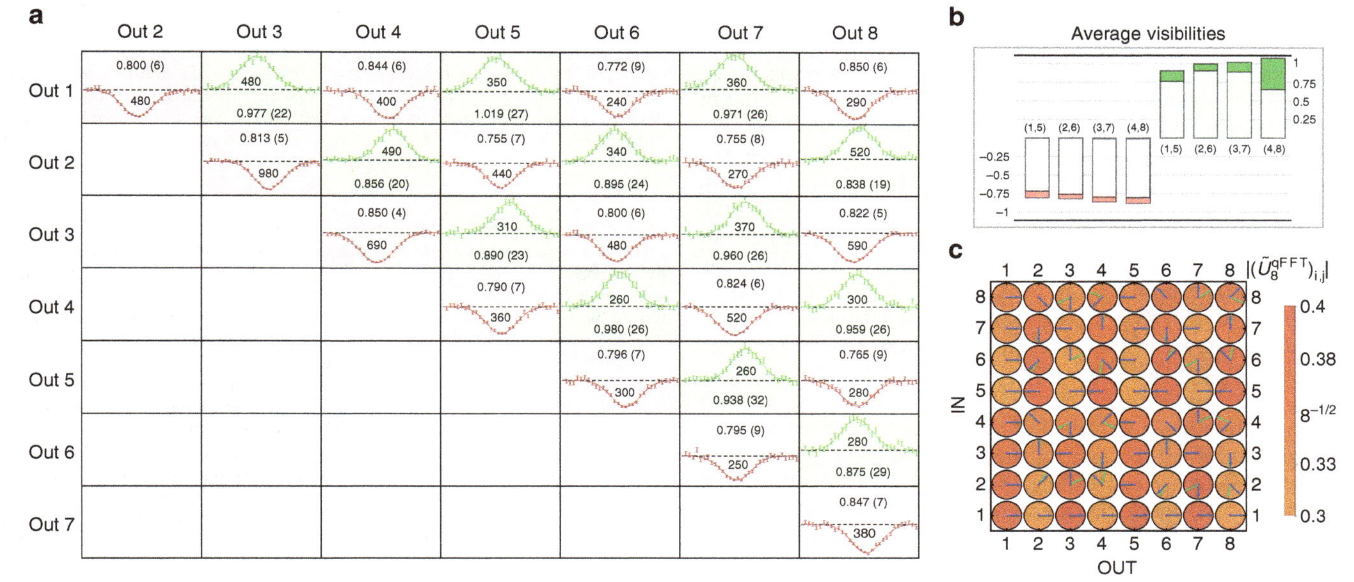

Figure 5 | Suppression law in a eight-mode qFFT integrated chip. (**a**) Set of 28 measured coincidence patterns (raw experimental data), corresponding to all collision-free output combinations for the input (2,6) of the eight-mode interferometer. For each output combination, the measured coincidence pattern as a function of the time delay is shown (points: experimental data, lines: best-fit curves). Red or green backgrounds correspond to dips and peaks, respectively. For all points, error bars are due to the Poissonian statistics of the events. In each subplot the measured visibility with corresponding error and the sample size are reported. For each visibility, the error is obtained through a Monte Carlo simulation by averaging over 3,000 simulated data sets. In each subplot the zero level coincides with the baseline, while a dashed line represents the number of coincidence events in the distinguishable limit. (**b**) Average visibilities of dips (red bars) and peaks (green bars) observed for the four collision-free cyclic inputs [(1,5), (2,6), (3,7), (4,8)]. Darker regions correspond to error bars of ± 1 s.d. (**c**) Representation of the reconstructed experimental transformation $\tilde{U}_8^{\mathrm{qFFT}}$, and comparison with U_8^{F}. Coloured disks represent the moduli of the reconstructed matrix elements (all equal to $8^{-1/2}$ for U_8^{F}). Arrows represent the phases of the unitary matrix elements (green: reconstructed unitary, blue: Fourier matrix).

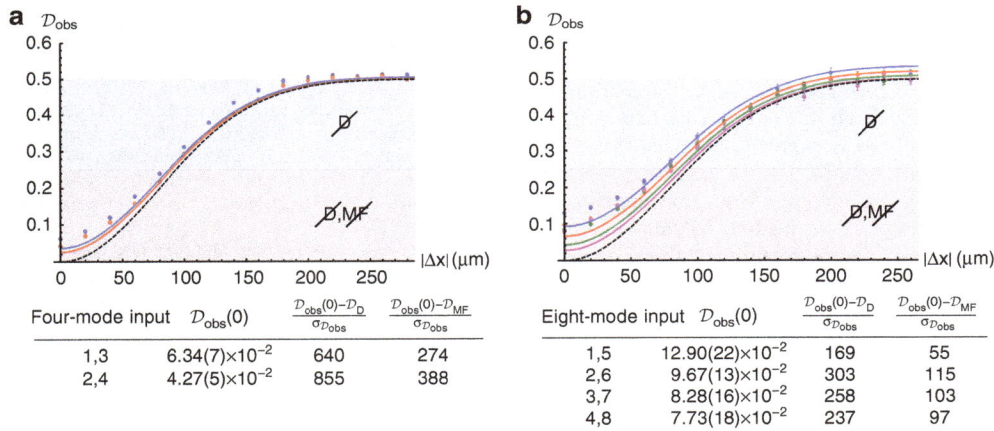

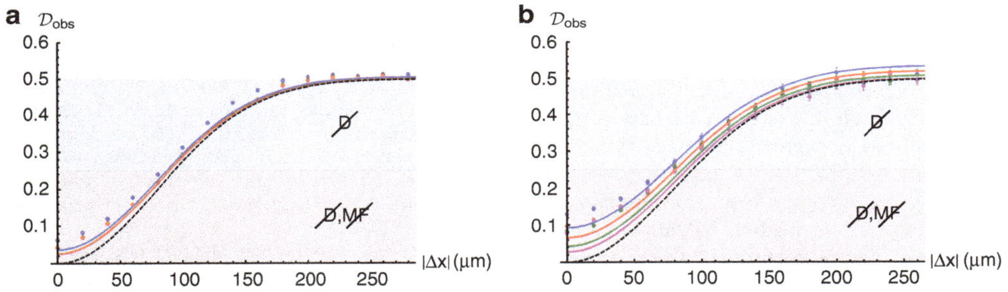

Four-mode input	$\mathcal{D}_{\mathrm{obs}}(0)$	$\frac{\mathcal{D}_{\mathrm{obs}}(0)-\mathcal{D}_{\mathrm{D}}}{\sigma_{\mathcal{D}_{\mathrm{obs}}}}$	$\frac{\mathcal{D}_{\mathrm{obs}}(0)-\mathcal{D}_{\mathrm{MF}}}{\sigma_{\mathcal{D}_{\mathrm{obs}}}}$
1,3	$6.34(7)\times10^{-2}$	640	274
2,4	$4.27(5)\times10^{-2}$	855	388

Eight-mode input	$\mathcal{D}_{\mathrm{obs}}(0)$	$\frac{\mathcal{D}_{\mathrm{obs}}(0)-\mathcal{D}_{\mathrm{D}}}{\sigma_{\mathcal{D}_{\mathrm{obs}}}}$	$\frac{\mathcal{D}_{\mathrm{obs}}(0)-\mathcal{D}_{\mathrm{MF}}}{\sigma_{\mathcal{D}_{\mathrm{obs}}}}$
1,5	$12.90(22)\times10^{-2}$	169	55
2,6	$9.67(13)\times10^{-2}$	303	115
3,7	$8.28(16)\times10^{-2}$	258	103
4,8	$7.73(18)\times10^{-2}$	237	97

Figure 6 | Measured violations. Observed violations $\mathcal{D}_{\mathrm{obs}}$ as a function of the path difference $|\Delta x| = c|\Delta\tau|$ between the two photons. Blue shaded regions in the plots correspond to the cases where the hypothesis of distinguishable particles can be ruled out. Red regions correspond to the cases when both the hypotheses of distinguishable particles and mean field state can be ruled out, and true two-particle interference is present. (**a**) Data for the four-mode interferometer. Blue points: input (1,3). Red points: input (2,4). Blue solid line: theoretical prediction for input (1,3). Red solid line: theoretical prediction for input (2,4). Black dashed line: theoretical prediction for a Fourier matrix. (**b**) Data for the eight-mode interferometer. Blue points: input (1,5). Red points: input (2,6). Green points: input (3,7). Magenta points: input (4,8). Coloured solid lines: corresponding theoretical predictions for the different inputs. Black dashed line: theoretical prediction for a Fourier matrix. Tables: violations $\mathcal{D}_{\mathrm{obs}}(0)$ at $\Delta x = 0$ and discrepancies (in sigmas) with the expected values for distinguishable particles ($\mathcal{D}_{\mathrm{D}}$) and MFs ($\mathcal{D}_{\mathrm{MF}}$), for the cyclic inputs of the two interferometers. $\mathcal{D}_{\mathrm{obs}}(0)$ are calculated following formula (5), while expected values for the other two cases are $\mathcal{D}_{\mathrm{D}} = 0.5$ and $\mathcal{D}_{\mathrm{MF}} = 0.25$. Error bars in all experimental quantities are due to the Poissonian statistics of measured events. All theoretical predictions in solid lines are calculated from the reconstructed unitaries, obtained from different sets of experimental data to ensure statistical independence. See Supplementary Note 2 for the modelling of imperfect state preparation.

For each of the cyclic input, we have evaluated here the violation degree $\mathcal{D}_{\mathrm{obs}}$ resulting from collected data. By measuring the coincidence pattern as a function of the path difference $\Delta x = c\Delta\tau$ between the two photons, and thus by tuning their degree of distinguishability, we could address the transition from distinguishable to indistinguishable particles. The value of $\mathcal{D}_{\mathrm{obs}}$ as a function of Δx has been obtained as $\sum_{(i,j)_{\mathrm{forbidden}}} P^{\mathrm{D}}_{i,j}(N^{\Delta x}_{i,j}/N^{\mathrm{D}}_{i,j})$, where $N^{\Delta x}_{i,j}$ and $N^{\mathrm{D}}_{i,j}$ are the number of measured coincidences for a given value of Δx and for distinguishable particles respectively. Two different regions can be identified. For intermediate values of Δx with respect to the coherence length of the photons, the measured data fall below the threshold $\mathcal{D}_{\mathrm{D}}$, and hence the hypothesis of distinguishable particles can be ruled out. Then, for smaller values of the path difference up to $\Delta x \to 0$, true two-photon interference can be certified since both hypothesis of distinguishable particles and MF state can be ruled out. The maximum violation occurring at $\Delta x = 0$ delay can be evaluated using equation (5). The experimental results retrieved from the protocol are shown in the tables of Fig. 6, in which we compare the values $\mathcal{D}_{\mathrm{obs}}(0)$ with the expected values for distinguishable particles $\mathcal{D}_{\mathrm{D}}$ and for a MF state $\mathcal{D}_{\mathrm{MF}}$. As shown for our implementation, the robustness of the protocol is ensured by the high number of s.d. separating the values in each comparison, thus unambiguously confirming the success of the certification protocol. In conclusion, the alternative hypotheses of distinguishable particles and of a MF state can be ruled out for all experiments.

Discussion

We have reported on the experimental observation of the suppression law on specific output combinations of a Fourier transformation due to quantum interference between photons. The observation of the suppression effect allowed us to rule out alternative hypotheses to the Fock state. The use of a novel implementation architecture, enabled by the 3D capabilities of femtosecond laser micromachining, extends the scalability of this technique to larger systems with lower experimental effort with respect to other techniques. While the presented architecture is designed to implement a Fourier matrix for a number of modes equal to $m = 2^p$, a generalization of the approach can be obtained by adopting a building block different from the beam splitter. For devices of odd dimension, for instance, such a tool can be provided by the tritter transformation[39]. At the same time, the universality of a generalized HOM effect with an arbitrary number of particles and modes is expected to make it a pivotal tool in the diagnostic and certification of quantum photonic platforms. Boson sampling represents a key example, since the scalability of the technique is expected to allow efficient certification of devices outperforming their classical counterparts. An interesting open problem is whether the computational hardness of boson sampling is maintained if the Haar-randomness condition is relaxed[46], and thus which is the minimal interferometer architecture required for an evidence of post-classical computation.

Fourier matrices can find application in different contexts. For instance, multiport beam splitters described by the Fourier matrix can be employed as building blocks for multiarm interferometers, which can be adopted for quantum-enhanced single and multiphase estimation protocols[47]. This would also allow the measurement of phase gradients with precision lower than the shot-noise limit[48]. Other fields where Fourier matrices are relevant include quantum communication scenarios[49], observation of two-photon correlations as a function of geometric phase[50], fundamental quantum information theory including mutually unbiased bases[51], as well as entanglement generation[52].

Methods

Waveguide device fabrication. Waveguide interferometers are fabricated in EAGLE2000 (Corning Inc.) glass chips. To inscribe the waveguides, laser pulses with 300 fs duration, 220 nJ energy and 1 MHz repetition rate from an Yb:KYW cavity dumped oscillator (wavelength 1,030 nm) are focused in the bulk of the glass, using a 0.6 NA microscope objective. The average depth of the waveguides, in the 3D interferometric structure, is 170 μm under the sample surface. The fabricated waveguides yield single-mode behaviour at 800 nm wavelength, with about 0.5 dB cm^{-1} propagation losses. The central part of the 3D interferometer, which includes all the relevant couplers, have a cross-section of about 50 μm × 50 μm (95 μm × 95 μm) for a length of 9.0 mm (14.7 mm) in the four-(eight-)modes case. The length of each fan-in and fan-out section, needed to bring the waveguides at 127 μm relative distance, is 7.8 mm (13.2 mm).

Photon generation and manipulation. The generation of two-photon states is performed by pumping a 2-mm long BBO crystal with a 392.5 nm wavelength Ti:Sa pulsed laser, with average power of 750 mW, which generates photons at 785 nm with a type II parametric downconversion process. The two photons are spectrally filtered by means of 3 nm interferential filters, and coupled into single-mode fibres. The indistinguishability of the photons is then ensured by a polarization compensation stage, and by propagation through independent delay lines (used to adjust the degree of temporal distinguishability) before injection within the interferometer via a single-mode fibre array. After the evolution through the integrated devices, photons are collected via a multimode fibre array. The detection system consists of four (8) single-photon avalanche photodiodes used for the four- (eight-) modes chip. An electronic data acquisition system allowed us to detect coincidences between all pairs of output modes. Typical coincidence rates for each collision-free output combination with distinguishable photons amounted to ~70–80 Hz (for the four-mode chip) and ~10–20 Hz (for the eight-mode chip).

References

1. Magyar, G. & Mandel, L. Interference fringes produced by superposition of two independent maser light beams. *Nature* **198,** 255–256 (1963).
2. Pfleegor, R. L. & Mandel, L. Interference of independent photon beams. *Phys. Rev. Lett.* **159,** 1084–1088 (1967).
3. Burnham, D. C. & Weinberg, D. L. Observation of simultaneity in parametric production of optical photon pairs. *Phys. Rev. Lett.* **25,** 84–87 (1970).
4. Hong, C. K., Ou, Z. Y. & Mandel, L. Measurement of subpicosecond time intervals between two photons by interference. *Phys. Rev. Lett.* **59,** 2046–2016 (1987).
5. Cosme, O., Padua, S., Bovino, F., Sciarrino, F. & De Martini, F. Hong-Ou-Mandel interferometer with one and two photon pairs. *Phys. Rev. A* **77,** 053822-1–053822-10 (2008).
6. Ou, Z. Y. & Mandel, L. Violation of Bell's inequality and classical probability in a two-photon correlation experiment. *Phys. Rev. Lett.* **61,** 50 (1988).
7. Walborn, S. P., Oliveira, A. N., Padua, S. & Monken, C. H. Multimode Hong-Ou-Mandel interference. *Phys. Rev. Lett.* **90,** 143601 (2003).
8. Sagioro, M. A., Olindo, C., Monken, C. H. & Padua, S. Time control of two-photon interference. *Phys. Rev. A* **69,** 053817 (2004).
9. Knill, E., Laflamme, R. & Milburn, G. J. A scheme for efficient quantum computation with linear optics. *Nature* **409,** 46–52 (2001).
10. O'Brien, J. L. Optical quantum computing. *Science* **318,** 1567–1570 (2007).
11. Walther, P. *et al.* Experimental one-way quantum computing. *Nature* **434,** 169–176 (2005).
12. Walmsley, I. A. & Raymer, M. G. Toward quantum-information processing with photons. *Science* **307,** 1733–1734 (2005).
13. Aaronson, S. & Arkhipov, A. in *Proceedings of the 43rd Annual ACM symposium on Theory of Computing* 333–342 (ACM Press, 2011).
14. Broome, M. A. *et al.* Photonic boson sampling in a tunable circuit. *Science* **339,** 794–798 (2013).
15. Spring, J. B. *et al.* Boson Sampling on a photonic chip. *Science* **339,** 798–801 (2013).
16. Tillmann, M. *et al.* Experimental boson sampling. *Nat. Photon.* **7,** 540–544 (2013).
17. Crespi, A. *et al.* Integrated multimode interferometers with arbitrary designs for photonic boson sampling. *Nat. Photon.* **7,** 545–549 (2013).
18. Spagnolo, N. *et al.* Experimental validation of photonic boson sampling. *Nat. Photon.* **8,** 615–620 (2014).
19. Carolan, J. *et al.* On the experimental verification of quantum complexity in linear optics. *Nat. Photon.* **8,** 621–626 (2014).
20. Bentivegna, M. *et al.* Experimental scattershot boson sampling. *Sci. Adv.* **1,** e1400255 (2015).
21. Rohde, P. P. & Ralph, T. C. Error tolerance of the boson sampling model for linear optics quantum computing. *Phys. Rev. A* **85,** 022332 (2012).
22. Leverrier, A. & Garcia-Patron, R. Analysis of circuit imperfections in boson sampling. *Quantum Inf. Comput.* **15,** 0489–0512 (2015).
23. Motes, K. R., Dowling, J. P. & Rohde, P. P. Spontaneous parametric down-conversion photon sources are scalable in the asymptotic limit for boson-sampling. *Phys. Rev. A* **88,** 063822 (2013).
24. Rohde, P. P., Moten, K. R. & Dowling, J. P. Evidence for the conjecture that sampling generalized cat states with linear optics is hard. *Phys. Rev. A* **91,** 012342 (2015).

25. Gogolin, C., Kliesch, M., Aolita, L. & Eisert, J. Boson-sampling in the light of sample complexity. Preprint at http://lanl.arxiv.org/abs/1306.3995 (2013).
26. Aaronson, S. & Arkhipov, A. Bosonsampling is far from uniform. *Quantum Info. Comput.* **14,** 1383–1423 (2014).
27. Tichy, M. C., Mayer, K., Buchleitner, A. & Molmer, K. Stringent and efficient assessment of boson-sampling devices. *Phys. Rev. Lett.* **113,** 020502 (2014).
28. Tichy, M. C., Tiersch, M., De Melo, F. & Mintert, F. and Buchleitner, A. Zero-transmission law for multiport beam splitters. *Phys. Rev. Lett.* **104,** 220405 (2010).
29. Cennini, G., Geckeler, C., Ritt, G. & Weitz, M. Interference of a variable number of coherent atomic sources. *Phys. Rev. A* **72,** 051601 (2005).
30. Hadzibabic, Z., Stock, S., Battelier, B., Bretin, V. & Dalibard, J. Interference of an array of independent Bose-Einstein condensates. *Phys. Rev. Lett.* **93,** 180403 (2004).
31. Tichy, M. C., Tiersch, M., Mintert, F. & Buchleitner, A. Many-particle interference beyond many-boson and many-fermion statistics. *New J. Phys.* **14,** 093015 (2012).
32. Peruzzo, A. *et al.* Quantum walks of correlated photons. *Science* **329,** 1500–1503 (2010).
33. Crespi, A. *et al.* Anderson localization of entangled photons in an integrated quantum walk. *Nat. Photon.* **7,** 322–328 (2013).
34. Carolan, J. *et al.* Universal linear optics. *Science* **349,** 711–716 (2015).
35. Barak, R. & Ben-Aryeh, Y. Quantum fast Fourier transform and quantum computation by linear optics. *J. Opt. Soc. Am. B* **24,** 231–240 (2007).
36. Osellame, R. *et al.* Femtosecond writing of active optical waveguides with astigmatically shaped beams. *J. Opt. Soc. Am. B* **20,** 1559–1567 (2003).
37. Gattass, R. & Mazur, E. Femtosecond laser micromachining in transparent materials. *Nat. Photon.* **2,** 219–225 (2008).
38. Meany, T. *et al.* Non-classical interference in integrated 3D multiports. *Opt. Express* **20,** 26895–26905 (2012).
39. Spagnolo, N. *et al.* Three-photon bosonic coalescence in an integrated tritter. *Nat. Commun.* **4,** 1606 (2013).
40. Poulios, K. *et al.* Quantum walks of correlated photon pairs in two-dimensional waveguide arrays. *Phys. Rev. Lett.* **112,** 143604 (2013).
41. Crespi, A. Suppression laws for multiparticle interference in Sylvester interferometers. *Phys. Rev. A* **91,** 013811 (2015).
42. Reck, M., Zeilinger, A., Bernstein, H. J. & Bertani, P. Experimental realization of any discrete unitary operator. *Phys. Rev. Lett.* **73,** 58–61 (1994).
43. Törmä, P. Beam splitter realizations of totally symmetric mode couplers. *J. Mod. Opt.* **43,** 245–251 (1996).
44. Cooley, J. W. & Tukey, W. An algorithm for the machine calculation of complex Fourier series. *Math. Comput.* **19,** 297–301 (1965).
45. Spagnolo, N. *et al.* General rules for bosonic bunching in multimode interferometers. *Phys. Rev. Lett.* **113,** 130503 (2013).
46. Matthews, J. C. F., Whittaker, R., O'Brien, J. L. & Turner, P. S. Testing randomness with photons by direct characterization of optical t-designs. *Phys. Rev. A* **91,** 020301(R) (2015).
47. Spagnolo, N. *et al.* Quantum interferometry with three-dimensional geometry. *Sci. Rep.* **2,** 862 (2012).
48. Motes, K. R. *et al.* Linear optical quantum metrology with single photons: exploiting spontaneously generated entanglement to beat the shot-noise limit. *Phys. Rev. Lett.* **114,** 170802 (2015).
49. Guha, S. Structured optical receivers to attain superadditive capacity and the Holevo limit. *Phys. Rev. Lett.* **106,** 240502 (2011).
50. Laing, A., Lawson, T., Martín López, E. & O'Brien, J. L. Observation of quantum interference as a function of Berry's phase in a complex Hadamard optical network. *Phys. Rev. Lett.* **108,** 260505 (2012).
51. Bengtsson, I. *et al.* Mutually unbiased bases and Hadamard matrices of order six. *J. Math. Phys.* **48,** 052106 (2007).
52. Lim, Y. L. & Beige, A. Multiphoton entanglement through a Bell-multiport beam splitter. *Phys. Rev. A* **71,** 062311 (2005).

Acknowledgements

We acknowledge technical support from Sandro Giacomini and Giorgio Milani. This work was supported by the ERC-Starting Grant 3D-QUEST (3D-Quantum Integrated Optical Simulation; grant agreement no. 307783): http://www.3dquest.eu, by the PRIN project Advanced Quantum Simulation and Metrology (AQUASIM) and by the H2020-FETPROACT-2014 Grant QUCHIP (Quantum Simulation on a Photonic Chip; grant agreement no. 641039). F.S. had full access to all the data in the study and takes responsibility for the integrity of the data and the accuracy of the data analysis.

Author contributions

A.C., R.O., M.B., F.F., N.S., N.V. and F.S. conceived the experimental approach for the observation of the suppression law. A.C. and R.O. developed the technique for three-dimensional circuits, and fabricated and characterized the integrated devices using classical optics. M.B., N.V., F.F., N.S., and F.S. carried out the quantum experiments, F.F., L.I., N.S., M.B., N.V. and F.S. elaborated the data. All authors discussed the experimental implementation and results, and contributed to writing the paper.

Additional information

Competing financial interests: The authors declare no competing financial interests.

16

Topological states in multi-orbital HgTe honeycomb lattices

W. Beugeling[1], E. Kalesaki[2,3], C. Delerue[2], Y.-M. Niquet[4,5], D. Vanmaekelbergh[6] & C. Morais Smith[7]

Research on graphene has revealed remarkable phenomena arising in the honeycomb lattice. However, the quantum spin Hall effect predicted at the *K* point could not be observed in graphene and other honeycomb structures of light elements due to an insufficiently strong spin-orbit coupling. Here we show theoretically that 2D honeycomb lattices of HgTe can combine the effects of the honeycomb geometry and strong spin-orbit coupling. The conduction bands, experimentally accessible via doping, can be described by a tight-binding lattice model as in graphene, but including multi-orbital degrees of freedom and spin-orbit coupling. This results in very large topological gaps (up to 35 meV) and a flattened band detached from the others. Owing to this flat band and the sizable Coulomb interaction, honeycomb structures of HgTe constitute a promising platform for the observation of a fractional Chern insulator or a fractional quantum spin Hall phase.

[1] Max-Planck-Institut für Physik komplexer Systeme, Nöthnitzer Straße 38, 01187 Dresden, Germany. [2] IEMN-Department ISEN, UMR CNRS 8520, 41 Boulevard Vauban, 59046 Lille, France. [3] Physics and Materials Science Research Unit, University of Luxembourg, 162A Avenue de la Faïencerie, L-1511 Luxembourg, Luxembourg. [4] Université Grenoble Alpes, INAC-SP2M, L_Sim, 17 avenue des Martyrs, 38054 Grenoble, France. [5] CEA, INAC-SP2M, L_Sim, 17 avenue des Martyrs, 38054 Grenoble, France. [6] Debye Institute for Nanomaterials Science, Utrecht University, Princetonplein 1, 3584 CC Utrecht, The Netherlands. [7] Institute for Theoretical Physics, Center for Extreme Matter and Emergent Phenomena, Utrecht University, Leuvenlaan 4, 3584 CE Utrecht, The Netherlands. Correspondence and requests for materials should be addressed to C.D. (email: christophe.delerue@isen.fr) or to C.M.S. (email: C.deMoraisSmith@uu.nl).

The discovery of graphene has confronted us with a material that exhibits fascinating electronic properties[1], such as zero-mass carriers, persisting conductivity at vanishing density at the Dirac point[2], Klein tunnelling[3] and an anomalous quantum Hall effect[4–6]. Nevertheless, the absence of a bandgap in its spectrum prevents its use as a field-effect transistor, and its weak spin–orbit coupling (SOC) hampers the possibility to realize the quantum spin Hall effect (QSHE)[7] and use it for quantum spintronics. The prospect of artificial graphene samples[8] that display the lacking properties has motivated research in various types of honeycomb lattice, such as arrays of ultracold atoms[9], molecular graphene[10], organometallic lattices[11] and two-dimensional (2D) electron gases subject to a geometric array of gates[12,13]. Recently, an alternative path came from the self-assembly of semiconductor nanocrystals forming atomically coherent 2D structures with a long-range honeycomb pattern; the thickness and honeycomb period are defined by the size of the nanocrystals, which is in the range of 5 nm (refs 14,15). Honeycomb lattices of PbSe and CdSe nanocrystals have been fabricated, with astonishing atomic coherence due to the oriented attachment of the nanocrystals. Theoretical investigations have shown that CdSe superlattices formed in such a way exhibit Dirac cones at two energies and nearly dispersionless bands. Unfortunately, these flat bands are connected to the nearby higher energy bands, and the SOC gaps in the conduction band are very small[16].

In this work, we propose a design for robust topological insulators that combine the properties of the honeycomb lattice and strong SOC. We consider three different types of HgTe layers with superimposed honeycomb geometry and present atomistic tight-binding (TB) calculations of their conduction band structure that we accurately describe by a 16-band effective model. Such lattices take advantage of the multi-orbital degrees of freedom in the honeycomb setup, allied to the strong SOC[17]. The access to multi-orbital degrees of freedom allows for further manipulation of the topological properties. While the Haldane[18] and Kane–Mele[7] models originally concerned honeycomb lattices characterized by a single orbital per site (for example, p_z in graphene) and isotropic nearest-neighbour (NN) hopping integrals between them, multi-orbital models have attracted much attention recently, in particular because they seem to be a paradigm to generate topologically non-trivial flat bands[19–23]. The flat band structure opens the way to the realization of interesting strongly correlated states, such as fractional QSHE, fractional Chern insulators or ferromagnetic fractional Chern insulators[20,22,24–31]. The currently proposed HgTe lattices exhibit conduction bands characterized by large topological gaps and an isolated flat band. We conclude that, depending on the position of the Fermi level, not only QSHE could be observed in these structures, but also fractional QSHE or fractional Chern insulator phases, as on-site and NN Coulomb-interaction parameters are found in the energy range required for their realization.

Results

Design of the HgTe honeycomb lattices. The three structures that we consider have in common that the [111] direction of the zincblende lattice is perpendicular to the plane. The first system is inspired by recent results obtained for nanocrystal self-assembly[14,15], that is, a graphene-type superlattice of truncated cubic nanocrystals attached via ⟨110⟩ facets. The second system consists of spheres connected by short cylinders, which allow us to vary the coupling between honeycomb lattice sites in a convenient way. The third system corresponds to a honeycomb array of cylinders. These two last systems may be experimentally realized by gas-phase deposition and lithography. We will show below that the multi-orbital topological effects are common to these three different structures.

The HgTe superlattices proposed here differ fundamentally from HgTe/CdTe quantum wells, where the QSHE has been predicted[32] and experimentally observed[33]. In the latter system, the appearance of the QSHE at the Γ point is connected to band inversion and vanishes for quantum wells of thickness below 6 nm. Here instead, the effect occurs at the K point, and it is driven by the honeycomb nanogeometry, allied to the strong SOC of the composing HgTe nanocrystals. This distinction is important because zero modes (Majoranas) bind to topological lattice defects when the bandgap opens at a non-Γ point in the Brillouin zone[34,35], and hence, the underlying topological order can be detected by measuring the structure of the topological defect[36,37].

Band structure of lattices of HgTe nanocrystals. To unveil the topological properties of these systems, we have performed atomistic TB band structure calculations. We use a basis of 20 atomic orbitals on each atom of the nanocrystals, including the spin degree of freedom. The methodology is described in ref. 38 and is summarized in the Methods section. In comparison with bulk HgTe, the electronic structure of HgTe superlattices is characterized by a large bandgap due to the strong quantum confinement. Conduction and valence bands close to this gap are composed of several minibands and minigaps due to the periodic scattering of the electronic waves in the honeycomb structure[16]. In the following, we only discuss the physics of the sixteen lowest conduction bands (Fig. 1).

A typical honeycomb lattice composed of HgTe nanocrystals is shown in Fig. 1a and the related conduction band structure is displayed in Fig. 1c. To visualize the effects of SOC, we compare it with the band structure of the same honeycomb structure composed of CdSe (Fig. 1b). The strong SOC in HgTe gives rise to effects which are absent in CdSe[16]. The overall behaviour of the band structure can be understood as follows. Each individual nanocrystal is characterized by two states with s envelope wavefunction and six p states at higher energy. In the honeycomb structure, strong coupling between the wavefunctions of neighbour nanocrystals leads to the formation of 16 bands grouped into two manifolds of four (s) and 12 (p) bands that are well separated. The s bands have the same type of dispersion as the π and π^* bands in real graphene[1]. In the case of CdSe (Fig. 1b), these bands are spin degenerate and are connected at the K and K' points of the Brillouin zone, where their dispersion is linear (Dirac points). In HgTe, instead, the s bands exhibit a small gap (5.7 meV) at the K points and have a quadratic dispersion (see Fig. 1c). In addition, they are characterized by a visible spin splitting at all points of the Brillouin zone, except at Γ and M. Among the 12 p bands of CdSe nanocrystal superlattices, eight have a small dispersion and the other four basically behave like the (Dirac) s bands. Four flat bands are built from the p_z states perpendicular to the lattice, which are not very dispersive because p_z–p_z (π) interactions are weak. Four other bands ($p_{x,y}$), respectively above and below the p Dirac band, are flat due to destructive interferences of electron hopping induced by the honeycomb geometry[19,20]. In honeycomb lattices of HgTe nanocrystals, the SOC induces spin splitting, opens a large gap at K in the p-like Dirac bands, and produces a considerable detachment of the lowest flat p band from the Dirac p band (Fig. 1c). The effects of the SOC are so strong that it is hardly possible to recognize the Dirac bands.

These unexpected features allow for the realization of several topological states of matter, by doping the system using a field-effect transistor or electrolyte gel gating[39]. At zero energy, the

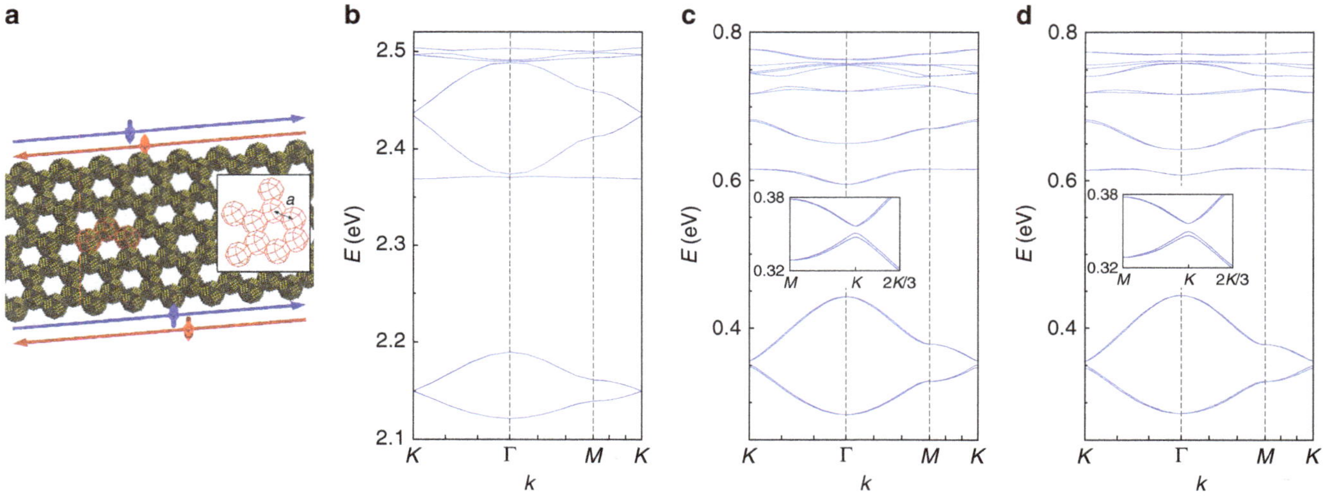

Figure 1 | Nanocrystal lattices and their conduction band dispersions. (**a**) honeycomb nanoribbon formed by the HgTe (CdSe) nanocrystals. Hg (Cd) atoms are in yellow, Te (Se) atoms are in grey. Each nanocrystal has a truncated nanocube shape, the vertices of which are given by all permutations of $[\pm 1, \pm(1-q), \pm(1-q)]l$, where q is the truncation factor and $2l$ is the size of the original nanocube before truncation. The honeycomb lattice spacing a, that is, the centre-to-centre distance between neighbor nanocrystals, is defined as $a=(2N+1)a_0/\sqrt{2}$ where N is an integer and a_0 is the cubic lattice parameter of HgTe (CdSe). The nanocrystals are attached via ⟨110⟩ facets ($a=\sqrt{2}(2-q)l$). The arrows along the ribbon indicate the electron propagation in the helical edge states present in the quantum spin Hall phase. Red and blue colours correspond to top and bottom edge for spin up, bottom and top edge for spin down, respectively. (**b**,**c**) Band dispersions for the bulk resulting from the atomistic TB calculations with $q=0.5$ ($a=5.0$ nm for HgTe, 4.7 nm for CdSe). (**d**) Same for the HgTe superlattice resulting from the effective model. Insets in **c**,**d** show the s bands in the gap region with higher magnification $\left[2K/3=\left(\frac{4\pi}{9a},\frac{4\pi}{9\sqrt{3}a}\right)\right]$.

undoped system is a semiconductor with a trivial gap of about 0.4 eV. On doping the material with one electron per nanocrystal, the small s-like topological gap may be reached, whereas for fillings between two and three electrons per nanocrystal, the fractional quantum (spin) Hall regime may be realized at the flat band. For a doping level of four electrons per nanocrystal, one reaches the QSHE gap. At this point, we should emphasize that doping of nanocrystals with up to 10 electrons has already been demonstrated experimentally[40]; therefore, all the interesting regimes that we discuss are at reach with the existing technology, at a simple switch of the doping level. Other examples of band structures for lattices of HgTe nanocrystals with different size or truncation factors are presented in Supplementary Fig. 1 and are discussed in Supplementary Note 1. They all show large topological gaps, especially in the p sector.

The topological properties of the bands are most transparently studied through an edge-state analysis in a one-dimensional nanoribbon[41]. We consider a zigzag ribbon composed of 16 nanocrystals (34,740 atoms) per unit cell. Figure 2a shows that edge states are crossing the three gaps between the p bands as well as the gap between the s bands. These results also hold for armchair ribbons. In Supplementary Fig. 2a and Supplementary Note 2, we present the band structure for another nanoribbon, which has two inequivalent edges. Still, helical edge states characteristic of the QSHE are found, as shown by the 2D plots of the wavefunctions (Supplementary Figs 2b–e).

Effective model. The band structures resulting from the atomistic TB calculations are accurately described by a 16-band effective model (Fig. 1d), where each nanocrystal is treated as one site on a honeycomb lattice. The effective TB model is written in the basis of the four aforementioned orbitals (s,p_x,p_y,p_z) per site as $H_{\mathrm{eff}}=H_\mu+H_{\mathrm{NN}}+H_{\mathrm{ISO}}+H_{\mathrm{RSO}}$. Here H_μ incorporates the on-site energies E_s, $E_{p_x}=E_{p_y}$, and E_{p_z}. The edge nanocrystals have a slightly different value of E_s compared with the bulk, to account for the different number of neighbours. The term

$$H_{\mathrm{NN}}=\sum_{\langle i,j\rangle}\sum_{\substack{\alpha\\ b,b'}} c^\dagger_{i,b,\alpha}V_{i,b;j,b'}c_{j,b',\alpha}, \tag{1}$$

encodes NN hopping, where $\langle i,j\rangle$ denotes NN sites, $\alpha=\uparrow,\downarrow$ denotes spin and b and b' the orbitals. The coupling coefficients $V_{i,b;j,b'}$ are expressed in terms of the hopping parameters $V_{ss\sigma}$, $V_{pp\sigma}$, $V_{pp\pi}$ and $V_{sp\sigma}$, following the notations of ref. 42.

The intrinsic SOC term H_{ISO} couples the electron orbital angular momentum $\mathbf{L}$ and spin $\mathbf{S}=\sigma/2$. In the p sector, it is encoded through the on-site term $\lambda^p_{\mathrm{ISO}}\mathbf{L}\cdot\sigma$. There is no on-site term in the s sector because the orbital angular momentum is 'frozen'. For the same reason, in graphene, the on-site intrinsic SOC term is absent because the sp^2 hybridization freezes the orbital momentum in the p_z state. However, as shown by Kane and Mele[7], the intrinsic SOC introduces a next-NN (NNN) hopping term, which is written as

$$H_{\mathrm{ISO}}=\mathrm{i}\lambda^s_{\mathrm{ISO}}\sum_{\langle\langle i,j\rangle\rangle}\sum_{\alpha} c^\dagger_{i,s,\alpha}\sigma^z_{\alpha\alpha}v_{ij}c_{j,s,\alpha}. \tag{2}$$

Here the summation is over NNNs, and $v_{ij}=\pm 1$, with the sign depending on the outer product of the two NN vectors that connect sites i and j. The Rashba SOC term, proportional to the cross product of momentum and spin, $\mathbf{p}\times\mathbf{S}$, is written as an NN hopping term

$$H_{\mathrm{RSO}}=\mathrm{i}\sum_{\langle i,j\rangle}\sum_{\substack{\alpha,\beta\\ b,b'}} c^\dagger_{i,b,\alpha}\gamma_{i,b;j,b'}\left[\hat{z}\cdot(\sigma\times\mathbf{r}_{ij})\right]_{\alpha\beta}c_{j,b',\beta}. \tag{3}$$

The coupling coefficients $\gamma_{i,b;j,b'}$ have the same structure as the $V_{i,b;j,b'}$ for the ordinary NN hopping and are expressed in terms of $\gamma_{ss\sigma}$, $\gamma_{pp\sigma}$ and $\gamma_{pp\pi}$. The sp term may be neglected due to the large energy separation between the s and p bands.

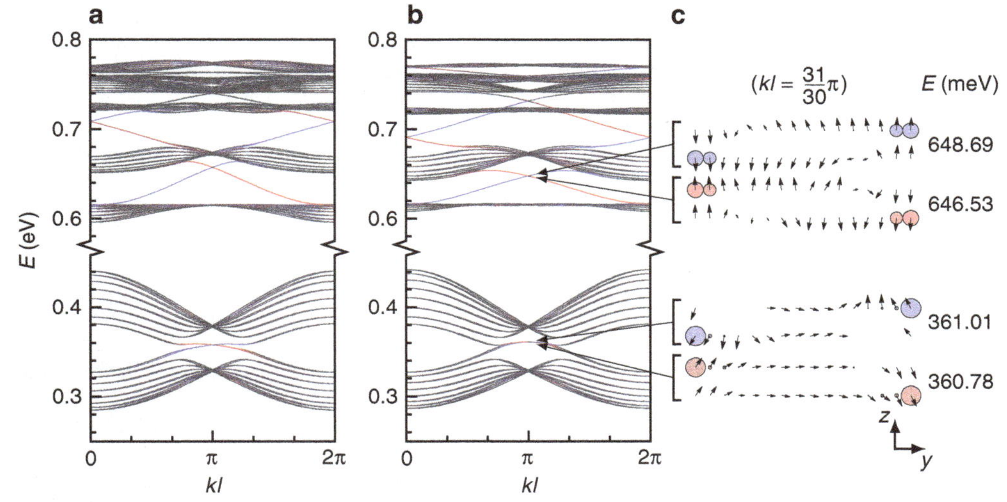

Figure 2 | Topological edge states and non-trivial gaps in honeycomb lattices of HgTe nanocrystals. (**a**) Conduction bands calculated using the atomistic TB method for a zigzag ribbon composed of 16 nanocrystals per unit cell ($q = 0.5$, body diagonal of 5.0 nm, cell length $l = 8.7$ nm). (**b**) Same but computed from the effective Hamiltonian. (**c**) Spin orientation on each site for a selection of states calculated at $k = 31\pi/(30l)$. A vertical arrow indicates that the spin is along the z direction, perpendicular to the lattice. The size of the circles represents the weight of the wavefunction on each site. In each figure, the colour indicates the expectation value $\langle y\sigma_z \rangle$, that is, red and blue correspond to top and bottom edge for spin up (bottom and top edge for spin down), respectively. At each energy E, there are two states that live on opposite edges with opposite spin; therefore, they are represented by the same colour. The bulk states are grey.

In Table 1, we present typical values for the parameters obtained numerically using least-squares fitting to the band structure of Fig. 1c. The band structure of the effective model is shown in Fig. 1d. As expected, the on-site term λ^p_{ISO} is much larger than the NNN term λ^s_{ISO}, explaining the opening of a very large gap at the K point and the detachment of the flat bands in the p sector. Using the effective model, the non-trivial topology of the bands is confirmed by the calculation of the Z_2 topological invariant, the spin Chern number (Methods).

The effective model yields a band structure for the ribbon (Fig. 2b) in excellent agreement with the atomistic TB calculations (Fig. 2a). The red and blue colours in Fig. 2a,b indicate the expectation value $\langle y\sigma_z \rangle = \langle \psi_i | \hat{y}\hat{\sigma}_z | \psi_i \rangle$, where y is the coordinate perpendicular to the ribbon edges. This expectation value allows us to identify helical edge states that come in pairs with identical dispersion, opposite spin, and live on opposite edges. Both atomistic and effective TB models show that the gaps in the s and p sectors exhibit helical edge states, characteristic of the QSHE.

When the Rashba coupling is neglected in the effective model, all states are spin degenerate. The Rashba term induces a small splitting in energy, and tilts the spins slightly away from the perpendicular direction. In Fig. 2c, we plot the spin direction on each site of the zigzag ribbon for four sets of four edge states (two in each edge). The spin direction is always perpendicular to the edge, that is, the spin lies in the yz plane if we choose the x direction to be parallel to the edge. The localization of the selected states on the edges is visible from the weight of the wavefunction, indicated by the size of the circles in the figure. The colours of the circles are determined by the local value of $y\sigma_z$, and correspond to the colours in the dispersion (Fig. 2a,b). The site dependence of the spin direction leads to interesting spin textures. For the edge states, the typical spin texture is almost smooth: Going from one edge to the other, the spin direction interpolates between (almost) up and (almost) down in a rotational manner. A slight tilt of $\sim 3°$ is observed for the edge states in the p bands. In the s bands, the tilt is stronger, similarly to graphene[41]. Here the spin vector at the edge site points $\sim 30°$ away from the vertical. This difference in tilt can be explained by the much larger intrinsic SOC in the p sector than in the s one, whereas the two Rashba couplings $\gamma_{ss\sigma}$ and $\gamma_{pp\sigma}$ are of similar order of magnitude. Strictly speaking, one should denote this state a Z_2 topological insulator, but since the Rashba SOC is extremely small, one can think of an approximate QSHE.

Table 1 | Parameters of the effective model.

On-site	NN hopping	Rashba SOC	Intrinsic SOC
$E^{bulk}_s = 0.365$ eV	$V_{ss\sigma} = -26.4$ meV	$\gamma_{ss\sigma} = 0.56$ meV	$\lambda^s_{ISO} = 0.71$ meV
$E^{edge}_s = 0.370$ eV	$V_{pp\sigma} = 45.6$ meV	$\gamma_{pp\sigma} = 1.50$ meV	$\lambda^p_{ISO} = 15.8$ meV
$E_{p_x} = 0.691$ eV	$V_{pp\pi} = -2.7$ meV	$\gamma_{pp\pi} = 0.80$ meV	
$E_{p_y} = 0.691$ eV	$V_{sp\sigma} = 31.1$ meV		
$E_{p_z} = 0.747$ eV			

Parameters derived for the lattice of HgTe nanocrystals described in Fig. 1. E_s, E_{p_x}, E_{p_y} and E_{p_z} are the on-site energies on the s, p_x, p_y and p_z orbitals, respectively. In ribbons, the edge nanocrystals have a slightly different value of E_s compared with the bulk ($E^{bulk}_s \neq E^{edge}_s$). $V_{ss\sigma}$, $V_{pp\sigma}$, $V_{pp\pi}$ and $V_{sp\sigma}$ are the hopping parameters, following the notations of ref. 42. $\gamma_{ss\sigma}$, $\gamma_{pp\sigma}$ and $\gamma_{pp\pi}$ are the terms describing the Rashba SOC, following the same notations. The intrinsic SOC is defined by λ^s_{ISO} and λ^p_{ISO} on s and p orbitals, respectively.

Flat band and Coulomb interactions. The large gaps in the p sector are mainly due to the intrinsic SOC, which, contrarily to the Kane-Mele model, is described by an on-site term ($\lambda^p_{ISO} = 15.8$ meV). In particular, the intrinsic SOC separates the lowest flat band from the other p bands, with large gaps, for example, 35 meV in the case of Fig. 1c. This gap ranges from 13 to 36 meV when we vary the nanocrystal size and shape (Supplementary Table 1).

Under partial filling of the lowest-energy flat p band by electrons, fascinating phenomena such as fractional QSHE are predicted in the presence of strong correlations, when the strength of Coulomb interactions between electrons is large compared with the bandwidth and smaller or comparable to the energy gap between the flat band and the next higher one (for a recent review, see ref. 31). In addition, it has been shown that also for large Coulomb interactions, a fractional Chern insulator phase cannot be ruled out[43]. We have thus calculated the on-site (U) and NN (V) electron–electron interaction energies for the honeycomb lattice of Fig. 1a, assuming two different dielectric environments around the nanocrystals (Methods). Table 2 shows that Coulomb energies are larger than the bandwidth and are comparable to the gap between p bands. Therefore, the gap sizes open the possibility to the experimental observation of strongly correlated quantum phases including the long sought fractional QSHE[20,24–27].

Band structure for other types of HgTe lattices. To understand the effects of the electronic coupling between honeycomb lattice sites, we have studied a second type of superlattices with a simpler geometry, consisting of tangent spheres connected by horizontal cylinders (Fig. 3a,b). The HgTe spheres have their [111] axis orthogonal to the lattice plane and the cylinders are oriented along ⟨110⟩ directions perpendicular to [111]. Figure 3c,d,e depicts the evolution of the band structure with the ratio between the diameters of cylinders (d) and spheres (D) (see also Supplementary Fig. 3). An increase of d/D induces larger NN hopping terms, broader s and p bands and stronger sp hybridization, as shown by the deformation of the s band. It also results in a larger NNN term λ^{s}_{ISO}, explaining why the non-trivial gap in the s sector only exists for $d/D > 0.3$. On the contrary, topological gaps are always present in the p sector, even for small values of d/D, because they are mainly determined by the on-site term λ^{p}_{ISO}. At high values of d/D (>0.7), the spin splitting of the bands in the entire Brillouin zone becomes particularly important due to increased Rashba couplings. In general, the lowest p band is rather flat and has a maximum separation from the next higher one for d/D close to 0.4–0.5. In that case, the energy separation can be as large as 35 meV (Fig. 3f). HgTe honeycomb structures with moderate coupling between lattice sites should provide the most suitable gap and band widths to observe the strongly correlated phases associated to the flatness of the bands.

Table 2 | Coulomb energies.

ε_{in}	ε_{out}	U	V	Bandwidth	Absolute gap	Gap at Γ
14	14	48 meV	23 meV	20 meV	35 meV	56 meV
14	6	76 meV	43 meV			

On-site (U) and nearest-neighbor (V) Coulomb-interaction energies calculated for a honeycomb lattice of HgTe nanocrystals (with lattice parameters $q = 0.5$, $a = 5$ nm) compared with the width of the lowest p band, the absolute gap between the lowest p bands (the gap between band extrema) and the vertical gap at Γ(k = 0) between them. ε_{in} and ε_{out} are the dielectric constants of the materials composing the lattice (HgTe) and its environment, respectively.

In Supplementary Fig. 4, we show further results for a third type of honeycomb structure made of overlapping HgTe cylinders parallel to each other. Once again, the band structure can be described by the effective model (Supplementary Note 3, Supplementary Table 2 and Supplementary Fig. 5). In that case, the NN couplings are even stronger, there is no gap in the s sector due to a large Rashba term, but the non-trivial gap above the lowest p band remains. We can conclude that the topological effects in the p sector are robust against changes in the electronic coupling and honeycomb period.

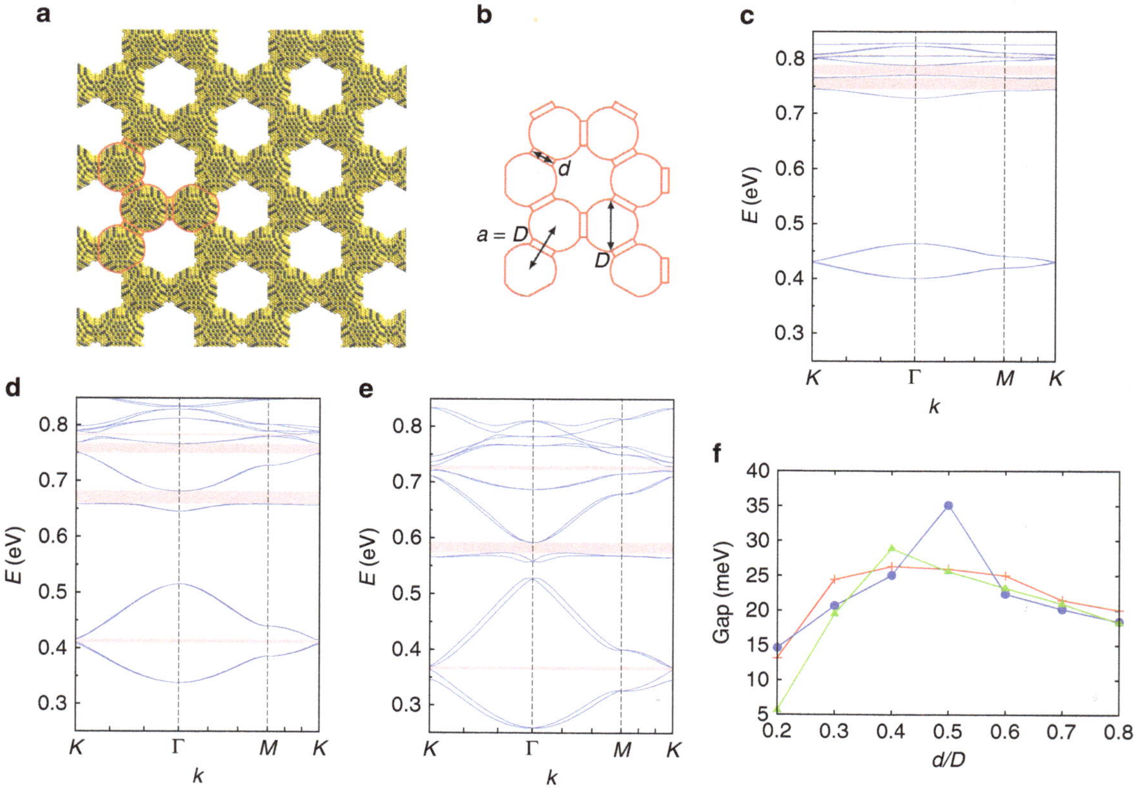

Figure 3 | Honeycomb lattice of HgTe spheres and its conduction band structure. (**a**,**b**) Top view of an assembly of spheres connected by cylinders, forming a honeycomb lattice of HgTe. Hg atoms are in yellow, Te atoms are in grey. The lattice spacing a, that is, the centre-to-centre distance between neighbour spheres, is equal to the diameter D of the spheres and d is the diameter of the cylinders. (**c**–**e**) Conduction band dispersions resulting from the atomistic TB calculation for $D = 5.0$ nm and $d/D = 0.3$ (**c**), $d/D = 0.6$ (**d**) or $d/D = 0.8$ (**e**). Non-trivial gaps are indicated by pink shaded regions. (**f**) Evolution of the gap between the lowest p bands versus d/D for three values of D (green triangles: $D = 5.9$ nm; blue circles: $D = 5.0$ nm; red crosses: $D = 4.1$ nm).

Discussion

In summary, we have performed atomistic TB calculations of the band structure of 2D honeycomb lattices of HgTe. We demonstrate that the strong SOC of HgTe combined with the honeycomb structure results in several topological phases. The calculated band structure can be described by a honeycomb lattice model as in graphene but including multi-orbital degrees of freedom that generate in particular a topologically non-trivial flat band. By taking advantage of these features, we show that in the same structure not only the QSHE, but potentially also the elusive fractional QSHE could be observed, just by varying the electron density. Both topological effects turn out to be protected by a gap as large as 35 meV, and can thus be observed at high temperatures. Honeycomb superlattices of HgTe are therefore platforms of high interest to study electrons on a multi-orbital honeycomb lattice under strong SOC. Such structures could be fabricated by nanocrystal self-assembly (in a similar way as for PbSe and CdSe[14,15]) or by a combination of gas-phase deposition and lithography. Our results open the path towards high-temperature quantum spintronics in artificial graphene.

Methods

Atomistic TB methodology. The electronic structure of HgTe superlattices is calculated within the TB approximation, as described in detail in ref. 38. The TB Hamiltonian matrix is written in a basis of atomic orbitals ($sp^3d^5s^*$ for each spin orientation) as function of parameters that have been obtained by fitting to two reference band structures: Close to the Fermi level, we use the $\mathbf{k}\cdot\mathbf{p}$ band structure of ref. 44, whereas elsewhere we use the band structure of ref. 45 obtained using a quasi-particle self-consistent GW approximation in a hybrid scheme. In the present work, we have used the TB parameters that give the band structure of HgTe at 300 K. The surfaces of the superlattices are saturated by pseudo-hydrogen atoms that push surface states far from the energy regions of interest in this study. Therefore surface states do not interfere with edge states predicted in ribbons. Owing to the large size of the systems that we have studied (up to $\approx 10^5$ atoms per unit cell), only near-gap eigenstates are calculated using the numerical methods described in ref. 46.

Calculation of the Z_2 topological invariant. The Z_2 topological invariant (spin Chern number C_s) for the bands of interest is calculated using the model Hamiltonian following the methodology proposed in ref. 47 and derived from ref. 48. This approach works even for systems without inversion symmetry, which is the case here. C_s is given by a sum of terms calculated on a regular lattice in the Brillouin zone. We have checked that the results converge for a mesh denser than 21×21 k vectors. In all cases, the invariants that we have computed for the bands are consistent with the number of edge states we observe in the bulk gaps.

Coulomb interactions. The Coulomb repulsion between electrons in honeycomb lattices of HgTe nanocrystals can be characterized as follows. For simplicity, we consider electrons in the s band, since in the case of individual nanocrystals, it was shown theoretically[49] and experimentally[50] that the Coulomb integrals are almost identical for states with s and p envelope functions. The Coulomb interaction associated to electrons on nanocrystals i and j is calculated as

$$\int |\Psi_i(\mathbf{r})|^2 |\Psi_j(\mathbf{r}')|^2 \mathcal{V}(\mathbf{r},\mathbf{r}')d\mathbf{r}d\mathbf{r}' \quad (4)$$

where $\Psi_i(\mathbf{r})$ is the s state on the nanocrystal i, and $\mathcal{V}(\mathbf{r},\mathbf{r}')$ is the screened Coulomb energy of two interacting electrons at $\mathbf{r}$ and $\mathbf{r}'$. Taking into account that there is just one (spin degenerate) s state per nanocrystal, Ψ_i and Ψ_j are simply defined as the components on nanocrystals i and j of the wavefunction calculated for the lowest s band at Γ (or another k vector), normalized on their respective nanocrystals. Coulomb matrix elements are decomposed in the basis of the atomic orbitals and are calculated following the methodology described in ref. 51, by using usual approximations in the TB description, that is, neglecting overlaps between atomic orbitals and considering atomic charges as point-like charges. For the potential $\mathcal{V}$, we consider two configurations: First, a dielectrically homogeneous system, for which $\mathcal{V}(\mathbf{r},\mathbf{r}') = e^2/(\varepsilon_{in}|\mathbf{r}-\mathbf{r}'|)$, where $\varepsilon_{in} = 14$ is the dielectric constant of HgTe; second, a dielectrically inhomogeneous system, for which $\mathcal{V}$ is calculated by solving the Poisson equation, with the dielectric constant inside (outside) the lattice equal to ε_{in} (ε_{out}). We have chosen $\varepsilon_{out} = 6$, a typical value taken to simulate the complex dielectric environment around semiconductor nanocrystals[49]. The on-site (U) and NN (V) terms are presented in Table 2. As expected, larger values are obtained for $\varepsilon_{out} = 6$ than for $\varepsilon_{out} = \varepsilon_{in}$. Longer-range Coulomb terms are expected to decay as the inverse of the distance between nanocrystals. However, it is important to note that these values do not take into account the extra screening induced by the electrons filling the bands. This could be computed, for example, in the random-phase approximation, but this is clearly beyond the scope of the present work. Long-range interactions will be strongly screened, while short-range ones will be only slightly reduced[52]. In this context, for a band filling of the order of 1/3, correlations will be mainly governed by short-range effects.

References

1. Castro Neto, A. H., Guinea, F., Peres, N. M. R., Novoselov, K. S. & Geim, A. K. The electronic properties of graphene. *Rev. Mod. Phys.* **81,** 109–162 (2009).
2. Nair, R. R. *et al.* Fine structure constant defines visual transparency of graphene. *Science* **320,** 1308 (2008).
3. Katsnelson, M. I., Novoselov, K. S. & Geim, A. K. Chiral tunnelling and the Klein paradox in graphene. *Nat. Phys.* **2,** 620–625 (2006).
4. Novoselov, K. S. *et al.* Two-dimensional gas of massless Dirac fermions in graphene. *Nature* **438,** 197–200 (2005).
5. Du, X., Skachko, I., Duerr, F., Luican, A. & Andrei, E. Y. Fractional quantum Hall effect and insulating phase of Dirac electrons in graphene. *Nature* **462,** 192–195 (2009).
6. Bolotin, K. I., Ghahari, F., Shulman, M. D., Stormer, H. L. & Kim, P. Observation of the fractional quantum Hall effect in graphene. *Nature* **462,** 196–199 (2009).
7. Kane, C. L. & Mele, E. J. Quantum spin Hall effect in graphene. *Phys. Rev. Lett.* **95,** 226801 (2005).
8. Polini, M., Guinea, F., Lewenstein, M., Manoharan, H. C. & Pellegrini, V. Artificial honeycomb lattices for electrons, atoms and photons. *Nat. Nanotechnol* **8,** 625–633 (2013).
9. Soltan-Panahi, P. *et al.* Multi-component quantum gases in spin-dependent hexagonal lattices. *Nat. Phys.* **7,** 434–440 (2011).
10. Gomes, K. K., Mar, W., Ko, W., Guinea, F. & Manoharan, H. C. Designer Dirac fermions and topological phases in molecular graphene. *Nature* **483,** 306–310 (2012).
11. Wang, Z. F., Su, N. & Liu, F. Prediction of a two-dimensional organic topological insulator. *Nano Lett.* **13,** 2842–2845 (2013).
12. Gibertini, M. *et al.* Engineering artificial graphene in a two-dimensional electron gas. *Phys. Rev. B* **79,** 241406 (2009).
13. Park, C.-H. & Louie, S. G. Making massless Dirac fermions from a patterned two-dimensional electron gas. *Nano Lett.* **9,** 1793–1797 (2009).
14. Evers, W. H. *et al.* Low-dimensional semiconductor superlattices formed by geometric control over nanocrystal attachment. *Nano Lett.* **13,** 2317–2323 (2013).
15. Boneschanscher, M. P. *et al.* Long-range orientation and atomic attachment of nanocrystals in 2D honeycomb superlattices. *Science* **344,** 1377–1380 (2014).
16. Kalesaki, E. *et al.* Dirac cones, topological edge states, and nontrivial flat bands in two-dimensional semiconductors with a honeycomb nanogeometry. *Phys. Rev. X* **4,** 011010 (2014).
17. Zhang, G.-F., Li, Y. & Wu, C. Honeycomb lattice with multiorbital structure: topological and quantum anomalous Hall insulators with large gaps. *Phys. Rev. B* **90,** 075114 (2014).
18. Haldane, F. D. M. Model for a quantum Hall effect without Landau levels: condensed-matter realization of the "parity anomaly". *Phys. Rev. Lett.* **61,** 2015–2018 (1988).
19. Wu, C., Bergman, D., Balents, L. & Das Sarma, S. Flat bands and Wigner crystallization in the honeycomb optical lattice. *Phys. Rev. Lett.* **99,** 070401 (2007).
20. Sun, K., Gu, Z., Katsura, H. & Das Sarma, S. Nearly flatbands with nontrivial topology. *Phys. Rev. Lett.* **106,** 236803 (2011).
21. Venderbos, J. W. F., Daghofer, M. & van den Brink, J. Narrowing of topological bands due to electronic orbital degrees of freedom. *Phys. Rev. Lett.* **107,** 116401 (2011).
22. Yang, S., Gu, Z.-C., Sun, K. & Das Sarma, S. Topological flat band models with arbitrary Chern numbers. *Phys. Rev. B* **86,** 241112 (2012).
23. Ölschläger, M. *et al.* Interaction-induced chiral $p_x \pm ip_y$ superfluid order of bosons in an optical lattice. *New J. Phys.* **15,** 083041 (2013).
24. Neupert, T., Santos, L., Chamon, C. & Mudry, C. Fractional quantum Hall states at zero magnetic field. *Phys. Rev. Lett.* **106,** 236804 (2011).
25. Hu, X., Kargarian, M. & Fiete, G. A. Topological insulators and fractional quantum Hall effect on the ruby lattice. *Phys. Rev. B* **84,** 155116 (2011).
26. Goerbig, M. O. From fractional Chern insulators to a fractional quantum spin Hall effect. *Eur. Phys. J. B* **85,** 1–8 (2012).
27. Regnault, N. & Bernevig, B. A. Fractional Chern insulator. *Phys. Rev. X* **1,** 021014 (2011).
28. Tang, E., Mei, J.-W. & Wen, X.-G. High-temperature fractional quantum Hall states. *Phys. Rev. Lett.* **106,** 236802 (2011).
29. Xiao, D., Zhu, W., Ran, Y., Nagaosa, N. & Okamoto, S. Interface engineering of quantum Hall effects in digital transition metal oxide heterostructures. *Nat. Commun.* **2,** 596 (2011).

30. He, J., Wang, B. & Kou, S.-P. Ferromagnetism and antiferromagnetism of a correlated topological insulator with a flat band. *Phys. Rev. B* **86,** 235146 (2012).
31. Bergholtz, E. J. & Liu, Z. Topological flat band models and fractional Chern insulators. *Int. J. Mod. Phys. B* **27,** 1330017 (2013).
32. Bernevig, B. A., Hughes, T. L. & Zhang, S.-C. Quantum spin Hall effect and topological phase transition in HgTe quantum wells. *Science* **314,** 1757–1761 (2006).
33. König, M. *et al.* Quantum spin Hall insulator state in HgTe quantum wells. *Science* **318,** 766–770 (2007).
34. Ran, Y., Zhang, Y. & Vishwanath, A. One-dimensional topologically protected modes in topological insulators with lattice dislocations. *Nat. Phys.* **5,** 298–303 (2009).
35. Juričić, V., Mesaros, A., Slager, R.-J. & Zaanen, J. Universal probes of two-dimensional topological insulators: dislocation and π flux. *Phys. Rev. Lett.* **108,** 106403 (2012).
36. Rüegg, A. & Lin, C. Bound states of conical singularities in graphene-based topological insulators. *Phys. Rev. Lett.* **110,** 046401 (2013).
37. Slager, R.-J., Mesaros, A., Juričić, V. & Zaanen, J. The space group classification of topological band-insulators. *Nat. Phys.* **9,** 98–102 (2013).
38. Allan, G. & Delerue, C. Tight-binding calculations of the optical properties of HgTe nanocrystals. *Phys. Rev. B* **86,** 165437 (2012).
39. Ye, J. T. *et al.* Liquid-gated interface superconductivity on an atomically flat film. *Nat. Mater.* **9,** 125–128 (2010).
40. Germeau, A. *et al.* Optical transitions in artificial few-electron atoms strongly confined inside ZnO nanocrystals. *Phys. Rev. Lett.* **90,** 097401 (2003).
41. Beugeling, W., Goldman, N. & Morais Smith, C. Topological phases in a two-dimensional lattice: magnetic field versus spin-orbit coupling. *Phys. Rev. B* **86,** 075118 (2012).
42. Slater, J. C. & Koster, G. F. Simplified LCAO method for the periodic potential problem. *Phys. Rev.* **94,** 1498–1524 (1954).
43. Kourtis, S., Neupert, T., Chamon, C. & Mudry, C. Fractional Chern insulators with strong interactions that far exceed band gaps. *Phys. Rev. Lett.* **112,** 126806 (2014).
44. Man, P. & Pan, D. S. Infrared absorption in HgTe. *Phys. Rev. B* **44,** 8745–8758 (1991).
45. Svane, A. *et al.* Quasiparticle band structures of β-HgS, HgSe, and HgTe. *Phys. Rev. B* **84,** 205205 (2011).
46. Niquet, Y. M., Delerue, C., Allan, G. & Lannoo, M. Method for tight-binding parametrization: Application to silicon nanostructures. *Phys. Rev. B* **62,** 5109–5116 (2000).
47. Fukui, T. & Hatsugai, Y. Quantum spin Hall effect in three dimensional materials: Lattice computation of Z_2 topological invariants and its application to Bi and Sb. *J. Phys. Soc. Jpn* **76,** 053702 (2007).
48. Fu, L. & Kane, C. L. Time reversal polarization and a Z_2 adiabatic spin pump. *Phys. Rev. B* **74,** 195312 (2006).
49. Niquet, Y. M., Delerue, C., Allan, G. & Lannoo, M. Interpretation and theory of tunneling experiments on single nanostructures. *Phys. Rev. B* **65,** 165334 (2002).
50. Banin, U., Cao, Y., Katz, D. & Millo, O. Identification of atomic-like electronic states in indium arsenide nanocrystal quantum dots. *Nature* **400,** 542–544 (1999).
51. Delerue, C., Lannoo, M. & Allan, G. Calculations of the electron-energy-loss spectra of silicon nanostructures and porous silicon. *Phys. Rev. B* **56,** 15306–15313 (1997).
52. Nozières, P. & Pines, D. *Theory Of Quantum Liquids* (Westview Press, 1999).

Acknowledgements

This work was supported by the French National Research Agency (ANR) project 'ETSFG' (ANR-09-BLAN-0421-01). W.B. is funded by the Max-Planck-Gesellschaft through the visitor's programme at MPI-PKS. E.K. acknowledges funding by the University of Luxembourg Research Office. D.V. and C.M.S. wish to acknowledge the Dutch FOM for financial support via the Programme 13DDC01 'Designing Dirac Carriers in Honeycomb Superlattices'. The work of C.M.S. is part of the D-ITP consortium, a programme of the Netherlands Organisation for Scientific Research (NWO) that is funded by the Dutch Ministry of Education, Culture and Science (OCW).

Author contributions

W.B., E.K. and C.D. performed the calculations. Y.-M.N. contributed to the development of the codes and methodologies. C.D., D.V. and C.M.S. supervised the project. All authors were involved in writing of the manuscript.

Additional information

Competing financial interests: The authors declare no competing financial interests.

17

Humans choose representatives who enforce cooperation in social dilemmas through extortion

Manfred Milinski[1], Christian Hilbe[2,3], Dirk Semmann[1], Ralf Sommerfeld[1] & Jochem Marotzke[4]

Social dilemmas force players to balance between personal and collective gain. In many dilemmas, such as elected governments negotiating climate-change mitigation measures, the decisions are made not by individual players but by their representatives. However, the behaviour of representatives in social dilemmas has not been investigated experimentally. Here inspired by the negotiations for greenhouse-gas emissions reductions, we experimentally study a collective-risk social dilemma that involves representatives deciding on behalf of their fellow group members. Representatives can be re-elected or voted out after each consecutive collective-risk game. Selfish players are preferentially elected and are hence found most frequently in the 'representatives' treatment. Across all treatments, we identify the selfish players as extortioners. As predicted by our mathematical model, their steadfast strategies enforce cooperation from fair players who finally compensate almost completely the deficit caused by the extortionate co-players. Everybody gains, but the extortionate representatives and their groups gain the most.

[1] Department of Evolutionary Ecology, Max-Planck-Institute for Evolutionary Biology, August-Thienemann-Strasse 2, 24306 Plön, Germany. [2] Department of Organismic and Evolutionary Biology, Department of Mathematics, Program for Evolutionary Dynamics, Harvard University, One Brattle Square, Cambridge, Massachusetts 02138, USA. [3] Institute of Science and Technology Austria, Am Campus 1, Klosterneuburg 3400, Austria. [4] Max Planck Institute for Meteorology, Department "The Ocean in the Earth System", 20146 Hamburg, Germany. Correspondence and requests for materials should be addressed to M.M. (email: milinski@evolbio.mpg.de) or to C.H. (email: hilbe@fas.harvard.edu) or to J.M. (email: jochem.marotzke@mpimet.mpg.de).

Although humans are regarded as champions of cooperation[1,2], there are social dilemmas that so far have defied solution—we have not yet collaborated successfully to stop the increase of global greenhouse-gas emissions[3,4], Europe continues to overexploit its marine fish stock[5] and the European Union has so far failed to reach an equitable solution to accommodating the large number of refugees arriving from Africa and the Middle East[6]. In these and other dilemmas, essential decisions are made not by individual social actors but by representatives such as officials from elected governments. Representatives have been shown to display a more competitive mindset than 'ordinary' group members[7]. However, the behaviour of representatives in a social dilemma has, to our knowledge, not been investigated experimentally. To fill this gap is the aim of our paper.

While we believe that our results apply to the role of representatives in social dilemmas more broadly, we have drawn our main inspiration and the concrete setting of our experiments from the challenge to prevent 'dangerous anthropogenic interference with the climate system'[8]. This challenge is now usually interpreted as limiting global warming to below 2 °C compared with the pre-industrial period. To prevent temperature from exceeding this limit, greenhouse-gas emissions should be reduced from about 2020 onwards; by 2050, emissions should fall to a level of $\leq 50\%$ of the year 2000 emissions[4,9–12]. However, as representatives attend climate summits to negotiate their country's share in reducing greenhouse-gas emissions, they are eagerly watched by their voters who might not re-elect their representatives when others negotiate a lower share[13]. Though everybody profits only if dangerous climate change is averted, none of the many climate summits has achieved sustained emissions reductions, the relative success of the Paris negotiations at COP21 notwithstanding.

The global emissions-reduction problem has been simulated experimentally in the 'collective-risk social dilemma' game[14–18]. A number of volunteers can invest anonymously from their individual endowments into a climate account in each of 10 consecutive rounds. If the group collectively reaches a specified target sum, everybody receives in cash what she has not invested from her endowment. However, if the group fails to reach the target, individuals risk losing all their remaining endowment with a high probability, mimicking the drastic economic losses that result from dangerous climate change. The social dilemma arises because all players benefit only if the collective target is reached, but individual payoff is maximised by lower-than-average contributions, spurred by the hope that others will compensate to reach the target[13].

In contrast to previous work, we have here assembled 15 groups of 18 players each where the groups are sub-divided into 6 'countries' of 3 players each who elect, re-elect or vote out their representative for the 6 representatives' 'summit'. For control, we have assembled 15 groups of 6 players each (as in ref. 14) and 15 groups of 18 players each. In 3 consecutive collective-risk games with 10 rounds each, each player in the 6-players and 18-players treatments contributes from her initial endowment of €40; in the 6-representatives treatment, each representative contributes from the combined endowments (€120) of her watching country mates and on their behalf (Fig. 1; see Methods). The target sum that must be collected by each group to prevent simulated dangerous climate change is €120 in the 6-players treatment and €360 both in the 18-players and the 6-representatives treatments.

We find that selfish players are preferentially elected and are hence found more frequently in the six-representatives treatment than in the other two treatments. Across all treatments, we identify the selfish players as extortioners. We develop a mathematical model and confirm its prediction that the extortioners' steadfast strategies enforce cooperation from fair players who finally compensate almost completely the deficit caused by the extortionate co-players.

Results

Simulated dangerous climate change. In the first game of the 18-players treatment and of the 6-representatives treatment, only 33% of the groups reach the target sum. By contrast, groups in the six-players treatment are almost twice as likely to collect sufficient contributions in the first game, with 60% of the groups reaching the target sum (Fig. 2a–c), similar to a previous study[14]. The percentage of groups reaching the target sum increases towards game 3 in the six-players and the six-representatives treatment, but the increase is not statistically significant. In game 3, the groups in the 18-players treatment are the least successful (Fig. 2a–c), but again differences are not statistically significant.

The total sums contributed per group do not differ among treatments in games 2 and 3 (Fig. 2e,f). In game 1, the six representatives contribute less than the six players ($P = 0.019$,

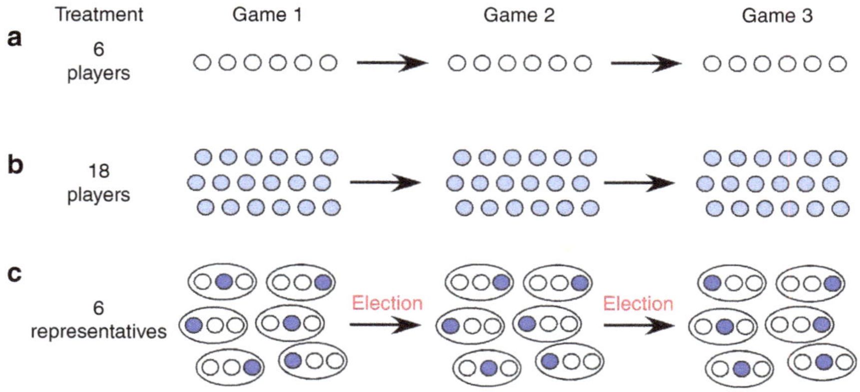

Figure 1 | Design of the three treatments. (**a**) The 6-players treatment, (**b**) the 18-players treatment, (**c**) the 6-representatives treatment. Each game consists of 10 rounds, during which players need to raise sufficient contributions to reach a specified target sum. Games 2 and 3 are replicates of game 1. The players remain the same in the 6-players and the 18-players treatment. In the 6-representatives treatment, representatives are randomly picked in game 1 and re-elected or voted out for games 2 and 3. Re-election of a representative may depend on the representatives' performance in previous games. In addition, except for the first four groups, after games 1 and 2 all players in the 6-representatives treatment are asked to write non-binding pledges about how they would contribute if elected. Players are only informed about the pledges of members of their own subgroup.

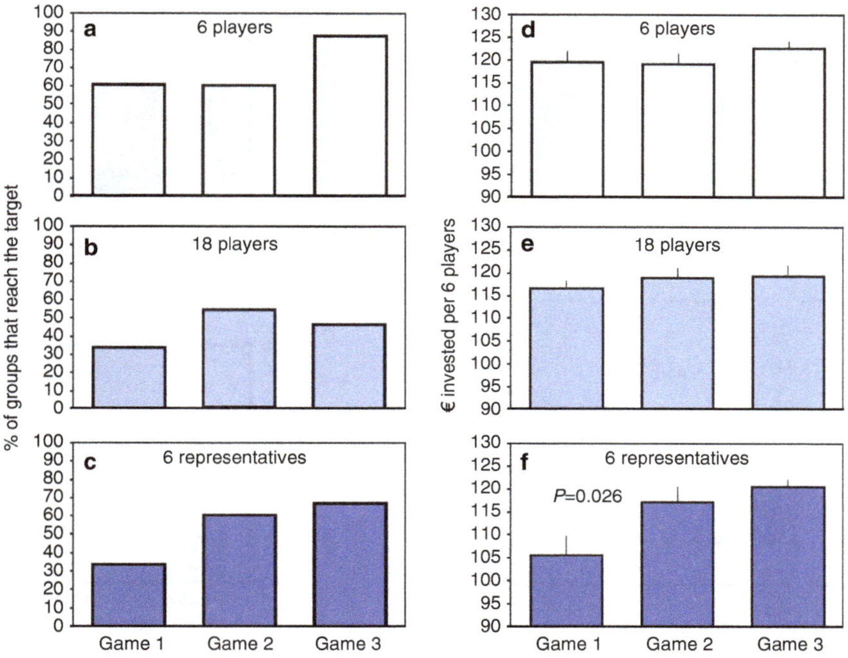

Figure 2 | Group success in reaching the target sum (left) and group investments (right). (**a**,**d**) Six-players treatment; (**b**,**e**) 18-players treatment; (**c**,**f**) 6-representatives treatment. In **f**, the sum invested is divided by 3 to allow comparison among treatments. Means ± s.e.m. of 15 groups per game and treatment are shown. See text for statistics.

$z = -2.341$, $n_1 = n_2 = 15$ groups, Mann–Whitney *U*-test, two-tailed; we use two-tailed tests throughout, with the group of six or 18 players as our statistical unit if not stated otherwise). Because in game 1, representatives are randomly picked from the group (see methods), the only difference between the two treatments is that representatives are contributing on behalf of their observing group. In such situations, representatives may have a more competitive mindset[7], which would explain why groups in the six-representatives treatment reach the target less often. Total contributions show a small increasing trend from the first to the third game in all treatments (Fig. 2e,f), but the differences are statistically significant only between games 1 and 2 in the six-representatives treatment ($P = 0.026$, $z = -2.230$, $n = 15$, Wilcoxon signed-rank matched pairs test). Summed up over all three games per group, contributions relative to the target sum are lowest in the six-representatives treatment, significantly lower than in the six-players treatment ($P = 0.0061$, $z = -2.742$, $n_1 = n_2 = 15$, Mann–Whitney *U*-test).

Fair and selfish players. For the group to reach the target sum, each player must on average contribute half of her total endowment—the 'fair share' of €20 (€60 per representative in the six-representatives treatment). Thus, whenever the target sum is not reached, one or several players must have contributed less than their fair share. We call these 'selfish players' to distinguish them from the 'fair players' who give at least their fair share. The percentage of selfish players is highest in the 6-representatives treatment (Fig. 3a), higher than in the 6-players treatment ($P = 0.01$, $z = -2.559$, $n_1 = n_2 = 15$, Mann–Whitney *U*-test) and almost significantly higher than in the 18-players treatment ($P = 0.06$, $z = -1.862$, $n_1 = n_2 = 15$, Mann–Whitney *U*-test). The average contribution of a selfish player (relative to the fair-share contribution) is lower in the 18-players treatment than in both the 6-players ($P = 0.02$, $z = -2.302$, $n_1 = n_2 = 15$, Mann–Whitney *U*-test; Fig. 3b) and the 6-representatives treatment ($P = 0.006$, $z = -2.739$, $n_1 = n_2 = 15$, Mann–Whitney *U*-test) (Fig. 3b). Over all three games, the net payoff (including trials where the group fails to collect the target sum and loses all remaining money) is higher for selfish than for fair players (Fig. 3c). Selfish players achieve a higher net payoff in the 6-players treatment, compared with both the 18-players treatment ($P = 0.024$, $z = -2.261$, $n_1 = n_2 = 15$, Mann–Whitney *U*-test) and the 6-representatives treatment ($P = 0.020$, $z = -2.325$, $n_1 = n_2 = 15$, Mann–Whitney *U*-test, shown per represented player; Fig. 3c).

Using a classification of players in a social dilemma proposed by Fischbacher and Gächter[19], the selfish representatives might be 'pessimistic conditional cooperators' who dislike that others contribute less than their fair share and thus stop contributing. However, all selfish representatives contribute more in the end than in the beginning ($P = 0.0002$, linear regression of contribution per selfish representative per group on rounds 1–10, analysed for game 3) and resemble 'imperfect conditional cooperators'[19]. By increasing their contribution during the 10 rounds as do fair representatives ($P = 0.002$), the selfish players help reaching the target, though they contribute much less than fair representatives.

Voters choose selfish representatives. After both games 1 and 2, representatives can be either re-elected or voted out. After game 1, those representatives who are re-elected have contributed significantly less in game 1 than those who are voted out (Fig. 4a) ($P = 0.01$, $z = -2.587$, $n = 15$, Wilcoxon signed-rank matched pairs test). While this is not the case after game 2, we still find a tendency that selfish representatives are preferentially re-elected, based on their past contributions. In addition, before each election the players formulate election pledges specifying their contribution strategy if elected. The percentage of selfish pledges (see Methods) is higher among the 6 elected representatives than among all 18 players of that treatment (Fig. 4b), although significantly so only after game 2 ($P = 0.0071$, $z = -2.692$, $n = 11$, Wilcoxon signed-rank matched pairs test). Thus,

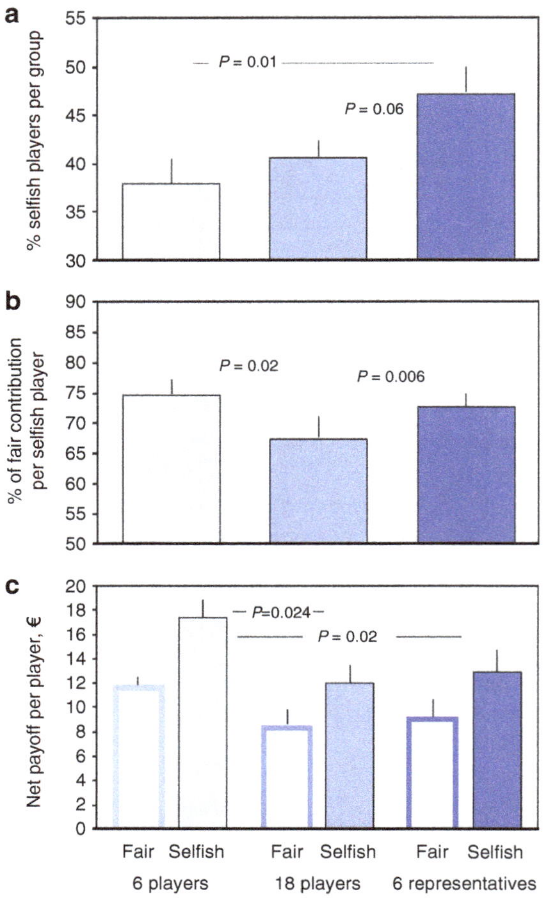

Figure 3 | Fair and selfish strategies. (**a**) The percentage of selfish players per group, (**b**) the average contribution of a selfish player (relative to the fair-share contribution), (**c**) the net payoff per fair and selfish player. Means ± s.e.m. of 15 groups per treatment are shown. See text for statistics.

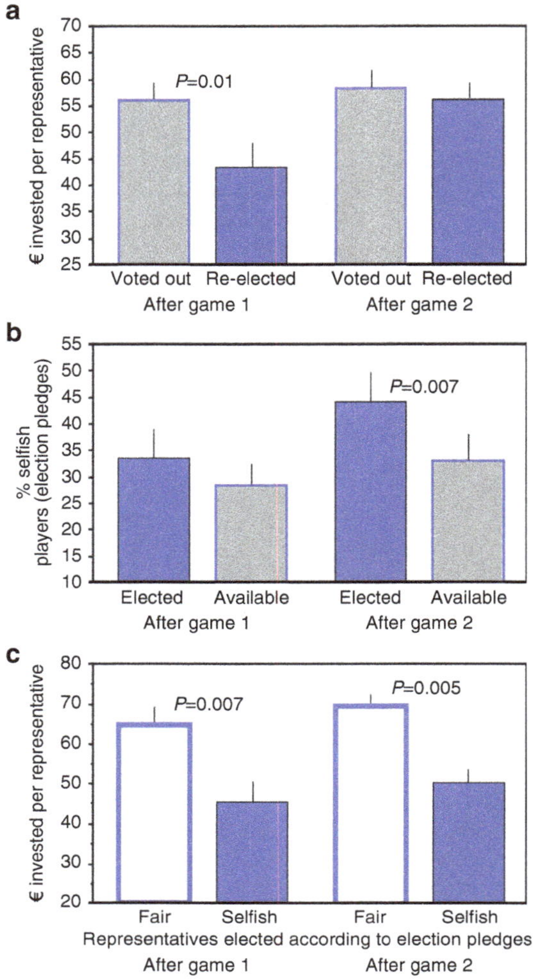

Figure 4 | Voting success and behaviour of selfish and fair representatives in the six-representatives treatment. (**a**) Previous investment of representatives who are either voted out or re-elected, (**b**) percentage of selfish players, according to their election pledges, available and elected, (**c**) future fulfilment of election pledges by selfish and fair players. Means ± s.e.m. groups are shown, for 15 groups in **a** and 11groups in **b** and 10 groups in **c**. See text for statistics.

representatives who act selfishly in game 1 are preferentially re-elected, and players who pledge to be selfish are preferentially elected after game 2.

Players classified as selfish according to their election pledges vote in 71.3% for classified selfish players and in 10.1% for classified fair representatives. Players classified as fair vote in 78.9% for classified fair players and in 14.6% for classified selfish players (the complement missing from 100% is due to players that could not be classified as either selfish or fair). Hence, selfish players want selfish representatives, and fair players want fair representatives.

Representatives who have pledged to be selfish contribute less in the following game than those who have pledged to be fair (Fig. 4c; after game 1: $P=0.007$, $z=-2.692$, $n=10$; after game 2: $P=0.0051$, $z=-2.803$, $n=10$, Wilcoxon signed-rank matched pairs test). Thus, players fulfil their pledges when acting as representatives.

Identification of selfish players as extortioners. Theorists have predicted for a long time that cooperative and fair strategies such as Tit-for-Tat would eventually succeed in social dilemmas[20–23]. Why then would subjects vote for representatives who mainly pursue the success of their own subgroup while disregarding the risks for the whole community? We hypothesize that the election procedure would favour representatives who motivate the other subgroups' representatives to reach the target, but at the same time ensure that the own subgroup contributes less than other subgroups. Individuals would like their representatives to be steadfast and to convince the other subgroups' representatives to compensate for any missing contributions. Such behaviour is reminiscent of the recently discovered class of extortionate ZD strategies for the repeated prisoner's dilemma[24–30], where extortionate players incentivize their opponents to cooperate although they themselves are not fully cooperative. In pairwise encounters, these extortionate players cannot be beaten by any other strategy, and they are predicted to perform well among adaptive co-players[24,25,27,29]. In the Methods section, we extend the theory of ZD strategies to the collective-risk social dilemma, and we prove that also in our experiment players may adopt extortionate strategies. Such players exhibit the following three characteristics: (i) Extortioners gain higher payoffs than their co-players by contributing less towards the climate account; that is, if x_i is the total contributions of an extortioner, and if x_{-i} is the average contribution of the other group members, then

$$x_i \leq x_{-i}. \quad (1)$$

(ii) Extortioners persuade their co-players to make up for the missing contributions; that is, the collective best response for the remaining $N-1$ group members is to choose x_{-i} such that the group reaches the target sum T,

$$x_i + (N-1)\cdot x_{-i} = T \quad (2)$$

(iii) Extortioners are consistent, meaning that the properties (i) and (ii) are not only satisfied in one particular instance of the game, but in every game the player participates in. We now test whether the selfish players in our experiment meet these three criteria.

Because we find both fair and selfish players in all three treatments, we perform a proof-of-principle with players of all treatments combined. To keep the group as statistical unit, we enter contribution averaged over all fair players of each group; contributions of representatives are divided by 3 to be comparable 'per player' to the other treatments. The contribution per fair player increases over the three games (Fig 5a; $P=0.0057$, $F_{2,130}=5.3788$, generalized linear model (GLM) with family = Gaussian). By contrast, the contribution per selfish player does not increase significantly ($P=0.66$, $F_{2,131}=0.4163$, GLM). We find a significant interaction between fair and selfish players' contributions over the three games ($P=0.032$, $F_{2,261}=3.4798$, GLM). Over the three games, as the contributions of fair players increase, so does the payoff of both fair players ($P=0.010$, $F_{2,132}=4.7574$, GLM, with family = gamma) and selfish players ($P=0.015$, $F_{2,132}=4.339$, GLM, with family = gamma; Fig. 5b). In each game, selfish players gain more than fair players; the difference increases from game 1 to game 3 (Fig. 5c) ($P=0.046$, $z=-1.995$, $n=45$, Wilcoxon matched pairs signed ranks test).

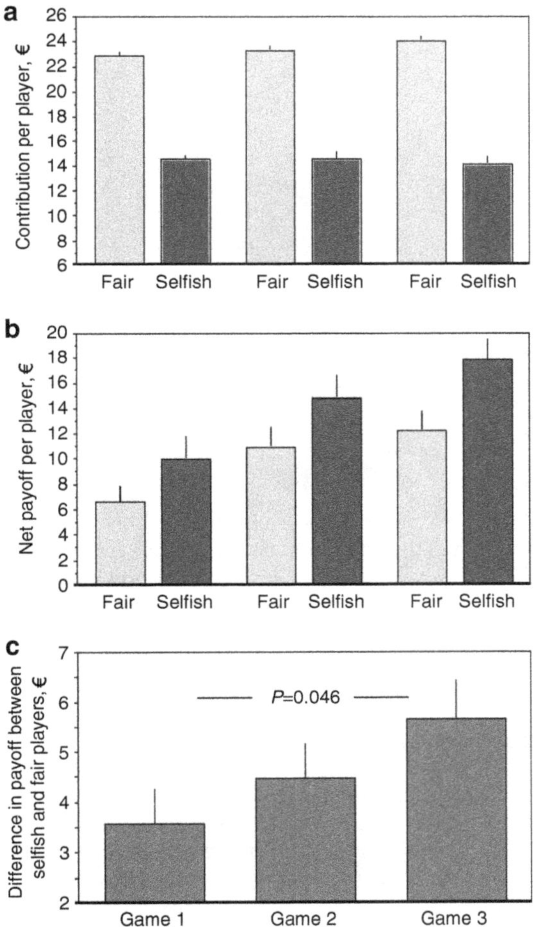

Figure 5 | Comparison of contributions and payoffs for fair players and selfish players across all three games. (**a**) Contribution of fair and selfish players; (**b**) net payoff of selfish and fair players; (**c**) difference in payoff between fair and selfish players. We enter contributions averaged over both all fair and all selfish players of each group. Contributions of representatives are divided by 3 to be comparable to other treatments. See text for statistics.

To test whether other group members are willing to compensate for missing contributions, we compare the contribution deficit of all selfish players in a group (the sum of all their negative deviations from the fair share) with the contribution surplus of all fair players (the sum of positive deviations of all the fair players; Fig. 6). For example, in game 1 in the six-players treatment, the dot most to the left (Fig. 6a) shows a group where the five selfish players contribute only €80 instead of the fair-share contribution of €100. The single fair player of that group contributes €22, €2 more than her fair share but not enough to compensate for the deficit of €20 caused by the selfish players. Hence the group misses the target sum of €120, and everybody loses the money not invested with 90% probability. As another example, the leftmost dot of those exactly on the red line depicts a group where the three selfish players invest €44 instead of €60, causing a deficit of €16, which is exactly compensated by the three remaining fair players. Thus the group meets the target of €120, but the selfish players receive a higher payoff than the fair players.

If selfish players were indeed able to persuade the remaining group members to compensate for missing contributions, we would expect the regression lines in Fig. 6 to have a significantly negative slope and to be close to the red lines marking exact (hypothetical) compensation. We see this compensation in the six-players treatment in game 2 (Fig. 6b, simple regression, F-test = 36.257, degree of freedom (DF) = 1, $P=0.0001$) and in game 3 (Fig. 6c, simple regression, F-test = 26.204, DF = 1, $P=0.0002$) and in the six-representatives treatment in game 3 (Fig. 6f, simple regression, F-test = 17.286, DF = 1, $P=0.0011$). By contrast, we find no significant compensation in the 18-players treatment.

In the 6-players treatment, fair players compensate or overcompensate the selfish players' deficit in 9 groups in games 1 and 2 (Fig. 6a,b) and in 13 groups in game 3 (Fig. 6c). In the 18-players treatment, fair players compensate or overcompensate the selfish players' deficit in 5 groups in game 1 (Fig. 6g) and in 7 groups in games 2 and 3 (Fig. 6h,i). In the 6-representatives treatment, the deficit of the selfish players is only compensated in 4 groups in game 1 (Fig. 6d) but in 9 groups in game 2 (Fig. 6e) and in 10 groups in game 3 (Fig. 6f). Over all treatments and games, selfish players or selfish representatives successfully drive their fair counterparts to compensation in 73 out of 135 individual games (54%). Moreover, groups become increasingly successful in reaching the target, improving from game 1 (40%) to game 2 (56%) and game 3 (67%). Because only fair players raise their contributions over the three games but not selfish players (see Fig. 5a), these results suggest that a considerable fraction of fair players learn to become even more cooperative in response to extortioners. The learning effect is demonstrated by the observation that the contribution per fair representative has no relation to the number of selfish representatives per group in game 1 but correlates significantly in game 3 (Supplementary Fig. 2).

Players behave consistently across the 3 games in the 6-players and the 18-players treatments, as witnessed by significant positive correlation of the contributions (see Supplementary Information for detailed analysis). For the six-representatives treatment, we have analysed the behaviour of representatives after being re-elected. In 34 out of the 42 cases in which a selfish representative is re-elected, the representative remains selfish in

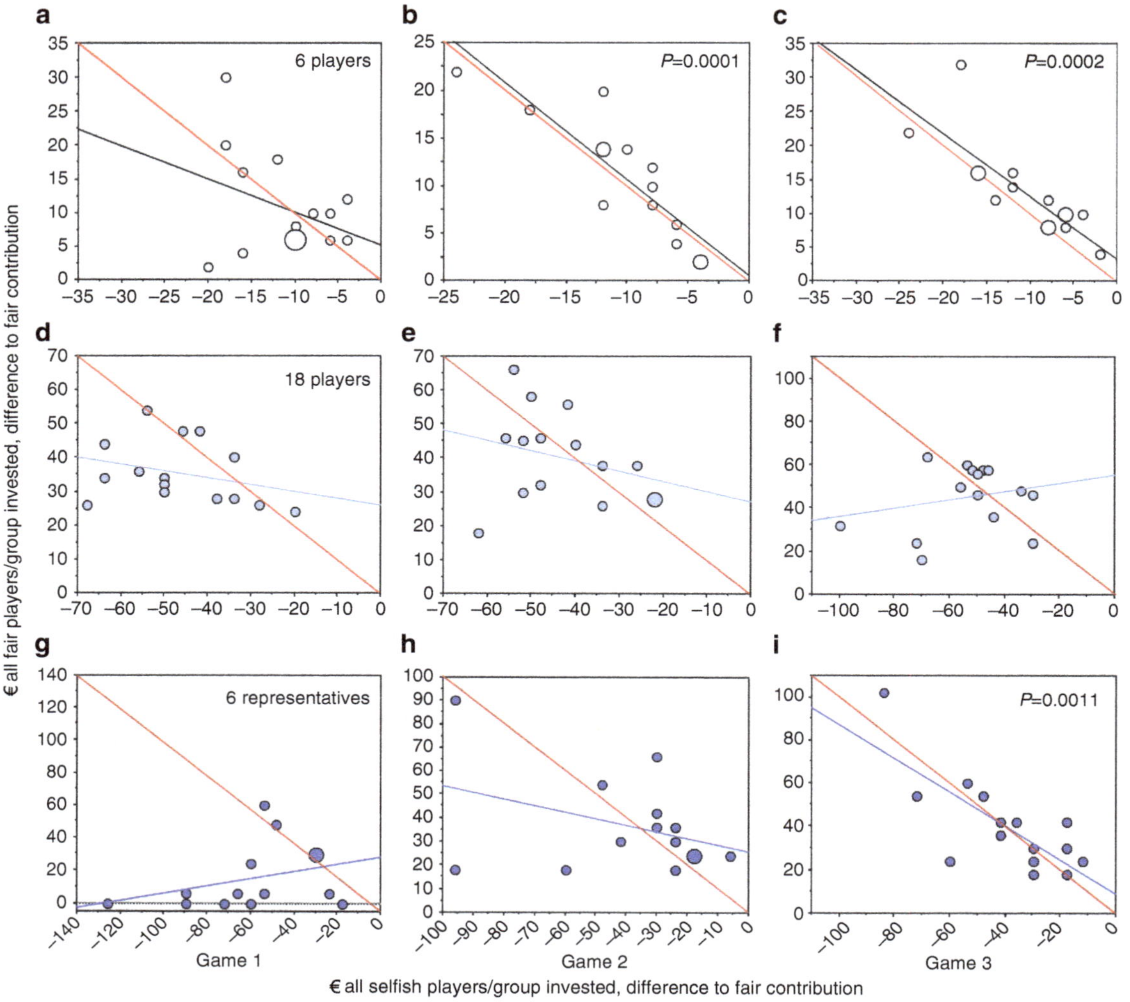

Figure 6 | Fair players' compensation of their selfish players' deficit. (**a**–**c**) Six-players treatment; (**d**–**f**) 18 players treatment; (**g**–**i**) 6-representatives treatment. Each dot represents a group; larger dots show overlaid results from two or three groups. Black and blue lines depict simple regressions. The red lines depict all combinations of hypothetical contributions in which fair players exactly compensate for the deficit caused by all selfish players of that group. Thus, dots on or above the red line correspond to groups that reach the target sum. See text for statistics.

the next game ($P = 0.005$, Fisher's exact test, two-tailed compared with 50%). Overall, we have thus established that selfish players gain much higher payoffs (Fig. 5); they are often successful in persuading their fair co-players to compensate for missing contributions (Fig. 6); and they are consistent across different games. Thus, selfish players show all three characteristics of extortionate behaviour.

Discussion

We have introduced into the collective-risk social dilemma the innovation that contributions into the climate account are decided on not by individual players but by representatives (six-representatives treatment). For control, we have assembled groups of 6 players and 18 players in 2 further treatments. We find selfish players in all treatments, but their concentration is highest in the 6-representatives treatment. Selfish representatives are preferentially elected or re-elected if they either contribute less than the fair share or pledge to do so. Having to cater to their electorates' preferences thus has the adverse effect that representatives risk losing the climate game to win elections. As a consequence, groups in the six-representatives treatment contribute less than groups in the six-players treatment (relative to the required target sum), and they receive lower average payoffs. On the other hand, in games 2 and 3 the groups tend to reach the target sum more often in the 6-representatives than in the 18-players treatment. While fair representatives compensate for missing contributions in game 3, 18 players do not achieve that compensation. Thus, our representatives tend to be more successful in preventing simulated dangerous climate change than 18 players deciding themselves. We speculate that this 'representatives' advantage' is much greater with much larger groups such as real countries.

The psychological consequences of acting as a representative of a group have been characterised as evoking both more competitive interaction goals and more competitive expectations of others[7]. A representative is faced with a powerful responsibility to provide good outcomes for her constituency and may face strong pressures by being monitored and evaluated[7]. The mindset that is activated by the role of representative shows up clearly in our experiments when we compare the behaviour of six players randomly selected to decide for themselves with the behaviour of six representatives randomly selected to decide for their group. Otherwise the players find themselves in exactly the same situation in both cases. As psychology predicts[7], the six players' groups are twice as successful as the six-representatives' groups in reaching the target sum. When voting, players preferentially choose representatives who either have displayed a competitive mindset as former representatives or have pledged to do so if elected.

In our experiments, selfish behaviour pays off only if others compensate any missing contributions. Selfish subjects apply an implicit form of extortion[24]—they contribute less than is needed on average, but in a way that makes it optimal for their peers to become even more cooperative. The effect of extortion in our experiments differs from that in the repeated prisoner's dilemma, in which subjects strongly oppose exploitation[31]. Here subjects in the six-players and six-representatives treatments eventually accept extortion up to a certain degree, especially in game 3, in which subjects have already gained some experience. We speculate that the higher tolerance towards extortioners in our experiment is due to the higher stakes involved—resisting extortion comes relatively cheap in the prisoner's dilemma, but it endangers the entire payoff in the collective-risk game. Only in the 18-players treatment was extortion unsuccessful in persuading others to cooperate, presumably because in larger groups it becomes more difficult to induce individuals to behave in a desired way.

Our identification of extortionate behaviour in the collective-risk social dilemma suggests two counteracting major effects when, with all due caution, we try to interpret the social dynamics of climate summits with our results in mind. On the one hand, the competitive advantage of selfish players in getting elected or re-elected appears to work against reaching a collective target such as preventing dangerous climate change—there might not be enough fair representatives around to support the target. On the other hand, selfish players, who are ubiquitous and show up in all but 1 of the 135 individual collective-risk games, consistently act as extortioners. Their steadfast strategies enhance the already-existing willingness of our fair players to contribute towards reaching the collective target. If we compare extortionate to hypothetical non-extortionate selfish players, we conclude—with more than just a hint of Machiavellian thinking—that extortion benefits the prevention of dangerous climate change.

Methods

Experimental procedures. A total of 630 undergraduate students from the Universities of Bonn, Hamburg, Göttingen, Kiel and Münster voluntarily participated in 45 experimental sessions with either 18 or 6 subjects each in a computerized experiment (for example, ref. 32). The subjects were separated by opaque partitions and each had a computer, on which they received the instructions for the experiment and with which they communicated their decisions. Throughout the whole experiment, subjects were anonymous, and they made their decisions under a neutral pseudonym.

There are three treatments (Fig. 1). For each treatment, we had 15 groups of subjects interacting in a variant of the 'collective-risk social dilemma' game[14]: subjects received an initial endowment, and they were asked, in each of 10 rounds, to contribute money from this endowment into a 'climate account'. At the end of round 10, the game software checked whether total contributions of all group members matched (or exceeded) a previously specified target sum. If that was the case, subjects received their remaining endowment in cash (in a way that maintained the subjects' anonymity). If the collective target was not reached, subjects lost their remaining endowment with 90% probability. Each game was repeated twice, such that every group played three games with all players keeping their pseudonyms (each time with a new endowment).

In the 6-players and the 18-players treatments, groups consisted of 6 and 18 subjects, respectively, and each subject had an initial endowment of €40. In each round, subjects could choose whether to contribute €0, €2 or €4 to the climate account, and the decisions of all subjects were shown to all subjects after each round. The target was reached if on average subjects contributed half their endowment (the target sum was €120 in the 6-players treatment, and it was €360 in the 18-players treatment).

In the 6-representatives treatment, groups of 18 subjects were sub-divided in 6 'countries' of 3 players. For game 1, the computer randomly determined six representatives, one from each country. Only the representatives were able to contribute money to the group's climate account: they had 3 times €40 at their disposal, for investing €0, €6 or €12 in each of 10 rounds. The decisions of all representatives were shown to all 18 subjects after each round. The target was to collect at least €360 in donations (€60 per representative or €20 per subject). If the target was reached, subjects received a third of their country's remaining endowment.

After games 1 and 2, the three subjects of each country could re-elect the previous representative or vote her out and elect a different member of their country with a majority vote. Except for the first four groups of this treatment, subjects could compose both after game 1 and game 2 election pledges of up to 500 characters on their laptop that could be seen only by the 3 subjects of a country. The pledges described how the person would decide if elected. We have blindly classified all pledges; those that promise to contribute 'less than the others', or 'less than the fair share' have been classified as 'selfish' and the others as 'fair'. When making the voting decision, each subject knew the observed decisions of her previous representative and those of the other representatives, and saw the three election pledges within her country (each with the respective pseudonym and a button for voting). In cases with no majority vote, the computer decided randomly for the next representative (in 9% of cases).

Subjects knew that the total sum of money in the climate account, accumulated from all participating groups, would be used to publish a press advertisement on climate protection in a daily German newspaper simultaneously with the publication of the present study. However, they received the 'little information' version from ref. 32 to explain the climate account, so that we could expect very weak motivation to invest in publishing the advertisement *per se*.

Theoretical model. Press and Dyson[24] describe a class of so-called ZD strategies for the repeated prisoner's dilemma, and they demonstrate that a subset of ZD strategies can be used to extort opponents. However, the collective-risk dilemma game used in our experiment is not a repeated two-player game. Herein, we thus extend the theory of ZD strategies to collective-risk dilemmas. As an application, we show the existence of extortionate strategies. Such strategies ensure that (i) a player gets at least the average payoff of the co-players; (ii) the collective best reply for the remaining group members is to reach the target; and (iii) the properties (i) and (ii) hold in any game the player participates in.

To this end, we consider a group of N individuals, with each group member having an initial endowment of E. The group engages in a collective-risk dilemma[12]: in each of R rounds, players can decide how much they want to contribute towards a common pool. We denote player i's contribution in round r by $x_i(r)$, and we assume that the minimum contribution per round is 0, whereas the maximum contribution is $x_{\max} = E/R$. To calculate the total contributions x_i of player i, we sum up over all rounds, $x_i = \sum x_i(r)$. The group's total contributions x are obtained by summing up over all individual contributions, $x = \sum x_i$. Payoffs for the collective-risk dilemma are defined as follows: if total contributions after round R exceed a threshold T, then all players receive their remaining endowment; that is, if $x \geq T$, then player i's expected payoff is $E - x_i$. Otherwise, if total contributions are below the threshold, all players risk losing their remaining endowment with some probability $p > 0$, and player i's expected payoff becomes $(1 - p)\ (E - x_i)$. Supplementary Table 1 gives a summary of all used variables.

In the experiment, players had to choose between three possible contribution levels in a given round, but for the model we assume for simplicity that players can contribute any amount $x_i(r) \in [0, x_{\max}]$. We note that the definition of ZD strategies given below can be extended to the case of discrete contribution levels. To achieve an arbitrary contribution level $y \in [0, x_{\max}]$, player i would need to randomize between the given discrete contribution levels such that the expected value satisfies $E[x_i(r)] = y$. Similar to Tit-for-Tat-like strategies in the Prisoner's Dilemma, we define ZD strategies in the collective-risk dilemma as behaviours that condition their contribution in the next round on the co-players' contributions in the previous round:

Definition (ZD strategies). Player i applies a ZD strategy for the collective-risk dilemma if i's contributions $x_i(r)$ in every round r satisfy

$$x_i(r) = s x_{-i}(r-1) + (1-s)\gamma E/R, \tag{3}$$

where $x_{-i}(r-1)$ is the average contribution of the other group members in the previous round, with $x_{-i}(0) := 0$, and s and γ are parameters that can be chosen by player i.

The parameter s is a measure for how a player reacts to the co-players' contributions of the previous round. The parameter γ, on the other hand, determines a player's baseline contribution level. These two parameters cannot be chosen arbitrarily—since player i's contribution needs to be in the interval $[0, E/R]$, the two parameters need to satisfy

$$\begin{aligned} -1 &\leq s \leq 1 \\ -s/(1-s) &\leq \gamma \leq 1/(1-s) \end{aligned} \tag{4}$$

It is the following property that makes ZD strategies interesting.

Proposition 1 (properties of ZD strategies). Suppose player i applies a ZD strategy with parameters s and γ.

1. If x_i denotes the total contributions of player i, and if x_{-i} denotes the average total contribution of i's co-players, then

$$|x_i - s x_{-i} - (1-s)\gamma E| \leq E/R \tag{5}$$

2. Similarly, if π_i and π_{-i} denote the corresponding realized payoffs, then payoffs either satisfy $\pi_i = \pi_{-i} = 0$ (if the group fails to reach the threshold and dangerous climate change occurs), or

$$|\pi_i - s\pi_{-i} - (1-s)(1-\gamma)E| \le E/R \tag{6}$$

Proof

1. By summing up Equation (3) over all rounds $1 \le r \le R$, we obtain

$$x_i = s[x_{-i} - x_{-i}(R)] + (1-s)\gamma E.$$

As a consequence,

$$|x_i - sx_{-i} - (1-s)\gamma E| \le |sx_{-i}(R)| \le E/R.$$

2. In case players do not lose their remaining endowment, Equation (6) follows directly from Equation (5) because $\pi_i = E - x_i$ and $\pi_{-i} = E - x_{-i}$.

For a collective-risk dilemma with sufficiently many rounds *R*, Proposition 1 thus implies that $x_i \approx sx_{-i} + \gamma(1-s)E$. That is, there is a linear relationship between the total contributions of player *i*, and the total contributions of *i*'s co-players. Similarly, it follows for the realized payoffs that either $\pi_i = \pi_{-i} = 0$ or $\pi_i \approx s\pi_{-i} + (1-\gamma)(1-s)E$. Therefore, unless payoffs are zero, there is also a linear relationship between the players' realized payoffs. This property makes strategies having the form of Equation (3) analogous to the ZD strategies described for the repeated prisoner's dilemma[24]. It is important to note that the above Proposition makes no restrictions on the strategies of *i*'s co-players—the stated results hold no matter what the other group members do. As a particular instance of ZD strategies, let us consider the following special case.

Definition (extortionate ZD strategies). A player applies an extortionate ZD if the parameters *s* and γ are chosen such that

$$\gamma = 0 \text{ and } \max[0, T/(pE) - (N-1)] \le s < 1. \tag{7}$$

If some player *i* applies such an extortionate strategy, it follows from Proposition 1 that approximately *i*'s total contribution only make up a fraction of the average contribution of the other group members, since $x_i \approx sx_{-i}$ (Supplementary Fig. 1 gives an illustration).

The following Proposition shows that the name 'extortionate ZD strategy' is justified: players with such a strategy show the typical characteristics of extortionate behaviour.

Proposition 2 (properties of extortionate ZD strategies). Suppose player *i* applies an extortionate ZD strategy. Then, irrespective of the strategies applied by the other group members (that is, in any game player *i* participates in),

1. Player *i*'s realized payoff is never below the mean payoff of the other group members, $\pi_i \ge \pi_{-i}$.
2. The collective best reply for the remaining group members is to reach the threshold *T*. In that case, player *i*'s payoff is strictly better than average, $\pi_i > \pi_{-i}$.

Proof. Because a player with an extortionate ZD strategy contributes strictly less than average, $x_i < x_{-i}$, it follows that either $\pi_i = \pi_{-i} = 0$ (if the group misses the target and players lose their remaining endowment) or $\pi_i > \pi_{-i}$ (otherwise). Moreover, for the other group members, it is collectively optimal to reach the target: by contributing nothing, their expected payoff becomes $(1-p)E$, whereas if they make the minimum contribution (in the first $R-1$ rounds) such that total contributions reach the target, then their payoff is $E - T/(N-1+s)$. Because $s \ge T/(pE) - (N-1)$, reaching the target is a collective best reply.

Proposition 2 is a proof-of-principle: there are strategies for the collective-risk dilemma that allow a player to extort the other group members. We note that the set of all extortionate strategies will typically be considerably bigger than the set of all extortionate ZD strategies. When we analyse experimental data, we therefore do not specifically look for strategies that have the functional form described in Equations (3) and (7); we rather look for all possible strategies that indicate extortionate behaviour (that is, we look whether players satisfy the conditions (i)–(iii) defined in the main text).

References

1. Fehr, E. & Fischbacher, U. The nature of human altruism. *Nature* **425,** 785–791 (2003).
2. Nowak, M. A. & Sigmund, K. Evolution of indirect reciprocity. *Nature* **437,** 1291–1298 (2005).
3. Peters, G. P. *et al.* The challenge to keep global warming below 2 degrees C. *Nat. Clim. Change* **3,** 4–6 (2013).
4. IPCC. *Climate Change 2014: Synthesis Report. Contribution of Working Groups I, II and III to the Fifth Assessment Report of the Intergovernmental Panel on Climate Change* (IPCC, 2014).
5. Froese, R. Fishery reform slips through the net. *Nature* **475,** 7–7 (2011).
6. Keep a welcome. *Nature* **525,** 157–157 (2015).
7. Reinders Folmer, C. P., Klapwijk, A., De Cremer, D. & Van Lange, P. A. M. One for all: what representing a group may do to us. *J. Exp. Soc. Psychol.* **48,** 1047–1056 (2012).
8. UNFCCC. *United Nations Framework Convention on Climate Change* (United Nations, 1992).
9. IPCC. in *Contribution of Working Group I to the Fourth Assessment Report of the Intergovernmental Panel on Climate Change.* (eds Solomon, Susan *et al.*) (Cambridge Univ. Press, 2007).
10. Meinshausen, M. *et al.* Greenhouse-gas emission targets for limiting global warming to 2°C. *Nature* **458,** 1158–1162 (2009).
11. Allen, M. R. *et al.* Warming caused by cumulative carbon emissions towards the trillionth tonne. *Nature* **458,** 1163–1166 (2009).
12. IPCC. in *Climate Change 2013: The Physical Science Basis. Contribution of Working Group I to the Fifth Assessment Report of the Intergovernmental Panel on Climate Change.* (eds Stocker, T. F. *et al.*) 3–29 (Cambridge Univ. Press, 2013).
13. Esslinger, D. There is currently no climate policy—because the voters do not want one (Es gibt derzeit keine Klimapolitik—weil die Wähler keine wollen). http://sz.de/1.1787481 (2013).
14. Milinski, M., Sommerfeld, R. D., Krambeck, H.-J., Reed, F. A. & Marotzke, J. The collective-risk social dilemma and the prevention of simulated dangerous climate change. *Proc. Natl Acad. Sci. USA* **105,** 2291–2294 (2008).
15. Milinski, M., Röhl, T. & Marotzke, J. Cooperative interaction of rich and poor can be catalyzed by intermediate climate targets. *Clim. Change* **109,** 807–814 (2011).
16. Tavoni, A., Dannenberg, A., Kallis, G. & Loeschel, A. Inequality, communication, and the avoidance of disastrous climate change in a public goods game. *Proc. Natl Acad. Sci. USA* **108,** 11825–11829 (2011).
17. Abou Chakra, M. & Traulsen, A. Evolutionary dynamics of strategic behavior in a collective-risk dilemma. *PLoS Comput. Biol.* **8,** e1002652 (2012).
18. Jacquet, J. *et al.* Intra- and intergenerational discounting in the climate game. *Nat. Clim. Change* **3,** 1025–1028 (2013).
19. Fischbacher, U. & Gächter, S. Social preferences, beliefs, and the dynamics of free riding in public goods experiments. *Am. Econ. Rev.* **100,** 541–556 (2010).
20. Axelrod, R. & Hamilton, W. D. The evolution of cooperation. *Science* **211,** 1390–1396 (1981).
21. Axelrod, R. The evolution of cooperation. (Basic Books, 1984).
22. Nowak, M. A. & Sigmund, K. Tit-for-tat in heterogeneous populations. *Nature* **355,** 250–253 (1992).
23. Nowak, M. A. & Sigmund, K. The alternating prisoner's dilemma. *J. Theor. Biol.* **168,** 219–226 (1994).
24. Press, W. H. & Dyson, F. J. Iterated Prisoner's Dilemma contains strategies that dominate any evolutionary opponent. *Proc. Natl Acad. Sci. USA* **109,** 10409–10413 (2012).
25. Hilbe, C., Nowak, M. A. & Sigmund, K. Evolution of extortion in iterated prisoner's dilemma games. *Proc. Natl Acad. Sci. USA* **110,** 6913–6918 (2013).
26. Stewart, A. J. & Plotkin, J. B. From extortion to generosity, evolution in the iterated prisoner's dilemma. *Proc. Natl Acad. Sci. USA* **110,** 15348–15353 (2013).
27. Chen, J. & Zinger, A. The robustness of zero-determinant strategies in iterated prisoner's dilemma games. *J. Theor. Biol.* **357,** 46–54 (2014).
28. Szolnoki, A. & Perc, M. Defection and extortion as unexpected catalysts of unconditional cooperation in structured populations. *Sci. Rep.* **4,** 5496 (2014).
29. Hilbe, C., Wu, B., Traulsen, A. & Nowak, M. A. Evolutionary performance of zero-determinant strategies in multiplayer games. *J. Theor. Biol.* **374,** 115–124 (2015).
30. Akin, E. What you gotta know to play good in the Iterated Prisoner's Dilemma. *Games* **6,** 175–190 (2015).
31. Hilbe, C., Roehl, T. & Milinski, M. Extortion subdues human players but is finally punished in the prisoner's dilemma. *Nat. Commun.* **5,** 3976 (2014).
32. Milinski, M., Semmann, D., Krambeck, H.-J. & Marotzke, J. Stabilizing the Earth's climate is not a losing game: supporting evidence from public goods experiments. *Proc. Natl Acad. Sci. USA* **103,** 3994–3998 (2006).

Acknowledgements

We thank the students for participation; H.-J. Krambeck for writing the software for the game; H. Arndt, T. Bakker, L. Becks, H. Brendelberger, S. Dobler and T. Reusch for support; and the Max Planck Society for the Advancement of Science for funding.

Author contributions

M.M. and J.M. designed the experiment; C.H. developed the mathematical model; M.M., D.S. and R.S. performed the experiments; M.M. analysed the data, M.M., J.M. and C.H. wrote the manuscript, and all authors revised the manuscript.

Additional information

Competing financial interests: The authors declare no competing financial interests.

18

Quantum algorithms for topological and geometric analysis of data

Seth Lloyd[1], Silvano Garnerone[2] & Paolo Zanardi[3]

Extracting useful information from large data sets can be a daunting task. Topological methods for analysing data sets provide a powerful technique for extracting such information. Persistent homology is a sophisticated tool for identifying topological features and for determining how such features persist as the data is viewed at different scales. Here we present quantum machine learning algorithms for calculating Betti numbers—the numbers of connected components, holes and voids—in persistent homology, and for finding eigenvectors and eigenvalues of the combinatorial Laplacian. The algorithms provide an exponential speed-up over the best currently known classical algorithms for topological data analysis.

[1] Department of Mechanical Engineering, Research Lab for Electronics, Massachusetts Institute of Technology, MIT 3-160, Cambridge, Massachusetts 02139, USA. [2] Institute for Quantum Computing, University of Waterloo, Waterloo, Ontario, Canada N2L 3G1. [3] Department of Physics and Astronomy, Center for Quantum Information Science & Technology, University of Southern California, Los Angeles, California 90089-0484, USA. Correspondence and requests for materials should be addressed to S.L. (email: slloyd@mit.edu).

Human society is currently generating on the order of Avogadro's number (6×10^{23}) of bits of data a year. Extracting useful information from even a small subset of such a huge data set is difficult. A wide variety of big data processing techniques have been developed to extract from large data sets the hidden information in which one is actually interested. Topological techniques for analysing big data represent a sophisticated and powerful tool[1–24]. By its very nature, topology reveals features of the data that robust to how the data were sampled, how it was represented and how it was corrupted by noise. Persistent homology is a particularly useful topological technique that analyses the data to extract topological features such as the number of connected components, holes, voids and so on (Betti numbers) of the underlying structure from which the data was generated. The length scale of analysis is then varied to see whether those topological features persist at different scales. A topological feature that persists over many length scales can be identified with a 'true' feature of the underlying structure.

Topological methods for analysis face challenges: a data consisting of n data points possesses 2^n possible subsets that could contribute to the topology. Performing methods of algebraic topology on simplicial complexes eventually requires matrix multiplication or diagonalization of matrices of dimension $O\binom{n}{k+1}$ to extract topological features at dimension k. For small k, such operations require time polynomial in n; however, to extract high-dimensional features, matrix multiplication and diagonalization lead to problem solution scalings that grow exponentially in the size of the complex. A variety of mathematical methods have been developed to cope with the resulting combinatorial explosion, notably mapping the complex to a smaller complex with the same homology, and then performing the matrix operations on the reduced complex[1–24]. Even in such cases, the initial reduction must identify all simplices in the original complex, and so can scale no better than linearly in the number of simplices. Consequently, even with only a few hundred data points, creating the persistent homology for Betti numbers at all orders of k is a difficult task. In particular, the most efficient classical algorithms for estimating Betti numbers at order k (the number of k-dimensional gaps, holes and so on), have computational complexity either exponential in k or exponential in n (refs 7–12), so that estimating Betti numbers to all orders scales exponentially in n, and algorithms for diagonalizing the combinatorial Laplacian (that reveal not only the Betti numbers but additional geometric structure) at order k have computational complexity as $O\left(\binom{n}{k}^2\right)$, where n is the number of vertices in the (possibly reduced) complex. That is, the best classical algorithms for estimating Betti numbers to all orders[9–12] and for diagonalizing the full combinatorial Laplacian grow exponentially in the number of vertices in the complex.

This paper investigates quantum algorithms for performing topological analysis of large data sets. We show that a quantum computer can find the eigenvectors and eigenvalues of the combinatorial Laplacian and estimate Betti numbers to all orders and to accuracy δ in time $O(n^5/\delta)$, thereby reducing a classical problem for which the best existing solutions have exponential computational complexity, to a polynomial-time quantum problem. Betti numbers can also be estimated by using a reduced, or 'witness' complex, that contains fewer points than the original complex[1–12]. Applied to such witness complexes, our method again yields a reduction in estimation time from $O(2^{2\tilde{n}})$ to $O(\tilde{n}^5)$, where $\tilde{n}$ is the number of points in the reduced complex.

Recently, quantum mechanical techniques have been proposed for machine learning and data analysis[25–34]. In particular, some quantum machine learning algorithms[31–33] provide exponential speed-ups over the best existing classical algorithms for supervised and unsupervised learning. Such 'big quantum data' algorithms use a quantum random access memory (qRAM)[35–37] to map an N-bit classical data set onto the quantum amplitudes of a ($\log_2 N$)-qubit quantum state, an exponential compression over the classical representation. The resulting state is then manipulated using quantum information processing in time poly($\log_2 N$) to reveal underlying features of the data set. That is, quantum computers that can perform 'quantum sampling' of data can perform certain machine learning tasks exponentially faster than classical computers performing classical sampling of data. A discussion of computational complexity in quantum machine learning can be found in ref. 34. Constructing a large-scale qRAM to access $N \sim 10^9 - 10^{12}$ pieces of data is a difficult task. By contrast, the topological and geometrical algorithms presented here do not require a large-scale qRAM: a qRAM with $O(n^2)$ bits suffices to store all pairwise distance information between the points of our data set. The algorithms presented here obtain their exponential speed-up over the best existing classical algorithms not by having quantum access to a large data set, but instead, by mapping a combinatorially large simplicial complex with $O(2^n)$ simplices to a quantum state with n qubits, and by using quantum information processing techniques such as matrix inversion and diagonalization to perform topological and geometrical analysis exponentially faster than classical algorithms. Essentially, our quantum algorithms operate by finding the eigenvectors and eigenvalues of the combinatorial Laplacian. But diagonalizing a 2^n by 2^n sparse matrix using a quantum computer takes time $O(n^2)$, compared with time $O(2^{2n})$ on a classical computer[38–40].

The algorithms given here are related to quantum matrix inversion algorithms[41]. The original matrix inversion algorithm[41] yielded as solution a quantum state, and left open the question of how to extract useful information from that state. The topological and geometric algorithms presented here answer that question: the algorithms yield as output not quantum states but rather topological invariants—Betti numbers—and do so in time exponentially faster than the best existing classical algorithms. The best classical algorithms for calculating the kth Betti number takes time $O(n^k)$, and estimating Betti numbers to all orders to accuracy δ takes time at least $O(2^n \log(1/\delta))$ (refs 7–12). Exact calculation of Betti numbers for some types of topological sets (algebraic varieties) is PSPACE hard[42]. By contrast, our algorithm provides approximate values of Betti numbers to all orders and to accuracy δ in time $O(n^5/\delta)$: although no polynomial classical algorithm for such approximate evaluation of topological invariants is known, the computational complexity of such approximation remains an open problem. We do not expect our quantum algorithms to solve a PSPACE-hard problem in polynomial time. We summarize the comparison between the amount of resources required by the classical and quantum algorithms in Table 1.

Results

The quantum pipeline. The quantum algorithm operates by mapping vectors, simplices, simplicial complexes and collections of simplicial complexes to quantum mechanical states, and reveals topology by performing linear operations on those states. The 2^n possible simplices of the simplicial complex are mapped onto an n-qubit quantum state. This state is then analysed using conventional quantum computational techniques of eigenvector and eigenvalue analysis, matrix inversion and so on. The quantum analysis reveals topological features of the data, and shows how those features arise and persist when the scale of analysis is varied. The resulting quantum algorithms provide an exponential speed-up over the best existing classical algorithms for topological data analysis.

In addition to revealing topological features such as Betti numbers, our algorithm uses the relationship between algebraic topology and Hodge theory[9–12,14–24] to reveal geometrical information about the data analysed at different scales. The algorithm operates by identifying the harmonic forms of the data, together with the other eigenvalues and eigenvectors of the combinatorial Laplacian—the quantities that famously allow one to 'hear the shape of a drum'[43]. The quantum algorithm reveals these geometric features exponentially faster than the corresponding classical algorithms. In particular, our quantum algorithm for finding all Betti numbers for the persistent homology for simplicial complexes over n points and for diagonalizing the combinatorial Laplacian takes time $O(n^5/\delta)$, where δ is the multiplicative accuracy to which Betti numbers and eigenvalues are determined. The best available classical algorithms to perform these tasks at all orders of k take time $O(2^{2n}\log(1/\delta))$.

The advantage of big quantum data techniques is that they provide exponential compression of the representation of the data. The challenge is to see if—and this is a big 'if'—it is still possible to process the highly compressed quantum data to reveal the desired hidden structure that underlies the original data set. Here we show that quantum information processing acting on large data sets encoded in a quantum form can indeed reveal topological features of the data set.

Classical algorithms for persistent homology have two steps (the 'pipeline'). First, one processes the data to allow the construction of a topological structure such as a simplicial complex that approximates the hidden structure from which the data was generated. The details of the topological structure depends on the scale at which data is grouped together. Second, one constructs topological invariants of that structure and analyses how those invariants behave as a function of the grouping scale. As above, topological invariants that persist over a wide range of scales are identified as features of the underlying hidden structure.

The quantum 'pipeline' for persistent homology also has two steps. First, one accesses the data in quantum parallel to construct quantum states that encode the desired topological structure: if the structure is a simplicial complex, for example, one constructs quantum states that are uniform superposition of descriptions of the simplices in the complex. Second, one uses the ability of quantum computing to reveal the ranks of linear maps to construct the topological invariants of the structure. The steps of the quantum pipeline are now described in more detail.

Constructing a simplicial complex. Classical persistent homology algorithms use the access to data and distances to construct a topological structure—typically a simplicial complex—that corresponds to the hidden structure whose topology one wishes to reveal. In the quantum algorithm, we use the ability to access data and to estimate distances in quantum parallel to construct quantum states that encode the simplicial complex. Each simplex in the complex consists of a fully connected set of vertices: a k-simplex s_k consists of $k+1$ vertices $j_0, j_1, \ldots, j_k$ (listed in ascending order, $j_0<j_1< \ldots <j_k$) together with the $k(k+1)/2$ edges connecting each vertex to all the other vertices in the simplex. Encode a k-simplex s_k as a string of n bits, for example, $0110 \ldots 1$, with $k+1$ 1s at locations $j_0, j_1, \ldots, j_k$ designating the vertices in the simplex. Removing the ℓth vertex and its associated edges from a k-simplex yields a $k-1$ simplex. The $k+1$ simplices $s_{k-1}(\ell)$ with vertices $j_0 \ldots \hat{j}_\ell \ldots j_k$ obtained by removing the ℓth vertex j_ℓ from s_k form the boundary of the original simplex. The number of potential simplices in a simplicial complex is equal to 2^n, the number of possible subsets of the n points in the graph. That is, every member of the power set is a potential simplex. If n is large, the resulting combinatorial explosion means that identifying large simplices can be difficult.

To define a simplicial complex, fix a grouping scale ϵ, and identify k simplices as subsets of $k+1$ points that are all within ϵ of each other. The resulting set of simplices S^ϵ is called the Vietoris–Rips complex. The form of the simplicial complex S^ϵ depends on the scale ϵ at which its points are grouped together: persistent homology investigates how topological invariants of the simplicial complex depend on the scale ϵ. The collection of simplicial complexes $\{S^\epsilon\}$ for different values of the grouping scale ϵ is called a filtration. Note that if a simplex belongs to the complex S^ϵ, then it also belongs to $S^{\epsilon'}$, $\epsilon' > \epsilon$. That is, the filtration consists of a sequence of nested simplicial complexes. When ϵ is sufficiently small, only the zero-simplices (points) lie in the complex. As ϵ increases, one and two simplices (edges and triangles) enter the complex, followed by higher order simplices. As ϵ continues to increase, topological features such as holes, gaps and voids come into existence, and then are eventually filled in. For sufficiently large ϵ, all possible simplices are contained in the complex.

Now construct quantum states that correspond to the simplicial complex. Encode simplices as quantum states over n qubits with 1 s at the positions of the vertices. We designate the k-simplex s_k by the n-qubit basis vector $|s_k\rangle \in C^{2^n}$. Denote the $\binom{n}{k+1}$ dimensional Hilbert space corresponding to all possible k simplices by W_k. Let $\mathcal{H}_k^\epsilon$ be the subspace of W_k spanned by $|s_k\rangle$ where $s_k \in S_k^\epsilon$, the set of k simplices in S^ϵ. The full simplex space at scale ϵ is defined to be $\mathcal{H}^\epsilon = \oplus_k H_k^\epsilon$. Assume that the distances between pairs of points are either given by a quantum algorithm or stored in qRAM (see Methods section). The ability to evaluate distances translates onto the ability to apply the projector P_k^ϵ that projects onto the k-simplex space $\mathcal{H}_k^\epsilon$ and the projector P^ϵ that projects onto the full simplex space $\mathcal{H}^\epsilon$.

Table 1 | Computational cost comparison.

Procedural steps	Classical cost	Quantum cost
Input pairwise distances, n points	$O(n^2)$ bits	$O(n^2)$ bits
Construct simplicial complex	$O(2^n)$ ops	$O(n^2)$ ops on $O(n)$ qubits
Diagonalize Laplacian/find Betti numbers	$O(2^{2n}\log(1/\delta))$ ops	$O(n^5/\delta)$ quantum ops

δ is the multiplicative accuracy to which the Betti numbers and the eigenvalues of the combinatorial Laplacian are determined. Note the trade-off between the exponential quantum speed-up and accuracy: the quantum algorithms obtain an exponential speed-up over classical algorithms but provide an accuracy that scales polynomially in $1/\delta$ rather than exponentially. This feature arises from the nature of the quantum phase estimation/matrix inversion algorithms, which obtain their exponential speed-up by estimating eigenvectors and eigenvalues using a 'pointer-variable' measurement interaction[38–40]. By contrast, classical algorithms need only keep $O(\log(1/\delta))$ bits of precision, but must perform $O(2^{2n})$ steps to diagonalize $2^n \times 2^n$ sparse matrices.

Grover's algorithm can then be used to construct the k-simplex state

$$|\psi\rangle_k^\epsilon = \frac{1}{\sqrt{|S_k^\epsilon|}} \sum_{s_k \in S_k^\epsilon} |s_k\rangle, \tag{1}$$

where as above S_k^ϵ is the set of k simplices in the complex at scale ϵ. That is, $|\psi\rangle_k^\epsilon$ is the uniform superposition of the quantum states corresponding to k simplices in the complex. For each simplex s_k, we can verify whether $s_k \in S_k^\epsilon$ in $O(k^2)$ steps. That is, we can implement a membership function $f_k^\epsilon(s_k) = 1$ of $s_k \in S_k^\epsilon$ in $O(k^2)$ steps. The multi-solution version of Grover's algorithm then allows us to construct the k-simplex state of equation (1).

The construction of the k-simplex state via Grover's algorithm reveals the number of k simplices $|S_k^\epsilon| = \dim H_k^\epsilon$ in the complex at scale ϵ, and takes time $O\left(n^2 (\zeta_k^\epsilon)^{-1/2}\right)$, where $\zeta_k^\epsilon = |S_k^\epsilon| / \binom{n}{k+1}$ is the fraction of possible k simplices that are actually n the complex at scale ϵ. When this fraction is too small, the quantum search procedure will fail to find the simplices. For $k \ll n$, we have $\binom{n}{k+1} = O(n^{k+1}/k!)$, and ζ_k^ϵ is only polynomially small in n. By contrast, for $k \approx n$, ζ_k^ϵ can be exponentially small in n: if only an exponentially small set of possible simplices actually lie in the complex, quantum search will fail to find them. For the purposes of performing the quantum algorithm, we fix a parameter ζ that determines the accuracy to which we wish to determine the simplex state, and run the simplex finding algorithm for a time $\zeta^{-1/2}$. At each grouping scale ϵ, the algorithm will find k simplices when $\zeta_k^\epsilon > \zeta$, and estimate the number of k simplices to accuracy $\zeta_k^\epsilon \pm \zeta$. As ϵ increases, more and more simplices enter into the complex; ζ_k^ϵ increases; and quantum search will succeed in constructing the simplex state to greater and greater accuracy. When ϵ becomes larger than the maximum distance between vectors, all simplices are in the complex.

Below, it will prove useful to have, in addition to the simplex state $|\psi\rangle_k^\epsilon$ the state $\rho_k^\epsilon = (1/|S_k^\epsilon|) \sum_{s_k \in S_k^\epsilon} |s_k\rangle\langle s_k|$, which is the uniform mixture of all k-simplex states in the complex at grouping scale ϵ. ρ_k^ϵ can be constructed in a straightforward fashion from the simplex state $|\psi\rangle_k^\epsilon$ by adding an ancilla and copying the simplex label to construct the state $\frac{1}{\sqrt{|S_k^\epsilon|}} \sum_{s_k \in S_k^\epsilon} |s_k\rangle \otimes |s_k\rangle$. Tracing out the ancilla then yields the desired uniform mixture over all k simplices.

In summary, we can represent the the simplicial complex in quantum mechanical form using exponentially fewer bits than that are required classically. Indeed, the quantum search method for constructing simplicial states works best when ζ_k^ϵ is not too small, so that a substantial fraction of simplices that could be in the complex are actually in the complex. But this regime is exactly the regime where the classical algorithms require an exponentially large amount of memory space bits merely to record which simplices are in the complex. Now we show how to act on this quantum mechanical representation of the filtration to reveal persistent homology.

Topological analysis. Having constructed a quantum state that represents the simplicial complex S^ϵ at scale ϵ, we use quantum information processing to analyse its topological properties. In algebraic topology in general, and in persistent homology in particular, this analysis is performed by investigating the properties of linear maps on the space of simplices. As above, let $\mathcal{H}_k^\epsilon$ be the Hilbert space spanned by vectors corresponding to k simplices in the complex at level ϵ. We identify the vector space $\mathcal{H}_k^\epsilon$ with the abelian group C_k (the kth chain group) under addition of vectors in the space. Let $j_0 \ldots j_k$ be the vertices of s_k. Define the boundary map ∂_k on the space of k simplices by

$$\partial_k |s_k\rangle = \sum_\ell (-1)^\ell |s_{k-1}(\ell)\rangle \tag{2}$$

where as above $s_{k-1}(\ell)$ is the $k-1$ simplex on the boundary of s_k with vertices $j_0 \ldots \hat{j}_\ell \ldots j_k$ obtained by omitting the ℓth vertex j_ℓ from s_k. The boundary map maps each simplex to the oriented sum of its boundary simplices. ∂_k is a $\binom{n}{k} \times \binom{n}{k+1}$ matrix with $n-k$ non-zero entries ± 1 in each row and $k+1$ non-zero entries ± 1 per column. Note that $\partial_k \partial_{k+1} = 0$: the boundary of a boundary is zero. As defined, ∂_k acts on the space of all k simplices. We define the boundary map restricted to operate from $\mathcal{H}_k^\epsilon$ to H_{k-1}^ϵ to be $\tilde{\partial}_k = \partial_k P_k^\epsilon$, where as above P_k^ϵ is the projector onto the space of k simplices in the complex at scale ϵ.

The kth homology group $\mathbf{H}_k$ is the quotient group, Ker $\tilde{\partial}_k/\text{Image}_{k+1}\tilde{\partial}_{k+1}$, the kernel of $\tilde{\partial}_k$ divided by the image of $\tilde{\partial}_{k+1}$ acting on $\mathcal{H}_{k+1}$at grouping scale ϵ. The kth Betti number β_k is equal to the dimension of $\mathbf{H}_k$, which in turn is equal to the dimension of the kernel of $\tilde{\partial}_k$ minus the dimension of the image of $\tilde{\partial}_{k+1}$.

The strategy that we use to identify persistent topological features operates by identifying the singular values and singular vectors of the boundary map. Connected components, holes, voids and so on, correspond to structures—chains of simplices—that have no boundary, but that are not themselves a boundary. That is, we are looking for the set of states that lie within the kernel of $\tilde{\partial}_k$, but that do not lie within the image of $\tilde{\partial}_{k+1}$. The ability to decompose arbitrary vectors in $\mathcal{H}_k^\epsilon$ in terms of these kernels and images allows us to identify Betti numbers at different grouping scales ϵ.

The quantum phase algorithm[38–40] allows one to decompose states in terms of the eigenvectors of an Hermitian matrix and to find the associated eigenvalues. Once the k-simplex states $|\psi\rangle_k^\epsilon$ have been constructed, the quantum phase algorithm allows one to decompose those states in terms of eigenvectors and eigenvalues of the boundary map. The boundary map is not Hermitian. We embed the boundary map $\tilde{\partial}_k$ into a Hermitian matrix B_k^ϵ defined by

$$B_k^\epsilon = \begin{pmatrix} 0 & \tilde{\partial}_k \\ \tilde{\partial}_k^\dagger & 0 \end{pmatrix}. \tag{3}$$

B_k^ϵ acts on the space $\mathcal{H}_{k-1}^\epsilon \oplus \mathcal{H}_k^\epsilon$. Note that B_k^ϵ is n-sparse: there are either k or $n-k$ entries per row. Similarly, define the full Hermitian boundary map to be

$$B^\epsilon = B_1^\epsilon \oplus B_2^\epsilon \oplus \ldots \oplus B_n^\epsilon. \tag{4}$$

B^ϵ is also n-sparse. Because $\tilde{\partial}_k \tilde{\partial}_{k+1} = 0$, we have $B^{\epsilon 2} = \Delta_0 \oplus \Delta_1 \oplus \ldots \oplus \Delta_n$, where $\Delta_k = \tilde{\partial}_k^\dagger \tilde{\partial}_k + \tilde{\partial}_{k+1} \tilde{\partial}_{k+1}^\dagger$ is the combinatorial Laplacian of the kth simplicial complex[22–24]. Because $(B^\epsilon)^2$ is the sum of the combinatorial Laplacians, B^ϵ is sometimes called the 'Dirac operator', since the original Dirac operator was the square root of the Laplacian. Explicit matrix forms of the Dirac operator and the combinatorial Laplacian are given in the Methods section. Hodge theory[9–12,14–24] implies that the kth homology group satisfies $\mathbf{H}_k = \text{Ker } \tilde{\partial}_k/\text{Image}_{k+1}\tilde{\partial}_{k+1} \cong \text{Ker } \Delta_k$. The dimension of this kernel is the kth Betti number.

To find the dimension of the kernel, apply the quantum phase algorithm[38–40] to B^ϵ starting from the uniform mixture of simplices ρ^ϵ. The quantum phase algorithm decomposes this state into the eigenvectors of the combinatorial Laplacian, and identifies the corresponding eigenvalues. The probability of yielding a particular eigenvalue is proportional to the dimension of the corresponding eigenspace. As above, classical algorithms for finding the eigenvalues and eigenvectors of the combinatorial Laplacians Δ_k, and calculating the dimension of the eigenspaces takes $O\left(\binom{n}{k}^2\right) \sim O(2^{2n})$ computational steps using sparse matrix diagonalization via Gaussian elimination or the Lanczos algorithm. On a quantum computer, however, the quantum phase algorithm[38–40] can project the simplex states $|\psi\rangle_k^\epsilon$ onto the eigenspaces of the Dirac operator B^ϵ and find corresponding eigenvalues to accuracy δ in time $O(n^5\delta^{-1}\zeta^{-1/2})$, where as above ζ is the accuracy to which we choose to construct the simplex state. The factor of n^5 arises because the quantum phase algorithm applied to an n-sparse matrix requires time n^3/δ^{-1}: the extra factor of n^2 arises because it takes time $O(k^2)$ to evaluate the projector P_k^ϵ onto the subspace of k simplices.

The algorithm also identifies the dimension of the eigenspaces of the Dirac operator and combinatorial Laplacian in time $O(n^5\delta^{-1}\zeta^{-1/2}\eta_\ell^{-1/2})$, where η_ℓ is equal to the dimension d_ℓ of the ℓth eigenspace divided by $|S|_k$, the dimension of the k-simplex space. The kth Betti number β_k is equal to the dimension of the kernel of Δ_k. The algorithm allows us to construct the full decomposition of the simplicial complex in terms of eigenvectors and eigenvalues of the combinatorial Laplacian, yielding useful geometric information such as harmonic forms. Monitoring how the eigenvalues and eigenspaces of the combinatorial Laplacian change as ϵ changes provides geometric information about how various topological features such as connected components, holes and voids come into existence and disappear as the grouping scale changes[16,17,44].

Discussion

This paper extended methods of quantum machine learning to topological data analysis. Homology is a powerful topological tool. The representatives of the homology classes for different k define the connected components of the simplicial complex, holes, voids and so on. The Betti numbers count the number of connected components, holes, voids and so on. Varying the simplicial scale ϵ and tracking how Betti numbers change as function of ϵ reveals how topological features come into existence and go away as the data is analysed at different length scales. Our algorithm also reveals how the structure of the eigenspaces and eigenvalues of the combinatorial Laplacian changes as a function of ϵ. This 'persistent geometry' reveals features of the data such as rate of change of harmonic forms over different simplicial scales.

The underlying methods of our quantum algorithms are similar to those in other big quantum data algorithms[19–21]. The primary difference between the topological and geometrical algorithms presented here, and algorithms for, for example, constructing clusters[19], principal components[20], and support vector machines[21], is that our topological algorithms require only a small qRAM of size $O(n^2)$. Consequently, even when the full qRAM resources are included in the accounting of the computational complexity of the algorithms, the topological algorithms require only an amount of computational resources polynomical in the number of data points, while the best existing classical algorithms for answering the same questions require exponential resources.

To recapitulate the steps of the algorithm: First, the quantum data is processed using standard techniques of quantum computation: distances between points are evaluated, simplices of neighbouring points are identified, and a simplicial complex is constructed. The simplicial complex depends on the grouping scale ϵ. We construct a quantum state that represents the filtration of the complex—the set of simplicial complexes, related by inclusion, for different ϵ. This quantum state contains exponentially fewer qubits than the number of bits required to describe the classical filtration of the complex. Second, we use the quantum phase algorithm[38–40] to calculate the eigenvalues and to construct the eigenspaces of the combinatorial Laplacian at each scale ϵ. The dimension of the kernel of the combinatorial Laplacian for k simplices is the kth Betti number. In addition, this construction gives us geometric information about the data set.

Classical algorithms for performing the full persistent homology over a space with n points over all scales k take time $O(2^{2n})$: there are 2^n possible simplices, and evaluating kernels and images of the boundary map via Gaussian elimination for sparse matrices takes time that goes as the square of the dimension of the space of simplices. By contrast, the quantum algorithm for constructing the Betti numbers and for decomposing the simplicial complex in terms of eigenvalues and eigenvectors of the combinatorial Laplacian takes time $O(n^5)$, compared with $O(2^{2n})$ for classical algorithms. The eigenvectors of the kernels of the combinatorial Laplacian are related to the representatives of the kth homology class via a boundary term. How to extend the quantum algorithms given here to construct the full barcode of persistent homology and to construct the representatives of the homology class directly is an open question. It would also be interesting to extend the quantum algorithmic methods developed here to further algebraic and combinatorial problems, for example, Morse theory.

Methods

Overview. In this section we provide further details of distance evaluation, simplex state construction, and the form of the Dirac operator and the combinatorial Laplacian.

State preparation and distance evaluation. Topological analysis of the data requires distances between data points. Assume that the data set contains n points together with the $n(n-1)/2$ distances between them. The data is stored in qRAM or qRAM[35–37], so that the algorithm can access the data in quantum parallel. The essential feature of a qRAM is that it preserves quantum coherence: the qRAM maps a quantum superposition of inputs $\sum_j \alpha_j|j\rangle|0\rangle$ to a quantum superposition of outputs $\sum_j \alpha_j|j\rangle|v_j\rangle$. Note that a quantum RAM is potentially significantly easier to construct than a full-blown quantum computer. The storage medium of a quantum RAM can be essentially classical: indeed, a single photon reflected off a compact disk encodes in its quantum state all the bits of information stored in the mirrors on the disk. In addition to a classical storage medium such as a CD, a qRAM contains quantum switches that can be opened in quantum superposition to access that information in quantum parallel. Each call to an N-bit qRAM requires $\log_2 N$ quantum operations. Quantum RAMS have been designed, and prototypes have been constructed[35–37]. In contrast to other big quantum data algorithms[31–33], the size of the qRAM required to perform topological and geometric analysis is relatively small: because the computational complexity of classical algorithms for persistent homology scales as $O(2^{2n})$, while the quantum algorithms require only $O(n^2)$ bits worth of qRAM, a significant quantum advantage could be obtained by a qRAM with hundreds to thousands of bits.

As an alternative to being presented with the pre-calculated distances, the data set could consist of n d-dimensional vectors $\{\vec{v}_j\}$ over the complex numbers, and we can use the qRAM to construct the distances $|\vec{v}_i - \vec{v}_j|$ between the ith and jth vectors[31]. Finally, the distances can be presented as the output of a quantum computation. In all cases, our quantum algorithms for topological and geometric analysis operate by accessing the distances in quantum parallel. Big quantum data analysis works by mapping each vector $\vec{v}_j$ to a quantum state $|v_j\rangle \in C^d$, and the entire database to a quantum state $(1/\sqrt{n})\sum_j |j\rangle|v_j\rangle \in C^n \otimes C^d$. A quantum RAM can be queried in quantum parallel: given an input state $|j\rangle|0\rangle$, it produces the output state $|j\rangle|v_j\rangle$, where $|v_j\rangle$ is normalized quantum state proportional to the vector $\vec{v}_j$. Such a quantum state can be encoded using $O(\log_2(nd))$ quantum bits, and $|\vec{v}_j|$ is the norm of the vector.

If we have not been given the $n(n-1)/2$ distances directly in qRAM, the next ingredient of the quantum algorithm is the ability to evaluate inner products and distances between vectors. In refs 20,31–33 it is shown how the access to vectors in quantum superposition: the ability to create the quantum states corresponding to the vectors translates into the ability to estimate

$\left|\vec{v}_i - \vec{v}_j\right|^2 = 2 - \vec{v}_i^\dagger \vec{v}_j - \vec{v}_j^\dagger \vec{v}_i$. That is, we can construct a quantum circuit that takes as input the state $|i\rangle|j\rangle|0\rangle$ and produces as output the state $|i\rangle|j\rangle\left|\left|\vec{v}_i - \vec{v}_j\right|^2\right\rangle$, where the third register contains an estimate of the distance between $\vec{v}_i$ and $\vec{v}_j$. To estimate the distance to accuracy δ takes $O(\delta^{-1})$ quantum memory calls and $O(\delta^{-1}(\log_2(nd))^2)$ quantum operations. As with the qRAM, the circuit to evaluate distances operates in quantum parallel.

Simplex state construction. To elucidate the construction of the k-simplex states (1), we look more closely into the implementation of Grover's algorithm to understand when it succeeds in constructing the k-simplex state, and how it fails. Start from a superposition $n^{-1/2}\sum_k |k\rangle$ over all values of k. Performing simplex construction in parallel via Grover's algorithm with the membership function f_k^ϵ yields the full simplex state at scale ϵ:

$$|\Psi\rangle^\epsilon = \frac{1}{\sqrt{n}}\sum_k |k\rangle|\psi\rangle_k^\epsilon. \tag{5}$$

By adding ancillae as above, we can also construct the uniform mixture over all values of k and all k simplices: $\rho^\epsilon = (1/n)\sum_k |k\rangle\langle k| \otimes \rho_k^\epsilon$. More precisely, if we run the quantum search procedure for a time $\zeta^{-1/2}$, we will obtain the state

$$|\Psi\rangle_\zeta^\epsilon = \frac{1}{\sqrt{n}}\left(\sum_{k:\zeta_k^\epsilon \geq \zeta} |k\rangle|\psi\rangle_k^\epsilon + \sum_{k:\zeta_k^\epsilon < \zeta} |k\rangle|0\rangle\right) \tag{6}$$

that contains the simplex states $|\psi\rangle_k^\epsilon$ for which $\zeta_k^\epsilon \geq \zeta$ and which returns a null result $|0\rangle$ for the simplex states for which $\zeta_k^\epsilon < \zeta$. For small ϵ—where only a small fraction of all possible simplices lie within the complex—and fixed ζ, the simplex state $|\Psi\rangle_\zeta^\epsilon$ will contain the actual simplex states $|\psi\rangle_k^\epsilon$ only for small k. As ϵ becomes larger and larger, higher and higher k-simplex states enter the filtration and $|\Psi\rangle_\zeta^\epsilon$ will contain more and more of the k-simplex states.

Constructing the simplex state in quantum parallel at m different grouping scales ϵ_i yields the filtration state

$$|\Phi\rangle_\zeta = \frac{1}{\sqrt{mn}}\sum_i |\epsilon_i\rangle|\Psi\rangle_\zeta^\epsilon. \tag{7}$$

The filtration state $|\Phi\rangle_\zeta$ contains the entire filtration of the simplicial complex in quantum superposition. The quantum filtration state contains exponentially fewer quantum bits than the number of classical bits required to describe the classical filtration of the complex: $\log m$ qubits are required to register the grouping scale ϵ, and n qubits are required to label the simplices. $|\Phi\rangle_\zeta$ takes time $O(\zeta^{-1/2}n^2\log(m))$ to construct. By contrast, a classical description of the filtration of the simplicial complex requires $O(2^n)$ bits.

Explicit form of the Dirac operator and simplicial Laplacian. Here we present the full matrix form of the Dirac operator B^ϵ and the combinatorial Laplacian $(B^\epsilon)^2$. The Dirac operator is

$$B^\epsilon = \begin{pmatrix} 0 & \tilde{\partial}_1 & 0 & & & & \\ \tilde{\partial}_1^\dagger & 0 & \tilde{\partial}_2 & & \cdots & & \\ 0 & \tilde{\partial}_2^\dagger & 0 & & & & \\ & & & \cdots & & & \\ & & & 0 & \tilde{\partial}_{n-1} & 0 \\ & \cdots & & \tilde{\partial}_{n-1}^\dagger & 0 & \tilde{\partial}_n \\ & & & 0 & \tilde{\partial}_n^\dagger & 0 \end{pmatrix}, \tag{8}$$

where as above $\tilde{\partial}_k = P_{k-1}^\epsilon \partial_k P_k^\epsilon$ is the boundary map confined to the simplicial subspace $\mathcal{H}^\epsilon$. It is straightforward to verify that the Dirac operator is n-sparse.

The combinatorial Laplacian is obtained by squaring the Dirac operator:

$$(B^\epsilon)^2 = \begin{pmatrix} \tilde{\partial}_1\tilde{\partial}_1^\dagger & 0 & 0 & & & \\ 0 & \tilde{\partial}_1^\dagger\tilde{\partial}_1 + \tilde{\partial}_2\tilde{\partial}_2^\dagger & 0 & & \cdots & \\ 0 & 0 & \tilde{\partial}_2^\dagger\tilde{\partial}_2 + \tilde{\partial}_3\tilde{\partial}_3^\dagger & & & \\ & & \cdots & & & \\ & \cdots & & \tilde{\partial}_{n-1}^\dagger\tilde{\partial}_{n-1} + \tilde{\partial}_n\tilde{\partial}^\dagger & 0 \\ & & & 0 & \tilde{\partial}_n^\dagger\tilde{\partial}_n \end{pmatrix}. \tag{9}$$

The quantum algorithm operates by diagonalizing the Dirac operator.

References

1. Zomorodian, A. & Carlsson, G. Computing persistent homology. *Discret. Comput. Geom.* **33,** 249–274 (2005).
2. Robins, V. Towards computing homology from finite approximations. *Topol. Proc.* **24,** 503–532 (1999).
3. Frosini, P. & Landi, C. Size theory as a topological tool for computer vision. *Pattern Recognit. Image Anal.* **9,** 596–603 (1999).
4. Carlsson, G., Zomorodian, A., Collins, A. & Guibas, L. Persistence barcodes for shapes. *Int. J. Shape Model.* **11,** 149–188 (2005).
5. Edelsbrunner, H., Letscher, D. & Zomorodian, A. Topological persistence and simplification. *Discret. Comput. Geom.* **28,** 511–533 (2002).
6. Zomorodian, A. in *Algorithms and Theory of Computation Handbook* 2nd edn Ch. 3, section 2 (Chapman and Hall/CRC, 2009).
7. Chazal, F. & Lieutier, A. Stability and computation of topological invariants of solids in R^n. *Discret. Comput. Geom.* **37,** 601–617 (2007).
8. Cohen-Steiner, D., Edelsbrunner, H. & Harer, J. Stability of persistence diagrams. *Discret. Comput. Geom.* **37,** 103–120 (2007).
9. Basu, S. On bounding the Betti numbers and computing the euler characteristic of semi-algebraic sets. *Discret. Comput. Geom.* **22,** 1–18 (1999).
10. Basu, S. Different bounds on the different Betti numbers of semi-algebraic sets. *Discret. Comput. Geom.* **30,** 65–85 (2003).
11. Basu, S. Computing the top Betti numbers of semi-algebraic sets defined by quadratic inequalities in polynomial time. *Found. Comput. Math.* **8,** 45–80 (2008).
12. Basu, S. Algorithms in real algebraic geometry: a survey. Preprint at http://arxiv.org/abs/1409.1534 (2014).
13. Friedman, J. Computing Betti numbers via combinatorial Laplacians. in *Proceedings of the 28th Annual ACM Symposium on Theory of Computing,* 386–391 (Atlanta, Georgia, 1996).
14. Hodge, W. V. D. *The Theory and Applications of Harmonic Integrals* (Cambridge University Press, 1941).
15. Munkrees, J. R. *Elements of Algebraic Topology* (Benjamin/Cummings, 1984).
16. Butler, S. & Chung, F. Small spectral gap in the combinatorial Laplacian implies Hamiltonian. *Ann. Comb.* **13,** 403–412 (2010).
17. Maletić, S. & Rjković, M. Combinatorial Laplacian and entropy of simplicial complexes associated with complex networks. *Eur. Phys. J. Spec. Top.* **212,** 77–97 (2012).
18. Niyogi, P., Smale, S. & Weinberger, S. A topological view of unsupervised learning from noisy data. *SIAM J. Comput.* **40,** 646–663 (2011).
19. Kozlov, D. *Algorithms and Computation in Mathematics* Vol. 21 (Springer, 2008).
20. Ghrist, R. Barcodes: the persistent topology of data. *Bull. Am. Math. Soc.* **45,** 61–75 (2008).
21. Harker, S., Mischaikow, K., Mrozek, M. & Nanda, V. Discrete Morse theoretic algorithms for computing homology of complexes and maps. *Found. Comput. Math.* **14,** 151–184 (2014).
22. Mischaikow, K. & Nanda, V. Morse theory for filtrations and efficient computation of persistent homology. *Discret. Comput. Geom.* **50,** 330–353 (2013).
23. CHOMP. Computational homology project. http://chomp.rutgers.edu.
24. CAPD::RedHom: Reduction homology algorithms. http://redhom.ii.uj.edu.pl/.
25. Servedio, R. A. & Gortler, S. J. Equivalences and separations between quantum and classical learnability. *SIAM J. Comput.* **33,** 1067 (2004).
26. Hentschel, A. & Sanders, B. C. Machine learning for precise quantum measurement. *Phys. Rev. Lett.* **104,** 063603 (2010).
27. Neven, H., Denchev, V. S., Rose, G. & Macready, W. G. Training a large scale classifier with the quantum adiabatic algorithm. Preprint at http://arxiv.org/abs/0912.0779 (2009).
28. Pudenz, K. L. & Lidar, D. A. Quantum adiabatic machine learning. *Quantum Inf. Process* **12,** 2027 (2013).
29. Anguita, D., Ridella, S., Rivieccion, F. & Zunino, R. Quantum optimization for training support vector machines. *Neural Netw.* **16,** 763–770 (2003).
30. Aïmeur, E., Brassard, G. & Gambs, S. Quantum speed-up for unsupervised learning. *Mach. Lear.* **90,** 261–287 (2013).
31. Lloyd, S., Mohseni, M. & Rebentrost, P. Quantum algorithms for supervised and unsupervised machine learning. Preprint at http://arxiv.org/abs/1307.0411 (2013).
32. Rebentrost, P., Mohseni, M. & Lloyd, S. Quantum support vector machine for big feature and big data classification. *Phys. Rev. Lett.* **113,** 130503 (2014).
33. Lloyd, S., Mohseni, M. & Rebentrost, P. Quantum principal component analysis. *Nat. Phys.* **10,** 631–633 (2014).
34. Aaronson, S. Read the fine print. *Nat. Phys.* **11,** 291–293 (2015).
35. Giovannetti, V., Lloyd, S. & Maccone, L. Quantum random access memory. *Phys. Rev. Lett.* **100,** 160501 (2008).
36. Giovannetti, V., Lloyd, S. & Maccone, L. Architectures for a quantum random access memory. *Phys. Rev. A* **78,** 052310 (2008).
37. De Martini, F. *et al.* Experimental quantum private queries with linear optics. *Phys. Rev. A* **80,** 010302 (2009).
38. Yu. Kitaev, A., Shen, A. H. & Vyalyi, M. N. *Classical and Quantum Computation, Graduate Studies in Mathematics* Vol. 47 (publications of the American Mathematical Society, 2004).
39. Abrams, D. S. & Lloyd, S. A quantum algorithm providing exponential speed increase for finding eigenvalues and eigenvectors. *Phys. Rev. Lett.* **83,** 5162–5165 (1999).

40. Nielsen, M. S. & Chuang, I. L. *Quantum Computation and Quantum Information* (Cambridge University Press, 2000).
41. Harrow, A. W., Hassidim, A. & Lloyd, S. Quantum algorithm for solving linear systems of equations. *Phys. Rev. Lett.* **15,** 150502 (2009).
42. Scheiblechner, P. On the complexity of deciding connectedness and computing Betti numbers of a complex algebraic variety. *J. Complex.* **23,** 359–379 (2007).
43. Kac, M. Can one hear the shape of a drum? *Am. Math. Mon.* **73,** 1–23 (1966).
44. Sadakane, K., Sugawara, N. & Tokuyama, T. Quantum computation in computational geometry. *Interdisc. Inf. Sci.* **8,** 129–136 (2002).

Acknowledgements

We thank Mario Rasetti for suggesting the topic of topological analysis of big data. We acknowledge helpful conversations with Patrick Rebentrost, Barbara Terhal and Francesco Vaccarino. S.L. was supported by ARO, AFOSR, DARPA and Jeffrey Epstein. P.Z. was supported by ARO MURI grant W911NF-11-1-0268 and by NSF grant PHY-969969.

Author contributions

All authors contributed to the problem formulation, quantum algorithm design and error analysis.

Additional information

Competing financial interests: The authors declare no competing financial interests.

19

Dynamic information routing in complex networks

Christoph Kirst[1,2,3,4,5], Marc Timme[1,3,4] & Demian Battaglia[6]

Flexible information routing fundamentally underlies the function of many biological and artificial networks. Yet, how such systems may specifically communicate and dynamically route information is not well understood. Here we identify a generic mechanism to route information on top of collective dynamical reference states in complex networks. Switching between collective dynamics induces flexible reorganization of information sharing and routing patterns, as quantified by delayed mutual information and transfer entropy measures between activities of a network's units. We demonstrate the power of this mechanism specifically for oscillatory dynamics and analyse how individual unit properties, the network topology and external inputs co-act to systematically organize information routing. For multi-scale, modular architectures, we resolve routing patterns at all levels. Interestingly, local interventions within one sub-network may remotely determine nonlocal network-wide communication. These results help understanding and designing information routing patterns across systems where collective dynamics co-occurs with a communication function.

[1] Network Dynamics, Max Planck Institute for Dynamics and Self-Organization (MPIDS), Göttingen 37077, Germany. [2] Nonlinear Dynamics, Max Planck Institute for Dynamics and Self-Organization (MPIDS), Göttingen 37077, Germany. [3] Institute for Nonlinear Dynamics, Georg-August University Göttingen, Göttingen 37077, Germany. [4] Bernstein Center for Computational Neuroscience (BCCN), Göttingen 37077, Germany. [5] Center for Physics and Biology, The Rockefeller University, New York, New York 10065, USA. [6] Université Aix-Marseille, INSERM UMR 1106, Institut de Neurosciences des Systémes, Marseille 13005, France. Correspondence and requests for materials should be addressed to C.K. (email: ckirst@rockefeller.edu).

Attuned function of many biological or technological networks relies on the precise yet dynamic communication between their subsystems. For instance, the behaviour of cells depends on the coordinated information transfer within gene-regulatory networks[1,2] and flexible integration of information is conveyed by the activity of several neural populations during brain function[3]. Identifying general mechanisms for the routing of information across complex networks thus constitutes a key theoretical challenge with applications across fields, from systems biology to the engineering of smart distributed technology[4–6].

Complex systems with a communication function often show characteristic dynamics, such as oscillatory or synchronous collective dynamics with a stochastic component[7–11]. Information is carried in the presence of these dynamics within and between neural circuits[12,13], living cells[14,15], ecological or social groups[16,17] as well as technical communication systems, such as *ad hoc* sensor networks[18,19]. While such dynamics could simply reflect the properties of the interacting unit's, emergent collective dynamical states in biological networks can actually contribute to the system's function. For example, it has been hypothesized that the widely observed oscillatory phenomena in biological networks enable emergent and flexible information routing[12]. Yet, what are the precise mechanism by which collective dynamics contribute to organizing communication in networks? Moreover, how do the intrinsic dynamics of units, their interaction topology and function as well as external driving signals and noise create specific patterns of information routing?

In this article, we derive a theory for information routing in complex networked systems, revealing the joint impact of all these elements. We identify a generic mechanism to dynamically route information in complex networked systems by conveying information in fluctuations around a collective dynamical reference state. Propagation of information then depends on the underlying reference dynamics and switching between multiple stable states induces flexible rerouting of information, even if the physical network stays unchanged. For oscillatory dynamics, analytic predictions show precisely how the physical coupling structure, the units' properties and the dynamical state of the network co-act to generate a specific communication pattern, quantified by time-delayed mutual information (dMI)[20,21] and transfer entropy[22] curves between time-series of the network's units. Resorting to a collective phase description[23], our theory further resolves communication patterns at all levels of multi-scale, modular topologies[24,25], as ubiquitous, for example, in the brain connectome and biochemical regulatory networks[26–29]. Interestingly, local interventions within one sub-network may remotely modify information transfer between other seemingly unrelated sub-networks. Finally, a combinatorial number of information routing patterns (IRPs) emerge if several multi-stable subsystems are combined into a larger modular network. These relations between multi-scale connectivity, collective network dynamics and flexible information routing have potential applications in the reconstruction and design of gene-regulatory circuits[15,30], wireless communication networks[4,19] or to the analysis of cognitive functions[31–35], among others. Moreover, these results offer generic insights into mechanisms for flexible and self-organized information routing in complex networked systems.

Results

Information routing via collective dynamics. To better understand how collective dynamics may contribute to specifically distribute bits of information from external or locally computed signals through a network or to it s downstream components we first consider a generic stochastic dynamical system. It evolves in time t according to

$$\frac{d}{dt}\boldsymbol{x} = \boldsymbol{f}(\boldsymbol{x}) + \boldsymbol{\xi} \qquad (1)$$

where $\boldsymbol{x} = (x_1, \ldots, x_N)$ denotes the variables of the network nodes, $\boldsymbol{f}$ describes the intrinsic dynamics of the network. The key premise is that the information to be routed through the network is carried in the stochastic external input $\boldsymbol{\xi} = (\xi_1, \ldots, \xi_N)$ driving instantaneous state variable fluctuations. To access the role of collective dynamics in routing this information we consider a deterministic intrinsic reference state $\boldsymbol{x}^{(\mathrm{ref})}(t)$ solving (2) in the absence of signals ($\boldsymbol{\xi} = 0$).

We use information theoretic measures that quantify the amount of information shared and transferred between nodes, independent of how this information is encoded or decoded. More precisely, we measure information sharing between signal $x_i(t)$ and the time d lagged signal $x_j(t+d)$ of nodes i and j in the network via the time-delayed mutual information (dMI)[20,21]

$$\mathrm{dMI}_{i,j}(d) = \iint p_{i,j^{(d)}}(t) log\left(\frac{p_{i,j^{(d)}}(t)}{p_i(t)p_j(t)}\right) \mathrm{d}x_i(t)\mathrm{d}x_j(t+d) \qquad (2)$$

Here $p_i(t)$ is the probability distribution of the variable $x_i(t)$ of unit i at time t and $p_{i,j^{(d)}}(t)$ the joint distribution of $x_i(t)$ and the variable $x_j(t+d)$ lagged by d. As a second measure, we use the delayed transfer entropy (dTE)[22] (cf. Methods) that genuinely measures information transfer between pairs of units[36]. Asymmetries in the curves $\mathrm{dMI}_{i,j}(d)$ and $\mathrm{dTE}_{i\to j}(d)$ then indicate the dominant direction in which information is shared or transferred between nodes (cf. Supplementary Note 1).

To identify the role of the underlying reference dynamical state $\boldsymbol{x}^{(\mathrm{ref})}(t)$ for network communication a small-noise expansion in the signals $\boldsymbol{\xi}$ turns a out to be ideally suited: while this expansion limits the analysis to the vicinity of a specific reference state which is usually regarded as a weakness of this technique, in the context of our study, this property is highly advantageous as it directly conditions the calculations on a particular dynamical state and enables us to extract it s role for the emergent pattern of information routing within the network. For white noise sources $\boldsymbol{\xi}$ this method yields general expressions for the conditional probabilities $p(\boldsymbol{x}(t+d)|\boldsymbol{x}(t))$ that depend on $\boldsymbol{x}^{(\mathrm{ref})}(t)$. Using this result the expressions for the dMI (2) and dTE (7) $\mathrm{dMI}_{i,j}(d)$ and $\mathrm{dTE}_{i\to j}(d)$ become a function of the underlying collective reference dynamical state (cf. Methods and Supplementary Note 2). The dependency on this reference state then provides a generic mechanism to change communication in networks by manipulation the underlying collective dynamics. In the following we show how this general principle gives rise to a variety of mechanisms to flexibly change information routing in networks. We focus on oscillatory phenomena widely observed in networks with a communication function[32,34,35,37,38].

Information exchange in phase signals. Oscillatory synchronization and phase locking[8,10] provide a natural way for the temporal coordination between communicating units. Key variables in oscillator systems are the phases $\phi_i(t)$ at time t of the individual units i. In fact, a wide range of oscillating systems display similar phase dynamics[8,11] (cf. Supplementary Note 3) and phase-based encoding schemes are common, for example, in the brain[32,34,35], genetic circuits[37] and artificial systems[38].

We first focus on systems in a stationary state with a stationary distribution for which the expressions for the dMI and dTE become independent of the starting time t and only depend on the lag d and reference state $\boldsymbol{\phi}^{(\mathrm{ref})}(t)$. To assess the dominant direction of the shared information between two nodes, we quantify asymmetries in the dMI curve by using the difference $\delta \mathrm{MI}_{i,j} = \mathrm{MI}_{i\to j} - \mathrm{MI}_{j\to j}$ between the integrated mutual informations $\mathrm{MI}_{i\to j} = \int_0^\infty \mathrm{dMI}_{i,j}(\tau)\mathrm{d}\tau$ and $\mathrm{MI}_{j\to i}$. If this is positive, information is shared predominantly from unit i to j, while negative values indicate the opposite direction. Analogously, we compute the differences in dTE as $\delta \mathrm{TE}_{i,j}$ (cf. Methods and Supplementary Note 1). The set of pairs $\{\delta \mathrm{MI}_{i,j}\}$ or $\{\delta \mathrm{TE}_{i,j}\}$ for all i, j then capture strength and directionality of information routing in the network akin to a functional connectivity analysis in neuroscience[39]. We refer to them as IRPs.

A range of networks of oscillatory units, with disparate physical interactions, connection topologies and external input signals support multiple IRPs. For instance, in a model of a gene-regulatory network with two oscillatory sub-networks

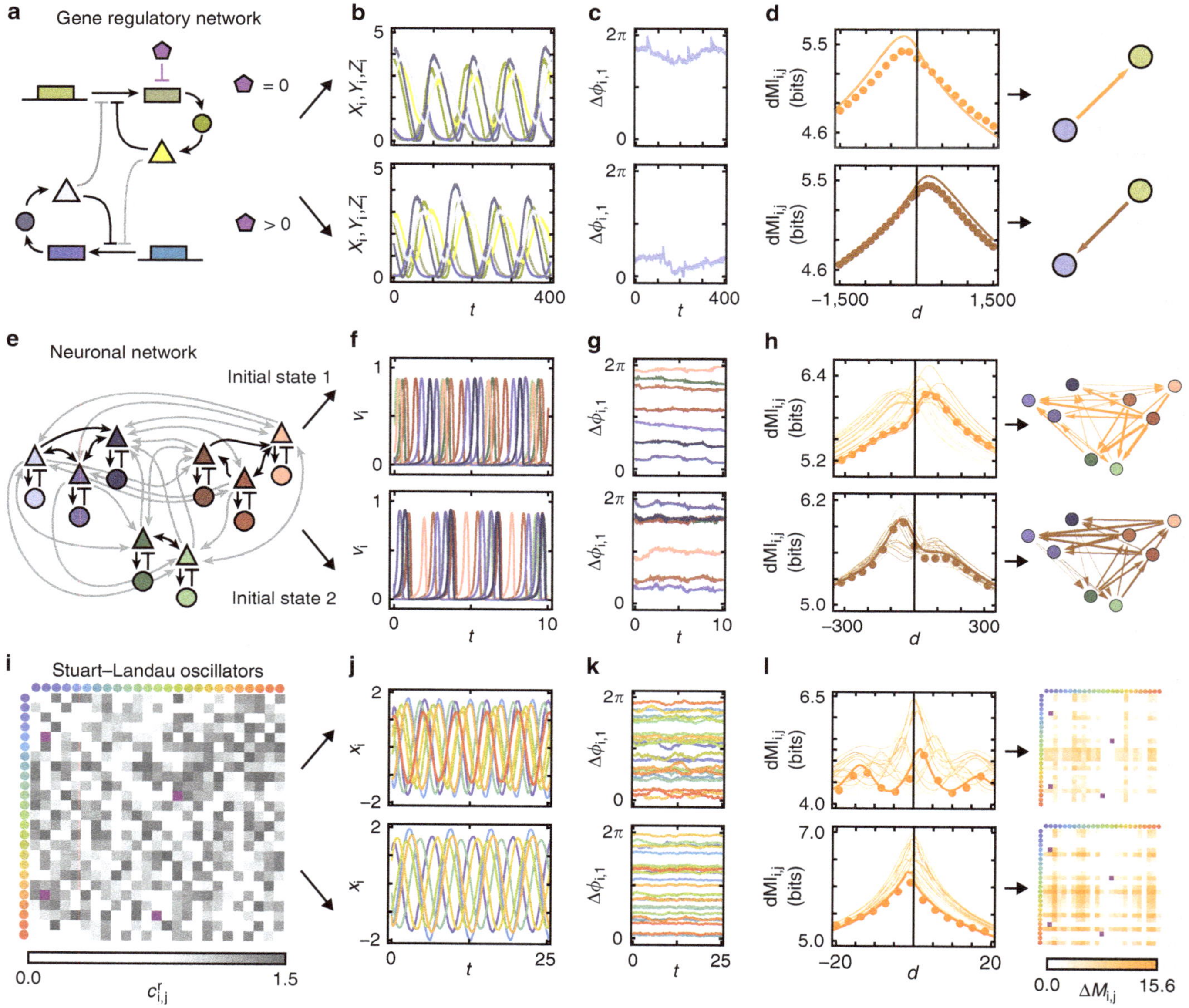

Figure 1 | Flexible information routing across networks. (**a**) Simple model of a gene-regulatory network of two coupled biochemical oscillators of Goodwin type (yellow and blue). An additional molecule (purple) degrades the transcribed mRNA in one of the oscillators and thereby changes its intrinsic frequency. Coupling strengths are gray coded (darker colour indicates stronger coupling), sharp arrows indicate activating and blunt arrows inhibiting influences (cf. Methods). (**b**) Stochastic oscillatory dynamics of the concentrations of the systems components x_i (mRNA), y_i (enzyme), and z_i (protein), $i \in \{1, 2\}$. (**c**) Fluctuations of the phases extracted from the full dynamics relative to a reference unit. (**d**) dMI ($\mathrm{dMI}_{1,2}$) between the phase signals. The numerical data (dots) agrees well with the theoretical prediction (4) (solid lines). The asymmetry in the dMI curves around $d = 0$ indicates a directed information sharing pattern summarized in the graphs (right). Arrow thickness indicates the strength of directed information sharing $\Delta \mathrm{MI}_{i,j}$ measured by the positively rectified differences of the areas below the $\mathrm{dMI}_{i,j}(d)$ curve for $d < 0$ and $d > 0$. (**e**-**h**) Same as in **a**-**d** but for a modular network of coupled neuronal sub-populations consisting each of excitatory (triangle) and inhibitory (disk) populations (Wilson-Cowan-type dynamics) that undergo neuronal oscillatory activity (cf. Methods). For the same network two different collective dynamical states accessed by different initial conditions give rise to two different information sharing patterns (**f**-**h** top versus bottom). (**f**) Oscillatory activities v_i of the $N = 8$ excitatory populations i. (**i**-**l**) As in **a**-**d** but for generic oscillators close to a Hopf bifurcation (Stuart-Landau oscillators) each described by two-dimensional normal-form coordinates (x_i, y_i) connected to a larger network of $N = 25$ oscillators with coupling coefficients $c^r_{i,j}$ (cf. Methods). In **i** and **l**, connectivity matrices are shown instead of graphs. Two different network-wide IRPs arise (top versus bottom in **j**-**l**) by changing a small number of connection weights (purple entries in **i** and **l**).

(Fig. 1a) dMI analysis reveals IRPs with different dominant directions (Fig. 1b–d, upper versus lower). The change is triggered by adding an external factor that degrades the transcribed mRNA in one of the oscillators and thereby changes its intrinsic frequency (see Methods). More complex changes in IRPs emerge in larger networks, possibly with modular architecture. In a network of interacting neuronal populations (Fig. 1e), different initial conditions lead to different underlying collective dynamical states. Switching between them induces complicated but specific changes in the IRPs (Fig. 1f–h). Different IRPs also emerge by changing a small number of connections in larger networks. Fig. 1i–l illustrates this for a generic system of coupled oscillators each close to a Hopf bifurcation.

In general, several qualitatively different options for modifying network-wide IRPs exist, all of which are relevant in natural and artificial systems: (i) changing the intrinsic properties of individual units (Fig. 1a–d); (ii) modifying the system connectivity (Fig. 1i–l); and (iii) selecting distinct dynamical states of structurally the same system (Fig. 1e–h).

Theory of phase information routing. To reveal how different IRPs arise and how they depend on the network properties and dynamics, we derive analytic expressions for the dMI and dTE between all pairs of oscillators in a network. We determine the phase of each oscillator i in isolation by extending its phase description to the full basin of attraction of the stable limit cycle[8,40]. For weak coupling, the effective phase evolution becomes

$$\frac{\mathrm{d}}{\mathrm{d}t}\phi_i = \omega_i + \sum_{j=1}^{N}\gamma_{i,j}(\phi_i - \phi_j) + \sum_{k=1}^{N}\varsigma_{i,k}\xi_k \qquad (3)$$

where ω_i is the intrinsic oscillation frequencies of node i and the coupling functions $\gamma_{i,j}(\cdot)$ depend on the phase differences only. The final sum in (3) models external signals as independent Gaussian white noise processes ξ_k and a covariance matrix $\varsigma_{i,k}$. The precise forms of $\gamma_{i,j}(\cdot)$ and $\varsigma_{i,k}$ generally depend on the specific system (Supplementary Note 3).

As visible from Fig. 1e–h, the IRP strongly depends on the underlying collective dynamical state. We therefore decompose the dynamics into a deterministic reference part $\phi_i^{(\mathrm{ref})}$ and a fluctuating component $\phi_i^{(\mathrm{fluct})}$. We focus on phase-locked configurations for the deterministic dynamics with constant phase offsets $\Delta\phi_{i,j}^{(\mathrm{ref})} = \phi_i^{(\mathrm{ref})} - \phi_j^{(\mathrm{ref})}$. We estimate the stochastic part $\phi_i^{(\mathrm{fluct})}$ via a small-noise expansion (Methods and Supplementary Note 4, Theorem 1) yielding a first-order approximation for the joint probabilities $p_{i,j^{(d)}}$. Using (2) together with the periodicity of the phase variables, we obtain the dMI

$$\mathrm{dMI}_{i,j}(d) = \frac{k_{i,j^{(d)}} I_1\left(k_{i,j^{(d)}}\right)}{I_0\left(k_{i,j^{(d)}}\right)} - \log\left(I_0\left(k_{i,j^{(d)}}\right)\right) \qquad (4)$$

between phase signals in coupled oscillatory networks; here, $I_n(k)$ is the nth modified Bessel function of the first kind, and $k_{i,j^{(d)}}$ is the inverse variance of a von Mises distributions ansatz for $p_{i,j^{(d)}}$. The system's parameter dependencies, including different inputs, local unit dynamics, coupling functions and interaction topologies are contained in $k_{i,j^{(d)}}$. By similar calculations we obtain analytical expressions for $\mathrm{dTE}_{i\to j}$ (Methods and Supplementary Note 4, Theorem 2). Our theoretical predictions well match the numerical estimates (Fig. 1d,h,l, see also Fig. 2c,d below and Supplementary Figs 1 and 2). For independent input signals ($\varsigma_{i,k} = 0$ for $i \neq k$) we typically obtain similar IRPs determined either by the dMI or the transfer entropy (Supplementary Fig. 1). Further, the results remain valid qualitatively when the noise level increases (Supplementary Fig. 2).

Mechanism of anisotropic information routing. To better understand how a collective state gives rise to a specific routing pattern with directed information sharing and transfer, consider a network of two symmetrically coupled identical neural population models (Fig. 2a). Because of permutation symmetry, the coupling functions $\gamma_{i,j}$, obtained from the phase reduction of the original Wilson–Cowan-type equations[41] (Methods, Supplementary Note 3), are identical. For biologically plausible parameters this network in the noiseless-limit has two stable phase-locked reference states (α and β). The fixed phase differences $\Delta\phi_{1,2}^{[\alpha]}$ and $\Delta\phi_{1,2}^{[\beta]}$ are determined by the zeros of the anti-symmetric coupling $\bar{\gamma}(\Delta\phi) = \gamma(\Delta\phi) - \gamma(-\Delta\phi)$ with negative slope (Fig. 2e). For a given level of (sufficiently weak) noise, the system shows fluctuations around either one of these states (Fig. 2b) each giving rise to a different IRP. Sufficiently strong external signals can trigger state switching and thereby effectively invert the dominant communication direction visible from the dMI (Fig. 2c) and even more pronounced from the dTE (Fig. 2d) without changing any structural properties of the network.

The anisotropy in information transfer in the fully symmetric network is due to symmetry broken dynamical states. For independent noise inputs, $\varsigma_{i,k} = \varsigma_i\delta_{i,k}$, that are moreover small, the evolution of $\phi_i^{(\mathrm{fluct})}$, $i \in \{1,2\}$, near the reference state α reduces to

$$\frac{\mathrm{d}}{\mathrm{d}t}\phi_i^{(\mathrm{fluct})} = g_i^{[\alpha]}\left(\phi_i^{(\mathrm{fluct})} - \phi_j^{(\mathrm{fluct})}\right) + \varsigma_i\xi_i \qquad (5)$$

with coupling constants $g_1^{[\alpha]} = \gamma'(\Delta\phi_{1,2}^{[\alpha]})$, $g_2^{[\alpha]} = -\gamma'(2\pi - \Delta\phi_{1,2}^{[\alpha]})$ (Methods). As $g_2^{[\alpha]} \approx 0$ (Fig. 2e), the phase $\phi_2^{(\mathrm{fluct})}$ essentially freely fluctuates driven by the noise input $\varsigma_2\xi_2$. This causes the system to deviate from the equilibrium phase difference $\Delta\phi_{1,2}^{[\alpha]}$. At the same time, the strongly negative coupling $g_1^{[\alpha]}$ dominates over the noise term $\varsigma_1\xi_1$ and unit 1 is driven to restore the phase difference by reducing $\left|\phi_1^{(\mathrm{fluct})} - \phi_2^{(\mathrm{fluct})}\right|$. Thus, $\phi_1^{(\mathrm{fluct})}$ is effectively enslaved to track $\phi_2^{(\mathrm{fluct})}$ and information is routed from unit 2 to 1, reflected in the dMI and dTE curves. The same mechanism accounts for the reversed anisotropy in communication when the system is near state β as the roles of units 1 and 2 are exchanged. Calculating the peak of the dMI curve in this example also provides a time scale $d_*^{[\alpha]} \approx -\log(2)/g_1^{[\alpha]}$ at which maximal information sharing is observed (Methods, equation (10), see also Supplementary Note 4). It furthermore becomes clear that the directionality of the information transfer in general need not be related to the order in which the oscillators phase-lock because the phase-advanced oscillator can either effectively pull the lagging one, or, as in this example, the lagging oscillator can push the leading one to restore the equilibrium phase-difference.

In summary, effective interactions local in state space and controlled by the underlying reference state together with the noise characteristics determine the IRPs of the network. Symmetry broken dynamical states then induce anisotropic and switchable routing patterns without the need to change the physical network structure.

Information routing in networks of networks. For networks with modular interaction topology[24–28], our theory relating topology, collective dynamics and IRPs between individual units can be generalized to predict routing between entire modules. Assuming that each sub-network X in the noiseless limit has a stable phase-locked reference state, a second-phase reduction[23] generalized to stochastic dynamics characterizes each module by a single meta-oscillator with collective phase Φ_X and frequency Ω_X, driven by effective noise sources Ξ_X with covariances $\Sigma_{X,Y}$. The collective phase dynamics of a network with M modules then satisfies

$$\frac{\mathrm{d}}{dt}\Phi_X = \Omega_X + \sum_{Y=1}^{M}\Gamma_{X,Y}(\Phi_X - \Phi_Y) + \sum_{Y=1}^{M}\Sigma_{X,Y}\Xi_Y \quad (6)$$

where $\Gamma_{X,Y}$ are the effective inter-community couplings (Supplementary Note 5). The structure of equation (6) is formally identical to equation (3) so that the expressions for inter-node information routing ($\mathrm{dMI}_{i,j}$ and $\mathrm{dTE}_{i\to j}$) can be lifted to expressions on the inter-community level ($\mathrm{dMI}_{X,Y}$ and $\mathrm{dTE}_{X\to Y}$) by replacing node- with community-related quantities (that is, ω_i with Ω_X or γ_{ik} with $\Gamma_{X,K}$ and so on; Supplementary Note 5, Corollaries 3 and 4). Importantly, this process can be further iterated to networks of networks and so on. Figure 3 shows examples of information flow patterns resolved at two scales. The information routing direction on the larger scale reflects the majority and relative strengths of IRPs on the finer scale.

Nonlocal information rerouting via local interventions. The collective quantities in the system (6) are intricate functions of the network properties at the lower scales. Intriguingly, the coupling functions $\Gamma_{X,Y}$ not only depend on the nonlocal interactions $\gamma_{i_X j_Y}$ between units i_X of module X and j_Y of cluster Y but also on purely local properties of the individual clusters. In particular, the form of $\Gamma_{X,Y}$ is a function of the intrinsic local dynamical states $\mathcal{D}_X$ and $\mathcal{D}_Y$ of both clusters as well as the phase response Z_X of sub-network X (see Methods and Supplementary Note 5). Thus IRPs on the entire network level depend on local community properties. This establishes several generic mechanisms to globally change information routing in networks via local changes of modular properties, local connectivity or via switching of local dynamical states.

In a network consisting of two sub-networks (Fig. 3a) the local change of the frequency of a single Hopf-oscillator in sub-network A induces a nonlocal inversion of the information routing between cluster A and B (Fig. 3b–d). In Fig. 3e–f, the direction in which information is routed between two sub-networks B and C of coupled phase oscillators is remotely changed by increasing the strength of a local link in module A. The origin in both examples is a non-trivial combination of several factors: the (small) manipulations alter the collective cluster frequency Ω_A and the local dynamical state $\mathcal{D}_A$ which in turn changes the collective phase response Z_A and the

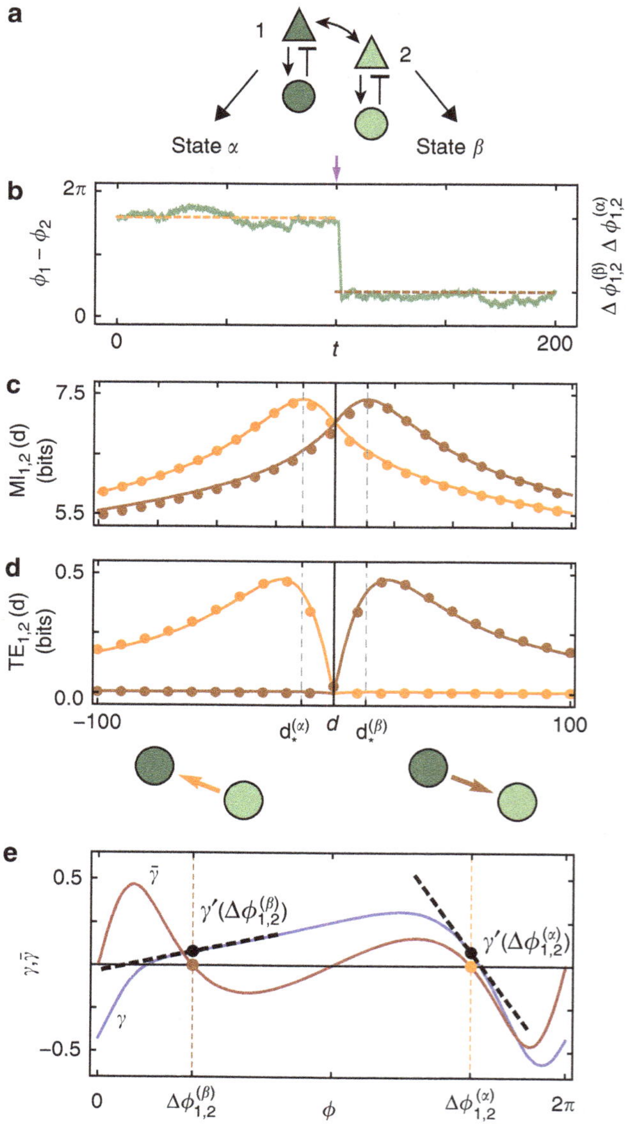

Figure 2 | Multi-stable dynamics and anisotropic information routing. (**a**) Two identical and symmetrically coupled neuronal circuits of Wilson-Cowan-type (dark and light green, modular sub-network in Fig. 1e). The noise free (that is, input free) network displays two different stable oscillatory dynamical states α and β. (**b**) Phase difference $\Delta\phi_{1,2}(t) := \phi_1(t) - \phi_2(t)$ between the extracted phases of the two neuronal populations is fluctuating around a locked value $\Delta\phi^{[\alpha]}_{1,2}$ of a the stable collective state α of the deterministic system (orange); a strong external perturbation (purple arrow) induces a switch to stochastic dynamics around the second stable deterministic reference state β (brown) with phase difference $\Delta\phi^{[\beta]}_{1,2}$. (**c**) Delayed mutual information $\mathrm{dMI}_{1,2}$ and (**d**) transfer entropy $\mathrm{dTE}_{1\to 2}$ curves between the phase signals in states α (orange) and β (brown) for numerical data (dots) and theory (lines) as a function of the time delay d between the stochastic phase signals $\phi_1(t)$ and $\phi_2(t+d)$. The change in peak latencies form $d^{[\alpha]}_* < 0$ to $d^{[\beta]}_* > 0$ in the $\mathrm{dMI}_{1,2}$ curves and the asymmetry of the $\mathrm{dTE}_{1\to 2}$ curves show anisotropic information routing for the two different states. Switching between the two dynamical states reverses the effective information routing pattern (IRP) (graphs, bottom). (**e**) Phase coupling function $\gamma(\Delta\phi) = \gamma_{1,2}(\Delta\phi) = \gamma_{2,1}(\Delta\phi)$ (blue) between the two neuronal oscillators and its anti-symmetric part $\bar{\gamma}(\Delta\phi) = \gamma(\Delta\phi) - \gamma(-\Delta\phi)$ (red). The two zeros of $\bar{\gamma}(\Delta\phi)$ with negative slope indicate the stable deterministic equilibrium phase differences $\Delta\phi^{[\alpha]}_{1,2}$ and $\Delta\phi^{[\beta]}_{1,2}$ of the dynamical states α and β, receptively. The directionality in the IRP arises due to symmetry breaking in the dynamics reflected in the different slopes of $\gamma(\Delta\phi)$ (dashed lines): In the state α, oscillator 1 receives inputs from oscillator 2 proportional to $\gamma'\left(\Delta\phi^{[\alpha]}_{1,2}\right)$, while oscillator 2 is coupled to 1 proportional to $\gamma'\left(2\pi - \Delta\phi^{[\alpha]}_{1,2}\right) = \gamma'\left(\Delta\phi^{[\beta]}_{1,2}\right)$ in linear small-noise approximation (cf. equation (5)). As $\left|\gamma'\left(\Delta\phi^{[\alpha]}_{1,2}\right)\right|$ is large deviations from the phase-locked state of oscillator 2 due to the noise inputs are strongly propagated to oscillator 1 to restore the phase-locking. Information injected to oscillator 2 is thus transmitted to oscillator 1. In contrast, inputs to oscillator 1 only weakly impact oscillator 2 as $\gamma'\left(\Delta\phi^{[\beta]}_{1,2}\right)$ is small. In total, the information is thus dominantly routed from 2 to 1. Switching to the dynamical state β reverses the roles of the oscillators and thus also the directionality of the information routing motive.

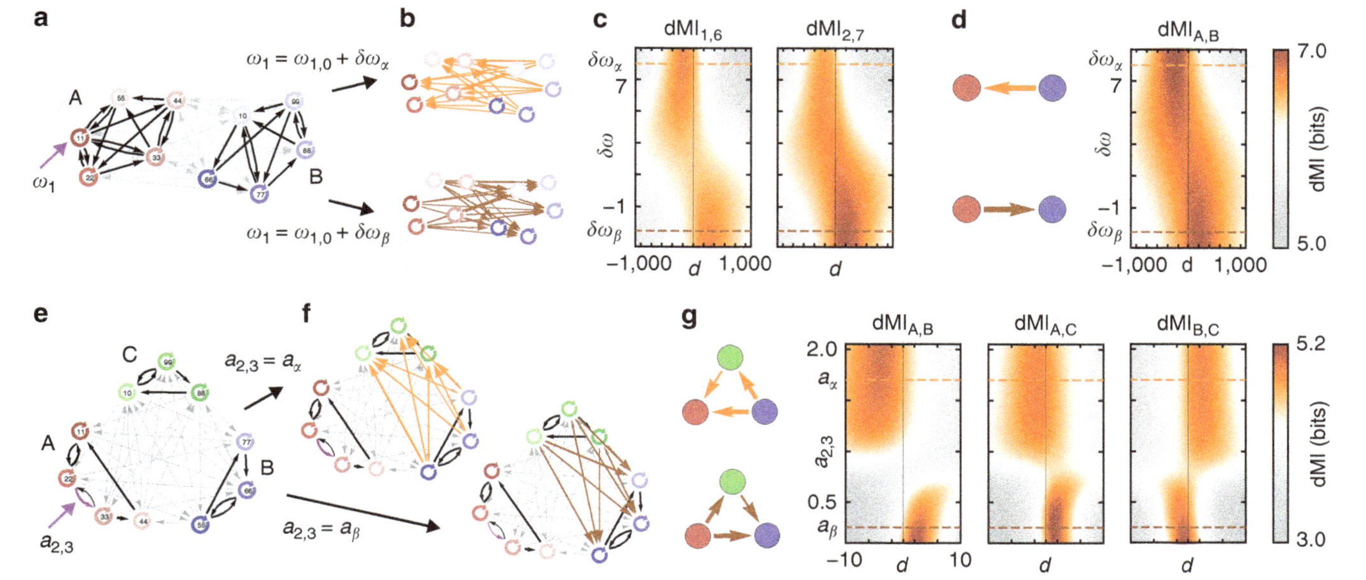

Figure 3 | Remotely induced rerouting of information in modular networks. (**a**) Network with two coupled communities A and B (red and blue) of oscillators close to a Hopf bifurcation. Changing the intrinsic frequency of a single node $i=1$ from $\omega_1+\delta\omega_\alpha$ to $\omega_1+\delta\omega_\beta$ induces a collective reorganization of equilibrium phase differences, that result in **b** oppositely directed information sharing patterns between the individual nodes of the two modules (top versus bottom). (**c**) dMI between two pairs of nodes from the two different clusters as a function of the time delay d and frequency change $\delta\omega_1$ of oscillator 1. The change of the peak from positive to negative delays reflecting the inversion of the information routing is visible. (**d**) Information routing patterns (IRPs) calculated from the hierarchically reduced system for the two configurations in **b** (left) and as a function of $\delta\omega_1$ (right) reflect the inversion on the finer scale (**b**,**c**). (**e**) Hierarchical network of three coupled modules of phase oscillators. (**f**) A change in the connection strength $a_{2,3}$ from a_α to a_β between two nodes ($3_A \rightarrow 2_A$) in sub-network A induces an inversion of information routing direction between the remote sub-networks B and C. (**g**) Full IRPs calculated form the hierarchical reduced system for $a_{2,3}=a_\alpha$ and $a_{2,3}=a_\alpha$ (left) and as a function of $a_{2,3}$ for all pairs of modules (density plots, right). The transition is not continuous but rather switch like.

effective noise strength Ξ_A of cluster A. These changes all contribute to changes in the effective couplings $\Gamma_{X,Y}$ as well as in the inter-cluster phase-locking values $\Delta\Phi_{X,Y}=\Phi_X-\Phi_Y$ (cf. Supplementary Note 5 and Supplementary Fig. 3). The changes in these properties, which enter the expressions for the information sharing and routing measures (Fig. 2 and Supplementary Note 4) then cause the observed changes in information routing direction. Interestingly, the transition in information routing has a switch-like dependency on the changed parameter (Fig. 3c,d,g) promoting digital like changes of communication modes. Combinatorially many information routing patterns.

Combinatorial many information routing patterns. As an alternative to interventions on local properties, also switching between multi-stable local dynamical states $\mathcal{D}_X$ can induce global information rerouting. In the example in Fig. 4, each of the $M=3$ modules $X\in\{A, B, C\}$ exhibits $\mathcal{N}_X=2$ alternative phase-locked states (labelled α_X and β_X, Supplementary Section 6 and Supplementary Fig. 4). For sufficiently weak coupling, this local multi-stability is preserved in the dynamics of the entire modular network. Consequently each choice of the $\mathcal{N}_A\times\mathcal{N}_B\times\mathcal{N}_C$ possible combinations of 'local' states gives rise to at least one network-wide collective state. Certain combinations of local states can give rise to one or even multiple globally phase-locked states (for example, $[\alpha_A\beta_B\alpha_C]$ in Fig. 4). Others support non-phase-locked dynamics that gives rise to time-dependent IRPs (cf. Fig. 4c and below). Thus, varying local dynamical states in a hierarchical network flexibly produces a combinatorial number $\mathcal{N}\geq\prod_X\mathcal{N}_X$ of different IRPs in the same physical network.

Time-dependent information routing. General reference states, including periodic or transient dynamics, are not stationary and hence the expressions for the dMI and dTE become dependent on the time t. For example, Fig. 4c shows IRPs that undergo cyclic changes due to an underlying periodic reference state (cf. also Supplementary Note 6 and Supplementary Fig. 5a-c). In systems with a global fixed point, systematic displacements to different starting positions in state space give rise to different stochastic transients with different and time-dependent IRPs (Supplementary Fig. 5d). Similarly, switching dynamics along heteroclinic orbits constitute another way of generating specific progressions of reference dynamics. Thus information 'surfing' on top of non-stationary reference dynamical configurations naturally yield temporally structured sequences of IRPs, resolvable also by other measures of instantaneous information flow[36,42,43].

Discussion

The above results establish a theoretical basis for the emergence of information routing capabilities in complex networks when signals are communicated on top of collective reference states. We show how information sharing (dMI) and transfer (dTE) emerge through the joint action of local unit features, global interaction topology and choice of the collective dynamical state. We find that IRPs self-organize according to general principles (cf. Figs 2–4) and can thus be systematically manipulated. Employing formal identity of our approach at every scale in oscillatory modular networks (equations (3) versus (6)) we identify local paradigms that are capable of regulating information routing at the nonlocal level across the whole network (Figs 3 and 4).

In contrast to self-organized technological routing protocols where local nodes use local routing information to locally propagate signals, such as in peer-to-peer networks[44], in the mechanism studied here the information routing modality is set

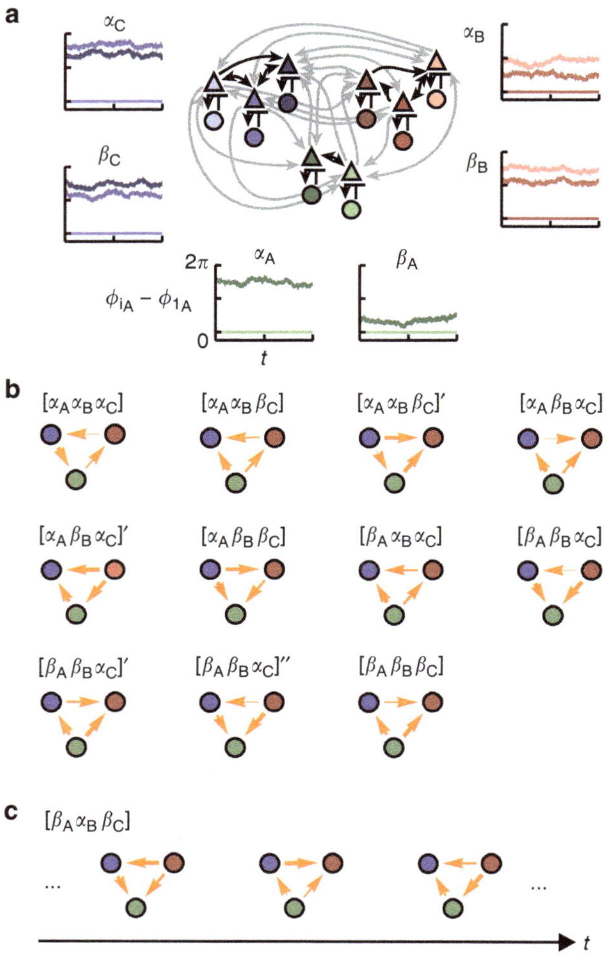

Figure 4 | Combinatorially many IRPs in networks with locally multi-stable dynamics. (**a**) Modular circuit as in Fig. 1e. Without inter-module coupling (grey arrows), each of the $M=3$ communities $X \in \{A, B, C\}$ exhibits multi-stability between two phase-locked configurations, denoted as states α_X and β_X (insets). The characteristics of these states remain present in the full network once weak coupling between the modules is introduced. We refer to these states as local dynamical states $\mathcal{D}_X$. (**b**) Information routing patterns (IRPs) between the hierarchically reduced sub-networks for different combinations of the local dynamical states $[\mathcal{D}_A\mathcal{D}_B\mathcal{D}_C]$, $\mathcal{D}_X \in \{\alpha_X, \beta_X\}$ that give rise to globally phase-locked dynamics. Arrows between nodes X and Y indicate strength (line width) and sign (arrow direction) of the information routing quantified via the difference in integrated dTE curves between the nodes (see Methods). The same combination of local dynamical states ($[\mathcal{D}_A\mathcal{D}_B\mathcal{D}_C]$) can give rise to more than one globally phase-locked collective state marked with dashes, i.e. $[\mathcal{D}_A\mathcal{D}_B\mathcal{D}_C]$, $[\mathcal{D}_A\mathcal{D}_B\mathcal{D}_C]'$,.... each giving rise to a separate IRP: For fixed local dynamical states the global network is multi-stable and choosing between those components is changing the information routing similarly as in Fig. 2. (**c**) The local dynamical state configuration $[\beta_A\alpha_B\beta_C]$ generates a periodic global dynamical state (cf. Supplementary Fig. 3) in which the hierarchically reduced IRP (graphs) becomes time-dependent (cf. also Supplementary Note 6).

by the entire network's collective dynamics. This collective reference state typically evolves on a slower timescale than the information carrying fluctuations that surf on top of it and is thus different from signal propagation in cascades[45] or avalanches[46] that dominate on shorter time scales.

We derived theoretical results based on information sharing and transfer obtained via delayed mutual information and transfer entropy curves. Using these abstract measures our results are independent of any particular implementation of a communication protocol and thus generically demonstrate how collective dynamics can have a functional role in information routing. For example, in the network in Fig. 2 externally injected streams of information are automatically encoded in fluctuations of the rotation frequency of the individual oscillators. The injected signals are then transmitted through the network and decodable from the fluctuating phase velocity of a target unit precisely along those pathways predicted by the current state-dependent IRP (Supplementary Note 7 and Supplementary Figs 6 and 7).

Our theory is based on a small-noise approximation that conditions the analysis onto a specific underlying dynamical state. In this way, we extracted the precise role of such a reference state for the network's information routing abilities. For larger signal amplitudes or in highly recurrent networks, in which higher order interactions can play an important role the expansion can be carried out systematically to higher orders using diagrammatic approaches[47] or numerically to accounting for better accuracy and non-Gaussian correlations (cf. also Supplementary Note 4).

In systems with multi-stable states two signal types need to be discriminated: those that encode the information to be routed and those that indicate a switch in the reference dynamics and consequently the IRPs. If the second type of stimuli is amplified appropriately a switch between multi-stable states can be induced that moves the network into the appropriate IRP state for the signals that follow. For example, in the network of Fig. 2a switch from states α to β can be induced by a strong positive pulse to oscillator 2 (and vice versa). If such pulses are part of the input a switch to the appropriate IRP state will automatically be triggered and the network auto-regulates its IRP function (Supplementary Fig. 6). More generally a separate part of the network that effectively filters out relevant signatures indicating the need for a different IRP could provide such pulses. Using the fact that local interventions are capable to switch IRPs in the network also the outcomes of local computations can be used to trigger changes in the global information routing and thereby enable context-dependent processing in a self-organized way.

When information surfs on top of dynamical reference states the control of IRPs is shifted towards controlling collective network dynamics making methods from control theory of dynamical systems available to the control of information routing. For example, changing the interaction function in coupled oscillators systems[18] or providing control signals to a subset of nodes[48,49] are capable of manipulating the network dynamics. Moreover, switch-like changes (cf. Fig. 3) can be triggered by crossing bifurcation points and the control of IRPs then gets linked to bifurcation theory of network dynamical systems.

While the mathematical part of our analysis focused on phase signals, including additional amplitude degrees of freedom into the theoretical framework can help to explore neural or cell signalling codes that simultaneously use activity- and phase-based representations to convey information[50]. Moreover, separating IRP generation, for example, via phase configurations, from actual information transfer, for instance in amplitude degrees of freedom, might be useful for the design of systems with a flexible communication function.

The role of self-organized collective dynamics in information routing in biological systems is still speculative. Our theoretical study, identifying natural mechanisms for state-dependent sharing and transfer of information, may thus foster further experimental explorations that seek for conclusive proofs and exploring the nonlocal effects of local system manipulations. While recent experimental studies point towards functional roles of collective dynamics[31,35,51–54] the predicted phenomena, including nonlocal changes of information routing by local interventions, could be directly experimentally verified using methods available to date, such as synthetic patterned neuronal cultures[55], electrochemical arrays[18] or synthetic gene-regulatory networks[5] (Supplementary Note 8). In addition, our results are applicable to the inverse problem: Unknown network characteristics may be inferred by fitting theoretically expected dMI and dTE patterns to experimentally observed data. For example, inferring state-dependent coupling strengths could further the analysis of neuronal dynamics during context-dependent processing[33,35,39,53,54,56].

Modifying inputs, initial conditions or system-intrinsic properties may well be viable in many biological and artificial systems whose function requires particular information routing. For instance, on long time scales, evolutionary pressure may select a particular IRP by biasing a particular collective state in gene-regulatory and cell-signalling networks[2,15,57]; on intermediate time scales, local changes in neuronal responses due to adaptation or varying synaptic coupling strength during learning processes[13] can impact information routing paths in entire neuronal circuits; on fast time scales, defined control inputs to biological networks or engineered communication systems that switch the underlying collective state, can dynamically modulate IRPs without any physical change to the network.

Methods

Transfer entropy. The dTE[22] from a time-series $x_i(t)$ to a time-series $x_j(t)$ is defined as

$$\mathrm{dTE}_{i\to j}(d)=\iiint p_{i,j,j^{(d)}}\log\left(\frac{p_{i,j,j^{(d)}}p_j}{p_{i,j}p_{j,j^{(d)}}}\right)\mathrm{d}x_i\mathrm{d}x_j\mathrm{d}x_{j^{(d)}} \tag{7}$$

with joint probability $p_{i,j,j^{(d)}}=p(x_j(t+d),x_i(t),x_j(t))$ and $p_{j,j^{(d)}}=p(x_j(t+d),x_j(t))$. This expression is not invariant under permutation of i and j, implying the directionality of TE. For a more direct comparison with dMI in Fig. 2, we define $\mathrm{dTE}_{i,j}(d)$ by $\mathrm{dTE}_{i\to j}(d)$ for $d>0$ and by $\mathrm{dTE}_{j\to i}(-d)$ for $d<0$ (cf. Supplementary Note 1 for additional details).

Dynamic information routing via dynamical states. For a dynamical system (2) the reference deterministic solution $\boldsymbol{x}^{(\mathrm{ref})}(t+s)$ starting at $\boldsymbol{x}(t)$ is given by the deterministic flow $\boldsymbol{x}^{(\mathrm{ref})}(t+s)=\boldsymbol{F}^{(\mathrm{ref})}(\boldsymbol{x}(t),s)$. The small-noise approximation for white noise $\boldsymbol{\xi}$ yields

$$p(\boldsymbol{x}(t+d)\mid\boldsymbol{x}(t))=\mathcal{N}_{\boldsymbol{x}^{(\mathrm{ref})}(t+d),\boldsymbol{Q}_d(\boldsymbol{x}(t))}(\boldsymbol{x}(t+s)) \tag{8}$$

where $\mathcal{N}_{\boldsymbol{x},\boldsymbol{\Sigma}}$ denotes the normal distribution with mean $\boldsymbol{x}$ and covariance matrix $\boldsymbol{\Sigma}$, $\boldsymbol{Q}_d(\boldsymbol{x})=\int_0^d e^{\int_s^d \boldsymbol{G}(r,\boldsymbol{x})dr}\varsigma\varsigma^{\mathrm{T}}e^{\int_s^d \boldsymbol{G}^{\mathrm{T}}(r,\boldsymbol{x})dr}ds$ and $\boldsymbol{G}(s,\boldsymbol{x})=\boldsymbol{D}\boldsymbol{f}(\boldsymbol{F}(\boldsymbol{x},s))$. We assumed $\boldsymbol{G}(s,\boldsymbol{x})\boldsymbol{G}(t,\boldsymbol{x})=\boldsymbol{G}(t,\boldsymbol{x})\boldsymbol{G}(s,\boldsymbol{x})$. From this and the initial distribution $p(\boldsymbol{x}(t))$, the dMI and transfer entropy $\mathrm{dMI}_{i,j}(d,t)$ and $\mathrm{dTE}_{i\to j}(d,t)$ are obtained via (2) and (7). The result depends on time t, lag d and the reference state $\boldsymbol{x}^{(\mathrm{ref})}$ (cf. Supplementary Note 2 for additional details).

Oscillator networks. In Fig. 1a, we consider a network of two coupled biochemical Goodwin oscillators[14,58]. Oscillations in the expression levels of the molecular products arise because of a nonlinear repressive feedback loop in successive transcription, translation and catalytic reactions. The oscillators are coupled via mutual repression of the translation process[59]. In addition, in one oscillator changes in concentration of an external enzyme regulate the speed of degradation of mRNAs, thus affecting the translation reaction, and, ultimately, the oscillation frequency. In Figs 1e, 2 and 4 we consider networks of Wilson–Cowan-type neural masses (population signals)[41]. Each neural mass intrinsically oscillates because of antagonistic interactions between local excitatory and inhibitory populations. Different neural masses interact, within and between communities, via excitatory synapses. In the generic networks in Figs 1i and 3a each unit is modelled by the normal form of a Hopf-bifurcation in the oscillatory regime together with linear coupling. Finally, the modular networks analysed in Fig. 3a,b are directly cast as phase-reduced models with freely chosen coupling functions. See the Supplementary Note 8 and Supplementary Figs 4, 8 and 9 for additional details, model equations and parameters and phase estimation.

Analytic derivation of the dMI and dTE curves. In the small-noise expansion[60], both dMI and dTE curves have an analytic approximation: For stochastic fluctuations around some phase-locked collective state with constant reference phase offsets $\Delta\phi_{i,j}=\phi_i-\phi_j$ the phases evolve as $\phi_i^{(\mathrm{ref})}(t)=\Omega t+\Delta\phi_{i,1}$ in the deterministic limit, where $\Omega=\omega_i+\sum_k\gamma_{i,k}(\Delta\phi_{i,k})$ is the collective network frequency and the $\gamma_{i,j}(\cdot)$ are the coupling functions from equation (3). In presence of noise, the phase dynamics have stochastic components $\phi_i^{(\mathrm{fluct})}(t)=\phi_i(t)-\phi_i^{(\mathrm{ref})}(t)$. In first-order approximation, independent noise inputs $\varsigma_{i,j}=\varsigma_i\delta_{i,j}$ yield coupled Ornstein–Uhlenbeck processes

$$\frac{\mathrm{d}}{\mathrm{d}t}\phi_i^{(\mathrm{fluct})}=\sum_k g_{i,k}\phi_k^{(\mathrm{fluct})}+\varsigma_i\xi_i \tag{9}$$

with linearized, state-dependent couplings given by the Laplacian matrix entries $g_{i,j}=-\gamma'_{i,j}(\Delta\phi_{i,j})$ and $g_{i,i}=\sum_k\gamma'_{i,k}(\Delta\phi_{i,k})$. The analytic solution to the stochastic equation (9) provides an estimate of the probability distributions, p_i, $p_{i,j^{(d)}}$ and $p_{i,j,j^{(d)}}$. Via (2) this results in a prediction for $\mathrm{dMI}_{i,j}(d)$, equation (4), as a function of the matrix elements $k_{i,j^{(d)}}$ specifying the inverse variance of a von Mises distribution ansatz for $p_{i,j^{(d)}}$. Similarly via (7) an expression for $\mathrm{dTE}_{i\to j}(d)$ is obtained. For the dependency of $k_{i,j^{(d)}}$, and $\mathrm{dTE}_{i\to j}(d)$ on network parameters and further details, see the derivation of the theorems 1 and 2 in Supplementary Note 4.

Time scale for information sharing. For a network of two oscillators as in Fig. (2) with linearized coupling strengths $g_1^{[\alpha]}$ and $g_2^{[\alpha]}$ and $g_1^{[\alpha]}<g_2^{[\alpha]}$, maximizing $\mathrm{dMI}_{1,2}(d)$ (see Supplementary Note 4 for full analytic expressions of dMI and dTE in two oscillator networks) yields

$$d^*=\left(g_1^{[\alpha]}+g_2^{[\alpha]}\right)^{-1}\log\left(\frac{1}{2}\left(1+\left(\frac{g_2^{[\alpha]}}{g_1^{[\alpha]}}\right)^2\right)\right) \tag{10}$$

Collective phase reduction. Suppose that each node $i=i_X$ belongs to a specific network module X out of $M\leq N$ non-overlapping modules of a network. Then equation (3) can be simplified to equation (6) under the assumption that in the absence of noise every community X has a stable internally phase-locked state $\phi_{i_X}^{(\mathrm{ref})}(t)=\Phi_X(t)+\Delta\phi_{i_X}$, where $\Delta\phi_{i_X}$ are constant phase offsets of individual nodes i_X. Every community can then be regarded as a single meta-oscillator with a collective phase $\Phi_X(t)$ and a collective frequency $\Omega_X=\omega_{i_X}+\sum_{j_X}\gamma_{i_X,j_X}(\Delta\phi_{i_X}-\Delta\phi_{j_X})$. The vector components of the collective phase response $\boldsymbol{Z}_X$, the effective couplings $\Gamma_{X,Y}$ and the noise parameters $\Sigma_{X,Y}$ and Ξ_X are obtained through collective phase reduction and depend on the respective quantities $(\omega_{i_X},\gamma_{i_X,j_X},\ldots)$ on the single-unit scale (see Supplementary Note 5 for a full derivation).

References

1. Tyson, J. J., Chen, K. & Novak, B. Network dynamics and cell physiology. *Nat. Rev. Mol. Cell Biol.* **2**, 908–916 (2001).
2. Tkacik, G., Callan, C. G. & Bialek, W. Information flow and optimization in transcriptional regulation. *Proc. Natl Acad. Sci. USA* **105**, 12265–12270 (2008).
3. Tononi, G., Edelman, G. M. & Sporns, O. Complexity and coherency: integrating information in the brain. *Trends Cogn. Sci.* **2**, 474–484 (1998).
4. Weber, W., Rabaey, J. M. & Aarts, E. (eds.) *Ambient Intelligence* (Springer, 2005).
5. Stricker, J. *et al.* A fast, robust and tunable synthetic gene oscillator. *Nature* **456**, 516–519 (2008).
6. Varela, F., Lachaux, J.-P., Rodriguez, E. & Martinerie, J. The brainweb: phase synchronization and large-scale integration. *Nat. Rev. Neurosci.* **2**, 229–239 (2001).
7. Winfree, A. T. *The Geometry of Biological Time* (Springer, 1980).
8. Kuramoto, Y. *Chemical Oscillations, Waves, and Turbulence* (Springer, 1984).
9. Strogatz, S. H. Exploring complex networks. *Nature* **410**, 268–276 (2001).
10. Pikovsky, A. S., Rosenblum, M. & Kurths, J. *Synchronization—A Universal Concept in Nonlinear Sciences* (Cambridge University Press, 2001).
11. Acebrón, J., Bonilla, L. & Vicente, C. The Kuramoto model: a simple paradigm for synchronization phenomena. *Rev. Mod. Phys.* **77**, 137–185 (2005).
12. Fries, P. A mechanism for cognitive dynamics: neuronal communication through neuronal coherence. *Trends. Cogn. Sci.* **9**, 474–480 (2005).
13. Salazar, R. F., Dotson, N. M., Bressler, S. L. & Gray, C. M. Content-specific fronto-parietal synchronization during visual working memory. *Science* **338**, 1097–1100 (2012).
14. Goldbeter, A. Computational approaches to cellular rhythms. *Nature* **420**, 238–245 (2002).
15. Feillet, C. *et al.* Phase locking and multiple oscillating attractors for the coupled mammalian clock and cell cycle. *Proc. Natl Acad. Sci. USA* **111**, 9828–9833 (2014).
16. Blasius, B., Huppert, A. & Stone, L. Complex dynamics and phase synchronization in spatially extended ecological systems. *Nature* **399**, 354–359 (1999).
17. Néda, Z., Ravasz, E., Brechet, Y., Vicsek, T. & Barabási, A. L. The sound of many hands clapping. *Nature* **403**, 849–850 (2000).
18. Kiss, I. Z., Rusin, C. G., Kori, H. & Hudson, J. L. Engineering complex dynamical structures: sequential patterns and desynchronization. *Science* **316**, 1886–1889 (2007).

19. Klinglmayr, J., Kirst, C., Bettstetter, C. & Timme, M. Guaranteeing global synchronization in networks with stochastic interactions. *New. J. Phys.* **14,** 073031 (2012).
20. Shaw, R. S. Strange attractors, chaotic behavior, and information flow. *Zeitschr. f. Naturforsch.* **36,** 80–112 (1981).
21. Vastano, J. A. & Swinney, H. L. Information transport in spatiotemporal systems. *Phys. Rev. Lett.* **18,** 1773–1776 (1988).
22. Schreiber, T. Measuring information transfer. *Phys. Rev. Lett.* **85,** 461–464 (2000).
23. Kawamura, Y., Nakao, H., Arai, K., Kori, H. & Kuramoto, Y. Collective phase sensitivity. *Phys. Rev. Lett.* **101,** 024101 (2008).
24. Newman, M. E. J. Communities, modules and large-scale structure in networks. *Nat. Phys.* **7,** 25–31 (2011).
25. Pajevic, S. & Plenz, D. The organization of strong links in complex networks. *Nat. Phys.* **8,** 429–436 (2012).
26. Song, S., Sjöström, P. J., Reigl, M., Nelson, S. & Chklovskii, D. B. Highly nonrandom features of synaptic connectivity in local cortical circuits. *PLoS Biol.* **3,** e68 (2005).
27. Perin, R., Berger, T. K. & Markram, H. A synaptic organizing principle for cortical neuronal groups. *Proc. Natl Acad. Sci. USA* **108,** 5419–5424 (2011).
28. Hagmann, P. *et al.* Mapping the structural core of human cerebral cortex. *PLoS Biol.* **6,** e159 (2008).
29. Ravasz, E., Somera, A. L., Mongru, D. A., Oltvai, Z. N. & Barabási, A.-L. Hierarchical Organization of modularity in metabolic networks. *Science* **297,** 1551–1555 (2002).
30. McMillen, D., Kopell, N., Hasty, J. & Collins, J. J. Synchronizing genetic relaxation oscillators by intercell signaling. *Proc. Natl Acad. Sci. USA* **99,** 679–684 (2002).
31. Cabral, J., Hugues, E., Sporns, O. & Deco, G. Role of local network oscillations in resting-state functional connectivity. *Neuroimage* **57,** 130–139 (2011).
32. Koepsell, K. & Sommer, F. T. Information transmission in oscillatory neural activity. *Biol. Cybern.* **99,** 403–416 (2008).
33. Kopell, N. J., Gritton, J. G., Whittington, A. W. & Kramer, M. A. Beyond the connectome: the dynome. *Neuron* **83,** 1319–1327 (2014).
34. Belitski, A. *et al.* Low-frequency local field potentials and spikes in primary visual cortex convey independent visual information. *J. Neurosci.* **28,** 5696–5709 (2008).
35. Agarwal, G. *et al.* Spatially distributed local fields in the hippocampus encode rat position. *Science* **344,** 626–630 (2014).
36. Lizier, J. T., Prokopenko, M. & Zomaya, A. Y. Local information transfer as a spatiotemporal filter for complex systems. *Phys. Rev. E* **77,** 026110 (2008).
37. Lauschke, V. M., Charisios, D. T., François, P. & Aulehla, A. Scaling of embryonic patterning based on phase-gradient encoding. *Nature* **7430,** 101–105 (2013).
38. Hoppensteadt, F. & Izhikevich, E. Synchronization of laser oscillators, associative memory, and optical neurocomputing. *Phys. Rev. E* **62,** 4010–4013 (2000).
39. Friston, K. J. Functional and effective connectivity: a review. *Brain Connect.* **1,** 13–36 (2011).
40. Teramae, J.-N., Nakao, H. & Ermentrout, G. B. Stochastic phase reduction for a general class of noisy limit cycle oscillators. *Phys. Rev. Lett.* **102,** 194102 (2009).
41. Wilson, H. R. & Cowan, J. D. Excitatory and inhibitory interactions in localized populations of model neurons. *Biophys. J.* **12,** 1–24 (1972).
42. Liang, X. & Kleeman, R. Information transfer between dynamical system components. *Phys. Rev. Lett.* **95,** 244101 (2005).
43. Majda, A. J. & Harlim, J. Information flow between subspaces of complex dynamical systems. *Proc. Natl Acad. Sci. USA* **104,** 9558–9563 (2007).
44. Lua, E. K., Crowcroft, J., Pias, M., Sharma, R. & Lim, S. A survey and comparison of peer-to-peer overlay network schemes. *IEEE Commun. Surveys Tutorials* **7,** 72–93 (2005).
45. Watts, D. J. A simple model of global cascades on random networks. *Proc. Natl Acad. Sci. USA* **99,** 5766–5771 (2002).
46. Beggs, J. M. & Plenz, D. Neuronal avalanches in neocortical circuits. *J. Neurosci.* **23,** 11167–11177 (2003).
47. Korutcheva, E. & Del Prete, V. A diagrammatic approach to study the information transfer in weakly non-linear channels. *Int. J. Mod. Phys. B* **16,** 3527–3544 (2002).
48. Cornelius, S. P., Kath, W. L. & Motter, A. E. Realistic control of network dynamics. *Nat. Commun.* **4,** 1942 (2013).
49. Liu, Y.-Y., Slotine, J.-J. & Barabási, A.-L. Controllability of complex networks. *Nature* **473,** 167–173 (2011).
50. Akam, T. & Kullmann, D. M. Oscillatory multiplexing of population codes for selective communication in the mammalian brain. *Nat. Rev. Neurosci.* **15,** 111–122 (2014).
51. Selimkhanov, J. *et al.* Accurate information transmission through dynamic biochemical signaling networks. *Science* **346,** 1370–1373 (2015).
52. Bastos, A. M. *et al.* Visual areas exert feedforward and feedback influences through distinct frequency channels. *Neuron* **85,** 390–401 (2015).
53. Canolty, R. T. *et al.* Oscillatory phase coupling coordinates anatomically dispersed functional cell assemblies. *Proc. Natl Acad. Sci. USA* **107,** 17356–17361 (2010).
54. Cole, M. W., Bassett, D. S., Power, J. D. & Braver, T. S. Intrinsic and task-evoked network architectures of the human brain. *Neuron* **83,** 238–251 (2014).
55. Feinerman, O., Rotem, A. & Moses, E. Reliable neuronal logic devices from patterned hippocampal cultures. *Nat. Phys.* **4,** 967–973 (2008).
56. Battaglia, D., Witt, A., Wolf, F. & Geisel, T. Dynamic effective connectivity of inter-areal brain circuits. *PLoS Comput. Biol.* **8,** e1002438 (2012).
57. Purvis, J. E. & Lahav, G. Encoding and decoding cellular information through signaling dynamics. *Cell* **152,** 945–956 (2013).
58. Goodwin, B. C. An entrainment model for timed enzyme synthesis in bacteria. *Nature* **209,** 479–481 (1966).
59. Wagner, A. Circuit topology and the evolution of robustness in two-gene circadian oscillators. *Proc. Natl Acad. Sci. USA* **102,** 11775–11780 (2005).
60. Gardiner, C. W. *Handbook of Stochastic Methods* 3rd edn (Springer Verlag, 2004).

Acknowledgements

We thank T. Geisel for valuable discussions. Partially supported by the Federal Ministry for Education and Research (BMBF) under grants no. 01GQ1005B (C.K., D.B., M.T.) and 03SF0472E (M.T.), by the NVIDIA Corp., Santa Clara, USA (M.T.), a grant by the Max Planck Society (M.T.), by the FP7 Marie Curie career development fellowship IEF 330792 (DynViB) (D.B.) and an independent postdoctoral fellowship by the Rockefeller University, New York, USA (C.K.).

Author contributions

All authors designed research. C.K. worked out the theory and derived the analytical results, developed analysis tools and carried out the numerical experiments. All authors analysed the data, interpreted the results and wrote the manuscript.

Additional information

Competing financial interests: The authors declare no competing financial interests.

20

Quantum coding with finite resources

Marco Tomamichel[1], Mario Berta[2] & Joseph M. Renes[3]

The quantum capacity of a memoryless channel determines the maximal rate at which we can communicate reliably over asymptotically many uses of the channel. Here we illustrate that this asymptotic characterization is insufficient in practical scenarios where decoherence severely limits our ability to manipulate large quantum systems in the encoder and decoder. In practical settings, we should instead focus on the optimal trade-off between three parameters: the rate of the code, the size of the quantum devices at the encoder and decoder, and the fidelity of the transmission. We find approximate and exact characterizations of this trade-off for various channels of interest, including dephasing, depolarizing and erasure channels. In each case, the trade-off is parameterized by the capacity and a second channel parameter, the quantum channel dispersion. In the process, we develop several bounds that are valid for general quantum channels and can be computed for small instances.

[1]School of Physics, University of Sydney, Sydney, New South Wales 2006, Australia. [2]Institute for Quantum Information and Matter, Division of Physics, Mathematics and Astronomy, California Institute of Technology, Pasadena, California 91125, USA. [3]Department of Physics, Institute for Theoretical Physics, ETH Zurich, 8093 Zürich, Switzerland. Correspondence and requests for materials should be addressed to M.T. (email: marco.tomamichel@sydney.edu.au).

One of the quintessential topics in quantum information theory is the study of reliable quantum information transmission over noisy quantum channels. Here 'channel' simply refers to a description of a physical evolution. In the standard formulation, one considers communication between two points connected by a memoryless channel that can be used many times in sequence. In this case, the sender first encodes a quantum state into a sequence of registers and then sends them one by one through the channel to the receiver. The receiver collects these registers and then attempts to decode the quantum state. Equivalently, one considers a collection of physical qubits that are exposed to independent noise. The goal is then to encode quantum information (logical qubits) into this system (physical qubits) so that the quantum information can be retrieved with high fidelity after a given time. One of the primary goals of information theory is to find fundamental limits imposed on any coding scheme that attempts to accomplish this task.

Following a tradition going back to Shannon's groundbreaking work[1], this problem is usually studied asymptotically: the quantum capacity of a channel[2–7] is defined as the optimal rate (in qubits per use of the channel) at which one can transmit quantum information with vanishing error as the number of sequential channel uses increases to infinity. In the context of information storage, the rate simply corresponds to the ratio of logical to physical qubits, and the number of physical qubits is taken to be asymptotically large. Such an asymptotic analysis has proven to be pertinent in the analysis of classical communication (cc) systems—but is it also satisfactory in the quantum setting?

Achieving (or approximately achieving) the quantum capacity generally requires both the receiver and sender to coherently manipulate an array of qubits that grows proportionally with the number of channel uses. More precisely, the sender is required to prepare arbitrary states that are entangled between all channel inputs and the receiver needs to perform a joint measurement on all channel outputs. While classical computers can readily operate on very large amounts of data, at least for the near future it appears unrealistic to expect that encoding and decoding circuits can store or coherently manipulate large numbers of qubits. Thus, it is natural to ask how well quantum coding schemes perform when we restrict the size of the quantum devices used for encoding the channel inputs and decoding its outputs. This is equivalent to considering communication with only a fixed number of channel uses.

In this work, following the footsteps of recent progress in classical information theory[8–11], we investigate how well one can transmit quantum information in a realistic scenario where the number of channel uses is limited. The quantum capacity is at most a proxy for the answer to this question, and we show with concrete examples that it is often not a very good one. For example, we find that in the order of a 1,000 qubits are required to get within 90% of the quantum capacity of a typical qubit dephasing channel. To overcome this issue, we develop a more precise approximate characterization of the performance of optimal coding schemes that takes into account finite size effects. We find that these effects are succinctly described by a second channel parameter (besides its capacity), which we name quantum channel dispersion. As such, our work generalizes recent progress in the study of cc over quantum channels[12,13].

Results

Model for quantum communication. In this work, we focus on codes enabling a state entangled with a reference system to be reliably transmitted through the channel. This is a strong requirement: reliable entanglement transmission implies reliable transmission, on average, of all pure input states. The coding scheme is depicted in Fig. 1. We are given a quantum channel $\mathcal{N} \equiv \mathcal{N}_{A\to B}$ and denote by $\mathcal{N}^{\otimes n}$ the n-fold parallel repetition of this channel. An entanglement-transmission code for $\mathcal{N}^{\otimes n}$ is given by a triplet $\{|M|, \mathcal{E}, \mathcal{D}\}$, where $|M|$ is the local dimension of a maximally entangled state $|\phi\rangle_{MM'}$ that is to be transmitted over $\mathcal{N}^{\otimes n}$. The quantum channels $\mathcal{E} \equiv \mathcal{E}_{M'\to A^n}$ and $\mathcal{D} \equiv \mathcal{D}_{B^n\to M''}$ are encoding and decoding operations, respectively. With this in hand, we now say that a triplet $\{R, n, \varepsilon\}$ is achievable on the channel $\mathcal{N}$ if there exists an entanglement-transmission code satisfying

$$\tfrac{1}{n}\log|M| \geq R \quad \text{and} \quad F(\phi_{MM''}, (\mathcal{D}\circ\mathcal{N}^{\otimes n}\circ\mathcal{E})(\phi_{MM'})) \geq 1-\varepsilon. \quad (1)$$

Here R is the rate of the code, n is the number of channel uses and ε is the tolerated error or infidelity, measured in terms of Uhlmann's fidelity[14], $F(\rho,\sigma) := \left\|\sqrt{\rho}\sqrt{\sigma}\right\|_1^2$.

The non-asymptotic achievable region of a quantum channel $\mathcal{N}$ is then given by the union of all achievable triplets $\{R, n, \varepsilon\}$. The goal of (non-asymptotic) information theory is to find tight bounds on this achievable region, in particular to determine if certain triplets are outside the achievable region and thus forbidden. For this purpose, we define its boundary

$$\hat{R}_{\mathcal{N}}(n;\varepsilon) := \max\{R : (R, n, \varepsilon) \text{is achievable on} \mathcal{N}\}, \quad (2)$$

and investigate it as a function of n for a fixed value of ε. We will often drop the subscript $\mathcal{N}$ if it is clear which channel is considered. An alternative approach would be to investigate the boundary $\hat{\varepsilon}_{\mathcal{N}}(n;R) := \min\{\varepsilon : (R, n, \varepsilon) \text{ is achievable}\}$, as in ref. 15. This leads to the study of error exponents (and the reliability function), as well as strong converse exponents. We will not discuss this here since such an analysis usually does not yield good approximations for small values of n.

To begin, let us rephrase the seminal capacity results in this language. The quantum capacity is defined as the asymptotic limit of $\hat{R}_{\mathcal{N}}(n;\varepsilon)$ when n (first) goes to infinity and ε vanishes. The capacity can be expressed in terms of a regularized coherent information[2,3,5–7,16]:

$$Q(\mathcal{N}) := \lim_{\varepsilon\to 0}\lim_{n\to\infty}\hat{R}_{\mathcal{N}}(n;\varepsilon) = \sup_{\ell\in\mathbb{N}}\frac{I_c\left(\mathcal{N}^{\otimes \ell}\right)}{\ell} \geq I_c(\mathcal{N}), \quad (3)$$

where the coherent information I_c is an entropic functional defined in Methods. This result is highly unsatisfactory, not least because the regularization makes its computation intractable. (The supremum in equation (3) is necessary in the following sense: there does not exist a universal constant ℓ_0 such that

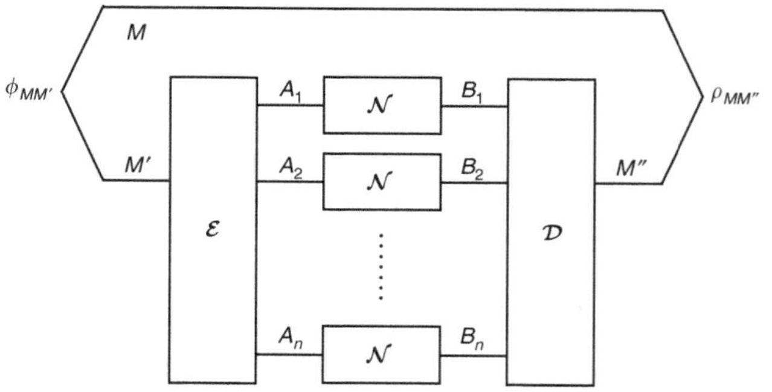

Figure 1 | Coding scheme for entanglement transmission. Coding scheme for entanglement transmission over n uses of a channel $\mathcal{N} \equiv \mathcal{N}_{A\to B}$. The systems M, M' and M'' are isomorphic. The encoder $\mathcal{E} \equiv \mathcal{E}_{M'\to A^n}$ encodes the part M' of the maximally entangled state $\phi_{MM'}$ into the channel input systems. Later, the decoder $\mathcal{D} \equiv \mathcal{D}_{B^n\to M''}$ recovers the state from the channel output systems. The performance of the code is measured using the fidelity $F(\phi_{MM''}, \rho_{MM''})$.

$Q(\mathcal{N}) \leq \frac{1}{\ell_0} I_c(\mathcal{N}^{\otimes \ell_0})$ for all channels $\mathcal{N}$[17].) Worse, the statement is not as strong as we would like it to be because it does not give any indication of the fundamental limits for finite ε or finite n.

For example, even sticking to the asymptotic limit for now, we might be willing to admit a small but nonzero error in our recovery. Formally, instead of requiring that the error vanishes asymptotically, we only require that it does not exceed a certain threshold, ε. Can we then achieve a higher asymptotic rate in the above sense? For cc this is ruled out by Wolfowitz's strong converse theorem[18]. However, surprisingly, the answer to this question is not known for general quantum channels. Recent work[19] at least settles the question in the negative for a class of generalized dephasing channels and in particular for the qubit dephasing channel

$$\mathcal{Z}_\gamma : \rho \mapsto (1-\gamma)\rho + \gamma Z \rho Z, \tag{4}$$

where $\gamma \in [0, 1]$ is a parameter and Z is the Pauli Z operator. Dephasing channels are particularly interesting examples because dephasing noise is dominant in many physical implementations of qubits. The results of ref. 19 thus allow us to fully characterize the achievable region in the limit $n \to \infty$ for such channels, and in particular ensure that

$$\lim_{n\to\infty} \hat{R}_{\mathcal{Z}_\gamma}(n;\varepsilon) = I_c(\mathcal{Z}_\gamma), \tag{5}$$

independent of the value of $\varepsilon \in (0, 1)$. Note also that the regularization is not required here since dephasing channels are degradable[20].

Here we go beyond studying the problem in the asymptotic limit and develop characterizations of the achievable region for finite values of n. We find inner (achievability) and outer (converse) bounds on the boundary of the achievable region. We first discuss these bounds for three important example channels, the qubit dephasing, erasure and depolarizing channel, and then present bounds for general channels.

Qubit dephasing channel. We show that the non-asymptotic achievable region of the qubit dephasing channel is equivalent to the corresponding region of a (classical) binary symmetric channel. This allows us to employ results from classical information theory[10,21,22] to establish the following characterization of the achievable region for the qubit dephasing channel.

Theorem 1. For the qubit dephasing channel $\mathcal{Z}_\gamma$ with $\gamma \in [0, 1]$, the boundary $\hat{R}(n;\varepsilon)$ satisfies

$$\hat{R}(n;\varepsilon) = 1 - h(\gamma) + \sqrt{\frac{v(\gamma)}{n}}\Phi^{-1}(\varepsilon) + \frac{\log n}{2n} + O\left(\frac{1}{n}\right), \tag{6}$$

where Φ is the cumulative normal distribution function, Φ^{-1} its inverse, $h(\cdot)$ denotes the binary entropy, and $v(\cdot)$ the corresponding variance, $v(\gamma) := \gamma(\log\gamma + h(\gamma))^2 + (1-\gamma)(\log(1-\gamma) + h(\gamma))^2$.

The expression without the remainder term $O(\frac{1}{n})$ is called the third order approximation of the (boundary of the) non-asymptotic achievable region. The quantity $v(\gamma)$ is the quantum channel dispersion and characterizes the finite size effects for quantum communication over the qubit dephasing channel. The approximation is visualized in Fig. 2 for an example channel with $\gamma = 0.1$. In Fig. 2a, we plot the smallest achievable error ε as a function of the rate R. Here we use the second order expansion without the term $\frac{1}{2n}\log n$ since it can conveniently be solved for ε. In the limit $n \to \infty$, we see an instantaneous transition of ε from 0 to 1, the signature of a strong converse: coding below the capacity $Q(\mathcal{Z}_\gamma) = 1 - h(\gamma)$ is possible with perfect fidelity, whereas coding above the capacity will necessarily result in a vanishing fidelity.

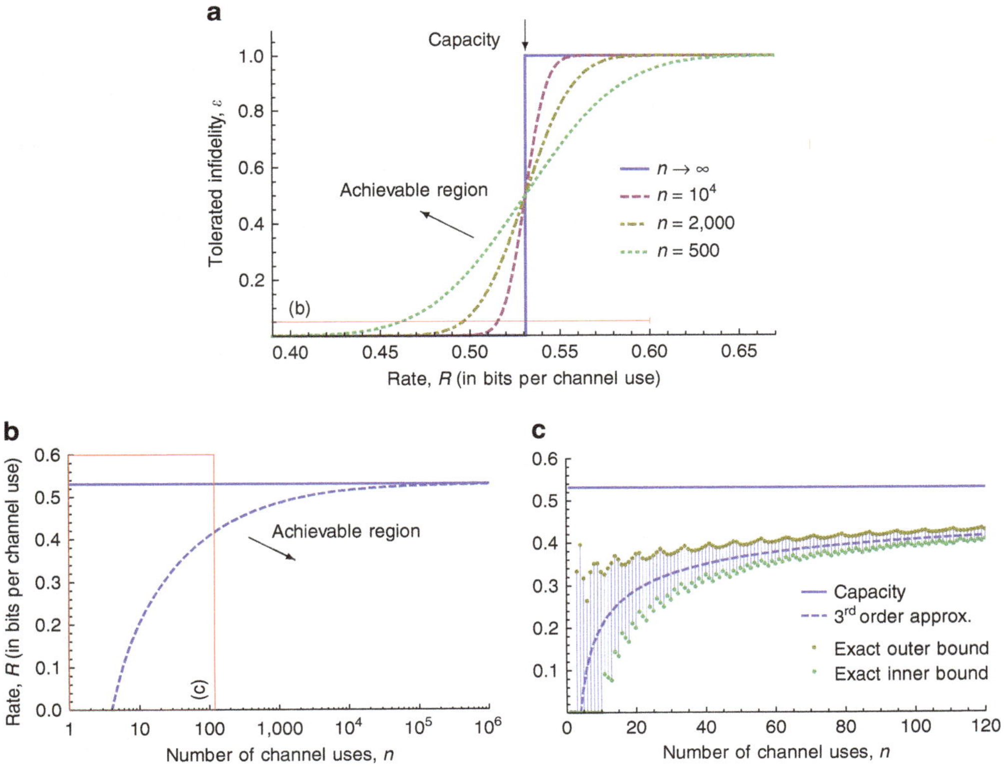

Figure 2 | Example 1—qubit dephasing channel. Approximation of the non-asymptotic achievable rate region of a qubit dephasing channel with $\gamma = 0.1$ (see Theorem 1). All numerical results are evaluated using the binary logarithm, that is, log≡log₂. **(a)** Boundary of the achievable region for fixed n with different values (second order approximation). **(b)** Boundary of the achievable region for fixed infidelity $\varepsilon = 5\%$ (third order approximation) in equation (6). **(c)** Comparison of strict bounds with third order approximation for fixed $\varepsilon = 5\%$.

In Fig. 2b, we plot the third order approximation in equation (6) for the highest achievable rate, $\hat{R}(n;\varepsilon)$, as a function of n for a fixed fidelity of 95% (that is, we set $\varepsilon = 5\%$). For example, this allows us to calculate how many times we need to use the channel to approximately achieve the quantum capacity. The third order approximation shows that we need ~ 850 channel uses to achieve 90% of the quantum capacity. Note that a coding scheme achieving this would probably require us to coherently manipulate 850 qubits in the decoder, which appears to be a quite challenging task. This example shows that the capacity does not suffice to characterize the ability of a quantum channel to transmit information, and further motivates the study of the achievable region for finite n.

Finally, we remark that the third order approximation is quite strong even for small n. To prove this, we compare it to exact upper and lower bounds on $\hat{R}(n;\varepsilon)$ in Fig. 2c and see that the remainder term $O(\frac{1}{n})$ becomes negligible for fairly small $n \approx 100$ for the present values of γ and ε.

Qubit erasure channel. Another channel we can analyse in this manner is the qubit erasure channel, given by the map

$$\mathcal{E}_\beta : \rho \mapsto (1-\beta)\rho + \beta |e\rangle\langle e|, \tag{7}$$

where $\beta \in [0, 1]$ is the probability of erasure and $|e\rangle\langle e|$ is a pure state orthogonal to ρ that indicates erasure. Here we investigate coding schemes that allow free cc assistance between the sender and receiver in both directions, in parallel to the quantum transmission. This setting is quite natural because we can often assume that cc is considerably easier to implement than quantum communication (see Fig. 5 in Methods for a description of such codes). We denote the corresponding boundary of the achievable region by $\hat{R}^{cc}(n;\varepsilon)$. Since this includes all codes that do not take advantage of cc, we clearly have $\hat{R}(n;\varepsilon) \le \hat{R}^{cc}(n;\varepsilon)$ for all channels. This inequality is strict for the erasure channel but for the dephasing channel we find that the asymptotic expansion in equation (6) holds for both $\hat{R}(n;\varepsilon)$ and $\hat{R}^{cc}(n;\varepsilon)$, that is, cc assistance does not help asymptotically (up to third order).

For the qubit erasure channel, we can determine the boundary $\hat{R}^{cc}(n;\varepsilon)$ exactly, again by generalizing[19] and relating the problem to that of the classical erasure channel.

Theorem 2. For the qubit erasure channel $\mathcal{E}_\beta$ with $\beta \in [0, 1]$, the boundary $\hat{R}^{cc}(n;\varepsilon)$ satisfies

$$\varepsilon = \sum_{l=n-k+1}^{n} \binom{n}{l} \beta^l (1-\beta)^{n-l} \left(1 - 2^{n\left(1-\hat{R}^{cc}(n;\varepsilon)\right)-l}\right). \tag{8}$$

Moreover, for large n, we have the expansion

$$\hat{R}^{cc}(n;\varepsilon) = 1 - \beta + \sqrt{\frac{\beta(1-\beta)}{n}}\,\Phi^{-1}(\varepsilon) + O\left(\frac{1}{n}\right). \tag{9}$$

The latter expression is a third order approximation of the achievable region, where $1-\beta$ is the quantum capacity and $\beta(1-\beta)$ is the quantum channel dispersion of the qubit erasure channel. In Fig. 3, we show this approximation for a qubit erasure channel with $\beta = 0.25$ and fidelity 99%. In Fig. 3a, we see that the non-asymptotic achievable region reaches 90% of the channel capacity for $n \approx 180$. Again, this confirms that the non-asymptotic treatment is crucial in the quantum setting. In Fig. 3b, we compare the third order approximation with the exact boundary of the achievable region in equation (8). We see that the approximation is already very precise (and the term $O(\frac{1}{n})$ thus negligible) for fairly small $n \approx 50$.

Qubit depolarizing channel. Another prominent channel is the qubit depolarizing channel. It is given by the map

$$\mathcal{D}_\alpha : \rho \mapsto (1-\alpha)\rho + \frac{\alpha}{3}(X\rho X + Y\rho Y + Z\rho Z), \tag{10}$$

where $\alpha \in [0, 1]$ is a parameter and X, Y, Z are the Pauli operators. For this channel, no closed formula for the quantum capacity $Q(\mathcal{D}_\alpha)$ is known, and the coherent information

$$I_c(\mathcal{D}_\alpha) = 1 - h(\alpha) - \alpha \log 3 \tag{11}$$

is only a strict lower bound on it[23]. However, various upper bounds on the quantum capacity of the qubit depolarizing channel have been established[24–28]. For example, in (ref. 24, Theorem 2) it is essentially shown that $Q(\mathcal{D}_\alpha) \le Q(\mathcal{Z}_\alpha) = 1 - h(\alpha)$, the quantum capacity of the qubit dephasing channel with dephasing parameter α. Here we extend this result to the non-asymptotic setting and find the following outer (converse) bound for the achievable rate region that holds even with cc assistance.

Theorem 3. For the qubit depolarizing channel $\mathcal{D}_\alpha$ with $\alpha \in [0, 1]$, the boundary $\hat{R}^{cc}(n;\varepsilon)$ satisfies

$$\hat{R}^{cc}(n;\varepsilon) \le 1 - h(\alpha) + \sqrt{\frac{v(\alpha)}{n}}\,\Phi^{-1}(\varepsilon) + \frac{\log n}{2n} + O\left(\frac{1}{n}\right), \tag{12}$$

where the right-hand side is simply the asymptotic expansion of the boundary of the achievable rate region for the qubit dephasing channel $\mathcal{Z}_\alpha$ with dephasing parameter α as in Theorem 1.

In Fig. 4a, we plot the second order approximation of the outer bound for a depolarizing channel with $\alpha = 0.05$ and 99% fidelity.

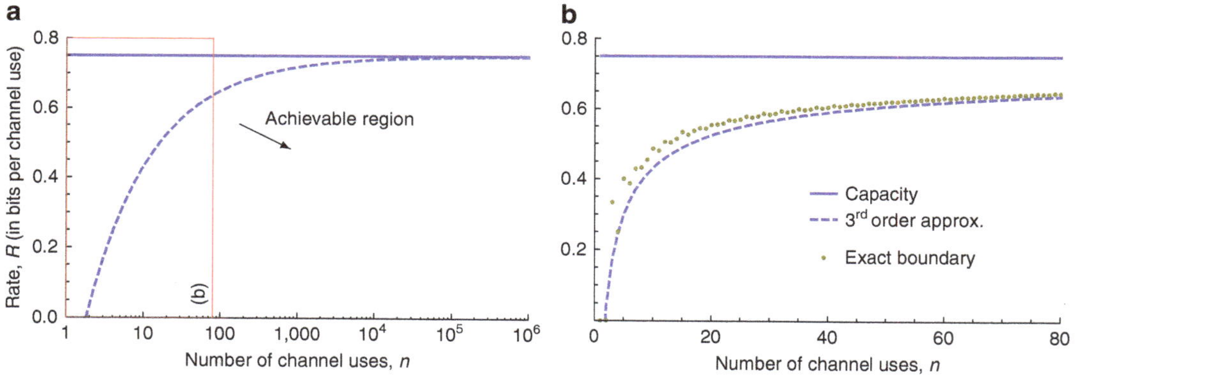

Figure 3 | Example 2—qubit erasure channel. Approximation of the non-asymptotic achievable rate region with classical communication assistance of a qubit erasure channel with $\beta = 0.25$ and fixed infidelity $\varepsilon = 1\%$ (see Theorem 2). (**a**) Boundary of the achievable region. (**b**) Comparison of exact bounds with third order approximation for small values of n.

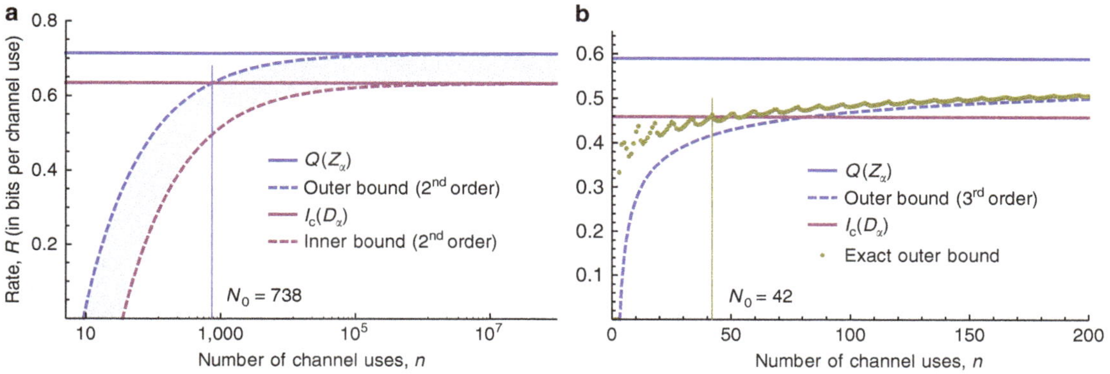

Figure 4 | Example 3—qubit depolarizing channel. Approximate inner and outer bounds on the non-asymptotic achievable rate region for the depolarizing channel (see Theorems 3 and 5) for fixed tolerated infidelity ε. The outer bounds apply to codes with classical communication assistance, whereas the inner bounds consider only unassisted codes. (**a**) Inner and outer bounds for $\alpha=0.05$ and $\varepsilon=1\%$. (**b**) Exact outer bound for $\alpha=0.0825$ and $\varepsilon=5.5\%$.

We see that to implement a code with a communication rate that exceeds the coherent information equation (3), we will need a quantum device that can process at least $N_0=738$ qubits coherently. Moreover, this statement remains true even if we allow for codes with cc assistance. This indicates that the question of whether the coherent information is a good or bad lower bound on the asymptotic quantum capacity is not of immediate practical relevance as long as we do not have a quantum computer that is able to perform a decoding operation on many hundreds of qubits.

In Fig. 4b, we examine a qubit depolarizing channel with parameters $\alpha=0.0825$ and $\varepsilon=5.5\%$. Instead of using an approximation for the outer bound, we use the exact outer bound to give the answer (it is 42) to the question of how many channel uses we need at minimum to exceed the coherent information. However, note that this does not give us any indication of what code (in particular if it is assisted or not), if any, can achieve this point.

General outer and inner bounds. We have so far focused our attention on three specific (albeit very important) examples of channels. However, many of the results derived in this article also hold more generally. For example, we find the following outer (converse) bound.

Theorem 4. For any quantum channel $\mathcal{N}$, the boundary $\hat{R}(n;\varepsilon)$ satisfies

$$\hat{R}(n;\varepsilon) \leq -\log f(\mathcal{N}^{\otimes n},\varepsilon), \tag{13}$$

where $f(\mathcal{N},\varepsilon)$ is the solution to a semidefinite optimization programme defined in equation 24 and Methods. Moreover, if $\mathcal{N}$ is covariant, we find the asymptotic expansion

$$\hat{R}^{cc}(n;\varepsilon) \leq \hat{R}^{cc}_{outer}(n;\varepsilon), \quad \text{with}$$
$$\hat{R}^{cc}_{outer}(n;\varepsilon) = I_R(\mathcal{N}) + \sqrt{\frac{V^{\varepsilon}_R(\mathcal{N})}{n}}\Phi^{-1}(\varepsilon) + O\left(\frac{\log n}{n}\right), \tag{14}$$

where the Rains information, $I_R(\mathcal{N})$, and its variance, $V^{\varepsilon}_R(\mathcal{N})$, are entropic functionals defined in equation (28) and equation (29) and Methods.

In fact, the bound in equation (13) holds also for codes that allow classical post-processing (cpp), as discussed in the Supplementary Notes. Covariant channels are discussed in Methods, and include the dephasing, erasure and depolarizing channels treated above. The semidefinite optimization programme $f(\mathcal{N},\varepsilon)$ is similar in spirit to the metaconverse for classical coding[10,29,30]. For quantum coding, alternative semidefinite optimization programme lower bounds on the

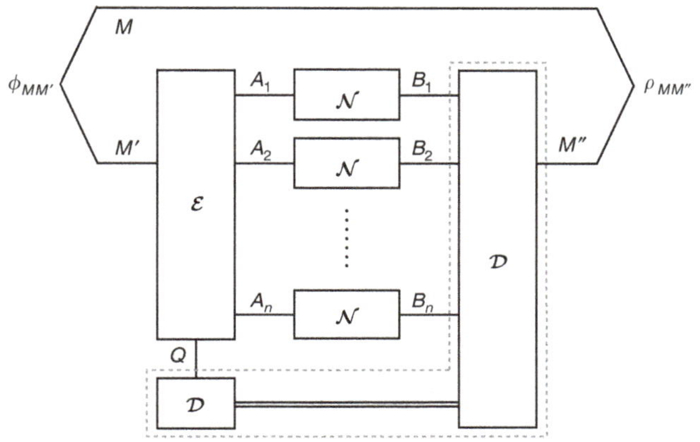

Figure 5 | Coding scheme for entanglement transmission with classical post-processing. Coding scheme for entanglement transmission over n uses of a channel $\mathcal{N}_{A\to B}$ with classical post-processing. The encoder $\mathcal{E}\equiv\mathcal{E}_{M'\to A^nQ}$ encodes M' into the channel input systems and a local memory Q. Later, the decoder $\mathcal{D}\equiv\mathcal{D}_{QB^n\to M''}$ recovers the maximally entangled state from the channel output systems and the memory Q using classical communication and local operations. The performance of the code is measured using the fidelity $F(\phi_{MM'},\rho_{MM''})$.

error boundary $\hat{\varepsilon}_{\mathcal{N}}(n;R) := \max\{\varepsilon : (R,n,\varepsilon) \text{is achievable}\}$ for fixed rate R have been derived in ref. 15. Note that our bound equation (14) is tight up to the second order asymptotically for the qubit dephasing channel (Theorem 1) and the erasure channel with cc assistance (Theorem 2). However, in the generic covariant case the bound is not expected to be tight. Moreover, if the channel is not covariant we cannot asymptotically expand our outer bounds on the achievable rate region in a closed form as above.

Finally, an inner (achievability) bound of the form shown in Theorem 1 also holds generally for all quantum channels.

Theorem 5. For any quantum channel $\mathcal{N}$, the boundary $\hat{R}(n;\varepsilon)$ satisfies

$$\hat{R}(n;\varepsilon) \geq R_{inner}(n;\varepsilon), \quad \text{with}$$
$$\hat{R}_{inner}(n;\varepsilon) = I_c(\mathcal{N}) + \sqrt{\frac{V^{\varepsilon}_c(\mathcal{N})}{n}}\Phi^{-1}(\varepsilon) + O\left(\frac{\log n}{n}\right), \tag{15}$$

where the coherent information, $I_c(\mathcal{N})$, and its variance, $V^{\varepsilon}_c(\mathcal{N})$, are entropic functionals defined in equation 35 and equation 36 and Methods.

Note that the bound equation (15) is tight up to the second order asymptotically for the qubit dephasing channel (Theorem 1). For the erasure channel, this bound does not match the outer bound since it does not take into account cc assistance. For general channels, the bound does not tightly characterize the achievable region. In particular, for $n \to \infty$, it converges to the coherent information and not the regularized coherent information, which can be strictly larger[23]. However, we have reasons to conjecture that the bound is tight for degradable channels[20,31].

The same inner bound has been shown independently and concurrently in ref. 32 using a different decoder.

Discussion

The main contributions of this work can be summarized as follows. We showed—both analytically and quantitatively—that the quantum channel capacity is insufficient to characterize achievable communication rates in the finite resource setting. We provided a remedy, showing that the capacity and quantum channel dispersion together provide a very good characterization, in particular for the practically relevant qubit dephasing, depolarization and erasure channels. This is crucial for practical considerations where one would like to rely on a simple and easy to evaluate formula to estimate the achievable rate region. For instance, one can use the estimated optimal rate region to benchmark explicit codes, for example, in designing a quantum repeater.

More precisely, for general channels, we gave inner (achievability) and outer (converse) bounds on the boundary of the achievable region for quantum communication with finite resources (*cf.*, Theorems 5 and 4). These bounds can be formulated as semidefinite programmes and thus evaluated for small instances. For larger instances, we show that the bounds admit a second order approximation featuring the dispersion (for the converse bound this requires the assumption of channel covariance) which can be evaluated efficiently. We then showed that the inner and outer bounds agree for the qubit dephasing channel (*cf.*, Fig. 2) and qubit erasure channel with cc assistance (*cf.*, Fig. 3) up to the third order asymptotically. For the qubit depolarizing channel (*cf.*, Fig. 4), we gave separate second order approximations for the inner and outer bounds. Closing the gap between these bounds (see shaded area in Fig. 4a), even asymptotically, remains one of the most tantalizing open questions in quantum information theory[26].

For general channels, many questions remain open. For example, we would like to understand if the inner bound in Theorem 5 characterizes the achievable region for all degradable channels[20] (*cf.*, the open questions in ref. 19). Also it would be interesting to explore higher order refinements for channels with zero quantum capacity (for example, for the erasure channel with $\beta \geq 1/2$ and no assistance). This might lead to a better understanding of superactivation of the quantum capacity[33]. Taking a broader view, convex relaxation, such as our semidefinite programme, provides a promising approach to better understand the rate region beyond studying entropic properties. For practical applications, the most important channel not addressed here is the qubit amplitude damping channel, and it is an important open question to analyse it in the finite resource regime.

Finally, we note that our analysis can be extended to the case of entanglement-assisted quantum communication. A short exhibition of this extension is provided in Supplementary Note 1.

Methods

General notation and codes. Here we sketch the main ideas of the proofs of Theorems 4 and 5, and a more detailed exposition is given in Supplementary Note 2. A detailed analysis of the example channels in Theorems 1–3 can be found in Supplementary Note 3.

We denote finite-dimensional Hilbert spaces corresponding to individual quantum systems by capital letters. In particular, we use A and B to model the channel input and output space, respevtively, whereas M and the isomorphic spaces M' and M'' are used to model the quantum systems containing the maximally entangled state to be transmitted. We also use A^n to denote the n-fold tensor product of A for any $n \in \mathbb{N}$. We use $\mathcal{P}(A)$ to denote the set of positive semidefinite operators on A, and $\mathcal{S}(A) := \{\rho_A \in \mathcal{P}(A)\colon \operatorname{tr}(\rho_A) = 1\}$ to denote quantum states with unit trace on A. We denote the dimension of A by $|A|$. Pure states are of the form $\rho_A = |\phi\rangle\langle\phi|$, where $|\phi\rangle$ is a vector in A and $\langle\phi|$ its dual functional. The marginals of a bipartite quantum state $\rho_{AB} \in \mathcal{S}(AB)$ on A and B are denoted by ρ_A and ρ_B, respectively. A quantum channel $\mathcal{N}_{A\to B}$ is a completely positive trace-preserving map from states on A to states on B. For any state ρ_A, we define the canonical purification $|\psi^\rho\rangle_{AA'} := |A|\sqrt{\rho_A} \otimes 1_{A'}|\phi\rangle_{AA'}$, where A' is isomorphic to A and $\phi_{AA'}$ is the maximally entangled state. To express our results, we use Umegaki's quantum relative entropy[34], $D(\rho\|\sigma) := \operatorname{tr}[\rho(\log\rho - \log\sigma)]$ and the quantum relative entropy variance[35,36], $V(\rho\|\sigma) := \operatorname{tr}\left[\rho(\log\rho - \log\sigma - D(\rho\|\sigma))^2\right]$. The coherent information and the coherent information variance[35] of a bipartite state ρ_{AB} are given as

$$I(A\rangle B)_\rho := D(\rho_{AB}\|1_A \otimes \rho_B) \quad \text{and} \quad V(A\rangle B)_\rho := V(\rho_{AB}\|1_A \otimes \rho_B). \tag{16}$$

We have defined unassisted entanglement-transmission codes in Results. Let us reintroduce them in the context of codes assisted by cpp. For this, we consider any quantum channel $\mathcal{N} \equiv \mathcal{N}_{A\to B}$ and its n-fold extension $\mathcal{N}^{\otimes n}$ that maps states on A^n to states on B^n. An entanglement-transmission code assisted by cpp for $\mathcal{N}^{\otimes n}$ is given by a triplet $\{|M|, \mathcal{E}, \mathcal{D}\}$, as depicted in Fig. 5. Here $|M|$ is the local dimension of a maximally entangled state $|\phi\rangle_{MM'}$ that is to be transmitted over $\mathcal{N}^{\otimes n}$. The encoder $\mathcal{E}_{M'\to A^nQ}$ is a completely positive trace-preserving map that prepares the channel inputs $A_1, A_2, \ldots A_n$ and a local memory system, which we denote by Q. The decoder $\mathcal{D}_{QB^n\to M''}$ is a completely positive trace-preserving map that is restricted to local operations and cc with regard to the bipartition $Q{:}B^n$ and outputs M'' on the receiver's side.

The boundary of the achievable rate region for these codes is denoted by $\hat{R}^{\mathrm{cpp}}_{\mathcal{N}}(n,\varepsilon)$. Finally, we note that unassisted codes are recovered if we choose Q to be trivial. Hence, unassisted codes are contained in the set of assisted codes and we have $\hat{R}(n,\varepsilon) \leq \hat{R}^{\mathrm{cpp}}(n,\varepsilon)$. Moreover, for covariant channels we will see later that $\hat{R}^{\mathrm{cpp}}(n,\varepsilon) = \hat{R}^{\mathrm{cc}}(n,\varepsilon)$ since all cc can be postponed to after the quantum communication. Hence, while we will in the following derive our converse bounds for $\hat{R}^{\mathrm{cpp}}(n,\varepsilon)$, they are also valid for $\hat{R}^{\mathrm{cc}}(n,\varepsilon)$ when the channel is covariant.

Outer bounds on the achievable rate region. Our converse results are inspired by the strong converse results for generalized dephasing channels and the meta-converse for classical channel coding[10]. They are expressed in terms of the channel hypothesis testing Rains relative entropy, which is defined following the generalized divergence framework discussed in ref. 19. First, let us introduce the Rains set[25,37], which is a superset of the set of positive partial transpose (PPT) states. It is defined as $\mathrm{PPT}^*(A{:}B) := \{\tau_{AB} \in \mathcal{P}(AB) \mid \|\mathcal{T}_B(\tau_{AB})\|_1 \leq 1\}$, where $\mathcal{T}_B$ denotes the partial transpose map on B. We have the following crucial inequality (ref. 38, Lemma 2): for every $\sigma_{AB} \in \mathrm{PPT}^*(A{:}B)$, we have

$$\langle\phi|\sigma_{AB}|\phi\rangle_{AB} \leq \frac{1}{|M|} \tag{17}$$

for all maximally entangled states ϕ_{AB} of local dimension $|M|$. The set is closed under local quantum operations on A and B supported by cc between A and B. Finally, we employ the hypothesis testing relative entropy[39], (in the form of ref. 40)

$$D_H^\varepsilon(\rho\|\sigma) := -\log\min\{\operatorname{tr}[\Lambda\sigma] \mid 0 \leq \Lambda \leq 1 \wedge \operatorname{tr}[\Lambda\rho] \geq 1-\varepsilon\}. \tag{18}$$

We first formulate a general metaconverse bounding possible rates R given a tolerated infidelity ε for a single use ($n = 1$) of a fixed channel $\mathcal{N} \equiv \mathcal{N}_{A\to B}$. For this purpose, consider any state $\rho_{MM''} = \mathcal{D} \circ \mathcal{N} \circ \mathcal{E}(\phi_{MM'})$ at the output of a code achieving fidelity $1-\varepsilon$ and any state $\sigma_{MM''} \in \mathrm{PPT}^*(M{:}M'')$. These must satisfy, according to equation (17),

$$\operatorname{tr}[\sigma_{MM'}\phi_{MM'}] \leq \frac{1}{|M|} \quad \text{and} \quad \operatorname{tr}[\rho_{MM'}\phi_{MM'}] \geq 1-\varepsilon. \tag{19}$$

From this, we can conclude that $D_H^\varepsilon(\rho_{MM'}\|\sigma_{MM'}) \geq \log|M|$ by using the projection $\Lambda = \phi_{MM'}$ as our hypothesis test in equation (18). At this stage, we can use the data-processing inequality of the hypothesis testing divergence[40] to remove the decoder from the picture. Minimizing over all auxiliary states $\sigma_{MQB} \in \mathrm{PPT}^*(MQ{:}B)$, this yields

$$\min_{\sigma_{MQB}} D_H^\varepsilon(\mathcal{N}_{A\to B}(\rho_{MQA})\|\sigma_{MQB}) \geq \log|M|, \quad \text{where} \quad \rho_{MQA} = \mathcal{E}(\phi_{MM'}). \tag{20}$$

Crucially, we rely on the fact that $\mathrm{PPT}^*(MQ{:}B)$ gets mapped into $\mathrm{PPT}^*(M{:}M'')$ by the action of the decoder. Now we observe that by choosing the register Q sufficiently large, we can assume that the encoder is an isometry without loss of generality. Hence, for a fixed marginal $\rho_A = \operatorname{tr}_{QM}(\rho_{MQA})$, we can rewrite the above inequality using the substitutions $A \to A'$ and $MQ \to A$ as

$$\min_{\sigma_{AB}\in\mathrm{PPT}^*(A:B)} D_H^\varepsilon(\mathcal{N}_{A'\to B}(\psi^\rho_{AA'})\|\sigma_{AB}) \geq \log|M|. \tag{21}$$

Optimized over all codes (and thus marginals ρ_A), we find that

$$\hat{R}^{\mathrm{cpp}}(1;\varepsilon) \leq I_{\mathrm{R}}^{\varepsilon}(\mathcal{N}) \tag{22}$$

with the channel hypothesis testing Rains relative entropy defined as

$$I_{\mathrm{R}}^{\varepsilon}(\mathcal{N}) := \max_{\rho_A\in\mathcal{S}(A)} \min_{\sigma_{AB}\in\mathrm{PPT}^*(A:B)} D_H^{\varepsilon}\left(\mathcal{N}_{A'\to B}(\psi_{AA'}^{\rho})\|\sigma_{AB}\right). \tag{23}$$

Note that this outer bound also holds for coding schemes with (unphysical) PPT assistance including classical pre- and post-processing assistance (see ref. 15 for a more comprehensive discussion of PPT assisted codes). The bound can be further relaxed to $\hat{R}^{\mathrm{cpp}}(1;\varepsilon) \leq -\log f(\mathcal{N},\varepsilon)$, where $f(\mathcal{N},\varepsilon)$ is a semidefinite programme given below. This semidefinite optimization is discussed in more detail in Supplementary Note 4.

$$\begin{aligned} f(\mathcal{N},\varepsilon) \quad = \quad & \text{minimum} \quad \mathrm{tr}[\xi_A] \\ & \text{subject to} \quad \xi_A, \Theta_{AB}, \Lambda_{AB}, \rho_A \geq 0 \\ & \xi_A \otimes 1_B \geq \Lambda_{AB} + \mathcal{T}_A(\Theta_{AB}) \\ & \Lambda_{AB} \leq \rho_A \otimes 1_B \\ & |A|\mathrm{tr}[\Lambda_{AB}\mathcal{N}_{A'\to B}(\phi_{AA'})] \geq 1-\varepsilon \\ & \mathrm{tr}[\rho_A] = 1 \end{aligned} \tag{24}$$

Moreover, the bound in equation (22) has the useful property that channel symmetries can be used to simplify its form, as we will see next. Suppose G is a group represented by unitary operators U_g on A and V_g on B. A quantum channel $\mathcal{N}_{A\to B}$ is covariant with respect to this group (and its representations) when

$$V_g\mathcal{N}(\cdot)V_g^{\dagger} = \mathcal{N}\left(U_g\cdot U_g^{\dagger}\right), \quad \forall g\in G. \tag{25}$$

Now the main workhorse to simplify our outer bounds for channels with symmetries is (ref. 19, Proposition 2), which states that we may restrict the optimization in equation (23) to input states that are invariant under the rotations $U_g\cdot U_g^{\dagger}$ for any $g\in G$. For channels of the form $\mathcal{N}^{\otimes n}$ which are invariant under permutation of the input and output systems, this allows us to restrict attention to input states that are permutation invariant.

Moreover, we call a channel covariant if it is covariant with respect to a group which has a representation U_g on A that is a one-design, that is, the map $\sum_{g\in G}\frac{1}{|G|}U_g(\cdot)U_g^{\dagger}$ always outputs the fully mixed state. In this case, the channel input state can be chosen to be fully mixed (respectively its purification is maximally entangled). Moreover, any such group allows for a corresponding teleportation protocol[41] (see the construction in ref. 42), and thus all interactive cc can be postponed until after the quantum communication is completed by the argument given in refs 43,44. From this, we can conclude that $\hat{R}^{\mathrm{cpp}}(n,\varepsilon)=\hat{R}^{\mathrm{cc}}(n,\varepsilon)$ for all covariant channels.

Now let $\mathcal{N}$ be a covariant quantum channel and $\phi_{AA'}$ a maximally entangled state. Then, our bound in equation (22) applied to the channel $\mathcal{N}^{\otimes n}$ yields

$$\hat{R}^{\mathrm{cc}}(n,\varepsilon) = \hat{R}^{\mathrm{cpp}}(n;\varepsilon) \leq \min_{\sigma_{AB}\in\mathrm{PPT}^*(A:B)} \frac{1}{n} D_H^{\varepsilon}\left(\mathcal{N}_{A'\to B}(\phi_{AA'})^{\otimes n}\|\sigma_{AB}^{\otimes n}\right), \tag{26}$$

where we voluntarily restricted the minimization to product states of the form $\sigma_{AB}^{\otimes n}$ for some $\sigma_{AB}\in\mathrm{PPT}^*(A{:}B)$. Moreover, since these states have tensor power structure, the outer bound can be expanded using[35,36]

$$\frac{1}{n}D_H^{\varepsilon}(\rho^{\otimes n}\|\sigma^{\otimes n}) = D(\rho\|\sigma) + \sqrt{\frac{V(\rho\|\sigma)}{n}}\Phi^{-1}(\varepsilon) + O\left(\frac{\log n}{n}\right). \tag{27}$$

This leads to the formal statement of Theorem 4.

Formal Theorem 4. Let $\mathcal{N} \equiv \mathcal{N}_{A\to B}$ be a covariant quantum channel and let $\phi_{AA'}$ be maximally entangled. We define the Rains information of $\mathcal{N}$ as

$$I_{\mathrm{R}}(\mathcal{N}) := \min_{\sigma_{AB}\in\mathrm{PPT}^*(A:B)} D(\mathcal{N}_{A'\to B}(\phi_{AA'})\|\sigma_{AB}), \tag{28}$$

where we let $\Pi\subset\mathrm{PPT}^*(A{:}B)$ be the set of states that achieve the minimum. The variance of the channel Rains information is

$$V_{\mathrm{R}}^{\varepsilon}(\mathcal{N}) := \begin{cases} \max_{\sigma_{AB}\in\Pi} V(\mathcal{N}_{A'\to B}(\phi_{AA'})\|\sigma_{AB}) & \text{for} \quad \varepsilon<\frac{1}{2} \\ \min_{\sigma_{AB}\in\Pi} V(\mathcal{N}_{A'\to B}(\phi_{AA'})\|\sigma_{AB}) & \text{for} \quad \varepsilon\geq\frac{1}{2} \end{cases}. \tag{29}$$

For any fixed $\varepsilon\in(0,1)$, the achievable region with classsical communication assistance satisfies

$$\hat{R}^{\mathrm{cc}}(n;\varepsilon) \leq \hat{R}_{\mathrm{outer}}^{\mathrm{cc}}(n;\varepsilon), \quad \text{with}$$
$$\hat{R}_{\mathrm{outer}}^{\mathrm{cc}}(n;\varepsilon) = I_{\mathrm{R}}(\mathcal{N}) + \sqrt{\frac{V_{\mathrm{R}}^{\varepsilon}(\mathcal{N})}{n}}\Phi^{-1}(\varepsilon) + O\left(\frac{\log n}{n}\right). \tag{30}$$

Inner bounds on the achievable rate region. We use the decoupling approach[45–47], and in particular a one-shot bound[31] which is a tighter version of previous bounds[48–50]. To reproduce their result, we need the following additional notation. Sub-normalized quantum states are collected in the set $\mathcal{S}_{\bullet}(A):=\{\rho_A\in\mathcal{P}(A)\colon \mathrm{tr}(\rho_A)\leq 1\}$. The purified distance[51] ε-ball around $\rho\in\mathcal{S}(A)$ is then defined as $\mathcal{B}^{\varepsilon}(\rho_A):=\{\bar{\rho}_A\in\mathcal{S}_{\bullet}(A)|F(\bar{\rho}_A,\rho_A)\geq(1-\varepsilon)^2\}$. Finally, for $\rho_{AB}\in\mathcal{S}(AB)$ and $\varepsilon\geq 0$ the smooth conditional min-entropy[51–53] is defined as

$$H_{\min}^{\varepsilon}(A|B)_{\rho} := \sup_{\bar{\rho}_{AB}\in\mathcal{B}^{\varepsilon}(\rho_{AB})} \sup_{\sigma_B\in\mathcal{S}(B)} \sup\{\lambda\in\mathbb{R}|\bar{\rho}_{AB}\leq 2^{-\lambda}\cdot 1_A\otimes\sigma_B\}. \tag{31}$$

Let us now restate (Proposition 20 in ref. 31) expressed in terms of the non-asymptotic achievable region as introduced in the Results. Let $\mathcal{N}_{A\to B}$ be a quantum channel with complementary channel $\mathcal{N}_{A\to E}^{c}$. Then $\{R,1,\varepsilon\}$ is achievable if, for some $\eta\in(0,\varepsilon]$ and some state $\rho_A\in\mathcal{S}(A)$, we have

$$R \leq H_{\min}^{\sqrt{\varepsilon}-\eta}(A|E)_{\omega} - 4\log\frac{1}{\eta}, \tag{32}$$

where $\omega_{AE}=(\mathcal{I}_A\otimes\mathcal{N}_{A'\to E}^{c})(\psi_{AA'}^{\rho})$. This leads immediately to the following inner bound on the achievable region. Using $\omega_{A^nE^n}=(\mathcal{I}_{A^n}\otimes(\mathcal{N}_{A'\to E}^{c})^{\otimes n})(\psi_{A^nA'^n})$, we have

$$\hat{R}(n;\varepsilon) \geq \sup_{\eta\in(0,\varepsilon)} \sup_{\rho_{A^n}\in\mathcal{S}(A^n)} \frac{1}{n}\left(H_{\min}^{\sqrt{\varepsilon}-\eta}(A^n|E^n)_{\omega} - 4\log\frac{1}{\eta} - 1\right). \tag{33}$$

The problem with this bound is that it is generally hard to evaluate, even for moderately large values of n. Hence we are interested to further simplify the expression on the right-hand side in this regime. To do so, we choose $\eta=1/\sqrt{n}$ and use input states of the form $\rho_A^{\otimes n}$. This yields the following relaxation, which holds if $n>\frac{1}{\varepsilon}$:

$$\hat{R}(n;\varepsilon) \geq \sup_{\rho_A\in\mathcal{S}(A)} \frac{1}{n}\left(H_{\min}^{\varepsilon_n}(A^n|E^n)_{\omega^{\otimes n}} - 2\log n - 1\right). \tag{34}$$

Here we introduced $\varepsilon_n=\sqrt{\varepsilon}-1/\sqrt{n}$ and ω_{AE} as in equation (32). Using a second order expansion[35] similar to the one in equation (27), we give an asymptotic expansion of the expression on the right-hand side of equation (34). This yields Theorem 5.

Formal Theorem 5. Let $\mathcal{N} \equiv \mathcal{N}_{A\to B}$ be a quantum channel. We define its coherent information as

$$I_c(\mathcal{N}) := \max_{\rho_A\in\mathcal{S}(A)} I(A\rangle B)_{\omega}, \quad \text{with} \quad \omega_{AB} = (\mathcal{I}_A\otimes\mathcal{N}_{A'\to B})(\psi_{AA'}^{\rho}) \tag{35}$$

and let $\Pi\subset\mathcal{S}(A)$ be the set of states that achieve the maximum. Define

$$V_c^{\varepsilon}(\mathcal{N}) := \begin{cases} \min_{\rho_A\in\Pi} V(A\rangle B)_{\omega} & \text{for} \quad \varepsilon<\frac{1}{2} \\ \max_{\rho_A\in\Pi} V(A\rangle B)_{\omega} & \text{for} \quad \varepsilon\geq\frac{1}{2} \end{cases}. \tag{36}$$

Then, for any fixed $\varepsilon\in(0,1)$, the achievable region satisfies

$$\hat{R}(n;\varepsilon) \geq \hat{R}_{\mathrm{inner}}(n;\varepsilon), \quad \text{with}$$
$$\hat{R}_{\mathrm{inner}}(n;\varepsilon) = I_c(\mathcal{N}) + \sqrt{\frac{V_c^{\varepsilon}(\mathcal{N})}{n}}\Phi^{-1}(\varepsilon) + O\left(\frac{\log n}{n}\right). \tag{37}$$

References

1. Shannon, C. A mathematical theory of communication. *Bell Syst. Tech. J.* **27,** 379–423 (1948).
2. Barnum, H., Knill, E. & Nielsen, M. A. On quantum fidelities and channel capacities. *IEEE Trans. Inf. Theory* **46,** 1317–1329 (2000).
3. Barnum, H., Nielsen, M. A. & Schumacher, B. Information transmission through a noisy quantum channel. *Phys. Rev. A* **57,** 4153 (1998).
4. Devetak, I. The private classical capacity and quantum capacity of a quantum channel. *IEEE Trans. Inf. Theory* **51,** 44–55 (2005).
5. Lloyd, S. The capacity of the noisy quantum channel. *Phys. Rev. A* **55,** 1613–1622 (1996).
6. Schumacher, B. & Nielsen, M. A. Quantum data processing and error correction. *Phys. Rev. A* **54,** 2629 (1996).
7. Shor, P. W. *Lectures Notes, MSRI Workshop on Quantum Computation* (2002).
8. Strassen., V. in *Trans. Third Prague Conference on Information Theory* 689–723 (Prague, Czech Republic, 1962).
9. Hayashi, M. Information spectrum approach to second-order coding rate in channel coding. *IEEE Trans. Inf. Theory* **55,** 4947–4966 (2009).
10. Polyanskiy, Y., Poor, H. V. & Verdú, S. Channel coding rate in the finite blocklength regime. *IEEE Trans. Inf. Theory* **56,** 2307–2359 (2010).
11. Tan, V. Y. F. Asymptotic Estimates in Information Theory with Non-Vanishing Error Probabilities. *Found. Trends Commun. Inf. Theory* **11** (2014).
12. Tomamichel, M. & Tan, V. Y. F. Second-order asymptotics for the classical capacity of image-additive quantum channels. *Commun. Math. Phys.* **338,** 103–137 (2015).
13. Datta, N., Tomamichel, M. & Wilde., M. M. *On the second-order coding rates for entanglement-assisted communication.* Preprint at http://arxiv.org/abs/1405.1797 (2016).
14. Uhlmann., A. The transition probability for states of star-algebras. *Ann. Phys.* **497,** 524–532 (1985).
15. Leung, D. & Matthews, W. On the power of PPT-preserving and non-signalling codes. *IEEE Trans. Inf. Theory* **61,** 4486–4499 (2015).

16. Devetak, I. & Winter, A. Distillation of secret key and entanglement from quantum states. *Proc. R. Soc. A* **461,** 207–235 (2005).
17. Cubitt, T. *et al.* Unbounded number of channel uses are required to see quantum capacity. *Nat. Commun.* **6,** 6739 (2015).
18. Wolfowitz, J. A note on the strong converse of the coding theorem for the general discrete finite-memory channel. *Inform. Control* **3,** 89–93 (1960).
19. Tomamichel, M., Wilde, M. M. & Winter, A. Strong converse rates for quantum communication. Preprint at http://arxiv.org/abs/1406.2946 (2014).
20. Devetak, I. & Shor, P. W. The capacity of a quantum channel for simultaneous transmission of classical and quantum information. *Commun. Math. Phys.* **256,** 287–303 (2005).
21. Gallager, R. G. *Information Theory and Reliable Communication* (Wiley, 1968).
22. Poltyrev, G. Bounds on the decoding error probability of binary linear codes via their spectra. *IEEE Trans. Inf. Theory* **40,** 1284–1292 (1994).
23. DiVincenzo, D., Shor, P. & Smolin, J. Quantum-channel capacity of very noisy channels. *Phys. Rev. A* **57,** 830–839 (1998).
24. Rains, E. M. Entanglement purification via separable superoperators. Preprint at http://arxiv.org/abs/quant-ph/9707002 (1997).
25. Rains, E. M. A semidefinite program for distillable entanglement. *IEEE Trans. Inf. Theory* **47,** 2921–2933 (2001).
26. Smith, G. & Smolin, J. A. in *IEEE Information Theory Workshop Proceedings* 368–372 (2008).
27. Smith, G., Smolin, J. A. & Winter., A. The quantum capacity with symmetric side channels. *IEEE Trans. Inf. Theory* **54,** 4208–4217 (2008).
28. Sutter, D., Scholz, V. B. & Renner, R. Approximate degradable quantum channels. Preprint at http://arxiv.org/abs/1412.0980 (2014).
29. Matthews, W. A linear program for the finite block length converse of Polyanskiy-Poor-Verdú via nonsignaling codes. *IEEE Trans. Inf. Theory* **58,** 7036–7044 (2012).
30. Matthews, W. & Wehner, S. Finite blocklength converse bounds for quantum channels. *IEEE Trans. Inf. Theory* **60,** 7317–7329 (2014).
31. Morgan, C. & Winter, A. 'Pretty strong' converse for the quantum capacity of degradable channels. *IEEE Trans. Inf. Theory* **60,** 317–333 (2014).
32. Beigi, S., Datta, N. & Leditzky, F. Decoding quantum information via the petz recovery map. Preprint at http://arxiv.org/abs/1504.04449 (2015).
33. Smith, G. & Yard, J. T. Quantum communication with zero-capacity channels. *Science* **321,** 1812–1815 (2008).
34. Umegaki, H. Conditional expectation in an operator algebra. *Kodai Math. Sem. Rep.* **14,** 59–85 (1962).
35. Tomamichel, M. & Hayashi, M. A hierarchy of information quantities for finite block length analysis of quantum tasks. *IEEE Trans. Inf. Theory* **59,** 7693–7710 (2013).
36. Li, K. Second-order asymptotics for quantum hypothesis testing. *Ann. Stat.* **42,** 171–189 (2014).
37. Audenaert, K., De Moor, B., Vollbrecht, K. & Werner, R. F. Asymptotic relative entropy of entanglement for orthogonally invariant states. *Phys. Rev. A* **66,** 032310 (2002).
38. Rains, E. M. Bound on distillable entanglement. *Phys. Rev. A* **60,** 179–184 (1999).
39. Hiai, F. & Petz, D. The proper formula for relative entropy and its asymptotics in quantum probability. *Commun. Math. Phys.* **143,** 99–114 (1991).
40. Wang, L. & Renner, R. One-shot classical-quantum capacity and hypothesis testing. *Phys. Rev. Lett.* **108,** 200501 (2012).
41. Bennett, C. H. *et al.* Teleporting an unknown quantum state via dual classical and Einstein-Podolsky-Rosen Channels. *Phys. Rev. Lett.* **70,** 1895–1899 (1993).
42. Werner, R. F. All teleportation and dense coding schemes. *J. Phys. A* **34,** 7081–7094 (2001).
43. Bennett, C. H., DiVincenzo, D. P., Smolin, J. A. & Wootters, W. K. Mixed-state entanglement and quantum error correction. *Phys. Rev. A* **54,** 3824–3851 (1996).
44. Pirandola, S., Laurenza, R., Ottaviani, C. & Banchi, L. The ultimate rate of quantum communications. Preprint at http://arxiv.org/abs/1510.08863 (2015).
45. Dupuis, F. *The Decoupling Approach to Quantum Information Theory.* PhD thesis, Univ. Montréal. Available at http://arxiv.org/abs/1004.1641 (2009).
46. Dupuis, F., Berta, M., Wullschleger, J. & Renner, R. One-shot decoupling. *Commun. Math. Phys.* **328,** 251–284 (2014).
47. Hayden, P., Horodecki, M., Yard, J. & Winter, A. A decoupling approach to the quantum capacity. *Open Syst. Inf. Dyn.* **15,** 7–19 (2008).
48. Berta., M. Single-Shot Quantum State Merging. Diploma thesis, ETH Zurich. Available at http://arxiv.org/abs/0912.4495 (2008).
49. Buscemi, F. & Datta, N. The quantum capacity of channels with arbitrarily correlated noise. *IEEE Trans. Inf. Theory* **56,** 1447–1460 (2010).
50. Datta, N. & Hsieh, M.-H. The apex of the family tree of protocols: optimal rates and resource inequalities. *N. J. Phys.* **13,** 093042 (2011).
51. Tomamichel, M., Colbeck, R. & Renner, R. Duality between smooth min- and max-entropies. *IEEE Trans. Inf. Theory* **56,** 4674–4681 (2010).
52. Renner., R. *Security of Quantum Key Distribution.* PhD thesis, ETH Zurich. Available at http://arxiv.org/abs/quant-ph/0512258 (2005).
53. Tomamichel, M. *Quantum Information Processing with Finite Resources—Mathematical Foundations,* Vol. 5 (Springer International Publishing, 2016).

Acknowledgements

We thank Chris Ferrie, Chris Granade, William Matthews, David Sutter and Mark Wilde for helpful discussions. M.T. is funded by an ARC Discovery Early Career Researcher Award Fellowship and acknowledges support from the ARC Centre of Excellence for Engineered Quantum Systems (EQUS). M.B. acknowledges funding provided by the Institute for Quantum Information and Matter. J.M.R. was supported by the Swiss National Science Foundation (through the National Centre of Competence in Research 'Quantum Science and Technology').

Author contributions

M.T., M.B. and J.R. developed the main ideas and technical results. M.T. wrote the manuscript with the help of M.B. and J.R.

Additional information

Competing financial interests: The authors declare no competing financial interests.

21

Almost quantum correlations

Miguel Navascués[1,2,3], Yelena Guryanova[3], Matty J. Hoban[4,5] & Antonio Acín[4,6]

Quantum theory is not only successfully tested in laboratories every day but also constitutes a robust theoretical framework: small variations usually lead to implausible consequences, such as faster-than-light communication. It has even been argued that quantum theory may be special among possible theories. Here we report that, at the level of correlations among different systems, quantum theory is not so special. We define a set of correlations, dubbed 'almost quantum', and prove that it strictly contains the set of quantum correlations but satisfies all-but-one of the proposed principles to capture quantum correlations. We present numerical evidence that the remaining principle is satisfied too.

[1] Department of Physics, Bilkent University, Ankara 06800, Turkey. [2] Universitat Autònoma de Barcelona, 08193 Bellaterra (Barcelona), Spain. [3] H.H. Wills Physics Laboratory, University of Bristol, Tyndall Avenue, Bristol BS8 1TL, UK. [4] ICFO-Institut de Ciencies Fotoniques, 08860 Castelldefels (Barcelona), Spain. [5] Department of Computer Science, University of Oxford, Parks Road, Oxford OX1 3QD, UK. [6] ICREA—Institucio Catalana de Recerca i Estudis Avançats, 08010 Barcelona, Spain. Correspondence and requests for materials should be addressed to A.A. (email: antonio.acin@icfo.es).

Ever since its conception, quantum mechanics has eluded consensual understanding. This has motivated the search for models alternative to quantum theory that could offer a more intuitive explanation of the observed phenomena. The study of modifications of quantum theory has also been motivated by the difficulty in merging it with relativity, the hope being that alternative models may facilitate this task. However, up to now, the existing alterations of quantum theory, even if *a priori* simple, have been problematic for fundamental reasons. For instance, they have been shown to imply a significant increase in the computational power of the theory[1] or violations of the no-signalling principle[2,3]: one of the pillars of relativity that states that superluminal communication is impossible. These results, among others, have led researchers to question whether 'quantum mechanics is special'[4] and 'an island in theory space'[5].

The possibility that quantum mechanics could indeed be the ultimate physical theory has motivated many attempts to reconstruct it from physically intuitive axioms within the framework of generalized probabilistic theories[6–9]. All such works, however, rely on a separation between states and observables, implicit in the formalism of generalized probabilistic theories. Furthermore, most axioms in this research programme are formulated in terms of abstract entities, such as states, which are not directly testable in an experiment.

A different approach was introduced in ref. 10, where the authors propose to classify and study physical theories according to the correlations that they allow to generate between different systems. In this approach, the goal is to understand quantum correlations from physical principles stated only in terms of correlations. The advantage of this approach comes from the fact that correlations among systems can be experimentally estimated in a 'device-independent' way, that is, just from the observed statistics and without any theoretical modelling of the systems. Thus, contrary to the axioms in the reconstruction paradigm, the validity of device-independent principles can be put to test without further considerations.

Popescu and Rohrlich[10] initiated this research line by considering whether quantum theory is the most nonlocal theory compatible with special relativity. They initially conjectured that, if Nature were more nonlocal, then we could violate special relativity just as observed in the aforementioned modifications to quantum theory. Remarkably, they proved that the answer to this question is negative and one can posit the existence of non-quantum but also non-signalling correlations[10].

Even though the principle of no-signalling fails to capture quantum correlations, the work of Popescu and Rohrlich motivated the quest to find a principle that does. As a result, several device-independent principles have been proposed so far that have been partly successful at capturing the nonlocality of quantum physics[10–15]. As above, the hope is to prove that the presence of correlations beyond quantum theory would have implausible consequences, in the sense of violating a reasonable principle for physical correlations, providing an intuitive explanation for the degree of nonlocality in quantum theory.

It is worth mentioning that the programme of understanding quantum correlations is vast and has been considered from different perspectives. For instance, in contextuality scenarios, quantum correlations refer to those correlations observed among different projective (or sharp) measurements subject to some compatibility constraints (see, for instance, refs 16–18 for different formalizations of contextuality scenarios). Here, we follow Popescu and Rohrlich and focus on correlations among outputs of measurements performed on separate systems. Note also that we do not consider sequences of measurements; therefore, measurements produce one single classical output. It is then assumed that no communication takes place among the systems while the measurements are performed, which can be enforced by arranging the measurements so that they define space-like separated events. As mentioned, and contrary to other approaches, the virtue of the considered scenario is that correlations can be estimated in a device-independent manner.

In this work, we show that quantum theory is, at the level of correlations among systems, not as special as expected. In particular, we identify a set of non-signalling correlations, which we name 'almost quantum', and prove that this set (i) is strictly larger than the set of quantum correlations but (ii) satisfies all-but-one of the device-independent principles proposed for quantum correlations. Moreover, we offer strong evidence that it satisfies the remaining principle too. Our results demonstrate that it is possible to modify the structure of quantum correlations without running into any operational contradiction with quantum predictions. They also shed doubts on the research programme of recovering quantum correlations from 'natural' device-independent physical principles. Finally, invoking arguments from the history approach introduced in the context of quantum gravity, we motivate the study of almost quantum correlations as a possible avenue for generalizations of quantum theory.

Results

The set of almost quantum correlations. The considered scenario consists of n different parties each having access to a system, or box in what follows, see Fig. 1. Each jth party, for $j \in \{1, 2,..., n\}$, inputs a classical symbol x_j from a set of M possible inputs into his box and as a result receives one output $a_j \in \{0, 1,..., d-1\}$. Thus, every scenario is labelled by these possible values as an (n, M, d) scenario. The correlations observed in this scenario are described by the $(dM)^n$ conditional probabilities $p(a_1, a_2, \ldots, a_n | x_1, x_2, \ldots, x_n) \equiv p(\bar{a}|\bar{x})$ of observing outputs $(a_1, a_2, \ldots, a_n) \equiv \bar{a}$ when using inputs $(x_1, x_2, \ldots, x_n) \equiv \bar{x}$. They can be collected into a vector of $(dM)^n$ positive numbers, which satisfy normalization over all d^n outputs (for each of the M^n choices of inputs). Since it is assumed that the different processes are local, that is, that the systems do not exchange any communication when producing the outputs given the inputs, the observed correlations must satisfy the no-signalling principle.

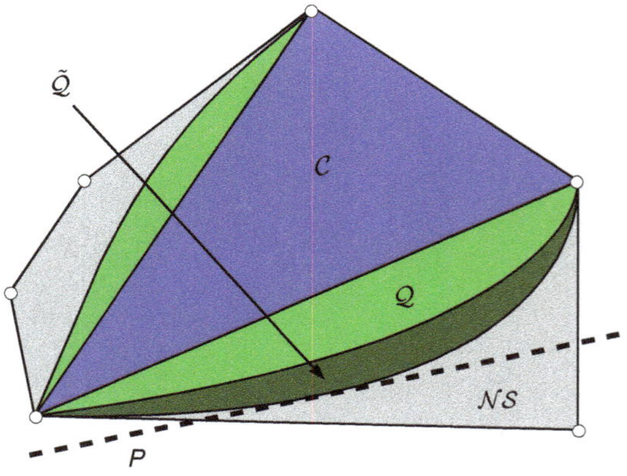

Figure 1 | Sets of correlations in a device-independent setting. In this work we study the sets of correlations produced by space-like separated boxes (inset) each with an input a_j and output x_j. All sets of correlations outside of the space of classical correlations labelled $\mathcal{C}$ are nonlocal correlations and all correlations within the set labelled $\mathcal{NS}$ satisfy special relativity. The set of quantum correlations is then $\mathcal{Q}$. We introduce a set that is an extension to quantum mechanics: the set of 'almost quantum' correlations $\tilde{\mathcal{Q}}$.

This principle implies that for all permutations (and relabelling of the parties after permutation) of n parties

$$\sum_{a_j, j\in\{1,2,\ldots,m\}} p(a_1, a_2, \ldots, a_n | x_1, x_2, \ldots, x_n) = p(a_{m+1}, a_{m+2}, \ldots, a_n | x_{m+1}, x_{m+2}, \ldots, x_n), \quad (1)$$

where the left-hand-side is a sum over $m \leq n$ parties. In other words, the marginal conditional probability distributions are well defined for all possible subsets of parties. The linear constraints (1) define the set of non-signalling correlations, denoted by $\mathcal{NS}$, see Fig. 1. The no-signalling constraints can, for instance, be enforced by arranging the input–output processes so that they define space-like separated events.

The set of quantum correlations, denoted by $\mathcal{Q}$, represents a strict subset of non-signalling correlations[10,19]. They are defined as those conditional probability distributions that can be generated by the Born rule as

$$p(a_1, a_2, \ldots, a_n | x_1, x_2, \ldots, x_n) = \langle\psi| \prod_{j=1}^{n} E_j^{a_j, x_j} |\psi\rangle \quad (2)$$

where:

(1) $|\psi\rangle$ is a normalized vector in a finite dimensional Hilbert space $\mathcal{H}$;

(2) $E_j^{a_j,x_j}$ are projectors acting on $\mathcal{H}$ and defining the measurements by each party, that is, $E_j^{a'_j,x_j} E_j^{a_j,x_j} = \delta_{a'_j}^{a_j} E_j^{a_j,x_j}$ and $\sum_{a_j} E_j^{a_j,x_j} = \mathbb{I}$, the identity matrix;

(3) any two measurement operators of two different parties commute, $\left[E_j^{a_j,x_j}, E_{j'}^{a_{j'},x_{j'}}\right] = 0$.

Note that in this work quantum correlations are defined using the commuting measurement paradigm rather than the tensor product of local measurements. This is because in the commuting measurement paradigm the comparison with almost quantum correlations (defined in what follows) becomes more transparent. Note that the two paradigms are known to be equivalent for finite dimensional systems, while the equivalence for infinite dimensions remains open[20]. Hence, by demanding $\mathcal{H}$ to be finite dimensional, all our results remain unchanged when switching to the tensor product representation.

After these preliminaries, we are now in position to define the set of almost quantum correlations, denoted by $\tilde{\mathcal{Q}}$. This set is defined by those correlations which can be written as

$$p(a_1, a_2, \ldots, a_n | x_1, x_2, \ldots, x_n) = \langle\psi| \prod_{j=1}^{n} E_j^{a_j, x_j} |\psi\rangle \quad (3)$$

where

(1) $|\psi\rangle$ is a normalized vector in a Hilbert space $\mathcal{H}$;

(2) $E_j^{a_j,x_j}$ are projectors acting on $\mathcal{H}$ and defining the measurements by each party, that is, $E_j^{a'_j,x_j} E_j^{a_j,x_j} = \delta_{a'_j}^{a_j} E_j^{a_j,x_j}$ and $\sum_{a_j} E_j^{a_j,x_j} = \mathbb{I}$, the identity matrix;

(3) one has $E_1^{a_1,x_1} \ldots E_n^{a_n,x_n}|\psi\rangle = E_{\pi(1)}^{a_{\pi(1)},x_{\pi(1)}} \ldots E_{\pi(n)}^{a_{\pi(n)},x_{\pi(n)}}|\psi\rangle$ for any arbitrary permutation π of the n parties.

We provide an alternative definition of the set of almost quantum correlations in terms of efficiently solvable semi-definite programmes (see Methods and Supplementary Note 1). For two parties the set of almost quantum correlations coincides with the set of correlations satisfying the $1+AB$ level of the Navascués–Pironio–Acin (NPA) hierarchy for quantum correlations[20], known as Q^{1+AB}. Correlations in Q^{1+AB} have already been shown as useful as quantum correlations for certain information-processing tasks[21,22].

When defining a set of non-signalling correlations, the first natural requirement to be met is that the set remains stable under post-processing. That is, it should be impossible to generate correlations outside the set by combining correlations within the set. The natural set of operations when dealing with correlated systems, or boxes, is given by post-selection, wirings of the boxes and distribution of boxes among parties (and the combinations of all of them), see Fig. 2. While this consistency requirement may seem trivial at first sight, and in fact it holds for classical, quantum and non-signalling correlations, it is not satisfied in general[23]. We prove that the set of almost quantum correlations is stable under classical post-processing (see Supplementary Note 2).

Quantum versus almost quantum correlations. It is already clear from the definitions that the set of quantum correlations is included in the set of almost quantum correlations. In fact, this inclusion can be proven to be strict; thus, $\mathcal{Q} \subsetneq \tilde{\mathcal{Q}}$. First of all, there already exist numerical results proving the existence of bipartite supraquantum correlations $p(\mathrm{a}|\mathrm{x}) \notin \mathcal{Q}$, which satisfy the step $1+AB$ of the NPA hierarchy, see for instance refs 20,24. In what follows, we provide an analytical proof of this strict inclusion already in the simplest possible case of two observers performing two measurements of two outputs.

Consider a Bell scenario as in Fig. 1 with two inputs $a_k \in \{0, 1\}$, two outputs $x_k \in \{0, 1\}$ and two parties $(k=1, 2)$. Owing to normalization and no-signalling constraints, any probability distribution in this scenario can be written as an eight-dimensional vector

$$\bar{p} \equiv (p_1(1|0), p_1(1|1), p_2(1|0), p_2(1|1), p(1,1|0,0), p(1,1|1,0), p(1,1|0,1), p(1,1|1,1)), \quad (4)$$

where $p_i(\alpha|\beta)$ is the probability $p(a_i|x_i)$ where $a_i = \alpha$ and $x_i = \beta$. In this scenario, one can define the following Bell-like expression built from a linear combination of the observed probabilities $B(\bar{p}) \equiv \bar{b} \cdot \bar{p}$, with

$$\bar{b} = \left(-\frac{30}{31}, \frac{167}{9}, \frac{167}{9}, -\frac{30}{31}, -\frac{174}{11}, -\frac{244}{23}, \frac{74}{11}, -\frac{174}{11} \right). \quad (5)$$

The minimum of this quantity over quantum correlations is slightly larger than -1 (see Supplementary Note 3), that is, $B(\bar{p}) > -1$ is an inequality satisfied by quantum correlations. This inequality can be violated by almost quantum correlations. We can provide an example of almost quantum correlations $p_{\tilde{\mathcal{Q}}}$ such that $B(\bar{p}_{\tilde{\mathcal{Q}}}) \approx -1.0029$ (see Supplementary Note 3). This value is smaller than the quantum minimum, and thus $\mathcal{Q} \subsetneq \tilde{\mathcal{Q}}$.

The discrepancy between quantum and almost quantum correlations is in principle experimentally testable. For instance, the previous Bell-type inequality for quantum correlations $B(\bar{p}) \geq -1$ can be tested as the standard Clauser–Horne–Shimony–Holt Bell inequality[25] in an experiment involving two measurements of two outputs on two different systems. More in general, any Bell inequality whose maximal quantum violation is not obtained at the $1+AB$ level of the NPA hierarchy provide experimental tests between quantum and almost quantum correlations. As mentioned, examples of these inequalities can be found in refs 20,24.

Of course, what is not possible at the moment is to think of a concrete experimental set-up in which this test could in principle be performed. Note that $\tilde{\mathcal{Q}}$ just defines a set of correlations, and not a complete physical theory, with well-defined state space, dynamics and measurements, which one can use to describe physical set-ups and whose correlation structure coincides with the set of almost quantum correlations. Thus, although from $\tilde{\mathcal{Q}}$ we can assert that, for some nonlocality experiments, the almost

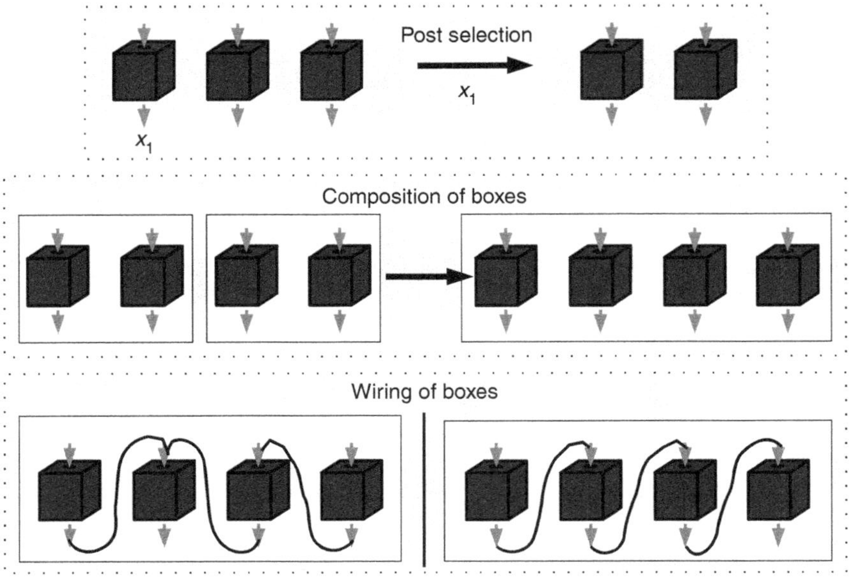

Figure 2 | Closure under classical post-processing operations. In order for a set of non-signalling correlations to be consistent, they need to be stable under classical post-processing. That is, given some boxes that produce correlations in a given set we should not be able to leave this set after classical post processing (resulting in new correlations) on the measurement data. All classical post-processing can be reduced to: (1) post selection (looking at correlations conditioned on a particular output of another box); (2) grouping of two sets of boxes to make a new set of boxes; and (3) wiring the boxes together so that inputs to some boxes are dependent on the outputs of others. We can also perform logical operations on the classical data to choose which form of post processing to perform. We show that if correlations belong to the almost quantum set $\tilde{\mathcal{Q}}$ then any classical post processing on these correlations will result in new correlations that are also inside $\tilde{\mathcal{Q}}$ (Supplementary Note 2). Therefore, the almost quantum set is stable under classical post processing.

quantum set predicts violations of the quantum bounds, the lack of a physical theory prevents us from proposing an explicit experiment in which this discrepancy could be observed.

Device-independent principles. Comparing the definitions of quantum and almost quantum correlations, it is clear that the only difference between the two sets appears in the third condition. In the case of almost quantum correlations, rather than demanding full commutation of the measurement operators by the different parties, one imposes the weaker condition that these measurements must commute when acting on the measured state $|\psi\rangle$. The intuition behind this definition is that in order to appreciate an operational difference between the two sets, one should probe the non-commutativity of the measurement operators in $\tilde{\mathcal{Q}}$ when acting on states different from $|\psi\rangle$. However, it is not clear how to get this information when one has only access to the correlations $p(\bar{a}|\bar{x})$. On the basis of this intuition, one may expect that the two sets are equivalent for many operational tasks.

Our next results show that this intuition is correct: despite the set of almost quantum correlations is strictly larger than the quantum one, almost quantum correlations can be proven to satisfy almost all the device-independent information principles that aim at singling out the set of quantum correlations. This implies that none of these principles suffice to characterize quantum correlations. To our knowledge, there have been five of such principles: Non-trivial Communication Complexity (NCC)[11], No Advantage for Nonlocal Computation[12], Information Causality (IC)[13], Macroscopic Locality (ML)[14] and Local Orthogonality (LO)[15].

The first three principles are defined in communication scenarios where two parties, call them Alice and Bob, must evaluate a bivariate function under limited communication or no communication at all. For NCC, Alice and Bob receive, respectively, the n-bit strings $\bar{x}, \bar{y}$. Their task is to compute the

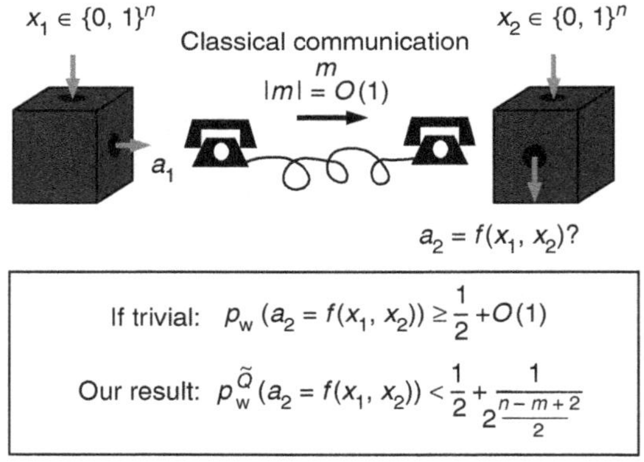

Figure 3 | The communication complexity scenario. The possibility of correlations more nonlocal than quantum, allowing for trivial communication complexity, was first highlighted in ref. 39. This observation was turned into the principle of NCC[11] and applies to the scenario of probabilistic communication complexity tasks. In this scenario, two parties, labelled 1 and 2, are each given a (generally different) input bit string x_1 and x_2, respectively, of length n and after receiving these inputs party 1 receives an output a_1 from their resources and communicates some number of bits to party 2 and party 2 outputs a_2. We assume that the number of bits is a constant $|m| = O(1)$. The most general statement we can make of the NCC principle says that no set of correlations will give a worst-case probability of success (for all possible inputs and all possible functions) bounded away from a half by a constant. An important result in our work is that for all correlations in the set $\tilde{\mathcal{Q}}$, this worst-case probability of success is not bounded away from a half by a constant.

binary function $f(\bar{x}, \bar{y}) \in \{0, 1\}$ with just one bit of communication. In this context, NCC postulates that there does *not* exist a universal number $1 \geq p > \frac{1}{2}$ such that any function $f(\bar{x}, \bar{y})$ can be computed with worst-case probability greater than p, see Fig. 3. For NANLC, the scenario is as follows: a referee randomly generates two bit strings $\bar{x}, \bar{z}$, according to the probability distributions $P_{\text{unif}}(\bar{x})$ (uniform) and $P(\bar{z})$, respectively. The referee then sends Alice (Bob) the string $\bar{x}(\bar{x} \oplus \bar{z})$, and Alice and Bob's task is to produce a pair of bits a, b such that $a \oplus b = f(\bar{z})$. Alice and Bob are thus carrying out a particular type of distributed computation. Their performance is limited by the principle of NANLC, which states that, no matter what type of nonlocality resources Alice and Bob share, their average success probability for computing function f should not beat the classical value. IC requires Alice (Bob) to be distributed a random n-bit string $\bar{x}$ (a random number $k \in \{1, ..., n\}$). Here both parties must cooperate so that Bob ouputs a bit b with $b = x_k$, and Alice is allowed to send Bob m bits of communication. The principle of IC claims that the efficiency of the above scheme is subject to the constraint

$$\sum_{k=1}^{n} I(b : x_k | k) \leq m, \tag{6}$$

where $I(A{:}B)$ denotes the mutual information between the random variables A and B[26].

The last two principles, ML and LO, have a somewhat different flavour. For ML, an experimental set-up where two parties carry out extensive measurements over a macroscopic number of independent particle pairs is considered. ML postulates that, under low-resolution detectors (of precision, $O(\sqrt{N})$), the statistics of such macroscopic experiments should admit a classical—local—description. LO is an intrinsically multipartite physical principle that states that, for any set E of pairwise locally orthogonal, $\sum_{e \in E} p(e) \leq 1$. Two measurement events $(a_1, \ldots, a_n | x_1, \ldots, x_n), (a'_1, \ldots, a'_n | x'_1, \ldots, x'_n)$ are called 'locally orthogonal' if $\exists\, k \in \{1, ..., n\}$ such that $x_k = x'_k, a_k \neq a'_k$.

Before this work, it was known that correlations satisy ML if and only if they satisfy the first step of the NPA hierarchy[14], the corresponding set being denoted by Q^1 (see Supplementary Note 4). Since Q^{1+AB} is included in Q^1, almost quantum correlations satisfy ML. The proofs that almost quantum correlations have non-trivial communication complexity provide no advantage for nonlocal computation and satisfy local orthogonality are much more involved (see Supplementary Notes 5–7).

At the moment, we are unable to prove that all almost quantum correlations satisfy IC. We believe that the difficulty in the proof comes from the fact that the entropic-like figure of merit (6) is hard to be bound without appealing to notions more sophisticated than correlation limits alone, such as the von Neumann entropy. However, up to now a series of works has explored the power of IC in different correlation scenarios[13,27–29]. We tested almost quantum correlations against all the known bounds and in all the cases we could reproduce or improve them (Supplementary Note 8). Put together, these numerical results strongly suggest that almost quantum correlations satisfy IC too.

Most of the existing results on IC have been derived in the case of two parties, as IC was originally defined in this scenario. However, it is possible to extend IC to scenarios involving more than two parties by simply imposing that correlations among N parties satisfy IC whenever they satisfy inequality (6) for any splitting of the parties into two groups. This generalization is analogous to what is done for the no-signalling principle. Adopting this generalization of IC, and using the methods described in ref. 30, we can show (Supplementary Note 9) that certain three-party almost quantum correlations beyond the quantum theory satisfy IC too (and thus all currently proposed principles). This second result implies that, even when taken together, all these principles are unable to capture quantum correlations, at least in the multipartite setting. This, however, does not exclude the possibility that IC is enough to single out bipartite quantum nonlocality.

Almost quantum correlations and the history approach. All our previous results have shown that the set of almost quantum correlations is, in terms of the known device-independent principles, as plausible as the quantum set. Thus, so far, we do not have any good reason justifying why correlations inside $\tilde{Q}$ should not be realizable in Nature. Now, what about correlations outside $\tilde{Q}$? Could we enlarge $\tilde{Q}$ further without violating basic physical laws? Or, in other words, is there a way of singling out the set of almost quantum correlations?

Our last goal is to discuss how it is possible to identify the set of almost quantum correlation using arguments arising from the history approach to quantum gravity. In ref. 31, Sorkin proves that the dynamics of any history-based physical theory with no third-order interference effects is described by a so-called 'decoherence functional', a concept that originated in the decoherent history approach to quantum mechanics (see, for example, refs 32–34). The decoherence functional is a matrix whose rows and columns are labelled by histories, and determines which partitions of the set of all possible histories are physically meaningful, together with the probability that the actual history of the universe falls into one of those subsets. In ref. 35, this general decoherence functional approach is applied to describe bipartite Bell experiments. Interestingly, the authors conclude that, demanding the decoherence functional to satisfy the extra condition of strong positivity (SP), which just means that the decoherence functional is demanded to be a semidefinite positive matrix, the set of accessible two-party correlations precisely corresponds to the $1 + AB$ level of the NPA hierarchy. As shown here, in the case of two parties, this level coincides with the set of almost quantum correlations.

While this connection is intriguing, the authors of ref. 35 do not give a strong justification for the SP condition used in its derivation. However, recently, the SP condition has been shown to emerge naturally from the requirement that any history-based theory, possibly supraquantum, should be able to describe single-site quantum experiments. The above results not only imply that almost quantum correlations are compatible with Sorkin's lack of third-order interference[31], see also ref. 36, but also that this principle, together with the need to accommodate single-site quantum experiments in our theory, precludes the existence of correlations outside the set of almost quantum correlations $\tilde{Q}$.

Discussion

In this study we have identified a set of correlations, named almost quantum, that, despite being strictly larger than the quantum set, seems to satisfy all device-independent principles proposed so far to single out quantum correlations. This set is consistent and has an efficient numerical characterization. Our results hence suggest that it is possible to modify the correlation structure of the quantum theory without running into any operationally relevant consequence. They also pose a convincing barrier to the research programme that aims to characterize quantum nonlocality, utilizing fundamental principles motivated by physical and information theory perspectives. The set of almost quantum correlations is thus a powerful tool for understanding nonlocality.

In view of all these points, it would be interesting to derive a physical theory underpinning $\tilde{Q}$. On the one hand, this theory could provide the first reasonable extension of quantum mechanics. As mentioned, so far, any such extension has often led to violations of the no-signalling principle or significant increases of the information power of the theory. A theory explaining almost quantum correlations would not suffer from these problems. On the other hand, this theory could be used to propose concrete experimental set-ups to test those Bell-like inequalities for which quantum and almost quantum correlations predict different bounds.

In this regard, it would be tempting to interpret the state and projection operators appearing in the definition of $\tilde{\mathcal{Q}}$, not as some auxiliary mathematical entities, but as the physical state and measurements of an underlying theory generating the correlations. However, this is just a possibility as valid as any other. In fact, it is unclear to us how to define a reasonable dynamics in this context. Another, possibly more promising, alternative is to search for a theory within the history approach. All these possibilities deserve further investigation but are beyond the scope of the present work.

We would like to conclude with an interesting speculation. An important conceptual difference between quantum and almost quantum correlations appears when considering the computational complexity of their characterization. Given some correlations $p(\mathrm{a}|\mathrm{x})$, determining whether they are quantum, $p(\mathrm{a}\,|\,\mathrm{x}) \in \mathcal{Q}$, or almost quantum, $p(\mathrm{a}\,|\,\mathrm{x}) \in \tilde{\mathcal{Q}}$, defines two decision problems. As indicated, the almost quantum problem can be efficiently solved using semidefinite programming techniques. On the contrary, the quantum problem appears to be much harder, as it has been shown that approximating the quantum violation of bipartite Bell inequalities is NP-hard[37]. Actually, it is not even known if the quantum problem is decidable. This raises a very provoking final thought: assume the quantum problem was undecidable, does one expect correlations in Nature to be undecidable?

Methods

Semidefinite programming and almost quantum correlations. In this section we re-express the definition of the set of almost quantum correlations as a semidefinite programme.

Let us first introduce some useful terminologies. Given $m \leq n$ parties, the pair $(\bar{a}\,|\,\bar{x})$, with the components of $\bar{a}$, $\bar{x}$ labelled by each of the m parties will be called an 'event'. For example, for $n=3$, the event $(a_1, a_3|x_1, x_3)$ represents the physical situation in which parties 1 and 3 have measured x_1, x_3, and obtained, respectively, the outcomes a_1, a_3. Following ref. 15, we say that two events $e \equiv (\bar{a}\,|\,\bar{x})$, $e' \equiv (\bar{a}'\,|\,\bar{x}')$ are 'locally orthogonal' (represented $e \perp e'$) if there is a common party k such that $x_k = x'_k$, and $a_k \neq a'_k$. We now present another useful characterization of the set $\tilde{\mathcal{Q}}$ of all almost quantum distributions.

Lemma 1. $P(a_1, \ldots, a_n\,|\,x_1, \ldots, x_n) \in \tilde{\mathcal{Q}}$ if, for any event $(\bar{a}\,|\,\bar{x})$ with $a_k \neq 0$ for all parties k involved, there exists a vector $|\bar{a}, \bar{x}\rangle \in \mathcal{H}$ with the properties

(1) $\langle \bar{a}', \bar{x}'\,|\,\bar{a}, \bar{x}\rangle = 0$, if $(\bar{a}'\,|\,\bar{x}') \perp (\bar{a}\,|\,\bar{x})$.

(2) $P(\bar{a}\,|\,\bar{x}) = \langle \phi\,|\,\bar{a}, \bar{x}\rangle$, where $|\phi\rangle$ is the vector corresponding to the null event, that is, none of the party measures. Note that $1 = P(\phi) = \langle \phi|\phi\rangle$, that is, $|\phi\rangle$ is normalized.

(3) $\langle \bar{a}, \bar{a}', \bar{x}, \bar{x}'\,|\,\bar{a}, \bar{a}'', \bar{x}, \bar{x}''\rangle = \langle \bar{a}', \bar{x}'\,|\,\bar{a}, \bar{a}'', \bar{x}, \bar{x}''\rangle = \langle \bar{a}, \bar{a}', \bar{x}, \bar{x}'\,|\,\bar{a}'', \bar{x}''\rangle$, where $(\bar{a}\,|\,\bar{x})$ is any event that doesnot involve the measuring parties in the events $(\bar{a}'\,|\,\bar{x}')$ and $(\bar{a}''\,|\,\bar{x}'')$.

Note that any set of complex vectors subject to restrictions (1), (2), (3) implies the existence of real vectors subject to the same constraints. It follows that, in the above semidefinite programme, all free variables can be taken to be real.

Proof. The right implication follows by defining $|\bar{a}, \bar{x}\rangle \equiv \Pi_k E^{a_k, x_k}|\phi\rangle$.

Regarding the left implication, consider the subspaces $V_k^{a,x} = \mathrm{span}\{|a_k, \bar{a}', x_k, \bar{x}'\rangle\}$. From condition (1), we have that $V_k^{a,x} \perp V_k^{a',x}$, for $a \neq a'$. It follows that the projectors on the subspaces $V_k^{a,x}$, that is, $\tilde{E}_k^{a,x} \equiv \mathrm{proj}\,(V_k^{a,x})$ satisfy $\tilde{E}_k^{a,x}\tilde{E}_k^{a',x} = \tilde{E}_k^{a,x}\delta_{a,a'}$. Now, the action of $\tilde{E}_k^{a_k,x_k}$ on the vector $|\phi\rangle$ is given by:

$$\tilde{E}_k^{a_k,x_k}|\phi\rangle = \tilde{E}_k^{a_k,x_k}|a_k, x_k\rangle + \tilde{E}_k^{a_k,x_k}(|\phi\rangle - |a_k, x_k\rangle) = |a_k, x_k\rangle, \qquad (7)$$

where we have used that $|a_k, x_k\rangle \in V_k^{a_k,x_k}$ in order to conclude $\tilde{E}_k^{a_k,x_k}|a_k, x_k\rangle = |a_k, x_k\rangle$. The second term $\tilde{E}_k^{a_k,x_k}(|\phi\rangle - |a_k, x_k\rangle)$ vanishes, since $|a_k, \bar{a}', x_k, \bar{x}'\,|\,\phi\rangle = |a_k, \bar{a}', x_k, \bar{x}'\,|\,a_k, x_k\rangle$ by virtue of relation (3). Note that, also by condition (3), the last equality holds when we replace $|\phi\rangle$, $|a_k, x_k\rangle$ by $|\bar{a}, \bar{x}\rangle$, $|a_k, \bar{a}, x_k, \bar{x}\rangle$, where $(\bar{a}\,|\,\bar{x})$ is any event where party k does not intervene. It follows that $\tilde{E}_k^{a_k,x_k}\;|\bar{a}, \bar{x}\rangle = |a_k, \bar{a}, x_k, \bar{x}\rangle$. By induction, we thus arrive at

$$\prod_k \tilde{E}_k^{a_k,x_k}\,|\phi\rangle = |\bar{a}, \bar{x}\rangle, \qquad (8)$$

for $a_k \neq 0$, where the product is taken in whatever order. Finally, define

$$\tilde{E}_k^{0,x_k} \equiv \mathbb{I} - \sum_{a \neq 0} \tilde{E}_k^{a,x_k}. \qquad (9)$$

From equation (8) and relation (2) it thus follows that the state $|\phi\rangle$ and the operators $\{\tilde{E}_k^{a,x}\}$ satisfy the conditions required to produce almost quantum correlations.

Given this lemma, we are now in a position to give the semidefinite programming version of almost quantum correlations. By the Gram decomposition[38], the existence of a set of vectors satisfying the conditions of Lemma 1 is equivalent to the existence of a positive semidefinite matrix Γ, with rows and columns labelled by events $(\bar{a}\,|\,\bar{x})$ with $a_k \neq 0$ for all parties involved, and such that

- $\Gamma_{(\bar{a}'|\bar{x}'),(\bar{a}\,|\,\bar{x})} = 0$, if $(\bar{a}'\,|\,\bar{x}') \perp (\bar{a}\,|\,\bar{x})$.
- $\Gamma_{\phi,\phi} = 1$.
- $P(\bar{a}\,|\,\bar{x}) = \Gamma_{\phi,(\bar{a}\,|\,\bar{x})}$.
- $\Gamma_{(\bar{a},\bar{a}'\,|\,\bar{x},\bar{x}'),(\bar{a},\bar{a}''\,|\,\bar{x},\bar{x}'')} = \Gamma_{(\bar{a}'\,|\,\bar{x}'),(\bar{a},\bar{a}''\,|\,\bar{x},\bar{x}'')} = \Gamma_{(\bar{a},\bar{a}'\,|\,\bar{x},\bar{x}'),(\bar{a}''\,|\,\bar{x}'')}$, where $(\bar{a}\,|\,\bar{x})$ is any event not involving the measuring parties in the events $(\bar{a}'\,|\,\bar{x}')$ and $(\bar{a}''\,|\,\bar{x}'')$.

We call any such matrix Γ an 'almost quantum certificate' for $P(\bar{a}\,|\,\bar{x})$.

References

1. Abrams, D. S. & Lloyd, S. Nonlinear quantum mechanics implies polynomial-time solution for *NP*-complete and *#P* problems. *Phys. Rev. Lett.* **81,** 3992–3995 (1998).
2. Gisin, N. Stochastic quantum dynamics and relativity. *Helv. Phys. Acta* **62,** 363–371 (1989).
3. Bancal, J.-D. *et al.* Quantum non-locality based on finite-speed causal influences leads to superluminal signalling. *Nat. Phys.* **8,** 867–870 (2012).
4. Popescu, S. Nonlocality beyond quantum mechanics. *Nat. Phys.* **10,** 264–270 (2014).
5. Aaronson, S. Quantum computing, postselection, and probabilistic polynomial-time. *Proc. R. Soc. A* **461,** 2063 (2005).
6. Hardy, L. Quantum theory from five reasonable axioms. Preprint at http://arxiv.org/abs/quant-ph/0101012 (2001).
7. Dakic, B. & Brukner, C. Quantum theory and beyond: is entanglement special? Preprint at http://arxiv.org/abs/0911.0695 (2009).
8. Masanes, L. & Mueller, M. P. A derivation of quantum theory from physical requirements. *New J. Phys.* **13,** 063001 (2011).
9. Chiribella, G., D'Ariano, G. M. & Perinotti, P. Informational derivation of quantum theory. *Phys. Rev. A* **84,** 012311 (2011).
10. Popescu, S. & Rohrlich, D. Quantum nonlocality as an axiom. *Foundation Phys.* **24,** 379–385 (1994).
11. Brassard, G. *et al.* Limit on nonlocality in any world in which communication complexity is not trivial. *Phys. Rev. Lett.* **96,** 250401 (2006).
12. Linden, N., Popescu, S., Short, A. J. & Winter, A. Quantum nonlocality and beyond: limits from nonlocal computation. *Phys. Rev. Lett.* **99,** 180502 (2007).
13. Pawlowski, M. *et al.* Information causality as a physical principle. *Nature* **461,** 1101 (2009).
14. Navascués, M. & Wunderlich, H. A glance beyond the quantum model. *Proc. R. Soc. A* **466,** 881–890 (2009).
15. Fritz, T. *et al.* Local orthogonality as a multipartite principle for quantum correlations. *Nat. Commun.* **4,** 2263 (2013).
16. Cabello, A., Severini, S. & Winter, A. Graph-theoretic approach to quantum correlations. *Phys. Rev. Lett.* **112,** 040401 (2014).
17. Abramsky, S. & Brandenburger, A. The sheaf-theoretic structure of non-locality and contextuality. *New J. Phys.* **13,** 113036 (2011).
18. Acín, A., Fritz, T., Leverrier, A. & Sainz, A. B. A combinatorial approach to nonlocality and contextuality. Preprint at http://arxiv.org/abs/1212.4084 (2012).
19. Khalfin, L. A. & Tsirelson, B. S. *Symposium on the Foundations of Modern Physics* 441–460 (1985).
20. Navascués, M., Pironio, S. & Acín, A. A convergent hierarchy of semidefinite programs characterizing the set of quantum correlations. *New J. Phys.* **10,** 073013 (2008).
21. de la Torre, G., Hoban, M. J., Dhara, C., Prettico, G. & Acín, A. Maximally nonlocal theories cannot be maximally random. Preprint at http://arxiv.org/abs/1403.3357 (2014).
22. Pironio *et al.* Random numbers certified by bells theorem. *Nature* **464,** 1021 (2010).
23. Allcock, J. *et al.* Closed sets of nonlocal correlations. *Phys. Rev. A* **80,** 062107 (2009).
24. Pál, K. F. & Vértesi, T. Quantum bounds on bell inequalities. *Phys. Rev. A* **79,** 022120 (2009).

25. Clauser, J. F., Horne, M. A., Shimony, A. & Holt, R. A. Proposed experiment to test local hidden-variable theories. *Phys. Rev. Lett.* **23,** 880–884 (1969).
26. Cover, T. M. & Thomas, J. A. *Elements of Information Theory,* 2nd edn (Wiley-Interscience, 2006).
27. Allcock, J., Brunner, N., Pawlowski, M. & Scarani, V. Recovering part of the boundary between quantum and nonquantum correlations from information causality. *Phys. Rev. A* **80,** 040103 (2009).
28. Cavalcanti, D., Salles, A. & Scarani, V. Macroscopically local correlations can violate information causality. *Nat. Commun.* **1,** 136 (2010).
29. Pawlowski, M. & Żukowski, M. Entanglement-assisted random access codes. *Phys. Rev. A* **81,** 042326 (2010).
30. Gallego, R., Wurflinger, L. E., Acín, A. & Navascués, M. Quantum correlations require multipartite information principles. *Phys. Rev. Lett.* **107,** 210403 (2011).
31. Craig, F. D. *et al.* A bell inequality analog in quantum measure theory. *J. Phys. A* **40,** 501–523 (2007).
32. Dowker, H. F. & Halliwell, J. J. Quantum mechanics of history: the decoherence functional in quantum mechanics. *Phys. Rev. D* **46,** 1580–1609 (1992).
33. Gell-Mann, M. & Hartle, J. B. Classical equations for quantum systems. *Phys. Rev. D* **47,** 3345–3382 (1993).
34. Griffiths, R. B. Consistent histories and quantum reasoning. *Phys. Rev. A* **54,** 2759–2774 (1996).
35. Dowker, F., Henson, J. & Wallden, P. A histories perspective on characterizing quantum non-locality. *New J. Phys.* **16,** 033033 (2014).
36. Henson, J. Bounding quantum contextuality with lack of third-order intereference. Preprint at http://arxiv.org/abs/1406.3281 (2014).
37. Ito, T., Kobayashi, H. & Matsumoto, K. Oracularization and two-prover one-round interactive proofs against nonlocal strategies. Preprint at http://arxiv.org/abs/0810.0693 (2008).
38. Horn, R. A. & Johnson, C. R. *Matrix Analysis* (Cambridge Univ. Press, 1990).
39. van Dam, W. Implausible consequences of superstrong nonlocality. Preprint at http://arxiv.org/abs/quant-ph/0501159 (2005).

Acknowledgements

This work is supported by the ERC CoG QITBOX and AdG NLST, EU RAQUEL and SIQS projects, the Spanish MINECO FIS2008-01236, with the support of FEDER funds, FOQUS and Chist-Era DIQIP, and the John Templeton Foundation.

Author contributions

M.N., Y.G., M.J.H. and A.A. contributed to developing the ideas, performing the calculations, analysing the results and writing the manuscript.

Additional information

Competing financial interests: The authors declare no competing financial interest.

22

Semiconductor-inspired design principles for superconducting quantum computing

Yun-Pil Shim[1,2] & Charles Tahan[1]

Superconducting circuits offer tremendous design flexibility in the quantum regime culminating most recently in the demonstration of few qubit systems supposedly approaching the threshold for fault-tolerant quantum information processing. Competition in the solid-state comes from semiconductor qubits, where nature has bestowed some very useful properties which can be utilized for spin qubit-based quantum computing. Here we begin to explore how selective design principles deduced from spin-based systems could be used to advance superconducting qubit science. We take an initial step along this path proposing an encoded qubit approach realizable with state-of-the-art tunable Josephson junction qubits. Our results show that this design philosophy holds promise, enables microwave-free control, and offers a pathway to future qubit designs with new capabilities such as with higher fidelity or, perhaps, operation at higher temperature. The approach is also especially suited to qubits on the basis of variable super-semi junctions.

[1]Laboratory for Physical Sciences, College Park, Maryland 20740, USA. [2]Department of Physics, University of Maryland, College Park, Maryland 20742, USA. Correspondence and requests for materials should be addressed to Y.-P.S. (email: ypshim@lps.umd.edu) or to C.T. (email: charlie@tahan.com).

Spin qubits[1] are based on the fundamental and intrinsic properties of semiconductor systems, such as electron spins trapped in the potential of a quantum dot[2] or a chemical impurity[3]. Spins can be naturally protected from charge noise due to weak spin–orbit coupling. In fact, the tiny matrix element between spin qubit states can allow spin qubits to operate at temperatures above the Zeeman splitting[4,5]. While a benefit to qubit coherence, this property of spins also leads to relatively slow single-qubit gates via, for example, a microwave pulse. It turns out that nature provides a solution: a very fast and robust two-qubit gate via the exchange interaction. This has led to "encoded" qubit schemes where the qubit is embedded logically in two to four physical spin qubits[6–8]. The fact that electrons are real particles can be used for fast initialization and readout techniques. Exchange-only qubits[7,9] allow all electrical implementation of qubit-gate operations, and enable universal quantum computation (QC) while providing some immunity to global field and timing fluctuations via a decoherence free subsystem, at the cost of more physical qubits and extra operations per encoded gate.

This work investigates how superconducting Josephson junction quantum circuits[10], whose properties can be engineered, can be improved by mimicking some of best properties of spin qubit systems. We propose a first step: an encoded superconducting qubit approach, which does not require microwave control, and thus divorces qubit frequency from control electronics. In analogy to the exchange-only qubit in semiconductor spin qubit systems, encoded qubits enable microwave-free control of the qubit states via fast DC-like voltage or flux pulses. In contrast to the exchange-only qubits, logical gate operations of this encoded superconducting (SC) qubit can be done with minimal overhead (zero overhead in physical two-qubit gates) in terms of control operations, a surprising result. We describe how to initialize the encoded qubit and implement single- and two-qubit logical gates using only *z*-control pulse sequences (via tunable frequency qubits). In the process we also lay out possible opportunities for future research on the basis of other insights from spin-based QC. To encourage implementation, we give an explicit protocol on the basis of qubits in operation today.

Small systems of superconducting qubits based on variations of the transmon qubit[11,12] have already demonstrated gates with fidelities approaching 99.99% along with rudimentary quantum algorithms including error correction cycles[13–19]. Note that because these architectures rely on single-qubit gates via microwaves, the future design space is constrained by the availability and convenience of microwave generators.

An alternative approach to combining the best properties of semiconductor and superconducting quantum systems is to take advantage of true superconducting-semiconductor systems. The appearance of superconductivity in conventional semiconductors[20,21] such as silicon[22–25] or germanium[26,27] could potentially allow for a new type of fully epitaxial super-semi devices[28,29]. And epitaxial super-semi Josephson juction devices based on the proximity effect have already led to new superconducting circuits[30,31]. Epitaxial super-semi systems may improve noise properties, but perhaps more importantly they enable gate-tunable Josephson junctions, which we can also take advantage of in our proposal introduced below.

Results

From encoded spins to tunable qubits. Spins in quantum dots, say in silicon, are typically assumed to have equivalent g-factors, so that in a magnetic field the frequency of each qubit is the same. Thus, to achieve universal QC (an ability to do arbitrary rotations around the Bloch sphere plus a two-qubit entangling gate), one needs at minimum three spins. In this case, an encoded one-qubit gate requires around 3 pulses and a CNOT gate requires roughly 20 pulses[7], a hefty overhead. Two-qubit encodings are possible, but require the complication of a magnetic field gradient (via for example a micromagnet). Superconducting qubits, on the other hand, can be man-made such that the qubit frequency is tunable. This allows arbitrary one-qubit rotations with just two-physical qubits, in theory.

In this work, we consider a qubit encoded in a system of two capacitively coupled-SC qubits. We take tunable transmons[16,32] like xmons[13] or gatemons[30] as our prototypical SC qubits (see Fig. 1a) and suggest one possible implementation following the capacitively-coupled xmon architecture of Martinis *et al.*[14] to encourage near-term realization. Although we explicitly chose the xmon geometry to be more specific about our proposed protocol, the general idea can easily be applied to other types of SC qubits, such as traditional transmons or capacitively-shunted flux qubits[33,34], which we will discuss later. A transmon qubit[11] is

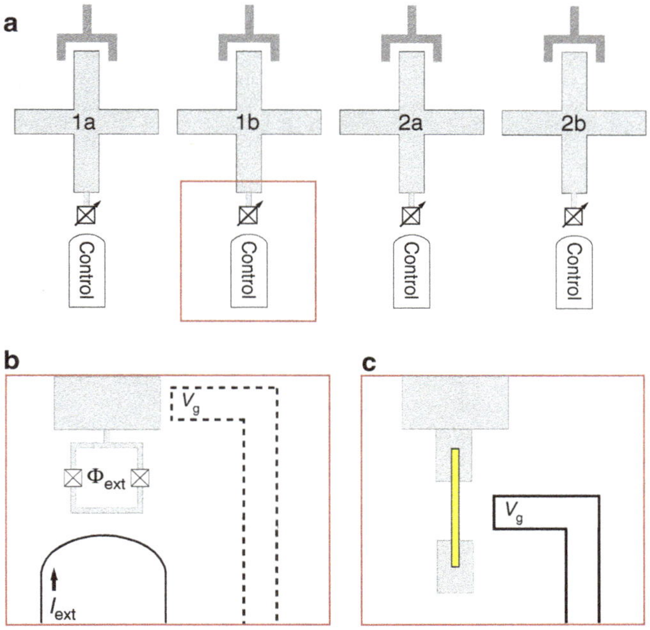

Figure 1 | Encoded superconducting qubits and tunable Josephson junctions. **(a)** Schematic diagram of a possible encoded superconducting qubit scheme as described in the text. An encoded qubit consists of two tunable physical SC qubits (for example, tunable transmons such as the xmon depicted here), with the encoded qubit states $|0\rangle_Q = |01\rangle$ and $|1\rangle_Q = |10\rangle$. In this picture, two encoded qubits are shown (for example, physical qubits 1a and 1b form an encoded qubit) and more encoded qubits can be introduced in a straightforward manner. Each SC qubit has a *z*-control line which tunes the Josephson energy E_J, and there are no additional microwave *xy*-control lines. All manipulation of the qubit states are done by the *z*-control pulses. Each transmon is capacitively-coupled to neighbouring transmons and also coupled to, as an example, a transmission line resonator for readout. **(b)** Double JJs in a loop act as a tunable JJ, controlled by an externally applied magnetic flux. In the SQUID tunable approach, one of the transmons in each encoded qubit needs a separate voltage control to tune the gate charge number n_g which may be needed to initialize the encoded qubit state. **(c)** electrostatically tunable JJ on the basis of a proximitized superconducting-semiconductor nanowire connecting two superconductors[30] used for gatemons. The nanowire is coated with SC and a portion is lifted off to form a semiconductor nanowire weak link. The JJ energy E_J is tuned by a side-gate voltage V_G, which can also serve as a capacitive tuning for initialization.

described by the charge qubit Hamiltonian:

$$H_X = 4E_C(\hat{n} - n_g)^2 - E_J \cos\hat{\theta}, \tag{1}$$

where $E_C = e^2/2C_\Sigma$ is the electron charging energy for total capacitance C_Σ and E_J is the Josephson energy. $\hat{n}$ and $\hat{\theta}$ are the number and phase operators, respectively, and n_g is the gate charge number that can be tuned by a capacitively-coupled external voltage. The qubit frequency $f_Q = \varepsilon/h$, where ε is the energy difference between the first excited state and the ground state, and $f_Q \simeq \sqrt{8E_C E_J}/h$ in the transmon regime, $E_J \gg E_C$. The Josephson energy of a JJ is determined by the material properties and geometry of the JJ, but a double JJ can be considered as a tunable JJ[35] where an externally applied magnetic flux through the double JJ loop can tune the effective coupling energy $E_J = E_{J0}\cos(\pi\Phi_{ext}/\Phi_0)$ (see Fig. 1b). Φ_{ext} is the external magnetic flux and Φ_0 is the SC flux quantum. Individual transmon qubits are typically controlled by tuning the qubit frequency with tunable E_J for z-control and by applying microwaves for x control.

Recently, there has been progress in an alternative approach for a tunable JJ using a superconductor proximitized semiconductor weak-link junction[30,31]. In ref. 30, an InAs nanowire was used to connect two superconductors (Al). The nanowire was epitaxially coated with Al and a small portion of the wire was etched off to form a semiconductor nanowire bridging two SCs (Fig. 1c). A side-gate voltage was used to tune the carrier density under the exposed portion of the wire and thus the Josephson energy of this SNS JJ. The gatemon, a tunable transmon based on this gate-tunable JJ, has several advantages. It requires only a single JJ that can be quickly tuned by a electrostatic voltage. It removes the need for external flux and hence reduces dissipation by a resistive control line and allows the device to operate in a magnetic field. The epitaxial growth of the nanowire JJ and its clean material properties[36,37] demonstrate the potential of a bottom-up approach for SC quantum devices[28,29].

Our encoded qubit is defined in a two-transmon system. The Hamiltonian for two transmons with the capacitive xx coupling is:

$$\begin{aligned} H_{2X} &= \sum_{k=a,b}\left[4E_C^{(k)}\left(\hat{N}_k - n_g^{(k)}\right)^2 - E_J^{(k)}\cos\hat{\theta}_k\right] \\ &\quad + E_{cc}\left(\hat{N}_a - n_g^{(a)}\right)\left(\hat{N}_b - n_g^{(b)}\right) \\ &= \varepsilon_a\tilde{\sigma}_a^z + \varepsilon_b\tilde{\sigma}_b^z + \varepsilon'\tilde{\sigma}_a^x \otimes \tilde{\sigma}_b^x, \end{aligned} \tag{2}$$

where E_{cc} is the capacitive coupling energy and $\tilde{\sigma}_k^i$ $(i = x, y, z)$ is the Pauli operator for k-th transmon in a reduced subspace of transmon qubit states. ε_k is the qubit energy of the k-th transmon, and $\varepsilon' = E_{cc}\alpha_a\alpha_b$ with $\alpha_k = \langle 1|\hat{N}_k|0\rangle$ where $|0\rangle$ and $|1\rangle$ are the two lowest energy states of individual transmons. In transmon qubit systems the capacitive coupling is usually turned on (off) by tuning the qubit frequencies to on (off) resonance. The capacitive xx coupling conserves the parity $\tilde{\sigma}_a^z \otimes \tilde{\sigma}_b^z$ of the two-transmon system and the Hamiltonian (equation (2)) is block-diagonal in the basis of $\{|00\rangle, |01\rangle, |10\rangle, |11\rangle\}$. We define our encoded qubit in the subspace of $\langle\{|01\rangle, |10\rangle\}\rangle$, since the other subspace $\langle\{|00\rangle, |11\rangle\}\rangle$ has states with a very large energy difference (much larger than the capacitive coupling), effectively turning off the capacitive coupling all the time.

In the encoded qubit basis $\{|0\rangle_Q, |1\rangle_Q\}$ where

$$|0\rangle_Q = |01\rangle, \quad |1\rangle_Q = |10\rangle, \tag{3}$$

the single qubit Hamiltonian is

$$H_Q = \begin{pmatrix} -\varepsilon_a + \varepsilon_b & \varepsilon' \\ \varepsilon' & \varepsilon_a - \varepsilon_b \end{pmatrix} = \frac{\varepsilon_a + \varepsilon_b}{2}1 + \Delta\varepsilon\hat{\sigma}^z + \varepsilon'\hat{\sigma}^x, \tag{4}$$

where $\Delta\varepsilon = (\varepsilon_b - \varepsilon_a)/2$ and $\hat{\sigma}^i$ $(i = x, y, z)$ is the Pauli operator for the encoded qubit. The qubit energies ε_a and ε_b can be controlled by the tunable JJ of each tunable transmon or gatemon, enabling logical gate operations with only fast DC-like voltage or flux pulses. In the following we will describe the logical gate operations, initialization and measurement schemes for this encoded qubit architecture.

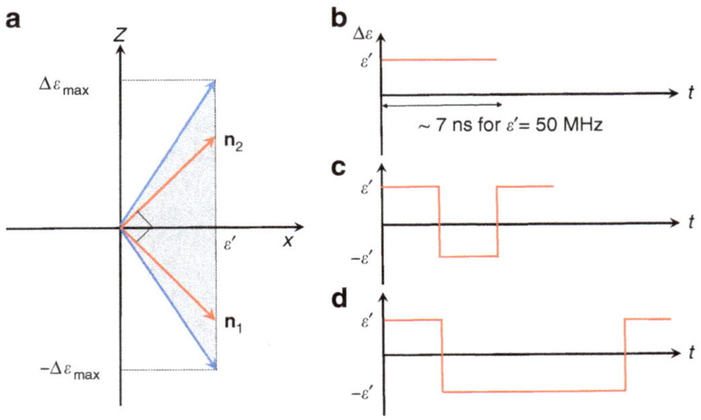

Figure 2 | Encoded single qubit operations. (**a**) Possible rotation axes in xz plane. The shaded grey region depicts the range of the direction of the possible rotation axis. The two red directions indicate a set of two orthogonal rotation axes, which can be used to implement any arbitrary single qubit gates in three steps. (**b–d**) Schematically shows an implementation of some logical gates in terms of rotations around $\hat{\mathbf{n}}_1 = \mathbf{n}_1/|\mathbf{n}_1|$ and $\hat{\mathbf{n}}_2 = \mathbf{n}_2/|\mathbf{n}_2|$. (**b**) Pulse shape for Hadamard gate. (**c**) Pulse shape for X gate. (**d**) Pulse shape for Z gate.

Single-qubit operations. The Hamiltonian for an encoded qubit is given by equation (4). For a fixed capacitive coupling between SC qubits, ε' is fixed, and the single-qubit operations can be implemented by pulsing the qubit energy ε through the z-control of individual transmons, in at most three rotations. Since the tunable range of $\Delta\varepsilon$ (order of GHz) is much greater than ε' (tens or hundreds of MHz), the rotation axis can be in almost any direction in the right half of the xz plane (see Fig. 2a), and most logical single-qubit gates can be implemented in two rotations[38]. In general, all single-qubit-gate operations can be implemented as a three-step Euler angle rotations around two orthogonal rotation axes (for example, see the two red axes in Fig. 2a).

We now provide implementations for a few representative single-qubit gates. The Hadamard gate, $H = ((1,1),(1,-1))/\sqrt{2}$, is a single-qubit gate that is almost ubiquitous in quantum circuits. Figure 2b shows an implementation of H gate as a single rotation $H = iR(\hat{\mathbf{n}}_2, \pi)$ around $\hat{\mathbf{n}}_2 = (1,0,1)/\sqrt{2}$. It can be achieved by tuning $\delta\varepsilon = \varepsilon'$. Here $R(\hat{\mathbf{n}}, \phi)$ is a rotation by angle ϕ around $\hat{\mathbf{n}}$ axis. Pauli X gate can be realized as a single rotation by tuning the two xmons on resonance ($\Delta\varepsilon = 0$), or three-step rotations such as $X = iR(\hat{\mathbf{n}}_2, \pi/2)R(\hat{\mathbf{n}}_1, \pi/2)R(\hat{\mathbf{n}}_2, \pi/2)$, where $\hat{\mathbf{n}}_1 = (1,0,-1)/\sqrt{2}$ and $\hat{\mathbf{n}}_2 = (1,0,1)/\sqrt{2}$, as was shown in Fig. 2c. Z gate requires three-step rotations: $Z = -iR(\hat{\mathbf{n}}_2, \pi/2)R(\hat{\mathbf{n}}_1, 3\pi/2)R(\hat{\mathbf{n}}_2, \pi/2)$. The above examples are for ideal systems with precise control over the system parameters. In real systems with fluctuating parameters, recently developed dynamical error-cancelling pulse sequences[39,40] could be useful for gate operations with higher fidelity.

Given that single-qubit gates in transmon systems through z-control have already demonstrated fidelities better than 0.999 (ref. 14), we expect the logical single-qubit gates (which require at most three rotation steps through z-control of transmons) will be

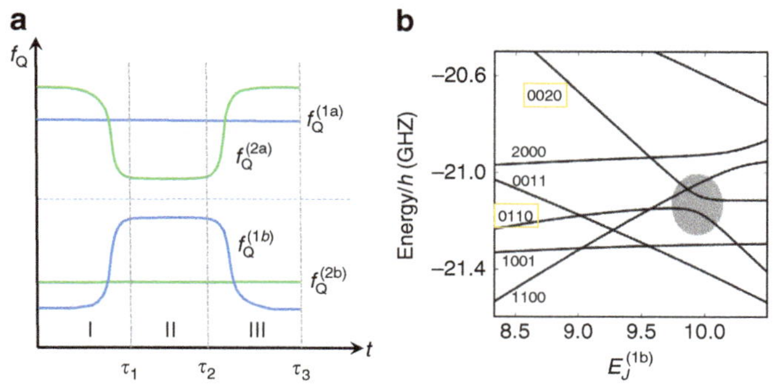

Figure 3 | Two-qubit gate operation. (**a**) Pulse scheme for two-qubit gate operations. *y* axis (f_Q) is the qubit frequencies of each transmon in Fig. 1a. $f_k^{(0)}$ is the idle qubit frequency of *k*-th transmon. The two blue curves (transmons 1a and 1b) form an encoded qubit, and the two green curves (transmons 2a and 2b) form the other encoded qubit. The transmons 1b and 2a are brought close to resonance while still far from being resonant with other transmons (transmon 1a and 2b), then are brought back to respective idle frequencies. (**b**) Energy spectrum for the process. The system is brought to the shaded area where (0110) and (0020) states are mixed. (0110) state accumulates non-trivial phase during this process, which leads to a CPHASE gate between transmon 1b and transmon 2a. This provides a non-trivial encoded two-qubit gate necessary for universal QC.

able to reach a fidelity better than $F_1 \geq F_z^3 = 0.999^3 = 0.997$ using currently available experimental techniques.

Two-qubit operations. For a scalable qubit architecture, we need to plan for the transmon qubit frequencies such that unnecessary resonances are avoided, especially if the two-qubit interaction cannot be completely shut off via, for example, a tunable coupler[41]. An encoded qubit has two transmons with idle frequency difference much larger than the capacitive coupling, so we can effectively turn the coupling off. In the two-encoded qubit system (four transmon system), we set the idle frequencies of next-nearest neighbour transmons to be different by more than the direct capacitive coupling between them, which is order of MHz[15]. We also set the encoded qubit frequencies $\Delta\varepsilon$ of the neighbouring encoded qubits to be different so we can mitigate some unintended resonances. For the calculations in this section, we set the four transmon qubit idle frequencies $f_Q^{(k)}$ as 5.6, 4.6, 5.9, 4.8 GHz for $k=$1a, 1b, 2a, 2b, respectively (see Fig. 1a. In this section and the following, we set $E_C^{(k)}/h = 375$ MHz and $E_{cc} = 30$ MHz for all transmons. Transmon qubit frequencies are controlled by tuning $E_J^{(k)}$.

Two-qubit operations can be implemented by adopting the adiabatic two-qubit CPHASE operations[14,42] between two-transmon qubits. By tuning the qubit frequencies of two-transmon qubits such that (11) and (02) states become resonant and then bringing them back to their idle frequencies, a unitary gate equivalent to the CPHASE gate between two qubits up to single-qubit unitary gates can be achieved[43]. This scheme has already been used in experiments and achieved reported fidelity better than 0.99 (ref. 14). In a similar manner, we can implement the CPHASE gate between two encoded qubits up to single-qubit unitary gates. Figure 3a shows schematically the pulse sequence of the transmon qubit frequencies, changing the qubit frequencies of transmon 1b and transmon 2a in Fig. 1a. First, we bring the transmons 1b and 2a closer during time τ_1 such that (0110) and (0020) states are on resonance in step (I). Then, in step (II), they stay there for a time period $\tau_{12}=\tau_2-\tau_1$, and finally we bring them back to initial point at time $\tau_3=\tau_2+\tau_1$ in step (III). The (0110) state gets mixed with (0020) due to the capacitive coupling during the pulse sequence with strength $\sqrt{2}\varepsilon'$. During this process the (0110) state obtains some non-trivial phase due to the interaction with (0020) while the other qubit states, (0101), (1001) and (1010), obtain only trivial phases since they do not get close to any other states that can mix. This process results in a unitary operation in the encoded qubit space, up to a global phase,

$$U = \begin{pmatrix} e^{i\phi_2} & 0 & 0 & 0 \\ 0 & e^{i(\phi_2+\phi_3+\delta\phi)} & 0 & 0 \\ 0 & 0 & 1 & 0 \\ 0 & 0 & 0 & e^{i\phi_3} \end{pmatrix}. \quad (5)$$

This is equivalent to the CPHASE gate $(1, 1, 1, e^{i\delta\phi})^T$ up to single-qubit operations.

$$\begin{aligned} \text{CPHASE} \;=\; & \left[\begin{pmatrix} 0 & 1 \\ e^{-i\phi_2} & 0 \end{pmatrix} \otimes \begin{pmatrix} 1 & 0 \\ 0 & 1 \end{pmatrix}\right] \\ & \times U \times \left[\begin{pmatrix} 0 & 1 \\ 1 & 0 \end{pmatrix} \otimes \begin{pmatrix} 1 & 0 \\ 0 & e^{-i\phi_3} \end{pmatrix}\right]. \end{aligned} \quad (6)$$

Note that, unlike refs 14,42, we tune both transmons 1b and 2a instead of tuning only one of them. If we only tuned transmon 2a to bring the (0110) state close to the (0020) state, then transmon 2a and transmon 2b would be close to resonance. Because the transmon–transmon interaction through capacitive coupling can be turned on and off by bringing the transmons on and off resonance, this will result in a complicated, unintended operation as well as leakage. So it is necessary to tune transmons 1b and 2a simultaneously so that transmons 1a and 2b do not come into play during the process. The resonance between next-nearest neighbours can also lead to some small anti-crossing, but these resonances only occur during the fast ramping up and down steps and thus can be negligible. This scheme is preferable to directly using the *xx* coupling between transmons 1b and 2a, since *xx* coupling drives the system outside of the encoded qubit space and hence leads to leakage, requiring a rather long sequence of pulse gates to implement a two-qubit logical operation[7,9]. The physical CPHASE gate has been successfully implemented for xmon qubits with gate time of ~40 ns (ref. 14), which can be directly applied for logical two-qubit gate here, too.

Figure 4 shows simulated numerical results of this physical two-qubit interaction between transmons 1b and 2a. We use an error function shape ramping up and down, similar to ref. 44,

$$E_J^{(1b)}(t) = \begin{cases} E_{J0}^{(1b)} + \frac{E_{res}^{(1b)}-E_{J0}^{(1b)}}{2}\left(1+\mathrm{erf}\left(\frac{t-\tau_1/2}{\sqrt{2}\sigma}\right)\right) & \text{(I)} \\ E_{res}^{(1b)} & \text{(II)} \\ E_{res}^{(1b)} - \frac{E_{res}^{(1b)}-E_{J0}^{(1b)}}{2}\left(1+\mathrm{erf}\left(\frac{t-\tau_1/2-\tau_2}{\sqrt{2}\sigma}\right)\right) & \text{(III)} \end{cases} \quad (7)$$

and $E_J^{(2a)}(t) = E_{J0}^{(1b)} + E_{J0}^{(2a)} - E_J^{(1b)}(t)$. E_{J0} is the idle value and E_{res} is for resonant (11) and (02) states. To find optimal solutions of this form, we change τ_1 and choose $\sigma=\tau_1/4\sqrt{2}$. $\tau_{12}=\tau_2-\tau_1$ is calculated analytically using a perturbative expression such that the whole process will result in the U with desired $\delta\phi$. Figure 4a shows τ_{12} and the total time τ_3 needed to implement a CZ gate ($\delta\phi=\pi$).

Due to the mixing with higher energy states which are out of the encoded qubit space, leakage error could pose a problem. We can compute the leakage error as follows. The full unitary operation matrix U can be written in a block-form

$$U = \begin{pmatrix} U_{AA} & U_{AB} \\ U_{BA} & U_{BB} \end{pmatrix}, \quad (8)$$

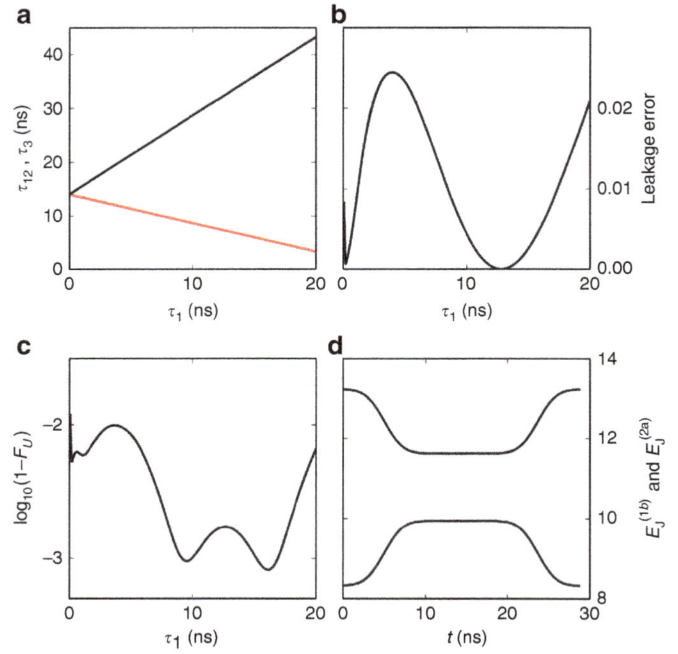

Figure 4 | Fidelity of adiabatic CZ interaction operation. (**a**) Operation time. The red curve is the staying time $\tau_{12}=\tau_2-\tau_1$ and the black curve is the total time τ_3. (**b**) Leakage error during the process. (**c**) Gate fidelity in terms of Makhlin invariants. This gives a measure of how close the unitary gate is to the CZ gate up to single qubit operations. (**d**) Pulse shape for $\tau_1=10$ ns. We used error function to model a smooth pulse shapes for $E_J^{(1b)}$ and $E_J^{(2a)}$.

where A is the encoded qubit subspace and B is the complementary subspace. For any qubit state $|\psi_A\rangle$ in the encoded qubit space, the leaked portion is $U_{BA}|\psi_A\rangle$ and $\|U_{BA}|\psi_A\rangle\|^2=\langle\psi_A|U_{BA}^\dagger U_{BA}|\psi_A\rangle\equiv\langle\psi_A|W_{AA}|\psi_A\rangle$. $W_{AA}=U_{BA}^\dagger U_{BA}$ is positive definite and the leakage error E_{leak} can be defined as $\max\langle\psi_A|W_{AA}|\psi_A\rangle=\max_\lambda\{w_\lambda\}$ where w_λ are the eigenvalues of W_{AA}. The leakage error (Fig. 4b) can be a few percent, but if we choose optimal τ_1, it can be significantly reduced, well below 1%. Note too that leakage can be dealt with algorithmically[45,46]; such circuit-based leakage reduction algorithms will likely be required in any quantum computing implementation.

Figure 4c shows the fidelity of this two-qubit unitary gate U from numerical simulation of the procedure. The fidelity of the unitary gate was defined as:

$$F_U=1-[f_1(U_{CZ})-f_1(U)]^2-[f_2(U_{CZ})-f_2(U)]^2 \qquad (9)$$

where f_1 and f_2 are the two Makhlin invariants[47] for two-qubit gates. Makhlin invariants are identical for different two-qubit unitary gates if they are equivalent up to single-qubit operations. We find that fidelity better than 99% is achievable for $\tau_1\simeq 10$ ns, which also leads to very small leakage. Figure 4d shows the pulse shape of $E_J^{(1b)}$ and $E_J^{(2a)}$ for $\tau_1=10$ ns. The total time duration for the whole process is about 30 ns. In real devices, the fidelity can be lower due to other sources of noise, but here we use only a simple form for the pulse shapes which are not fully optimized as in refs 44,48, so there is some room for improvement. We also considered Gaussian shape pulses and obtained similar results.

We can estimate the realistic fidelity of the encoded CPHASE gate constructed here from the fidelities of the z-control pulses and the adiabatic process. Since any single-qubit logical gate involves at most three rotations (that is, three pulse steps), the encoded CPHASE gate requires at most 12 pulse steps. Assuming the z-pulse fidelity of 0.999 and a fidelity of the adiabatic gate U in equation (6) betweeen two transmons of about 0.99, the fidelity of the total process can be estimated to be better than $F_2>0.999^{12}\times 0.99\simeq 0.978$. Better optimization or different sequences may improve the fidelity. Of critical comparison, the already demonstrated physical CPHASE gate fidelity of 0.99 (ref. 14) also includes a single-adiabatic operation and single-qubit corrective operations, so the encoded CPHASE gate should be achievable with a similar fidelity. The encoded CNOT gate can be implemented with CPHASE gate and single-qubit gates, and we can expect similar fidelity for CNOT gate.

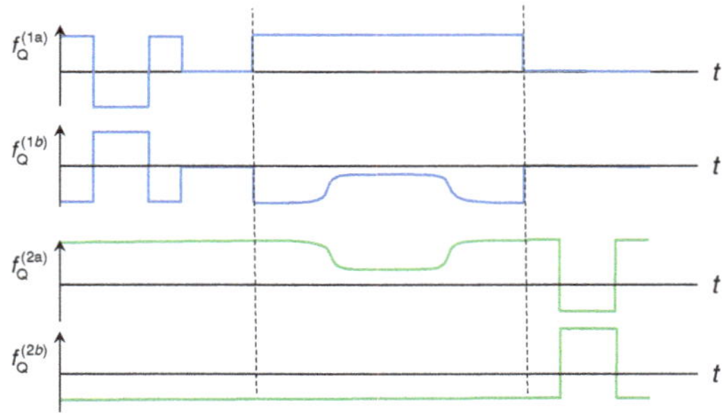

Figure 5 | Pulse sequence for encoded CPHASE gate. This schematically shows a sequence for CPHASE gate in equation (6). Single qubit phase gate is implemented with three-step Euler rotations, and Pauli X and identity gates are implemented as a single rotation. The two vertical dashed lines separate single qubit gates and two-qubit adiabatic operation.

Figure 5 schematically depicts a sequence of DC pulses for the logical CPHASE gate, using the expression in equation (6). The first three pulses in encoded qubit 1 implement a phase gate and the next resonant pulse realizes a Pauli X gate. The second encoded qubit is pulsed to qubit frequencies such that the encoded qubit 2 rotates by $2n\pi$ to implement the identity operation. Then the two-qubit adiabatic gate between transmons 1b and 2a is applied. After that, an X gate is applied to encoded qubit 1 as a single-resonant pulse step and a phase gate is applied to encoded qubit 2 in three rotations. This particular implementation of CPHASE contains only 9 single-qubit operations, better than the general 12 single-qubit gates we discussed above.

Our choice of encoded qubit is for the sake of simplicity and straightforward incorporation of physical qubit operations into logical gate operations. We also considered an alternative choice, $(|01\rangle\pm|10\rangle)/\sqrt{2}$ in the same subspace, which more closely resembles choice for encoded spin qubits. With this encoded qubit, the constant capacitive coupling leads to a constant energy gap between encoded qubit states and the z-control of each physical qubit allows tunable $\hat{\sigma}^x$ operation. Single-qubit logical gates can be implemented in a similar way, and the adiabatic two-qubit operation will need additional single-qubit unitary gates to transform to the CPHASE gate due to the basis change of the encoded qubit.

The capacitive coupling between transmons is typically constant and determined solely by the geometry of the SC islands. This coupling is effectively turned on and off by the qubit frequency differences. With more complicated control circuits as in the gmon architecture[41,49], the capacitive coupling can also be tunable and completely turned off, giving a very large on/off ratio. The tunable capacitive coupling removes the need to detune each transmon to avoid unwanted resonances, hence significantly simplifying the qubit frequency controls during the CPHASE operation. This also allows rotating the encoded qubit around any axis in the full xz plane, reducing the necessary rotation steps to two for any single-qubit logical gates[38].

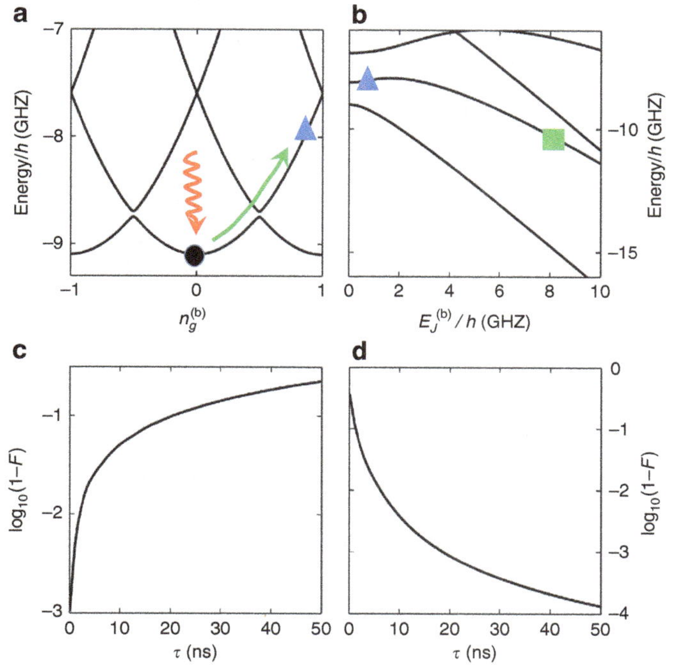

Figure 6 | Initialization scheme for the encoded qubit. (**a**) The energy spectrum of an encoded qubit with the second transmon in the charge qubit regime. After thermalizing the qubit into the ground state (black dot), $n_g^{(b)}$ is tuned from 0 to some value between 0.5 and 1. If this change is done fast enough, the qubit is in the first excited state (blue triangle). Then the qubit is adiabatically moved into the transmon regime (green square) by increasing $E_J^{(b)}$ as shown in (**b**). (**c**,**d**) Fidelity of these processes as a function of total time duration.

Initialization. In spin systems the encoded qubit can be initialized fast and with high fidelity by loading pairs of electrons in the singlet state directly from the Fermi sea provided by the leads supplying the quantum dots, then adiabatically separating the singlet into two dots[6]. Electrons' fermionic and particle nature enables this—a quantum property that may be emulated with engineered many-body photonic systems (for example, (refs 50,51)), but which is in no way practical in the near-term. One could also engineer a two-qubit system where the ground state is a singlet, for example, by making the coupling between the two qubits much greater than the qubit splittings (and, for example, waiting for relaxation to the ground state). Here, although, one would want to quickly move out of this regime to do gates at an implementable speed in addition to turning off as much as possible qubit–qubit couplings, which would be very challenging. Here, we provide an alternative initialization scheme that only requires fast DC pulses.

The ground state of the two-transmon system is $|00\rangle$, which is not in the encoded qubit subspace defined by equation (3). To initialize the system into $|0\rangle_Q = |01\rangle$ without microwave control, we propose using a process analogous to the Landau–Zener (LZ) tunnelling[52,53]. For this procedure, we need tunability of the gate charge $n_g^{(b)}$ of the second transmon, which can be provided by connecting a capacitor with a voltage control to the transmon (see Fig. 1b) or by using the side-gate for gatemons. The initialization procedure is as follows. First, we tune the transmon qubit into the charge qubit regime where $E_J^{(b)}$ is much smaller than $E_C^{(b)}$ by tuning Φ_{ext} (or V_g for a variable super-semi JJ) with $n_g^{(b)} \simeq 0$. Then, via thermalization (by waiting the relaxation time or by coupling to a dissipative reservoir) the qubit reaches the ground state (black dot in Fig. 6a). (The thermalization could instead be done before tuning to the charge qubit regime.) In this charge qubit regime, the two lowest energy states anti-cross at the sweet spot $n_g^{(b)} = 0.5$. By changing the gate charge $n_g^{(b)}$ from 0 to a value larger than 0.5, we can induce the LZ tunnelling to prepare the charge qubit in the first excited state (blue triangle). Then, we can tune $E_J^{(b)}$ back to the operating transmon regime ($E_J^{(b)} \gg E_C^{(b)}$) (green square in Fig. 6b). If we tune $E_J^{(b)}$ exactly to be zero, then there is a crossing instead of anti-crossing, and the fidelity will be much better. But some finite value will be allowable as long as we can change $n_g^{(b)}$ fast enough.

Figure 6c shows the calculated fidelity of the LZ tunnelling in the charge qubit regime of Fig. 6a as a function of the total time τ taken to change $n_g^{(b)}$. Here fidelity is defined as $F = |\langle\Psi_{target}|\Psi_{final}\rangle|$. We have used system parameters easily available in real systems, $E_C^{(a)}/h = E_C^{(b)}/h = 375$ MHz, $E_J^{(a)}/h = 12$ GHz, $E_J^{(b)}/h = 50$ MHz, $E_{cc} = 30$ MHz. $n_g^{(b)}$ was changed from 0 to 0.8. As is the case for typical LZ tunnelings, the fidelity is better with faster change of the parameter. We expect to see fidelity better than 99% for a LZ process of a few nanoseconds. Tuning back to the transmon regime is essentially an adiabatic process, and the fidelity increases with slower change (Fig. 6d). We changed $E_J^{(b)}/h$ from 50 MHz to 8.33 GHz, and the fidelity is better than 99% for a process of a few tens of nanoseconds. So this initialization process will take $\sim$20 ns to prepare the logical qubit state with fidelity of $\sim$99%. The effect of charge and quasiparticle noise during this process is a concern that should be investigated experimentally, but charge qubits have been shown to have T_1 times up to 0.2 ms (ref. 54). Variants of the flux qubit are especially stable to quasiparticle and charge noise fluctuations even at small qubit splittings[34].

Measurement of the qubit states. Since an encoded qubit is in a state

$$|\Psi\rangle = \alpha|0\rangle_Q + \beta|1\rangle_Q = \alpha|01\rangle + \beta|10\rangle, \qquad (10)$$

the encoded qubit can be measured by measuring either of the physical qubits using a standard method, such as dispersive measurement[55–58] (which can be multiplexed). The choice of our encoded qubit in equation (3) allows us to translate the single-qubit state into the encoded qubit states. With a choice of a singlet-triplet-like encoded qubit states, $(|01\rangle \pm |10\rangle)/\sqrt{2}$, the encoded qubit state can also be measured after some single-qubit gates are applied to turn them into the encoded qubit states as above, or they could be measured directly by dispersive measurement since these states correspond to different resonator frequencies[59,60].

Unlike the spin system where measurement of a singlet can be done electrostatically using a projective measurement[6], the dispersive measurement of SC qubits using a transmission line resonator still requires a microwave carrier, which is fine as a proof of concept. We would prefer a measurement approach that takes full advantage of our encoded qubit architecture, with qubit energy completely separated from microwave source. One possibility is to convert the encoded qubit to another quantum system (or measurement qubit) that is long-lived classically, but can be read out digitally or with fast base-band pulses (in other words a latched readout), for example,[61]. A compromise option is to do dispersive measurement but still utilize lower bandwidth lines: we can either tune E_J directly or swap the qubit with another one with a different frequency such that it can be readily measured.

Discussion

We proposed a scheme for a "dual rail" superconducting quantum computer where each qubit consists of two tunable-physical qubits. Encoded two-qubit operations are found to require only a single physical two-qubit gate and single-qubit pulses. Since physical two-qubit gates are typically much more costly in time and fidelity, this means that the overhead of encoded operations as proposed here is not significant, especially compared to spin qubits.

In this encoded qubit architecture all qubit manipulations are achieved solely by the z-control pulse sequences of individual qubits. This removes the requirement of microwave xy-control lines necessary in conventional transmon or similar qubit devices, simplifying classical control circuitry significantly. In addition, the encoded approach may allow lower requirements for available bandwidth per line, the potential for less crosstalk, and a reduction in needed timing accuracy as the encoded qubit states are nearly degenerate. Removing the need for microwave control frees the choice of qubit frequency from the cost and availability of microwave electronics. One is then able to design physical qubits with higher (or much lower) frequency that might enable higher temperature qubit operation (which may benefit from work already underway to enable high magnetic field compatible circuits for Majorana experiments[62] in higher-T_c materials) or qubits made from degenerate quantum circuits as in symmetry protected approaches[63–65], of which there is a natural connection to how spin qubits are inherently protected.

Encoding a qubit in a two-dimensional subspace in a larger Hilbert space introduces leakage error. For our encoded qubits, the relaxation process of individual transmons will lead to leakage out of the encoded qubit space. For a single-gate operation such as CNOT of duration τ, the leakage error due to the T_1 process would be $1-\exp(-\tau/T_1) \simeq 0.04$ % for $\tau = 40$ ns and $T_1 = 100\,\mu$s, which would slightly reduce the error threshold for quantum error correction[66]. While a single-gate operation of a few tens of ns does not lead to significant leakage errors, a long sequence of gate operations in a large system can be a problem. Particularly, a single-logical qubit for fault-tolerant quantum computing such as the surface code will consist of many encoded qubits and a logical operation will be a sequence of operations on those encoded qubits. Therefore, leakage reduction units[67] will likely be essential. For example, a full-leakage reduction unit on the basis of one-bit teleportation[66] would require an ancilla qubit for each encoded qubit and additional CNOT operations, and measurements after each logical CNOT operation. Qubits especially designed for large relaxation times, such as variants of fluxonium[68], may be particularly promising for our approach (for example, a T_1 time of 1 ms would lead to a leakage error per CNOT of 4×10^{-5}) and would reduce the overhead for leakage mitigation dramatically.

The recently demonstrated capacitively-shunted flux qubits[33,34] (or "fluxmon") may also provide a promising alternative. They have comparable coherence times and a larger anharmonicity than transmons. Qubit–qubit coupling through mutual inductance would also provide transversal xx coupling like the capacitive coupling between transmon qubits, so the formalism used in this work should be applicable as well. They also offer benefits for initialization as they can be tuned to the flux qubit regime down to very small qubit splitting while being protected to T_1 processes that flux qubits offer, and readout can also be done by using a DC SQUID[69,70] without a transmission line.

In the next phase of this design philosophy one can consider how to mimic other beneficial properties of spin qubits: very weak coupling between qubit states to charge noise and phonons, a fast and selective two-qubit gate via a Pauli exclusion like mechanism or an interaction that mimics it, very large ON/OFF ratios and fast initialization via some as yet unknown method.

References

1. Awschalom, D. D., Bassett, L. C., Dzurak, A. S., Hu, E. L. & Petta, J. R. Quantum spintronics: engineering and manipulating atom-like spins in semiconductors. *Science* **339,** 1174–1179 (2013).
2. Loss, D. & DiVincenzo, D. P. Quantum computation with quantum dots. *Phys. Rev. A* **57,** 120–126 (1998).
3. Kane, B. E. A silicon-based nuclear spin quantum computer. *Nature* **393,** 133–137 (1998).
4. Tyryshkin, A. M. *et al.* Electron spin coherence exceeding seconds in high-purity silicon. *Nat. Mater.* **11,** 143–147 (2012).
5. Saeedi, K. *et al.* Room-temperature quantum bit storage exceeding 39 minutes using ionized donors in silicon-28. *Science* **342,** 830–833 (2013).
6. Petta, J. R. *et al.* Coherent manipulation of coupled electron spins in semiconductor quantum dots. *Science* **309,** 2180–2184 (2005).
7. DiVincenzo, D. P., Bacon, D., Kempe, J., Burkard, G. & Whaley, K. B. Universal quantum computation with the exchange interaction. *Nature* **408,** 339–342 (2000).
8. Bacon, D., Kempe, J., Lidar, D. A. & Whaley, K. B. Universal fault-tolerant quantum computation on decoherence-free subspaces. *Phys. Rev. Lett.* **85,** 1758–1761 (2000).
9. Fong, B. H. & Wandzura, S. M. Universal quantum computation and leakage reduction in the 3-qubit decoherence free subsystem. *Quant. Inf. Comput.* **11,** 1003–1018 (2011).
10. Devoret, M. H. & Schoelkopf, R. J. Superconducting circuits for quantum information: an outlook. *Science* **339,** 1169–1174 (2013).
11. Koch, J. *et al.* Charge-insensitive qubit design derived from the Cooper pair box. *Phys. Rev. A* **76,** 042319 (2007).
12. Schreier, J. A. *et al.* Suppressing charge noise decoherence in superconducting charge qubits. *Phys. Rev. B* **77,** 180502 (R) (2008).
13. Barends, R. *et al.* Coherent Josephson qubit suitable for scalable quantum integrated circuits. *Phys. Rev. Lett.* **111,** 080502 (2013).
14. Barends, R. *et al.* Superconducting quantum circuits at the surface code threshold for fault tolerance. *Nature* **508,** 500–503 (2014).
15. Kelly, J. *et al.* State preservation by repetitive error detection in a superconducting quantum circuit. *Nature* **519,** 66–69 (2015).
16. Reed, M. D. *et al.* Realization of three-qubit quantum error correction with superconducting circuits. *Nature* **482,** 382–385 (2012).
17. Sun, L. *et al.* Tracking photon jumps with repeated quantum non-demolition parity measurements. *Nature* **511,** 444–448 (2014).
18. Chow, J. M. *et al.* Implementing a strand of a scalable fault-tolerant quantum computing fabric. *Nat. Commun.* **4,** 4015 (2014).
19. Córcoles, A. D. *et al.* Demonstration of a quantum error detection code using a square lattice of four superconducting qubits. *Nat. Commun.* **6,** 6979 (2015).
20. Blase, X., Bustarret, E., Chapelier, C., Klein, T. & Marcenat, C. superconducting group-IV semiconductors. *Nat. Mater.* **8,** 375–382 (2009).
21. Bustarret, E. Superconductivity in doped semiconductors. *Physica C* **514,** 36–45 (2015).
22. Bustarret, E. *et al.* Superconductivity in doped cubic silicon. *Nature* **444,** 465–468 (2006).
23. Marcenat, C. *et al.* Low-temperature transition to a superconducting phase in boron-doped silicon films grown on (001)-oriented silicon wafers. *Phys. Rev. B* **81,** 020501 (R) (2010).
24. Dahlem, F. *et al.* Subkelvin tunneling spectroscopy showing Bardeen–Cooper–Schrieffer superconductivity in heavily boron-doped silicon epilayers. *Phys. Rev. B* **82,** 140505 (R) (2010).
25. Grockowiak, A. *et al.* Superconducting properties of laser annealed implanted Si:B epilayers. *Supercond. Sci. Technol.* **26,** 045009 (2013).
26. Herrmannsdörfer, T. *et al.* Superconducting state in a gallium-doped germanium layer at low temperatures. *Phys. Rev. Lett.* **102,** 217003 (2009).
27. Skrotzki, R. *et al.* The impact of heavy Ga doping on superconductivity in germanium. *Low Temp. Phys.* **37,** 877–883 (2011).
28. Shim, Y.-P. & Tahan, C. Bottom-up superconducting and Josephson junction device inside a group-IV semiconductor. *Nat. Commun.* **5,** 4225 (2014).
29. Shim, Y.-P. & Tahan, C. Superconducting-semiconductor quantum devices: from qubits to particle detectors. *IEEE J. Sel. Top. Quant. Electron.* **21,** 9100209 (2015).
30. Larsen, T. W. *et al.* Semiconductor-nanowire-based superconducting qubit. *Phys. Rev. Lett.* **115,** 127001 (2015).
31. deLange, G. *et al.* Realization of microwave quantum circuits using hybrid superconducting-semiconducting nanowire Josephson elements. *Phys. Rev. Lett.* **115,** 127002 (2015).
32. Ristè, D. *et al.* Detecting bit-flip errors in a logical qubit using stabilizer measurements. *Nat. Commun.* **6,** 6983 (2015).
33. You, J. Q., Hu, X., Ashhab, S. & Nori, F. Low-decoherence flux qubit. *Phys. Rev. B* **75,** 140515 (2007).
34. Yan, F. *et al.* The flux qubit revisited. Preprint at http://arxiv.org/abs/1508.06299 (2015).

35. Makhlin, Y., Schön, G. & Shnirman, A. Josephson-junction qubits with controlled couplings. *Nature* **398,** 305–307 (1999).
36. Chang, W. *et al.* Hard gap in epitaxial semiconductorsuperconductor nanowires. *Nat. Nanotechonol.* **10,** 232–236 (2015).
37. Krogstrup, P. *et al.* Epitaxy of semiconductorsuperconductor nanowires. *Nat. Mater.* **14,** 400–406 (2015).
38. Shim, Y.-P., Fei, J., Oh, S., Hu, X. & Friesen, M. Single-qubit gates in two steps with rotation axes in a single plane. Preprint at http://arxiv.org/abs/1303.0297 (2013).
39. Wang, X. *et al.* Composite pulses for robust universal control of singlettriplet qubits. *Nat. Commun.* **3,** 997 (2012).
40. Wang, X., Bishop, L. S., Barnes, E., Kestner, J. P. & Sarma, S. D. Robust quantum gates for singlet-triplet spin qubits using composite pulses. *Phys. Rev. A* **89,** 022310 (2014).
41. Chen, Y. *et al.* Qubit architecture with high coherence and fast tunable coupling. *Phys. Rev. Lett.* **113,** 220502 (2014).
42. DiCarlo, L. *et al.* Demonstration of two-qubit algorithms with a superconducting quantum processor. *Nature* **460,** 240–244 (2009).
43. Strauch, F. W. *et al.* Quantum logic gates for coupled superconducting phase qubits. *Phys. Rev. Lett.* **91,** 167005 (2003).
44. Ghosh, J. *et al.* High-fidelity controlled-σ^z gate for resonator-based superconducting quantum computers. *Phys. Rev. A* **87,** 022309 (2013).
45. Wu, L.-A., Byrd, M. S. & Lidar, D. A. Efficient universal leakage elimination for physical and encoded qubits. *Phys. Rev. Lett.* **89,** 127901 (2002).
46. Fowler, A. G. Coping with qubit leakage in topological codes. *Phys. Rev. A* **88,** 042308 (2013).
47. Makhlin, Y. Nonlocal properties of two-qubit gates and mixed states, and the optimization of quantum computations. *Quant. Inf. Proc* **1,** 243–252 (2002).
48. Martinis, J. M. & Geller, M. R. Fast adiabatic qubit gates using only σ_z control. *Phys. Rev. A* **90,** 022307 (2014).
49. Geller, M. R. *et al.* Tunable coupler for superconducting Xmon qubits: perturbative nonlinear model. *Phys. Rev. A* **92,** 012320 (2015).
50. Greentree, A. D., Tahan, C., Cole, J. H. & Hollenberg, L. C. L. Quantum phase transitions of light. *Nat. Phys.* **2,** 856–861 (2006).
51. Hartmann, M., Brandão, F. & Plenio, M. Quantum many-body phenomena in coupled cavity arrays. *Laser Photon. Rev.* **2,** 527–556 (2008).
52. Landau, L. Zur Theorie der Energieubertragung. II. *Phys. Z. Sowjetunion* **2,** 46–51 (1932).
53. Zener, C. Non-Adiabatic Crossing of Energy Levels. *Proc. R. Soc. Lond. A* **137,** 696–702 (1932).
54. Kim, Z. *et al.* Decoupling a Cooper-pair box to enhance the lifetime to 0.2 ms. *Phys. Rev. Lett.* **106,** 120501 (2011).
55. Blais, A., Huang, R.-S., Wallraff, A., Girvin, S. M. & Schoelkopf, R. J. Cavity quantum electrodynamics for superconducting electrical circuits: an architecture for quantum computation. *Phys. Rev. A* **69,** 062320 (2003).
56. Wallraff, A. *et al.* Strong coupling of a single photon to a superconducting qubit using circuit quantum electrodynamics. *Nature* **431,** 162–167 (2004).
57. Schuster, D. I. *et al.* ac stark shift and dephasing of a superconducting qubit strongly coupled to a cavity field. *Phys. Rev. Lett.* **94,** 123602 (2005).
58. Wallraff, A. *et al.* Approaching unit visibility for control of a superconducting qubit with dispersive readout. *Phys. Rev. Lett.* **95,** 060501 (2005).
59. Gambetta, J. M., Houck, A. A. & Blais, A. Superconducting qubit with purcell protection and tunable coupling. *Phys. Rev. Lett.* **106,** 030502 (2011).
60. Srinivasan, S. J., Hoffman, A. J., Gambetta, J. M. & Houck, A. A. Tunable coupling in circuit quantum electrodynamics using a superconducting charge qubit with a V-shaped energy level diagram. *Phys. Rev. Lett.* **106,** 083601 (2011).
61. Berkley, A. J. *et al.* A scalable readout system for a superconducting adiabatic quantum optimization system. *Supercond. Sci. Technol.* **23,** 105014 (2010).
62. van Woerkom, D. J., Geresdi, A. & Kouwenhoven, L. P. One minute parity lifetime of a NbTiN Cooper-pair transistor. *Nat. Phys.* **11,** 547550 (2015).
63. Douçot, B. & Ioffe, L. B. Physical implementation of protected qubits. *Rep. Prog. Phys.* **75,** 072001 (2012).
64. Kitaev, A. Y. Fault-tolerant quantum computation by anyons. *Ann. Phys.* **303,** 2–30 (2003).
65. Brooks, P., Kitaev, A. & Preskill, J. Protected gates for superconducting qubits. *Phys. Rev. A* **87,** 052306 (2013).
66. Suchara, M., Cross, A. W. & Gambetta, J. M. Leakage suppression in the toric code. *Quant. Inf. Comp.* **15,** 997–1016 (2015).
67. Aliferis, P. & Terhal, B. M. Fault-tolerant quantum computation for local leakage faults. *Quant. Inf. Comp.* **7,** 139–156 (2007).
68. Manucharyan, V. E., Koch, J., Glazman, L. I. & Devoret, M. H. Fluxonium: single Cooper-pair circuit free of charge offsets. *Science* **326,** 113–116 (2009).
69. Bylander, J. *et al.* Noise spectroscopy through dynamical decoupling with a superconducting flux qubit. *Nat. Phys.* **7,** 565–570 (2011).
70. Jin, X. Y. *et al.* Z-gate operation on a superconducting flux qubit via its readout SQUID. *Phys. Rev. Applied* **3,** 034004 (2015).

Acknowledgements

We thank C.M. Marcus, A. Mizel, W.D. Oliver, B. Palmer and K.D. Petersson for useful discussions.

Author contributions

All the authors contributed to the planning of the project, interpretation of the results, discussions and writing of the manuscript. Y.-P.S performed the theoretical and numerical calculations.

Additional information

Competing financial interests: The authors declare no competing financial interests.

23

Signatures of the Adler–Bell–Jackiw chiral anomaly in a Weyl fermion semimetal

Cheng-Long Zhang[1,*], Su-Yang Xu[2,*], Ilya Belopolski[2,*], Zhujun Yuan[1,*], Ziquan Lin[3], Bingbing Tong[1], Guang Bian[2], Nasser Alidoust[2], Chi-Cheng Lee[4,5], Shin-Ming Huang[4,5], Tay-Rong Chang[2,6], Guoqing Chang[4,5], Chuang-Han Hsu[4,5], Horng-Tay Jeng[6,7], Madhab Neupane[2,8,9], Daniel S. Sanchez[2], Hao Zheng[2], Junfeng Wang[3], Hsin Lin[4,5], Chi Zhang[1,10], Hai-Zhou Lu[11], Shun-Qing Shen[12], Titus Neupert[13], M. Zahid Hasan[2] & Shuang Jia[1,10]

Weyl semimetals provide the realization of Weyl fermions in solid-state physics. Among all the physical phenomena that are enabled by Weyl semimetals, the chiral anomaly is the most unusual one. Here, we report signatures of the chiral anomaly in the magneto-transport measurements on the first Weyl semimetal TaAs. We show negative magnetoresistance under parallel electric and magnetic fields, that is, unlike most metals whose resistivity increases under an external magnetic field, we observe that our high mobility TaAs samples become more conductive as a magnetic field is applied along the direction of the current for certain ranges of the field strength. We present systematically detailed data and careful analyses, which allow us to exclude other possible origins of the observed negative magnetoresistance. Our transport data, corroborated by photoemission measurements, first-principles calculations and theoretical analyses, collectively demonstrate signatures of the Weyl fermion chiral anomaly in the magneto-transport of TaAs.

[1] International Center for Quantum Materials, School of Physics, Peking University, Beijing, China. [2] Laboratory for Topological Quantum Matter and Spectroscopy (B7), Department of Physics, Princeton University, Princeton, New Jersey 08544, USA. [3] Wuhan National High Magnetic Field Center, Huazhong University of Science and Technology, Wuhan 430074, China. [4] Centre for Advanced 2D Materials and Graphene Research Centre, National University of Singapore, Singapore 117546, Singapore. [5] Department of Physics, National University of Singapore, Singapore 117542, Singapore. [6] Department of Physics, National Tsing Hua University, Hsinchu 30013, Taiwan. [7] Institute of Physics, Academia Sinica, Taipei 11529, Taiwan. [8] Condensed Matter and Magnet Science Group, Los Alamos National Laboratory, Los Alamos, New Mexico 87545, USA. [9] Department of Physics, University of Central Florida, Orlando, Florida 32816, USA. [10] Collaborative Innovation Center of Quantum Matter, Beijing 100871, China. [11] Department of Physics, South University of Science and Technology of China, Shenzhen, China. [12] Department of Physics, The University of Hong Kong, Pokfulam Road, Hong Kong, China. [13] Princeton Center for Theoretical Science, Princeton University, Princeton, New Jersey 08544, USA. * These authors contributed equally to this work. Correspondence and requests for materials should be addressed to M.Z.H. (email: mzhasan@princeton.edu) or to S.J. (email: gwljiashuang@pku.edu.cn).

The principles of physics rest crucially on symmetries and their associated conservation laws. Over the past century, physicists have repeatedly observed the violations of apparent conservation laws in particle physics, each time leading to new insights and a refinement of our understanding of nature. One of the most interesting phenomena of this type is the breaking of a conservation law of classical physics by quantum-mechanical effects, a so-called anomaly in quantum field theory[1]. Perhaps the most primitive example is the so-called chiral anomaly associated with Weyl fermions[2–6]. A Weyl fermion is a massless fermion that carries a definite chirality. Due to the chiral anomaly, the chiral charge of Weyl fermions is not conserved by the full quantum-mechanical theory. Historically, the chiral anomaly was crucial in understanding a number of important aspects of the standard model of particle physics. The most well-known case is the triangle anomaly associated with the decay of the neutral pion π^0 (refs 3,4). Despite having been discovered more than 40 years ago, it remained solely in the realm of high-energy physics.

Recently, there has been considerable progress in understanding the correspondence between high-energy and condensed matter physics, which has led to deeper knowledge of important topics in physics such as spontaneous symmetry breaking, phase transitions and renormalization. Such knowledge has, in turn, greatly helped physicists and materials scientists to better understand magnets, superconductors and other novel materials, leading to important practical device applications. Here, we present the signatures of the chiral anomaly in a low-energy condensed matter Weyl system. In order to measure the chiral anomaly in a solid-state system, one needs to find a perturbation that couples differently to the two Weyl fermions of opposite chiralities. This is most naturally realized in a Weyl semimetal, in which the two Weyl cones are separated in momentum space. Recent theoretical and experimental advances have shown that Weyl fermions can arise in the bulk of certain novel semimetals with nontrivial topology[7–16]. A Weyl semimetal is a bulk crystal whose low-energy excitations satisfy the Weyl equation. Therefore, the conduction and valence bands touch at discrete points, the Weyl nodes, with a linear dispersion relation in all three momentum space directions moving away from the Weyl node. The nontrivial topological nature of a Weyl semimetal guarantees that Weyl fermions with opposite chiralities are separated in momentum space (Fig. 1a), and host a monopole and an antimonopole of Berry flux in momentum space, respectively (Fig. 1b). In this situation, parallel magnetic and electric fields can pump electrons between Weyl cones of opposite chirality that are separated in momentum space (Fig. 1a). This process violates the conservation of the the chiral charge, meaning that the number of particles of left and right chirality are not separately conserved[5,17–26], giving rise to an analogue of the chiral anomaly in a condensed matter system. Apart from this elegant analogy and correspondence between condensed matter and high-energy physics, the chiral anomaly also serves as a crucial transport signature for Weyl fermions in a Weyl semimetal phase. Furthermore, theoretical studies have recently suggested that it has potential applications[27].

In this paper, we perform magneto-transport experiments on the Weyl semimetal TaAs[12–14,16]. We observe a negative longitudinal magnetoresistance (LMR) in the presence of parallel magnetic and electric fields, which is indicative of the chiral anomaly due to Weyl fermions. On the other hand, due to the complicated nature of the magnetoresistence[28–38], an unambiguous demonstration of the chiral anomaly remains lacking despite the volume of works reporting negative LMR[39–44]. Our data and careful analyses, which go beyond a simple observation of a negative LMR, allow us to systematically exclude other possible origins for the observed negative LMR. These data strongly support the chiral anomaly due to Weyl fermions in TaAs. Our studies demonstrate a low-energy platform where the fundamental physics of Weyl fermions and quantum anomalies can be studied in a piece of solid metal[17–27].

Results

ARPES band structure. We start by presenting the key aspects of the bulk band structure of TaAs both in theory and in experiment. According to our first-principles calculation[13,14], in total there are 24 bulk Weyl cones. We denote the 8 Weyl nodes that are located on the $k_z = \frac{2\pi}{c}$ as W1 and the other 16 nodes that are away from this plane as W2 (Fig. 1c). There is a 13 meV offset between the energies of the W1 and W2 Weyl nodes (Fig. 1e). The pockets that arise from the Weyl fermions are shown in blue in Fig. 1d. Apart from the Weyl cones, there are additional (non-Weyl) hole-like bands crossing the Fermi level shown by the red ring-shaped contours in Fig. 1d.

We independently study the bulk electronic structure via angle-resolved photoemission spectroscopy (ARPES). This is important because relying entirely on numerical band structure calculations is not conclusive. Particularly, numerical band calculations have little power in predicting the position of the chemical potential of real samples, which is crucial for transport experiments. Figure 1f shows an $E - k_{||}$ dispersion map that cuts across the two nearby W2 Weyl cones. The dispersion map reveals two linearly dispersive bands. The k-space distance between the two crossing points is about 0.08 Å^{-1}, consistent with the calculated results. More importantly, our ARPES measurement shows that the native chemical potential of the samples is very close to the energy of the W2 Weyl nodes. Our data also reveals the W1 Weyl cones. As shown in Fig. 1g, the energy of the W1 Weyl node is below that of the W2 Weyl node (Fig. 1f), which is consistent with band calculation results. Systematic ARPES data can be found in the Supplementary Fig. 1 and Supplementary Note 1. We also observe the trivial hole bands in ARPES. The essential observations are listed as follows. There are three types of bands at the Fermi level, the W1 and W2 Weyl nodes and a trivial hole-like band. The native chemical potential is close to the energy of the W2 Weyl nodes, which is 13 meV higher than that of the W1 Weyl nodes. Therefore, the W1 Weyl cones form electron-like pockets, the trivial hole-like bands form hole-like pockets, and the W2 Weyl cones have low-carrier concentration, which can be electron- or hole-like depending on the specific position of the native chemical potential with respect to the W2 Weyl node in each sample batch.

Quantum oscillation data. We have performed magneto-transport measurements on our TaAs samples, in order to probe the band structure at the Fermi level (Supplementary Fig. 2 and Supplementary Note 2). Our Hall data indeed reveal a coexistence of electron and hole carriers. We obtained critical band parameters, such as Fermi wavevector, Fermi velocity, chemical potential, carrier mobilities and so on, from the Shubnikov de-Haas (SdH) oscillation data. All band parameters obtained from the SdH oscillations are consistent with first-principles calculation and ARPES results. Most importantly, this enables us to determine the position of the chemical potential with respect to the Weyl nodes, as shown in Fig. 2i. We name the samples by a letter (a or c) followed by a number (1–5). The letter 'a' or 'c' means that the electrical current is along the crystallographic a- or c-axes. The number '1–5' refers to a sequence of the samples' chemical potential with respect to the energy of the W2 Weyl node from below the node to above the node (Fig. 2i). We also provide the Fermi energy of the TaAs samples determined

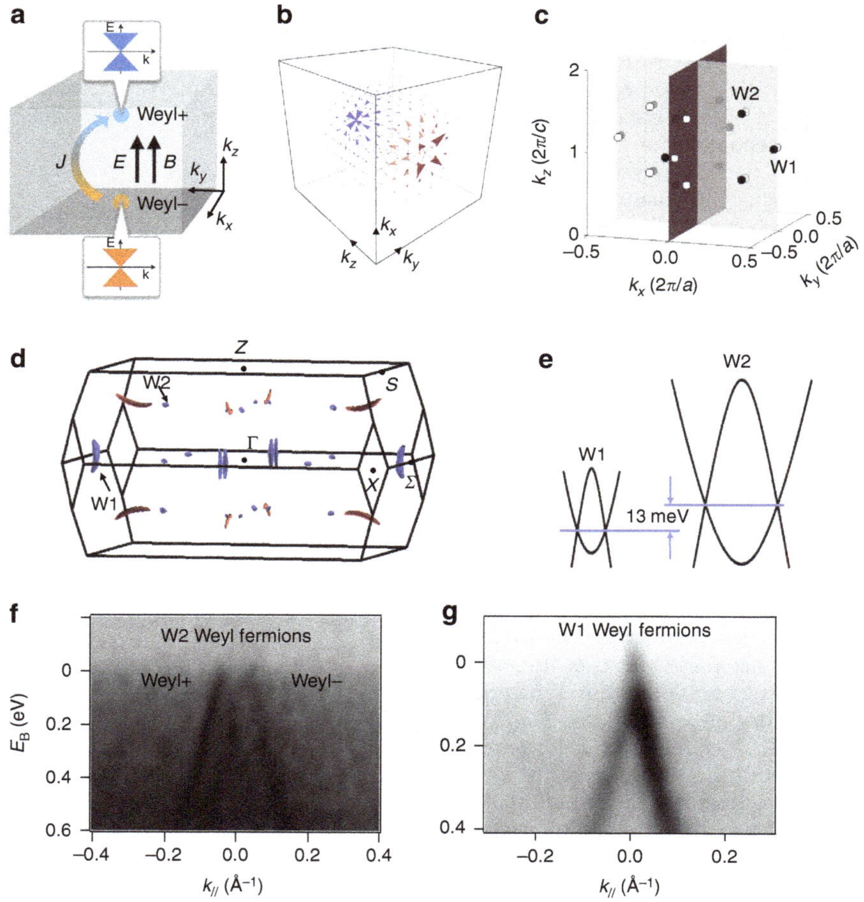

Figure 1 | Electronic band structure of the Weyl semimetal TaAs. (**a**) Schematics of the separation of the pairs of Weyl fermions in a Weyl semimetal with opposite chiralities in momentum space, which is a direct consequence of its nontrivial topological nature. (**b**) Distribution of the Berry curvature near two Weyl nodes in momentum space with the opposite chiralities. (**c**) The location of the Weyl nodes in the first Brillouin zone. (**d**) First-principles calculated constant energy contour of TaAs. The energy is set at about 5 meV above the energy of the W2 Weyl nodes. (**e**) Schematic energy dispersions of the W1 and the W2 Weyl cones. (**f**) ARPES measured energy dispersions of the W2 Weyl cones. (**g**) ARPES measured energy dispersions of the W1 Weyl cones.

by different approaches in Supplementary Note 3 and Supplementary Table 1.

Longitudinal magnetoresistance. We now present our LMR data, without a pre-biased assumption of their origin. Figure 2a–e show the LMR data of five different batches of samples. The LMR data show three main features as a function of the magnetic field, as schematically drawn in Fig. 2f. At very small fields close to $B=0$, we observe a sharp increase of the LMR. Following the sharp increase, the LMR is found to decrease in an intermediate B field range. This is the negative LMR. While further increasing the B field, the LMR starts to increase again. We note that because features I and II likely have independent origins, the LMR is not necessarily absolutely negative (we do, however, observe absolute negative LMR in samples c2 and c4). Hence, more precisely, we speak of negative LMR in this paper, if the resistivity decreases with an increasing B field. In addition to these general features, we observe other more sample-dependent features: for sample c4, our data show clear quantum oscillations at a quite wide B field range of $0.5\,\mathrm{T} \leq B \leq 8\,\mathrm{T}$. For other samples, the quantum oscillations are much weaker but they are still visible. For sample a5, the LMR increases monotonically as a function of the B field. No negative LMR is observed.

We study the systematic dependence of the LMR on different parameters, including temperature, the angle between the **E** and **B** fields, and the direction of the current with respect to the crystallographic axis. The temperature-dependent data are shown in Fig. 3a for sample a1. Most notably, the negative LMR (feature II) shows a strong temperature dependence. At higher temperatures, for example, $T \geq 50\,\mathrm{K}$, the negative LMR vanishes. The dependence on the angle between the **E** and **B** fields are shown in Fig. 3b–e. Our data show that the negative MR exhibits a very strong angular dependence. It becomes quickly suppressed as one varies the direction of the magnetic **B** field away from that of the electric **E** field. The dependence on the direction of the current with respect to the crystallographic direction is presented in Fig. 2. The measurements were performed with current along the crystallographic a axis for samples a1, a3 and a5, and with current along the c axis for samples c2 and c4. In both cases, the negative LMR is observed except for sample a5, whose chemical potential is far away from the energy of the Weyl nodes (Fig. 2i).

Origins of the negative LMR. We now use these observations to understand the origin of the negative LMR. First, it is well-known that a negative LMR can arise in magnetic materials[28]. This obviously does not apply to our non-magnetic TaAs samples. The second possible origin is more classical due to geometry or size effects of the samples, such as the current-jetting effect[29,30]. These geometrical MR effects are also not consistent with our data, because they do not vanish quickly as one raises temperature[30], and furthermore we have carefully shaped our samples to exclude the geometrical effects (Supplementary Fig. 3 and Supplementary

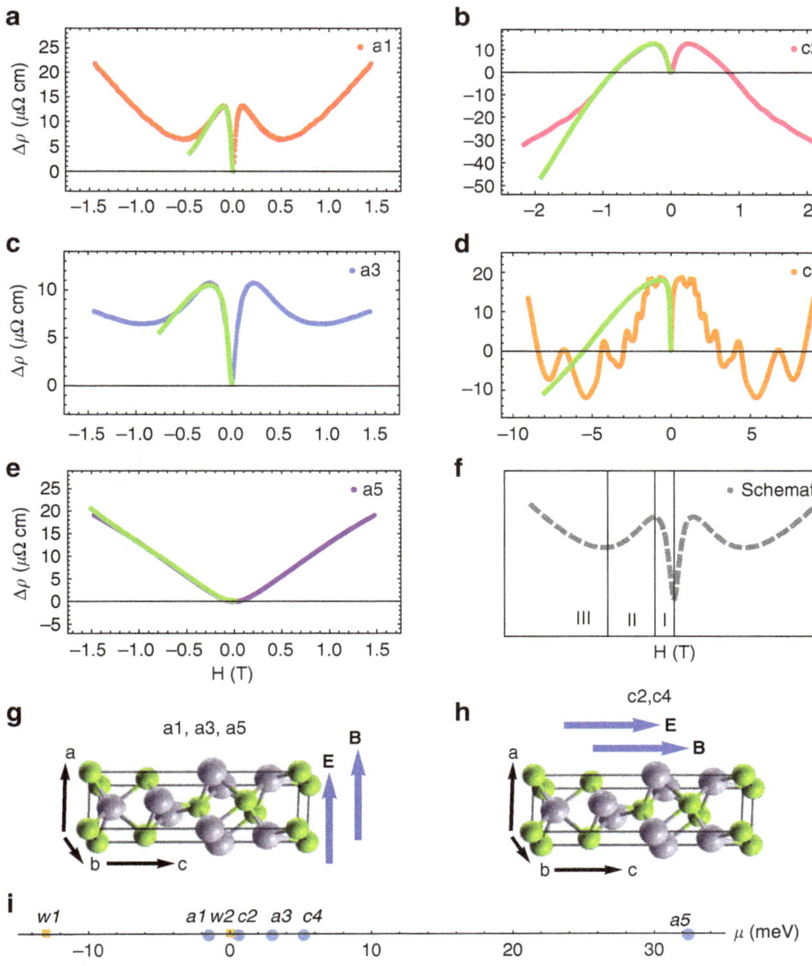

Figure 2 | Observation of negative longitudinal magneto-resistances. (**a–e**) LMR data at $T = 2$ K for samples a1, c2, a3, c4 and a5, respectively. The green curves are the fits to the LMR data in the semiclassical regime. The *y* axes of **a–e** are the change of the resistivity with respect to the zero-field resistivity, $\Delta\rho = \rho(B) - \rho(B = 0)$. (**h**) A schematic drawing of the LMR data to show the three important features (I–III) observed in our data as a function of the magnetic field. (**g**) Measurement geometry for samples a1, a3 and a5. (**h**) Measurement geometry for samples c2 and c4. (**f**) A schematic illustration of the LMR data. The data consists of three sections as a function of the magnetic field, which are labled as I–III. (**i**) Position of the samples' chemical potential with respect to the energy of the Weyl nodes obtained from SdH oscillation measurements.

Note 4). Third, we observe the negative LMR with current flowing both along the crystallographic *a*- and *c*-axes. We note that TaAs has a tetragonal lattice. Hence the *a*- and *c*-axes represent the largest anisotropy that the system could offer. The fact that the negative LMR is observed along both *a*- and *c*-axes proves that anisotropies in the system cannot explain our data[32]. Fourth, in the quantum limit, negative LMR can arise from the chiral, quasi one-dimensional character of the Landau levels that are formed by the band structure under magnetic fields. Essentially, it was predicted[31,33] that a negative LMR can arise in any 3D metal irrespective of its band structure if the sample is in the ultra-quantum limit, which means that one has $\omega\tau \gg 1$ (ω is the cyclotron frequency and τ is the transport life time) and that the chemical potential only crosses the lowest Landau level (the Landau level index $N = 0$, see Supplementary Fig. 4). This has been observed in doped semiconductor samples[35]. We have carefully checked whether our negative LMR is due to this mechanism. Particularly, one needs to be careful about the trivial hole-like bands in TaAs, because if they were in the ultra-quantum limit then it would have been entirely possible that the observed negative LMR were due to these trivial bands, rather than due to Weyl fermions in our samples. We note that the negative LMR are observed at small magnetic fields (for example, $0.1\,\mathrm{T} \leq B \leq 0.5\,\mathrm{T}$ for sample a1). We have checked the $\omega\tau$ and the Landau level index N of our samples quantitatively (Supplementary Table 2), and our results show that all samples are always in the semiclassical limit at the small magnetic fields where the negative LMR are observed. Therefore, our data are inconsistent with this origin[31,33]. Fifth, a recent theoretical work has predicted a linear B-dependent magneto-conductivity in small fields[45]. However, this is also inconsistent with our data because predicted linear B-dependent magneto-conductivity requires the system to lie in the ultra-quantum limit. That is, only the lowest Landau band crosses the Fermi level, which is clearly not the case for our systems under study. Finally, in the semiclassical limit, nonzero LMR can arise from finite Berry curvature, as follows from the semiclassical equations of motion. Having excluded all other possibilities, we are led to conclude that our observed negative LMR must have this origin. However, as suggested in ref. 36 in addition to nonzero Berry curvature, an approximately conserved chiral charge density with long relaxation time—as found in Dirac and Weyl semimetals—is required to yield a negative LMR. The observed negative LMR is most likely to be attributed to Weyl nodes, in accordance with the theoretical analysis of (refs 24,26 and 36). To confirm this picture independent of the assumption of effective low-energy Weyl

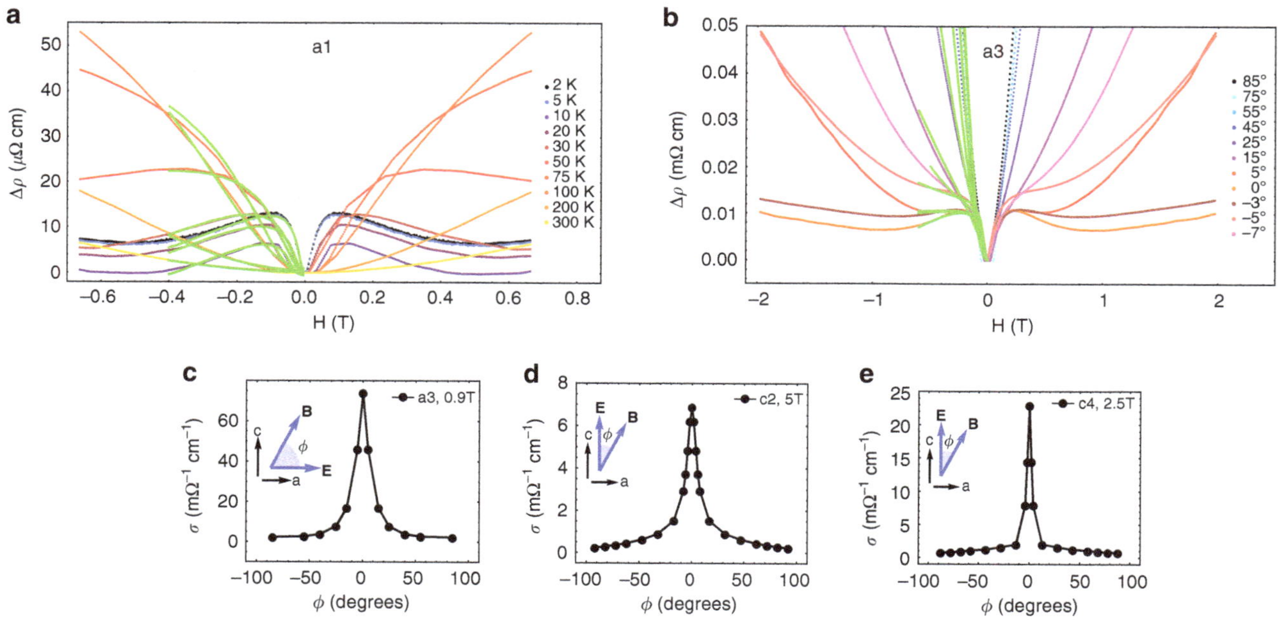

Figure 3 | Systematic dependence of the negative longitudinal magneto-resistances. (**a**) Temperature-dependent LMR data for sample a1. (**b**) Magnetoresistance data as a function of the angle between the $\vec{E}$ and $\vec{B}$ fields. The green curves of **a**,**b** are the fits to the LMR data in the semiclassical regime. The *y* axes are the change of the resistivity with respect to the zero-field resistivity, $\Delta\rho = \rho(B) - \rho(B=0)$. (**c**-**e**) The magnetoresistance as a function of the angle for samples c2, a3 and c4 at a fixed field.

Hamiltonians, we have studied the contribution of Berry curvature from each band carefully in a first-principles-derived model for TaAs (see Fig. 4c,d and Supplementary Fig. 5 and Supplementary Note 5). Our results show that in our TaAs system the Berry curvature almost entirely arises from the Weyl cones.

The chiral anomaly. With such a conclusion, we are entitled to fit our LMR data with a semiclassical magnetoconductance formula that includes the contribution from Weyl nodes due to their Berry curvature. Specifically, we use the following equation.

$$\sigma_{xx}(B) = 8C_W B^2 - C_{WAL}\left(\sqrt{B}\frac{B^2}{B^2+B_c^2} + \gamma B^2 \frac{B_c^2}{B^2+B_c^2}\right) + \sigma_0 \quad (1)$$

All coefficients are positive. The first term $\sigma^{chiral} = C_W B^2$ is due to the Weyl fermions and will lead to a B^2-dependent negative LMR. This term was systematically studied by transport theories in refs 24,26. The chiral coefficient is $C_W = \frac{e^4 \tau_a}{4\pi^4 \hbar^4 g(E_F)}$ (refs 24,26), where $g(E_F)$ is the density of states at the Fermi level, τ_a is the axial charge relaxation time and the additional factor of 8 is because we have 8 pairs of W2 Weyl nodes. All remaining terms contribute to positive LMR in the semiclassical regime. The C_{WAL} term arises from the 3D weak anti-localization (WAL) effect of the Weyl cones, which accounts for the initial steep uprise of the LMR at small magnetic fields. The 3D WAL is known to have a $-B^2$ dependence near zero field and $-\sqrt{B}$ dependence at higher fields[46]. So we include a critical field B_c that characterizes a crossover. For the four samples a1, c2, a3, c4, the increase of the LMR at small magnetic fields are 230, 5, 156 and 47% compared to the zero-field resistance. Particularly, the increase for samples a1 and a3 is larger than 100%, which is usually not expected from the WAL scenario. On the other hand, we do notice that the increase is quite sample dependent, and that a similarly large increase ($\sim$100%) of the magnetoresistance has also been reported in a concurrent transport work on TaAs[40]. In this work, we fit this initial uprise of the LMR by the WAL effect, but the anomalously large increase in samples a1, a3 and also in ref. 40 remains an theoretically open question that needs further investigation, which does not affect our main conclusion, that is, signatures of the chiral anomaly. Finally, the σ_0 term is the positive LMR that arises from the Drude conductivity of conventional charge carriers present in TaAs. In parallel fields, the Lorentz force is zero so the Drude conductivity is a constant. More systematic details regarding the fitting are presented in Supplementary Fig. 6 and Supplementary Note 6.

The fitting results are shown by the green curves in Figs 2a–e and 3a,b. It can be seen that the fitting works well for the small B field region which includes the negative LMR. This is reasonable because the fitting formula is derived in the semiclassical limit. The angle dependence of the chiral coefficient C_W is shown in Fig. 4b for sample a3, which demonstrates that C_W is only significant in the presence of parallel electric and magnetic fields. The sharp angular dependence is an open theoretical problem. More importantly, we study the chemical-potential dependence of the LMR data. Our fitting captures quantitatively the relative size of the low-field positive LMR and the higher-field negative LMR as a function of chemical potential. We plot this ratio as a dimensionless quantity in Fig. 4a. We find that despite the simple form of the fitting formula, the different measurement geometries for the different samples, the presence of large quantum oscillations in sample c4 and large differences in the absolute resistivities of different samples, the chiral anomaly ratio scales as $1/E_F^2$. It is remarkable that this fitting result matches the simplest theoretical model for a Weyl point, where the Berry curvature $\Omega \propto 1/E_F^2$. We emphasize that this provides powerful evidence that the negative LMR is due to the Weyl fermions. Note that the specific expression of the chiral coefficient, $C_W = \frac{e^4 \tau_a}{4\pi^4 \hbar^4 g(E_F)} \propto \frac{1}{E_F^2}$, is a result of the linear dispersion and the specific Berry curvature distribution of the Weyl cones (see Fig. 1b). Especially, at the energy of the W2 Weyl nodes, the trivial hole bands do not have

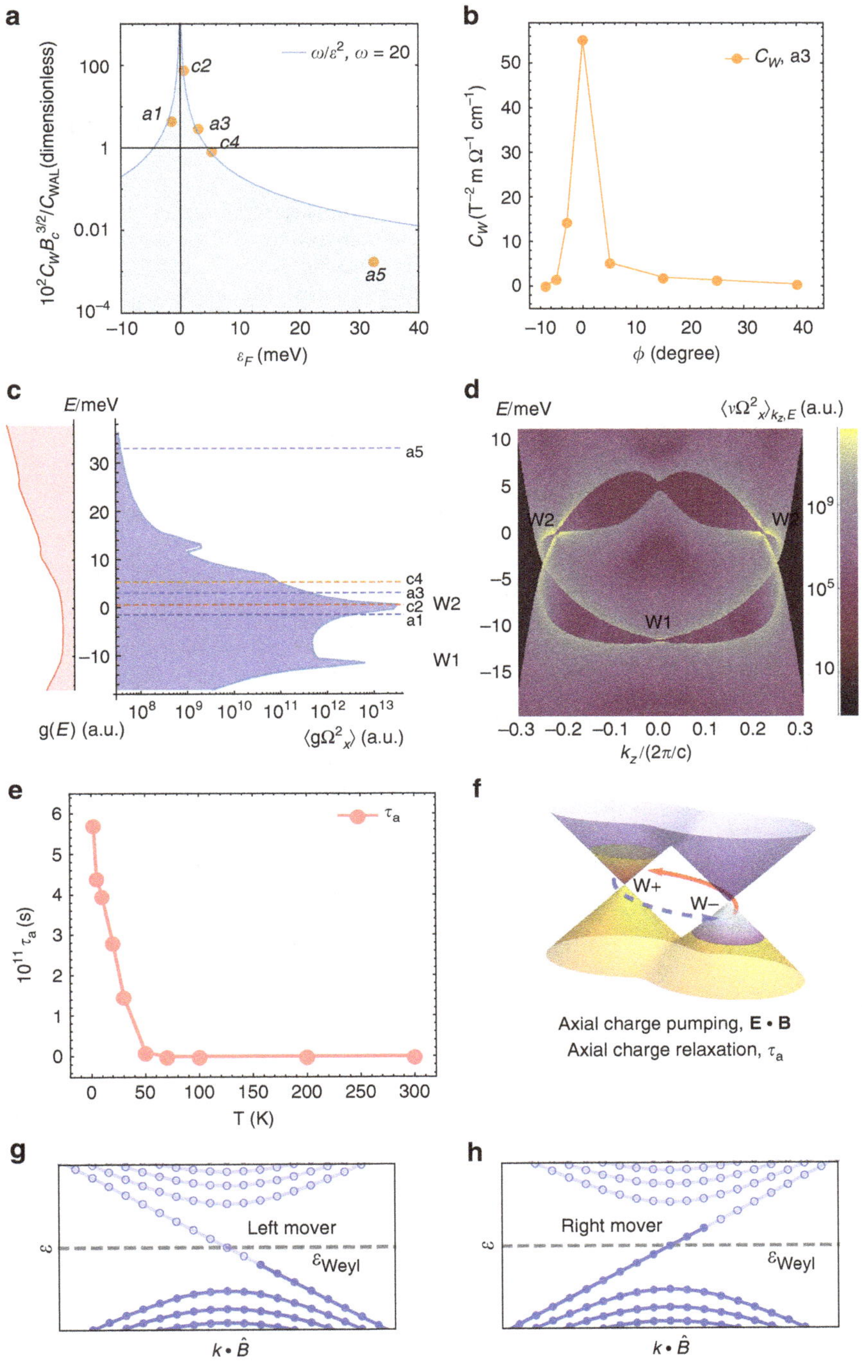

Figure 4 | Signatures of the chiral anomaly due to Berry curvature of the Weyl fermions. (**a**) Chemical potential E_F dependence of the chiral coefficient C_W. We expect the chiral coefficient C_W to decay as a function of $1/E_F^2$. (**b**) Angle (**E** versus **B**) dependence of the chiral coefficient C_W. (**c**) Density of states ($g(E)$) of the bulk electronic structure of TaAs shows a slow variation as a function of energy. The Berry curvature $\langle\Omega_x^2\rangle$ increases markedly at the energy close to the Weyl nodes. (**d**) Distribution of the square of the Berry curvature as a function of k_z and energy E, evidencing that the Weyl points are the dominant source of Berry curvature. The plot is integrated with respect to kx and ky over the whole Brillouin zone. (**e**)Temperature dependence of the axial charge relaxation time τ_a for sample a1. (**f**) A cartoon illustrating the chiral anomaly based on our LMR data. The chiral anomaly leads to the axial charge pumping, **E** · **B**. This causes a population imbalance difference between the Weyl cones with the opposite chiralities. The charge-pumping effect is balanced by the axial charge relaxation, characterized by the time scale τ_a (refs 24,26,36). Note that the axial charge relaxation time τ_a can be directly obtained from the observed negative LMR data through the chiral coefficient $C_W = \frac{e^4\tau_a}{4\pi^4\hbar^4 g(E_F)}$. We also note that this is a cartoon that assumes the Fermi level at zero B field is exactly at the Fermi level. (**g**,**h**) Landau energy spectra of the left- and right-handed Weyl fermions in the presence of parallel electric and magnetic fields.

any singularity. Thus if the negative LMR arose from the hole bands, then the chiral coefficient C_W would not have increased markedly as the chemical potential approaches the energy of the W2 Weyl nodes (Fig. 4a). Therefore, the obtained $\frac{1}{E_F^2}$ dependence of the chiral coefficient (Fig. 4a) provides a unique demonstration that our negative LMR is due to the Weyl fermions, because the $\frac{1}{E_F^2}$

dependence reveals the details of the band dispersion and Berry curvature distribution of the Weyl cones, not just the fact that the bands have some nonzero Berry curvature. We use the above data and analyses to further exclude the possibility of negative LMR due to weak localization arising from the intervalley scattering. Our data is not due to the weak localization for the following reasons: First, weak localization does not have a strong E_F dependence, let alone the marked $\frac{1}{E_F^2}$ dependence observed in our data. Second, it has been theoretically shown that the magneto-conductivity without the chiral anomaly is always monotonic, even though the intervalley scattering can induce a negative LMR arising from the weak localization[46]. This is not consistent with our data, which means that, without the chiral anomaly, only weak localization/anti-localization cannot explain our data.

Up to here, we have demonstrated that the negative LMR arises from the nonzero Berry curvature of the Weyl fermions in our TaAs samples. We now establish the connection between our data and the chiral anomaly, the non-conservation of the electron quasi-particle number of the Weyl cones with a given chirality. In a real Weyl semimetal sample, this can be understood by two crucial components, the axial charge-pumping effect and the axial charge relaxation, as schematically shown in Fig. 4f. The charge-pumping effect means that a nonzero $\mathbf{E}\cdot\mathbf{B}$ can pump charges from one Weyl cone to the other, leading to an imbalance of the quasi-particle number of the Weyl cones with the opposite chiralities. This effect is well-established to occur between Weyl nodes of different chiral charge[24,26,36], which are monopoles of Berry field strength in momentum space. We have directly shown the nontrivial Berry curvature monopoles associated with the Weyl fermions via our LMR transport data. The axial charge pumping creates an out-of-equilibrium quasi-particle distribution between the Weyl cones with opposite chiralities. To form a steady state, it is counteracted by the relaxation of the axial charge disproportionation through scattering between the Weyl nodes. The relaxation is characterized by a time scale, the axial charge relaxation time τ_a. From our negative LMR data, we directly obtain the axial charge relaxation time τ_a (Fig. 4e). The nonzero axial charge relaxation time τ_a not only directly demonstrates the axial charge relaxation, but also confirms the existence of the axial charge pumping because these two are directly coupled, which means that one cannot exist alone if the other is absent.

We can directly obtain the axial charge relaxation time τ_a, which serves as the critical physical quantity that characterizes the chiral anomaly, from the chiral coefficient C_W using t he relationship $C_W = \frac{e^4 \tau_a}{4\pi^4 \hbar^4 g(E_F)}$ (refs 24,26). In Fig. 3a, we present fitting results as a function of temperature for sample s1. We use the fitting coefficients C_W to obtain the axial charge relaxation time as a function of temperature, presented in Fig. 4e. We find that the τ_a rapidly decays to zero with increasing temperature. This decay of τ_a corresponds to the decay of the negative LMR with increasing temperature in the raw data and is expected because scattering typically increases with temperature. We obtain an axial charge relaxation time $\tau_a = 5.96 \times 10^{-11}$ s for sample a1 at $T = 2$ K (Fig. 4e). Note that this τ_a is associated with the W2 Weyl cones because the Fermi level is very close to the W2 nodes. On the other hand, it is difficult to obtain the transport life time of the W2 Weyl cones because the density of states at the Fermi level is dominated by contributions from the W1 Weyl cones and the trivial hole bands (Fig. 1d). Therefore, we estimate the quasi-particle life time associated with the W2 Weyl cones via $\tau \simeq \hbar/E_F = 7.04 \times 10^{-13}$ s for sample a1. We see that the axial charge relaxation time τ_a is much longer than the quasi-particle life time τ. The imbalance of population due to the axial charge pumping can be also estimated by the uncertainty principle $\Delta\mu = \hbar/\tau_a \simeq 0.011$ meV. At $E_F = -1.5$ meV, the density of states per W2 Weyl cone is $g(E_F) = 1.6 \times 10^{16}$ states/(eV cm^{-3}). Therefore, we estimate the chiral charge, the non-conservation of the quasi-particle number of the Weyl cone with a given chirality, to be $\Delta\mu \times g(E_F) = 1.6 \times 10^{14}$. This directly characterizes the chiral anomaly in our Weyl semimetal TaAs sample.

Discussion

We emphasize the critical logical sequences that are key to our demonstration. Unlike previous studies, we do not assume that the negative LMR arises from the chiral anomaly[40–44]. To demonstrate the chiral anomaly, it is critically important to consider all possible origins for a negative LMR and to discuss how one can distinguish each of the other origins from the chiral anomaly. We first excluded the geometry and spin(magnetic) effects. Then we show that our observed LMR is not in the quantum (large B field) limit, in which the Fermi energy crosses only the lowest Landau level. This is important because the LMR in the quantum (large B field) limit can be negative or positive depending on specific scenarios, such as the band dispersion and nature of the impurities[31,33,34]. In fact, it is even theoretically shown that the Weyl cones that respect time-reversal symmetry can contribute a positive (not a negative) LMR in the quantum limit if the field dependence of the scattering time and Fermi velocity of the Landau bands is fully respected[34]. Therefore, observing a negative LMR in the large-field quantum limit may not be a compelling signature of Weyl fermions. In the semiclassical (small B field) limit, after excluding the geometry and magnetic effects, one can avoid ambiguities in the physical interpretation since a negative LMR can only arise from a nonzero Berry curvature[24,26,36,37]. In fact, it has been shown that the LMR from a band with zero Berry curvature will always be positive[37]. However, we emphasize that, at a qualitative level, the negative LMR in the semiclassical limit is only a signature of the Berry curvature but it is not unique to Weyl fermions. In order to uniquely attribute the negative LMR to Weyl fermions, we discovered here that it is crucial to obtain comprehensive information about the band structure. Specifically, first we have shown that the Berry curvature in our TaAs is dominated by the Weyl cones. Second, the chiral coefficient has a $\frac{1}{E_F^2}$ dependence. These two pieces of evidence, together with the full systematics of the data sets uniquely presented here, provides strong signatures of the chiral anomaly of Weyl fermions.

Methods

Sample growth and electrical transport. High-quality single crystals of TaAs were grown by the standard chemical vapour transport method as described in ref. 47. TaAs crystals were structurally characterized by powder X-ray diffraction to confirm bulk quality, and to determine (001) crystal face. A small portion of the obtained samples were ground into fine powders for X-ray diffraction measurements on Rigaku MiniFlex 600 with Cu K_α (40 kV, 15 mA; $\lambda = 0.15405$ nm) at room temperature, and then refined by a Rietica Rietveld program. Magneto-transport measurements were performed using a Quantum Design Physical Property Measurement System. High-field electrical transport measurements were carried out using a pulsed magnet of 50 ms in Wuhan National High Magnetic Field Center. All the measurements were carried out from -9 to 9 T or -56 to 56 T.

Angle-resolved photoemission spectroscopy. The soft X-ray ARPES (SX-ARPES) measurements were performed at the ADRESS Beamline at the Swiss Light Source in the Paul Scherrer Institut in Villigen, Switzerland using photon energies ranging from 300 to 1,000 eV (ref. 48). The sample was cooled down to 12 K to quench the electron–phonon interaction effects reducing the k-resolved spectral fraction. The energy and angle resolution was better than 80 meV and 0.07°, respectively. Vacuum ultraviolet ARPES measurements were performed at beamlines 4.0.3, 10.0.1 and 12.0.1 of the Advanced Light Source at the Lawrence Berkeley National Laboratory in Berkeley, California, USA, Beamline 5–4 of the Stanford Synchrotron Radiation Light source at the Stanford Linear Accelerator Center in Palo Alto, California, USA and Beamline I05 of the Diamond Light Source in Didcot, UK, with the photon energy ranging from 15 to 100 eV. The energy and momentum resolution was better than 30 meV and 1% of the surface Brillouin zone.

Theoretical calculations. First-principles calculations were performed by the OPENMX code within the framework of the generalized gradient approximation of density-functional theory[49]. Experimental lattice parameters were used[47], and the details for the computations can be found in our previous work in ref. 13. A real-space tight-binding Hamiltonian was obtained by constructing symmetry-respecting Wannier functions for the As *p* and Ta *d* orbitals without performing the procedure for maximizing localization, similar for calculations done for topological insulators[50,51].

References

1. Bertlmann, R. A. *Anomalies in Quantum Field Theory* (International Series of Monographs on Physics, 2011).
2. Weyl, H. Elektron und gravitation. *I. Z. Phys.* **56,** 330–352 (1929).
3. Adler, S. Axial-vector vertex in spinor electrodynamics. *Phys. Rev.* **177,** 2426–2438 (1969).
4. Bell, J. S. & Jackiw, R. A PCAC puzzle: $\pi^0 \rightarrow \gamma\gamma$ in the σ-model. *Il. Nuovo. Cimento. A* **60,** 47–61 (1969).
5. Nielsen, H. B. & Ninomiya, M. The Adler-Bell-Jackiw anomaly and Weyl fermions in a crystal. *Phys. Lett. B* **130,** 389–396 (1983).
6. Volovik, G. E. *The Universe in a Helium Droplet* (Clarendon Press, 2003).
7. Balents, L. Weyl electrons kiss. *Physics* **4,** 36 (2011).
8. Wan, X. G., Turner, A. M., Vishwanath, A. & Savrasov, S. Y. Topological semimetal and fermi-arc surface states in the electronic structure of pyrochlore iridates. *Phys. Rev. B* **83,** 205101 (2011).
9. Hasan, M. Z., Xu, S.-Y. & Neupane, M. in *Topological Insulators: Fundamentals and Perspectives*. (eds Ortmann, F., Roche, S. & Valenzuela, S. O.) (John Wiley & Sons, 2015).
10. Turner, A. M. & Vishwanath, A. in *Topological Insulators*. (eds Franz, M. & Molenkamp, L.) (Elsevier, 2013).
11. Murakami, S. Phase transition between the quantum spin Hall and insulator phases in 3D: emergence of a topological gapless phase. *New. J. Phys.* **9,** 356 (2007).
12. Xu, S.-Y. *et al.* Discovery of a Weyl Fermion semimetal and topological Fermi arcs. *Science* **349,** 613–617 (2015).
13. Huang, S. M. *et al.* A Weyl Fermion semimetal with surface Fermi arcs in the transition metal monopnictide TaAs class. *Nat. Commun.* **6,** 7373 (2015).
14. Weng, H. *et al.* Weyl semimetal phase in non-centrosymmetric transition metal monophosphides. *Phys. Rev. X* **5,** 011029 (2015).
15. Lu, L. *et al.* Experimental observation of Weyl points. *Science* **349,** 622–624 (2015).
16. Lv, B. Q. *et al.* Experimental discovery of Weyl semimetal TaAs. *Phys. Rev. X* **5,** 031013 (2015).
17. Duval, C., Horvath, Z., Horvathy, P. A., Martina, L. & Stichel, P. C. Berry phase correction to electron density in solids and "exotic" dynamics. *Mod. Phys. Lett. B* **20,** 373 (2006).
18. Fukushima, K., Kharzeev, D. E. & Warringa, H. J. Chiral magnetic effect. *Phys. Rev. D* **78,** 074033 (2008).
19. Zyuzin, A. A. & Burkov, A. A. Topological response in Weyl semimetals and the chiral anomaly. *Phys. Rev. B* **86,** 115133 (2012).
20. Aji, V. Adler-Bell-Jackiw anomaly in Weyl semimetals: application to pyrochlore iridates. *Phys. Rev. B* **85,** 241101 (2012).
21. Zyuzin, A. A., Wu, S. & Burkov, A. A. Weyl semimetal with broken time reversal and inversion symmetries. *Phys. Rev. B* **85,** 165110 (2012).
22. Grushin, A. G. Consequence of a condensed matter realization of Lorentz-violating QED in Weyl semi-metals. *Phys. Rev. D* **86,** 045001 (2012).
23. Hosur, P. & Qi, X.-L. Recent developments in Weyl semimetals. *Comptes. Rendus. Physique.* **14,** 857–870 (2013).
24. Son, D. T. & Spivak, B. Z. Chiral anomaly and classical negative magnetoresistance of Weyl metals. *Phys.Rev. B* **88,** 104412 (2013).
25. Jho, Y.-S. & Kim, K.-S. Interplay between interaction and chiral anomaly: anisotropy in the electrical resistivity of interacting Weyl metals. *Phys. Rev. B* **87,** 205133 (2013).
26. Burkov, A. A. Chiral anomaly and diffusive magnetotransport in Weyl metals. *Phys. Rev. Lett.* **113,** 247203 (2014).
27. Parameswaran, S. A., Grover, T., Abanin, D. A., Pesin, D. A. & Vishwanath, A. Probing the chiral anomaly with nonlocal transport in three-dimensional topological semimetals. *Phys. Rev. X* **4,** 031035 (2014).
28. Ritchie, L. *et al.* Magnetic, structural, and transport properties of the Heusler alloys Co2MnSi and NiMnSb. *Phys. Rev. B* **68,** 104430 (2003).
29. Pippard, A. B. *Magnetoresistance in Metals* (Cambridge University Press, 1989).
30. Hu, J., Rosenbaum, T. F. & Betts, J. B. Current jets, disorder, and linear magnetoresistance in the silver chalcogenides. *Phys. Rev. Lett.* **95,** 186603 (2005).
31. Argyres, P. N. & Adams, E. N. Longitudinal magnetoresistance in the quantum limit. *Phys. Rev.* **104,** 900–908 (1956).
32. Kikugawa *et al.* Realization of the axial anomaly in a quasi-two-dimensional metal. Preprint at http://arxiv.org/abs/1412.5168 (2014).
33. Goswami, P., Pixley, J. H. & Das Sarma, S. Axial anomaly and longitudinal magnetoresistance of a generic three dimensional metal. *Phys. Rev. B* **92,** 075205 (2015).
34. Lu, H.-Z., Zhang, S.-B. & Shen, S.-Q. High-field magnetoconductivity of topological semimetals with short-range potential. *Phys. Rev. B* **92,** 045203 (2015).
35. Sugihara, K., Tokumoto, M., Yamanouchi, C. & Yoshihiro, K. Longitudinal magnetoresistance of n-InSb in the quantum limit. *J. Phys. Soc. Jpn* **41,** 109–115 (1976).
36. Burkov, A. A. Negative longitudinal magnetoresistance in Dirac and Weyl metals. *Phys. Rev. B* **91,** 245157 (2015).
37. Pal, H. K. & Maslov, D. L. Necessary and sufficient condition for longitudinal magnetoresistance. *Phys. Rev. B* **81,** 214438 (2010).
38. Spivak, B. Z. & Andreev, A. V. Magneto-transport phenomena related to the chiral anomaly in Weyl semimetals. Preprint at http://arxiv/abs/1510.01817 (2015).
39. Kim, H.-J. *et al.* Dirac versus Weyl Fermions in topological insulators: Adler-Bell-Jackiw anomaly in transport phenomena. *Phys. Rev. Lett.* **111,** 246603 (2013).
40. Huang, X. *et al.* Observation of the chiral anomaly induced negative magneto-resistance in 3D Weyl semi-metal TaAs. *Phys. Rev. X* **5,** 031023 (2015).
41. Zhang, C. *et al.* Observation of the Adler-Bell-Jackiw chiral anomaly in a Weyl semimetal. Preprint at http://arxiv.org/abs/1503.02630 (2015).
42. Xiong, J. *et al.* Sinature of the chiral anomaly in a Dirac semimetal: a current plume steered by a magnetic field. Preprint at http://arxiv.org/abs/1503.08179 (2015).
43. Zhang, C. *et al.* Detection of chiral anomaly and valley transport in Dirac semimetals. Preprint at http://arxiv.org/abs/1504.07698 (2015).
44. Shekhar, C. *et al.* Large and unsaturated negative magnetoresistance induced by the chiral anomaly in the Weyl semimetal TaP. Preprint at http://arxiv.org/abs/1506.06577 (2015).
45. Zhang, S.-B., Lu, H.-Z. & Shen, S.-Q. Chiral anomaly and linear magnetoconductivity in a topological Weyl semimetal. Preprint at http://arxiv.org/abs/1509.02001 (2015).
46. Lu, H.-Z. & Shen, S.-Q. Weak antilocalization and localization in disordered and interacting Weyl semimetals. *Phys. Rev. B* **92,** 035203(2015).
47. Murray, J. J. *et al.* Phase relationships and thermodynamics of refractory metal pnictides: the metal-rich tantalum arsenides. *J. Less Common Met.* **46,** 311–320 (1976).
48. Strocov, V. N. *et al.* oft-X-ray ARPES facility at the ADRESS beamline of the SLS: concepts, technical realisation and scientific applications. *J. Synchrotron Rad.* **21,** 32–44 (2014).
49. Perdew, J. P., Burke, K. & Ernzerhof, M. Generalized gradient approximation made simple. *Phys. Rev. Lett.* **77,** 3865–3868 (1996).
50. Hasan, M. Z. & Kane, C. L. Topological insulators. *Rev. Mod. Phys.* **82,** 3045–3067 (2010).
51. Weng, H., Ozaki, T. & Terakura, K. Revisiting magnetic coupling in transition-metal-benzene complexes with maximally localized Wannier functions. *Phys. Rev. B* **79,** 235118 (2009).

Acknowledgements

M.Z.H., S.-Y.X. and I.B. thank I. Klebanov, A. Polyakov and H. Verlinde for theoretical discussions. T.N. thanks A.G. Grushin for discussions. S.J. thanks J. Xiong and F. Wang for valuable discussions, and C.-L.Z. and Z.Y. thank Y. Li and J. Feng for using instruments in their groups. The work at Princeton and Princeton-led synchrotron-based measurements were supported by Gordon and Betty Moore Foundation through Grant GBMF4547 (Hasan). S.J. was supported by the National Basic Research Program of China (Grant Nos. 2013CB921901 and 2014CB239302) and by the Opening Project of Wuhan National High Magnetic Field Center (Grant No.PHMFF2015001), Huazhong University of Science and Technology. C.Z. was supported by the National Science Foundation of China (Grant No.11374020). H.-Z.L. acknowledges the Singapore National Research Foundation for the support under NRF Award No. NRF-NRFF2013-03. S.-Q.S. was supported by the Research Grant Council, University Grants Committee, Hong Kong under Grant No. 17303714. M.N. was supported by the start-up funds from University of Central Florida and Los Alamos National Laboratory Laboratory Directed Research & Development (LDRD) program. H.L. was supported by the Natural Science Foundation of China under Grant No. 11574127. We gratefully acknowledge J.D. Denlinger, S.K. Mo, A.V. Fedorov, M. Hashimoto, M. Hoesch, T. Kim and V.N. Strocov for their beamline assistance at the Advanced Light Source, the Stanford Synchrotron Radiation Lightsource, the Diamond Light Source and the Swiss Light Source. Visits to Princeton University by S.-M.H., G.C., T.-R.C and H.L. were partially funded by the U.S. Department of Energy (DOE), Office of Science, Basic Energy Sciences (BES) under the funding number DE-FG-02-05ER46200.

Author contributions

C.-L.Z., performed the electrical transport experiments with help from Z.Y., Z.L., B.T., J.W., C.Z. and S.J.; S.-Y.X., I.B., G.B. conducted the ARPES experiments with assistance from N.A., M.N., D.S.S., H.Z. and M.Z.H.; C.-L.Z., Z.Y. and S.J. grew the single-crystal samples; C.-C.L., S.-M.H., T.-R.C., G.C., C.-H.H., H.-T.J. and H.L. performed first-principles band structure calculations; H.-Z.L., S.-Q.S. and T.N. did theoretical analyses; H.-Z.L. and S.-Q.S. proposed the fitting formula for the weak anti-localization. I.B. performed the fitting to the magnetoresistance data. S.-Y.X., M.Z.H. and S.J. were responsible for the overall direction, planning and integration among different research units.

Additional information

Competing financial interests: The authors declare no competing financial interests.

Structure and inference in annotated networks

M.E.J. Newman[1,2,3] & Aaron Clauset[3,4,5]

For many networks of scientific interest we know both the connections of the network and information about the network nodes, such as the age or gender of individuals in a social network. Here we demonstrate how this 'metadata' can be used to improve our understanding of network structure. We focus in particular on the problem of community detection in networks and develop a mathematically principled approach that combines a network and its metadata to detect communities more accurately than can be done with either alone. Crucially, the method does not assume that the metadata are correlated with the communities we are trying to find. Instead, the method learns whether a correlation exists and correctly uses or ignores the metadata depending on whether they contain useful information. We demonstrate our method on synthetic networks with known structure and on real-world networks, large and small, drawn from social, biological and technological domains.

[1] Department of Physics, University of Michigan, 450 Church Street, Ann Arbor, Michigan 48109, USA. [2] Center for the Study of Complex Systems, University of Michigan, 450 Church Street, Ann Arbor, Michigan 48109, USA. [3] Santa Fe Institute, 1399 Hyde Park Road, Santa Fe, New Mexico 87501, USA. [4] Department of Computer Science, University of Colorado, 430 UCB, Boulder, Colorado 80309, USA. [5] BioFrontiers Institute, University of Colorado, 596 UCB, Boulder, Colorado 80309, USA. Correspondence and requests for materials should be addressed to M.E.J.N. (email: mejn@umich.edu) or to A.C. (email: aaron.clauset@colorado.edu).

Networks arise in many fields and provide a powerful and compact representation of the internal structure of a wide range of complex systems[1]. Examples include social networks of interactions among people, technological and information networks such as the Internet or the World Wide Web, and biological networks of molecules, cells, or entire species. The last two decades have witnessed rapid growth both in the availability of network data and in the number and sophistication of network analysis techniques. Borrowing ideas from graph theory, statistical physics, computer science, statistics and other areas, network analysis typically aims to characterize a network's structural features in a way that sheds light on the behaviour of the system the network describes. Studies of social networks, for instance, might identify the most influential or central individuals in a population. Studies of road networks can shed light on traffic flows or bottlenecks within a city or country. Studies of pathways in metabolic networks can lead to a more complete understanding of the molecular machinery of the cell.

Most research in this area treats networks as objects of pure topology, unadorned sets of nodes and their interactions. Most network data, however, are accompanied by annotations or metadata that describe properties of nodes such as a person's age, gender or ethnicity in a social network, feeding mode or body mass of species in a food web, data capacity or location of nodes on the Internet and so forth. (There can be metadata on the edges of a network as well as on the nodes[2], but our focus here is on the node case.) In this paper, we consider how to extend the analysis of networks to directly incorporate such metadata. Our approach is based on methods of statistical inference and can in principle be applied to a range of different network analysis tasks. Here we focus specifically on one of the most widely studied tasks, the community detection problem. Community detection, also called node clustering or classification, searches for a good division of a network's nodes into groups or classes[3]. Typically, one searches for assortative structure, groupings of nodes such that connections are denser within groups than between them. This structure is common in social networks, for example, where groups might correspond to sets of friends or co-workers, but it also occurs in other cases, including biological and ecological networks, the Web, transportation and distribution networks, and others. Less common, but no less important, is disassortative structure, in which network connections are sparser within groups than between them, and mixtures of assortative and disassortative structure can also occur, where different groups may have varying propensities for within- or between-group connections.

In some cases, the groups identified by community detection correlate meaningfully with other network properties or functions, such as allegiances or personal interests in social networks[3,4] or biological function in metabolic networks[5,6]. Some recent research, however, has suggested that these cases may be the exception rather than the rule[7,8], an important point that we address later in this paper.

A large number of methods have been proposed for detecting communities in unannotated networks[3]. Among these, some of the most powerful, both in terms of rigorously provable performance and of raw speed, are those based on statistical inference. Here we build on these methods to incorporate node metadata—either categorical or real-valued—into the community detection problem in a principled and flexible manner. (For real-valued metadata we restrict ourselves to the scalar or one-dimensional case, but multi-dimensional metadata, such as locations in physical or latent space[9–11], would be a natural focus for future extensions of our approach.) The resulting methods have several attractive features. First, they can make use of metadata in arbitrary format to improve the accuracy of community detection. Second, and crucially for our goals, they do not assume *a priori* that the metadata correlate with the communities we seek to find. Instead, they detect and quantify the relationship between metadata and community, if one exists, then exploit that relationship to improve the results. Even if the correlation is imperfect or noisy, the method can still use what information is present to return improved results. Conversely, if no correlation exists the method will automatically ignore the metadata, returning results based on network structure alone.

Third, our methods allow us to select between competing divisions of a network. Many networks have a number of different possible divisions[12]. For example, a social network of acquaintances may have meaningful divisions along lines of age, gender, race, religion, language, politics or many other variables. By incorporating metadata that correlate with a particular division of interest, we can favour that division over others, steering the analysis in a desired direction. (Approaches like this are sometimes referred to as supervised learning techniques, particularly in the statistics and machine-learning literature.) Thus, if we are interested for instance in a division of a social network along lines of age, and we have age data for some fraction of the nodes, we can use those data to steer the algorithm towards age-correlated divisions. Even if the metadata are incomplete or noisy, the algorithm can still use them to guide its analysis. However, if we hand the algorithm metadata that do not correlate with any good division of the network, the method will decline to follow along blindly, and will inform us that no good correlation exists.

Finally, the correlation between metadata and network structure learned by the algorithm (if one exists) is interesting in its own right. Once found, it allows us to quantify the agreement between network communities and metadata, and to predict community membership for nodes for which we lack network data and have only metadata. If we have learned, for example, that age is a good predictor of social groupings, then we can make quantitative predictions of group membership for individuals about whom we know their age and nothing else.

A number of other researchers have investigated ways to incorporate metadata into community detection calculations[13–19], though they have typically made stronger assumptions about the nature of the communities or metadata, assuming, for example, that communities are always assortative, or that the metadata represent locations in physical space. Perhaps closest to our approach are semi-supervised learning methods[17,20–22], where it is assumed that we are given the exact community assignments of some fraction of the nodes and the goal is to deduce the reminder. A variant of this approach is active learning, in which the community membership of some nodes is given, but the known nodes are not specified *a priori*, being instead chosen by the algorithm itself as it runs[23,24]. Another vein of research, somewhat further from our approach, considers the case where we are told some pairs of nodes that are either definitely in or definitely not in the same community, and then assigns communities subject to these constraints[25,26].

Our approach, which is described in detail in the Methods section, takes as input a network accompanied by a set of node metadata, which may be, for instance, numerical values or arbitrary textual or alphanumeric labels, and produces as output a division of the nodes of the network into a specified number k of groups or communities. The method does not (as some methods do) assume a particular pattern of connections among communities—such as denser connections within groups than between them—and it is numerically efficient, making use of a so-called belief propagation scheme to perform rapid inference of the optimal group assignments making possible applications to very large networks. The largest network we have analysed using the method has over 1.4 million nodes.

In the following sections we give results showing that our method is able to recover known communities in benchmark data sets with higher accuracy than algorithms based on network structure alone, that we can select between competing community divisions in both real and synthetic tests, that the method is able accurately to divine correlations between network structure and metadata, or determine that no such correlation exists, and that learned correlations between structure and metadata can be used to predict community membership based on metadata alone.

Results

Synthetic networks. Our first tests are on computer-generated (synthetic) networks that have known community structure embedded within them. These networks were created using the stochastic block model, a standard model of network structure in which n nodes are assigned to groups then edges are placed between them independently with probabilities that are a function of group membership only[27,28]. After the networks are created, we generate discrete-valued node metadata that match the true community assignments of nodes a given fraction of the time, and are chosen randomly from the non-matching values otherwise. This allows us to control the extent to which the metadata correlate with the community structure and hence test the algorithm's ability to make use of metadata of varying quality.

Figure 1a shows results for a set of such networks having two communities of equal size, with edge probabilities $p_{\text{in}} = c_{\text{in}}/n$ and $p_{\text{out}} = c_{\text{out}}/n$ for within-group and between-group edges, respectively, where n is the number of nodes as before and c_{in} and c_{out} are constants whose values we choose. When c_{in} is much greater than c_{out} the communities are easy to detect from network structure alone, but as c_{in} approaches c_{out} the structure becomes weaker and harder to detect. Each curve in the figure shows the fraction of nodes classified into their correct groups by our algorithm, as we vary the strength of the community structure,

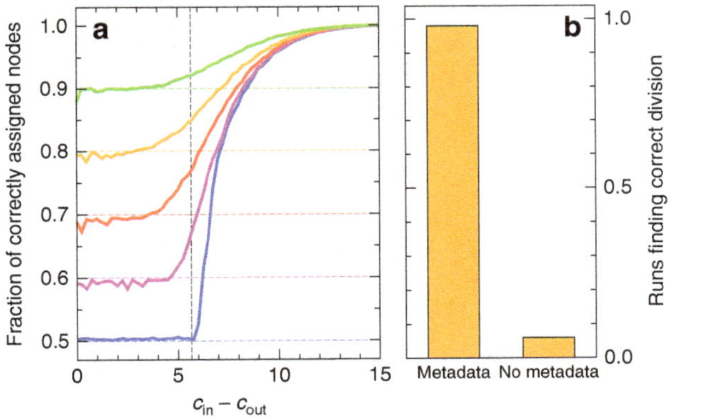

Figure 1 | Tests on synthetic benchmark networks with $n = 10{,}000$ nodes. (**a**) Fraction of correctly assigned nodes for networks with two planted communities with mean degree $c = 8$, as a function of the difference between the numbers of within- and between-group connections. The five curves show results for networks with a match between metadata and planted communities on a fraction 0.5, 0.6, 0.7, 0.8 and 0.9 of nodes (bottom to top). The vertical dashed line indicates the theoretical detectability threshold, below which no algorithm without metadata can detect the communities. (**b**) Fraction of 100 four-group test networks where the algorithm selects a particular two-way division, out of several competing possibilities, with and without the help of metadata that are weakly correlated with the desired division. A run is considered to find the correct division if the fraction of correctly classified nodes exceeds 85%. Network parameters are $c_{\text{out}} = 4$ and $c_{\text{in}} = 20$.

measured by the difference $c_{\text{in}} - c_{\text{out}}$. Individual curves show results for different levels of correlation between communities and metadata.

When metadata and community agree for exactly half of the nodes (bottom curve) there is no correlation between the two, and the metadata cannot help in community detection. It thus comes as no surprise that this curve shows the lowest success rate. At higher levels of correlation the metadata contain useful information and the algorithm's performance improves accordingly.

Examining the figure, a clear pattern emerges. For large $c_{\text{in}} - c_{\text{out}}$ the network contains strong community structure and the algorithm reliably classifies essentially all nodes into the correct groups, as we would expect of any effective algorithm. As the structure weakens the fraction of correct nodes declines, but it remains higher in all the cases where the metadata are useful than in the lowest curve where they are not. Moreover, the algorithm's success rate appears to improve monotonically with the level of correlation between metadata and communities.

When there are no metadata, it is known that the belief propagation algorithm we use gives optimal answers to the community detection problem in the sense that no other algorithm will classify a higher fraction of nodes correctly on average[29]. The fact that our algorithm does better when there are metadata thus implies that the algorithm with metadata does better than any possible algorithm without metadata.

Furthermore, it has previously been shown that below the so-called detectability threshold, which occurs at $c_{\text{in}} - c_{\text{out}} = \sqrt{2(c_{\text{in}} + c_{\text{out}})}$ (indicated by the vertical dashed line in the figure, and aligning with the sharp transition in the bottom curve), community structure becomes so weak as to be undetectable by any algorithm that relies on network structure alone[29,30]. Well below this threshold, however, our algorithm still correctly classifies a fraction of the nodes roughly equal to the fraction of metadata that match the communities, meaning that the algorithm does better with metadata than without it even below the threshold. Figure 1a also shows that the fraction of correctly classified nodes beats this baseline level for values of $c_{\text{in}} - c_{\text{out}}$ somewhat below the threshold, suggesting that the use of the metadata shifts the threshold downward or perhaps eliminates it altogether.

In short, our method automatically combines the available information from network structure and metadata to do a better job of community detection than any algorithm based on network structure alone. And when either the network or the metadata contain no information about community structure the algorithm correctly ignores them and returns an estimate based only on the other.

Figure 1b shows a different synthetic test, of the algorithm's ability to select between competing divisions of a network. In this test, networks were generated with four equally sized communities but the algorithm was tasked with finding a division into just two communities. There are eight ways of dividing such a network in two if we are to keep the four underlying groups undivided. We imagine a situation in which we are interested in finding a particular one out of these eight. A conventional community detection algorithm might find a reasonable division of these networks, but there is no guarantee it would find the 'correct' one—some fraction of the time we can expect it to find one of the competing divisions. But if our algorithm is given a set of metadata that correlate with the division of interest, even if the correlation is poor, then that division will be favoured over the others.

In our tests the desired division was one that places two of the underlying four groups in one community and the remaining two in the other. Two-valued metadata were generated that agree with this division 65% of the time, a relatively weak level of correlation, not far above the 50% of completely uncorrelated data. Nonetheless, as shown in Fig. 1b, this is enough for the algorithm to reliably find the correct division of the network in almost every case—98% of the time in our tests. Without the metadata, by contrast, we succeed only 6% of the time. Some practical applications of this ability to select among competing divisions are given in the next section.

Real-world networks. In this section we describe applications of our method to a range of real-world networks, drawn from social, biological and technological domains.

For our first application we analyse a network of school students, drawn from the US National Longitudinal Study of Adolescent Health. The network represents patterns of friendship, established by survey, among the 795 students in a medium-sized American high school (US grades 9–12, ages 14–18 years) and its feeder middle school (grades 7 and 8, ages 12–14 years).

Given that this network combines middle and high schools, it comes as no surprise that there is a clear division (previously documented) into two network communities corresponding roughly to the two schools. Previous work, however, has also shown the presence of divisions by ethnicity[31]. Our method allows us to select between divisions by using metadata that correlate with the one we are interested in.

Figure 2 shows the results of applying our algorithm to the network three times. Each time, we asked the algorithm to divide the network into two communities. In Fig. 2a, we used the six school grades as metadata and the algorithm readily identifies a division into grades 7 and 8 on the one hand and grades 9–12 on the other—that is, the division into middle school and high school. In Fig. 2b, by contrast, we used the students' self-identified ethnicity as metadata, which in this data set takes one of four values: white, black, hispanic, or other (plus a small number of nodes with missing data). Now the algorithm finds a completely different division into two groups, one group consisting principally of black students and one of white. (The small number of remaining students are distributed roughly evenly between the groups.)

One might be concerned that in these examples the algorithm is mainly following the metadata to determine community membership, and ignoring the network structure. To test for this possibility, we performed a third analysis, using gender as metadata. When we do this, as shown in Fig. 2c, the algorithm does not find a division into male and female groups. Instead, it finds a new division that is a hybrid of the grade and ethnicity divisions (white high-school students in one group and everyone else in the other). That is, the algorithm has ignored the gender metadata, because there was no good network division that correlated with it, and instead found a division based on the network structure alone. The algorithm makes use of the metadata only when doing so improves the quality of the network division (in the sense of the maximum-likelihood fit described in the Methods section).

The extent to which the communities found by our algorithm match the metadata (or any other 'ground truth' variable) can be quantified by calculating a normalized mutual information (NMI)[32,33], as described in the Methods section. NMI ranges in value from 0 when the metadata are uninformative about the communities to 1 when the metadata specify the communities completely. The divisions shown in Fig. 2a,b have NMI scores of 0.881 and 0.820, respectively, indicating that the metadata are strongly though not perfectly correlated with community membership. By contrast, the division in Fig. 2c, where gender was used as metadata, has an NMI score of 0.003, indicating that

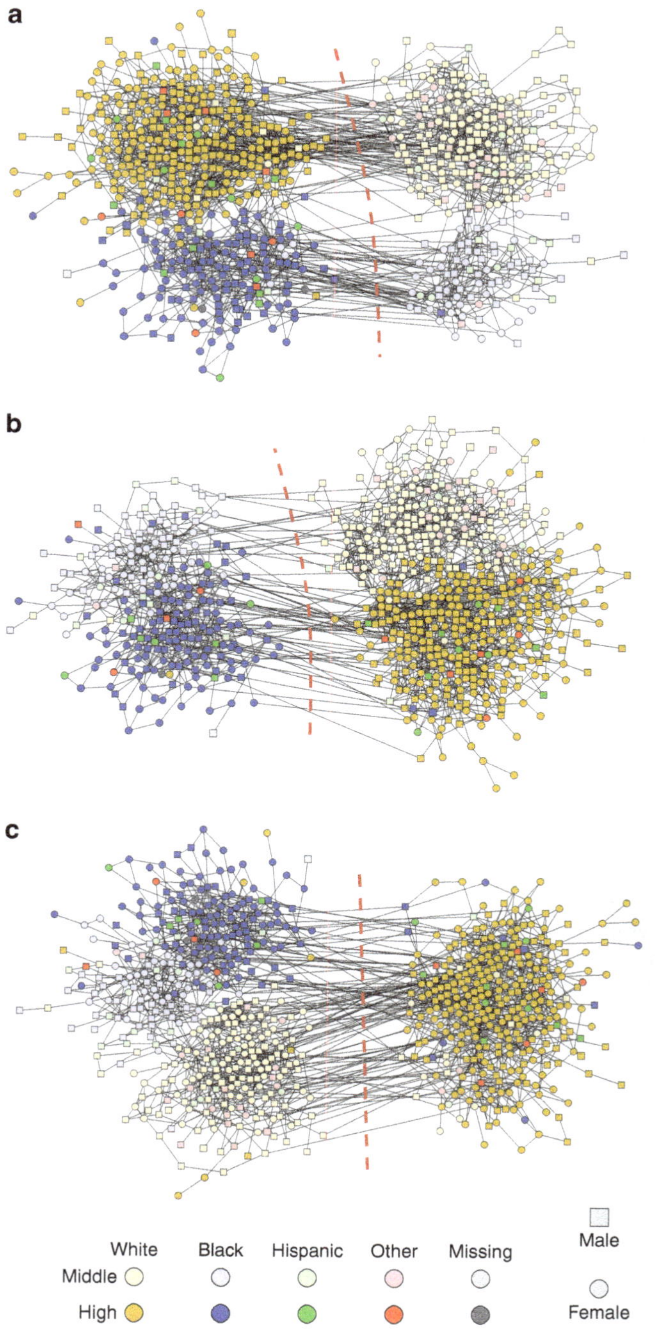

Figure 2 | Communities found in a high school friendship network with various types of metadata. Three divisions of a school friendship network, using as metadata (**a**) school grade, (**b**) ethnicity and (**c**) gender.

the metadata contain essentially zero information about the communities.

Our next application is to an ecological network, a food web of predator–prey interactions between 488 marine species living in the Weddell Sea, a large bay off the coast of Antarctica[34,35]. A number of different metadata are available for these species, including feeding mode (deposit feeder, suspension feeder, scavenger and so on), zone within the ocean (benthic, pelagic and so on) and others. In our analysis, however, we focus on one in particular, the average adult body mass. Body masses of species in this ecosystem have a wide range, from microorganisms weighing nanograms or less to hundreds of tonnes for the largest whales.

Conventionally, in such cases one often works with the logarithm of mass, which makes the range more manageable, and we do so here. Then we perform k-way community decompositions using this log-mass as metadata, for various values of k.

Figure 3a shows the results for $k=3$. Nodes are coloured according to their role in the ecosystem—carnivores, herbivores, primary producers and so forth. The division found by the algorithm appears to match these roles quite closely, with one group composed almost entirely of primary producers and herbivores, one of omnivores and a third that contains most of the carnivores. Node sizes in the figure are proportional to log-mass, which increases as we go up the figure, indicating that the algorithm has recovered from the network structure the well-known correlation between body mass and ecosystem role[36]. This point is further emphasized by the probabilities of membership in the three groups, which are an incidental, but often useful, additional output of the algorithm we use (see Methods). These probabilities, plotted as a function of body mass in Fig. 3b, show that low-mass organisms are overwhelmingly likely to be in the first group, and high-mass ones in the third group. Organisms of intermediate mass have a broader distribution, but are particularly concentrated in the second group.

The membership probabilities are also of interest in their own right. If, for instance, we were to learn of a new species, previously unrepresented in our food-web data set, then even without knowing its pattern of network connections we can make a statement about its probability of belonging to each of the communities, as well as its probability of interaction with other species, so long as we know its body mass. For instance, a low body mass of 10^{-12} g would put a species with high probability in group 1 in Fig. 3, meaning it is almost certainly a primary producer or a herbivore, with the interaction patterns that implies.

Community detection is widely studied precisely because network communities are believed to be correlated with network function. More specifically, it is commonly assumed that communities correlate with some underlying functional variable, which may or may not be observed. This assumption, however, has been challenged by recent work that compared communities in real-world networks against 'ground truth' metadata variables and found little correlation between the two[7,8]. This is a striking discovery, but there is a caveat. As we have seen, there are often multiple meaningful community divisions of a network (as in the school friendship network of Fig. 2, for example), and the fact that one division is uncorrelated with a given metadata variable does not rule out the possibility that another could be.

Our third real-world example application illustrates these issues using one of the same networks studied in ref. 8, a 46,676-node representation of the peering structure of the Internet at the level of autonomous systems. The 'ground truth' variable for this network is the country in which each autonomous system is located. The analysis of ref. 8 found there to be little correlation between community structure and countries.

We first analyse this network without metadata, performing a traditional 'blind' community division, into five groups using standard methods. We then repeat the analysis using the algorithm of this paper, with the countries as metadata. Recall that, in doing this, we do not force the algorithm to find a community division that aligns with the metadata if no such division exists, but if a division does exist it will be favoured over competing divisions that do not align with the metadata. There are 173 distinct countries in the data set, a significantly larger number of metadata values than for any of the other networks we have considered, but by no means beyond the capabilities of our method.

As before, we assess the results using the normalized mutual information. If indeed there are many competing divisions of the network, only some of which correlate with the particular metadata we are given, then we would expect our blind analysis to return a range of NMI values on different runs, some low and (maybe) some higher. This is indeed what we see, with the NMI in our calculations ranging from a high of 0.626 to a relatively low 0.398, the latter being in agreement with results quoted in ref. 8.

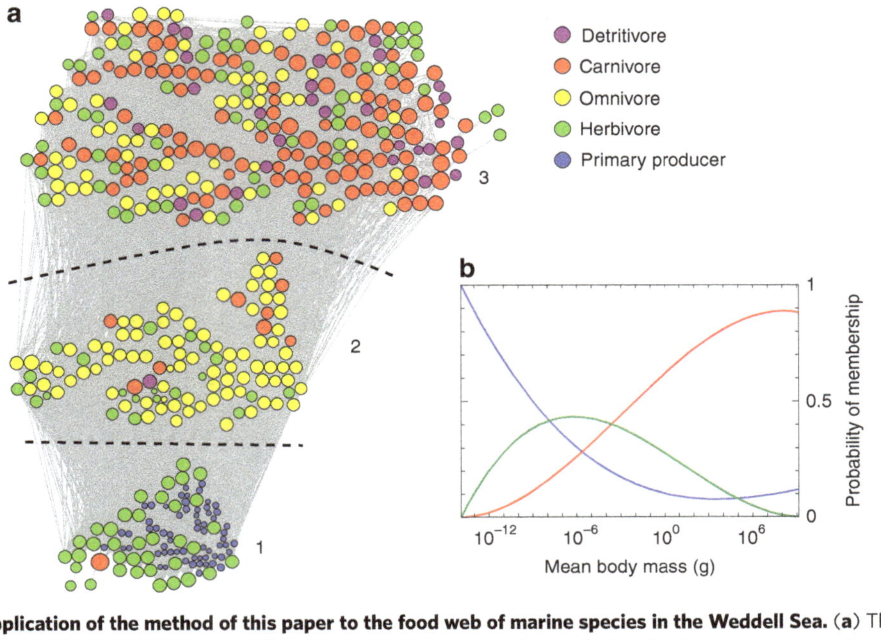

Figure 3 | Results of the application of the method of this paper to the food web of marine species in the Weddell Sea. (**a**) Three-way decomposition of the marine food web described in the text, with the logarithm of mean body mass used as metadata. Node sizes are proportional to log-mass, and colours indicate species role within the ecosystem. (**b**) Learned probabilities of belonging to each of the communities as a function of body mass. We use log mass as the metadata variable in our calculations, but the horizontal axis here is calibrated to read in terms of the original mass in grams using a logarithmic scale. The blue, green and red curves correspond, respectively, to the communities labelled 1, 2 and 3 in **a**.

Conversely, when the algorithm of this paper is applied with countries as metadata, we find an NMI score significantly higher than any of these figures, at 0.870, which would conventionally be interpreted as an indication of strong correlation.

These results emphasize that an apparent lack of correlation between network communities and metadata could be the result of the presence of competing network divisions, which are not correlated with the particular metadata we have at hand. The algorithm of this paper allows us to select among divisions and hence find ones that correlate with the variable of interest.

Our fourth example is drawn from the FB100 data set of Traud *et al.*[37], which is a set of friendship networks among college students at the US universities compiled from friend relations on the social networking website Facebook. The networks date from the early days of Facebook when its services were available only to universities and each university formed a separate and unconnected subgraph in the larger network. The nodes in these networks represent the participants, who are mainly though not exclusively students, the edges represent friend relations on Facebook, and in addition to the network structure there are metadata of several types, including gender, college year (that is, year of college graduation), major (that is, principal subject of students' study, if known) and a numerical code indicating which dorm students lived in.

The primary divisions in these networks appear to be by age, or more specifically by college year. For instance, we have looked in some detail at the network for Harvard University, the birthplace of Facebook, which has 15,126 nodes. Most of these represent undergraduate students, who span college years 2003–2009, but there are also a small number of alumni (that is, former students), primarily those recently graduated (graduation years 2000–2002), as well as grad students, summer students, and some faculty and staff.

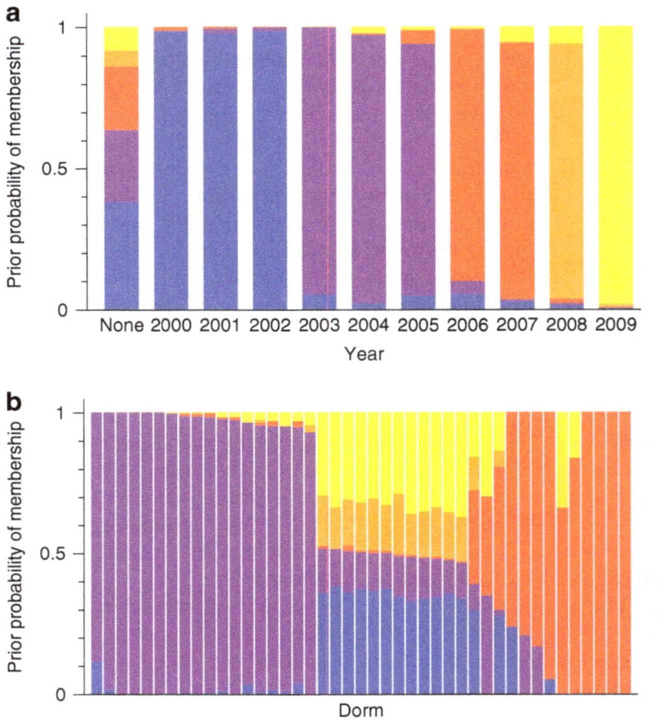

Figure 4 | Learned prior probability of community membership for two five-way divisions of the Harvard Facebook friendship network described in the text. The horizontal axis is (**a**) year of graduation and (**b**) dormitory, and the colours represent the learned prior probabilities of membership in each of the communities.

Figure 4a shows results from a five-way division of the network using our algorithm with year as metadata. This calculation provides another example of the usefulness of the learned probabilities of group membership in shedding light on the structure of the network. The figure shows a visualization of the probabilities as a function of year, with the colours showing the relative probability of belonging to each of the communities. Each of the bars in the plot has the same height of 1 since the probabilities are required to sum to 1, while the balance of colours shows the distribution over communities. Examination of the top panel in the figure shows clearly a division of the network along age lines. Two groups, in orange and yellow at the right of the plot, correspond to the most recent two years of students at the time of the study (graduation years 2008 and 2009) and the next, in red, account for the two years before that (2006 and 2007). The purple community corresponds to the next three years, 2003–2005, while the sixth group, shown in blue, corresponds to the alumni. Finally, students for whom year was not recorded are shown in the column marked 'None,' which is a mixture of all five groups.

These results align well with the original analysis of the same data by Traud *et al.*[37], who performed a traditional community division of the network and then carried out *post hoc* statistical tests to measure correlations between communities and metadata. They found strong correlations with college year metadata, in agreement with our results. With the benefit of hindsight the results may appear unsurprising—anyone who has been to college knows that a large number of your friends are in the same year as you—but one could certainly formulate competing hypotheses. One alternative that Traud *et al.* considered was that friendship might be influenced by where students live, with students living in the same dormitory more likely to be friends, regardless of what year they are in. Traud *et al.* found that there was some evidence for this hypothesis, but that the effect was weaker than that for age, and our analysis confirms this. The bottom panel in Fig. 4 shows a plot of the priors for a division with dorm as the metadata variable and there is a clear correlation between dorm and community membership, but it is not as clean as in the case of age. There appear to be two groups that align strongly with particular sets of dorms (coloured red and purple in the figure) while the rest of the dorms are a mix of different communities (the region in the middle of the figure). The impression that the community structure is more closely aligned with graduation year than with dormitory is also borne out by the normalized mutual information values for the two divisions, which are 0.668 for graduation year but 0.255 for dormitory.

Our final real-world network example is drawn from a gene recombination network for the human parasite *Plasmodium falciparum*, which causes malaria. Malaria is endemic in tropical regions and is responsible for roughly a million deaths annually, mostly children in sub-Saharan Africa[38]. During infection, parasites evade the host immune system and prolong the infection by repeatedly changing a protein camouflage displayed on the surface of an infected red blood cell. To enable this behaviour, each parasite has a repertoire of roughly 60 immunologically distinct proteins, each of which is encoded by a *var* gene in the parasite's genome[39]. These genes undergo frequent recombination, producing novel proteins by shuffling and splicing substrings from existing *var* genes.

The process of recombination induces a natural bipartite network with two types of nodes, *var* genes on the one hand and their constituent substrings on the other, where each gene node is connected by an edge to every substring it contains[40,41]. Recombination in these genes occurs mainly within a number of distinct highly variable regions (HVRs) and each HVR

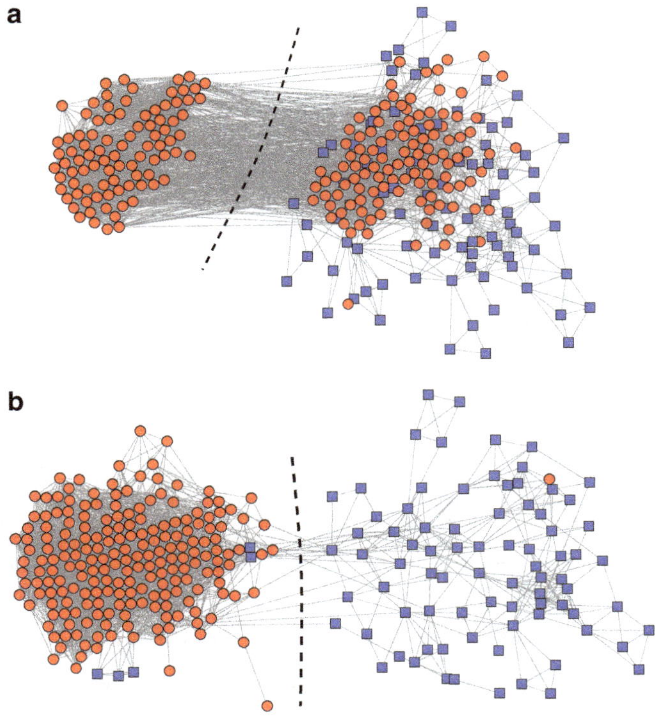

Figure 5 | Inferred communities for the malaria HVR 6 gene recombination network. Communities inferred (**a**) without metadata and (**b**) with metadata for the HVR 6 network of the human malaria parasite *P. falciparum*, where metadata values are the CP labels for the genes and nodes are coloured according to their biologically relevant Cys label.

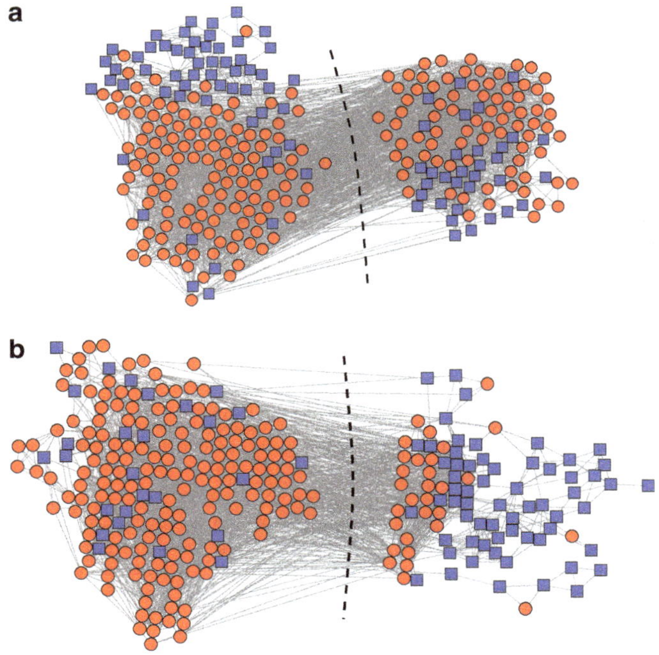

Figure 6 | Inferred communities for the malaria HVR 5 gene recombination network. Communities inferred (**a**) without metadata and (**b**) with metadata for the HVR 6 network of the human malaria parasite *P. falciparum*, where metadata values are the CP labels for the genes and nodes are coloured according to their biologically relevant Cys label.

represents a distinct set of edges among the same nodes. Here we focus on the one-mode gene–gene projections of the HVR 5 and HVR 6 subnetworks, which have previously been analysed using community detection methods without metadata[40,41]. Each of these one-mode networks consists of 297 genes.

We analyse these networks using as metadata the Cys labels derived from the HVR 6 sequence and the Cys-PoLV (CP) labels derived from the sequences adjacent to HVRs 5 and 6 (refs 39,42,43). Both types of labels depend only on the sequences' characteristics: Cys indicates the number of cysteines the HVR 6 sequence contains (2 or 4) while CP subdivides the Cys classifications into six groups depending on particular sequence motifs. Thus, each node has two metadata values, a Cys label and a CP label. The Cys labels are biologically important because cysteine counts have been implicated in severe disease phenotypes[39,42].

In our calculations we use the six CP labels as metadata for a two-way community division of the network and then evaluate the degree to which the inferred communities correlate with the Cys metadata. Figure 5 shows the results for the HVR 6 network with and without the CP labels as metadata. Without metadata, the Cys labels are mixed across the inferred groups (Fig. 5a), but with metadata we obtain a nearly perfect partition (Fig. 5b). This indicates that the CP label correlates well with the network's community structure, a fact that was obscured in the analysis without metadata. Furthermore, the inferred communities correlate strongly with the coarser Cys labels, which were not shown to the method: observing that a gene has two cysteines is highly predictive (96% probability) of that gene being in one group, while having four cysteines is modestly predictive (67% probability) of being in the other group. Thus, the method has discovered by itself that the motif sequences that define the CP labels, along with their corresponding network communities, correlate with cysteine counts and their associated severe disease phenotypes[39,42].

The communities in the HVR 6 network represent highly non-random patterns of recombination, which are thought to indicate functional constraints on protein structure. Previous work has conjectured that common constraints on recombination span distinct HVRs[40]. We can test this hypothesis using the methods described in this paper. There is no reason *a priori* to expect that the community structure of HVR 6 should correlate with that of HVR 5 because the Cys and CP labels are derived from outside the HVR 5 sequences—Cys labels reflect cysteine counts in HVR 6 while CP labels subdivide Cys labels based on sequence motifs adjacent to, but outside of, HVR 5. Applying our methods to HVR 5 without any metadata (Fig. 6a), we find mixing of the HVR 6 Cys labels across the HVR 5 communities. By contrast, using the CP labels as metadata for the HVR 5 network, our method finds a much cleaner partition (Fig. 6b), indicating that indeed the HVR 6 Cys labels correlate with the community structure of HVR 5.

Discussion

There are a number of possible extensions of this work. At the simplest level one could include more complex metadata types, such as combinations of discrete and continuous variables, or vector variables such as spatial coordinates. Metadata could also be incorporated into methods for detecting other types of structure, such as hierarchies[44], motifs[45], core-periphery structures[46], rankings[47] or latent-space structures[48]. And the resulting fits could form the starting point for a variety of additional applications, such as the prediction of missing links or missing metadata in incomplete data sets. These and other possibilities we leave for future work.

Methods

Our method makes use of techniques of Bayesian statistical inference in which we construct a generative network model possessing the specific features we hope to find in our data, namely community structure and a correlation between that structure and node metadata, then we fit the model to an observed network plus accompanying metadata and the parameters of the fit tell us about the structure of the network.

The model we use is a modified version of a stochastic block model. The original stochastic block model, proposed in 1983 by Holland *et al.*[27], is a simple model for generating random networks with community structure in which nodes are divided among some number of communities and edges are placed randomly and independently between them with probabilities that depend only on the communities to which the nodes belong. We modify this model in two ways. First, following ref. 28, we note that the standard stochastic block model does poorly at mimicking the structure of networks with highly heterogeneous degree sequences (which includes nearly all real-world networks), and so we include a 'degree-correction' term that matches node degrees (that is, the number of connections each node has) to those of the observed data. Second, we introduce a dependence on node metadata via a set of prior probabilities. The prior probability of a node belonging to a particular community becomes a function of the metadata, and it is this function that is learned by our algorithm to incorporate the metadata into the calculation.

Unordered data. Consider an undirected network with n nodes labelled by integers $u = 1 \ldots n$, divided among k communities, and denote the community to which node u belongs by $s_u \in 1 \ldots k$. In the simplest case, we consider metadata with a finite number K of discrete, unordered values and we denote node u's metadata by $x_u \in 1 \ldots K$. The choice of labels $1 \ldots K$ is arbitrary and does not imply an ordering for the metadata or that the metadata are one-dimensional. If a social network has two-dimensional metadata describing both language and race, for example, we simply encode each possible language/race combination as a different value of x: English/white, Spanish/white, English/black and so forth. If a network has nodes that are missing metadata values, we just let 'missing' be another metadata value.

Given metadata $\mathbf{x} = \{x_u\}$ and degree $\mathbf{d} = \{d_u\}$ for all nodes, a network is generated from the model as follows. First, each node u is assigned to a community s with a probability depending on u's metadata x_u. The probability of assignment we denote γ_{sx} for each combination s,x of community and metadata, so the full prior probability on community assignments is $P(\mathbf{s}|\boldsymbol{\Gamma},\mathbf{x}) = \prod_i \gamma_{s_i,x_i}$, where $\boldsymbol{\Gamma}$ denotes the $k \times K$ matrix of parameters γ_{sx}. (More complex forms of the prior are appropriate in other cases, as we will see.) Once every node has been assigned to a community, edges are placed independently at random between nodes, with the probability of an edge between nodes u and v being

$$p_{uv} = d_u d_v \theta_{s_u,s_v}. \tag{1}$$

where θ_{st} are parameters that we specify, with $\theta_{st} = \theta_{ts}$. The factor $d_u d_v$ allows the model to fit arbitrary degree sequences as described above. Models of this kind have been found to fit community structure in real networks well[28].

Community detection then consists of fitting the model to observed network data using the method of maximum likelihood. Given an observed network, we define its adjacency matrix $\mathbf{A}$ to be the $n \times n$ real symmetric matrix with elements $a_{uv} = 1$, if there is an edge between nodes u and v and 0 otherwise. Then the probability, or likelihood, that this network was generated by our model, given the parameters and metadata, is

$$\begin{aligned} P(\mathbf{A}|\boldsymbol{\Theta},\boldsymbol{\Gamma},\mathbf{x}) &= \sum_{\mathbf{s}} P(\mathbf{A}|\boldsymbol{\Theta},\mathbf{s}) P(\mathbf{s}|\boldsymbol{\Gamma},\mathbf{x}) \\ &= \sum_{\mathbf{s}} \prod_{u<v} p_{uv}^{a_{uv}} (1-p_{uv})^{1-a_{uv}} \prod_u \gamma_{s_u,x_u}, \end{aligned} \tag{2}$$

where $\boldsymbol{\Theta}$ is the $k \times k$ matrix with elements θ_{st} and the sum is over all possible community assignments $\mathbf{s}$.

Fitting the model involves maximizing this likelihood with respect to $\boldsymbol{\Theta}$ and $\boldsymbol{\Gamma}$ to determine the most likely values of the parameters, which we do using an expectation-maximization (EM) algorithm. Typically, rather than maximizing (2) itself, we maximize instead its logarithm,

$$\log P(\mathbf{A}|\boldsymbol{\Theta},\boldsymbol{\Gamma},\mathbf{x}) = \log \sum_{\mathbf{s}} P(\mathbf{A}|\boldsymbol{\Theta},\mathbf{s}) P(\mathbf{s}|\boldsymbol{\Gamma},\mathbf{x}), \tag{3}$$

which gives the same answer for $\boldsymbol{\Theta}$ and $\boldsymbol{\Gamma}$ but is often more convenient. The most obvious approach for performing the maximization would be simply to differentiate with respect to the parameters, set the result to zero, and solve the resulting equations. This, however, produces a complex set of implicit equations that have no easy solution. Instead, therefore, we make use of Jensen's inequality, which says that for any set of positive quantities x_i the log of their sum obeys

$$\log \sum_i x_i \geq \sum_i q_i \log \frac{x_i}{q_i}, \tag{4}$$

where q_i is any correctly normalized probability distribution such that $\sum_i q_i = 1$. Note that the exact equality is recovered by the particular choice

$$q_i = \frac{x_i}{\sum_i x_i}. \tag{5}$$

Applying Jensen's inequality to equation (3), we find that

$$\begin{aligned} \log P(\mathbf{A}|\boldsymbol{\Theta},\boldsymbol{\Gamma},\mathbf{x}) &\geq \sum_{\mathbf{s}} q(\mathbf{s}) \log \frac{P(\mathbf{A}|\boldsymbol{\Theta},\mathbf{s}) P(\mathbf{s}|\boldsymbol{\Gamma},\mathbf{x})}{q(\mathbf{s})} \\ &= \sum_{\mathbf{s}} q(\mathbf{s}) \log P(\mathbf{A}|\boldsymbol{\Theta},\mathbf{s}) + \sum_{\mathbf{s}} q(\mathbf{s}) \log P(\mathbf{s}|\boldsymbol{\Gamma},\mathbf{x}) - \sum_{\mathbf{s}} q(\mathbf{s}) \log q(\mathbf{s}), \end{aligned} \tag{6}$$

where $q(\mathbf{s})$ is any distribution over community assignments $\mathbf{s}$ such that $\sum_{\mathbf{s}} q(\mathbf{s}) = 1$. The maximum of the right-hand side of this inequality with respect to possible choices of the distribution $q(\mathbf{s})$ coincides with the exact equality, which, following equation (5), is when

$$q(\mathbf{s}) = \frac{P(\mathbf{A}|\boldsymbol{\Theta},\mathbf{s}) P(\mathbf{s}|\boldsymbol{\Gamma},\mathbf{x})}{\sum_{\mathbf{s}} P(\mathbf{A}|\boldsymbol{\Theta},\mathbf{s}) P(\mathbf{s}|\boldsymbol{\Gamma},\mathbf{x})}. \tag{7}$$

Thus, the maximization of the left-hand side of (6) with respect to $\boldsymbol{\Theta}, \boldsymbol{\Gamma}$ to give the optimal values of the parameters is equivalent to a maximization of the right-hand side both with respect to $q(\mathbf{s})$ (which makes it equal to the left-hand side) and with respect to $\boldsymbol{\Theta}, \boldsymbol{\Gamma}$. A simple algorithm for performing such a double maximization is to repeatedly maximize with respect to first $q(\mathbf{s})$ and then $\boldsymbol{\Theta}, \boldsymbol{\Gamma}$ until we converge to an answer. In other words:

1. Make an initial guess about the parameter values and use them to calculate the optimal $q(\mathbf{s})$ from equation (7).
2. Using that value, maximize the right-hand side of (6) with respect to the parameters, while holding $q(\mathbf{s})$ constant.
3. Repeat from step 1 until convergence is achieved.

Step 2 can be performed by differentiating with $q(\mathbf{s})$ fixed and subject to the normalization constraint $\sum_s \gamma_{sx} = 1$ for all x. Performing the derivatives and assuming that the network is large and sparse so that p_{uv} is small, we find to leading order in small quantities that

$$\theta_{st} = \frac{\sum_{uv} a_{uv} q_{st}^{uv}}{\sum_{uv} d_u d_v q_{st}^{uv}}, \qquad \gamma_{sx} = \frac{\sum_u \delta_{x,x_u} q_s^u}{\sum_u \delta_{x,x_u}}, \tag{8}$$

where

$$q_s^u = \sum_{\mathbf{s}} q(\mathbf{s}) \delta_{s_i,s}, \qquad q_{st}^{uv} = \sum_{\mathbf{s}} q(\mathbf{s}) \delta_{s_u,s} \delta_{s_v,t}. \tag{9}$$

In addition, for a large sparse network, the community assignments of distant nodes will be uncorrelated and hence we can write $q_{st}^{uv} \simeq q_s^u q_t^v$ in the denominator of (8) to get

$$\theta_{st} = \frac{\sum_{uv} a_{uv} q_{st}^{uv}}{\sum_u d_u q_s^u \sum_v d_v q_t^v}, \tag{10}$$

which reduces the denominator sums from n^2 terms to only n and considerably speeds the calculation. (We cannot make the same factorization in the numerator, since the terms in the numerator involve q_{st}^{uv} on adjacent nodes u, v only, so the nodes are not distant from one another.)

Equation (7) tells us that once the iteration converges, the value of $q(\mathbf{s})$ is

$$q(\mathbf{s}) = \frac{P(\mathbf{A}|\boldsymbol{\Theta},\mathbf{s}) P(\mathbf{s}|\boldsymbol{\Gamma},\mathbf{x})}{\sum_{\mathbf{s}} P(\mathbf{A}|\boldsymbol{\Theta},\mathbf{s}) P(\mathbf{s}|\boldsymbol{\Gamma},\mathbf{x})} = \frac{P(\mathbf{A},\mathbf{s}|\boldsymbol{\Theta},\boldsymbol{\Gamma},\mathbf{x})}{P(\mathbf{A}|\boldsymbol{\Theta},\boldsymbol{\Gamma},\mathbf{x})} = P(\mathbf{s}|\mathbf{A},\boldsymbol{\Theta},\boldsymbol{\Gamma},\mathbf{x}). \tag{11}$$

In other words $q(\mathbf{s})$ is the posterior distribution over community assignments $\mathbf{s}$, the probability of an assignment $\mathbf{s}$ given the inputs $\mathbf{A}$, $\boldsymbol{\Theta}$, $\boldsymbol{\Gamma}$, and $\mathbf{x}$, and q_s^u is the marginal posterior probability that node u belongs to community s. Normally, in fact, q_s^u is the object of primary interest in the calculation, as it tells us to which group each node belongs. That is, it tells us the optimal division of the network into communities. As discussed in the Results section, the prior probabilities γ_{sx} may also be of interest, since they tell us how and to what extent the metadata are correlated with the communities. If the metadata are uncorrelated with the network communities, the prior probabilities become constant, independent of the metadata, and thus have no impact on the posterior probabilities of the communities. Similarly, if the network is large and has strong community structure (as in the region on the right of Fig. 1a where $c_{in} - c_{out}$ is large), the prior probabilities will have little effect on the results and the algorithm will find the structure embedded in network with or without help from the metadata.

Computationally, the most demanding part of the EM algorithm is calculating the sum in the denominator of equation (7), which has an exponentially large number of terms, making its direct evaluation intractable on all but the smallest of networks. Traditionally one gets around this problem by approximating the full distribution $q(\mathbf{s})$ by Monte Carlo importance sampling. In our calculations, however, we instead use a recently proposed alternative method based on belief propagation[29], which is significantly faster, and fast enough in practice for applications to very large networks.

Final likelihood value. The EM algorithm always converges to a maximum of the likelihood but is not guaranteed to converge to the global maximum—it is possible for there to be one or more local maxima as well. To get around this problem we normally run the algorithm repeatedly with different random initial guesses for the parameters and from the results choose the one that finds the highest likelihood value. In the calculations presented in this paper we did at least 10 such 'random restarts' for each network. To determine which run has the highest final value of the likelihood we calculate the log-likelihood from the right-hand side of (6) using $P(\mathbf{A}|\mathbf{\Theta}, \mathbf{s})$ and $P(\mathbf{s}|\mathbf{\Gamma}, \mathbf{x})$ as in equation (2), the final fitted values of the parameters $\mathbf{\Theta}$ and $\mathbf{\Gamma}$ from the EM algorithm, and $q(\mathbf{s})$ as in equation (7). (As we have said, the right-hand side of (6) becomes equal to the left, and hence equal to the true log-likelihood, when $q(\mathbf{s})$ is given the value in equation (7).)

Putting it all together, our expression for the log-likelihood is

$$\log P(\mathbf{A}|\mathbf{\Theta},\mathbf{\Gamma},\mathbf{x}) = \sum_{\mathbf{s}} q(\mathbf{s}) \sum_{u<v} \left[a_{uv}\log\left(d_u d_v \theta_{s_u,s_v}\right) + (1-a_{uv})\log\left(1-d_u d_v \theta_{s_u,s_v}\right)\right] + \sum_{\mathbf{s}} q(\mathbf{s}) \sum_u \log \gamma_{s_u,x_u} - \sum_{\mathbf{s}} q(\mathbf{s}) \log q(\mathbf{s}). \tag{12}$$

Neglecting terms beyond first order in small quantities, the first sum can be rewritten as

$$\tfrac{1}{2}\sum_{uv}\sum_{st}\left[q_{st}^{uv} a_{uv}(\log d_u + \log d_v + \log \theta_{st}) - q_{st}^{uv} d_u d_v \theta_{st}\right] = \tfrac{1}{2}\left[\sum_u d_u \log d_u + \sum_v d_v \log d_v + \sum_{st} \log\theta_{st} \sum_{uv} a_{uv} q_{st}^{uv} - \sum_{st}\theta_{st}\sum_{uv} d_u d_v q_{st}^{uv}\right], \tag{13}$$

where we have made use of $\sum_{st} q_{st}^{uv}=1$ and $\sum a_{uv}=d_u$.

The first two terms in (13) are constant for any given network and hence can be neglected—they are irrelevant for comparing the likelihood values between different runs on the same network. The final term can be rewritten using equation (8) as

$$\sum_{st}\theta_{st}\sum_{uv} d_u d_v q_{st}^{uv} = \sum_{st}\sum_{uv} a_{uv} q_{st}^{uv} = \sum_{uv} a_{uv}, \tag{14}$$

which is also a constant and can be neglected. Thus, only the third term in (13) need be carried over.

The second sum in (12) is

$$\sum_{\mathbf{s}} q(\mathbf{s}) \sum_u \log \gamma_{s_u,x_u} = \sum_{su} q_s^u \log \gamma_{s,x_u} = \sum_{su} q_s^u \sum_x \delta_{x,x_u} \log \gamma_{sx} = \sum_{usx} \delta_{x,x_u} \gamma_{sx} \log \gamma_{sx} = \sum_{su} \gamma_{s,x_u} \log \gamma_{s,x_u}, \tag{15}$$

where we have used equation (8) again in the third equality.

The final sum in (12) is the entropy of the posterior distribution $q(\mathbf{s})$, which is harder to calculate because it requires not just the marginals of q but the entire distribution. We get around this by making the so-called Bethe approximation[49]:

$$q(\mathbf{s}) = \frac{\prod_{u<v}\left[q_{s_u,s_v}^{uv}\right]^{a_{uv}}}{\prod_u \left[q_s^u\right]^{d_u-1}}, \tag{16}$$

which is exact on trees and locally tree-like networks, and is considered to be a good working approximation on other networks. Substituting this form into the entropy term gives

$$\sum_{\mathbf{s}} q(\mathbf{s}) \log q(\mathbf{s}) = \tfrac{1}{2}\sum_{uv} a_{uv} \sum_{st} q_{st}^{uv} \log q_{st}^{uv} - \sum_u (d_u - 1) \sum_s q_s^u \log q_s^u. \tag{17}$$

Finally, combining equations (13)–(17) and substituting into equation (12), our complete expression for the log-likelihood, neglecting constants, is

$$\log P(\mathbf{A}|\mathbf{\Theta},\mathbf{\Gamma},\mathbf{x}) = \tfrac{1}{2}\sum_{st}\log\theta_{st}\sum_{uv} a_{uv} q_{st}^{uv} + \sum_u\sum_s \gamma_{s,x_u}\log\gamma_{s,x_u} - \tfrac{1}{2}\sum_{uv} a_{uv}\sum_{st} q_{st}^{uv}\log q_{st}^{uv} + \sum_u (d_u-1)\sum_s q_s^u \log q_s^u. \tag{18}$$

The run that returns the largest value of this quantity is the run with the highest likelihood and hence the best fit to the model.

Ordered metadata. We also consider cases in which the metadata are ordered and potentially continuous variables, such as age or income in a social network, which require a different algorithm. The prior probability $P(s|x)$ of belonging to community s given metadata value x now becomes a continuous function of x. In most cases the metadata have a finite range and for convenience we normalize them to fall in the range $x \in [0, 1]$. (In the rarer case of metadata with infinite range a transformation can be applied first to bring them into a finite range.) One immediate question that arises is what limitations should be placed on the form of the probability $P(s|x)$. We cannot allow it to take any functional form, such as ones that vary arbitrarily rapidly, for (at least) two reasons. First, it would be unphysical—there are good reasons in most cases to believe that nodes with infinitesimally different metadata x have only infinitesimally different probabilities of falling in a particular group. In other words, $P(s|x)$ should be smooth and slowly varying in some sense. Second, a function that can vary arbitrarily rapidly can have arbitrarily many degrees of freedom, which would lead to overfitting of the model.

To avoid of these problems, we enforce a slowly varying prior by writing the function $P(s|x)$ as an expansion in a finite set of suitably chosen basis functions. In our work we use the Bernstein polynomials of degree N:

$$B_j(x) = \binom{N}{j} x^j (1-x)^{N-j}, \quad j = 0 \ldots N. \tag{19}$$

(There is an interesting model selection problem inherent in the choice of the degree, which we do not tackle here but which would be a good topic for future research.)

Bernstein polynomials have three particular properties that make them useful for representing probabilities:

1. They form a complete basis set for polynomials of degree N.
2. They fall in the range $0 \le B_j(x) \le 1$ for all $x \in [0, 1]$ and all j.
3. They satisfy the sum rule

$$\sum_{j=0}^{N} B_j(x) = 1 \tag{20}$$

for all $x \in [0, 1]$.

The first of these implies that any degree-N representation of the probability $P(s|x)$ can be written in the form

$$P(s|x) = \sum_{j=0}^{N} \gamma_{sj} B_j(x) \tag{21}$$

for some choice of coefficients γ_{sj}. Moreover, if $\gamma_{sj} \in [0, 1]$ for all s,j then $P(s|x) \in [0, 1]$ for all $x \in [0, 1]$, meaning it is a well-defined probability within this domain. To see this observe first that $P(s|x) \ge 0$ when $\gamma_{sj} \ge 0$ since all $B_j(x) \ge 0$, and second that for $\gamma_{sj} \le 1$ we have

$$P(s|x) = \sum_{j=0}^{N} \gamma_{sj} B_j(x) \le \sum_{j=0}^{N} B_j(x) = 1, \tag{22}$$

where we have made use of equation (20).

Finally, the normalization condition $\sum_s P(s|x) = 1$ can be satisfied for all x by requiring that

$$\sum_s \gamma_{sj} = 1, \tag{23}$$

so that

$$\sum_s P(s|x) = \sum_s \sum_{j=0}^{N} \gamma_{sj} B_j(x) = \sum_{j=0}^{N} B_j(x) = 1. \tag{24}$$

We now employ the form (21) to represent the prior probabilities in our EM algorithm, writing

$$P(\mathbf{s}\,|\,\mathbf{\Gamma},\mathbf{x}) = \prod_u P(s_u\,|\,x_u). \tag{25}$$

The only change to the algorithm from the previous case arises when we maximize the right-hand side of equation (6). Instead of maximizing with respect to the prior probabilities directly, we now maximize with respect to the coefficients γ_{sj} of the expansion. The optimal values of the coefficients are given by

$$\gamma_{sj} = \underset{\{\gamma_{sj}\}}{\operatorname{argmax}} \sum_{ut} q_t^u \log \sum_k \gamma_{tk} B_k(x_u), \tag{26}$$

subject to the constraint (23). One can derive conditions for the maximum by direct differentiation, but the equations do not have a closed-form solution, so instead we once again employ Jensen's inequality (4) to write

$$\sum_{ut} q_t^u \log \sum_k \gamma_{tk} B_k(x_u) \ge \sum_{ut} q_t^u \sum_k Q_k^{tu} \log \frac{\gamma_{tk} B_k(x_u)}{Q_k^{tu}}, \tag{27}$$

which is true for any Q_j^{su} satisfying $\sum_j Q_j^{su}=1$ for all u, s. The exact equality is achieved when

$$Q_j^{su} = \frac{\gamma_{sj} B_j(x_u)}{\sum_k \gamma_{sk} B_k(x_u)}, \tag{28}$$

and the maximum of equation (26) can be computed by first maximizing over Q_j^{su} in this way and then over γ_{sj}. This leads to an iterative algorithm analogous to the EM algorithm in which one computes the Q_j^{su} from (28) and then, using those values, computes the maximum with respect to γ_{sj} by differentiating the right-hand side of (27) subject to the condition (23), which gives

$$\gamma_{sj} = \frac{\sum_u q_s^u Q_j^{su}}{\sum_{tu} q_t^u Q_j^{tu}}. \tag{29}$$

Iterating (28) and (29) alternately to convergence now gives us the coefficients γ_{sj} of the optimal degree-N polynomial prior. Note that (29) always gives γ_{sj} in the range from zero to one, so that, as discussed above, the resulting prior $P(s|x)$ also lies between zero and one and is thus a lawful probability.

Implementation. The calculations for this paper were implemented in the C programming language for speed. The code is included as a Supplementary Software file. We also used a number of additional techniques to improve speed and convergence. We find that the majority of the running time of the algorithm is taken up by the belief propagation calculations, and this time can be shortened by noting that highly converged values of the beliefs are pointless in early steps of the EM algorithm. The parameter values used to calculate the beliefs in these steps are, presumably, highly inaccurate since the EM algorithm has not converged yet, so there is little point waiting for the beliefs to converge to high accuracy when there are much bigger sources of error in the calculation. In the calculations of this paper, we limited the belief propagation to no more than 20 steps at any point. In the early stages of the EM algorithm this gives rather crude values for the beliefs, but these values would not be particularly good under any circumstances, no matter how many steps we used, because of the poor parameter values. In the later stages of the EM algorithm, 20 steps are enough to ensure good convergence (and indeed we often get good convergence after many fewer steps than this).

We also place a limit on the total number of iterations of the EM algorithm, discarding results that fail to converge within the allotted time. In the calculations in this paper, this second limit was set at either 20 or 100 steps. We have performed some runs with higher limits (up to 1,000 EM steps) but, paradoxically, we find this often gives poorer results, for instance in our tests on synthetic networks. This seems to be because the EM algorithm sometimes converges (as we have said) to the wrong solution and empirically when it does so it also often converges more slowly. By discarding runs that converge slowly, therefore, we tend to discard incorrect solutions and improve the average quality of our results.

Normalized mutual information. In our calculations we make use of normalized mutual information to measure the quality of our results. NMI is a widely used measure of the level of agreement between community divisions and 'ground truth' variables, proposed by Danon *et al.*[32]. Given a community division represented by an n-element vector $\mathbf{s}$ of group labels and discrete metadata represented by $\mathbf{x}$, the conditional entropy of the community division is[50]

$$H(\mathbf{s}|\mathbf{x}) = -\sum_x P(x) \sum_s P(s|x) \log P(s|x), \tag{30}$$

$P(x)$ is the fraction of nodes with metadata x and $P(s|x)$ is the probability that a node belongs to community s if it has metadata x. Traditionally the logarithm is taken in base 2, in which case the units of conditional entropy are bits. The conditional entropy is equal to the amount (in bits) of additional information one would need, on top of the metadata themselves, to specify the community membership of every node in the network. If the metadata are perfectly correlated with the communities, so that knowing the metadata tells us the community of every node, then the conditional entropy is zero. Conversely, if the metadata are worthless, telling us nothing at all about community membership, then the conditional entropy takes its maximum value, equal to the total entropy of the community assignment $H(\mathbf{s}) = -\sum_s P(s) \log P(s)$. In our case we already know the value of $P(s|x)$: it is equal to the prior probability γ_{sx} of belonging to community s, one of the outputs of our algorithm. Hence

$$\begin{aligned} H(\mathbf{s}|\mathbf{x}) &= -\sum_x P(x) \sum_s \gamma_{sx} \log \gamma_{sx} = -\sum_x \frac{n(x)}{n} \sum_s \gamma_{sx} \log \gamma_{sx} \\ &= -\frac{1}{n} \sum_{su} \gamma_{s,x_u} \log \gamma_{s,x_u}, \end{aligned} \tag{31}$$

where $n(x) = nP(x)$ is the number of nodes with metadata x and n is the total number of nodes in the network, as previously.

Alternatively, if we want a measure that increases (rather than decreases) with the amount of information the metadata give us, we can subtract $H(\mathbf{s}|\mathbf{x})$ from $H(\mathbf{s})$, which gives the (unnormalized) mutual information

$$I(\mathbf{s};\mathbf{x}) = H(\mathbf{s}) - H(\mathbf{s}|\mathbf{x}), \tag{32}$$

This quantity has a range from zero to $H(\mathbf{s})$, making it potentially hard to interpret, so commonly one normalizes it, creating the normalized mutual information. There are several different normalizations in use. As discussed by McDaid *et al.*[33], it is mathematically reasonable to normalize by the larger, the smaller or the mean of the entropies $H(\mathbf{s})$ and $H(\mathbf{x})$ of the communities and metadata. Danon *et al.*[32] originally used the mean, while Hric *et al.*[8] in their work on lack of correlation between communities and metadata (discussed in the Results section) used the maximum. In the present case, however, we contend that the best choice is the minimum.

The largest possible value of the mutual information is $H(\mathbf{s})$, which sets the scale on which the mutual information should be considered large or small. Thus, one might imagine the correct normalization would be achieved by simply dividing $I(\mathbf{s};\mathbf{x})$ by $H(\mathbf{s})$, yielding a value that runs from zero to one. This, however, would give a quantity that was asymmetric with respect to $\mathbf{s}$ and $\mathbf{x}$—if the values of the two vectors were reversed the value of the mutual information would change. Mutual information, by convention, is symmetric and we would prefer a symmetric definition. Dividing by $\min[H(\mathbf{s}), H(\mathbf{x})]$ achieves this. In all the examples we consider, the number of communities is less than the number of metadata values, in some cases by a wide margin. Assuming the values of both to be reasonably broadly distributed, this implies that the entropy $H(\mathbf{s})$ of the communities will be smaller than that of the metadata $H(\mathbf{x})$ and hence $\min[H(\mathbf{s}), H(\mathbf{x})] = H(\mathbf{s})$. Thus if we define

$$\mathrm{NMI} = \frac{I(\mathbf{s};\mathbf{x})}{\min[H(\mathbf{s}), H(\mathbf{x})]}, \tag{33}$$

we ensure that the normalized mutual information lies between zero and one, that it has a symmetric definition with respect to $\mathbf{s}$ and $\mathbf{x}$, and that it will achieve its maximum value of one when the metadata perfectly predict the community membership. Other definitions, normalized using the mean or maximum of the two entropies, satisfy the first two of these three conditions but not the third, giving values smaller than one by an unpredictable margin even when the metadata perfectly predict the communities. We use the definition (33) in the calculations presented in this paper.

References

1. Newman, M. E. J. *Networks: An Introduction* (Oxford Univ. Press, 2010).
2. Aicher, C., Jacobs, A. Z. & Clauset, A. Learning latent block structure in weighted networks. *J. Complex Networks* **3,** 221–248 (2015).
3. Fortunato, S. Community detection in graphs. *Phys. Rep.* **486,** 75–174 (2010).
4. Adamic, L. A. & Glance, N. The political blogosphere and the 2004 U.S. election: divided they blog. In *Proceedings of the 3rd International Workshop on Link Discovery,* 36–43 (2005).
5. Holme, P., Huss, M. & Jeong, H. Subnetwork hierarchies of biochemical pathways. *Bioinformatics* **19,** 532–538 (2003).
6. Guimerà, R. & Amaral, L. A. N. Functional cartography of complex metabolic networks. *Nature* **433,** 895–900 (2005).
7. Yang, J. & Leskovec, J. Community-affiliation graph model for overlapping community detection. In *Proceedings of the 12th IEEE International Conference on Data Mining (ICDM),* 1170–1175 (2012).
8. Hric, D., Darst, R. K. & Fortunato, S. Community detection in networks: structural communities versus ground truth. *Phys. Rev. E* **90,** 062805 (2014).
9. Barthélemy, M. Spatial networks. *Phys. Rep.* **499,** 1–101 (2011).
10. Jacobs, A. Z. & Clauset, A. A unified view of generative models for networks: models, methods, opportunities, and challenges. Preprint at http://arxiv.org/abs/1411.4070 (2014).
11. Zuev, K., Marián Boguñá, G. B. & Krioukov, D. Emergence of soft communities from geometric preferential attachment. *Sci. Rep.* **5,** 9421 (2015).
12. Good, B. H., de Montjoye, Y.-A. & Clauset, A. Performance of modularity maximization in practical contexts. *Phys. Rev. E* **81,** 046106 (2010).
13. Bothorel, C., Cruz, J. D., Magnani, M. & Micenková, B. Clustering attributed graphs: models, measures and methods. *Network Sci.* **3,** 408–444 (2015).
14. Yang, J., McAuley, J. & Leskovec, J. Community detection in networks with node attributes. In *Proceedings of the 13th IEEE International Conference On Data Mining (ICDM),* 1151–1156 (2013).
15. Binkiewicz, N., Vogelstein, J. T. & Rohe, K. Covariate assisted spectral clustering. Preprint at http://arxiv.org/abs/1411.2158 (2014).
16. Galbrun, E., Gionis, A. & Tatti, N. Overlapping community detection in labeled graphs. *Data Min. Knowl. Discovery* **28,** 1586–1610 (2014).
17. Hansen, T. J. & Mahoney, M. W. Semi-supervised eigenvectors for large-scale locally-biased learning. *J. Mach. Learn. Res.* **15,** 3871–3914 (2014).
18. Zhang, Y., Levina, E. & Zhu, J. Community detection in networks with node features. Preprint at https://arxiv.org/abs/1509.01173 (2015).
19. Expert, P., Evans, T. S., Blondel, V. D. & Lambiotte, R. Uncovering space-independent communities in spatial networks. *Proc. Natl Acad. Sci. USA* **108,** 7663–7668 (2011).
20. Peel, L. Supervised blockmodeling. *ECML/PKDD Workshop on Collective Learning and Inference on Structured Data.* http://arxiv.org/abs/1209.5561 (2012).
21. Eaton, E. & Mansbach, R. A spin-glass model for semi-supervised community detection. In *Proceedings of the 26th AAAI Conference on Artificial Intelligence (AAAI),* 900–906 (2012).
22. Zhang, P., Moore, C. & Zdeborová, L. Phase transitions in semisupervised clustering of sparse networks. *Phys. Rev. E* **90,** 052802 (2014).

23. Moore, C., Yan, X., Zhu, Y., Rouquier, J.-B. & Lane, T. Active learning for node classification in assortative and disassortative networks. In *Proceedings of the 17th ACM SIGKDD International Conference on Knowledge Discovery and Data Mining (KDD),* 841–849 (2011).
24. Leng, M., Yao, Y., Cheng, J., Lv, W. & Chen, X. in *Database Systems for Advanced Applications* (eds Meng, W., Feng, L., Bressan, S., Winiwarter, W. & Song, W.) Vol. 7826, 324–338 (Springer, 2013).
25. Maa, X., Gaoa, L., Yongb, X. & Fua, L. Semi-supervised clustering algorithm for community structure detection in complex networks. *Phys. A* **389,** 187–197 (2010).
26. Zhang, Z.-Y. Community structure detection in complex networks with partial background information. *Europhys. Lett.* **101,** 48005 (2013).
27. Holland, P. W., Laskey, K. B. & Leinhardt, S. Stochastic blockmodels: some first steps. *Social Networks* **5,** 109–137 (1983).
28. Karrer, B. & Newman, M. E. J. Stochastic blockmodels and community structure in networks. *Phys. Rev. E* **83,** 016107 (2011).
29. Decelle, A., Krzakala, F., Moore, C. & Zdeborová, L. Inference and phase transitions in the detection of modules in sparse networks. *Phys. Rev. Lett.* **107,** 065701 (2011).
30. Mossel, E., Neeman, J. & Sly, A. Reconstruction and estimation in the planted partition model. *Probab. Theory Related Fields* **162,** 431–461 (2015).
31. Moody, J. Race, school integration, and friendship segregation in America. *Am. J. Sociol.* **107,** 679–716 (2001).
32. Danon, L., Duch, J., Diaz-Guilera, A. & Arenas, A. Comparing community structure identification. *J. Stat. Mech.* **2005,** P09008 (2005).
33. McDaid, A. F., Greene, D. & Hurley, N. Normalized mutual information to evaluate overlapping community finding algorithms. Preprint at http://arxiv.org/abs/1110.2515 (2011).
34. Brose, U. *et al.* Body sizes of consumers and their resources. *Ecology* **86,** 2545–2545 (2005).
35. Jacob, U. *Trophic Dynamics of Antarctic Shelf Ecosystems Food Webs and Energy Flow Budgets* (PhD thesis, Univ. Bremen, 2005).
36. Woodward, G. *et al.* Body size in ecological networks. *Trends Ecol. Evol.* **20,** 402–409 (2005).
37. Traud, A. L., Mucha, P. J. & Porter, M. A. Social structure of Facebook networks. *Phys. A* **391,** 4165–4180 (2012).
38. Report, W. M. *World Malaria Report* (World Health Organization, 2012).
39. Bull, P. C. *et al. Plasmodium falciparum* variant surface antigen expression patterns during malaria. *PLOS Pathog.* **1,** e26 (2005).
40. Larremore, D. B., Clauset, A. & Buckee, C. Z. A network approach to analyzing highly recombinant malaria parasite genes. *PLOS Comput. Biol.* **9,** e1003268 (2013).
41. Larremore, D. B., Clauset, A. & Jacobs, A. Z. Efficiently inferring community structure in bipartite networks. *Phys. Rev. E* **90,** 012805 (2014).
42. Warimwe, G. M. *et al. Plasmodium falciparum var* gene expression is modified by host immunity. *Proc. Natl Acad. Sci. USA* **106,** 21801–21806 (2009).
43. Bull, P. C. *et al.* An approach to classifying sequence tags sampled from *Plasmodium falciparum var* genes. *Mol. Biochem. Parasitol.* **154,** 98–102 (2007).
44. Clauset, A., Moore, C. & Newman, M. E. J. Hierarchical structure and the prediction of missing links in networks. *Nature* **453,** 98–101 (2008).
45. Milo, R. *et al.* Network motifs: simple building blocks of complex networks. *Science* **298,** 824–827 (2002).
46. Borgatti, S. P. & Everett, M. G. Models of core/periphery structures. *Social Networks* **21,** 375–395 (1999).
47. Ball, B. & Newman, M. E. J. Friendship networks and social status. *Network Sci.* **1,** 16–30 (2013).
48. Hoff, P. D., Raferty, A. E. & Handcock, M. S. Latent space approaches to social network analysis. *J. Am. Stat. Assoc.* **97,** 1090–1098 (2002).
49. Yedidia, J. S., Freeman, W. T. & Weiss, Y. in *Exploring Artificial Intelligence in the New Millennium* (eds Lakemeyer, G. & Nebel, B.) 239–270 (Morgan Kaufmann, 2003).
50. Cover, T. M. & Thomas, J. A. *Elements of Information Theory* 2nd edn (Wiley, 2006).

Acknowledgements

We thank Daniel Larremore, Cristopher Moore, Leto Peel and Mason Porter for useful conversations, and Darko Hric, Richard Darst and Santo Fortunato for sharing the Internet data set and taking the time to explain it to us. This work uses data from Add Health, a programme project designed by J. Richard Udry, Peter S. Bearman and Kathleen Mullan Harris, and funded by a grant P01HD31921 from the National Institute of Child Health and Human Development, with cooperative funding from 17 other agencies. Special acknowledgment is due Ronald R. Rindfuss and Barbara Entwisle for assistance in the original design. This research was funded in part by the US National Science Foundation under grants DMS-1107796 and DMS-1407207 (M.E.J.N.) and IIS-1452718 (A.C.).

Author contributions

M.E.J.N. and A.C. conceived the research, designed the analyses, conducted the analyses and wrote the manuscript.

Additional information

Competing financial interests: The authors declare no competing financial interests.

25

Topological vacuum bubbles by anyon braiding

Cheolhee Han[1], Jinhong Park[1], Yuval Gefen[2] & H.-S. Sim[1]

According to a basic rule of fermionic and bosonic many-body physics, known as the linked cluster theorem, physical observables are not affected by vacuum bubbles, which represent virtual particles created from vacuum and self-annihilating without interacting with real particles. Here we show that this conventional knowledge must be revised for anyons, quasiparticles that obey fractional exchange statistics intermediate between fermions and bosons. We find that a certain class of vacuum bubbles of Abelian anyons does affect physical observables. They represent virtually excited anyons that wind around real anyonic excitations. These topological bubbles result in a temperature-dependent phase shift of Fabry-Perot interference patterns in the fractional quantum Hall regime accessible in current experiments, thus providing a tool for direct and unambiguous observation of elusive fractional statistics.

[1] Department of Physics, Korea Advanced Institute of Science and Technology, 291, Daehak-ro, Yuseong-gu, Daejeon 34141, Korea. [2] Department of Condensed Matter Physics, Weizmann Institute of Science, Rehovot 76100, Israel. Correspondence and requests for materials should be addressed to H.-S.S. (email: hssim@kaist.ac.kr).

When two identical particles adiabatically exchange their positions $\mathbf{r}_{i=1,2}$, their final state ψ (up to dynamical phase) is related to the initial one through an exchange statistics phase θ^*,

$$\psi(\mathbf{r}_2, \mathbf{r}_1) = e^{i\theta^*} \psi(\mathbf{r}_1, \mathbf{r}_2), \qquad (1)$$

with $\theta^* = 0$ (π) for bosons (fermions)[1].

Anyons[2–4] are quasiparticles in two dimensions, not belonging to the two classes of elementary particles, bosons and fermions. Abelian anyons appear in the fractional quantum Hall (FQH) system of filling factor $\nu = 1/(2n+1)$, $n = 1,2,\cdots$. They carry a fraction $e^* = \nu e$ of the electron charge e and obey fractional exchange statistics, satisfying equation (1) with $\theta^* = \pm\pi\nu$. Two anyons gain a phase $\pm 2\pi\nu$ from a braiding, whereby one winds around the other; $\pm$ depends on the winding direction. Although fractional charges have been detected[5–8], experimental measurement of statistics phase $\pi\nu$ has been so far elusive. Existing theoretical proposals for the measurement involve quantities inaccessible in current experiments or suffer from unintended change of a proposed setup with external parameters[9–18].

In many-body quantum theory[1], Feynman diagrams are used to compute the expectation value of observables. This approach invokes vacuum bubble diagrams, which describe virtual particles excited from vacuum and self-annihilating without interacting with real particles. According to the linked cluster theorem[1], each diagram possessing vacuum bubbles comes with, hence is exactly cancelled by, a partner diagram of the same magnitude but of the opposite sign. Consequently, vacuum bubbles do not contribute to physical observables.

In the following, we demonstrate that this common wisdom has to be revised for anyons: a certain class of vacuum bubbles of Abelian anyons does affect observables. These virtual particles, which we call topological vacuum bubbles, wind around a real anyonic excitation, gaining the braiding phase $\pm 2\pi\nu$. We propose a realistic setup for detecting them and $\theta^* = \pi\nu$.

Results

Topological vacuum bubble. We illustrate topological vacuum bubbles. In Fig. 1a, a Feynman diagram represents interference $a_1 a_2^*$ between processes a_1 and a_2 of propagation of a real particle. In a_1, a virtual particle-hole pair is excited then self-annihilates after the virtual particle winds around the real particle, forming a vacuum bubble, while it is not excited in a_2. The winding results in a braiding phase $2\pi\nu$ and an Aharonov–Bohm phase $2\pi\Phi/\Phi_0^*$ from the magnetic flux Φ enclosed by the winding path, contributing to the interference signal as $e^{i(2\pi\Phi/\Phi_0^* + 2\pi\nu)}$; $\Phi_0^* = h/e^*$ is the anyon flux quantum[9,19].

The limiting cases of bosons ($\nu = 0$) and fermions ($\nu = 1$) imply that this bubble diagram appears together with, and is cancelled by, a partner diagram in Fig. 1b. The partner diagram has a bubble not encircling the real particle and involves only $2\pi\Phi/\Phi_0^*$. The two diagrams (and their complex conjugates) yield

$$\text{Interference signal} \propto \text{Re}\left[e^{i(2\pi\Phi/\Phi_0^* + 2\pi\nu)} - e^{2\pi i\Phi/\Phi_0^*}\right]$$
$$= -\sin(\pi\nu)\sin\left(2\pi\Phi/\Phi_0^* + \pi\nu\right). \qquad (2)$$

For bosons and fermions, the two diagrams fully cancel each other with $\sin(\pi\nu) = 0$ in agreement with the linked cluster theorem; hence, the signal disappears. By contrast, for anyons they cancel only partially, producing the non-vanishing interference in an observable, and are topological as the braiding phase is involved.

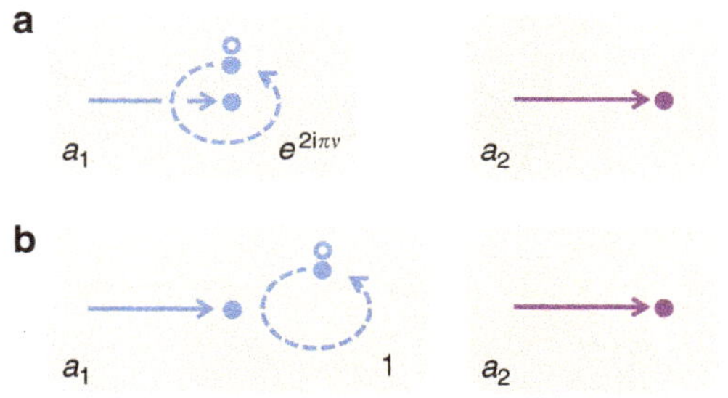

Figure 1 | Topological vacuum bubble. Feynman diagrams for interference involving a real particle and a virtual particle-hole excitation from vacuum. Full (empty) circles represent particles (holes). Solid (dashed) lines denote propagations of real (virtual) particles. (**a**) Diagram for the interference $a_1 a_2^*$ of two processes: (a_1, blue) A real particle propagates, a virtual particle-hole pair is excited, then the pair self-annihilates after the virtual particle winds around the real one. (a_2, magenta) A real particle propagates. The entire virtual process constitutes a vacuum bubble. For anyons, the bubble gains a topological braiding phase $2\pi\nu$ from the winding. (**b**) Partner diagram of **a**. Here a virtual particle, constituting another bubble, does not encircle a real one and hence gains no braiding phase. The diagrams in **a** and **b** contribute to observables for anyons, while they do not for bosons and fermions.

Interferometer setup. In Fig. 2a, we propose a minimal setup for observing topological vacuum bubbles. It is a Fabry–Perot interferometer[9,17,20–23] in the $\nu = 1/(2n+1)$ FQH regime, coupled to an additional edge channel (Edge 1) via a quantum point contact (QPC1). At QPCi, there occurs tunnelling of a single anyon (rather than anyon bunching), fulfilled[24] with $\gamma_i \ll k_B T$; γ_i is the tunnelling strength and T is the temperature. Gate voltage V_G is applied, to change the interferometer loop enclosing Aharonov–Bohm flux Φ. The interference part $I_{D_3}^{\text{int}}$ of charge current at drain D_3 is measured with bias voltage V applied to source S_1; the other S_i's and D_i's are grounded. Together with 'virtual' (thermal) anyon excitations in the interferometer, a voltage-biased 'real' anyon, dilutely injected at QPC1 from Edge 1 to the interferometer, forms topological vacuum bubbles, as shown below. The bubbles contribute to $I_{D_3}^{\text{int}}$ at the leading order ($I_{D_3}^{\text{int}} \propto \gamma_1^2\gamma_2\gamma_3$) in QPC tunnelling, as Edges 2 and 3 are unbiased. It is noteworthy that in the setups previously studied[9–18], topological bubbles do not contribute to current at the leading order.

We consider the regime of $e^*V \gg k_B T \gtrsim \hbar v_p/L$, where the size $L_V \equiv \hbar v_p/(e^*V)$ of the dilutely injected anyons is much smaller than interferometer size L and the injection of hole-like anyons at QPC1 is ignored; v_p is anyon velocity along the edges and e^*V should be much smaller than the FQH energy gap. Because of the dilute injection and $L_V \ll L$, anyon braiding is well defined in the interferometer. As shown below, the dependence of $I_{D_3}^{\text{int}}$ on Φ or on V_G provides a clear signature of the topological bubbles, consequently, $\theta^* = \pi\nu$ in both of the pure Aharonov–Bohm regime (where Coulomb interaction of the edge channels with bulk anyons localized inside the interferometer loop is negligible) and the Coulomb-dominated regime (where the interaction is strong)[22,25]. Below, we first ignore bulk anyons.

Interference current. Employing the chiral Luttinger liquid theory[26,27] for FQH edges and Keldysh Green's functions[10,12], we compute $I_{D_3}^{\text{int}} (\propto \gamma_1^2\gamma_2\gamma_3)$ at the leading order in γ. There are four types of the processes mainly contributing to $I_{D_3}^{\text{int}} \simeq I_{D_3}^{\text{I}-1} + I_{D_3}^{\text{I}-2} + I_{D_3}^{\text{II}} + I_{D_3}^{\text{III}}$; see Fig. 2b. For $e^*V \gg k_B T \gg \hbar v_p/L$, we obtain the analytical expression of $I_{D_3}^{\text{int}}$: The interference current contributed by Type I-1 processes is

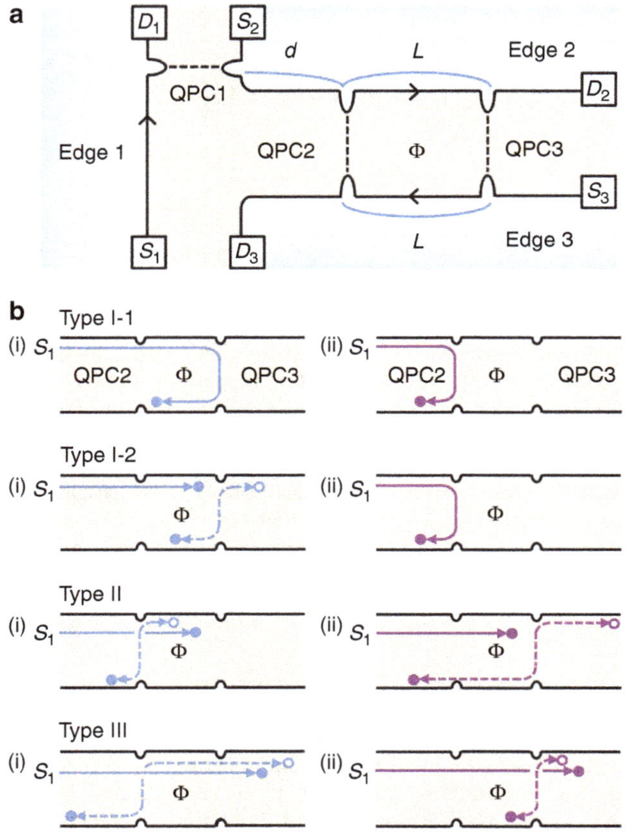

Figure 2 | Interferometry for detecting topological vacuum bubbles. (**a**) In the setup, anyons move (see arrows) along FQH edge channel $i=1,2,3$ that connects source S_i and drain D_i, and jump (dashed) between the channels via tunnelling at QPCs. The loop defined by Edges $i=2,3$, QPC2 and QPC3 encloses magnetic flux Φ, forming a Fabry–Perot interferometer. Distance between QPC2 and QPC3 (QPC1) is L (d). (**b**) Two interfering paths (i) and (ii) of each main interference process at $e^*V \gg k_B T \gtrsim \hbar v_p/L$. Following Fig. 1, filled (empty) circles represent particle-like (hole-like) anyons and solid (dashed) lines denote propagation of an anyon injected from S_1 (anyon pair excitation at QPCs). Type II and III processes involve a topological vacuum bubble.

$I_{D_3}^{\mathrm{I}-1} \propto V^{\nu-2}T^{\nu-1}g(V,T)\cos(2\pi\Phi/\Phi_0^* + e^*VL/\hbar v_\mathrm{p})$, which by Type I-2 is $I_{D_3}^{\mathrm{I}-2} \propto V^{-1}g(V,T)(\sin^2\pi\nu)\cos(2\pi\Phi/\Phi_0^*)$, and those by Type II and III are

$$I_{D_3}^{\mathrm{II}} \simeq g(V,T)\frac{2L+C(\nu)L_T}{\hbar v_\mathrm{p}}(\sin^2\pi\nu)\cos(2\pi\Phi/\Phi_0^*+\pi\nu), \quad (3)$$

$$I_{D_3}^{\mathrm{III}} \simeq g(V,T)\frac{C(\nu)L_T}{\hbar v_\mathrm{p}}(\sin^2\pi\nu)\cos(2\pi\Phi/\Phi_0^*-\pi\nu), \quad (4)$$

where $g(V,T) \propto \gamma_1^2\gamma_2\gamma_3(VT)^{2\nu-1}e^{-2L/L_T}$, e^{-2L/L_T} is a thermal suppression factor, thermal length $L_T \equiv \hbar v_\mathrm{p}/(\pi\nu k_B T)$ and $C(\nu=1/3) \simeq 0.43$; see Methods and the Supplementary Note 3.

Type I-1 processes describe interference between two paths of an anyon moving from S_1 to D_3 via (i) QPC2 and (ii) QPC3, respectively. They were previously studied[9].

In Type I-2, an anyon injected from S_1 interferes with a particle-like anyon excited at a QPC or annihilates a hole-like anyon; particle-like and hole-like anyons are pairwise excited thermally at QPCs. For example, consider the two following interfering histories: (i) an anyon is injected from S_1 to D_2, an anyon pair is excited at QPC3 and then the particle-like (hole-like) anyon of the pair moves to D_3 (D_2). (ii) An anyon is injected from S_1 to D_3 via QPC2 without any excitations. The hole-like anyon annihilates the injected anyon on Edge 2 in history (i) and the particle-like anyon of (i) interferes with the injected anyon of (ii) on Edge 3. The sum of such interference processes yields $I_{D_3}^{\mathrm{I}-2} \propto \sin^2\pi\nu\cos(2\pi\Phi/\Phi_0^*)$. The $\sin^2\pi\nu$ factor appears, because relative locations of anyons on Edge 2 or 3 differ between the processes, leading to an exchange phase $\pm\pi\nu$, and because a process with an excitation (of a particle-like anyon moving to D_2 and a hole-like one to D_3) yields charge current in the opposite direction to another with its particle-hole conjugated excitation (of a particle to D_3 and a hole to D_2).

In Types II and III, a real anyon injected from S_1 moves to D_2 and a virtual anyon pair excited at QPC2 interferes with another at QPC3. The interference path effectively encloses the real anyon, forming a topological vacuum bubble (*cf.* Fig. 1). In Type II, when the real anyon is located on Edge 2 between QPC2 and QPC3, a virtual pair is excited at QPC2 in history (i) and at QPC3 in history (ii). Next, the hole-like (particle-like) anyon of each pair moves, for example, to D_2 (D_3). The interference of the two histories corresponds to the winding of a virtual anyon around the real one and Φ, forming a topological bubble with interference phase $\pm(2\pi\Phi/\Phi_0^*+2\pi\nu)$; $\pm$ depends on whether the hole-like anyon moves to D_2 or D_3. In the interference, the winding of a virtual anyon around the real one effectively occurs through the exchanges of the positions of the anyons in each of Edges 2 and 3, as relative locations of anyons on Edges 2 and 3 differ between (i) and (ii) (see Supplementary Fig. 2 and Supplementary Note 6). This interference is accompanied by a partner process. The latter has a bubble that winds around Φ (gaining phase $\pm 2\pi\Phi/\Phi_0^*$), but not around a real anyon. The two partner processes partially cancel each other, yielding $I_{D_3}^{\mathrm{II}} \propto \sin\pi\nu\cos(2\pi\Phi/\Phi_0^*+\pi\nu)$; the remaining $\sin\pi\nu$ factor in equations (3) and (4) has a similar origin to the $\sin^2\pi\nu$ factor of $I_{D_3}^{\mathrm{I}-2}$.

In Type III, the two interfering histories are as follows: (i) a virtual pair is excited at QPC2 before a real anyon injected from S_1 arrives at QPC2 and (ii) another pair is excited at QPC3 after the real one arrives at QPC3. The ensuing chronological sequence on Edge 2 is opposite to Type II: an anyon excited at QPC2 arrives at QPC3; the real one arrives at QPC3; a pair is excited at QPC3. The resulting topological bubble effectively winds around the real anyon in the direction opposite to its winding around Φ, yielding a phase $\pm(2\pi\Phi/\Phi_0^*-2\pi\nu)$. Partial cancellation of the bubble and its partner leads to $I_{D_3}^{\mathrm{III}} \propto \sin\pi\nu\cos(2\pi\Phi/\Phi_0^*-\pi\nu)$. The factor $2L+CL_T$ of $I_{D_3}^{\mathrm{II}}$ (CL_T in $I_{D_3}^{\mathrm{III}}$) in equation (3) (equation (4)) comes from the time window compatible with the chronological sequence on Edge 2.

Type II and III processes of topological bubbles do not affect any observables at $\nu=1$ (fermions), due to full cancellation between partner bubbles (the linked cluster theorem). They are distinct from I-2. I-2 processes produce, for example, non-vanishing current noise $\langle (I_{D_3}^{\mathrm{int}} - \langle I_{D_3}^{\mathrm{int}}\rangle)^2\rangle$ at $\nu=1$, as the particle-hole conjugated excitations (mentioned before) equally contribute to the noise (although the contributions of the conjugations to $I_{D_3}^{\mathrm{I}-2}$ cancel each other, leading to $I_{D_3}^{\mathrm{I}-2}=0$).

In the more general regime of $e^*V \gg k_B T \gtrsim \hbar v_\mathrm{p}/L$, we employ the parametrization

$$I_{D_3}^{\mathrm{int}} \propto \cos(2\pi\Phi/\Phi_0^*+\theta). \quad (5)$$

The phase θ is determined by competition between the various contributions to $I_{D_3}^{\mathrm{int}}$ and contains information about statistics phase $\pi\nu$. At $e^*V \gg k_B T \gtrsim \hbar v_\mathrm{p}/L$, $I_{D_3}^{\mathrm{II}} + I_{D_3}^{\mathrm{III}}$ is much larger than $I_{D_3}^{\mathrm{I}-1} + I_{D_3}^{\mathrm{I}-2}$ and dominates $I_{D_3}^{\mathrm{int}}$, because the interfering anyon of Types I-1 and I-2 is voltage biased and has width $L_V \propto V^{-1}$ much narrower than the thermal anyon excitations (whose width $L_T \propto T^{-1}$) of II and III, showing much weaker interference. From

$I_{D_3}^{\mathrm{II}}$ and $I_{D_3}^{\mathrm{III}}$, we find

$$\theta \simeq \arctan\left[\frac{L}{L+C(\nu)L_T}\tan\left(\pi\nu+(1-2\nu)\arctan\frac{2L_V}{L_T}\right)\right]$$
$$\rightarrow \arctan\left[\frac{L}{L+C(\nu)L_T}\tan\pi\nu\right] \quad \text{as } T/V\rightarrow 0$$
$$\rightarrow \pi\nu \quad \text{as } L_T/L\rightarrow 0 \text{ and } T/V\rightarrow 0. \tag{6}$$

The arctan $2L_V/L_T$ term represents an error in the braiding phase $2\pi\nu$ of the topological bubbles. It occurs when the size L_V of a real anyon is not sufficiently smaller than the winding radius of a virtual anyon around the real one. It is negligible at $e^*V \gg k_BT$, as the radius is effectively large when $L_V \ll L_T$; those corresponding to the error are ignored in equations (3) and (4), and are shown in Supplementary Note 3. For $e^*V \gg k_BT \gg \hbar v_p/L$, $I_{D_3}^{\mathrm{II}}$ dominates $I_{D_3}^{\mathrm{int}}$ and $\theta\rightarrow\pi\nu$. Remarkably, θ depends on T, contrary to common practice in electron interferometry[28].

Coulomb-dominated regime. In experimental situations of a Fabry–Perot interferometer in the FQH regime, it is expected that there exist bulk anyons localized inside the interferometer loop. There are two regimes of Fabry–Perot interferometers, the pure Aharonov–Bohm regime and the Coulomb-dominated regime. In the former regime, Coulomb interaction between the bulk anyons and the edge of the interferometer is negligible, whereas it is crucial in the latter[25]. The Fabry–Perot interferometers of recent experiments[17,20–23] in the FQH regime are in the Coulomb-dominated regime. Below we compute the interference current I_{D_3} in the presence of the Coulomb interaction and show that equation (6) is applicable to both of the pure Aharonov–Bohm regime and the Coulomb-dominated regime.

For $e^*V \gg k_BT \gtrsim \hbar v_p/L$, we numerically compute $I_{D_3}^{\mathrm{int}}(\propto \gamma_1^2\gamma_2\gamma_3)$ in Fig. 3, combining our theory with the capacitive interaction model[25] that successfully describes thermally fluctuating bulk anyons and the interaction (see the Method and Supplementary Note 5). We find the gate-voltage dependence of $I_{D_3}^{\mathrm{int}}\propto\cos(2\pi V_G/V_{G,0}+\theta)$ with periodicity $V_{G,0}$ in the Coulomb-dominated regime and $I_{D_3}^{\mathrm{int}}\propto\cos(2\pi\Phi/\Phi_0+\theta)$ in the pure Aharonov–Bohm limit; here the periodicity of the Φ dependence is $\Phi_0\equiv h/e$ rather the period Φ_0^* of equation (5), because of the fluctuation of the number of bulk anyons[9,19]. In both the regimes, the interference processes discussed before (Fig. 2) appear in the same manner; hence, θ satisfies the analytic expression in equation (6) (*cf.* Fig. 3c).

How to measure the phase θ. Experimental measurements of θ can be affected by possible side effects, including the external-parameter (magnetic fields, gate voltages and bias voltages) dependence of the size, shape, QPC tunnelling and bulk anyon excitations. Below we propose how to detect θ with avoiding the side effects, using the setup in Fig. 2a.

The phase θ is experimentally measurable, by comparing $I_{D_3}^{\mathrm{int}}$ with a reference current $I_{\mathrm{Ref},D_3}^{\mathrm{int}}$. $I_{\mathrm{Ref},D_3}^{\mathrm{int}}$ is measured at D_3 in the same setup under the same external parameters (temperature, gate voltages, magnetic field and so on) with $I_{D_3}^{\mathrm{int}}$, but with applying infinitesimal bias voltage $V_{\mathrm{ref}}/2$ to S_2 and $-V_{\mathrm{ref}}/2$ to S_3 and keeping S_1 and all D_i's grounded[9] (*cf.* Supplementary Note 4). In any regimes $I_{D_3}^{\mathrm{int}}$ shows the same interference pattern with $I_{\mathrm{Ref},D_3}^{\mathrm{int}}$, but is phase-shifted from $I_{\mathrm{Ref},D_3}^{\mathrm{int}}$ by θ; $I_{\mathrm{Ref},D_3}^{\mathrm{int}}\propto\cos 2\pi V_G/V_{G,0}$ ($I_{\mathrm{Ref},D_3}^{\mathrm{int}}\propto\cos 2\pi\Phi/\Phi_0$) in the Coulomb-dominated (pure Aharonov–Bohm) limit. Importantly, the side effects modify $I_{D_3}^{\mathrm{int}}$ and $I_{\mathrm{Ref},D_3}^{\mathrm{int}}$ in the same manner; hence, the phase shift between the patterns remains as θ.

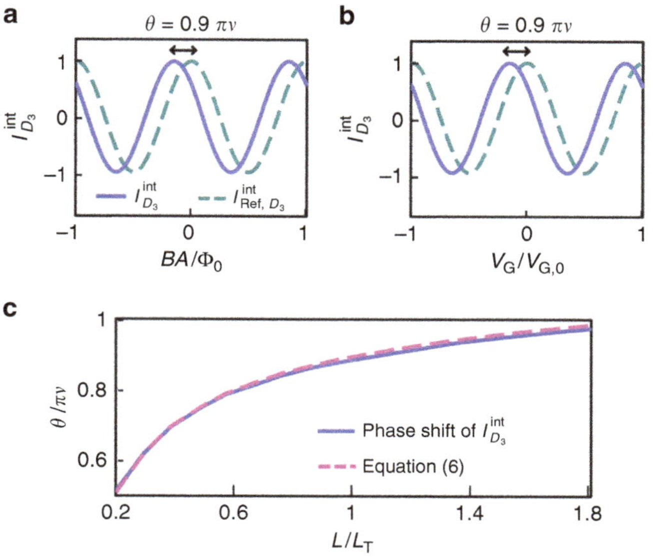

Figure 3 | Detection of anyon phase $\pi\nu$ from interference phase shift θ. (**a**) Dependence of $I_{D_3}^{\mathrm{int}}$ (blue, normalized) and $I_{\mathrm{Ref},D_3}^{\mathrm{int}}$ (cyan, normalized) on Φ in the pure Aharonov–Bohm regime and (**b**) their dependence on V_G in the Coulomb dominated regime. We choose $\nu=1/3$, $T=30$ mK, $e^*V=45\,\mu$eV, $L=3\,\mu$m and $v_p=10^4$ m s^{-1} ($L/L_T=1.2$ and $L/L_V=20$); see Supplementary Note 5 for the Coulomb interaction parameter of the regimes. For these parameters, the phase shift θ between $I_{D_3}^{\mathrm{int}}$ and $I_{\mathrm{Ref},D_3}^{\mathrm{int}}$ is $0.9\pi\nu$. (**c**) Dependence of θ on T. The same parameters (except T) with (**a**) and (**b**) are chosen. $\theta\rightarrow\pi\nu$ as T increases (yet $e^*V \gg k_BT \gg \hbar v_p/L$). In both the pure Aharonov–Bohm regime and the Coulomb-dominated regime, the same numerical result (blue curve) of $\theta(T)$, which agrees with equation (6) (magenta), is obtained.

The fractional statistics phase is directly and unambiguously identifiable in experiments, by observing $\theta\rightarrow\pi\nu$ at $e^*V \gg k_BT \gg \hbar v_p/L$ with excluding the side effects as above, or one applies the fit function of arctan $[A_1/(1+A_2/T)]$ with fit parameters A_1 and A_2 to measured data of $\theta(T)$ and extracts arctan $A_1=\pi\nu$ from the fit (*cf.* the second line of equation (6)). Observation of $\theta=\pi\nu$ or $\theta(T)$ will suggest a strong evidence of anyon braiding and topological bubbles.

The parameters in Fig. 3 are experimentally accessible[17,20–23]. For the QPCs, there are constraints (i) that the number of voltage-biased anyons injected through QPC1 is at most one in the interferometer loop at any instance (to ensure that the braiding phase of a topological vacuum bubble is $2\pi\nu$), (ii) that the anyon tunnelling probabilities at QPC2 and QPC3 are sufficiently small (to ensure that the double winding of an anyon along the interferometer loop is negligible) and (iii) that anyon tunnelling (rather than electron tunnelling) occurs at the QPCs. The constraint (i) is satisfied when the anyon tunnelling probability at QPC1 is $<\hbar v_p/(2Le^*V)$, which is ~ 0.05 under the parameters. To achieve the constraints (ii) and (iii), each tunnelling probability of QPC2 and QPC3 is typically set to be 0.4 in experiments[22,29]; then the amplitude of the double winding is smaller than that of the single winding by the factor 0.4 exp$(-2L/L_T)$, which is ~ 0.04 at 30 mK. With the constraints we estimate the amplitude of $I_{D_3}^{\mathrm{int}} \lesssim (\nu e^2V/h)(\hbar v_p/(2Le^*V))0.4\exp(-2L/L_T)$, which is 1.5 pA at 30 mK and 0.6 pA at 40 mK under the parameters. It is noteworthy that $\theta=0.9\pi\nu$ is reached at 30 mK, while $\theta=0.95\pi\nu$ at 40 mK under the parameters. The estimation is within a measurable range in experiments, where current $\gtrsim 0.5$ pA is well detectable[30].

We remark that the above strategy of detecting the fractional statistics phase is equally applicable to the more general quantum Hall regime[22] of filling factor $\nu' = \nu + \nu_0$, in which the edge channels from the integer filling ν_0 are fully transmitted through the QPCs, while the channel from the fractional filling ν forms the interferometry in Fig. 2. For this case we compute $I_{D_3}^{\rm int}$ and $I_{{\rm Ref},\,D_3}^{\rm int}$, and find that in both of the pure Aharonov–Bohm regime and the Coulomb-dominated regime the interference-pattern phase shift between them is identical to the phase θ of the $\nu_0 = 0$ case discussed in equation (6) and Fig. 3 (*cf.* Supplementary Note 5).

Certain anyonic vacuum bubbles involve topological braiding and affect physical observables surprisingly, contrary to vacuum bubbles of bosons and fermions. They can be detected with current experiment tools, which will provide an unambiguous evidence of anyonic fractional statistics. We expect that they are relevant also for other filling fractions $\nu = p/(2np+1)$, non-Abelian anyons[17,21] and topological quantum computation setups[31].

Methods

Hamiltonian for the interferometer. We present the Hamiltonian for the setup. We recall the chiral Luttinger liquid theory for FQH edges.

The Hamiltonian $H = \sum_i H_{{\rm edge},i} + H_{\rm tun}$ for the interferometer in Fig. 2a consists of $H_{{\rm edge},i}$ for edge channel i and $H_{\rm tun}$ for anyon tunnelling at QPCs. Edge channel 1 is biased by V and its Hamiltonian, employing the bosonization[26] for chiral Luttinger liquids, is given by

$$H_{{\rm edge},1} = \frac{\hbar v_{\rm p}}{4\pi\nu}\int_{-\infty}^{\infty} dx : (\partial_x\phi_1(x))^2 : + e^* V\hat{N}_1. \qquad (7)$$

For the other unbiased channels, $H_{{\rm edge},i=2,3} = (\hbar v_{\rm p}/4\pi\nu)\int_{-\infty}^{\infty} dx : (\partial_x\phi_i(x))^2 :$ Here, $e^* > 0$, $\hat{N}_i$ is the anyon number operator of channel i and $\phi_i(x)$ is the bosonic field of channel i at position x, which describes the plasmonic excitation of anyons. The tunnelling Hamiltonian is $H_{\rm tun} = T_1^{\rightarrow} + T_2^{\downarrow} + T_3^{\downarrow} + {\rm h.c.}$, $T_1^{\rightarrow}$ is the operator from Edge channel 1 to 2 at QPC 1, $T_2^{\downarrow}$ from Edge 2 to 3 at QPC2 and $T_3^{\downarrow}$ from Edge 2 to 3 at QPC3. These are written as

$$T_1^{\rightarrow}(t) = \gamma_1\Psi_2^{\dagger}(0,t)\Psi_1(0,t), \qquad (8)$$

$$T_2^{\downarrow}(t) = \gamma_2 e^{-i\pi\Phi/\Phi_0^*}\Psi_3^{\dagger}(L,t)\Psi_2(d,t), \qquad (9)$$

$$T_3^{\downarrow}(t) = \gamma_3 e^{i\pi\Phi/\Phi_0^*}\Psi_3^{\dagger}(0,t)\Psi_2(d+L,t), \qquad (10)$$

where $\Psi_i^{\dagger}(x,t) = F_i^{\dagger}(t)e^{-i\phi_i(x,t)}/\sqrt{2\pi a}$ creates an (particle-like) anyon at position x and time t on Edge i, a is the short-length cutoff, γ_i is the tunnelling strength at QPC i (chosen as real) and the Aharonov–Bohm flux Φ enclosed by Edges $i = 2,3$, QPC2 and QPC3 is attached to $T_2^{\downarrow}$ and $T_3^{\downarrow}$, respectively, under certain gauge transformation; the dynamical phase common to the all edge channels is absorbed to $2\pi\Phi/\Phi_0^*$. The Klein factor $F_i^{\dagger}$ increases the number of anyons on Edge i by 1 and satisfies $F_i^{\dagger}F_i = F_iF_i^{\dagger} = 1$, $[N_iF_j^{\dagger}] = \delta_{ij}F_i^{\dagger}$, $F_1(t) = F_1(0)e^{-ie^*Vt/\hbar}$ and $[\phi_i, F_j] = 0$.

The exchange rule in equation (1) is described by ϕ_i and F_i. On Edge i, it is satisfied by

$$[\phi_i(x_1), \phi_j(x_2)] = i\pi\nu\delta_{ij}{\rm sgn}(x_1 - x_2). \qquad (11)$$

The exchange rule between anyons on different edges is achieved with the commutators of F_i,

$$F_iF_j = F_jF_ie^{-i\pi\nu{\rm sgn}(i-j)}, \qquad F_i^{\dagger}F_j = F_jF_i^{\dagger}e^{i\pi\nu{\rm sgn}(i-j)}. \qquad (12)$$

A conventional way[32] for obtaining the commutators is to think of an extended edge connecting the different channel segments with no twist (*cf.* Fig. 4). The connection should preserve the chiral propagation direction of the channels. The exchange rule $\Psi(x_1)\Psi(x_2) = \Psi(x_2)\Psi(x_1)e^{-i\pi\nu{\rm sgn}(x_1-x_2)}$ of anyons of the extended edge agrees with equations (11) and (12).

We consider the regime of weak tunnelling of anyons and treat $H_{\rm tun}$ as a perturbation on $\sum_i H_{{\rm edge},i}$. Perturbation theory is applicable[24] in the renormalization group sense, when e^*V and $k_{\rm B}T$ are higher than $C\gamma^{1/(1-\nu)}$, C being a non-universal constant.

The current I_{D_3} is expressed as $I_{D_3} = ie^*[N_3, H] = ie^*(T_2^{\downarrow} - T_2^{\uparrow} + T_3^{\downarrow} - T_3^{\uparrow})$. I_{D_3} is decomposed, $I_{D_3} = I_{D_3}^{\rm dir} + I_{D_3}^{\rm int}$, into direct current $I_{D_3}^{\rm dir} \propto \gamma_1^2\gamma_2^2, \gamma_1^2\gamma_3^2$ and interference current $I_{D_3}^{\rm int} \propto \gamma_1^2\gamma_2\gamma_3$ depending on Φ (the leading-order contribution). Equations (3) and (4) are obtained by employing Keldysh Green's function technique with semiclassical approximation (see Supplementary Fig. 1 and Supplementary Notes 1, 2 and 3).

Coulomb interaction. In the presence of Coulomb interaction between bulk anyons and edge channels, we compute $I_{D_3}^{\rm int}$, combining our chiral Luttinger liquid theory with the capacitive interaction model[25] that successfully describes the Coulomb-dominated regime. The interferometer Hamiltonian $H = \sum_i H_{{\rm edge},i} + H_{\rm tun}$ is modified by the Coulomb interaction as

$$\begin{aligned} H \rightarrow & H + U_{\rm bulk}Q_{\rm bulk}^2 \\ & + U_{\rm int}Q_{\rm bulk}\int_0^L dx(:\partial_x\phi_2(x+d): + :\partial_x\phi_3(x):) \\ & = \sum_{i=1,2,3}\frac{\hbar v_{\rm p}}{4\pi\nu}\int_{-\infty}^{\infty} dx : (\partial_x\bar{\phi}_i(x))^2 : + H_{\rm bulk} + H_{\rm tun}. \end{aligned} \qquad (13)$$

Here, $Q_{\rm bulk} = \nu BA_{\rm area}/\Phi_0 + \nu N_{\rm L} - \bar{q}$ is the number of the net charges localized within the interferometer bulk (inside the interference loop), $A_{\rm area}$ is the area of the interferometer, $N_{\rm L}$ is the net number of quasiparticles minus quasiholes and $\bar{q}$ is the number of positive background charges induced by the gate voltage applied to the interferometer. $U_{\rm int}$ is the strength of Coulomb interaction between the charges of the interferometer edge and the charges localized in the interferometer bulk, and $U_{\rm bulk}$ is the strength of interaction between the bulk charges. In the second equality of equation (13), we introduce a boson field $\bar{\phi}_i$ for each Edge i, $\bar{\phi}_i(x) = \phi_i(x) + \frac{2\pi\nu}{\hbar v_{\rm p}}U_{\rm int}Q_{\rm bulk}\int_{-\infty}^{x} K_i(x')dx'$, where $K_2(x) = 1$ for $d < x < d+L$, $K_3(x) = 1$ for $0 < x < L$ and $K_i(x) = 0$ otherwise. The second term of $\bar{\phi}$ describes the charges $-\frac{2\pi\nu}{\hbar v_{\rm p}}U_{\rm int}Q_{\rm bulk}$ induced per unit length by the interaction. In equation (13), the Hamiltonian is quadratic in $\bar{\phi}_i$ and has $H_{\rm bulk} = (U_{\rm bulk} - \frac{2\pi\nu L}{\hbar v_{\rm p}}U_{\rm int}^2)Q_{\rm bulk}^2$. It is noteworthy that $U_{\rm bulk} - \frac{2\pi\nu L}{\hbar v_{\rm p}}U_{\rm int}^2 > 0$. The main interference signal $I_{D_3}^{\rm int}$ and the reference signal $I_{{\rm Ref},\,D_3}^{\rm int}$ are computed by taking ensemble average over the thermal fluctuations of $N_{\rm L}$ (see Supplementary Note 5).

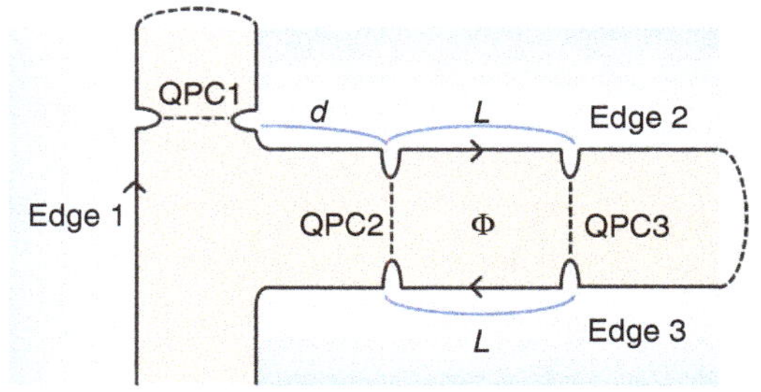

Figure 4 | Extended edge channel scheme. It is obtained by connecting the edge channels of the setup in Fig. 2a. The connection is represented by dashed arcs, whereas anyon propagation direction and anyon tunnelling at QPCs are represented by arrows and dashed lines, respectively.

References

1. Fetter, A. L. & Walecka, J. D. *Quantum Theory Of Many-Particle Systems* (McGraw-Hill, 1971).
2. Leinaas, J. M. & Myrheim, J. On the theory of identical particles. *Il Nuovo Cimento B Ser* **37**, 1–23 (1977).
3. Arovas, D., Schrieffer, J. R. & Wilczek, F. Fractional statistics and the quantum Hall effect. *Phys. Rev. Lett.* **53**, 722–723 (1984).
4. Stern, A. Anyons and the quantum Hall effect—a pedagogical review. *Ann. Phys.* **1**, 204–249 (2008).
5. Goldman, V. J. & Su, B. Resonant tunneling in the quantum Hall regime: measurement of fractional charge. *Science* **267**, 1010–1012 (1995).
6. De-Picciotto, R. *et al.* Direct observation of a fractional charge. *Nature* **389**, 162–164 (1997).
7. Saminadayar, L., Glattli, D. C., Jin, Y. & Etienne, B. Observation of the e/3 fractionally charged Laughlin quasiparticle. *Phys. Rev. Lett.* **79**, 2526–2529 (1997).
8. Dolev, M., Heiblum, M., Umansky, V., Stern, A. & Mahalu, D. Observation of a quarter of an electron charge at the $\nu = 5/2$ quantum Hall state. *Nature* **452**, 829–834 (2008).
9. Chamon, C. D. C., Freed, D. E., Kivelson, S. A., Sondhi, S. L. & Wen, X. G. Two point-contact interferometer for quantum Hall systems. *Phys. Rev. B* **55**, 2331–2343 (1997).
10. Safi, I., Devillard, P. & Martin, T. Partition noise and statistics in the fractional quantum Hall effect. *Phys. Rev. Lett.* **86**, 4628–4631 (2001).
11. Vishveshwara, S. Revisiting the Hanbury Brown—Twiss setup for fractional statistics. *Phys. Rev. Lett.* **91**, 196803 (2003).
12. Kim, E.-A., Lawler, M., Vishveshwara, S. & Fradkin, E. Measuring fractional charge and statistics in fractional quantum Hall fluids through noise experiments. *Phys. Rev. B* **74**, 155324 (2006).
13. Law, K. T., Feldman, D. E. & Gefen, Y. Electronic Mach-Zehnder interferometer as a tool to probe fractional statistic. *Phys. Rev. B* **74**, 045319 (2006).
14. Feldman, D. E., Gefen, Y., Kitaev, A., Law, K. T. & Stern, A. Shot noise in an anyonic Mach-Zehnder interferometer. *Phys. Rev. B* **76**, 085333 (2007).

15. Campagnano, G. *et al.* Hanbury Brown—Twiss interference of anyons. *Phys. Rev. Lett.* **109,** 106802 (2012).
16. Kane, C. L. Telegraph noise and fractional statistics in the quantum Hall effect. *Phys. Rev. Lett.* **90,** 226802 (2003).
17. An, S. *et al.* Braiding of abelian and non-abelian anyons in the fractional quantum Hall effect. Preprint at http://arXiv.org/abs/1112.3400 (2011).
18. Rosenow, B. & Simon, S. H. Telegraph noise and the Fabry-Perot quantum Hall interferometer. *Phys. Rev. B* **85,** 201302 (2012).
19. Kivelson, S. Semiclassical theory of localized many-anyon states. *Phys. Rev. Lett.* **65,** 3369–3372 (1990).
20. Camino, F. E., Zhou, W. & Goldman, V. J. e/3 Laughlin quasiparticle primary-filling = 1/3 interferometer. *Phys. Rev. Lett.* **98,** 076805 (2007).
21. Willett, R. L., Pfeiffer, L. N. & West, K. W. Measurement of filling factor 5/2 quasiparticle interference with observation of charge e/4 and e/2 period oscillations. *Proc. Natl Acad. Sci. USA* **106,** 8853–8858 (2009).
22. Ofek, N. *et al.* Role of interactions in an electronic Fabry Perot interferometer operating in the quantum Hall effect regime. *Proc. Natl Acad. Sci. USA* **107,** 5276–5281 (2010).
23. McClure, D. T., Chang, W., Marcus, C. M., Pfeiffer, L. N. & West, K. W. Fabry-Perot interferometry with fractional charges. *Phys. Rev. Lett.* **108,** 256804 (2012).
24. Kane, C. L. & Fisher, M. P. A. Transmission through barriers and resonant tunneling in an interacting one-dimensional electron gas. *Phys. Rev. B* **46,** 15233–15262 (1992).
25. Halperin, B., Stern, A., Neder, I. & Rosenow, B. Theory of the Fabry-Perot quantum Hall interferometer. *Phys. Rev. B* **83,** 155440 (2011).
26. von Delft, J. & Schoeller, H. Bosonization for beginners---refermionization for experts. *Ann. Phys. (Leipzig)* **7,** 225–306 (1998).
27. Wen, X. G. Chiral Luttinger liquid and the edge excitations in the fractional quantum Hall states. *Phys. Rev. B* **41,** 12838–12844 (1990).
28. Ji, Y. *et al.* An electronic Mach Zehnder interferometer. *Nature* **422,** 415–418 (2003).
29. Griffiths, T. G. *et al.* Evolution of quasiparticle charge in the fractional quantum Hall regime. *Phys. Rev. Lett.* **85,** 3918–3921 (2000).
30. Comforti, E. *et al.* Bunching of fractionally charged quasiparticles tunnelling through high-potential barriers. *Nature* **416,** 515–518 (2002).
31. Nayak, C., Simon, S. H., Stern, A., Freedman, M. & Sarma, S. D. Non-Abelian anyons and topological quantum computation. *Rev. Mod. Phys.* **80,** 1083–1159 (2008).
32. Guyon, R., Devillard, P., Martin, T. & Safi, I. Klein factors in multiple fractional quantum Hall edge tunneling. *Phys. Rev. B* **65,** 153304 (2002).

Acknowledgements

H.-S.S. thanks Eddy Ardonne, Hyungkook Choi, Yunchul Chung, Dmitri Feldman, Woowon Kang, Kirill Shtengel and Joost Slingerland for useful discussion. We thank the support by Korea NRF (grant numbers NRF-2010-00491 and NRF-2013R1A2A2A01007327 to H.-S.S.), by KAIST-HRHRP (C.H.) and by DFG (grant number RO 2247/8-1 to Y.G.).

Author contributions

C.H. has conceived the concept of topological vacuum bubbles, performed the detailed calculation, analysed the Coulomb-dominated regime and wrote the paper. J.P. has performed the analysis and wrote the paper. Y.G. has conceived the interferometry setup and wrote the paper. H.-S.S. has conceived the concept of topological vacuum bubbles and the interferometry setup, analysed the Coulomb-dominated regime, wrote the paper and supervised the project.

Additional information

Competing financial interests: The authors declare no competing financial interests.

26

Experimental perfect state transfer of an entangled photonic qubit

Robert J. Chapman[1,2], Matteo Santandrea[3,4], Zixin Huang[1,2], Giacomo Corrielli[3,4], Andrea Crespi[3,4], Man-Hong Yung[5], Roberto Osellame[3,4] & Alberto Peruzzo[1,2]

The transfer of data is a fundamental task in information systems. Microprocessors contain dedicated data buses that transmit bits across different locations and implement sophisticated routing protocols. Transferring quantum information with high fidelity is a challenging task, due to the intrinsic fragility of quantum states. Here we report on the implementation of the perfect state transfer protocol applied to a photonic qubit entangled with another qubit at a different location. On a single device we perform three routing procedures on entangled states, preserving the encoded quantum state with an average fidelity of 97.1%, measuring in the coincidence basis. Our protocol extends the regular perfect state transfer by maintaining quantum information encoded in the polarization state of the photonic qubit. Our results demonstrate the key principle of perfect state transfer, opening a route towards data transfer for quantum computing systems.

[1]Quantum Photonics Laboratory, School of Engineering, RMIT University, Melbourne, Victoria 3000, Australia. [2]School of Physics, The University of Sydney, Sydney, New South Wales 2006, Australia. [3]Istituto di Fotonica e Nanotecnologie, Consiglio Nazionale delle Ricerche, Piazza Leonardo da Vinci 32, Milano I-20133, Italy. [4]Dipartimento di Fisica, Politecnico di Milano, Piazza Leonardo da Vinci 32, Milano I-20133, Italy. [5]Department of Physics, South University of Science and Technology of China, Shenzhen 518055, China. Correspondence and requests for materials should be addressed to A.P. (email: alberto.peruzzo@rmit.edu.au).

Transferring quantum information between locations without disrupting the encoded information *en route* is crucial for future quantum technologies[1–8]. Routing quantum information is necessary for communication between quantum processors, addressing single qubits in topological surface architectures, and for quantum memories as well as many other applications.

Coupling between stationary qubits and mobile qubits via cavity and circuit quantum electrodynamics has been an active area of research with promise for long-distance quantum communication[9–12]; however, coupling between different quantum information platforms is challenging as unwanted degrees of freedom lead to increased decoherence[13]. Quantum teleportation between distant qubits allows long-distance quantum communication via shared entangled states[14–17]; however, in most quantum information platforms this would again require coupling between stationary and mobile qubits. Physically relocating trapped ion qubits has also been demonstrated[18,19], however, with additional decoherence incurred during transport.

By taking advantage of coupling between neighbouring qubits, it is possible to transport quantum information across a stationary lattice[2]. This has the benefits that one physical platform is being used and the lattice sites remain at fixed locations. The most basic method is to apply a series of SWAP operations between neighbouring sites such that, with enough iterations, the state of the first qubit is relocated to the last. This method requires a high level of active control on the coupling and is inherently weak as individual errors accumulate after each operation, leading to an exponential decay in fidelity as the number of operations increases[20].

The perfect state transfer (PST) protocol utilizes an engineered but fixed coupled lattice. Quantum states are transferred between sites through Hamiltonian evolution for a specified time[2–7]. For a one-dimensional system with N sites, the state intially at site n is transferred to site $N-n+1$ with 100% probability without need for active control on the coupling[21]. PST can be performed on any quantum computing architecture where coupling between sites can be engineered, such as ion traps[18] and quantum dots[22]. Figure 1 presents an illustration of the PST protocol. The encoded quantum state, initially at the first site, is recovered at the final site after a specific time. In the intermediate stages, the qubit is in a superposition across the lattice. Aside from qubit relocation, the PST framework can be applied to entangled W-state preparation[23], state amplification[24] and even quantum computation[25–29].

To date, most research on PST has been theoretical[2–7,20,21,23,24,28,30–42], with experiments[43,44] being limited to demonstrations where no quantum information is transferred, and do not incorporate entanglement, often considered the defining feature of quantum mechanics[45]. Here, we present the implementation of a protocol that extends PST for relocating a polarization-encoded photonic qubit across a one-dimensional lattice, realized as an array of 11 evanescently coupled waveguides[46–48]. We show that the entanglement between a photon propagating through the PST waveguide array and another photon at a different location is preserved.

Results

PST Hamiltonian. The Hamiltonian for our system in the nearest-neighbour approximation is given by the tight-binding formalism

$$\hat{H} = \sum_{\sigma\in\{H,V\}} \sum_{n=1}^{N-1} C_{n,n+1}\left(\hat{a}^{\dagger}_{n+1,\sigma}\hat{a}_{n,\sigma} + \hat{a}^{\dagger}_{n,\sigma}\hat{a}_{n+1,\sigma}\right), \quad (1)$$

where $C_{n,n+1}$ is the coupling coefficient between waveguides n and $n+1$, and $\hat{a}_{n,\sigma}$ $(\hat{a}^{\dagger}_{n,\sigma})$ is the annihilation (creation) operator applied to waveguide n and polarization σ (horizontal or vertical). Hamiltonian evolution of a state $|\psi_0\rangle$ for a time t is calculated via the Schrödinger equation, giving the final state $|\psi(t)\rangle = \exp(\frac{-i\hat{H}t}{\hbar})|\psi(0)\rangle$ (ref. 49). Equation (1) is constructed of independent tight-binding Hamiltonians acting on each orthogonal polarization. This requires there to be no cross-talk terms $\hat{a}^{\dagger}_{n,H}\hat{a}_{m,V}$ or $\hat{a}^{\dagger}_{n,V}\hat{a}_{m,H}$ $\forall\ m,n$. The spectrum of coupling coefficients $C_{n,n+1}$ is crucial for successful PST. Evolution of this Hamiltonian with a uniform coupling coefficient spectrum, equivalent to equally spaced waveguides, is not sufficient for PST with over three lattice sites as simulated in Fig. 2a. PST requires the coupling coefficient spectrum to follow the function

$$C_{n,n+1} = C_0\sqrt{n(N-n)}, \quad (2)$$

where C_0 is a constant, N is the total number of lattice sites and evolution is for a specific time $t_{\mathrm{PST}}=\frac{\pi}{2C_0}$ (refs 3, 4). This enables arbitrary-length PST as simulated in Fig. 2b for 11 sites. The coupling coefficient spectrum for each polarization must be equal and follow equation (2) for the qubit to be faithfully relocated and the polarization-encoded quantum information to be preserved. The distance between waveguides dictates the coupling coefficient; however, for planar systems, the coupling coefficient of each polarization will in general be unequal due to the waveguide birefringence. To achieve equal coupling between polarizations, the waveguide array is fabricated along a tilted plane in the substrate[50]. This is made possible by the unique three-dimensional capabilities of the femtosecond laser-writing technique (see Supplementary Note 1 for further fabrication and device details). We measure a total propagation loss of 1.8 ± 0.2 dB; however, our figure of merit is how well preserved the polarization quantum state is after the transfer protocol. Therefore we calculate fidelity without loss. Ideally the PST protocol exhibits unit fidelity and efficiency, where the quantum state is reliably transferred and the encoded state is preserved. Due to loss in our experiment, we have less than unit efficiency; however, this loss is largely unrelated to the PST Hamiltonian in equation (1). Further optimizing the fabrication process could

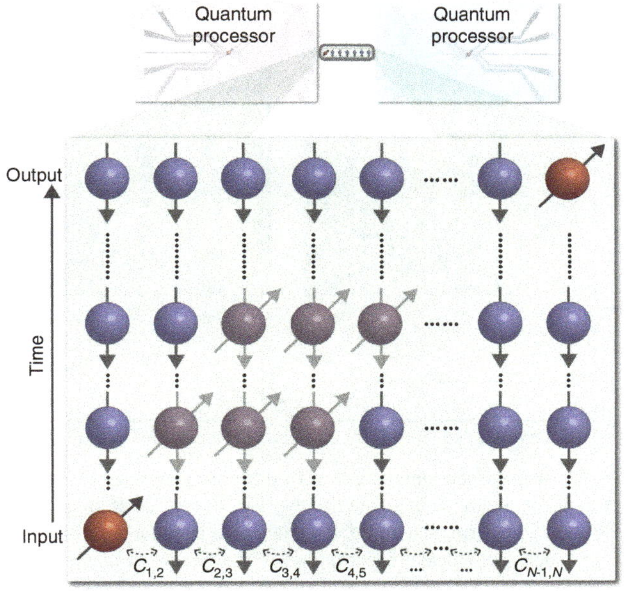

Figure 1 | Illustration of a one-dimensional perfect state transfer lattice connecting two quantum processors. By engineering the Hamiltonian of a lattice, the state at the first site is transferred to the last site after a specific time. This Hamiltonian defines the perfect state transfer protocol[3], which can be used for routing quantum information inside a quantum processor.

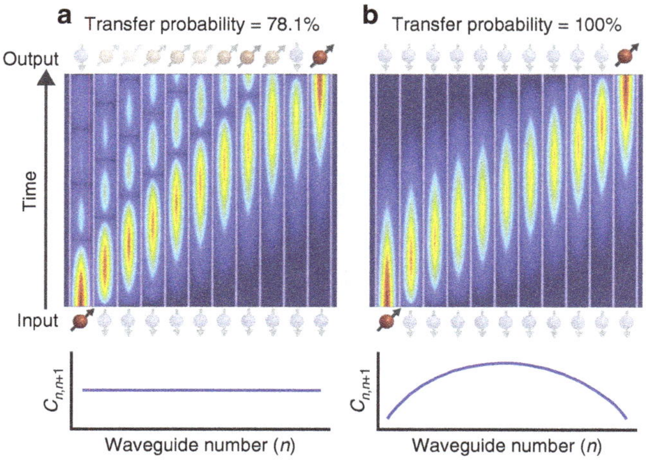

Figure 2 | Propagation simulations with different coupling coefficient spectra. (**a**) A photon is injected into the first waveguide of an array of eleven coupled waveguides with the Hamiltonian in equation (1) and a uniform coupling coefficient spectrum. With the constraint that reflections off boundaries are not allowed, we calculate a maximum probability of transferring the photon to waveguide 11 of 78.1% (ref. 2). (**b**) A photon is injected into the first waveguide of an array of eleven coupled waveguides, this time with the coupling coefficient spectrum of equation (2). After evolution for a pre-determined time, the photon is received at waveguide 11 with 100% probability[3-7].

reduce the level of propagation loss (see Methods and Supplementary Note 2 for further details on loss).

We inject photons into waveguides 1, 6 and 10 of the array, which after time t_{PST} transfer to waveguides 11, 6 and 2, respectively. Figure 3a–c presents propagation simulations for each transfer. Input waveguides extend to the end of the device to allow selective injection.

Transfer characterization. To characterize the coupling coefficient spectra, we inject horizontally and vertically polarized laser light at 808 nm into each input waveguide. Laser light is more robust to noise than single photons and we can monitor the output with a CCD camera to fast gather results. Using laser light at the same wavelength as our single photons will give an output intensity distribution equivalent to the output probability distribution for detecting single photons[46]. Ideally light injected into waveguide n will output the device only in waveguide $N-n+1$; however, this assumes an approximate model of nearest-neighbour coupling only. Taking into account coupling between further separated waveguides reduces the transfer probability. This decrease is greater for light injected closer to the centre of the array (see Supplementary Note 1). Figure 3d–f presents our measured output probability distribution for horizontally (P_n^H) and vertically (P_n^V) polarized laser light injected into each input waveguide, where n is the output waveguide number. Fidelity between the probability distributions for each polarization is given by $F_{\mathrm{distribution}} = \sum_n \sqrt{P_n^H P_n^V}$. This fidelity is closely related to how similar the two coupling coefficient spectra are. We measure an average probability distribution fidelity for all transfers of 0.976 ± 0.006 (see Supplementary Table 1 for all fidelity values). We encode quantum information in the polarization state of the photon and are interested in reliably relocating this qubit. We use a single optical fibre to capture photons from the designed output waveguide, which, in all cases, is the waveguide with the greatest output probability.

Quantum process tomography. We perform quantum process tomography to understand the operation performed on the single-photon polarization state during each PST transfer. We inject single-photon states $|\psi_{\mathrm{in}}\rangle = (\alpha\hat{a}^\dagger_{S,H} + \beta\hat{a}^\dagger_{S,V})|0\rangle$ into each input waveguide $S \in \{1,6,10\}$, where α (β) is the probability amplitude of the horizontal (vertical) component of the photon and $|\alpha|^2 + |\beta|^2 = 1$. From quantum process tomography on the output polarization states, we can generate a process matrix χ_{pol} for each transfer[1,51]. We aim to perform the identity operation so that the quantum information encoded in the polarization can be recovered after relocation. We measure a polarization phase shift associated with each transfer. This phase shift can be compensated for with a local polarization rotation applied before injection. Figure 3g–i presents our measured process matrix for each transfer. Across all transfers we demonstrate an average fidelity of the polarization process including compensation to an identity of 0.982 ± 0.003 (see Supplementary Note 3 for details of the compensation scheme and Supplementary Table 2 for all fidelities). Process fidelity is calculated as $F_{\mathrm{process}} = \mathrm{Tr}\{\chi_1 \chi_{\mathrm{pol+comp}}\}$ (ref. 52), where χ_1 is the process matrix for the identity operation and $\chi_{\mathrm{pol+comp}}$ is the combined polarization operation and compensation process matrix.

Ideally the output state for each transfer is $|\psi_{\mathrm{out}}\rangle = (\alpha\hat{a}^\dagger_{T,H} + \beta\hat{a}^\dagger_{T,V})|0\rangle$, where $T \in \{11,6,2\}$ and the probability amplitude of each polarization component remains equal to the input state. Our high-fidelity measurements on single-photon relocation demonstrate that we can route a polarization-encoded photonic qubit across our device and faithfully recover the encoded quantum information.

Entangled state transfer. Entanglement is likely to be a defining feature of quantum computing, and preserving entanglement is therefore critical to the success of any qubit relocation protocol. We prepare the Bell state $\frac{1}{\sqrt{2}}(|H_1V_2\rangle + |V_1H_2\rangle)$ using the spontaneous parametric downconversion process. The polarization is controlled using rotatable half and quarter waveplates (HWPs and QWPs), and polarizing beam splitters (PBSs) as shown in Fig. 4 (ref. 53) (see Methods for details). This set-up prepares a general state $\alpha|H_1V_2\rangle + \beta|V_1H_2\rangle$ when measuring in coincidence, where $|\alpha|^2 + |\beta|^2 = 1$. Photon 1 is injected into the waveguide array, while photon 2 propagates through polarization-maintaining fibre (PMF). In terms of waveguide occupancy, our input state is $|\psi_{\mathrm{in}}\rangle = \frac{1}{\sqrt{2}}(\hat{a}^\dagger_{S,H}\hat{a}^\dagger_{0,V} + \hat{a}^\dagger_{S,V}\hat{a}^\dagger_{0,H})|00\rangle$ for each input waveguide $S \in \{1,6,10\}$, where $\hat{a}^\dagger_{0,\sigma}$ denotes the creation operator acting on polarization σ in PMF. Full two-qubit polarization tomography[54] is performed on the output and the fidelity calculated as

$$F_{\mathrm{quantum}} = \left(\mathrm{Tr}\left\{\sqrt{\sqrt{\rho_{\mathrm{input}}}\rho_{\mathrm{output}}\sqrt{\rho_{\mathrm{input}}}}\right\}\right)^2, \quad (3)$$

where ρ_{output} is the density matrix after the PST protocol has been applied and ρ_{input} is the density matrix after propagation through a reference straight waveguide[55]. After all qubit relocations we measure an average polarization state fidelity of 0.971 ± 0.014. Fidelity is measured in the two-photon coincidence basis. This value is therefore the fidelity on the quantum state transferred without taking into account the loss (see Methods and Supplementary Note 2 for loss analysis). We can use the results from quantum process tomography to generate a characterized model of our device. We can now use this model to calculate the similarity between the predicted output state and our measured output state as

$$S_{\mathrm{quantum}} = \left(\mathrm{Tr}\left\{\sqrt{\sqrt{\rho_{\mathrm{predicted}}}\rho_{\mathrm{output}}\sqrt{\rho_{\mathrm{predicted}}}}\right\}\right)^2. \quad (4)$$

We calculate an average similarity of 0.987 ± 0.014 across all transfers (see Supplementary Table 3 for all fidelities

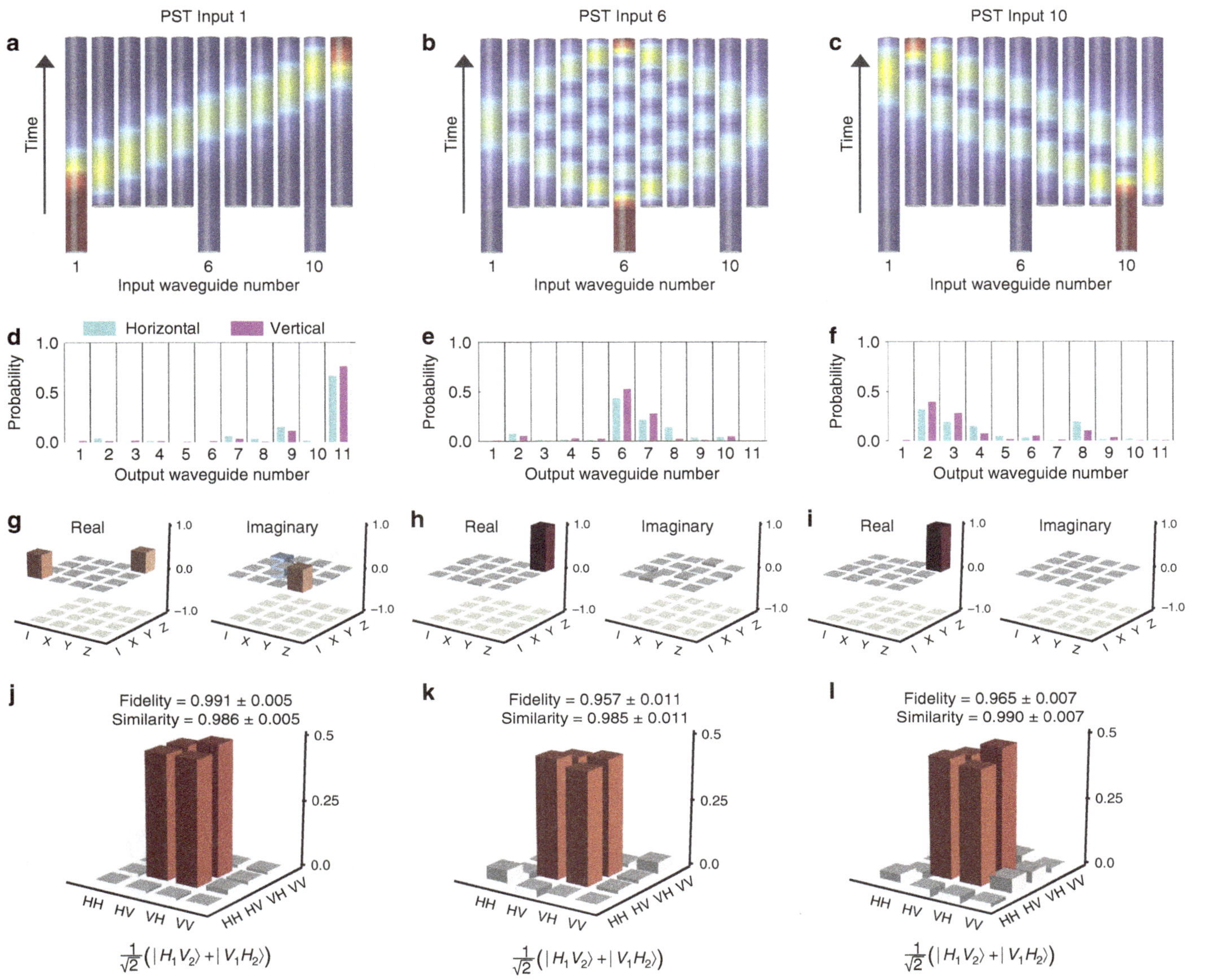

Figure 3 | Experimental data from the characterization and performance of perfect state transfer waveguide array. (**a–c**) Propagation simulations showing the device implementation to enable specific waveguide input. (**d–f**) Output probability distributions for each input of the PST array for horizontally and vertically polarized laser light. (**g–i**) Quantum process matrix for each transfer in the PST array measured with single-photon quantum process tomography. (**j–l**) Two-photon quantum state tomography is performed after photon 1 of the polarization entangled Bell state $\frac{1}{\sqrt{2}}(|H_1V_2\rangle + |V_1H_2\rangle)$ has been relocated. Results have had the small imaginary components removed for brevity.

and similarities). Figure 3j–l presents our measured density matrix after each entangled state transfer.

Ideally, the output state for each transfer is $|\Psi_{\text{out}}\rangle = \frac{1}{\sqrt{2}}(\hat{a}^\dagger_{T,H}\hat{a}^\dagger_{0,V} + \hat{a}^\dagger_{T,V}\hat{a}^\dagger_{0,H})|00\rangle$, where $T\in\{11,6,2\}$. With high fidelity the probability amplitude of each component is preserved and the state remains almost pure. This result demonstrates that with our device we can relocate a polarization qubit between distant sites and preserve entanglement with another qubit at a different location. In principle our device could route qubits from any waveguide n to waveguide $N-n+1$. Quantum error correction protocols require sophisticated interconnection to access individual qubits for control and measurement within large, highly entangled surface code geometries[56]. PST is a clear gateway towards accessing qubits in such systems without disrupting quantum states and entanglement throughout the surface code.

Decohered state transfer. Decoherence has applications in quantum simulation to emulate systems in nature[57], and it is therefore important to note that this approach for relocating quantum information can be applied to states of any purity[3]. We prepare decohered states by introducing a time delay between the horizontal and vertical components of the polarization qubit. We implement this delay by extending one arm of the source, which reduces the overlap of the photons after they are both incident on the PBS, as shown in Fig. 4. This delay extends the state into a time-bin basis, which we trace over on measurement, leading to a mixed state. The purity of the state can be calculated as the convolution of the horizontal and vertical components with a time delay τ:

$$\text{Purity}(\tau) \equiv \int_{-\infty}^{\infty} H(t)V(\tau - t)\mathrm{d}t, \qquad (5)$$

where τ is controlled by altering the path length of the vertical

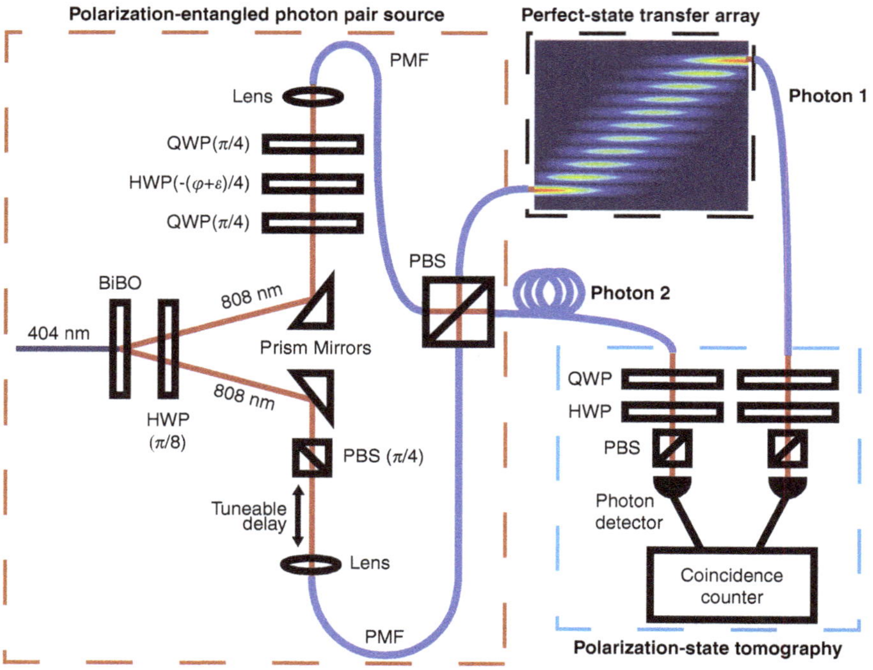

Figure 4 | Experimental set-up. Polarization entangled photons are generated in free space before coupling into PMF. Photon 1 is injected into the perfect state transfer array, while photon 2 travels through PMF. Full two-qubit polarization tomography is performed on the output. See Methods for experimental set-up details.

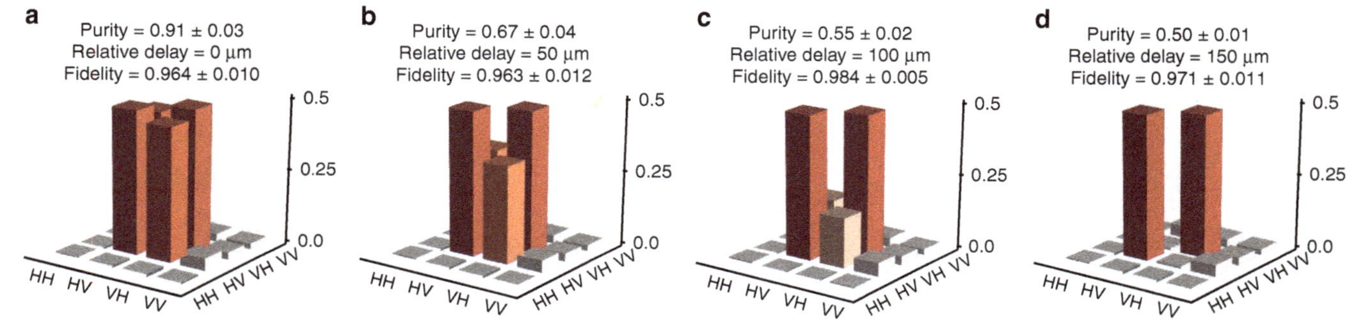

Figure 5 | Perfect state transfer of entangled states with varying purity. Photon 1 of the state $\frac{1}{\sqrt{2}}(|H_1V_2\rangle + |V_1H_2\rangle)$ is injected into waveguide 1 of the PST array. A delay is applied to the vertical component to control the purity of the state. (**a**) Relative delay of 0 µm, (**b**) 50 µm, (**c**) 100 µm and (**d**) 150 µm. Results have had the small imaginary components removed for brevity.

component of the state. $H(V)$ is the horizontal (vertical) component of the photon. Figure 5 presents density matrices for PST from waveguide 1 to waveguide 11 applied to entangled states of varying purity. The injected states are recovered with an average fidelity of 0.971 ± 0.019 and an average similarity of 0.978 ± 0.019 (see Supplementary Table 4 for all values).

Discussion

We have proposed and experimentally demonstrated a protocol for relocating a photonic qubit across eleven discrete sites, maintaining the quantum state with high fidelity and preserving entanglement with another qubit at a different location. We can aim to improve our fidelity by reducing next-nearest-neighbour coupling by further separating the waveguides and having a longer device. This would increase the contrast between nearest- and next-nearest-neighbour coupling to better fit the Hamiltonian in equation (1). A by-product of longer devices, however, is an increase in propagation loss. Depth-dependent spherical aberrations in the laser irradiation process may also affect the homogeneity of the three-dimensional waveguide array. Additional optics in the laser writing set-up could be employed to reduce this effect. Protocols for relocating quantum information across discrete sites are essential for future quantum technologies. Our protocol builds on the PST with extension to include an additional degree of freedom for encoding quantum information. This demonstration opens pathways towards faithful quantum state relocation in quantum computing systems.

Methods

Experimental set-up. Horizontally polarized photon pairs at 807.5 nm are generated via type 1 spontaneous parametric downconversion in a 1-mm-thick BiBO crystal, pumped by an 80-mW, 403.75-nm CW diode laser. Both photons are rotated into a diagonal state $\frac{1}{\sqrt{2}}(|H\rangle + |V\rangle)$ by a half waveplate (HWP) with fast axis at 22.5° from vertical. One photon has a phase applied by two 45° quarter waveplates (QWP) on either side of a HWP at θ°. The second photon has its diagonal state optimized with a PBS at $\sim$45°.

Each photon is collected in PMF and are incident on the two input faces of a fibre pigtailed PBS. When measuring in the coincidence basis, this post-selects the entangled state $\frac{1}{\sqrt{2}}(|H_1V_2\rangle + e^{i\phi}|V_1H_2\rangle)$, where $\phi = 4(\theta + \epsilon)$ and ϵ is the intrinsic phase applied by the whole system. The experimental set-up is illustrated in Fig. 4.

PMF is highly birefringent, resulting in full decoherence of the polarization state after $\sim$1 m of fibre giving a mixed state. To maintain polarization superposition over several metres of fibre, we use 90° connections to ensure that both polarizations propagate through equal proportions of fast- and slow-axis fibre. Slight length differences between fibres and temperature variations mean the whole system applies a residual phase ϵ to the state, which can be compensated for in the source using the phase-controlling HWP.

Polarization state tomography combines statistics from projection measurements to generate the density matrix of a state. Single-photon rotations are applied by a QWP and HWP before a PBS. Single-qubit tomography requires four measurements and two-qubit tomography requires 16. Accidental counts are removed by taking each reading with and without an electronic delay. This helps reduce noise in our measurements.

Photon count rate. In our experiment, we prepare polarization Bell states with a count rate of $\sim 2 \times 10^3$ s^{-1}. After the PST array we measure a count rate of $\sim 10^2$ s^{-1}. The propagation loss of the array is only 1.8 dB. Most of the total loss ($\sim$ 13 dB) is indeed due to mode mismatch between the waveguides and fibres, imperfect coupling, reflections at interfaces, and non-unit relocation efficiency. We integrate our measurements for 30 s to reduce the statistical noise due to the Poisson distribution of the photon count rate.

References

1. Nielsen, M. A. & Chuang, I. L. *Quantum Computation and Quantum Information* 10th anniversary edition (Cambridge University Press, 2000).
2. Bose, S. Quantum communication through an unmodulated spin chain. *Phys. Rev. Lett.* **91,** 207901 (2003).
3. Christandl, M., Datta, N., Ekert, A. & Landahl, A. J. Perfect state transfer in quantum spin networks. *Phys. Rev. Lett.* **92,** 187902 (2004).
4. Plenio, M. B., Hartley, J. & Eisert, J. Dynamics and manipulation of entanglement in coupled harmonic systems with many degrees of freedom. *New J. Phys.* **6,** 36 (2004).
5. Gordon, R. Harmonic oscillation in a spatially finite array waveguide. *Opt. Lett.* **29,** 2752 (2004).
6. Nikolopoulos, G. M., Petrosyan, D. & Lambropoulos, P. Electron wavepacket propagation in a chain of coupled quantum dots. *J. Phys. Condens. Matter* **16,** 4991 (2004).
7. Nikolopoulos, G. M., Petrosyan, D. & Lambropoulos, P. Coherent electron wavepacket propagation and entanglement in array of coupled quantum dots. *Europhys. Lett.* **65,** 297–303 (2004).
8. DiVincenzo, D. P. The physical implementation of quantum computation. *Fortschr. Phys.* **48,** 771–783 (2000).
9. Wallraff, A. *et al.* Strong coupling of a single photon to a superconducting qubit using circuit quantum electrodynamics. *Nature* **431,** 162–167 (2004).
10. Majer, J. *et al.* Coupling superconducting qubits via a cavity bus. *Nature* **449,** 443–447 (2007).
11. Herskind, P. F., Dantan, A., Marler, J. P., Albert, M. & Drewsen, M. Realization of collective strong coupling with ion Coulomb crystals in an optical cavity. *Nat. Phys.* **5,** 494–498 (2009).
12. Paik, H. *et al.* Observation of high coherence in Josephson junction qubits measured in a three-dimensional circuit QED architecture. *Phys. Rev. Lett.* **107,** 240501 (2011).
13. Schoelkopf, R. J. & Girvin, S. M. Wiring up quantum systems. *Nature* **451,** 664–669 (2008).
14. Bennett, C. H. *et al.* Teleporting an unknown quantum state via dual classical and Einstein-Podolsky-Rosen channels. *Phys. Rev. Lett.* **70,** 1895–1899 (1993).
15. Bouwmeester, D. *et al.* Experimental quantum teleportation. *Nature* **390,** 575–579 (1997).
16. Furusawa, A. *et al.* Unconditional quantum teleportation. *Science* **282,** 706–709 (1998).
17. Braunstein, S. L. & Kimble, H. J. Teleportation of continuous quantum variables. *Phys. Rev. Lett.* **80,** 869–872 (1998).
18. Kielpinski, D., Monroe, C. & Wineland, D. J. Architecture for a large-scale ion-trap quantum computer. *Nature* **417,** 709–711 (2002).
19. Seidelin, S. *et al.* Microfabricated surface-electrode ion trap for scalable quantum information processing. *Phys. Rev. Lett.* **96,** 253003 (2006).
20. Yung, M.-H. Quantum speed limit for perfect state transfer in one dimension. *Phys. Rev. A* **74,** 030303 (2006).
21. Christandl, M. *et al.* Perfect transfer of arbitrary states in quantum spin networks. *Phys. Rev. A* **71,** 032312 (2005).
22. Loss, D. & DiVincenzo, D. P. Quantum computation with quantum dots. *Phys. Rev. A* **57,** 120–126 (1998).
23. Kay, A. Perfect efficient, state transfer and its application as a constructive tool. *Int. J. Quantum Inf.* **08,** 641–676 (2010).
24. Kay, A. Unifying quantum state transfer and state amplification. *Phys. Rev. Lett.* **98,** 010501 (2007).
25. Raussendorf, R. & Briegel, H. J. A one-way quantum computer. *Phys. Rev. Lett.* **86,** 5188–5191 (2001).
26. Zhou, X., Zhou, Z.-W., Guo, G.-C. & Feldman, M. J. Quantum computation with untunable couplings. *Phys. Rev. Lett.* **89,** 197903 (2002).
27. Benjamin, S. C. & Bose, S. Quantum computing in arrays coupled by "always-on" interactions. *Phys. Rev. A* **70,** 032314 (2004).
28. Kay, A. Computational power of symmetric Hamiltonians. *Phys. Rev. A* **78,** 012346 (2008).
29. Mkrtchian, G. F. Universal quantum logic gates in a scalable Ising spin quantum computer. *Phys. Lett. A* **372,** 5270–5273 (2008).
30. Cook, R. J. & Shore, B. W. Coherent dynamics of N-level atoms and molecules. III. An analytically soluble periodic case. *Phys. Rev. A* **20,** 539–544 (1979).
31. Burgarth, D. & Bose, S. Conclusive and arbitrarily perfect quantum-state transfer using parallel spin-chain channels. *Phys. Rev. A* **71,** 052315 (2005).
32. Burgarth, D., Giovannetti, V. & Bose, S. Efficient and perfect state transfer in quantum chains. *J. Phys. A Math. Gen.* **38,** 6793 (2005).
33. Yung, M.-H. & Bose, S. Perfect state transfer, effective gates, and entanglement generation in engineered bosonic and fermionic networks. *Phys. Rev. A* **71,** 032310 (2005).
34. Plenio, M. B. & Semio, F. L. High efficiency transfer of quantum information and multiparticle entanglement generation in translation-invariant quantum chains. *New J. Phys.* **7,** 73 (2005).
35. Zhang, J. *et al.* Simulation of Heisenberg XY interactions and realization of a perfect state transfer in spin chains using liquid nuclear magnetic resonance. *Phys. Rev. A* **72,** 012331 (2005).
36. Kay, A. Perfect state transfer: beyond nearest-neighbor couplings. *Phys. Rev. A* **73,** 032306 (2006).
37. Bose, S. Quantum communication through spin chain dynamics: an introductory overview. *Contemp. Phys.* **48,** 13–30 (2007).
38. Kostak, V., Nikolopoulos, G. M. & Jex, I. Perfect state transfer in networks of arbitrary topology and coupling configuration. *Phys. Rev. A* **75,** 042319 (2007).
39. Di Franco, C., Paternostro, M. & Kim, M. S. Perfect state transfer on a spin chain without state initialization. *Phys. Rev. Lett.* **101,** 230502 (2008).
40. Gualdi, G., Kostak, V., Marzoli, I. & Tombesi, P. Perfect state transfer in long-range interacting spin chains. *Phys. Rev. A* **78,** 022325 (2008).
41. Paz-Silva, G. A., Rebic, S., Twamley, J. & Duty, T. Perfect mirror transport protocol with higher dimensional quantum chains. *Phys. Rev. Lett.* **102,** 020503 (2009).
42. Perez-Leija, A., Keil, R., Moya-Cessa, H., Szameit, A. & Christodoulides, D. N. Perfect transfer of path-entangled photons in Jx photonic lattices. *Phys. Rev. A* **87,** 022303 (2013).
43. Bellec, M., Nikolopoulos, G. M. & Tzortzakis, S. Faithful communication Hamiltonian in photonic lattices. *Opt. Lett.* **37,** 4504 (2012).
44. Perez-Leija, A. *et al.* Coherent quantum transport in photonic lattices. *Phys. Rev. A* **87,** 012309 (2013).
45. Einstein, A., Podolsky, B. & Rosen, N. Can quantum-mechanical description of physical reality be considered complete? *Phys. Rev.* **47,** 777–780 (1935).
46. Perets, H. B. *et al.* Realization of quantum walks with negligible decoherence in waveguide lattices. *Phys. Rev. Lett.* **100,** 170506 (2008).
47. Rai, A., Agarwal, G. S. & Perk, J. H. H. Transport and quantum walk of nonclassical light in coupled waveguides. *Phys. Rev. A* **78,** 042304 (2008).
48. Peruzzo, A. *et al.* Quantum walks of correlated photons. *Science* **329,** 1500–1503 (2010).
49. Bromberg, Y., Lahini, Y., Morandotti, R. & Silberberg, Y. Quantum and classical correlations in waveguide lattices. *Phys. Rev. Lett.* **102,** 253904 (2009).
50. Sansoni, L. *et al.* Two-particle bosonic-fermionic quantum walk via integrated photonics. *Phys. Rev. Lett.* **108,** 010502 (2012).
51. O'Brien, J. L. *et al.* Quantum process tomography of a controlled-NOT gate. *Phys. Rev. Lett.* **93,** 080502 (2004).
52. Gilchrist, A., Langford, N. K. & Nielsen, M. A. Distance measures to compare real and ideal quantum processes. *Phys. Rev. A* **71,** 062310 (2005).
53. Matthews, J. C. F. *et al.* Observing fermionic statistics with photons in arbitrary processes. *Sci. Rep.* **3,** 1539 (2013).
54. James, D. F. V., Kwiat, P. G., Munro, W. J. & White, A. G. Measurement of qubits. *Phys. Rev. A* **64,** 052312 (2001).
55. Jozsa, R. Fidelity for mixed quantum states. *J. Mod. Opt.* **41,** 2315–2323 (1994).
56. Devitt, S. J., Munro, W. J. & Nemoto, K. Quantum error correction for beginners. *Rep. Prog. Phys.* **76,** 076001 (2013).
57. Lloyd, S. Universal quantum simulators. *Science* **273,** 1073–1078 (1996).

Acknowledgements

G.C., A.C. and R.O. acknowledge support from the European Union through the projects FP7-ICT-2011-9-600838 (QWAD-Quantum Waveguides Application and Development; www.qwad-project.eu) and H2020-FETPROACT-2014-641039 (QUCHIP-Quantum Simulation on a Photonic Chip; www.quchip.eu). M.-H.Y. acknowledges support by the National Natural Science Foundation of China under Grants No. 11405093. A.P. acknowledges an Australian Research Council Discovery Early Career Researcher Award under project number DE140101700 and an RMIT University Vice-Chancellor's Senior Research Fellowship.

Author contributions
All authors contributed to all aspects of this work.

Additional information

Competing financial interests: The authors declare no competing financial interests.

27

Implementation of quantum and classical discrete fractional Fourier transforms

Steffen Weimann[1,*], Armando Perez-Leija[1,*], Maxime Lebugle[1,*], Robert Keil[2], Malte Tichy[3], Markus Gräfe[1], René Heilmann[1], Stefan Nolte[1], Hector Moya-Cessa[4], Gregor Weihs[2], Demetrios N. Christodoulides[5] & Alexander Szameit[1]

Fourier transforms, integer and fractional, are ubiquitous mathematical tools in basic and applied science. Certainly, since the ordinary Fourier transform is merely a particular case of a continuous set of fractional Fourier domains, every property and application of the ordinary Fourier transform becomes a special case of the fractional Fourier transform. Despite the great practical importance of the discrete Fourier transform, implementation of fractional orders of the corresponding discrete operation has been elusive. Here we report classical and quantum optical realizations of the discrete fractional Fourier transform. In the context of classical optics, we implement discrete fractional Fourier transforms of exemplary wave functions and experimentally demonstrate the shift theorem. Moreover, we apply this approach in the quantum realm to Fourier transform separable and path-entangled biphoton wave functions. The proposed approach is versatile and could find applications in various fields where Fourier transforms are essential tools.

[1]Institute of Applied Physics, Abbe School of Photonics, Friedrich-Schiller-Universität Jena, Max-Wien Platz 1, 07743 Jena, Germany. [2]Institut für Experimentalphysik, Universität Innsbruck, Technikerstraße 25, 6020 Innsbruck, Austria. [3]Department of Physics and Astronomy, University of Aarhus, 8000 Aarhus, Denmark. [4]INAOE, Coordinacion de Optica, Luis Enrique Erro No. 1, Tonantzintla, Puebla 72840, Mexico. [5]CREOL, The College of Optics & Photonics, University of Central Florida, Orlando, Florida 32816, USA. * These authors contributed equally to this work. Correspondence and requests for materials should be addressed to A.S. (email: alexander.szameit@uni-jena.de).

Two hundred years ago, Joseph Fourier introduced a major concept in mathematics, the so-called Fourier transform (FT). It was not until 1965, when Cooley and Tukey developed the 'fast Fourier transform' algorithm, that Fourier analysis became a standard tool in contemporary sciences[1]. Two crucial requirements in this algorithm are the discretization and truncation of the domain, where the signals to be transformed are defined. These requirements are always satisfiable, since observable quantities in physics must be well behaved and finite in extension and magnitude.

In 1980, Namias made another significant leap with the introduction of the fractional Fourier transform (FrFT), which contains the FT as a special case[2]. Several investigations quickly followed, leading to a more general theory of joint time-frequency signal representations[3] and fractional Fourier optics[4]. The vast scope of the FrFT has been demonstrated in areas such as wave propagation, signal processing and differential equations[3,5–7]. So far, the FT of fractional order was realized only by single-lens systems[8,9], although other theoretical suggestions, including multi-lens systems[10] or graded index fibres exist[11]. The aim to discretize this generalized FT led to the introduction of the discrete fractional Fourier transform (DFrFT) operating on a finite grid in a way similar to that of a discrete FT[12]. Along those lines, several versions of the DFrFT have been introduced[5], however, without any experimental realization, so far. In this work we focus on the optical implementation of the so-called Fourier–Kravchuk transform[12] that can be equally applied to the classical and quantum states. Throughout our paper, we simply refer to this transform as DFrFT whose application reaches from the demonstration of the Fourier suppression law[13], N00N-state generation[14,15] and qubit storage[16] to the realization of perfect discrete lenses for non-uniform input distributions.

In this work, we report on the realization of DFrFTs of one-dimensional optical signals based on an integrated lattice of evanescently coupled waveguides. In these photonic arrangements, the inter-channel couplings are designed in such a way that the system readily performs the DFrFT of any incoming signal. The signal evolution is governed by the Schrödinger equation and the associated Hamilton operator is known as the J_x-operator in the quantum theory of angular momentum or likewise as the Heisenberg XY model from the quantum theory of ferromagnetism. We foresee that the inherent versatility of this approach will make other realizations of the DFrFT, the FT and the fast Fourier transform recognizably simple and thus may open the door to many interesting applications in integrated quantum computation[17].

Results

Theoretical approach. Similarly to its continuous counterpart, the DFrFT can be interpreted physically as a continuous rotation of the associated wave functions through an angle Z in phase space (see Fig. 1a)[18]. The idea is thus to construct finite circuits that are capable of imprinting such rotations to any light field. In quantum mechanics, three-dimensional spatial rotations of complex state vectors are generated via operations of the angular momentum operators J_k ($k = x, y, z$) on the Hilbert space of the associated system[19]. In particular, the rotation imprinted by the J_x-operator turns out to be an elaborated definition of the DFrFT (see Methods section for discussion). These concepts can be readily translated to the optical domain by mapping the matrix elements of the J_x-operator over the inter-channel couplings of engineered waveguide arrays (Fig. 1b–e)[20]. The coupling matrix of such waveguide arrays is thus given by $(J_x)_{m,n} = \kappa_0\left(\sqrt{(j-m)(j+m+1)}\delta_{m+1,n} + \sqrt{(j+m)(j-m+1)}\delta_{m-1,n}\right)/2$ (ref. 19). Here, κ_0 is a scaling factor introduced for experimental reasons. The indices m and n range from $-j$ to j in unit steps.

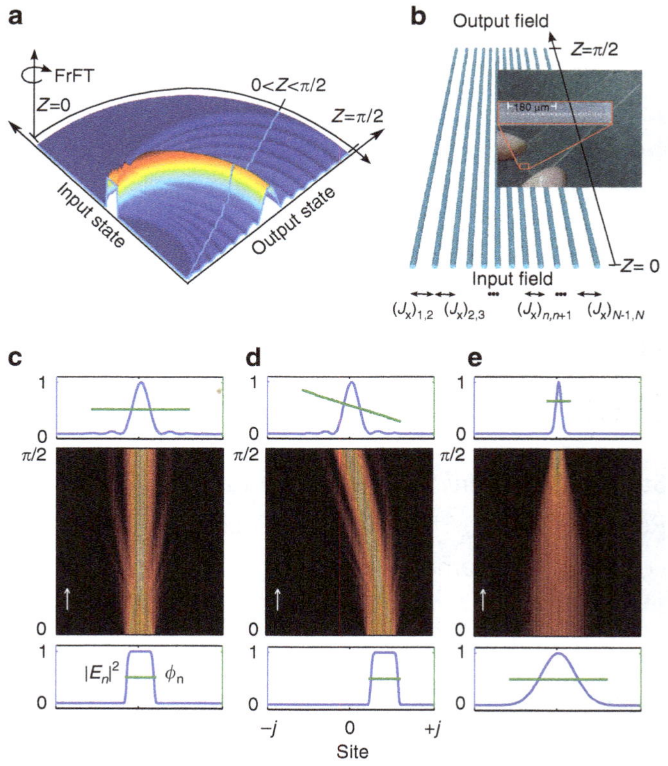

Figure 1 | Discretization of the FrFT. (**a**) Pictorial view of actual fractional Fourier transforms exemplified as continuous rotations in phase space. (**b**) Schematic representation of a pre-engineered J_x-array. (**c**-**e**) Top views of continuous 'rotations' of a rectangular (**c**), displaced rectangular (**d**) and Gaussian (**e**) optical wave functions in a J_x-array with $N = 151$. The bottom and top plots show the intensities $|E_n|^2$ and phases ϕ_n of the ingoing and outgoing wave packets, respectively. The green lines describe the magnitude of phase distributions of the optical fields, that is, the phase jumps of π due to a change of sign of the signal's amplitude are not shown.

Meanwhile, j represents an arbitrary positive integer or half-integer that determines the total number of waveguides via $N = 2j + 1$ (Fig. 1b).

Coupled mode theory states that the evolution of light in the J_x-waveguide array is governed by the following set of equations[20]

$$i\frac{\mathrm{d}}{\mathrm{d}Z}E_m(Z) = \frac{1}{\kappa_0}\sum_{n=-j}^{j}(J_x)_{m,n}E_n(Z). \tag{1}$$

Here, $E_n(Z)$ denotes the complex electric field amplitude at site n. In the quantum optics regime, single photons traversing such devices are governed by a set of Heisenberg equations that are isomorphic to equation (1). The only difference is that in the quantum case $E_n(Z)$ must be replaced by the photon creation operator $a_n^\dagger(Z)$. In a spintronic context, the evolution parameter Z is associated with time, whereas in the framework of integrated quantum optics, Z represents the propagation distance, see Fig. 1b–e. A spectral decomposition of the J_x-matrix yields the eigenvectors $u_n^{(m)} = 2^n\sqrt{\frac{(j+n)!(j-n)!}{(j+m)!(j-m)!}}P_{j+n}^{(m-n,-m-n)}(0)$[14,19], which in combination with the eigenvalues, $\beta_m = -j, \ldots, j$, render the closed-form point-spread function

$$G_{p,q}(Z) = i^{p-q}\sqrt{\frac{(j+p)!(j-p)!}{(j+q)!(j-q)!}}\left[\sin\left(\frac{Z}{2}\right)\right]^{q-p}\left[\cos\left(\frac{Z}{2}\right)\right]^{-q-p}P_{j+p}^{(q-p,-q-p)}(\cos(Z)). \tag{2}$$

Note that q and p represent the excited and observed sites, respectively, and $P_n^{(A,B)}(x)$ are the Jacobi polynomials of order n (see Methods section for discussion). Using equation (2), we can compute the response of the system to any input signal, which in turn gives the DFrFT[12]. Accordingly, DFrFT of any particular order arises at one specific propagation distance Z lying between 0 and $\pi/2$. In the limit $N \rightarrow \infty$, the eigenvectors of J_x, $u^{(m)}_{x\sqrt{2/N}}$, become the continuous Hermite–Gauss polynomials $H_m(x)$, which are known to be the eigenfunctions of the fractional Fourier operator[12]. As a result, in the continuous limit, the DFrFT described by equation (2) converges to the continuous FrFT[5,12]; and the standard FT is recovered at $Z=\pi/2$ (Fig. 1c–e). Note that in general, the DFrFT obtained in our devices and the usual DFT become equal only in the continuous limit $N \rightarrow \infty$ and at $Z=\pi/2$.

Experiments with classical light. To experimentally demonstrate the functionality of the suggested waveguide system, we use $N=21$ waveguides to perform FTs of simple wave packets. We first consider a Gaussian wave packet with a full-width at half-maximum (FWHM) covering the five central sites (Fig. 2a). The input signal is prepared by focusing a Gaussian beam from a HeNe laser onto the front facet of the sample. By exploiting the fluorescence from colour centres within the waveguides[21], we monitor the full intensity evolution from the input to the output plane. The fluorescence image, Fig. 2a, shows a gradual transition from an initially narrow Gaussian distribution at the input to a broader one at the Fourier plane (left and right panels Fig. 2a), demonstrating that narrow signals in space correspond to broad signals in Fourier space. For intermediate propagation distances ($Z \in [0, \pi/2]$) we extract other orders of the DFrFT, simultaneously. For comparison, we plot the continuous FrFT produced by the corresponding continuous Gaussian profile (red curves Fig. 2a). The agreement between the computed FrFT and the experimental DFrFT proves that for the considered Gaussian input signal, $N=21$ is sufficient to achieve the continuous limit. We now shift the input Gaussian beam by six channels towards the edge. Since the separations between adjacent waveguides at the edges are bigger than the separations between adjacent waveguides in the centre, the discretization grid is not perfectly homogeneous. Strictly speaking, the discretized shifted Gaussian just at the input plane covers slightly less than five waveguides FWHM. We observe that the well-approximated off-centre Gaussian travels to the centre at $Z=\pi/2$ (Fig. 2b), hereby showing the famous shift theorem. In additional experiments, extended signals, for example, a shifted top-hat function, are found to be well transformed according to equation (2) as well. However, we find that for this type of excitation $N>21$ would be required to discuss the continuous limit (see Supplementary Note 1 with Supplementary Fig. 1).

An unequivocal criterion, for the functionality of devices that perform the DFrFT, equation (2), can be formulated by evaluating $G_{p,q}\left(\frac{\pi}{2}\right)$. At this particular distance, point-like excitations will give rise to signal magnitudes that perfectly resemble the magnitudes of one of the eigenstates of the transform. More specifically, for transforms such as equation (2), one finds that an excitation of the qth site excites the qth system eigenstate up to local phases ($G_{p,q}\left(\frac{\pi}{2}\right)=i^{p-q}u_p^{(q)}$) (see the Methods section for explanations). The experimental demonstration of this intriguing

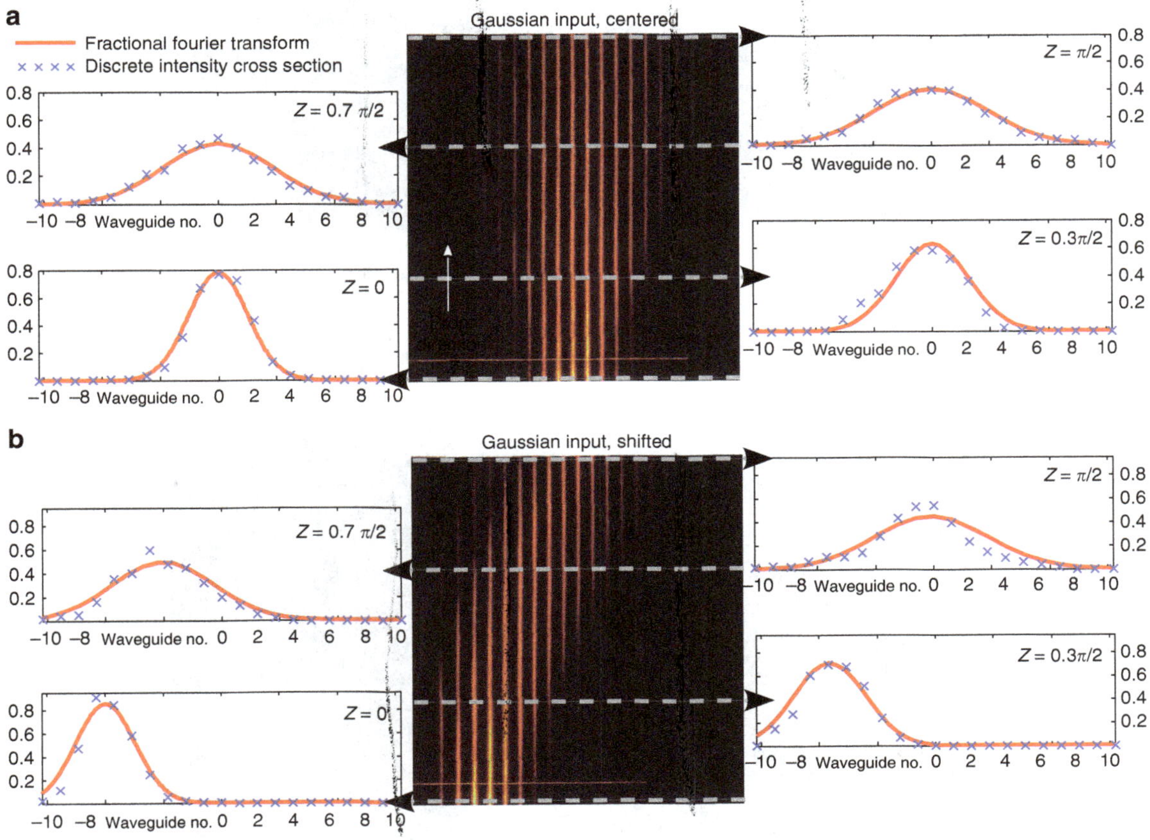

Figure 2 | DFrFT of classical light. (**a**) Transformation of a Gaussian input into a Gaussian profile of larger width along the evolution in the J_x-array. The FT is obtained at $Z=\pi/2$. The experimental data (blue crosses) is compared with the numeric FrFT (red curves). (**b**) A shifted input Gaussian profile evolves towards the centre of the array and acquires the same width as in **a**.

effect is shown in the subpanels of Fig. 3a–d along with the theoretical predictions. It can be argued that for any point-like excitation, the continuous limit cannot be met experimentally (see the Methods for discussion). Instead, equation (2) creates a non-uniform amplitude distribution with a phase difference of $\pi/2$ between adjacent sites. Nevertheless, in the continuous limit, $G_{p,q}\left(\frac{\pi}{2}\right)$ tends to the usual FT kernel[12]. At this point, it is worth emphasizing the formal equivalence to the quantum Heisenberg XY model in condensed matter physics[22,23]. In this respect, our observations demonstrate the capability of the here-presented systems to store quantum information in XY Hamiltonians by converting specific inputs into eigenstates of the system[16]. To our knowledge, this rather rare property has never been thoroughly investigated before.

Quantum experiments. To demonstrate the applicability of our approach in the quantum domain, we now analyse intensity correlations of separable and path-entangled photon pairs propagating through these Fourier transformers. To do so, we fabricated J_x-arrays involving $N=8$ channels. The importance of exploring FTs of such states has been highlighted in several investigations, demonstrating interesting effects such as suppression of states and portraying biphoton spatial correlations[24–26].

In this discrete quantum optical context, pure separable two-photon states are readily produced by coupling pairs of indistinguishable photons into two distinct lattice sites (m, n), this state is mathematically described by $|\Psi(0)\rangle = a_m^\dagger a_n^\dagger|0\rangle$. Conversely, path-entangled two-photon states are created by simultaneously launching both photons at either site m or n with exactly the same probability, that is, $|\Psi(0)\rangle = [(a_m^\dagger)^2 + (a_n^\dagger)^2]|0\rangle/2$. Furthermore, the probability of observing one of the photons at site k and its twin at site l is given by the intensity correlation matrix $\Gamma_{k,l}(Z) = \langle a_k^\dagger a_l^\dagger a_l a_k\rangle$ (ref. 27). An intriguing and unique property of the J_x-systems is that at $Z=\pi/2$ the correlation matrices are

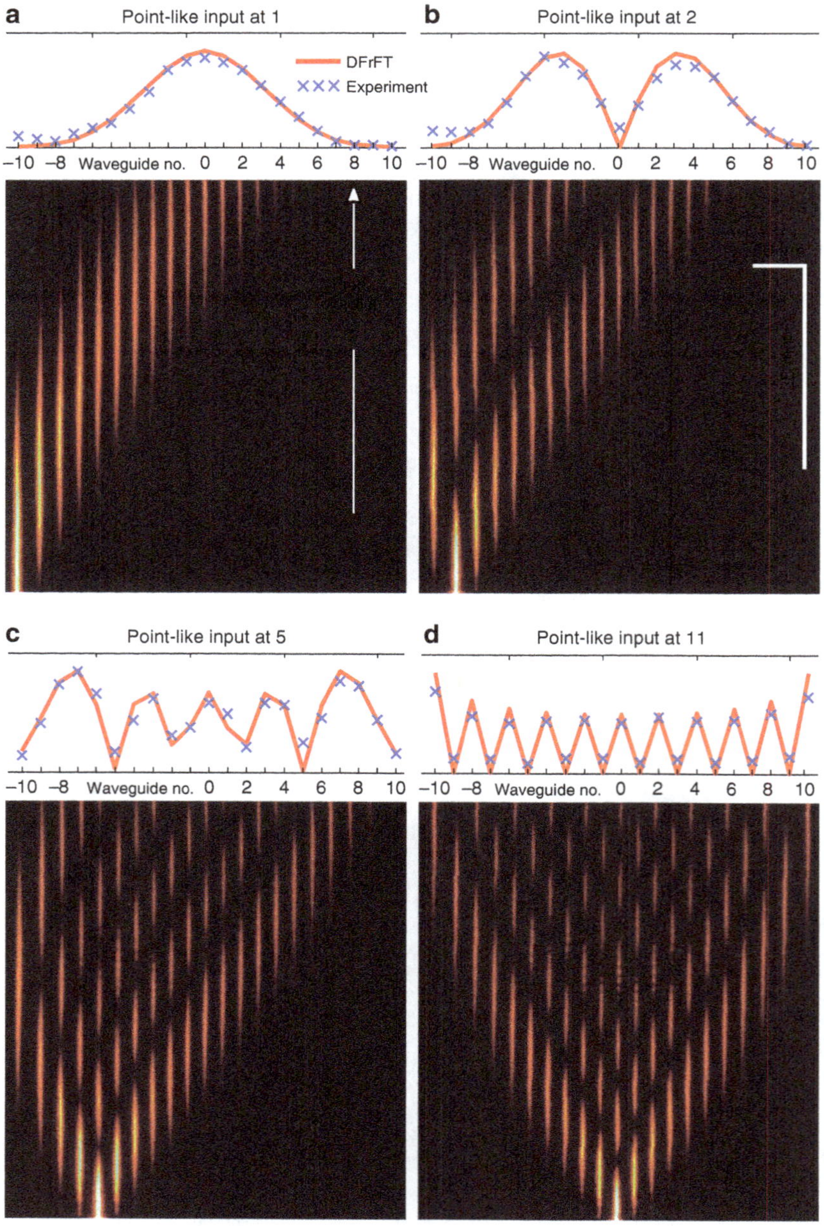

Figure 3 | Experimental visualization of the discrete Hermite-Gauss polynomials. (**a–d**) Evolution of single-site inputs into the magnitudes of the respective eigensolutions, as predicted theoretically (methods). The experimental data (blue crosses) is compared with the analytic DFrFT (red curves).

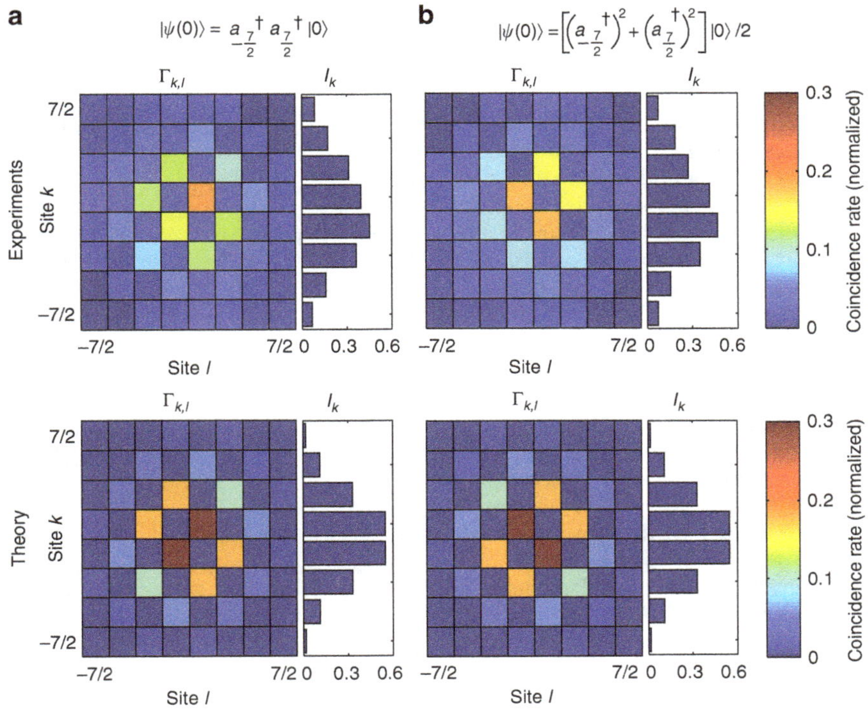

Figure 4 | DFrFT of quantum light. Correlation maps $\Gamma_{k,l}$ of a two-photon state either prepared (**a**) in a product state or (**b**) in a path-entangled state after propagating through a J_x-array. The photon density I_k at the output is shown on the right side of each map. The evaluation of the s.d. of the coincidence rates is presented in the Supplementary Note 2.

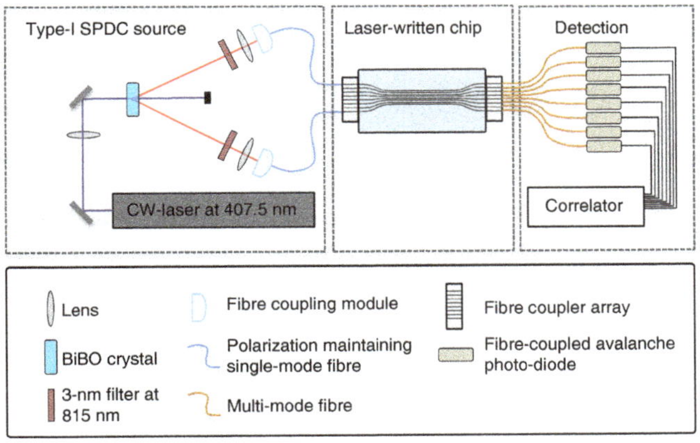

Figure 5 | Set-up used to carry out spatial correlation measurements of photonic quantum states. The set-up consists of three parts—the state preparation, the execution of the DFrFT and the correlation measurement.

given in terms of the eigenstates, as noticed above. Hence, for the separable case, $|\Psi(0)\rangle = a^\dagger_m a^\dagger_n |0\rangle$, the correlation matrices are given by $\Gamma_{k,l} = |u_k^{(m)} u_l^{(n)} + u_k^{(n)} u_l^{(m)}|^2$, whereas for the path-entangled state, $|\Psi(0)\rangle = [(a^\dagger_m)^2 + (a^\dagger_n)^2]|0\rangle/2$, we have $\Gamma_{k,l} = |(-i)^{2m} u_k^{(m)} u_l^{(m)} + (-i)^{2n} u_k^{(n)} u_l^{(n)}|^2$. Of particular interest is the separable case, where the photons are symmetrically coupled into the outermost waveguides, $|\Psi(0)\rangle = a^\dagger_j a^\dagger_{-j} |0\rangle$. In this scenario, only the correlation matrix elements for which $(k+l) = \text{odd}$ are nonzero, and are given by $\Gamma_{k,l} = 4^{k+l+1}(j+k)!(j-k)!(j+l)!(j-l)!(P_{j+k}^{(j-k,-j-k)}(0)P_{j+l}^{(-j-l,j-l)}(0))/[(N-1)!]^2$. These effects are demonstrated for the initial state $|\Psi(0)\rangle = a^\dagger_{-\frac{7}{2}} a^\dagger_{\frac{7}{2}} |0\rangle$ in Fig. 4a, where concentration and absence of probability in the correlation matrix clearly show that some states are completely suppressed—a hallmark of any Fourier unitary process[13]. An estimation of the statistical significance of the data set, along with a short discussion on incoherence effects can be found in the Supplementary Note 2 involving Supplementary Figs 2 and 3.

As a second case, we consider a fully symmetric path-entangled two-photon state of the form $|\Psi(0)\rangle = [(a^\dagger_j)^2 + (a^\dagger_{-j})^2]|0\rangle/2$. Physically, both photons are entering together into the array at either site j or $-j$ with equal probability[28–30]. The correlations are determined by $\Gamma_{k,l} = |u_k^{(j)} u_l^{(j)} + u_k^{(-j)} u_l^{(-j)}|^2$, from which we infer that the probability of measuring photon coincidences at coordinates (k, l) vanishes at sites where the sum $(k+l)$ is odd. In contrast, at coordinates where $(k+l)$ is even, the correlation function collapses to the expression $\Gamma_{k,l} = 4^{k+l+1}(j+k)!(j-k)!(j+l)!(j-l)!(P_{j+k}^{(j-k,-j-k)}(0)P_{j+l}^{(-j-l,j-l)}(0))^2/[(N-1)!]^2$. This indicates that in this path-entangled case the correlation map appears rotated by 90° with respect to the matrix obtained with separable two-photon states. We performed an experiment to demonstrate these predictions using states of the type $|\Psi(0)\rangle = [(a^\dagger_{-\frac{7}{2}})^2 + (a^\dagger_{\frac{7}{2}})^2]|0\rangle/2$, which were prepared using a 50:50 directional coupler acting as a beam splitter[29]. The whole experiment is achieved using a single chip containing both the state preparation stage followed by a J_x-system, yielding high interferometric control over the field dynamics (Fig. 5). The experimental measurements are presented in Fig. 4b. Similarly, suppression of states occurs as a result of destructive quantum interference. As predicted, a closer look into the correlation pattern reveals that indeed the correlation map appears rotated by 90° with respect to the matrix obtained with separable two-photon states.

Discussion

We emphasize that our quantum measurements feature interference fringes akin to the ones observed in quantum Young's two-slit experiments of biphoton wave functions in free space as demonstrated in ref. 24. In such free-space experiments, however, far-field observations were carried out using lenses and the two slits were emulated by optical fibres[24]. Along those lines, we have created a fully integrated quantum interferometer to observe fundamental quantum mechanical features[25]. This additionally suggests an effective way to generate quantum states containing only even (odd) non-vanishing inter-particle distance probabilities for the separable input state (symmetric path-entangled state). In addition, the eigenfunctions associated with the Hamiltonian system explored in our work are specific Jacobi polynomials, which are well known as the optimal basis for quantum phase-retrieval algorithms[31] and these eigenstates can be retrieved by limited phase operations. Knowing a quantum wavefunction and its FT, a phase-retrieval algorithm for signals that are a superposition of a finite number of Hermite–Gauss polynomials has been introduced[32]. This phase-retrieval algorithm might be implemented employing our system, since its discrete character automatically possesses a finite number of polynomials that are closely connected to the Hermite–Gauss polynomials. Another potential application is the realization of the Radon–Wigner transform given by the squared modulus of the FrFT[6,33]. The Radon–Wigner function is a basic tool for the reconstruction of Wigner quasi-probability distributions in quantum optics[34,35]. Also, FrFTs appear naturally in optics as free-space propagation between two spherical reference planes in general[4]. Like in our system, the order of the FrFT is proportional to the propagation distance. On that basis, complex spatial filtering involving several fractional Fourier planes was suggested[36]. In this description, the optical signal is discretized and thus described by a vector of field amplitudes at certain sites. In our approach this is inherently realized.

In conclusion, we have successfully demonstrated a universal discrete optical device capable of performing classical and quantum DFrFTs. Our studies might find applications in developing a more general quantum suppression law[13] and perhaps in the development of new quantum algorithms.

Methods

The fractional Fourier transform in quantum harmonic oscillations. In this section, we briefly describe the relation between the continuous FrFT operator and the Hamiltonian of the quantum harmonic oscillator[2].

The FrFT operator $\hat{F}_Z$ is defined by the following eigenvalue equation involving the Hermite–Gauss polynomials of order n and the eigenvalues $\lambda_n = \exp(inZ)$

$$\hat{F}_Z\{\exp(-x^2/2)H_n(x)\} = \exp(inZ)\exp(-x^2/2)H_n(x), \quad (3)$$

where $Z \in \mathbb{R}$. Concurrently, one can interpret equation (3) as quantum time evolution from time $t=0$ to $t=Z$. We write $\hat{F}_Z$ as $\exp(iZ\hat{A})$.

$$\exp(iZ\hat{A})\{\exp(-x^2/2)H_n(x)\} = \exp(inZ)\exp(-x^2/2)H_n(x). \quad (4)$$

To show that $\hat{A}$ is the Hamilton operator of the harmonic oscillator, we differentiate both sides of equation (4) with respect to Z and evaluate the result at $Z=0$. We obtain

$$\hat{A}\exp(-x^2/2)H_n(x) = n\exp(-x^2/2)H_n(x). \quad (5)$$

To find the spatial representation of $\hat{A}$, consider the differential equation

$$\left(-\frac{1}{2}\frac{\mathrm{d}^2}{\mathrm{d}x^2} + x\frac{\mathrm{d}}{\mathrm{d}x}\right)\mathrm{H}_n(x) = nH_n(x), \quad (6)$$

for the Hermite polynomials $H_n(x)$ of order n. Using the identities $1=\exp(x^2/2)\exp(-x^2/2)$ and $\exp(\xi x^2/2)\left(\frac{\mathrm{d}^n}{\mathrm{d}x^n}\right)\exp(-\xi x^2/2) = \left(\frac{\mathrm{d}}{\mathrm{d}x} - \xi x\right)^n$, one can show that equation (6) can be written as

$$\left(-\frac{1}{2}\frac{\mathrm{d}^2}{\mathrm{d}x^2} + \frac{1}{2}x^2 - \frac{1}{2}\right)\exp(-x^2/2)\mathrm{H}_n(x) = n\exp(-x^2/2)H_n(x), \quad (7)$$

where we have used the commutator $\left[x, \frac{\mathrm{d}}{\mathrm{d}x}\right] = -1$. Comparing equation (5) and equation (7), one can see that $\hat{A}$ becomes

$$\hat{A} = \left(-\frac{1}{2}\frac{\mathrm{d}^2}{\mathrm{d}x^2} + \frac{1}{2}x^2 - \frac{1}{2}\right). \quad (8)$$

In summary, the FrFT operator $\hat{F}_z$ can be written as

$$\hat{F}_z = \exp\left(\frac{iz}{2}\left(-\frac{\mathrm{d}^2}{\mathrm{d}x^2} + x^2 - 1\right)\right). \quad (9)$$

Because of this one-to-one correspondence between the dynamics of the quantum harmonic oscillator and the fractional Fourier operator, the implementation of such transform is immediate using harmonic oscillator systems[37].

J_x-photonic lattices as discrete harmonic oscillators. Our aim in this section is to show that in the continuous limit $N \to \infty$, the eigenvalue equation for J_x-arrays becomes the eigenvalue equation of the quantum harmonic oscillator. Consider the matrix representation of the J_x-operator again. For convenience we take, in this section only, $\kappa_0 = 1$.

$$\begin{aligned}(J_x)_{m,n} &= \tfrac{1}{2}\left(\sqrt{(j-m)(j+m+1)}\delta_{n,m+1} + \sqrt{(j+m)(j-m+1)}\delta_{n,m-1}\right) \\ &= \tfrac{1}{2}\left(\sqrt{j(j+1)-m(m+1)}\delta_{n,m+1} + \sqrt{j(j+1)-m(m-1)}\delta_{n,m-1}\right).\end{aligned} \quad (10)$$

The indices m and n range from $-j$ to j in unit steps and j is an arbitrary positive integer or half-integer. The dimension of the J_x-matrix is $N=2j+1$. We now introduce the variable $\gamma = j(j+1) = (N^2-1)/4$, which implies that $(J_x)_{m,n}$ can be written as

$$(J_x)_{m,n} = \frac{\sqrt{\gamma}}{2}\left(\sqrt{1-\frac{1}{\gamma}m(m+1)}\delta_{n,m+1} + \sqrt{1-\frac{1}{\gamma}m(m-1)}\delta_{n,m-1}\right). \quad (11)$$

Let us consider the eigenvalue equation for this matrix

$$\frac{\sqrt{\gamma}}{2}\left(\sqrt{1-\frac{1}{\gamma}m(m+1)}\psi_{m+1} + \sqrt{1-\frac{1}{\gamma}m(m-1)}\psi_{m-1}\right) = \beta_m\psi_m. \quad (12)$$

Considering the region $m \ll j$, since $\gamma \propto N^2$, in the limit $N \to \infty$, the terms $m(m \pm 1)/\gamma \ll 1$. Hence, in the domain far from the edge of the array a Taylor expansion yields

$$\sqrt{\gamma}\sqrt{1-\frac{m}{\gamma}(m \pm 1)} \approx \sqrt{\gamma}\left(1 - \frac{m}{2\gamma}(m \pm 1) - \frac{m^2}{8\gamma^2}(m \pm 1)^2\right). \quad (13)$$

By defining $m = x\gamma^{1/4}$ (or $x = m/\gamma^{1/4}$), we obtain

$$\sqrt{\gamma}\sqrt{1-\frac{m}{\gamma}(m \pm 1)} \approx \gamma^{1/2} - \frac{x^2}{2} \mp \frac{x}{2\gamma^{1/4}} - \frac{x^4}{8\gamma^{1/2}} \quad (14)$$

Plugging this expression into equation (12)

$$\left(\gamma^{1/2} - \frac{x^2}{2} - \frac{x}{2\gamma^{1/4}} - \frac{x^4}{8\gamma^{1/2}}\right)\psi_{m+1} + \left(\gamma^{1/2} - \frac{x^2}{2} + \frac{x}{2\gamma^{1/4}} - \frac{x^4}{8\gamma^{1/2}}\right)\psi_{m-1} \approx 2\beta_m\psi_m \quad (15)$$

We redefine the functions $\psi_m = \psi(x) = \psi\left(\frac{m}{\gamma^{1/4}}\right)$ and $\psi_{m+1} = \psi\left(x + \frac{1}{\gamma^{1/4}}\right) = \psi\left(\frac{m}{\gamma^{1/4}} + \frac{1}{\gamma^{1/4}}\right)$ such that we can introduce the Taylor series

$$\psi_{m\pm 1} = \psi\left(x \pm \frac{1}{\gamma^{1/4}}\right) = \psi(x) \pm \frac{1}{\gamma^{1/4}}\psi'(x) + \frac{1}{2\gamma^{1/2}}\psi''(x), \quad (16)$$

where, again, we have kept only terms up to second order in $1/\gamma^{1/4}$. Substituting equation (16) into equation (15), and using the limit

$$\lim_{N\to\infty}\sqrt[4]{\left(\frac{4}{N^2}\right)\left(\frac{1}{1-\frac{1}{N^2}}\right)} = 0.$$

We obtain the time-independent Schrödinger equation for the harmonic oscillator

$$\left(-\frac{1}{2}\frac{\mathrm{d}^2}{\mathrm{d}x^2} + \frac{1}{2}x^2\right)\psi(x) = (\sqrt{\gamma} - \beta)\psi(x). \quad (17)$$

Therefore, in the continuous limit $N \to \infty$, the difference equation describing J_x-photonic lattices becomes the time-independent Schrödinger equation for the quantum harmonic oscillator. Note, however, that due to the importance of the condition $m \ll j$ in this derivation, this statement is only valid when dealing with signals that are square integrable in the continuous limit. Thus, the operator equation (11) can be used to define the discrete version of the quantum harmonic oscillator and thus the DFrFT.

The point-spread function for J_x-photonic lattices. In this section, it is shown that at $Z=\pi/2$, the green function of J_x-systems becomes proportional to the amplitude of one of the eigenstates. The evolution of light in J_x-arrays is governed by the set of N coupled differential equations (equation (1)).

The normalized propagation coordinate Z is given by $Z = \kappa_0 z$, where z is the actual propagation distance and κ_0 is an arbitrary scale factor. The quantity $E_n(Z)$ denotes the mode field amplitude at site n. A spectral decomposition of the J_x-matrix yields the eigensolutions

$$u_n^{(m)} = (2)^n\sqrt{\frac{(j+n)!(j-n)!}{(j+m)!(j-m)!}}P_{j+n}^{(m-n,\,-m-n)}(0). \quad (18)$$

$P_n^{(A,B)}(x)$ are the Jacobi polynomials of order n. And the corresponding eigenvalues are integers or half-integers, $\beta_m = -j, \ldots, j$, depending on the parity of N[14,19]. Using the eigenvectors and eigenvalues we obtain the point-spread function

$$G_{p,q}(Z) = \sum_{r=-j}^{j} u_q^{(r)} u_p^{(r)} \exp(irZ). \quad (19)$$

$G_{p,q}(Z)$ represents the amplitude at site p after an excitation of site q. Using equation (18) and the properties of the Jacobi polynomials one can show that equation (19) reduces to the closed-form expression

$$G_{p,q}(Z) = (-i)^{q-p}\sqrt{\frac{(j+p)!\,(j-p)!}{(j+q)!\,(j-q)!}}\left[\sin\left(\frac{Z}{2}\right)\right]^{q-p} \left[\cos\left(\frac{Z}{2}\right)\right]^{-q-p} P_{j+p}^{(q-p,\,-q-p)}(\cos(Z)). \quad (20)$$

Evaluation of equation (20) at $Z = \pi/2$ yields

$$G_{p,q}\left(\frac{\pi}{2}\right) = (-i)^{q-p}(2)^{p}\sqrt{\frac{(j+p)!\,(j-p)!}{(j+q)!\,(j-q)!}}P_{j+p}^{(q-p,\,-q-p)}(0) = (-i)^{q-p}u_p^{(q)}. \quad (21)$$

Equation (21) shows that at $Z = \pi/2$ the point-spread function becomes proportional to the amplitude of the corresponding eigenstates depending on the excited site. In other words, there is a one-to-one correspondence between the excited site number and the eigenstates of the system: excitation of the qth site excites the qth eigenstate up to well-defined local phases.

Devices fabrication and specifications. Our devices are fabricated in bulk-fused silica samples (Corning 7980, ArF grade) using the femtosecond laser direct-write approach[21]. The transparent material is modified within the focal region due to nonlinear absorption resulting in a local increase of the refractive index. Effectively, the waveguides possess only the fundamental mode. The coupling between neighbouring waveguides depends on their separation within the glass chip. Regarding the theoretical description of the waveguide array, the validity of equation (1) is only given if the fundamental modes of neighbouring waveguides have a negligible overlap. For a given N and a maximum length of the glass wafers of Z/κ_0, this can only be ensured for a certain range of κ_0. Outside this range of κ_0, coupled mode theory will break down and errors are introduced to the implementation of the DFrFT. In our experiments these errors are kept small but impossibly perfectly zero.

For the fabrication of the devices used to transform classical light, we employed an Yb-doped fibre laser (Amplitude Systèmes) operating at a wavelength of 532 nm, a repetition rate of 200 kHz and a pulse length of 300 fs. Waveguides were written with 300 nJ pulses focused by a 20x objective. The sample was moved at a velocity of 200 mm min^{-1} by high-precision positioning stages (ALS 130, Aerotech Inc.) with a positioning error of ± 0.1 µm. From this random positioning error, the realized inter-channel couplings inherit a relative error of 2%. The mode field diameters of the guided mode were 4 µm × 7 µm at 632 nm. In the classical experiments, the Fourier plane lies at $Z/\kappa_0 = 7.48$ cm, that is, $\kappa_0 = 0.21$ cm^{-1}. The desired nearest-neighbor couplings $(J_x)_{n\pm1,n}$ determine the separations of the waveguides n and $n \pm 1$. The largest separation of 22.8 µm between adjacent waveguides occurs at the edges. In the centre, the separation is 17.6 µm.

For the samples illuminated with single-photon states of light, we used a RegA 9,000 seeded by a Mira Ti:Sa femtosecond laser oscillator. The amplifier produced 150 fs pulses centred at 800 nm at a repetition rate of 100 kHz, with energy of 450 nJ. The structures were permanently inscribed with a 20x objective while moving the sample at a constant speed of 60 mm min^{-1}, using the positioning system described above. The mode field diameters of the guided mode were 18 µm × 20 µm at 815 nm. All structures were designed with fan-in and fan-out sections arranged in a three-dimensional geometry, and located prior and after the J_x-lattice, as illustrated in Fig. 5. This effectively suppresses any unwanted crosstalk between the guides and permits easy coupling to fibre arrays with a standard spacing of 127 µm. In the presented device used to transform quantum states, we have $\kappa_0 = 0.6$ cm^{-1}, that is, the Fourier plane is located 2.62 cm after the beginning of the J_x-array.

Experiment on the characterization of two-photon correlations. A BiB_3O_6 nonlinear crystal was pumped with a 70 mW continuous wave pump laser emitting at 407.5 nm, which provided pairs of indistinguishable photons due to type-I spontaneous parametric down-conversion, see Fig. 5. Photon pairs with a central wavelength of 815 nm were filtered by 3 nm (FWHM) interference filters. They were further coupled to the chip via fibre arrays through polarization maintaining fibres, and subsequently fed into single-photon detectors (avalanche photodiodes). The two-photon correlation function was determined by analysing the twofold coincidences recorded between all output channels with the help of an electronic correlator card (Becker & Hickl: DPC230). The spatial correlation results presented in Fig. 4 were extracted from a data set with total integration time of 5 min. The coincidences were then analysed with a time window set at 5 ns and are corrected for detector efficiencies. To assess the statistical consistency of the results in Fig. 4, we discuss in the Supplementary Note 2 the data set presented in terms of correlation event numbers before normalization. For both measurements, the detector clicks data set initially consisted of $\sim 1.5 \times 10^6$ events in total, which after post selection was reduced to a set of $\sim 5 \times 10^4$ coincidence events in total, for both input states. Accidental coincidences due to simultaneous detection of two photons not coming from the same pair are estimated to occur with a negligible rate of $< 2 \times 10^{-6}$s^{-1}. Non-deterministic number-resolved photon detection was achieved using fibre beam splitters. This set-up thus allows the determination of all 36 two-photon coincidence events occurring in photonic lattices consisting of eight channels. Furthermore, at the wavelength of interest, propagation losses and birefringence are estimated to be 0.3 dB cm^{-1} and in the order of 10^{-7}, respectively. Additional discrepancies in between the measured correlation matrices and the theoretical ones may appear due to imperfect excitations, asymmetric output coupling losses (or detector efficiencies) or limited indistinguishability of the photons.

References

1. Cooley, J. & Tukey, J. W. An algorithm for the machine calculation of complex Fourier series. *Math. Comput.* **19,** 297–301 (1965).
2. Namias, V. The fractional order Fourier transform and its application to quantum mechanics. *J. Inst. Maths. Applics* **25,** 241–265 (1980).
3. Almeida, L. B. The fractional Fourier transform and time-frequency representations. *IEEE Trans. Signal Process.* **42,** 3084–3091 (1994).
4. Ozaktas, H. M. & Mendlovic, D. Fractional Fourier Optics. *J. Opt. Soc. Am. A* **12,** 743–751 (1995).
5. Ozaktas, H. M., Zalevsky, Z. & Kutay, M. A. *The Fractional Fourier Transform with Applications in Optics and Signal Processing* (Wiley, 2001).
6. Lohmann, A. W. & Soffer, B. H. Relationship between the Radon–Wigner and fractional Fourier transforms. *J. Opt. Soc. Am. A* **11,** 1798–1801 (1994).
7. Man'ko, M. A. Fractional Fourier transform in information processing, tomography of optical signal, and green function of harmonic oscillator. *J. of Russ. Laser Research* **20,** 226–228 (1999).
8. Dorsch, R. G., Lohmann, A. W., Bitran, Y., Mendlovic, D. & Ozaktas, H. M. Chirp filtering in the fractional Fourier domain. *Appl. Opt.* **33,** 7599–7602 (1994).
9. Lohmann, A. W. & Mendlovic, D. Fractional Fourier transform: photonic implementation. *Appl. Opt.* **33,** 7661–7664 (1994).
10. Lohmann, A. W. Image rotation, Wigner rotation, and the Fractional Fourier transform. *J. Opt. Soc. Am. A* **10,** 2181–2186 (1993).
11. Mendlovic, D. & Ozaktas, H. Fractional Fourier transforms and their optical implementation: I. *J. Opt. Soc. Am. A* **10,** 1875–1881 (1993).
12. Atakishiyev, N. M. & Wolf, K. B. Fractional Fourier-Kravchuk transform. *J. Opt. Soc. Am. A* **14,** 1467–1477 (1997).
13. Tichy, M. C., Mayer, K., Buchleitner, A. & Molmer, K. Stringent and efficient assessment of Boson Sampling Devices. *Phys. Rev Lett.* **113,** 020502–020506 (2014).
14. Perez-Leija, A., Keil, R. & Moya-Cessa, H. Perfect transfer of path-entangled photons in J_x-photonic lattices, A. Szameit, and D. N. Christodoulides. *Phys. Rev. A* **87,** 022303 (2013).
15. Humphreys, P. C., Barbieri, M., Datta, A. & Walmsley, I. A. Quantum enhanced multiple phase estimation. *Phys. Rev. Lett.* **111,** 070403 (2013).
16. Perez-Leija, A. *et al.* Eigenstate-assisted longitudinal quantum state transfer and qubit-storage in photonic and spin lattices, 45-th annual meeting of the APS Division of Atomic. *Molecular and optical Physics* **59,** J4.010 (2014).
17. Aspuru-Guzik, A., Dutoi, A. D., Love, P. J. & Head-Gordon, M. Simulated Quantum computation of molecular energies. *Science* **309,** 1704–1707 (2005).
18. Wolf, K. B. *Integral transforms in science and engineering* Vol. 11 (Mathematical Concepts and Methods for Science and Engineering, 1979).
19. Narducci, L. M. & Orzag, M. Eigenvalues and Eigenvectors of angular momentum operator J_x without theory of rotations. *Am. J. of Phys* **40,** 1811–1814 (1972).
20. Christodoulides, D. N., Lederer, F. & Silberberg, Y. Discretizing light behavior in linear and nonlinear waveguide lattices. *Nature* **424,** 817–823 (2003).
21. Szameit, A. & Nolte, S. Discrete optics in femtosecond-laser-written photonic structures. *J. Phys. B* **43,** 163001 (2010).
22. Lieb, E., Schultz, T. & Matts, D. Two soluble models of an antiferromagnetic chain. *Ann. Phys.* **16,** 407–466 (1961).
23. Cormick, C., Bermudez, A., Huelga, S. F. & Plenio, M. Preparation of the ground state of a spin chain by dissipation in a structured environment. *New J. Phys.* **15,** 073027 (2013).
24. Bobrov, I. B., Kalashnikov, D. A. & Krivitsky, L. A. Imaging of spatial correlations of two-photon states. *Phys. Rev. A* **89,** 043814 (2014).
25. Peeters, W. H., Renema, J. J. & van Exter, M. P. Engineering of two-photon spatial quantum correlations behind a double slit. *Phys. Rev. A* **79,** 043817 (2009).
26. Poem, E., Gilead, Y., Lahini, Y. & Silberberg, Y. Fourier processing of quantum light. *Phys. Rev. A* **86,** 023836 (2012).
27. Bromberg, Y., Lahini, Y., Morandotti, R. & Silberberg, Y. Quantum and classical correlations in waveguide lattices. *Phys. Rev. Lett.* **102,** 253904 (2009).

28. Marshall, G. D. *et al.* Laser written waveguide photonic quantum circuits. *Opt. Exp.* **17,** 12546–12554 (2009).
29. Lebugle, M. *et al.* Experimental observation of N00N state Bloch oscillations. *Nat. Commun.* **6,** 8273 (2015).
30. Hong, C. K., Ou, Z. Y. & Mandel, L. Measurement of subpicosecond time intervals between two photons by interference. *Phys. Rev. Lett.* **59,** 2044 (1987).
31. Wiseman, H. M. & Milburn, G. J. *Quantum Measurement and control* (Cambridge University Press, 2010).
32. Orlowski, A. & Paul, H. Phase retrieval in quantum mechanics. *Phys. Rev. A* **50,** R921–R924 (1994).
33. Wood, J. & Barry, D. T. Radon transformation of time-frequency distributions for analysis of multicomponent signals. *IEEE Transac. Signal Process.* **42,** 3166–3177 (1994).
34. Vogel, K. & Risken, H. Determination of quasiprobability distributions in terms of probability distributions for the rotated quadrature phase. *Phys. Rev. A* **40,** 2847 (1989).
35. Raymer, M. G., Beck, M. & McAlister, D. F. *Quantum Optics VI.* (eds Harvey, J. D. & Wall, D. F.) (Springer-Verlag, 1994).
36. Ozaktas, H. M. & Mendlovic, D. Fractional Fourier transforms and their optical implementations. *J. Opt. Soc. Am. A* **10,** 2522–2531 (1993).
37. Marhic, M. E. Roots of the identity operator and optics. *J. Opt. Soc. Am. A* **12,** 1448 (1995).

Acknowledgements

We gratefully acknowledge financial support by the German Ministry of Education and Research (Center for Innovation Competence programme, grant no. 03Z1HN31) and the Deutsche Forschungsgemeinschaft (grant no. NO462/6-1 and SZ276/7-1). M.L. thanks the Initial Training Network PICQUE (grant no. 608062) within the Seventh Framework Programme for Research of the European Commission for funding.

Author contributions

S.W. and A.P.-L. conceived the idea. S.W. and M.L. designed the samples and performed the measurements. A.P.-L. and S.W. developed the theory. S.W., M.L. and A.P.-L. analysed the data. A.S. supervised the project. All authors discussed the results and co-wrote the manuscript.

Additional information

Competing financial interests: The authors declare no competing financial interests.

28

Typical fast thermalization processes in closed many-body systems

Peter Reimann[1]

The lack of knowledge about the detailed many-particle motion on the microscopic scale is a key issue in any theoretical description of a macroscopic experiment. For systems at or close to thermal equilibrium, statistical mechanics provides a very successful general framework to cope with this problem. However, far from equilibrium, only very few quantitative and comparably universal results are known. Here a quantum mechanical prediction of this type is derived and verified against various experimental and numerical data from the literature. It quantitatively describes the entire temporal relaxation towards thermal equilibrium for a large class (in a mathematically precisely defined sense) of closed many-body systems, whose initial state may be arbitrarily far from equilibrium.

[1] Fakultät für Physik, Universität Bielefeld, 33615 Bielefeld, Germany. Correspondence and requests for materials should be addressed to P.R. (email: reimann@physik.uni-bielefeld.de).

In a macroscopic object, which is spatially confined and unperturbed by the rest of the world, every single atom exhibits an essentially unpredictable, chaotic motion *ad infinitum*, yet the system as a whole seems to approach in a predictable and often relatively simple manner some steady equilibrium state. Paradigmatic examples are compound systems, parts of which are initially hotter than others, or a simple gas in a box, streaming through a little hole into an empty second box. While such equilibration and thermalization phenomena are omnipresent in daily life and extensively observed in experiments, they entail some very challenging fundamental questions: why are the macroscopic phenomena reproducible though the microscopic details are irreproducible in any real experiment? How can the irreversible tendency towards macroscopic equilibrium be reconciled with the basic laws of physics, implying a perpetual and essentially reversible motion on the microscopic level?

Such fundamental issues are widely considered as still not satisfactorily understood[1–6]. Within the realm of classical mechanics, they go back to Maxwell, Boltzmann and many others[7]. Their quantum mechanical treatment was initiated by von Neumann[8] and is presently attracting renewed interest[9–12], for example, in the context of imitating thermal equilibrium by single pure states due to such fascinating phenomena as concentration of measure[2,13,14], canonical typicality[3,15–18] or eigenstate thermalization[4,19–26]. Numerically, scrutinizing ultracold atom experiments[27–30] and unravelling the relations between thermalization, integrability and many-body localization are among the current key issues[4,31–37]. Analytically, essential equilibration and thermalization properties of closed many-body systems or of subsystems thereof were deduced from first principles under increasingly weak assumptions about the initial disequilibrium, the system Hamiltonian and the observables[1,8–12,38–45]. In particular, groundbreaking results regarding pertinent relaxation timescales have been obtained in refs 45–51. Of foremost relevance for our present study is the work of the Bristol collaboration[49], showing, among others, that all two-outcome measurements, where one of the projectors is of low rank, equilibrate as fast as they possibly can without violating the time–energy uncertainty relation. A second recent key result is due to Goldstein, Hara and Tasaki[50,51], demonstrating that most systems closely approach an overwhelmingly large, so-called equilibrium Hilbert subspace on the extremely short Boltzmann timescale $t_{\rm B}:=h/k_{\rm B}T$. A more detailed account of pertinent previous works is provided as Supplementary Note 1.

Here we will further extend these findings in two essential respects: instead of upper bounds for some suitably defined characteristic timescale, as in refs 49–51, the entire temporal relaxation will be approximated in the form of an equality. As an even more decisive generalization of ref. 49–51, we will admit largely arbitrary observables. Finally, and actually for the first time within the realm of the above-mentioned analytical approaches[1,8–12,38–51], we will compare our predictions with various experimental as well as numerical data from the literature. In fact, most of those data have not been quantitatively explained by any other analytical theory before. Adopting a 'typicality approach' similar in spirit to random matrix theory[9–12], our result covers the vast majority (in a suitably defined mathematical sense) of initial conditions, observables and system Hamiltonians. On the other hand, many commonly considered observables and initial conditions actually seem to be rather special in that they are close to or governed by a hidden conserved quantity and therefore thermalize 'untypically slowly'.

Results

Set-up. Employing textbook quantum mechanics, we consider time-independent Hamiltonians H with eigenvalues E_n and eigenvectors $|n\rangle$ on a Hilbert space $\mathcal{H}$ of large (but finite) dimensionality $D \gg 1$. As usual, system states (pure or mixed) are described by density operators $\rho : \mathcal{H} \to \mathcal{H}$ and observables by Hermitian operators $A : \mathcal{H} \to \mathcal{H}$ with matrix elements $\rho_{mn} := \langle m|\rho|n\rangle$ and $A_{mn} := \langle m|A|n\rangle$, respectively. Expectation values are given by $\langle A\rangle_\rho := \mathrm{Tr}\{\rho A\}$ and the time evolution by $\rho(t) = \mathcal{U}_t\rho(0)\mathcal{U}_t^\dagger$ with propagator $\mathcal{U}_t := e^{-iHt/\hbar}$, yielding

$$\langle A\rangle_{\rho(t)} = \sum_{m,n=1}^{D} \rho_{mn}(0)A_{nm}\, e^{i(E_n - E_m)t/\hbar}. \tag{1}$$

The main examples are closed many-body systems with a macroscopically well-defined energy, that is, all relevant eigenvalues $E_1, \ldots, E_D$ are contained in some microcanonical energy window $[E-\Delta E, E]$, where ΔE is small on the macroscopic but large on the microscopic scale. For systems with $f \gg 1$ degrees of freedom, D is then exponentially large in f (refs 9,41). Accordingly, the relevant Hilbert space $\mathcal{H}$ is spanned by the eigenvectors $\{|n\rangle\}_{n=1}^{D}$ and is sometimes also named energy shell or active Hilbert space, see, for example, refs 8–12 and Supplementary Note 2 for more details.

Analytical results. Our main players are the three Hermitian operators H (Hamiltonian), A (observable) and $\rho(0)$ (initial state), each with its own eigenvalues (spectrum) and eigenvectors (basis of $\mathcal{H}$). In the following, the three spectra will be considered as arbitrary but fixed, while the eigenbases will be randomly varied relatively to each other. More precisely, all unitary transformations $U : \mathcal{H} \to \mathcal{H}$ between the eigenbases of H and A are considered as equally likely (Haar distributed[8–11]), while the basis of $\rho(0)$ relatively to that of A is arbitrary but fixed. (Equivalently, we could let 'rotate' H relatively to $\rho(0)$ while keeping A fixed relatively to $\rho(0)$.) In particular, the initial expectation value $\langle A\rangle_{\rho(0)}$ can be chosen arbitrary but then remains fixed (U independent). It is only for times $t>0$ that the randomness of the unitary U also randomizes (via H) the further temporal evolution of $\rho(t)$ and thus of $\langle A\rangle_{\rho(t)}$.

The basic idea behind this randomization of U is akin to random matrix theory[9–12], namely, to derive an approximation for $\langle A\rangle_{\rho(t)}$, which applies to the overwhelming majority of all those randomly sampled U's, hence it typically should apply also to the particular (non-random) U of the actual system of interest. A more detailed justification of this 'typicality approach' will be provided in the section 'Typicality of thermalization'.

Since A_{mn} refers to the basis of H, these matrix elements depend on U, and likewise for $\rho_{mn}(0)$ (the explicit formulae are provided in the section 'Basic matrices'). Indicating averages over U by the symbol $[\cdots]_U$ and exploiting that all basis transformations U are equally likely, it follows for symmetry reasons that $[\rho_{nn}(0)A_{nn}]_U$ must be independent of n. Likewise, $[\rho_{mn}(0)A_{nm}]_U$ must be independent of m and n for all $m \neq n$. We thus can conclude that for any n

$$D[\rho_{nn}(0)A_{nn}]_U = \left[\sum_{k=1}^{D} \rho_{kk}(0)A_{kk}\right]_U \tag{2}$$

and that for any $m \neq n$

$$\begin{aligned} D(D-1)[\rho_{mn}(0)A_{nm}]_U &= \left[\sum_{j\neq k} \rho_{jk}(0)A_{kj}\right]_U \\ &= \left[\sum_{j,k=1}^{D} \rho_{jk}(0)A_{kj}\right]_U - \left[\sum_{k=1}^{D} \rho_{kk}(0)A_{kk}\right]_U. \end{aligned} \tag{3}$$

Defining the auxiliary density operator ω via the matrix elements $\omega_{mn} := \delta_{mn}\rho_{nn}(0)$, equation (2) can be rewritten as $[\mathrm{Tr}\{\omega A\}]_U$. Working in a reference frame where only H (and thus ω) changes with U, but not A and $\rho(0)$, implies $[\mathrm{Tr}\{\omega A\}]_U = \mathrm{Tr}\{[\omega]_U A\}$. With $\rho_{\mathrm{av}} := [\omega]_U$ it follows that

$$[\rho_{nn}(0)A_{nn}]_U = \mathrm{Tr}\{\rho_{\mathrm{av}}A\}/D = \langle A\rangle_{\rho_{\mathrm{av}}}/D \tag{4}$$

for arbitrary n. Likewise, equation (3) yields

$$[\rho_{mn}(0)A_{nm}]_U = \frac{\langle A\rangle_{\rho(0)} - \langle A\rangle_{\rho_{\mathrm{av}}}}{D(D-1)} \tag{5}$$

for arbitrary $m \neq n$.

Upon separately averaging in equation (1), the summands with $m = n$ and those with $m \neq n$ over U, and then exploiting equations (4) and (5) one readily finds that

$$\left[\langle A\rangle_{\rho(t)}\right]_U = \langle A\rangle_{\rho_{\mathrm{av}}} + F(t)\left\{\langle A\rangle_{\rho(0)} - \langle A\rangle_{\rho_{\mathrm{av}}}\right\} \tag{6}$$

$$F(t) := \frac{D}{D-1}\left(|\phi(t)|^2 - \frac{1}{D}\right), \tag{7}$$

where $\phi(t)$ is the Fourier transform of the spectral density from ref. 46 (see also refs 51–53)

$$\phi(t) := \frac{1}{D}\sum_{n=1}^{D} e^{iE_n t/\hbar}. \tag{8}$$

The following results can be derived in principle along similar lines (symmetry arguments being one key ingredient), but since the actual details are quite tedious, they are postponed to Methods. As a first result, one obtains

$$\langle A\rangle_{\rho_{\mathrm{av}}} = \langle A\rangle_{\rho_{\mathrm{mc}}} + \frac{\langle A\rangle_{\rho(0)} - \langle A\rangle_{\rho_{\mathrm{mc}}}}{D+1}, \tag{9}$$

where $\rho_{\mathrm{mc}} := I/D$ is the microcanonical density operator and I the identity on $\mathcal{H}$. As a second result, one finds for the statistical fluctuations

$$\xi(t) := \langle A\rangle_{\rho(t)} - \left[\langle A\rangle_{\rho(t)}\right]_U \tag{10}$$

the estimate

$$[\xi^2(t)]_U = \mathcal{O}(\Delta_A^2 \mathrm{Tr}\{\rho^2(0)\}/D) \tag{11}$$

for arbitrary t, where Δ_A is the range of A, that is, the difference between the largest and smallest eigenvalues of A. Since averaging over U and integrating over t are commuting operations, equation (11) implies that

$$\left[\frac{1}{t_2 - t_1}\int_{t_1}^{t_2} \xi^2(t)dt\right]_U = \mathcal{O}\left(\frac{\Delta_A^2 \mathrm{Tr}\{\rho^2(0)\}}{D}\right) \tag{12}$$

for arbitrary $t_2 > t_1$.

Considering t in equation (11) as arbitrary but fixed, equation (10) and $D \gg 1$ imply (obviously or by exploiting Chebyshev's inequality[1,9,40,43,45]) that $\langle A\rangle_{\rho(t)}$ is practically indistinguishable from the average in equation (6) for the vast majority of all unitaries U. Indeed, the fraction (normalized Haar measure) of exceptional U's is unimaginably small for typical macroscopic systems with, say, $f \approx 10^{23}$ degrees of freedom, since D in equation (11) is exponentially large in f (see below equation (1)). Likewise, considering an arbitrary but fixed time interval $[t_1, t_2]$ in equation (12), it follows for all but a tiny fraction of U's that the time average over $\xi^2(t)$ on the left-hand side of equation (12) must be unimaginably small, and hence also the integrand $\xi^2(t)$ itself must be exceedingly small for the overwhelming majority of all $t \in [t_1, t_2]$. Accordingly, $\langle A\rangle_{\rho(t)}$ must remain extremely close to equation (6) simultaneously for all those $t \in [t_1, t_2]$.

Due to equation (9) and $D \gg 1$, we furthermore can safely approximate $\langle A\rangle_{\rho_{\mathrm{av}}}$ in equation (6) by $\langle A\rangle_{\rho_{\mathrm{mc}}}$. Altogether, we thus can conclude that in very good approximation

$$\langle A\rangle_{\rho(t)} = \langle A\rangle_{\rho_{\mathrm{mc}}} + F(t)\left\{\langle A\rangle_{\rho(0)} - \langle A\rangle_{\rho_{\mathrm{mc}}}\right\} \tag{13}$$

for the vast majority of unitaries U and times t. As detailed in Methods, the neglected corrections in equation (13) consist of a systematic (U independent) part, which is bounded in modulus by $\Delta_A/(D^2-1)$ for all t, and a random (U dependent) part (namely $\xi(t)$), whose typical order of magnitude is $\Delta_A\sqrt{\mathrm{Tr}\{\rho^2(0)\}/D}$ (for most U and t, cf. equations (11) and (12)), that is, $\xi(t)$ is dominating by far (note that $1 \geq \mathrm{Tr}\{\rho^2(0)\} \geq \mathrm{Tr}\{\rho_{\mathrm{mc}}^2\} = 1/D$). Moreover, the correlations of $\xi(t)$ decay on timescales comparable to those governing $F(t)$.

These are our main formal results. In the rest of the paper, we discuss their physical content.

Basic properties of $F(t)$. Equation (8) implies that $\phi(0) = 1$, $\phi(-t) = \phi^*(t)$ and $|\phi(t)| \leq 1$. With equation (7) and $D \gg 1$, it follows that in very good approximation

$$F(t) = |\phi(t)|^2, \tag{14}$$

and thus

$$F(0) = 1, 0 \leq F(t) \leq 1, F(-t) = F(t). \tag{15}$$

Indicating averages over all $t \geq 0$ by an overbar, one can infer from equations (8) and (14) that $\overline{F(t)} = \sum_k d_k^2/D^2$, where k labels the eigenspaces of H with mutually different eigenvalues and d_k denotes their dimensions. Since $\sum_k d_k = D$, we thus obtain $\overline{F(t)} \leq \max_k(d_k/D)$. Excluding extremely large multiplicities (degeneracies) of energy eigenvalues, it follows that the time average $\overline{F(t)}$ is negligibly small, and hence[1,9,40,43,45] that $F(t)$ itself must be negligibly small for the overwhelming majority of all sufficiently large t, symbolically indicated as

$$F(t \to \infty) \rightsquigarrow 0. \tag{16}$$

Note that there still exist arbitrarily large exceptional t's owing to the quasi-periodicity of $\phi(t)$ implied by equation (8). We also emphasize that our main result (13) itself admits arbitrary degeneracies of H.

As an example, we focus on the microcanonical set-up introduced below equation (1) and on not too large times, so that equation (8) is well approximated by

$$\phi(t) = \int_{E-\Delta E}^{E} \rho(x)e^{ixt/\hbar}dx, \tag{17}$$

where $\rho(x)$ represents the (smoothened and normalized) density of energy levels E_n in the vicinity of the reference energy x. If the level density is constant throughout the energy window $[E-\Delta E, E]$, we thus obtain with equation (14)

$$F(t) = \frac{\sin^2(\Delta E t/2\hbar)}{(\Delta E t/2\hbar)^2}. \tag{18}$$

Next, we recall Boltzmann's entropy formula $S(x) = k_{\mathrm{B}}\ln(\Omega(x))$, where $\Omega(x)$ counts the number of E_n's below x and k_{B} is Boltzmann's constant. Hence, $\Omega'(x)$ must be proportional to the level density $\rho(x)$ from above. Furthermore, $T := 1/S'(E)$ is the usual microcanonical temperature of a system with energy E at thermal equilibrium. A straightforward expansion then yields the approximation $\rho(E-y) = ce^{-y/k_{\mathrm{B}}T}$ for $y \geq 0$, where c is fixed via $\int_{E-\Delta E}^{E}\rho(x)dx = 1$. The omitted higher-order terms are safely negligible for all $y \geq 0$ and systems with $f \gg 1$ degrees of freedom, see also ref. 53. With equations (14) and (17), one thus finds

$$F(t) = \frac{1 - 2\alpha\cos(\Delta E\, t/\hbar) + \alpha^2}{(1-\alpha)^2\left[1 + (k_B T t/\hbar)^2\right]}, \quad (19)$$

where $\alpha := e^{-\Delta E/k_B T}$. For $\Delta E \ll k_B T$, one recovers equation (18), and for $\Delta E \gg k_B T$, one obtains

$$F(t) = \frac{1}{1 + (k_B T t/\hbar)^2}. \quad (20)$$

Typicality of thermalization. Equations (13) and (16) imply thermalization in the sense that the expectation value $\langle A\rangle_{\rho(t)}$ becomes (for most *U*) practically indistinguishable from the microcanonical average $\langle A\rangle_{\rho_{mc}}$ for the overwhelming majority of all sufficiently large *t*. Exceptional *t*'s are, for instance, due to quantum revivals, which, in turn, are apparently closely related to the quasi-periodicities of *F*(*t*).

Our assumption that energy eigenvalues must not be extremely highly degenerate (see above equation (16)) is similar to refs 46,47,49–51, but considerably weaker than the corresponding premises in most other related works[1,8–12,39–45].

The usual time inversion invariance on the fundamental, microscopic level[7] is maintained by equation (13) due to equation (15). Surprisingly, and in accordance with the second law of thermodynamics, the latter symmetry persists even if it is broken in the microscopic quantum dynamics, for example, by an external magnetic field.

By propagating $\rho(0)$ backward in time (with respect to one particular *U*) and taking the result as new initial state, one may easily tailor[41] examples of the very rare *U*'s and *t*'s, which notably deviate from the typical behaviour (13). Equivalently, one may back-propagate *A* instead of $\rho(0)$ (Heisenberg picture).

Note that *S* and *T* were introduced below equation (18) not in the sense of associating some entropy and temperature to the non-equilibrium states $\rho(t)$ but rather as a convenient level-counting tool. However, we now can identify them *a posteriori* with the pertinent entropy and temperature after thermalization.

The randomization via *U* (see the section 'Analytical results') can be viewed in two ways: either one considers $\rho(0)$, *A* and the spectrum of *H* as arbitrary but fixed, while the eigenbasis of *H* is sampled from a uniform distribution (Haar measure). Or one considers *H* and the spectra of $\rho(0)$ and *A* as arbitrary but fixed and randomizes the eigenvectors of *A* and $\rho(0)$. In doing so, a key point is that the relative orientation of the eigenbases of $\rho(0)$ and *A* can be chosen arbitrarily but then is kept fixed. Indeed, it is well known[12,49] that for 'most' such orientations the expectation values $\langle A\rangle_{\rho(0)}$ and $\langle A\rangle_{\rho_{mc}}$ are practically indistinguishable, that is, an initial $\langle A\rangle_{\rho(0)}$ far from equilibrium requires a careful fine-tuning of $\rho(0)$ relatively to *A*.

In reality, there is usually nothing random in the actual physical systems one has in mind. Hence, results such as equation (13), which (approximately) apply to the overwhelming majority of unitaries *U*, should be physically interpreted according to the common lore of random matrix theory[9,10,12], namely, as to apply practically for sure to a concrete system under consideration, unless there are particular reasons to the contrary.

Such reasons arise, for instance, when *A* is known to be a conserved quantity, implying a common eigenbasis of *A* and *H*, that is, the basis transformations *U* must indeed be very special. Furthermore, this non-typicality is structurally stable against sufficiently small perturbations of *A* and/or *H* so that the eigenvectors remain 'almost aligned' (each eigenvector of *A* mainly overlaps with one or a few eigenvectors of *H*), and hence *A* remains 'almost conserved' (almost commuting with *H*). Analogous non-typical *U*'s are expected when $\rho(0)$ is known to be (almost) conserved (commuting with *H*).

Further well-known exceptions are integrable systems, for which thermalization in the above sense may be absent for certain $\rho(0)$ and *A*[4,32] (but not for others[22]), systems exhibiting many-body localization[34,36] or trivial cases with non-interacting subsystems (Supplementary Note 2).

Our present focus is different: taking thermalization for granted is the temporal relaxation well approximated by equation (13)?

Typical fast relaxation and prethermalization. Equation (20) is governed by the Boltzmann time $t_B := h/k_B T$, amounting to $t_B \approx 10^{-13}$ s at room temperature. Equation (19) gives rise to comparably short timescales, unless the temperature is exceedingly low or the energy window ΔE is unusually small. Such relaxation times are much shorter than commonly observed in real systems[46,49–51]. Moreover, the temporal decay is typically non-exponential (see, for example, equations (18)–(20)), again in contrast to the usual findings.

This seems to imply that typical experiments correspond to non-typical unitaries *U*. Plausible explanations are as follows: to begin with, the above-predicted typical relaxation times are so short that they simply could not be observed in most experiments. Second (or as a consequence), the usual initial conditions and/or observables are indeed quite 'special' with respect to the prominent role of almost conserved quantities (see previous section), in particular, 'local descendants' of globally conserved quantities such as energy, charge, particle numbers and so on: examples are the amount of energy, charge and so on within some subdomain of the total system or, more generally, local densities, whose content within a given volume can only change via transport currents through the boundaries of that volume. As a consequence, the global relaxation process becomes 'unusually slow' if the densities between macroscopically separated places need to equilibrate (small surface-to-volume ratio) or if there exists a natural 'bottleneck' for their exchange (weakly interacting subsystems).

Put differently, our present theory is meant to describe the very rapid relaxation towards local equilibrium, but not any subsequent global equilibration. Only if there exists a clear-cut timescale separation between these two relaxation steps (or if there is no second step at all), can we hope to quantitatively capture the first step by our results. Conversely, the timescale separation usually admits some Markovian approximation for the second step, yielding an exponential decay, whose timescale still depends on many details of the system.

Natural further generalizations include the closely related concepts of hindered equilibrium, quasi-equilibrium (metastability) and, above all, prethermalization[29,54,55], referring, for example, to a fast partial thermalization within a certain subset of modes, (quasi-)particles or other generalized degrees of freedom. (Like in ref. 54, we do not adopt here the additional requirement[55] that the almost conserved quantities originate from a weak perturbation of an integrable system.)

In short, our working hypothesis is that the theory (13) describes the temporal relaxation of $\langle A\rangle_{\rho(t)}$ for any given pair $(\rho(0), A)$ unless one of them is exceptionally close to or in some other way slowed down by an (almost) conserved quantity.

Comparison with experimental results. We focus on experiments in closed many-body systems in accordance with the above general requirements. In comparing them with our theory (13), we furthermore assume that the (pre-)thermalized system occupies a microcanonical energy window with some (effective)

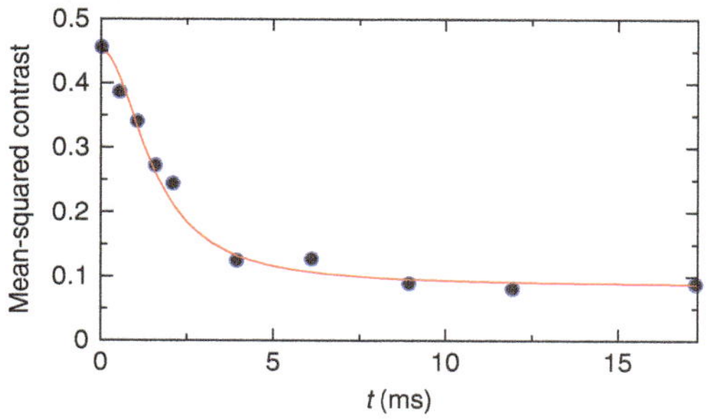

Figure 1 | Prethermalization of ultracold atoms. The considered observable 'mean-squared contrast' quantifies the spatial correlation of the matter-wave interference pattern after coherently splitting a Bose gas into two quasi-condensates (see ref. 29 for more details). Symbols: experimental data from Fig. 2a of ref. 29. Line: theoretical prediction (13) and (20) with $T=5$ nK. The pertinent effective temperature has also been roughly estimated in ref. 29 (see Fig. 2b therein) and is still compatible with our present fit $T=5$ nK. As discussed at the end of the section 'Typical fast relaxation and prethermalization', the depicted prethermalization is followed by a much slower, global thermalization[29], which is omitted in the present figure.

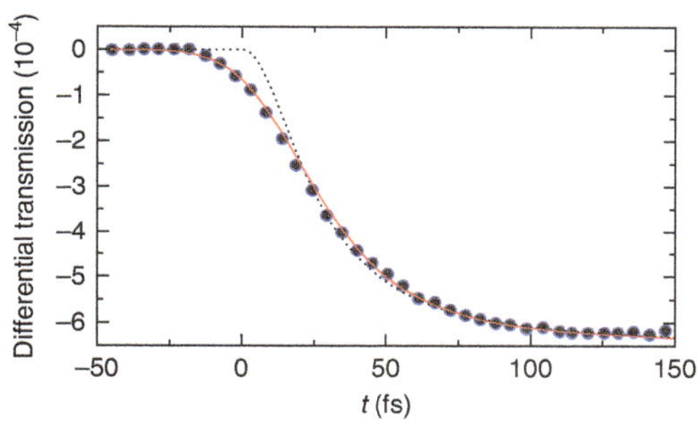

Figure 2 | Ultrafast relaxation of hot electrons. A first laser pulse (at $t=0$) 'heats up' the electron gas in a thin ferromagnetic film, whose rethermalization is then probed by means of a second laser pulse. As detailed in ref. 56, the considered observable 'differential transmission' quantifies the magneto-optical polarization rotation of the probe laser light. Symbols: experimental data from Fig. 2a of ref. 56. Dotted: theoretical prediction (13) and (20) with $T=310$ K and $F(t<0):=1$. Solid: convolution of the dotted line with a Gaussian of 35 fs FWHM, accounting for the finite widths of the pump and probe laser pulses (see also main text). Similarly as in Fig. 1, on larger timescales than covered by the present figure, the prethermalized electrons also exhibit non-negligible interactions with the lattice phonons and magnons, resulting in a much slower global relaxation of the compound electron-lattice system[56]. Concerning the pertinent temperature T, a direct experimental estimate is not available for the set-up from ref. 56 (I contacted one of the authors), but it has been provided for a similar experiment by the same group in ref. 65, except that the fluence (energy per spot area of the pump laser pulse) was 70 times larger than in ref. 56. Taking all this into account, the estimate $T=310$ K adopted in the present figure seems very reasonable.

temperature T and $\Delta E \gg k_B T$, so that equation (20) applies. Finally, the asymptotic values $\langle A\rangle_{\rho(0)}$ and $\langle A\rangle_{\rho_{mc}}$ in equation (13) are either obvious or will be estimated from the measurements, hence no further knowledge about the often quite involved details of the experimental observables will be needed!

Figure 1 demonstrates the very good agreement of the theory with the rapid initial prethermalization of a coherently split Bose gas, observed by the Schmiedmayer group in ref. 29.

In Fig. 2, the theory is compared with the pump-probe experiment by the Bigot group from ref. 56. The finite widths of the pump and the probe laser pulses are roughly accounted for by convoluting equation (13) with a Gaussian of 35 fs FWHM (full width at half maximum). In ref. 56, the FWHM of the pump pulse is estimated as 20 fs and the combined FWHM for both pulses as 22 fs, implying a FWHM of 9 fs for the probe pulse. The latter value seem quite optimistic to us. A second 'excuse' for our slightly larger FWHM value of 35 fs is that the tails of the experimental pulse shape may be considerably broader than those of a Gaussian with the same FWHM (see, for example, Fig. 2c in the Supplementary Material of ref. 57). Finally, the convolution of equation (13) with a Gaussian represents a rather poor 'effective description' in the first place: our entire theoretical approach becomes strictly speaking invalid when the duration of the perturbation becomes comparable to the thermalization time.

A similar comparison with the pump-probe experiments from ref. 58 is presented in Fig. 3. As before, we adopted a slightly larger FWHM of 100 fs than the estimate of 76 fs in ref. 58. Due to the above-mentioned fundamental limitations of our theory for such rather large FWHM values, the temperatures adopted in Fig. 3 should still be considered as quite crude estimates. Apart from that, Fig. 3 nicely confirms the predicted temperature dependence from equation (20).

We close with three remarks: first, refs 56,58 also implicitly confirm our prediction that the essential temporal relaxation (encapsulated by $F(t)$ in equation (13)) is generically the same for different observables. Second, similar pump-probe experiments abound in the literature, but usually the pulse widths are too large for our purposes. Third, the temporal relaxation in Figs 1–3 has also been investigated numerically, but closed analytical results have not been available before[29,58].

Comparison with numerical results. Figure 4 illustrates the very good agreement of our theory with Rigol's numerical findings from ref. 32, both for an integrable and an non-integrable example. A similar agreement is found for all other parameters and also for an analogous hardcore boson model examined in refs 31,32. On the other hand, a second observable considered in ref. 32, deriving from the momentum distribution function, exhibits in all cases a significantly slower and also qualitatively different temporal relaxation. According to the discussion in the section 'Typical fast relaxation and prethermalization', it is quite plausible that the latter observable is indeed 'non-typical' in view of the fact that it represents a conserved quantity for fermions with $V=\tau'=V'=0$ (ref. 32).

In Fig. 5, we compare our theory with the simulations of a different one-dimensional electron model from ref. 59. In doing so, the pertinent temperature T has been estimated as follows: the textbook Sommerfeld expansion for N electrons in a one-dimensional box yields $E=E_0[1+(3\pi^2/8)(k_B T/E_F)^2]$, where E is their total energy, $E_0=(1/3)NE_F$ the ground-state energy, $E_F=(\pi\hbar N/gL)^2/2m$ the Fermi energy, L the box length, m the electron mass and $g:=2s+1=2$ ($s=1/2$ for electrons). Assuming that the pulse acts solely on the small well implies $N=16$, $L\simeq 15$ nm (ref. 59), and $E-E_0\simeq 0.045$ eV (see Fig. 8a in ref. 59). Altogether, we thus obtain $T\simeq 170$ K.

The remnant 'fluctuations' of the numerical data in Figs 4 and 5 can be readily explained as finite particle number effects (see Fig. 4 in ref. 32 and Fig. 10 in ref. 59), and their temporal correlations are as predicted below equation (13). The seemingly rather strong fluctuations in Fig. 5 are a fallacy since the systematic changes themselves are very small.

Next, we turn to the numerical findings for a qubit in contact with a spin bath by the Trauzettel group from ref. 60. The agreement with our theory in Fig. 6 is as good as it possibly can be for such a rather small dimensionality of $D=2^7$. Indeed, the remaining differences nicely confirm the predictions below equation (13), regarding both their typical order of magnitude

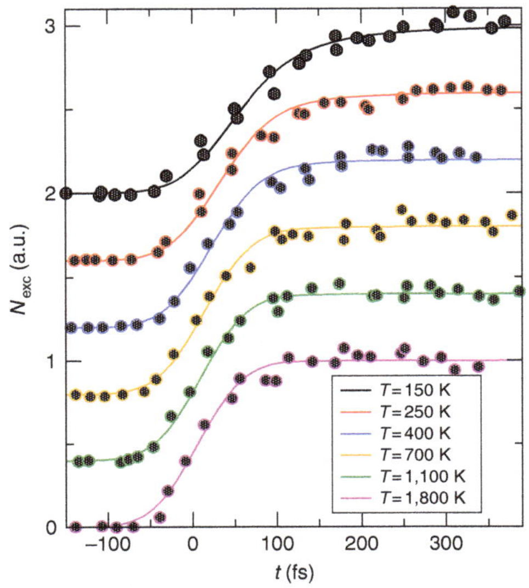

Figure 3 | Temperature-dependent relaxation of hot electrons. Symbols: similar pump-probe experiments as in Fig. 2, but now conducted on bismuth and for six different fluences (energy per spot area of the pump laser pulses). As detailed in ref. 58, the considered observable N_{exc} quantifies (in a.u.) the number of excited electrons above the Fermi level. The depicted data are from Fig. 5b of ref. 58 for fluences (top-down) 0.12, 0.2, 0.36, 0.52, 0.68, and 0.84 mJ cm^{-2}. Lines: theoretical prediction (13) and (20) with temperatures as indicated and convoluted with a Gaussian of 100 fs FWHM (see also main text). The conversion of a given fluence into a temperature change of the electron gas is not obvious. In particular, the estimates provided in ref. 58 seem not very reliable to us: first of all, Fig. 6 in ref. 58 indicates a temperature of ca. 250 K at four different time points ~200 fs before the pump pulse, while the actual temperature of the unperturbed system is known to be 130 K. Second, the temperature error bars in Fig. 6b of ref. 58 are quite large. Third, a key premise of those estimates in ref. 58 is that the 'renormalized' curves in Supplementary Fig.3b of ref 66 should coincide, while their actual agreement is only moderately better than for the 'bare' curves in Supplementary Fig. 3a. For all these reasons, we used the temperature as a fit parameter in the present figure.

$\Delta_A\sqrt{\mathrm{Tr}\{\rho^2(0)\}/D} = 1\sqrt{2^{-6}/2^7} \simeq 0.01$ and their temporal correlations (where we exploited that $\mathrm{Tr}\{\rho^2(0)\} = 2^{-6}$ for the particular initial condition $\rho(0)$ adopted in Fig. 6).

Our final example is Bartsch and Gemmer's random matrix model from ref. 17. Referring to the notation and definitions in the caption of Fig. 7, one readily sees that the considered observable A is a conserved quantity for the unperturbed Hamiltonian ($\lambda = 0$). In agreement with our discussion in the section 'Typical fast relaxation and prethermalization', A is therefore still 'almost conserved' for small λ and indeed exhibits a slow, exponential decay towards $\langle A\rangle_{\rho_{\mathrm{mc}}} = 0$ (see Fig. 1a in ref. 17). Upon increasing λ, one recovers the much faster, non-exponential decay of our present theory (see Fig. 1b in ref. 17). Unfortunately, the λ-value 1.77×10^{-3} from Fig. 1b of ref. 17 is still somewhat too small and the eigenvalues E_1, ..., $E_{6,000}$ are not any more available (I asked the authors). Therefore, we repeated the numerics from ref. 17 on our own for $\lambda = 7\times10^{-3}$. The resulting agreement with equation (13) in Fig. 7 is very good, and the temporal correlations of the deviations as well as their typical order of magnitude $\Delta_A\sqrt{\mathrm{Tr}\{\rho^2(0)\}/D} = 2\sqrt{1/6,000} \simeq 0.03$ are as predicted below equation (13).

We close with two remarks: first, there is no fit parameter in any of the above examples apart from $\langle A\rangle_{\rho(0)}$ in Fig. 4 and $\langle A\rangle_{\rho_{\mathrm{mc}}}$ in Figs 4 and 5. Second, especially in the case of the integrable

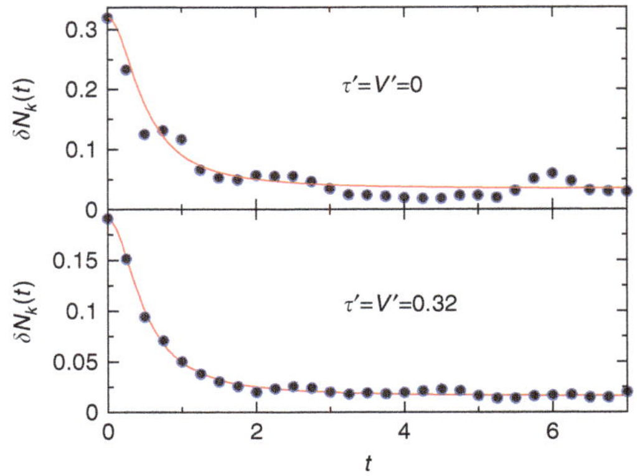

Figure 4 | Relaxation of an integrable and a non-integrable fermionic model. The upper part of the plot refers to an integrable model, the lower to a nonintegrable one. Symbols: numerical results from ref. 32 for eight strongly correlated fermions on a one-dimensional lattice with 24 sites, described in terms of an extended Hubbard model with nearest- and next-nearest-neighbour hopping and interaction parameters τ, τ', V and V', respectively. Working in units with $\hbar = k_B = \tau = V = 1$ and focusing on parameters $\tau' = V'$, the model is integrable if $\tau' = V' = 0$ and non-integrable otherwise. A quantum quench generates an initial pure state out of equilibrium, whose energy corresponds to that of a canonical ensemble with temperature $T = 2$. As detailed in ref. 32, the considered observable $\delta N_k(t)$ is a dimensionless descendant of the density-density structure factor. The depicted data are from Fig. 1g,j of ref. 32. Lines: theoretical predictions (13) and (20) with $T = 2$.

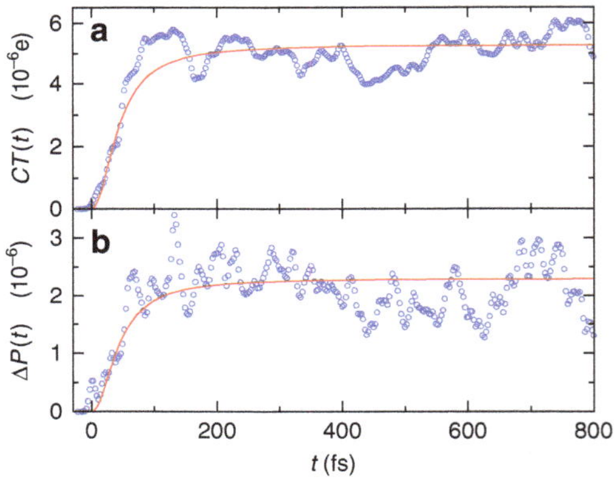

Figure 5 | Prethermalization in a one-dimensional electron gas. Symbols: numerical results from ref. 59 for a one-dimensional model of the many-electron dynamics in an asymmetric double-well potential (emulating a metal-insulator-metal junction). Starting with 44 electrons in the ground state, a laser pulse-like electrical perturbation acts predominantly on the 16 electrons in the smaller, box-shaped well, and then their rethermalization is monitored via the charge transfer into the larger well (denoted in (a) as $CT(t)$), and via the change of the ground state population (denoted in (b) as $\Delta P(t)$). Depicted are the numerical results from Fig. 8 of ref. 59. For further details regarding the simulations, we refer to refs 59,67. Lines: theoretical predictions (13) and (20), exploiting the estimate $T = 170$ K from the main text, and neglecting the finite temporal width (20 fs) of the pulse. As in Figs 1-3, we are actually dealing with a prethermalization process within the smaller well. The subsequent global thermalization is much slower due to the high barrier between the wells. Considering that $\langle A\rangle_{\rho_{\mathrm{mc}}}$ is the only remaining fit parameter in the theory from equations (13) and (20), the agreement with the simulations is remarkably good. In particular, the two very different observables $CT(t)$ and $\Delta P(t)$ are indeed governed by the same $F(t)$, as predicted by equations (13) and (20).

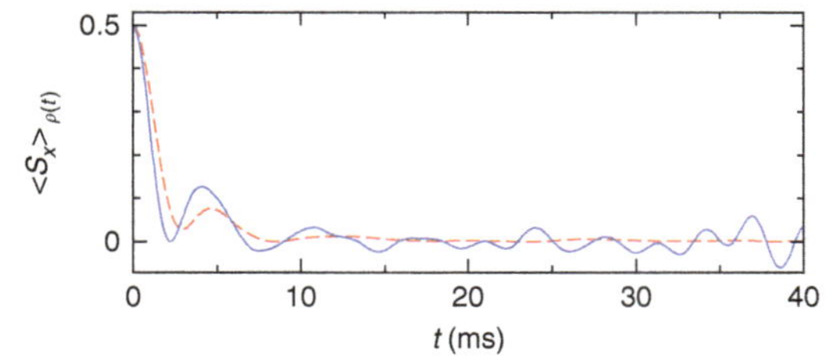

Figure 6 | Thermalization of a spin qubit coupled with a bath. Solid: numerical results for the model with seven spin-1/2 degrees of freedom in an external magnetic field from ref. 60: a central spin (qubit) is randomly (and reasonably weakly) coupled with a bath of six spins. The initial state $\rho(0)$ is the product of a totally mixed bath state and an eigenstate of the central spin component S_x. Depicted are the data from Fig. 2 of ref. 60 for the central spin component S_x. Dashed: theoretical prediction (13), (14) and (8). Due to the above-mentioned initial condition and the quite small dimension $D=2^7$, the approximations (18)-(20) are not very well satisfied by the actual energy eigenvalues $E_1, \ldots, E_{128}$ (kindly provided by the authors of ref. 60). Hence, we have evaluated $F(t)$ in equation (13) directly via equations (14) and (8).

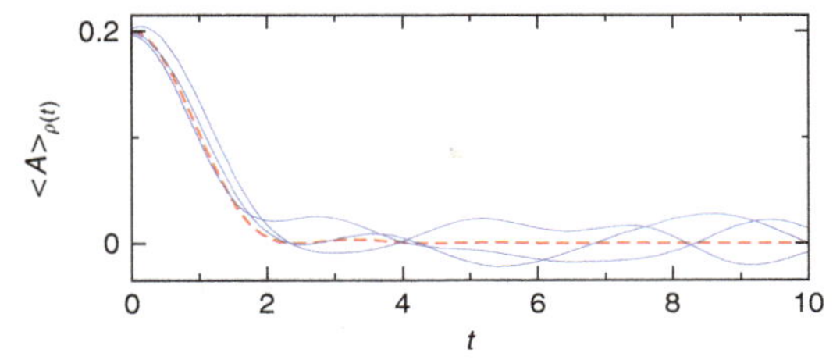

Figure 7 | Thermalization in a random matrix model. Solid: numerical results for the random matrix model of the form $H=H_0+\lambda V$ from ref. 17. Adopting dimensionless units with $\hbar=1$, the $D=6{,}000$ eigenvalues of H_0 are chosen equidistant with level spacing 8.33×10^{-5} (ref. 17). The matrix elements of A (observable) and V (perturbation) in the basis of H_0 satisfy $A_{ik}=(-1)^k\delta_{ik}$ and $V_{ki}=V^*_{ik}$. Apart from the latter constraint, the real and imaginary parts of V_{ik} are independent, normally distributed random numbers. The initial state is $\rho(0)=|\psi\rangle\langle\psi|$, where $|\psi\rangle$ is randomly sampled from the energy shell $\mathcal{H}$ under the constraint $\langle A\rangle_{\rho(0)}\simeq 0.2$ (ref. 17). Depicted are three representative numerical realizations for $\lambda=7\times10^{-3}$ akin to Fig. 1b of ref. 17 (in dimesionless units). Dashed: theoretical prediction (13), (14) and (8). Similarly as in Fig. 6, the numerically obtained energies $E_1, \ldots, E_{6{,}000}$ were found to satisfy equations (18)-(20) not very well, hence we have directly evaluated equations (8) and (14).

model in Fig. 4, one may question whether the considered system exhibits thermalization in the first place, as is tacitly assumed in equation (13). In Supplementary Note 2, we argue that equation (13) indeed is expected to still remain valid in such cases if $\langle A\rangle_{\rho_{\mathrm{mc}}}$ is replaced by the pertinent non-thermal long-time asymptotics (which, in turn, is estimated from the numerical data in Fig. 4).

Discussion

Our main result (13) implies thermalization in the sense that a generic non-equilibrium system with a macroscopically well-defined energy becomes practically indistinguishable from the corresponding microcanonical ensemble for the overwhelming majority of all sufficiently late times. Apart from the concrete initial and long-time expectation values (that is, $\langle A\rangle_{\rho(0)}$ and $\langle A\rangle_{\rho_{\mathrm{mc}}}$ in equation (13)), the temporal relaxation (that is, $F(t)$ in equation (13)) depends only on the spectrum of the Hamiltonian within the pertinent interval of non-negligibly populated energy eigenstates, but not on any further details of the initial condition or the observable. This represents one of the rare instances of a general quantitative statement about systems far from equilibrium.

The theory agrees very well with a wide variety of experimental and numerical results from the literature (though none of them was originally conceived for the purpose of such a comparison). We are in fact not aware of any other quantitative analytical explanation of those data comparable to ours. Indeed, the usual paradigm to identify and then analytically quantify the main physical mechanisms seems almost hopeless here. In a sense, our present approach thus amounts to a different paradigm: there is no need of any further 'explanations', since the observed behaviour is expected with overwhelming likelihood from the very beginning, that is, unless there are special *a priori* reasons to the contrary.

Similarly as in refs 46,49–51, generic thermalization is found to happen extremely quickly (unless the system's energy or temperature is exceedingly low). Moreover, the temporal decay is typically non-exponential. A main prediction of our theory is that these features should in fact be very common (at least in the form of prethermalization), but often they are unmeasurably fast or they have simply not been looked for so far. Conversely, most of the usually considered observables and initial conditions are actually quite 'special', namely, exceptionally slow, 'almost conserved' quantities. A better understanding of those principally untypical but practically very common thermalization processes remains an open problem[49–51].

Methods

Basic matrices. According to the section 'Analytical results', the unitary U represents the basis transformation between the eigenvectors $|n\rangle$ ($n=1, \ldots, D$) of the Hamiltonian H and those of the observable A. Denoting the eigenvalues of A by λ_ν and the eigenvectors by $|\psi_\nu\rangle$ ($\nu=1, \ldots, D$), the matrix elements of U are thus $U_{n\nu} := \langle n \mid \psi_\nu\rangle$. Accordingly, the matrix elements of $\rho(0)$ in the basis of H are related to those in the basis of A via

$$\rho_{mn}(0)=\sum_{\mu,\nu=1}^{D} U_{m\mu}\rho_{\mu\nu}U^*_{n\nu}, \tag{21}$$

where $\rho_{\mu\nu} := \langle\psi_u|\rho(0)|\psi_\nu\rangle$. Similarly, the matrix elements of A satisfy

$$A_{mn}=\sum_{\xi=1}^{D} U_{m\xi}\lambda_\xi U^*_{n\xi}, \tag{22}$$

and hence

$$\rho_{mn}(0)A_{nm}=\sum_{\mu,\nu,\xi=1}^{D}\rho_{\mu\nu}\lambda_\xi U_{m\mu}U^*_{n\nu}U_{n\xi}U^*_{m\xi}. \tag{23}$$

As announced below equation (3), we work (without loss of generality) in a reference frame (or reference basis of $\mathcal{H}$) so that only H (and thus $|n\rangle$) depends on U, while A and $\rho(0)$ (and thus $|\psi_\nu\rangle$) are independent of U. Hence, $\rho_{\mu\nu}$ and λ_ξ on the right-hand side of equations (21)–(23) are independent of U.

Derivation of equation (9). As a simple first exercise, let us average equation (23) over all uniformly (Haar) distributed unitaries U, as specified in the section 'Analytical results'. Since the factors $\rho_{\mu\nu}\lambda_\xi$ on the right-hand side are independent of U, we are left with averages over the U matrix elements. Such averages have been evaluated repeatedly and often independently of each other in the literature, see, for example, refs 5,61–63, a key ingredient being symmetry arguments due to the invariance of the Haar measure under arbitrary unitary transformations. Particularly convenient for our present purposes is the formalism adopted by Brouwer and Beenakker, see ref. 63, and further references therein. The general structure of such averages is provided by equation (2.2) in ref. 63, reading

$$\left[U_{a_1b_1}\cdots U_{a_mb_m}U^*_{\alpha_1\beta_1}\cdots U^*_{\alpha_n\beta_n}\right]_U=\delta_{mn}\sum_{P,P'}V_{P,P'}\prod_{j=1}^{n}\delta_{a_j\alpha_{P(j)}}\delta_{b_j\beta_{P'(j)}}. \tag{24}$$

Quoting verbatim from ref. 63, 'the summation is over all permutations P and P' of the numbers $1, \ldots, n$. The coefficients $V_{P,P'}$ depend only on the cycle structure of the permutation $P^{-1}P'$. Recall that each permutation of $1, \ldots, n$ has a unique factorization in disjoint cyclic permutations ('cycles') of lengths $c_1, \ldots, c_k$ (where $n=\sum_{j=1}^{k}c_j$). The statement that $V_{P,P'}$ depends only on the cycle structure of $P^{-1}P'$ means that $V_{P,P'}$ depends only on the lengths $c_1, \ldots, c_k$ of the cycles in the factorization of $P^{-1}P'$. One may therefore write $V_{c_1,\ldots,c_k}$ instead of $V_{P,P'}$.' The explicit numerical values of all $V_{c_1,\ldots,c_k}$ with $n \le 5$ are provided by the columns 'CUE' of Tables II and IV in ref. 63. Further remarks: the labels m and n in equation (24) have nothing to do with those in equation (23). Equation (24) equals zero unless $m = n$. Every label a_j must have a 'partner', that is, its value must coincide with one of the α_j's and vice versa, since otherwise the product over the Kronecker delta's $\delta_{a_j\alpha_{P(j)}}$ in equation (24) would be zero for all P's. Note that some a_j's may assume the same value, but then an equal number of α_j's also must assume that value. Likewise, every b_j needs a 'partner' among the β_j's and vice versa.

Adopting the abbreviation

$$X_{mn} := [\rho_{mn}(0)A_{nm}]_U \tag{25}$$

and the renamings $a_1 := m$, $a_2 := n$, $b_1 := \mu$, $b_2 := \xi$ and $b_3 := \nu$, equation (23) yields

$$X_{a_1a_2} = \sum_{b_1,b_2,b_3} \rho_{b_1b_3}\lambda_{b_2}\left[U_{a_1b_1}U_{a_2b_2}U^*_{a_1b_2}U^*_{a_2b_3}\right]_U . \tag{26}$$

The connection with equation (24) is established via the identifications $\alpha_1 := a_1$, $\alpha_2 := a_2$, $\beta_1 := b_2$ and $\beta_2 := b_3$. Therefore, if $b_1 \neq b_2$, then the only potential 'partner' of b_1 is β_2, and only if their values coincide, that is, $b_3 = b_1$, the corresponding summands may be non-zero. The same conclusion can be drawn if $b_1 = b_2$. We thus can rewrite equation (26) with equation (24) as

$$X_{a_1a_2} = \sum_{b_1,b_2} \rho_{b_1b_1}\lambda_{b_2}\sum_{P,P'} V_{P,P'}\prod_{j=1}^{2}\delta_{a_ja_{P(j)}}\delta_{b_j\beta_{P'(j)}}, \tag{27}$$

where $\beta_1 = b_2$ and $\beta_2 = b_1$.

There are two permutations of the numbers 1 and 2, namely, the identity and one, which exchanges 1 and 2. Denoting them as P_1 and P_2, respectively, and observing that $\beta_j = b_{P_2(j)}$, equation (27) can be rewritten as

$$X_{a_1a_2} = \sum_{k=1}^{2}\prod_{j=1}^{2}\delta_{a_ja_{P_k(j)}}\sum_{l=1}^{2}V_{P_k,P_l}S_l \tag{28}$$

$$S_l := \sum_{b_1,b_2}\rho_{b_1b_1}\lambda_{b_2}\prod_{j=1}^{2}\delta_{b_jb_{P_2(P_l(j))}} \tag{29}$$

For $l = 1$, the two Kronecker delta's in equation (29) both require that $b_1 = b_2$ and hence

$$S_1 = \sum_{b_1}\rho_{b_1b_1}\lambda_{b_1} = \mathrm{Tr}\{\rho(0)A\}. \tag{30}$$

The last equality can be verified by evaluating the trace in the eigenbasis of A, see above equation (21). In the same way, one finds that

$$S_2 = \sum_{b_1,b_2}\rho_{b_1b_1}\lambda_{b_2} = \mathrm{Tr}\{\rho(0)\}\mathrm{Tr}\{A\} = D\mathrm{Tr}\{\rho_{\mathrm{mc}}A\}. \tag{31}$$

In the last equation, we exploited that $\mathrm{Tr}\{\rho(0)\} = 1$ and $\rho_{\mathrm{mc}} := I/D$, see below equation (9). Observing that the two Kronecker delta's in equation (28) equal one if $k = 1$ or if $k = 2$ and $a_1 = a_2$, the overall result is

$$\begin{aligned} X_{a_1a_2} =& \langle A\rangle_{\rho(0)}\left(V_{P_1,P_1} + \delta_{a_1a_2}V_{P_2,P_1}\right)\\ &+ D\langle A\rangle_{\rho_{\mathrm{mc}}}\left(V_{P_1,P_2} + \delta_{a_1a_2}V_{P_2,P_2}\right), \end{aligned} \tag{32}$$

where, as usual, $\langle A\rangle_{\rho(0)} := \mathrm{Tr}\{\rho(0)A\}$ and $\langle A\rangle_{\rho_{\mathrm{mc}}} := \mathrm{Tr}\{\rho_{\mathrm{mc}}A\}$.

Finally, the coefficients V_{P_k,P_l} are evaluated as explained below equation (24): if $k = l$, then $P_l^{-1}P_k = P_1$ factorizes in two cycles of lengths $c_1 = c_2 = 1$, that is, $V_{P_k,P_l} = V_{c_1,c_2} = V_{1,1}$. Likewise, if $k \neq l$, then $P_l^{-1}P_k = P_2$ consists of one cycle with $c_1 = 2$, that is, $V_{P_k,P_l} = V_2$. Referring to columns 'CUE' and rows '$n = 2$' of Tables II and IV in ref. 63 yields $V_{1,1} = 1/(D^2-1)$ and $V_2 = -1/[D(D^2-1)]$. Returning to the original labels m and n in equation (25), we thus can rewrite equation (32) as

$$X_{mn} = \langle A\rangle_{\rho(0)}\frac{D-\delta_{mn}}{D(D^2-1)} + \langle A\rangle_{\rho_{\mathrm{mc}}}\frac{D\delta_{mn}-1}{D^2-1}. \tag{33}$$

As a consequence, we can infer from equations (4) and (25) that $\langle A\rangle_{\rho_{\mathrm{av}}} = DX_{nn}$ and with equation (33) that

$$\langle A\rangle_{\rho_{\mathrm{av}}} = \langle A\rangle_{\rho(0)}\frac{1}{D+1} + \langle A\rangle_{\rho_{\mathrm{mc}}}\frac{D}{D+1}. \tag{34}$$

Hence, one readily recovers equation (9).

A relation remarkably similar to our present equation (9), albeit in a quite different physical context, has been previously obtained also in ref. 64 (see equation (2) therein).

Derivation of equation (11). Without any doubt, there are much faster ways to obtain equations (33) or (34). The advantage of our present way is that it can be readily adopted without any conceptual differences (albeit the actual calculations become more lengthy) to more demanding cases like

$$\left[\xi^2(t)\right]_U = \left[\langle A\rangle^2_{\rho(t)}\right]_U - \left[\langle A\rangle_{\rho(t)}\right]^2_U, \tag{35}$$

see equation (10).

To evaluate the last term in equation (35), we recast equation (6) with equations (7) and (9) into the form

$$\left[\langle A\rangle_{\rho(t)}\right]_U = F_0(t)\langle A\rangle_{\rho(0)} + \bar{F}_0(t)\langle A\rangle_{\rho_{\mathrm{mc}}} + R_1(t) \tag{36}$$

$$R_1(t) := \bar{F}_0(t)\frac{\langle A\rangle_{\rho_{\mathrm{mc}}} - \langle A\rangle_{\rho(0)}}{D^2-1} \tag{37}$$

$$\bar{F}_0(t) := 1 - F_0(t) \tag{38}$$

$$F_0(t) := \frac{1}{D^2}\sum_{m,n=1}^{D} e^{i(E_n-E_m)t/\hbar} = |\phi(t)|^2, \tag{39}$$

where $\phi(t)$ is defined in equation (8). Similarly as in equation (15), one sees that $F_0(t), \bar{F}_0(t) \in [0,1]$ for all t. Denoting by $\lambda_{\max}$ and $\lambda_{\min}$ the largest and smallest among the eigenvalues $\lambda_1, \ldots, \lambda_D$ of A, the range of A is defined as $\Delta_A := \lambda_{\max} - \lambda_{\min}$. Furthermore, we can and will add a constant to A so that $\lambda_{\min} = -\lambda_{\max}$ without any change in the final conclusions below. It readily follows that $|\lambda_\nu| \le \Delta_A/2$ for all ν and hence that

$$\left|\langle A^\kappa\rangle_\rho\right| \le (\Delta_A/2)^\kappa \tag{40}$$

for arbitrary density operators ρ and $\kappa \in \mathbb{N}$. We thus can infer from equation (37) that

$$|R_1(t)| \le \Delta_A/(D^2-1). \tag{41}$$

Likewise, one finds upon squaring equation (36) that

$$\left[\langle A\rangle_{\rho(t)}\right]^2_U = \left(F_0(t)\langle A\rangle_{\rho(0)} + \bar{F}_0(t)\langle A\rangle_{\rho_{\mathrm{mc}}}\right)^2 + R_2(t) \tag{42}$$

$$|R_2(t)| \le 3\Delta_A^2/(D^2-1). \tag{43}$$

Turning to the first term on the right-hand side of (35), one can infer, similarly as in equations (25) and (26), from (1) and (23) that

$$\left[\langle A\rangle^2_{\rho(t)}\right]_U = \sum_{a_1,\ldots,a_4} e^{i(E_{a_1}-E_{a_2}+E_{a_3}-E_{a_4})t/\hbar}X_{a_1,\ldots,a_4} \tag{44}$$

$$X_{a_1,\ldots,a_4} := \sum_{b_1,\ldots,b_6}\rho_{b_1b_5}\lambda_{b_2}\rho_{b_3b_6}\lambda_{b_4}\left[U_{a_1b_1}\ldots U_{a_4b_4}U^*_{a_1\beta_1}\ldots U^*_{a_4\beta_4}\right]_U, \tag{45}$$

with $\beta_1 := b_2$, $\beta_2 := b_5$, $\beta_3 := b_4$ and $\beta_4 := b_6$. Similarly as below equation (26), it follows that only those summands may be non-zero, for which b_1 and b_3 have 'partners' among β_2 and β_4, and vice versa. This condition can be satisfied in two ways: (i) $b_5 = b_1$ and $b_6 = b_3$; and (ii) $b_5 = b_3$, $b_6 = b_1$ and $b_1 \neq b_3$. The latter condition is due to the fact that the case $b_1 = b_3$ is already covered by (i). Exploiting equation (24) and with the abbreviation $\vec{a} := (a_1, \ldots, a_4)$ and likewise for $\vec{b}$, $\vec{\beta}$ and so on, we thus obtain

$$X_{\vec{a}} = X^{(i)}_{\vec{a}} + X^{(ii)}_{\vec{a}} \tag{46}$$

$$X^{(i)}_{\vec{a}} := \sum_{\vec{b}}\rho_{b_1b_1}\lambda_{b_2}\rho_{b_3b_3}\lambda_{b_4}\sum_{P,P'}V_{P,P'}\prod_{j=1}^{4}\delta_{a_ja_{P(j)}}\delta_{b_j\beta^{(i)}_{P'(j)}} \tag{47}$$

$$X^{(ii)}_{\vec{a}} := \sum_{\vec{b},b_1\neq b_3}\rho_{b_1b_3}\lambda_{b_2}\rho_{b_3b_1}\lambda_{b_4}\sum_{P,P'}V_{P,P'}\prod_{j=1}^{4}\delta_{a_ja_{P(j)}}\delta_{b_j\beta^{(ii)}_{P'(j)}}, \tag{48}$$

where $\vec{\beta}^{(i)} := (b_2, b_1, b_4, b_3)$ and $\vec{\beta}^{(ii)} := (b_2, b_3, b_4, b_1)$.

There are $4! = 24$ permutations P of the numbers 1, 2, 3 and 4. Adopting the shorthand notation $[P(1)P(2)P(3)P(4)]$ to explicitly specify a given P, these 24 permutations are:

$$P_1 = [1234], P_2 = [2134], P_3 = [3214], P_4 = [4231],$$

$$P_5 = [1324], P_6 = [1432], P_7 = [1243], P_8 = [2143],$$

$$P_9 = [3412], P_{10} = [4321], P_{11} = [1342], P_{12} = [1423],$$

$$P_{13} = [3241], P_{14} = [4213], P_{15} = [2431], P_{16} = [4132],$$

$$P_{17} = [2314], P_{18} = [3124], P_{19} = [2341], P_{20} = [2413],$$

$$P_{21} = [3421], P_{22} = [3142], P_{23} = [4312], P_{24} = [4123].$$

Observing that $\beta^{(i)}_j = b_{P_8(j)}$ and $\beta^{(ii)}_j = b_{P_{19}(j)}$ for all $j = 1, \ldots, 4$, it is quite straightforward but very arduous to explicitly carry out the sums over P' and $\vec{b}$ in equations (47) and (48), and the sum over $\vec{a}$ in equation (44), yielding

$$\left[\langle A\rangle^2_{\rho(t)}\right]_U = \sum_{k=1}^{24} f_k(t)T(P_k), \tag{49}$$

where the functions $f_k(t)$ are given by

$$\begin{aligned} f_1(t) &= D^4 F_0^2(t), \\ f_2(t) &= f_4(t) = f_5(t) = f_7(t) = D^3 F_0(t), \\ f_3(t) &= f_6^*(t) = D^3[\phi(t)]^2[\phi(2t)]^*, \\ f_8(t) &= f_{10}(t) = D^2, \\ f_9(t) &= D^2 F_0(2t), \\ f_k(t) &= D^2 F_0(t) \text{ for } k = 11, \ldots, 18, \\ f_k(t) &= D \text{ for } k = 19, \ldots, 24, \end{aligned} \tag{50}$$

and the coefficients $T(P)$ are given by

$$\begin{aligned} T(P) = {} & D^2\langle A\rangle_{\rho_{mc}}^2 (V_{P,P_8} + V_{P,P_{24}} \mathrm{Tr}\{\rho^2(0)\}) \\ & + D\langle A^2\rangle_{\rho_{mc}} (V_{P,P_{10}} \mathrm{Tr}\{\rho^2(0)\} + V_{P,P_{19}}) \\ & + D\langle A\rangle_{\rho_{mc}} \langle A\rangle_{\rho(0)} (V_{P,P_2} + V_{P,P_7} + V_{P,P_{20}} + V_{P,P_{22}}) \\ & + D\langle A\rangle_{\rho_{mc}} \mathrm{Tr}\{\rho^2(0)A\} (V_{P,P_{12}} + V_{P,P_{14}} + V_{P,P_{16}} + V_{P,P_{18}}) \\ & + \langle A\rangle_{\rho(0)}^2 (V_{P,P_1} + V_{P,P_9}) \\ & + \mathrm{Tr}\{[\rho(0)A]^2\} (V_{P,P_3} + V_{P,P_6}) \\ & + \langle A^2\rangle_{\rho(0)} (V_{P,P_{11}} + V_{P,P_{13}} + V_{P,P_{15}} + V_{P,P_{17}}) \\ & + \mathrm{Tr}\{\rho^2(0)A^2\} (V_{P,P_4} + V_{P,P_5} + V_{P,P_{21}} + V_{P,P_{23}}) \, . \end{aligned} \tag{51}$$

To explicitly evaluate equations (49)–(51), we still need the coefficients V_{P_k,P_l} for all $k,l \in \{1, \ldots, 24\}$. They are obtained as explained below equation (24): defining $j = j(k,l)$ implicitly via $P_j = P_l^{-1} P_k$, one finds by factorizing each P_j into its disjoint cycles and exploiting Tables II and IV of ref. 63 that V_{P_k,P_l} is given by

$$\begin{aligned} V_{1,1,1,1} &= D^{-4} \text{ for } j = 1, \\ V_{2,1,1} &= -D^{-5} \text{ for } j = 2, \ldots, 7, \\ V_{2,2} &= D^{-6} \text{ for } j = 8, \ldots, 10, \\ V_{3,1} &= 2D^{-6} \text{ for } j = 11, \ldots, 18, \\ V_4 &= -5D^{-7} \text{ for } j = 19, \ldots, 24, \end{aligned} \tag{52}$$

up to correction factors of the form $1 + \mathcal{O}(D^{-2})$ on the right-hand side of each of those relations. One thus is left with finding $P_j = P_l^{-1} P_k$ for all 24^2 pairs (k,l). To mitigate this daunting task, we have restricted ourselves to those summands in equation (49), which are at least of the order D^{-1}. Along these lines, one finally recovers with equations (35), (40) and (42) the result (11).

Derivation of equation (13). While the essential steps in deriving equation (13) have been outlined already in the main text, we still have to provide the details of the statements below equation (13): our first observation is that $R_1(t)$ in equation (36) amounts to the systematic (U independent) part of the omitted corrections in equation (13), and equation (41) to the bound announced below equation (13).

By means of a straightforward (but again very tedious) generalization of the calculations from the preceding subsection, one finds that

$$[\xi(t)\xi(s)]_U = C(t,s)\frac{\Delta_A^2 \mathrm{Tr}\{\rho^2(0)\}}{D} + \mathcal{O}\left(\frac{\Delta_A^2}{D^2}\right), \tag{53}$$

where $C(t, s)$ has the following six properties: first, $C(t, s) = C(s, t) = C(-t, -s)$ for all t, s. Second, $|C(t, s)| \leq 9$ for all t,s. Third, $C(t, 0) = 0$ for all t. Fourth, $C(t, s) \rightsquigarrow 0$ for $|t - s| \to \infty$, cf. equation (16). Fifth, $C(t,s)F(t-s)\langle(A - \langle A\rangle_{\rho_{mc}})^2\rangle_{\rho_{mc}}$ for $t, s \to \infty$. Sixth, given s, the behaviour of $C(t, s)$ as a function of t is roughly comparable to that of $F(t - s)$ for most t.

Though we did not explicitly evaluate the last term in equation (53), closer inspection of its general structure shows that it can be bounded in modulus by $c\Delta_A^2/D^2$ for some c, which is independent of t, s, D, A, $\rho(0)$ and H. Moreover, there is no indication of any fundamental structural differences in comparison with the leading and next-to-leading order terms, which we did evaluate. In other words, the last term in equation (53) is expected to satisfy properties analogous to those mentioned below equation (53). Recalling that the purity $\mathrm{Tr}\{\rho^2(0)\}$ satisfies the usual bounds $1 \geq \mathrm{Tr}\{\rho^2(0)\} \geq \mathrm{Tr}\{\rho_{mc}^2\} = 1/D$, we thus recover the properties of $\xi(t)$ announced below equation (13).

References

1. Tasaki, H. From quantum dynamics to the canonical distribution: general picture and rigorous example. *Phys. Rev. Lett.* **80,** 1373–1376 (1998).
2. Popescu, S., Short, A. J. & Winter, A. Entanglement and the foundations of statistical mechanics. *Nat. Phys.* **2,** 754–758 (2006).
3. Goldstein, S., Lebowitz, J. L., Tumulka, R. & Zanghì, N. Canonical typicality. *Phys. Rev. Lett.* **96,** 050403 (2006).
4. Rigol, M., Dunjko, V. & Olshanii, M. Thermalization and its mechanism for generic isolated quantum systems. *Nature* **452,** 854–858 (2008).
5. Gemmer, J., Michel, M. & Mahler, G. *Quantum Thermodynamics* 2nd edn (Springer, 2009).
6. Eisert, J., Friesdorf, M. & Gogolin, C. Quantum many-body systems out of equilibrium. *Nat. Phys.* **11,** 124–130 (2015).
7. Sklar, L. *Physics and Chance* (Cambridge Univ. Press, 1993).
8. von Neumann, J. Beweis des Ergodensatzes und des H-Theorems in der neuen Mechanik. *Z. Phys.* **57,** 30–70 (1929) [English translation by Tumulka, R. Proof of the ergodic theorem and the H-theorem in quantum mechanics. *Eur. Phys. J. H* **35,** 201–237 (2010)].
9. Goldstein, S., Lebowitz, J. L., Tumulka, R. & Zanghì, N. Long-time behavior of macroscopic quantum systems: commentary accompanying the english translation of John von Neumann's 1929 article on the quantum ergodic theorem. *Eur. Phys. J. H* **35,** 173–200 (2010).
10. Goldstein, S., Lebowitz, J. L., Mastrodonato, C., Tumulka, R. & Zanghì, N. Approach to thermal equilibrium of macroscopic quantum systems. *Phys. Rev. E* **81,** 011109 (2010).
11. Goldstein, S., Lebowitz, J. L., Mastrodonato, C., Tumulka, R. & Zanghì, N. Normal typicality and von Neumann's quantum ergodic theorem. *Proc. R. Soc. A* **466,** 3203–3224 (2010).
12. Reimann, P. Generalization of von Neumann's approach to thermalization. *Phys. Rev. Lett.* **115,** 010403 (2015).
13. Popescu, S., Short, A. J. & Winter, A. The foundations of statistical mechanics from entanglement: Individual states vs. averages. Preprint at http://arxiv.org/abs/quant-ph/0511225 (2005).
14. Müller, M. P., Gross, D. & Eisert, J. Concentration of measure for quantum states with a fixed expectation value. *Commun. Math. Phys.* **303,** 785–824 (2011).
15. Sugita, A. On the basis of quantum statistical mechanics. *Nonlinear Phenom. Complex Syst.* **10,** 192–195 (2007).
16. Reimann, P. Typicality for generalized microcanonical ensembles. *Phys. Rev. Lett.* **99,** 160404 (2007).
17. Bartsch, C. & Gemmer, J. Dynamical typicality of quantum expectation values. *Phys. Rev. Lett.* **102,** 110403 (2009).
18. Sugiura, S. & Shimizu, A. Thermal pure quantum states at finite temperature. *Phys. Rev. Lett.* **108,** 240401 (2012).
19. Deutsch, J. M. Quantum statistical mechanics in a closed system. *Phys. Rev. A* **43,** 2046–2049 (1991).
20. Srednicki, M. Chaos and quantum thermalization. *Phys. Rev. E* **50,** 888–901 (1994).
21. Neuenhahn, C. & Marquardt, F. Thermalization of interacting fermions and delocalization in Fock space. *Phys. Rev. E* **85,** 060101(R) (2012).
22. Rigol, M. & Srednicki, M. Alternatives to eigenstate thermalization. *Phys. Rev. Lett.* **12,** 110601 (2012).
23. Ikeda, T. N., Watanabe, Y. & Ueda, M. Finite-size scaling analysis of the eigenstate thermalization hypothesis in a one-dimensional interacting Bose gas. *Phys. Rev. E* **87,** 012125 (2013).
24. Beugeling, W., Moessner, R. & Haque, M. Finite-size scaling of eigenstate thermalization. *Phys. Rev. E* **89,** 042112 (2014).
25. Steinigeweg, R., Khodja, A., Niemeyer, H., Gogolin, C. & Gemmer, J. Pushing the limits of the eigenstate thermalization hypothesis towards mesoscopic quantum systems. *Phys. Rev. Lett.* **112,** 130403 (2014).
26. Goldstein, S., Huse, D. A., Lebowitz, J. L. & Tumulka, R. Thermal equilibrium of a macroscopic quantum system in a pure state. *Phys. Rev. Lett.* **115,** 100402 (2015).
27. Cramer, M., Flesch, A., McCulloch, I. P., Schollwöck, U. & Eisert, J. Exploring local quantum many-body relaxation by atoms in optical superlattices. *Phys. Rev. Lett.* **101,** 063001 (2008).
28. Trotzky, S. *et al.* Probing the relaxation towards equilibrium in an isolated strongly correlated one-dimensional Bose gas. *Nat. Phys.* **8,** 325–330 (2012).
29. Gring, M. *et al.* Relaxation and prethermalization in an isolated quantum system. *Science* **337,** 1318–1322 (2012).
30. Pertot, D. *et al.* Relaxation dynamics of a Fermi gas in an optical superlattice. *Phys. Rev. Lett.* **113,** 170403 (2014).
31. Rigol, M. Breakdown of thermalization in finite one-dimensional systems. *Phys. Rev. Lett.* **103,** 100403 (2009).
32. Rigol, M. Quantum quenches and thermalization in one-dimensional fermionic systems. *Phys. Rev. A* **80,** 053607 (2009).
33. Santos, L. F. & Rigol, M. Onset of quantum chaos in one-dimensional bosonic and fermionic systems and its relation to thermalization. *Phys. Rev. E* **81,** 036206 (2010).
34. Pal, A. & Huse, D. A. Many-body localization phase transitions. *Phys. Rev. E* **82,** 174411 (2010).
35. Brioli, G., Kollath, C. & Läuchli, A. Effect of rare fluctuations on the thermalization of isolated quantum systems. *Phys. Rev. Lett.* **105,** 250401 (2010).
36. Gogolin, C., Müller, M. & Eisert, J. Absence of thermalization in nonintegrable systems. *Phys. Rev. Lett.* **106,** 040401 (2011).
37. Banuls, M. C., Cirac, J. I. & Hastings, M. B. Strong and weak thermalization of infinite non-integrable quantum systems. *Phys. Rev. Lett.* **106,** 050405 (2011).

38. Cramer, M., Dawson, C. M., Eisert, J. & Osborne, T. J. Exact relaxation in a class of non-equilibrium quantum lattice systems. *Phys. Rev. Lett.* **100,** 030602 (2008).
39. Reimann, P. Foundation of statistical mechanics under experimentally realistic conditions. *Phys. Rev. Lett.* **101,** 190403 (2008).
40. Linden, N., Popescu, S., Short, A. J. & Winter, A. Quantum mechanical evolution towards equilibrium. *Phys. Rev. E* **79,** 061103 (2009).
41. Reimann, P. Canonical thermalization. *New J. Phys.* **12,** 055027 (2010).
42. Short, A. J. Equilibration of quantum systems and subsystems. *New J. Phys.* **13,** 053009 (2011).
43. Reimann, P. & Kastner, M. Equilibration of macroscopic quantum systems. *New J. Phys.* **14,** 043020 (2012).
44. Reimann, P. Equilibration of isolated macroscopic quantum systems under experimentally realistic conditions. *Phys. Scr.* **86,** 058512 (2012).
45. Short, A. J. & Farrelly, T. C. Quantum equilibration in finite time. *New J. Phys.* **14,** 013063 (2012).
46. Cramer, M. Thermalization under randomized local Hamiltonians. *New. J. Phys.* **14,** 053051 (2012).
47. Goldstein, S., Hara, T. & Tasaki, H. Time scales in the approach to equilibrium of macroscopic quantum systems. *Phys. Rev. Lett.* **111,** 140401 (2013).
48. Monnai, T. Generic evaluation of relaxation time for quantum many body systems: analysis of system size dependence. *J. Phys. Soc. Jpn* **82,** 044006 (2013).
49. Malabarba, A. S. L., Garcia-Pintos, L. P., Linden, N., Farrelly, T. C. & Short, A. J. Quantum systems equilibrate rapidly for most observables. *Phys. Rev. E* **90,** 012121 (2014).
50. Goldstein, S., Hara, T. & Tasaki, H. Extremely quick thermalization in a macroscopic quantum system for a typical nonequilibrium subspace. *New. J. Phys.* **17,** 045002 (2015).
51. Goldstein, S., Hara, T. & Tasaki, H. The approach to equilibrium in a macroscopic quantum system for a typical nonequilibrium subspace. Preprint at http://arxiv.org/abs/1402.3380 (2014).
52. Znidaric, M., Pineda, C. & Garcia-Mata, I. Non-Markovian behavior of small and large complex quantum systems. *Phys. Rev. Lett.* **107,** 080404 (2011).
53. Monnai, T. General relaxation time of the fidelity for isolated quantum thermodynamic systems. *J. Phys. Soc. Jpn* **82,** 044006 (2014).
54. Berges, J., Borsányi, Sz. & Wetterich, C. Prethermalization. *Phys. Rev. Lett.* **93,** 142002 (2004).
55. Moeckel, M. & Kehrein, S. Interaction quench in the Hubbard model. *Phys. Rev. Lett.* **100,** 175702 (2008).
56. Guidoni, L., Beaurepaire, E. & Bigot, J.-Y. Magneto-optics in the ultrafast regime: Thermalization of spin populations in ferromagnetic films. *Phys. Rev. Lett.* **89,** 017401 (2002).
57. Gierz, I. *et al.* Tracking primary thermalization events in graphene with photoemission at extreme time scales. *Phys. Rev. Lett.* **115,** 086803 (2015).
58. Faure, J. *et al.* Direct observation of electron thermalization and electron-phonon coupling in photoexcited bismuth. *Phys. Rev. B* **88,** 075120 (2013).
59. Thon, A. *et al.* Photon-assisted tunneling versus tunneling of excited electrons in metal-insulator-metal junctions. *Appl. Phys. A* **78,** 189–199 (2004).
60. Hetterich, D., Fuchs, M. & Trauzettel, B. Equilibration in closed quantum systems: Application to spin qubits. *Phys. Rev. B* **92,** 155314 (2015).
61. Brody, T. A. *et al.* Random-matrix physics: spectrum and strength fluctuations. *Rev. Mod. Phys.* **53,** 385–480 (1981).
62. Collins, B. & Sniady, P. Integration with respect to the Haar measure on unitary, orthogonal and symplectic group. *Commun. Math. Phys.* **264,** 773–795 (2006).
63. Brouwer, P. W. & Beenakker, C. W. J. Diagrammatic method of integration over the unitary group, with applications to quantum transport in mesoscopic systems. *J. Math. Phys.* **37,** 4904–4934 (1996).
64. Olshanii, M. *et al.* An exactly solvable model for the integrability-chaos transition in rough quantum billiards. *Nat. Commun.* **3,** 641 (2012).
65. Beaurepaire, E., Merle, J.-C., Daunois, A. & Bigot, J.-Y. Ultrafast spin dynamics in ferromagnetic Nickel. *Phys. Rev. Lett.* **76,** 4250–4253 (1996).
66. Papalazarou, E. *et al.* Supplemental Material of *Phys. Rev. Lett.* **108,** 256808 (2012).
67. Klamroth, T. Laser-driven electron transfer through metal-insulator-metal contacts: Time-dependent configuration interaction singles calculations for a jellium model. *Phys. Rev. B* **68,** 245421 (2003).

Acknowledgements

I am indebted to Walter Pfeiffer and Thomas Dahm for numerous enlightening discussions. I also thank all authors of refs. 17,32,59,60 for providing the raw data of their published works, in particular Christian Bartsch, Marcos Rigol, Tillmann Klamroth and Daniel Hetterich. This project was supported by DFG-Grant RE1344/7-1.

Author contributions

This work was carried out by P.R.

Additional information

Competing financial interests: The authors declare no competing financial interests.

29

Realistic noise-tolerant randomness amplification using finite number of devices

Fernando G.S.L. Brandão[1,2], Ravishankar Ramanathan[3], Andrzej Grudka[4], Karol Horodecki[5], Michał Horodecki[3], Paweł Horodecki[6], Tomasz Szarek[7] & Hanna Wojewódka[8]

Randomness is a fundamental concept, with implications from security of modern data systems, to fundamental laws of nature and even the philosophy of science. Randomness is called certified if it describes events that cannot be pre-determined by an external adversary. It is known that weak certified randomness can be amplified to nearly ideal randomness using quantum-mechanical systems. However, so far, it was unclear whether randomness amplification is a realistic task, as the existing proposals either do not tolerate noise or require an unbounded number of different devices. Here we provide an error-tolerant protocol using a finite number of devices for amplifying arbitrary weak randomness into nearly perfect random bits, which are secure against a no-signalling adversary. The correctness of the protocol is assessed by violating a Bell inequality, with the degree of violation determining the noise tolerance threshold. An experimental realization of the protocol is within reach of current technology.

[1] Quantum Architectures and Computation Group, Microsoft Research, Redmond, Washington 98052, USA. [2] Department of Computer Science, University College London, WC1E 6BT London, UK. [3] Faculty of Mathematics, Physics and Informatics, Institute of Theoretical Physics and Astrophysics and National Quantum Information Centre, University of Gdańsk, 80-309 Gdańsk, Poland. [4] Faculty of Physics, Adam Mickiewicz University, 61-614 Poznań, Poland. [5] Faculty of Mathematics, Physics and Informatics, Institute of Informatics and National Quantum Information Centre, University of Gdańsk, 80-309 Gdańsk, Poland. [6] Faculty of Applied Physics and Mathematics, National Quantum Information Centre, Gdańsk University of Technology, 80-233 Gdańsk, Poland. [7] Faculty of Mathematics, Physics and Informatics, Institute of Mathematics and National Quantum Information Centre, University of Gdańsk, 80-309 Gdańsk, Poland. [8] Faculty of Mathematics, Physics and Informatics, Institute of Theoretical Physics and Astrophysics, Instutute of Mathematics and National Quantum Information Centre, University of Gdańsk, 80-309 Gdańsk, Poland. Correspondence and requests for materials should be addressed to H.W. (email: hwojewod@mat.ug.edu.pl).

The simplest cryptographic resource is a random bit, unknown to any adversary. It is a basic brick of virtually any cryptographic protocol. However, traditional random number generators (RNGs) are based on classical physics, which is deterministic. Therefore, the output randomness cannot be trusted without further assumptions. For instance, a pseudo-random number generator based on a deterministic algorithm might rely on a small set of initial values (the seed) being unknown to any adversary or on certain mathematical hardness assumptions. Hardware RNGs based on physical phenomena such as thermal or atmospheric noise, radioactive decay or unstable dynamical systems are susceptible to attacks by an adversary feeding known signals into the supposedly random source. Indeed, it is not possible to create randomness out of nothing, and random numbers generated by any sort of software are in principle vulnerable to hacking.

In view of such pervasiveness of determinism, computer scientists have considered the weaker task of amplifying imperfect randomness. The goal here is to extract nearly perfect random bits given many samples of somewhat random, but potentially almost deterministic bits. Unfortunately, even this weaker task was proven to be impossible by Santha and Vazirani in 1984 (ref. 1). They introduced a model of the sources of weak randomness (described in detail later) and showed that randomness amplification from a single such source through any deterministic method is impossible. Classical information processing only allows for randomness amplification when one has access to at least two independent weak sources of randomness.

However, for almost a century, we have been aware that Nature is not ruled by deterministic classical laws. It is described by quantum theory, which is intrinsically non-deterministic, because of the famous rule postulated by Max Born in 1927, stating that the quantum-mechanical wave-function only describes probabilities of events. This peculiar feature of Nature has generated conflict among generations of scientists, but was shown to be an indispensable part of quantum theory by John Bell[2].

In recent decades, the information revolution has influenced almost every aspect of present-day life and stimulated a huge effort towards establishing cryptographic security. Quantum indeterminism has now been promoted from a weird peculiarity of quantum theory to a potentially important resource. New RNGs, based on quantum principles, have been built and are now commercially available (for example, the ones offered by IdQuantique[3]). However, to trust the randomness produced by such devices, one has to either trust that the device works as per its specification or verify its internal construction by direct inspection. As a typical user would not be able to make such a verification, a basic question arises: can we build a device to produce certified randomness, in a device-independent way?

In view of the requirement of using untrusted devices, we still face the same limitation as in traditional cryptography: randomness cannot be created out of nothing. Fortunately, recently it turned out that randomness amplification, while inaccessible classically, can be accomplished in a device-independent manner by exploiting quantum-mechanical systems[4]. Earlier work has also shown that the similar task of randomness expansion (in which a small fully random seed is stretched into a longer random string) is possible using quantum-mechanical correlations[5]. The central role in both these processes is played by the so-called Bell inequalities[2]. The test of violation of Bell inequalities, which certifies true randomness, is performed solely on the statistics, and therein lies its device-independent potential. This idea first appeared in a seminal paper by Ekert[6], followed by Barrett *et al.*[7], as well as others.

There is, however, a caveat to this concept. Namely, there is no way to guarantee the random nature of the world, because of the possibility of super-determinism, that is, all events, including the choices of the measurement settings in any experiment, may be predetermined (no free-will). Indeed, a test of violation of Bell inequalities allows to uncover true randomness only if we can choose the measurement settings at random. We thus face a sort of vicious circle.

Recently, Colbeck and Renner, building on a breakthrough results in refs 5–7, showed that one can amplify randomness: even though the inputs are only weakly random, the outputs provide almost ideal randomness. Subsequent protocols have further developed this original result[8,9].

The existing protocols that implement the task of amplifying randomness, while ingenious and conceptually important, have some drawbacks that make them impossible to implement in reality. Namely, after each single round of measurement, the device has to be discarded and new devices have to be used (equivalently, a large number of devices may be used in parallel). To imagine how limited such a protocol would be in practice, consider a typical implementation in which we need to create a kilo-byte of secure randomness. Then, even if we used a single photon per output random bit, 8,000 devices, shielded from each other, would be required.

Only one protocol proposed so far does not have this unfeasible requirement of many devices—the original Colbeck and Renner's protocol[4]. However, it also suffers from many disadvantages—namely, the protocol does not tolerate noise, besides using a large number of settings, being therefore impractical. Thus, the existing protocols are trapped between Scylla of many devices and Charybdis of fragility to noise. There have been other proposals to solve this problem in refs 10,11 for an adversary obeying quantum-mechanical laws. However, full proofs of security of these protocols are still missing.

In contrast, good protocols have been developed for the task of randomness expansion[12]. Therefore, it has been a pressing open question in the field whether randomness amplification can realistically be implemented.

In this paper, we provide protocols that escape this dual restrictive alternative, being therefore directly implementable in practice. We propose protocols that use a small constant number of devices and produce randomness out of an arbitrarily weak SV-source while being able to tolerate a constant noise rate. Moreover, in the protocols the security can be tested without referring to quantum mechanics.

Results

Santha–Vazirani ε-SV sources. As an illustration of the SV-source model, assume that an adversary has two different coins, one biased towards heads (for example, probability of heads is 2/3) and the other biased towards tails (probability of heads is 1/3, say). The adversary, in each time step, chooses one of the two coins and tosses it; the choice of coin may depend (probabilistically) on the outcomes of the prior tosses. The sequence of random outcomes of these coin tosses then gives an SV-source.

In general, Santha and Vazirani[1] considered an ε-SV source, where $\varepsilon \in [0,0.5]$ is a parameter, which indicates how much the source bits deviate from fully random (in the above example, $\varepsilon = 1/6$). In the most general case, an ε-SV source is given by a probability distribution $p(\varphi_0,\ldots,\varphi_n)$ over bit strings such that

$$0.5-\varepsilon \le P(\varphi_0) \le 0.5+\varepsilon,$$
$$0.5-\varepsilon \le P(\varphi_i \mid \varphi_0,\ldots,\varphi_{i-1}) \le 0.5+\varepsilon \quad \text{for } 1 \le i \le n. \tag{1}$$

Note that, when $\varepsilon=0$, the bits are fully random, whereas for $\varepsilon=0.5$, they are fully deterministic.

The Santha–Vazirani no-go result holds for any $\varepsilon\neq 0$ and says that any bit extracted by a deterministic procedure from a single ε-SV source will always have an ε bias.

Bell inequalities. In the Nature, there exist correlations that cannot be described by any deterministic theory, that is, the correlations cannot be explained by any model where all the randomness is due to lack of knowledge. Such correlations are manifested operationally through the violation of Bell inequalities. This immediately brings to mind applications in security: were the probabilistic description simply due to lack of one's knowledge, an eavesdropper could potentially have this knowledge. Thus, the security would be compromised and there would be no randomness whatsoever. On the contrary, violation of Bell inequalities makes room for the possibility for true randomness.

Main results. We propose here two protocols, which are both discussed in detail, as well as compared with similar protocols for randomness amplification, in Supplementary Note 1. The Bell inequality and randomness extractors used in the protocols are presented in Supplementary Note 2. Assumptions are summarized in Supplementary Note 3, while the main mathematical tools, essential to establish the security of the protocols, are summarized in Supplementary Note 4.

Protocol I is the more basic one and employs just four devices, however, it needs an extractor that so far is only known to exist *implicitly* in the full range $0<\varepsilon<1/2$ (alternatively, there is an explicit extractor that can be employed in the protocol, but then it can produce just one bit of randomness). This is overcome in our Protocol II, which is an extended version of Protocol I. It requires eight devices, but works with a fully explicit extractor for the whole range of ε, that is, even for arbitrarily weak sources. Protocol I is depicted in Fig. 1, whereas Protocol II is illustrated in Fig. 2. More precisely, our results can be formulated as the following two (informal) theorems, whose formal versions are given in the Supplementary Notes 5 and 6 (see Supplementary Theorems 20 and 27 for details).

Theorem 1 [informal]: For every $\varepsilon<\frac{1}{2}$, there is a protocol using an ε-SV source and four no-signalling devices with the following properties:

- Using the devices $(n, \log(1/\delta))$ times, the protocol either aborts or produces n bits, which are δ-close to uniform and independent of any side information (for example, held by an adversary).
- Local measurements on many copies of a four-partite entangled state, with $(1-2\varepsilon)$ error rate, give rise to devices that do not abort the protocol with probability larger than $1-2^{-\Omega(n)}$.

The protocol is non-explicit and runs in $(n, \log(1/\delta))$ time. Alternatively, it can use an explicit extractor to produce a single bit of randomness that is δ-close to uniform in $(\log(1/\delta))$ time.

Theorem 2 [informal]: For every $\varepsilon<\frac{1}{2}$, there is a protocol using an ε-SV source and eight no-signalling devices with the following properties:

- Using the devices $2^{\mathrm{poly}(n,\ \log(1/\delta))}$ times, the protocol either aborts or produces n bits, which are δ-close to uniform and independent of any side information (for example, held by an adversary).
- Local measurements on many copies of a four-partite entangled state, with $(1-2\varepsilon)$ error rate, give rise to devices that do not abort the protocol with probability larger than $1-2^{-\Omega(n)}$.

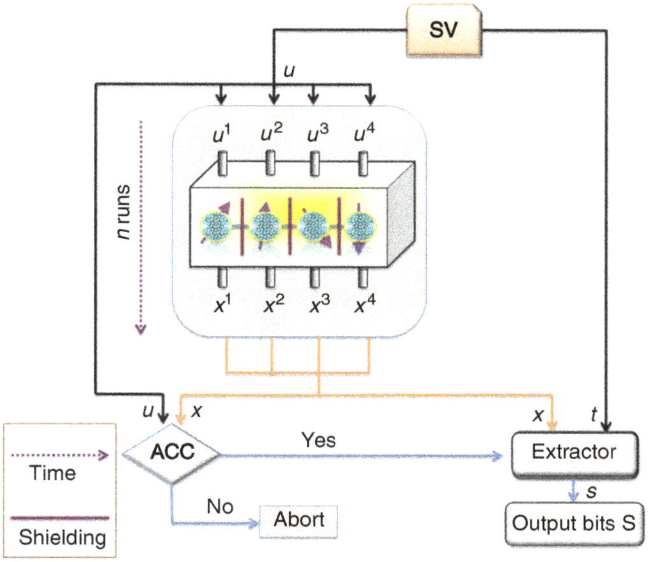

Figure 1 | Protocol I—designed for randomness amplification with four devices. Devices are shielded one from another and used in a sequence of n runs. $\boldsymbol{u}^1,\ldots\boldsymbol{u}^4$ denote binary inputs in each run and they determine which measurement is made in the given run, whereas $\boldsymbol{x}^1,\ldots,\boldsymbol{x}^4$ are binary outputs of the measurements. If the test, performed on the inputs and outputs of the device, is passed (denoted by ACC—Yes), then the outputs and another set of bits from the SV-source (denoted by t) are introduced into an extractor in order to obtain final output bits S. Black arrows mark the directions of bits from the SV-source; orange ones indicate where output bits are introduced and blue arrows show possible alternatives: accepting the protocol and obtaining (supposedly random) bits or aborting the protocol.

The protocol is fully explicit and runs in $2^{\mathrm{poly}(n,\ \log(1/\delta))}$ time.

General setup. Let us first outline our general setup. We divide the bits from the SV-source into two parts. One part of the weakly random bits is fed into the devices, either as inputs or to choose some particular runs. Then, a test is performed on the inputs and outputs of the device. If the test is passed (denoted by an 'ACC'), then the outputs and the other part of bits from the SV-source are fed into a randomness extractor—a deterministic function, whose output constitutes the almost ideal randomness. If the test is failed, the protocol is aborted.

The devices work as follows. A source produces four particle entangled states (of photons, atoms and so on). Each device performs a measurement on one of the four particles produced in each run. One of two measurements can be performed, which is decided by a binary input to each device. When the input is zero, one type of measurement is performed, whereas the other type is performed when the input is one. The measurements have binary outputs.

Description of the protocols. Protocol I works as follows. A portion of bits from the SV-source is used as an input to the devices. A test is then performed, which amounts to checking whether a particular Bell inequality is violated to a certain specified level. The test consists of computing the following quantity

$$B_{\mathrm{est}}=\frac{1}{n}\sum_{k=1}^{n}B(x_k,u_k) \qquad (2)$$

from n 4-tuples of inputs and outputs obtained in n runs (every x_k and u_k is of the form $(\boldsymbol{x}^1,\ldots\boldsymbol{x}^4)$ and $(\boldsymbol{u}^1,\ldots\boldsymbol{u}^4)$, respectively), and

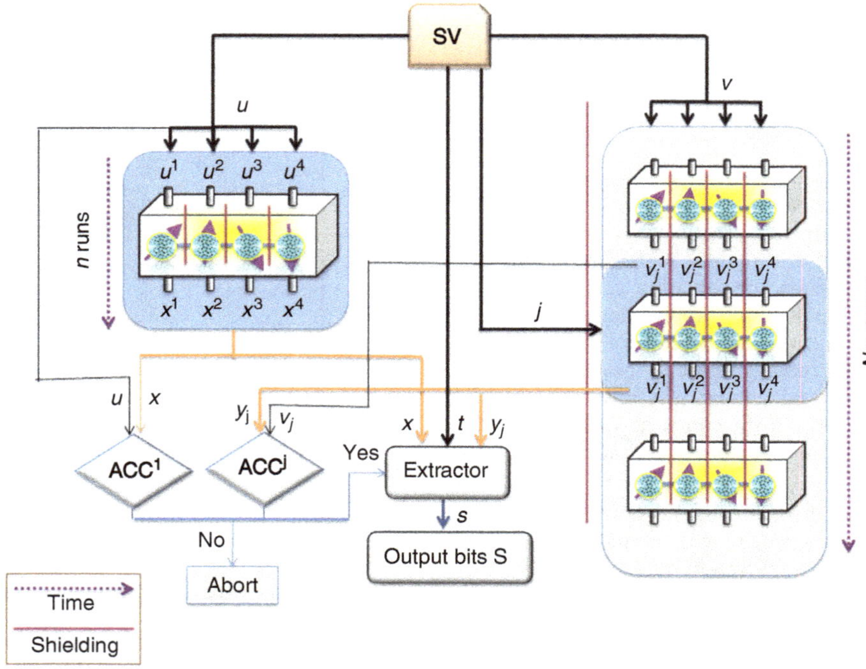

Figure 2 | Protocol II—designed for randomness amplification from eight devices. The eight devices are shielded from each other and one block of n runs is performed sequentially on the first four devices while N blocks of n runs are performed sequentially on the second four devices. One of these N blocks, marked in blue, is selected using some portion of bits (denoted by j) from the SV-source. $\boldsymbol{u}^1, \ldots \boldsymbol{u}^4, \boldsymbol{v}_j^1, \ldots, \boldsymbol{v}_j^4$ denote binary inputs in each run and they determine which measurement is made in the given run, whereas $\boldsymbol{x}^1, \ldots, \boldsymbol{x}^4, \boldsymbol{y}_j^1, \ldots, \boldsymbol{y}_j^4$ are binary outputs of the measurements. Black arrows indicate where bits from SV source are directed, whereas orange ones mark where output bits are introduced. The test is performed separately on inputs and outputs of the first four devices and the chosen block from the second four devices. Blue arrows show possible alternatives: either the tests are passed (ACC^1—Yes, ACC^j—Yes), which enables further action, or the protocol is aborted. If both tests are passed, then output bits together with further bits from the SV-source (denoted by t) are introduced into an extractor to obtain final bits S.

the explicit expression for $B(x_k, u_k)$ is given in equation (7) of the Supplementary Note 2. The test is passed if $B_{est} \leq \delta$. The parameter $\delta \geq 0$ can be interpreted as the noise level, that is, in the absence of noise we would observe $B_{est} = 0$. For the explicit form of the quantum state and measurements that achieve $B_{est} = 0$, see equations (8) and (9) of the Supplementary Note 2. If the test is failed, the protocol is aborted. If it is passed, the outputs of the devices and a second part of bits from SV-source are fed into an extractor, designed to extract randomness from two independent sources.

For Protocol II, we add a second group of four devices, operating in the same way as the original group. The runs (uses) of the devices from this second group are divided into blocks and a portion of bits from the SV-source is used to choose a block. The same test as in Protocol I is performed twice: first on all runs from the first group of devices, and then on the chosen block of runs from the second group. If the tests are passed, an extractor designed to extract randomness from three independent weak sources is applied to the three groups of variables: one from the SV-source, another formed by outputs from the first four devices and a third formed by the outputs from the chosen block of runs from the second group of four devices.

The merit of Protocol II is that it offers amplification of arbitrarily weak sources under a constant noise rate and with just a few devices. The probability of failure (failure occurs if the test was passed, but the output of the protocol is not random) scales as an inverse polynomial in the total number of runs. On the contrary, Protocol I has probability of failure exponentially small in the number of runs. Moreover, since for $\varepsilon < \frac{\sqrt{2}-1}{2}$ there exist explicit two source extractors[13], Protocol I also gives exponential security for this range of ε. As a matter of fact, the field of extractors is being constantly developed. For instance, in a very recent development[14] an extractor was found that, if used in our Protocol I, allows to draw one bit of randomness for an arbitrarily weak SV-source with exponential security.

Let us emphasize, that our protocols exhibit a strong security criterion called composable security[15,16]. This means that the obtained randomness can be securely used as an input to any other protocol. It also means that if an adversary Eve would in future learn part of the random bits output by the protocols (for example, by some espionage), the remaining bits would still be completely secure.

Last but not least, the security of our protocols relies on quantum-mechanical predictions, but can be verified by a person that either does not know or does not trust the quantum mechanical theory. Indeed, the security of our protocols is based on the very statistics of the outcomes of the device and the quantum mechanics is needed only to produce the required statistics. Moreover, Protocol I offers exponential security within such a paradigm.

In the proof sketched in the Methods section, we combine results from the classical theory of extractors obtained in refs 14,17,18, the recently discovered information-theoretic approach to the de Finetti theorem[19] and the Azuma–Hoeffding inequality[20].

Additional remarks on assumptions. In the present work, we have also uncovered all the assumptions that were not necessarily explicit in the literature so far. Clearly, the minimal assumption one has to make is that of shielding devices, which means that the devices do not send signals to each other or to the external world and vice-versa. Indeed, if an eavesdropper can monitor our device instantaneously or if they can force it to behave in various ways, then any output of the device will be insecure. In addition, the device might have a transmitter inside that reports everything to the adversary. The shielding assumption is thus mandatory. Apart from this minimal assumption, we also make the constant device assumption, which has been either implicitly or explicitly used in almost all previous papers in this subject. This imposes that the device's behaviour is not correlated with the source of weak randomness (see Supplementary Note 3 for details). The only work that does not use this assumption is ref. 9, which, however, requires the use of many devices, thus falling into the above-described unpleasant alternative. Moreover, it also assumes that the adversary is restricted by the laws of quantum theory rather than just the shielding assumption.

Let us emphasize that in the classical world, the above assumptions are not enough to amplify randomness, as classical correlations are not strong enough to allow the verification procedure outlined above. Thus, the functioning of our protocols hinges on the intrinsic indeterminacy in the quantum description of nature.

Discussion

We have presented realistic protocols for obtaining secure random bits from an arbitrarily weak Santha–Vazirani source. Both protocols use a finite number of devices, work even with correlations attainable by noisy quantum-mechanical resources and are composably secure against general no-signalling adversaries. The Protocol I uses four devices, and under the requirement of explicit extractors, can be either applied for a restricted range of epsilon to produce a non-zero rate of output randomness in polynomial time or for the entire range of epsilon to produce one bit of output randomness in polynomial time. The Protocol II uses eight devices and works for the entire range of epsilon to produce a non-zero rate of output randomness using an explicit extractor, however, it has the drawback of an exponential runtime. Important open questions for future research concern the relaxation of the constant device assumption (of independence between source and devices), and the development of protocols that can amplify general min-entropy sources of randomness in a secure manner against no-signalling adversaries. More open questions are raised in Supplementary Note 7.

Methods

Min-entropy sources and extractors. Before we sketch the proof of our result, let us describe one of the important ingredients of the proof—the min-entropy sources and randomness extractors. For given probability distribution $\{p_i\}$, its min-entropy is defined as $H_{\min} = -\log(\max_i p_i)$, and a source which produces an n-bit distribution with min-entropy k is called an (n,k) min-entropy source. An example is the SV-source, whose min-entropy per bit is no smaller than $log(\frac{1}{2}+\varepsilon)$. Although the randomness of a single min-entropy source cannot be amplified, it is known that one can classically amplify randomness from two or more independent min-entropy sources. The (deterministic) functions that do the job are called randomness extractors. We shall use the following results on extractors:

- There exists a (non-explicit) deterministic extractor that, given two independent sources of min-entropy larger than h, outputs $O(h)$ bits $2^{-O(h)}$-close to uniform[17].
- There exists an explicit extractor that, given three independent sources, one having min-entropy larger than τn (for any $\tau>0$) and the other two larger than $h \geq log^c(n)$ (with $c>0$ being an absolute constant), outputs $O(h)$ bits $2^{-h^{O(1)}}$-close to uniform[21]. The extractor can be implemented in time (n,h).
- There exists an explicit extractor that, given two independent sources of min-entropy at least $\log^C(h)$ for large enough constant C outputs 1 bit with error $h^{-\Omega(1)}$ (ref. 14).

Security of Protocol I: proof sketch. Let us now sketch the security proof of Protocol I (the full proof is provided in the Supplementary Note 5). The idea is the following: we prove that with high probability either, when conditioned on the inputs and upon acceptance (passing the test), the outputs of the devices form a min-entropy source or else the probability of acceptance is small. Thus, up to a small probability of failure (failure means that one accepts, but at the same time the devices do not constitute a min-entropy source), we have at our disposal two sources of weak randomness: the SV-source and the min-entropy source from the outputs of the devices. Because of the constant device assumption, when conditioned on the inputs, the two sources are independent, and we can apply a classical randomness extractor. Thus, the crux of the proof is to show that the outputs of devices constitute a min-entropy source. Here is where the Bell inequality comes into play.

Before going into more detail, let us introduce some notation. Consider n uses of each device, and let x, u be the outputs and inputs to the devices, respectively (the u come from the SV-source). Note that we have $x=(x_1,\ldots,x_n)$, $u=(u_1,\ldots,u_n)$, where every x_k and u_k, for $k\in\{1,\ldots,n\}$, is of the form $\boldsymbol{x}=(\boldsymbol{x}^1,\boldsymbol{x}^2,\boldsymbol{x}^3,\boldsymbol{x}^4)$ and $\boldsymbol{u}=(\boldsymbol{u}^1,\boldsymbol{u}^2,\boldsymbol{u}^3,\boldsymbol{u}^4)$, respectively. Let w and z be the input and output of the eavesdropper's device, respectively, and let e denote the side information possessed by the eavesdropper. Finally, we denote further bits drawn from the SV-source (apart from u) as t.

For any given e, the eavesdropper chooses optimally her input, which finally gives rise to a probability distribution $q(x,u,z,e,t)$ satisfying certain conditions, given by equations (80)–(83) in Supplementary Note 4. The conditions come from our basic assumptions: shielding (equation (80) in Supplementary Note 5), constant devices (equation (81) in Supplementary Note 5) and causality (time-ordered no-signalling equation (82) in Supplementary Note 5). Let us denote by ACC the event of acceptance. We are thus interested in the min-entropy of the distribution $q(x|u, z, e, t, \text{ACC})$. We will actually prove that, with probability greater than $1-\sqrt{\alpha/q(\text{ACC})}$, we have

$$\max_x q(x|z,u,e,t,\text{ACC}) \leq \sqrt{\frac{\alpha}{q(\text{ACC})}}, \tag{3}$$

where $\alpha=2^{-Cn}$ with C being a constant, depending only on the SV-source parameter ε and the noise level δ (for details, see Proposition 19 in Supplementary Note 5). To this end, note that passing the test assures that the estimated value of the Bell quantity satisfies $B_{\text{est}}\leq\delta$. Using the Azuma–Hoeffding theorem, we prove that, if the estimated Bell value is no greater than δ, then with probability $1-O(e^{-\delta^2 n})$, in a linear fraction of runs μn, with $\mu=1-\sqrt{2\delta}$, the 'true' value of the SV-Bell quantity (conditioned on the history) is also small, that is, it is bounded by $\sqrt{2\delta}$ (see Lemma 9 in Supplementary Note 4 for details). The true value B_i^{SV} in the i-th run is here meant to be the average of $B(\boldsymbol{x},\boldsymbol{u})$ over the probability distribution in the run, conditioned on the history—the previous inputs and outputs, as well as Eve's variables e, z:

$$B_i^{SV} = \sum_{x_i,u_i} q(x_i,u_i|u_{<i},x_{<i},z,e,t)B(x_i,u_i). \tag{4}$$

We then prove, using linear programming (see Lemma 10 and 11 in Supplementary Note 4 for details), that for any arbitrary distribution $p(\boldsymbol{x},\boldsymbol{u})$ of a single run, we have

$$p(\boldsymbol{x}|\boldsymbol{u}) \leq \frac{1}{3}\left(1+2\frac{B^{SV}}{(\frac{1}{2}-\varepsilon)^4}\right) \quad \text{with} \quad B^{SV}=\sum_{\boldsymbol{x},\boldsymbol{u}} p(\boldsymbol{x},\boldsymbol{u})B(\boldsymbol{x},\boldsymbol{u}), \tag{5}$$

where $\boldsymbol{u}$ comes from the SV-source, so that $(\frac{1}{2}-\varepsilon)^4 \leq p(\boldsymbol{u}) \leq (\frac{1}{2}+\varepsilon)^4$. Applying this general relation to our case and using the Bayes rule, we obtain that for $(x,u)\in\text{ACC}$,

$$q(x|u,z,e,t) \leq \max\{\gamma^{\mu n}, \epsilon_{Az}\} \tag{6}$$

with $\gamma=\frac{1}{3}\left(1+\frac{2(2\delta)^{1-c}}{(\frac{1}{2}-\varepsilon)^4}\right)$ for some constant $c>0$ (see Lemma 18 in Supplementary Note 5). Here ϵ_{Az} denotes the (small) failure probability of the estimation of the true SV-Bell value. For any given value of SV-source parameter ε, we see that there is a small enough value of the noise parameter δ, which ensures that $\gamma<1$, so that the probability $q(x|u, z, e, t)$ in equation (6) is bounded by an exponentially (in n) small parameter, which in turn ensures that in the numerator of the right-hand-side of equation (3) there appears an exponentially decaying parameter α. The denominator therein appears due to the fact that we are conditioning on the event ACC in equation (3), unlike in equation (6). The technical details (some of them along the lines of ref. 20) that finally lead to equation (6) are presented in the proof of Theorem 20 in Supplementary Note 5 (ref. 22). This ends the sketch of the proof that our quantum device gives rise to a min-entropy source.

Now, as stated before, the constant device assumption assures that the min-entropy source obtained from the outputs of the device is independent of the bits from the SV-source that were not used as inputs. The application of a randomness extractor then gives the final output bits, although as noted earlier an explicit extractor is not known for all values of ε, except one that outputs a single bit of randomness[14].

Security of Protocol II: comment on the proof. In Protocol II, we manage to create three independent min-entropy sources, for which explicit extractors are known. To do this, we prove a version of the de Finetti theorem (see Lemma 13 in Supplementary Note 4 for details), which ensures that the outputs from the first 4-tuple of devices and the outputs belonging to the selected block of runs from the second 4-tuple are independent. As in Protocol I, each of the two sets of outputs constitutes a min-entropy source (conditioned on the two acceptances, as we perform separate tests for both sets). These, together with an unused portion of bits from the SV-source constitute three independent min-entropy sources, for which explicit extractors exist. The full proof of security of this protocol is provided in the Supplementary Note 6.

References

1. Santha, M. & Vazirani, U. V. Generating quasi-random sequences from slightly-random sources. *Proc. 25th IEEE Symp. Found. Comput. Sci. (FOCS'84)* 434–440 (1984).
2. Bell, J. S. On the Einstein-Podolsky-Rosen Paradox. *Physics* **1,** 195–200 (1964).
3. ID Quantique (IDQ) http://www.idquantique.com.
4. Colbeck, R. & Renner, R. Free randomness can be amplified. *Nat. Phys.* **8,** 450–453 (2012).
5. Pironio, S. *et al.* Random numbers certified by Bell's theorem. *Nature* **464,** 1021–1024 (2010).
6. Ekert, A. Quantum cryptography based on Bell's theorem. *Phys. Rev. Lett* **67,** 661–663 (1991).
7. Barrett, J., Hardy, L. & Kent, A. No signaling and quantum key distribution. *Phys. Rev. Lett.* **95,** 010503 (2005).
8. Gallego, R. *et al.* Full randomness from arbitrarily deterministic events. *Nat. Commun.* **4,** 2654 (2013).
9. Chung, K. M., Shi, Y. & Wu, X. Physical randomness extractors: generating random numbers with minimal assumptions, Preprint at http://arxiv.org/abs/1402.4797 (2014).
10. Mironowicz, P., Gallego, R. & Pawłowski, M. Amplification of arbitrarily weak randomness. *Phys. Rev. A* **91,** 032317 (2015).
11. Bouda, J., Pawłowski, M., Pivoluska, M. & Plesch, M. Device-independent randomness extraction for arbitrarily weak min-entropy source. *Phys. Rev. A* **90,** 032313 (2014).
12. Miller, C. A. & Shi, Y. Robust protocols for securely expanding randomness and distributing keys using untrusted quantum devices. *Proc. 46th Annu. ACM Symp. Theory Comput. (STOC'14)* 417–426 (2014).
13. Raz, R. Extractors with weak random seeds. *Proc. 37th Annu. ACM Symp. Theory Comput. (STOC'05)* 11–20 (2005).
14. Chattopadhyay, E. & Zuckerman, D. Explicit two-source extractors and resilient functions, Electronic colloquium on computational complexity, Revision 1 of Report No **119** (2015).
15. Ben-Or, M., Horodecki, M., Leung, D. W., Mayers, D. & Oppenheim, J. in *Proc. TCC 2005, LNCS*, vol. 3378, 386–406 (Springer, 2005).
16. Renner, R. & Köenig, R. in *Proc. of TCC 2005, LNCS*, vol. 3378, 407–425 (Springer, 2005).
17. Chor, B. & Goldreich, O. Unbiased bits from sources of weak randomness and probabilistic communication complexity. *IEEE 26th Annu. Symp. Found. Comput. Sci.* 429–442 (1985).
18. Xin-Li Extractors for a constant number of independent sources with polylogarithmic min-entropy. *IEEE 54th Annu. Symp. Found. Comput. Sci. (FOCS)* 100–109 (2013).
19. Brandão, F. G. S. L. & Harrow, A. W. Quantum de Finetti theorems under local measurements with applications. *Proc. 45th Annu. ACM Symp. Theory Comput. (STOC'13)* 861–870 (2013).
20. Pironio, S. & Massar, S. Security of practical private randomness generation. *Phys. Rev. A* **87,** 012336 (2013).
21. Rao, A. Extractors for a constant number of polynomially small min-entropy independent sources. *Proc. 38th Annu. ACM Symp. Theory Comput. (STOC'06)* 497–506 (2006).
22. Brandão, F.G.S.L. *et al. Robust device-independent randomness amplification with few devices.* Preprint at http://arxiv.org/abs/1310.4544 (2015).

Acknowledgements

We thank Rotem Arnon-Friedman for discussions. The work is supported by ERC AdG grant QOLAPS, EC grant RAQUEL and by Foundation for Polish Science TEAM project co-financed by the EU European Regional Development Fund. F.G.S.L.B. acknowledges support from EPSRC and Polish Ministry of Science and Higher Education Grant no. IdP2011 000361. Part of this work was done in the National Quantum Information Center of Gdańsk. Part of this work was done when F.G.S.L.B., R.R., K.H. and M.H. attended the programme 'Mathematical Challenges in Quantum Information' at the Isaac Newton Institute for Mathematical Sciences in the University of Cambridge. Another part was done in the programme 'Quantum Hamiltonian Complexity' in the Simons Institute for the Theory of Computing. Finally, M.H. thanks the Department of Physics and Astronomy and the Department of Computer Science of UCL, where part of this work was also performed, for hospitality.

Author contributions

F.G.S.L.B. and M.H. conceptualized central ideas, all authors contributed extensively to the work presented in the paper.

Additional information

Competing financial interests: The authors declare no competing financial interests.

Permissions

All chapters in this book were first published in NC, by Nature Publishing Group; hereby published with permission under the Creative Commons Attribution License or equivalent. Every chapter published in this book has been scrutinized by our experts. Their significance has been extensively debated. The topics covered herein carry significant findings which will fuel the growth of the discipline. They may even be implemented as practical applications or may be referred to as a beginning point for another development.

The contributors of this book come from diverse backgrounds, making this book a truly international effort. This book will bring forth new frontiers with its revolutionizing research information and detailed analysis of the nascent developments around the world.

We would like to thank all the contributing authors for lending their expertise to make the book truly unique. They have played a crucial role in the development of this book. Without their invaluable contributions this book wouldn't have been possible. They have made vital efforts to compile up to date information on the varied aspects of this subject to make this book a valuable addition to the collection of many professionals and students.

This book was conceptualized with the vision of imparting up-to-date information and advanced data in this field. To ensure the same, a matchless editorial board was set up. Every individual on the board went through rigorous rounds of assessment to prove their worth. After which they invested a large part of their time researching and compiling the most relevant data for our readers.

The editorial board has been involved in producing this book since its inception. They have spent rigorous hours researching and exploring the diverse topics which have resulted in the successful publishing of this book. They have passed on their knowledge of decades through this book. To expedite this challenging task, the publisher supported the team at every step. A small team of assistant editors was also appointed to further simplify the editing procedure and attain best results for the readers.

Apart from the editorial board, the designing team has also invested a significant amount of their time in understanding the subject and creating the most relevant covers. They scrutinized every image to scout for the most suitable representation of the subject and create an appropriate cover for the book.

The publishing team has been an ardent support to the editorial, designing and production team. Their endless efforts to recruit the best for this project, has resulted in the accomplishment of this book. They are a veteran in the field of academics and their pool of knowledge is as vast as their experience in printing. Their expertise and guidance has proved useful at every step. Their uncompromising quality standards have made this book an exceptional effort. Their encouragement from time to time has been an inspiration for everyone.

The publisher and the editorial board hope that this book will prove to be a valuable piece of knowledge for researchers, students, practitioners and scholars across the globe.

List of Contributors

E. Tjhung, A. Tiribocchi, D. Marenduzzo and M.E. Cates
SUPA, School of Physics and Astronomy, University of Edinburgh, JCMB Kings Buildings, Mayfield Road, Edinburgh EH9 3JZ, UK

Vittorio Peano and Christian Brendel
Institute for Theoretical Physics, University of Erlangen-Nürnberg, Staudtstr. 7, 91058 Erlangen, Germany

Martin Houde and Aashish A. Clerk
Department of Physics, McGill University, 3600 rue University, Montreal, Quebec, Canada H3A 2T8

Florian Marquardt
Institute for Theoretical Physics, University of Erlangen-Nürnberg, Staudtstr. 7, 91058 Erlangen, Germany
Max Planck Institute for the Science of Light, Günther-Scharowsky-Straβe 1/Bau 24, 91058 Erlangen, Germany

Matteo Lostaglio, David Jennings and Terry Rudolph
Department of Physics, Imperial College London, London SW7 2AZ, UK

Linyuan Lü
Alibaba Research Center for Complexity Sciences, Alibaba Business College, Hangzhou Normal University, Hangzhou 311121, China

Tao Zhou
CompleX Lab, Web Sciences Center, University of Electronic Science and Technology of China, Chengdu 611731, China
Big Data Research Center, Univesrsity of Electronic Science and Technology of China, Chengdu 611731, China

Qian-Ming Zhang
CompleX Lab, Web Sciences Center, University of Electronic Science and Technology of China, Chengdu 611731, China
Department of Physics and Center for Polymer Studies, Boston University, Boston, Massachusetts 02215, USA

H. Eugene Stanley
Alibaba Research Center for Complexity Sciences, Alibaba Business College, Hangzhou Normal University, Hangzhou 311121, China
Department of Physics and Center for Polymer Studies, Boston University, Boston, Massachusetts 02215, USA

Eliot Bolduc, Genevieve Gariepy and Jonathan Leach
Institute of Photonics and Quantum Sciences, School of Engineering & Physical Sciences, Heriot-Watt University, David Brewster Building, Edinburgh EH14 4AS, UK

Xiaoming Mao
Department of Physics, University of Michigan, Ann Arbor, Michigan 48109, USA

Anton Souslov
School of Physics, Georgia Institute of Technology, Atlanta, Georgia 30332, USA

Carlos I. Mendoza
Instituto de Investigaciones en Materiales, Universidad Nacional Autónoma de México, Apartado Postal 70-360, 04510 México, D.F., Mexico

T.C. Lubensky
Department of Physics and Astronomy, University of Pennsylvania, Philadelphia, Pennsylvania 19104, USA

Anatoly Zlotnik
Center for Nonlinear Studies, MS B258, Los Alamos National Laboratory, Los Alamos, New Mexico 87545, USA

Raphael Nagao and István Z. Kiss
Department of Chemistry, Saint Louis University, 3501 Laclede Ave., St Louis, Missouri 63103, USA

Jr-Shin Li
Department of Electrical and Systems Engineering, Washington University in St Louis, CB
1042, 1 Brookings Drive, St Louis, Missouri 63130, USA

Sergio Boixo, Alireza Shabani, Sergei V. Isakov, Vasil S. Denchev, Masoud Mohseni and Hartmut Neven
Google, Venice, California 90291, USA

Vadim N. Smelyanskiy
Google, Venice, California 90291, USA
NASA Ames Research Center, Moffett Field, California 94035, USA

Mark Dykman
Department of Physics and Astronomy, Michigan State University, East Lansing, Michigan 48824, USA

Anatoly Yu. Smirnov
D-Wave Systems Inc., Burnaby, British Columbia, Canada V5C 6G9

Mohammad H. Amin
D-Wave Systems Inc., Burnaby, British Columbia, Canada V5C 6G9
Department of Physics, Simon Fraser University, Burnaby, British Columbia, Canada V5A 1S6

John W. Simpson-Porco
Department of Electrical and Computer Engineering, Engineering Building 5, University of Waterloo, Waterloo, Ontario, Canada N2L 3G1

Florian Dörfler
Automatic Control Laboratory, Swiss Federal Institute of Technology (ETH), Physikstrasse 3, CH-8092 Zürich, Switzerland

Francesco Bullo
Department of Mechanical Engineering, Center for Control, Dynamical Systems and Computation, Engineering Building II, University of California at Santa Barbara, Santa Barbara, 93106-9560 California, USA

Michiel Hermans
OPERA photonique, Université Libre de Bruxelles, Avenue F. Roosevelt 50, 1050 Brussels, Belgium

Michaël Burm and Joni Dambre
ELIS Department, Ghent University, Sint Pietersnieuwstraat 41, 9000 Ghent, Belgium

Thomas Van Vaerenbergh and Peter Bienstman
INTEC Department, Ghent University, Sint Pietersnieuwstraat 41, 9000 Ghent, Belgium

Rafael Chaves and David Gross
Institute for Physics, University of Freiburg, Rheinstrasse 10, D-79104 Freiburg, Germany

Christian Majenz
Department of Mathematical Sciences, University of Copenhagen, Universitetsparken 5, DK-2100 Copenhagen Ø, Denmark

Serdar Çolak
Department of Civil and Environmental Engineering, MIT, Cambridge, Massachusetts 02139, USA

Antonio Lima
Department of Civil and Environmental Engineering, MIT, Cambridge, Massachusetts 02139, USA
School of Computer Science, University of Birmingham, Edgbaston B15 2TT, UK

Marta C. González
Department of Civil and Environmental Engineering, MIT, Cambridge, Massachusetts 02139, USA
Engineering Systems Division, MIT, Cambridge, Massachusetts 02139, USA

Jichang Zhao
School of Economics and Management, Beihang University, Beijing 100191, China

Daqing Li
School of Reliability and Systems Engineering, Beihang University, Beijing 100191, China
Science and Technology on Reliability and Environmental Engineering Laboratory, Beijing 100191, China

Hillel Sanhedrai and Shlomo Havlin
Department of Physics, Bar-Ilan University, Ramat Gan 5290002, Israel

Reuven Cohen
Department of Mathematics, Bar-Ilan University, Ramat Gan 5290002, Israe

Yusuke Goto and Hajime Tanaka
Institute of Industrial Science, University of Tokyo, 4-6-1 Komaba, Meguro-ku, Tokyo 153-8505, Japan

Andrea Crespi, Roberto Osellame and Roberta Ramponi
Istituto di Fotonica e Nanotecnologie, Consiglio Nazionale delle Ricerche (IFN-CNR), Piazza Leonardo da Vinci, 32, I-20133 Milano, Italy
Dipartimento di Fisica, Politecnico di Milano, Piazza Leonardo da Vinci, 32, I-20133 Milano, Italy

Marco Bentivegna, Fulvio Flamini, Nicolò Spagnolo, Niko Viggianiello, Paolo Mataloni and Fabio Sciarrino
Dipartimento di Fisica, Sapienza Universita` di Roma, Piazzale Aldo Moro 5, I-00185 Roma, Italy

Luca Innocenti
Dipartimento di Fisica, Sapienza Universita` di Roma, Piazzale Aldo Moro 5, I-00185 Roma, Italy
Università di Roma Tor Vergata, Via della ricerca scientifica 1, I-00133 Roma, Italy

W. Beugeling
Max-Planck-Institut für Physik komplexer Systeme, Nöthnitzer Straβe 38, 01187 Dresden, Germany

C. Delerue
IEMN-Department ISEN, UMR CNRS 8520, 41 Boulevard Vauban, 59046 Lille, France

E. Kalesaki
IEMN-Department ISEN, UMR CNRS 8520, 41 Boulevard Vauban, 59046 Lille, France
Physics and Materials Science Research Unit, University of Luxembourg, 162A Avenue de la Faïencerie, L-1511 Luxembourg, Luxembourg

Y.-M. Niquet
Université Grenoble Alpes, INAC-SP2M, L_Sim, 17 avenue des Martyrs, 38054 Grenoble, France
CEA, INAC-SP2M, L_Sim, 17 avenue des Martyrs, 38054 Grenoble, France

D. Vanmaekelbergh
Debye Institute for Nanomaterials Science, Utrecht University, Princetonplein 1, 3584 CC Utrecht, The Netherlands

C. Morais Smith
Institute for Theoretical Physics, Center for Extreme Matter and Emergent Phenomena, Utrecht University, Leuvenlaan 4, 3584 CE Utrecht, The Netherlands

Manfred Milinski, Dirk Semmann and Ralf Sommerfeld
Department of Evolutionary Ecology, Max-Planck-Institute for Evolutionary Biology, August-Thienemann-Strasse 2, 24306 Plön, Germany

Christian Hilbe
Department of Organismic and Evolutionary Biology, Department of Mathematics, Program for Evolutionary Dynamics, Harvard University, One Brattle Square, Cambridge, Massachusetts 02138, USA
Institute of Science and Technology Austria, Am Campus 1, Klosterneuburg 3400, Austria

Jochem Marotzke
Max Planck Institute for Meteorology, Department "The Ocean in the Earth System", 20146 Hamburg, Germany

Seth Lloyd
Department of Mechanical Engineering, Research Lab for Electronics, Massachusetts Institute of Technology, MIT 3-160, Cambridge, Massachusetts 02139, USA

Silvano Garnerone
Institute for Quantum Computing, University of Waterloo, Waterloo, Ontario, Canada N2L 3G1

Paolo Zanardi
Department of Physics and Astronomy, Center for Quantum Information Science & Technology, University of Southern California, Los Angeles, California 90089-0484, USA

Christoph Kirst
Network Dynamics, Max Planck Institute for Dynamics and Self-Organization (MPIDS), Göttingen 37077, Germany
Nonlinear Dynamics, Max Planck Institute for Dynamics and Self-Organization (MPIDS), Göttingen 37077, Germany
Institute for Nonlinear Dynamics, Georg-August University Göttingen, Göttingen 37077, Germany
Bernstein Center for Computational Neuroscience (BCCN), Göttingen 37077, Germany
Center for Physics and Biology, The Rockefeller University, New York, New York 10065, USA

Marc Timme
Network Dynamics, Max Planck Institute for Dynamics and Self-Organization (MPIDS), Göttingen 37077, Germany
Institute for Nonlinear Dynamics, Georg-August University Göttingen, Göttingen 37077, Germany
Bernstein Center for Computational Neuroscience (BCCN), Göttingen 37077, Germany

Demian Battaglia
Université Aix-Marseille, INSERM UMR 1106, Institut de Neurosciences des Systémes, Marseille 13005, France

Marco Tomamichel
School of Physics, University of Sydney, Sydney, New South Wales 2006, Australia

Mario Berta
Institute for Quantum Information and Matter, Division of Physics, Mathematics and Astronomy, California Institute of Technology, Pasadena, California 91125, USA

Joseph M. Renes
Department of Physics, Institute for Theoretical Physics, ETH Zurich, 8093 Zürich, Switzerland

Miguel Navascués
Department of Physics, Bilkent University, Ankara 06800, Turkey
Universitat Autònoma de Barcelona, 08193 Bellaterra (Barcelona), Spain
H.H. Wills Physics Laboratory, University of Bristol, Tyndall Avenue, Bristol BS8 1TL, UK

Yelena Guryanova
H.H. Wills Physics Laboratory, University of Bristol, Tyndall Avenue, Bristol BS8 1TL, UK

Matty J. Hoban
ICFO-Institut de Ciencies Fotoniques, 08860 Castelldefels (Barcelona), Spain
Department of Computer Science, University of Oxford, Parks Road, Oxford OX1 3QD, UK

Antonio Acín
ICFO-Institut de Ciencies Fotoniques, 08860 Castelldefels (Barcelona), Spain
ICREA—Institucio Catalana de Recerca i Estudis Avançats, 08010 Barcelona, Spain

Charles Tahan
Laboratory for Physical Sciences, College Park, Maryland 20740, USA

Yun-Pil Shim
Laboratory for Physical Sciences, College Park, Maryland 20740, USA
Department of Physics, University of Maryland, College Park, Maryland 20742, USA

Cheng-Long Zhang, Zhujun Yuan and Bingbing Tong
International Center for Quantum Materials, School of Physics, Peking University, Beijing, China

Guang Bian, Su-Yang Xu, Ilya Belopolski, M. Zahid Hasan, Nasser Alidoust, Daniel S. Sanchez and Hao Zheng
Laboratory for Topological Quantum Matter and Spectroscopy (B7), Department of Physics, Princeton University, Princeton, New Jersey 08544, USA

Tay-Rong Chang
Department of Physics, National Tsing Hua University, Hsinchu 30013, Taiwan
Laboratory for Topological Quantum Matter and Spectroscopy (B7), Department of Physics, Princeton University, Princeton, New Jersey 08544, USA

Ziquan Lin and JunfengWang
Wuhan National High Magnetic Field Center, Huazhong University of Science and Technology, Wuhan 430074, China

Guoqing Chang, Chuang-Han Hsu, Chi-Cheng Lee, Shin-Ming Huang and Hsin Lin
Centre for Advanced 2D Materials and Graphene Research Centre, National University of Singapore, Singapore 117546, Singapore
Department of Physics, National University of Singapore, Singapore 117542, Singapore

Horng-Tay Jeng
Department of Physics, National Tsing Hua University, Hsinchu 30013, Taiwan
Institute of Physics, Academia Sinica, Taipei 11529, Taiwan

Madhab Neupane
Laboratory for Topological Quantum Matter and Spectroscopy (B7), Department of Physics, Princeton University, Princeton, New Jersey 08544, USA
Condensed Matter and Magnet Science Group, Los Alamos National Laboratory, Los Alamos, New Mexico 87545, USA
Department of Physics, University of Central Florida, Orlando, Florida 32816, USA

Shuang Jia and Chi Zhang
International Center for Quantum Materials, School of Physics, Peking University, Beijing, China
Collaborative Innovation Center of Quantum Matter, Beijing 100871, China

Hai-Zhou Lu
Department of Physics, South University of Science and Technology of China, Shenzhen, China

Shun-Qing Shen
Department of Physics, The University of Hong Kong, Pokfulam Road, Hong Kong, China

Titus Neupert
Princeton Center for Theoretical Science, Princeton University, Princeton, New Jersey 08544, USA

M.E.J. Newman
Department of Physics, University of Michigan, 450 Church Street, Ann Arbor, Michigan 48109, USA
Center for the Study of Complex Systems, University of Michigan, 450 Church Street, Ann Arbor, Michigan 48109, USA
Santa Fe Institute, 1399 Hyde Park Road, Santa Fe, New Mexico 87501, USA

Aaron Clauset
Santa Fe Institute, 1399 Hyde Park Road, Santa Fe, New Mexico 87501, USA
Department of Computer Science, University of Colorado, 430 UCB, Boulder, Colorado 80309, USA
BioFrontiers Institute, University of Colorado, 596 UCB, Boulder, Colorado 80309, USA

Cheolhee Han, Jinhong Park and H.-S. Sim
Department of Physics, Korea Advanced Institute of Science and Technology, 291, Daehak-ro, Yuseong-gu, Daejeon 34141, Korea

Yuval Gefen
Department of Condensed Matter Physics, Weizmann Institute of Science, Rehovot 76100, Israel

Robert J. Chapman, Zixin Huang and Alberto Peruzzo
Quantum Photonics Laboratory, School of Engineering, RMIT University, Melbourne, Victoria 3000, Australia
School of Physics, The University of Sydney, Sydney, New South Wales 2006, Australia

Roberto Osellame, Giacomo Corrielli, Andrea Crespi and Matteo Santandrea
Istituto di Fotonica e Nanotecnologie, Consiglio Nazionale delle Ricerche, Piazza Leonardo da Vinci 32, Milano I-20133, Italy
Dipartimento di Fisica, Politecnico di Milano, Piazza Leonardo da Vinci 32, Milano I-20133, Italy

Man-Hong Yung
Department of Physics, South University of Science and Technology of China, Shenzhen 518055, China

Steffen Weimann, Armando Perez-Leija, Maxime Lebugle, Markus Gräfe, René Heilmann, Stefan Nolte and Alexander Szameit
Institute of Applied Physics, Abbe School of Photonics, Friedrich-Schiller-Universität Jena, Max-Wien Platz 1, 07743 Jena, Germany

Robert Keil and Gregor Weihs
Institut für Experimentalphysik, Universität Innsbruck, Technikerstraβe 25, 6020 Innsbruck, Austria

Malte Tichy
Department of Physics and Astronomy, University of Aarhus, 8000 Aarhus, Denmark

Hector Moya-Cessa
INAOE, Coordinacion de Optica, Luis Enrique Erro No. 1, Tonantzintla, Puebla 72840, Mexico

Demetrios N. Christodoulides
CREOL, The College of Optics & Photonics, University of Central Florida, Orlando, Florida 32816, USA

Peter Reimann
Fakultät für Physik, Universität Bielefeld, 33615 Bielefeld, Germany

Fernando G.S.L. Brandão
Quantum Architectures and Computation Group, Microsoft Research, Redmond, Washington 98052, USA
Department of Computer Science, University College London, WC1E 6BT London, UK

Ravishankar Ramanathan and Michał Horodecki
Faculty of Mathematics, Physics and Informatics, Institute of Theoretical Physics and Astrophysics and National Quantum Information Centre, University of Gdańsk, 80-309 Gdańsk, Poland

Andrzej Grudka
Faculty of Physics, Adam Mickiewicz University, 61-614 Poznań, Poland

Karol Horodecki
Faculty of Mathematics, Physics and Informatics, Institute of Informatics and National Quantum Information Centre, University of Gdańsk, 80-309 Gdańsk, Poland

Paweł Horodecki
Faculty of Applied Physics and Mathematics, National Quantum Information Centre, Gdańsk University of Technology, 80-233 Gdańsk, Poland

Tomasz Szarek
Faculty of Mathematics, Physics and Informatics, Institute of Mathematics and National Quantum Information Centre, University of Gdańsk, 80-309 Gdańsk, Poland

Hanna Wojewódka
Faculty of Mathematics, Physics and Informatics, Institute of Theoretical Physics and Astrophysics, Instutute of Mathematics and National Quantum Information Centre, University of Gdańsk, 80-309 Gdańsk, Poland

Index

www.ingramcontent.com/pod-product-compliance
Lightning Source LLC
Chambersburg PA
CBHW080517200326
41458CB00012B/4236
9781632385468